# Lecture Notes in Artificial Intelligence 4629

Edited by J. G. Carbonell and J. Siekmann

Subseries of Lecture Notes in Computer Science

Václav Matoušek    Pavel Mautner (Eds.)

# Text, Speech and Dialogue

10th International Conference, TSD 2007
Pilsen, Czech Republic, September 3-7, 2007
Proceedings

 Springer

Series Editors

Jaime G. Carbonell, Carnegie Mellon University, Pittsburgh, PA, USA
Jörg Siekmann, University of Saarland, Saarbrücken, Germany

Volume Editors

Václav Matoušek
Pavel Mautner
University of West Bohemia
Department of Computer Science and Engineering
Univerzitni 8, CZ - 306 14 Plzen, Czech Republic
E-mail: {matousek; mautner}@kiv.zcu.cz

Library of Congress Control Number: 2007933813

CR Subject Classification (1998): I.2.7, I.2, H.3, H.4, I.7

LNCS Sublibrary: SL 7 – Artificial Intelligence

ISSN       0302-9743
ISBN-10    3-540-74627-7 Springer Berlin Heidelberg New York
ISBN-13    978-3-540-74627-0 Springer Berlin Heidelberg New York

Springer is a part of Springer Science+Business Media

springer.com

© Springer-Verlag Berlin Heidelberg 2007

Typesetting: Camera-ready by author, data conversion by Scientific Publishing Services, Chennai, India
SPIN: 12115303       06/3180                           5 4 3 2 1 0

# Preface

The International Conference TSD 2007 presented state-of-the-art technology and recent achievements in the field of natural language processing. It declared its intent to be an interdisciplinary forum, intertwining research in speech and language processing with its applications in everyday practice. We feel that the mixture of different approaches and applications offered a great opportunity to get acquainted with the current activities in all aspects of language communication and to witness the amazing vitality of researchers from developing countries too. The financial support of the ISCA (International Speech Communication Association) enabled the wide attendance of researchers from all active regions of the world.

This year's conference was the 10<sup>th</sup> event in the series on Text, Speech, and Dialogue, originated in 1998, and it was partially oriented towards language modeling, which was chosen as the main topic of the conference with the aim to celebrate the forthcoming 75th. birthday of the general chair of the conference Frederick Jelinek. All other invited speakers (Eva Hajičová, Heinrich Niemann, Renato De Mori and David Nahamoo) read the lectures celebrating mainly the above mentioned jubilee of the founder of our conference. Also several appendant actions of the conference were dedicated to this celebrational aim.

This volume contains a collection of the papers presented at the conference organized by the Faculty of Applied Sciences of the University of West Bohemia in Pilsen in collaboration with the Faculty of Informatics, Masaryk University in Brno, and held in the new Primavera Conference Center in Pilsen, September 3–7, 2007. Theoretical and more general contributions were presented in common (plenary) sessions. Problem oriented sessions as well as panel discussions then brought together the specialists in limited problem areas with the aim of exchanging the knowledge and skills resulting from research projects of all kinds. Each of the submitted papers was thoroughly reviewed by three members of the conference reviewing team consisting of more than 40 top specialists in the conference topic areas. A total of 80 accepted papers out of 198 submitted, altogether contributed by 215 authors and co-authors, were selected for presentation at the conference by the program committee and then included in this book.

We would like to gratefully thank the invited speakers and the authors of the papers for their valuable contributions and the ISCA for its financial support. Last but not least, we would like to express our gratitude to the authors for providing their papers on time, to the members of the conference reviewing team and program committee for their careful reviews and paper selection, to the editors for their hard work in preparing this volume, and to the members of the local organizing committee for their enthusiasm in organizing the conference.

June 2007                                                           Václav Matoušek

# Organization

TSD 2007 was organized by the Faculty of Applied Sciences, University of West Bohemia in Plzeň (Pilsen), in cooperation with the Faculty of Informatics, Masaryk University in Brno, Czech Republic. The conference Web–page is located at URL:

http://www.kiv.zcu.cz/events/tsd2007/.

## Program Committee

Frederick Jelinek (USA), *General Chair*
Hynek Heřmanský (Switzerland), *Executive Chair*
Eneko Agirre (Spain)
Geneviève Baudoin (France)
Jan Černocký (Czech Republic)
Alexander Gelbukh (Mexico)
Louise Guthrie (UK)
Jan Hajič (Czech Republic)
Eva Hajičová (Czech Republic)
Patrick Hanks (UK and Czech Republic)
Ludwig Hitzenberger (Germany)
Jaroslava Hlaváčová (Czech Republic)
Aleš Horák (Czech Republic)
Eduard Hovy (USA)
Ivan Kopeček (Czech Republic)
Steven Krauwer (The Netherlands)
Siegfried Kunzmann (Germany)
Natalija Loukachevitch (Russia)
Václav Matoušek (Czech Republic)
Hermann Ney (Germany)
Elmar Nöth (Germany)
Karel Oliva (Austria)
Karel Pala (Czech Republic)
Nikola Pavešić, (Slovenia)
Vladimír Petkevič (Czech Republic)
Fabio Pianesi (Italy)
Josef Psutka (Czech Republic)
James Pustejovsky (USA)
Léon J. M. Rothkrantz (The Netherlands)
Ernst Günter Schukat-Talamazzini (Germany)
Pavel Skrelin (Russia)

Pavel Smrž  (Czech Republic)
Marko Tadić  (Croatia)
Tamás Varadi  (Hungary)
Zygmunt Vetulani  (Poland)
Taras Vintsiuk  (Ukraine)
Yorick Wilks  (UK)
Victor Zakharov  (Russia)

## Local Organizing Committee

Václav Matoušek  *(Chair)*
Štěpán Albrecht
Kamil Ekštein
Jana Hesová
Martin Hošna
Svatava Kindlová
Jana Klečková
Miloslav Konopík
Jana Krutišová
Pavel Mautner
Roman Mouček
Helena Ptáčková *(Secretary)*
Tomáš Pavelka

# About Plzeň (Pilsen)

The New Town of Pilsen was founded at the confluence of four rivers – Radbuza, Mže, Úhlava and Úslava – following a decree issued by the Czech king, Wenceslas II. He did so in 1295. From the very beginning, the town was a busy trade center located at the crossroads of two important trade routes. These linked the Czech lands with the German cities of Nuremberg and Regensburg.

In the 14$^{th}$ century, Pilsen was the third largest town after Prague and Kutna Hora. It comprised 290 houses on an area of 20 ha. Its population was 3,000 inhabitants. In the 16$^{th}$ century, after several fires that damaged the inner center of the town, Italian architects and builders contributed significantly to the changing character of the city. The most renowned among them was Giovanni de Statia. The Holy Roman Emperor, the Czech king Rudolf II, resided in Pilsen twice between 1599–1600. It was at the time of the Estates revolt. He fell in love with the city. He even bought two houses neighboring the town hall and had them reconstructed according to his taste.

Later, in 1618, Pilsen was besieged and captured by Count Mansfeld's army. Many Baroque style buildings dating to the end of the 17$^{th}$ century were designed by Jakub Auguston. Sculptures were done by Kristian Widman. The historical heart of the city – almost identical with the original Gothic layout – was declared protected historic city reserve in 1989.

Pilsen experienced a tremendous growth in the first half of the 19$^{th}$ century. The City Brewery was founded in 1842 and the Skoda Works in 1859. With the population of 175,038 inhabitants, Pilsen prides itself on being the seat of the University of West Bohemia and Bishopric.

The historical core of the city of Pilsen is limited by the line of the former town fortification walls. These gave way, in the middle of the 19$^{th}$ century, to a green belt of town parks. Entering the grounds of the historical center, you walk through streets that still respect the original Gothic urban layout, i.e., the unique developed chess ground plan.

You will certainly admire the architectonic dominant features of the city. These are mainly the Church of St. Bartholomew, the loftiness of which is accentuated by its slim church spire. The spire was reconstructed into its modern shape after a fire in 1835, when it had been hit by a lightening bolt during a night storm.

The placement of the church right within the grounds of the city square was also rather unique for its time. The church stands to right of the city hall. The latter is a Renaissance building decorated with graffiti in 1908–12. You will certainly also notice the Baroque spire of the Franciscan monastery.

All architecture lovers can also find more hidden jewels, objects appreciated for their artistic and historic value. These are burgher houses built by our ancestors in the styles of the Gothic, Renaissance or Baroque periods. The architecture of these sights was successfully modeled by the construction whirl of the end of the 19$^{th}$ century and the beginning of the 20$^{th}$ century.

Thanks to the generosity of the Gothic builders, the town of Pilsen was predestined for free architectonic development since its very coming to existence. The town has therefore become an example of a harmonious coexistence of architecture both historical and historicizing.

# Sponsoring Institutions

International Speech Communication Association (ISCA)
Czech Society for Cybernetics and Informatics (CSKI)

# Table of Contents

## Speech

# Dialog

# Language Modeling with Linguistic Cluster Constraints

Frederick Jelinek and Jia Cui

Center for Language and Speech Processing,
The Johns Hopkins University, Baltimore, USA

**Abstract.** In the past, Maximum Entropy based language models were constrained by training data n-gram counts, topic estimates, and triggers. We will investigate the obtainable gains from imposing additional constraints related to linguistic clusters, such as parts of speech, semantic/syntactic word clusters, and semantic labels. It will be shown that there substantial profit is available provided the estimates use Gaussian a priori statistics.

V. Matoušek and P. Mautner (Eds.): TSD 2007, LNAI 4629, p. 1, 2007.

# Some of Our Best Friends Are Statisticians

Jan Hajič and Eva Hajičová

Institute of Formal and Applied Linguistics, Charles University in Prague,
Malostranské nám. 25, CZ-11800 Prague, Czech Republic
{hajic,hajicova}@ufal.mff.cuni.cz

In his LREC 2004 invited talk when awarded by the first ever Antonio Zampolli prize for his essential contributions to the use of spoken and written language resources, Frederick Jelinek has used the title "Some of My Best Friends Are Linguists". He did so for many reasons, one of them being that he wanted to remove the perception that he dislikes linguists and linguistics after so many people used to cite his famous line from an old presentation at a Natural Language Processing Evaluation workshop in 1988, in which he said "Whenever I fire a linguist our system performance improves."

We linguists (at least in Prague) have never struck back with a similar line. However, counterexamples abound in cases where statistics as currently (and at that time) implemented simply "cannot work", and these counterexamples have been cited by many people (not just linguists) many times.

In any case, we have paraphrased his title today. It is intentional, since we realized that (punch lines aside) there were lessons to be learned from this seemingly "counter-linguistic war".

Of course, it was never really a war... but we have to mention here one real war, during which the history of computational linguistics as a "joint venture" between mathematics, computing and linguistics began: The World War II. During this war, we saw the first true, successful and to a certain extent, pretty decisive application of machines to linguistic problems: the "computing machine" (called "Bombe") helped Alan Turing and his colleagues in their Bletchley Park offices to break the German Military Enigma code (which was used to convey important command messages originally expressed in the German language, of course), leading to many successful Allied operations.[1] Moreover, despite using computing machinery, he has used some of the linguistic properties of language usage to help this otherwise computationally very complex task.

After the war, when focus shifted to the cold war realities, we can read in Warren Weaver's letter to Norbert Wiener in March 1947: "When I look at an article in Russian, I say: 'This is really written in English but it has been coded in some strange symbols. I will proceed to decode it.' " (quoted in [38], p. 18). These 31 words describe in an excellent abbreviation the prevailing approach in statistical modeling that was started "for real" (after computers gained enough power to really implement such things) in

---

[1] The first steps in breaking the Enigma code were in fact done by three Polish mathematicians who in 1939 handed their findings over to France and Britain. The code-breaking effort was later (mid-1943 onwards) greatly expanded in the US at the NCR Corporation to handle the more advanced version of the German Enigma, enabling to break the German Navy messages almost in "real time".

V. Matoušek and P. Mautner (Eds.): TSD 2007, LNAI 4629, pp. 2–10, 2007.

the early 80s at places like CMU and Fred Jelinek's team at IBM Research in Yorktown Heights. This approach simply turns the given problem around and looks at the problem rather as an information-theoretic decoding problem. After all, from this point of view there is obviously no difference in the "signal" that goes through the proverbial "noisy channel", being it texts, words, grammatical information, speech signal, or the now-abandoned Morse signal.

However, between the mid-50s and early 80s there was a totally different paradigm being used in Natural Language Processing, the one started by Chomsky in his Syntactic Structures book published in 1957. It was the formal linguistic approach that has led to many theories and many implementations of various tools and problems, without real solutions and fully working systems. To do justice to this "new paradigm" of linguistics, as it is sometimes labeled, we should admit that its initiators did not claim to offer theoretical background for computerized natural language systems. Quite on the contrary: Chomsky (and many of his followers throughout the time) even explicitly discouraged anyone from taking the statistical or information-theoretic approach by quite harshly dismissing any such attempts.

It thus required quite a courage to stand up against this prevailing view. It is no surprise that it has come not from the core NLP community, but from the Automatic Speech Recognition area, which had been typically separate from the NLP itself (sadly enough, it still is). IBM Research group under Fred Jelinek's leadership realized (and experimentally showed) that linguistic rules and Artificial Intelligence techniques had inferior results even when compared to very simplistic statistical techniques. This was first demonstrated on phonetic baseforms in the acoustic model for a speech recognition system, but later it became apparent that this can be safely assumed almost for every other problem in the field (e.g., Jelinek [19] and Jelinek, Bahl and Mercer [22]). Statistical learning mechanisms were apparently and clearly superior to any human-designed rules, especially those using any preference system, since humans are notoriously bad at estimating quantitative characteristics in a system with many parameters (such as a natural language). It was at that time when the linguists' work was deemed counterproductive–because supported by too simple performance evaluations– and when firing linguists became a popular matter.

Textual Natural Language Processing applications followed soon, started with an application of Hidden Markov Models for part-of-speech tagging (Bahl and Mercer [1], followed in 1988 by Ken Church's PARTS system [10]) and (again) within Fred Jelinek's group at IBM Research by the now famous Candide statistical machine translation system from French to English, which relied heavily (and at the beginning, almost exclusively) on statistical information derived from parallel texts (Brown et al. [7] and Berger et al. [4]).[2]

Even though fiercely opposed by many in the field (Peter Brown, the leading scientist in the Candide machine translation project, had an amazing Coling-1990 hand-written flameful review framed in his office, and Bob Mercer, another member of the Candide group, was called "the world's biggest dreamer" at a MT conference in Garmisch-Partenkirchen in Germany in 1987 by some leading MT specialists of that time), the numbers talk clearly: whereas there was only 1 statistical paper at the ACL conference

---

[2] The Canadian Hansards, transcripts of parliamentary debates in the Canadian Parliament.

in 1990 (out of 40 accepted papers), in Sapporo 2003 there were 48 such papers (over three fourth of the papers accepted). It's probably even more today–and the "Empirical Methods in NLP" conference, devoted exclusively to the use of statistical methods in NLP (which started 12 years ago as a workshop on using very large corpora), has now the same length and almost the same size as the main ACL conference itself.

This all might seem to lead to almost immediate (or near-future) death of "classical" rule- or "AI"-based computational linguistics and natural language processing. The more so, given that corpora grow larger and larger and some people are expressing even more dramatic view of the future in this field. Starting with citing Bob Mercer's "There are no data like more data" (said already in 1985), we can point to Eric Brill's talk at North American ACL in 2001 in Pittsburgh where he essentially claimed that linguistics is dead and that we "shouldn't think, but use machine learning on large corpora" and "if we really have to think, we should think about how to get more data" (Brill [6] and Banko and Brill [2], [3]). At Eurospeech in Geneva in 2003, Ken Church [11] went even further and said that in the area of part-of-speech tagging "if all the money spent on developing better [statistical] methods went to annotating corpora, we would almost surely have had better taggers today". And Franz Josef Och has demonstrated (at this year's North American ACL conference) that improvement in machine translation still continues if its language model training data size increases even from hundreds of billions of words to almost a trillion [28].

Not so. We all should read very carefully what all these visionaries and undoubtedly great minds of Speech and Natural Language Processing have said–either explicitly or between the lines. Fred Jelinek himself says (in the Award talk [21]) that "my colleagues and I always hoped that linguistics will eventually allow us to strike gold". The recent Franz Josef Och's presentation about the trillion-word language model also showed that the gains no longer follow the logarithmic growth we used to take for granted.[3] Eric Brill's provocative talk which was (when read only superficially) dismissing linguistics and even algorithm research almost absolutely only experimented with pseudoproblems that just need plain text data, which are easy to find. It is an unfortunate fact that problems that lead to using plain text data for statistical modeling and learning are only a minority in the portfolio of important computational linguistics issues. In the majority of problems, however, we need more or less profound linguistic expertise to at least divide the problem into manageable parts. For example, in machine translation today we need to lemmatize and identify (at least) "phrases"–indeed, very linguistic phenomena. For predicate extraction, we need to know at least some aspects of the syntactic structure of the sentence. For word sense disambiguation, we need to know which senses of the words in question to disambiguate. And even in speech recognition, where plain text language modeling solves half of the problem, we need transcripts of what has been said. Ken Church was comparing algorithms and methodology to annotating of data, again a linguists' task–or at least a process of linguistic interpretation of plain text, albeit (in his case) a relatively simple one (for English).

We are not disputing the vision that (possibly) machines will be able to simulate human mind perfectly (and perhaps better than humans themselves) and (re)discover

---

[3] "Logarithmic growth" means a constant improvement of performance when training data is doubled.

everything in linguistics (and other such fields of science) from plain data. We would like to argue "only" that this vision looks ahead to an extremely distant future and that it will not even come true if computational linguistics is not viewed as an experimental science and we act accordingly today. What does "experimental science" mean? Let's take physics: Newton and others in the Middle Ages observed and measured what they could observe and measure. And then, they tried to explain the data they observed by a set (the simplest possible set)[4] of mathematical equations and formulas. In a similar vein, astronomers observed the positions of stars relatively to the Earth and tried to come up with formulas that would explain their movements, in this case without a possibility to observably confirm their findings in the near future. Eventually they succeeded and their findings were even later confirmed by new observations achieved thanks to the possibility of outer space travel.

All well, but how can linguistics be ever looked at that way? What is an "observation" and "measurement" in linguistics? There are no objective criteria for anything but plain text corpora and digitized speech recordings. For the rest, we have to use intermediaries–and these intermediaries must be humans, humans with their intuition about the relationship between the observed data (text or speech) and its meaning through their *understanding* and *interpretation*. However, in order to provide "measurable" data these intermediaries must also be able to *formalize* their interpretation and understanding. We call such intermediaries *annotators* and the results of their work are annotated linguistic data. Annotators and those who work on the formalization of the problem and provide annotation support from both the linguistic and technical side thus should be in the center of the current methodology in computational linguistics.

It is interesting to note that first annotated corpora were created at least 10 years before the onset of using statistical methods in linguistics and intended for various purposes, last but not least in the area of quantitative linguistics, which tries to find and publish various statistics about language (such as word and letter counts, part-of-speech and morphological phenomena relative to the lexicon etc.) but makes no attempt to use those statistics for predictions and modeling. The first such known corpus is the Brown corpus[5] (of English), a one million token corpus published in the late 1960s, annotated on the level of part-of-speech tags. As far as we know, tied for the second place[6] is the Czech Academic Corpus[7] published in the early 1970. This corpus was, again as far as we know, the first large-scale corpus (with 540,000 real word tokens) with systematic annotation of most syntactic relations.[8]

---

[4] Based on the ages-old principle of "Occam's razor": that scientific explanations should be as simple as possible (but not simpler, obviously).

[5] Created at the Brown University, Providence, RI, USA, by a Czech emigrant Jindřich (later known as Henry) Kučera and Nelson Francis [14].

[6] Together with the Lancaster-Oslo-Bergen Corpus, published in 1970.

[7] Created by the group of Marie Těšitelová in the Institute of Czech Language of the Czechoslovak Academy of Sciences.

[8] The Czech Academic Corpus is now being restored and preserved, and its annotation is being brought up to current standards–not just for historical reasons, but also to provide additional compatible data for many morphological and syntactic experiments on Czech (see CAC 1.0, at http://ufal.mff.cuni.cz/rest and [37]; for some aspects of the restoration of CAC see Ribarov et al. [30]).

We have to again mention Fred Jelinek, since it was his group who commissioned in 1987 the first English large-scale syntactically annotated corpus–it was built in Lancaster in Britain after Geoffrey Leech and Geoffrey Sampson had built their first Treebank there. The most famous and the most frequently used corpus in this area is the one million word token Penn Treebank[9], which was the first annotation project of the then-established and now famous data creator and distributor, the Linguistic Data Consortium associated with the University of Pennsylvania in Philadelphia, PA in the U.S.

Again, the first Treebank conceived as a dependency one from scratch was the Czech-language Prague Dependency Treebank (PDT) published first in 2001 (For an outline of the project, see Hajič [15]; for the latest version of PDT, see [16]). This treebank follows the theory of the Functional Generative Description developed by Petr Sgall and his collaborators at Charles University in Prague since the early 1960s (for the first accounts see [33], [34]; for a more comprehensive treatment see Sgall et al. [35]).[10] Dependency style of annotation (obviously, also of parser output specification as used earlier) is seen as more compact and economic yet closer to what we really want parsing to be used for: information search, information extraction, predicate-argument structure etc. The German-language Negra and later the Tiger corpus followed soon (see Brants et al. [5], [27] and [36]), as well as many smaller treebanks of other languages. The PropBank (see Palmer, Gildea and Kingsbury [29]) and NomBank [26] projects for English, and the Prague Dependency Treebank Version 2.0 [16] with its associated valency lexicon [17] (containing, for each valency-capable word sense, its labeled arguments and possible surface form realizations)[11] both try to add more information on the meaning of the sentence to the underlying syntactic structure annotation, albeit both in a different way. The new version of the Czech-language Prague Dependency Treebank tries to interlink several annotation tasks (morphology, underlying syntax, certain aspects of semantics, co-reference, information structure) in a consistent way, again a world's first – no such treebank for English or another language exists yet.

Just to give an overall feeling about our vision of linguistic resources and without going into any detail (for a more detailed discussion see the web page referred to below and the references quoted there), we illustrate here on the example of the Prague Dependency Treebank how we believe a good start of a development should look like. PDT is an annotated collection of Czech texts, randomly chosen from the Czech National Corpus [12], with a mark-up on three layers: morphemic, surface shape ("analytical") and underlying ("tectogrammatical"). The current version [16], annotated on all three layers, contains 3165 documents (mainly from journalistic style) comprising almost 50,000 sentences with over 830,000 tokens. The main concern of our linguistically oriented research is the tectogrammatical level, on which every node of the tectogrammatical (underlying) representation of the sentence (Tectogrammatical Tree Structure, or TGTS) is assigned a label consisting of, e.g., the lexical value of the word, of its "grammatemes" (i.e. the values of semantic categories typically displayed at the

---

[9] Created by the group led by Mitch Marcus and published by LDC in 1993 [25].

[10] Though the origins of the dependency-based syntactic description are located in Europe, it should be noted that one of the fundamental contributions to this view of underlying syntactic structure is C. J. Fillmore's "case theory" [13]; this is well documented by Robinson [31], [32].

[11] For an independent and more elaborate Czech valency lexicon see [39].

morphological level, like (semantic) number, (semantic) tense, etc.), of its "functors" (with a more subtle differentiation of syntactic relations by means of "subfunctors" (e.g. *inside, above, on-the-surface* with the Locative functor), and the Topic-Focus Articulation attribute containing values for contextual boundness ($t$ for a contextually bound non-contrastive node, $c$ for a contextually bound contrastive node and $f$ for a contextually non-bound node). It should be noted that only nodes for autosemantic lexical units are included, the function words (appearing in the surface shape of the sentence) are being "hidden" and their contribution to the meaning of the sentence and to its underlying syntax is captured by indices of different kinds within the complex labels of the nodes (functors and grammatemes, see above).

In addition, some basic inter-sentential links are also added. It should be noted that TGTSs may contain nodes not present in the morphemic form of the sentence in case of surface deletions; TGTSs differ from the theoretically adequate rendering of tectogrammatical representation in that coordinating conjunctions are represented as head nodes of the coordinated structures, which makes it possible for the TGTSs to constitute two-dimensional trees.

It is our conviction that any modern linguistic theory has to be formulated in a way that it can be tested by some objective means; one of the ways how to test a theory is to use it as a basis for a consistent annotation of large language resources, i.e. of text corpora. As our experience with PDT documents, annotation may concern not only the surface and morphemic shape of sentences, but also (and first of all) the underlying sentence structure, which elucidates phenomena hidden on the surface although unavoidable for the representation of the meaning and functioning of the sentence, for modeling its comprehension and for studying its semantico-pragmatic interpretation.

Having annotated data at our disposal allows us to compare different implementations of our methods among themselves and to evaluate their results in terms of the degree of agreement with the "gold truth", or the "measured" reality (the formalized interpretation of the plain text or speech recordings). This is a crucial point which allows us to call computational linguistics an experimental science and put it among such sciences as physics or astronomy. Moreover, using (the larger) part of the annotated data for statistically-based machine learning takes computational linguistics a step further: getting rid of the ineffective AI-style estimation of various weights, preferences and such, which apparently can be much effectively estimated by the computer (for excellent textbooks on this topic, see Charniak [8], Jelinek [20], Manning and Schütze [24] and Jurafsky and Martin [23]). So one might ask if there is any place today for linguistics and linguists (apart from annotation)? The answer is a definite yes: machine learning today, especially for complex language applications, cannot be effectively solved without a prescribed and relatively fixed model structure. For example, to solve the problem of part of speech tagging, one must decide in advance (before we can ever start the statistical training process) which features are important: Should we use the tags assigned to the neighboring words in the process? Is it the word on the left or the word on the right? How far can we go in the context? Which linguistic features to take into consideration when designing the sets of tags? We cannot simply say to the machine learner "please use any context and any combination of features and tell me which are important" – there are simply too many. It is here where the statistician and programmer should really talk to the linguist(s) and come up with appropriate features. We know this is hard, and we believe that this is where we currently owe each other the most.[12]

All statistical modeling aside, let us conclude with one simple prediction. Regardless of what Natural Language Processing methods will be in place in 5, 10, 25 or 50 years time, the golden statistical computational linguistics era will be credited with at least one thing that has changed the field forever: objective evaluation. There will be new evaluation metrics and procedures, there will be even new attempts to dismiss the importance of evaluation as we know it today, there will be various re-incarnations of the idea, the standards will move towards the evaluation of more complex systems and even the simplest tools will be evaluated in the context of such systems, and not just in isolation. But "blind" and objective evaluation, as relentlessly explained and pushed forward by Fred Jelinek and some others already from the beginning of the 80s, will stay. Statisticians will not protest; some linguists still may. However, we view this as an achievement that has served well not just Computational Linguistics, but it is in fact in all linguists' and statisticians' interest alike.

Long live evaluation, annotated corpora, cooperation between linguists and statisticians, and first of all, long live Fred Jelinek!

## Acknowledgments

The authors gratefully acknowledge the support of the grants of the Ministry of Education of the Czech Republic No. MSM0021620838 and LC536.

## References

1. Bahl, L.R., Mercer, R.L.: Part-of-speech assignment by a statistical decision algorithm. In: Proceedings of the IEEE International Symposium on Information Theory, pp. 88–89. IEEE Computer Society Press, Los Alamitos (1976)
2. Banko, M., Brill, E.: Scaling to Very Very Large Corpora for Natural Language Disambiguation. In: Proceedings of the 39th Annual Meeting of the Association for Computational Linguistics. Toulouse, France (2001)
3. Banko, M., Brill, E.: Mitigating the Paucity-of-Data Problem: Exploring the Effect of Training Corpus Size on Classifier Performance for Natural Language Processing. In: Proceedings of the First International Conference on Human Language Technology. San Diego, California, pp. 1–5 (2001)
4. Berger, A.L., Brown, P.F., Della Pietra, S.A., Della Pietra, V.J., Gillett, J.R., Lafferty, J.D., Mercer, R.L., Printz, H., Ureš, L.: The Candide System for Machine Translation. In: Proceedings of the ARPA Conference on Human Language Technology. Plainsborough, New Jersey (1994)
5. Brants, S., Dipper, S., Hansen, S., Lezius, W., Smith, G.: TIGER treebank. In: Proceedings of the First Workshop on Treebanks and Linguistic Theories (TLT 2002), Sozopol, Bulgaria, pp. 24–42 (2002)
6. Brill, E.: Paucity Shmaucity–What Can We Do With A Trillion Words? In: Invited talk at EMNLP-NAACL 2001 Conference. Pittsburgh, PA, USA (2001)

---

[12] See Hajičová [18] for some suggestions pointing out what the "old", "traditional" linguistic theory can still offer to modern computational linguistics.

7. Brown, P.F., Cocke, J., Della Pietra, S.A., Della Pietra, V.J., Jelinek, F., Lafferty, J.D., Mercer, R.L., Roossin, P.S.: A Statistical Approach to Machine Translation. Computational Linguistics 16(2), 79–85 (1990)
8. Charniak, E.: Statistical Language Learning. The MIT Press, Cambridge, MA (1996)
9. Chomsky, N.: Syntactic Structures. Mouton, The Hague (1957)
10. Church, K.W.: A Stochastic PARTS Program and Noun Phrase Parser for Unrestricted Text. In: Proceedings of the Second Conference on Applied Natural Language Processing. 26th Annual Meeting of the ACL. Austin, Texas, pp. 136–143 (1988)
11. Church, K.W.: Speech and Language Processing: Where Have We Been and Where Are We Going? In: Proceedings of the 8th European Conference on Speech Communication and Technology (EUROSPEECH/INTERSPEECH-2003). Geneva, Switzerland (2003)
12. Czech National Corpus, http://ucnk.ff.cuni.cz
13. Fillmore, C.J.: The case for case. In: Bach, E., Harms, R. (eds.) Universals in Linguistic Theory. New York, pp. 1–90 (1968)
14. Francis, N. F.: Standard Corpus of Edited Present-day American English. College English 26, 267-273. Reprinted in Geoffrey Sampson and Diana McCarthy (eds.) Corpus Linguistics: Readings in a Widening Discipline. Continuum 2004, London/New York, pp. 27–34 (1965)
15. Hajič, J.: Building a syntactically annotated corpus: The Prague Dependency Treebank. In: Issues of Valency and Meaning. Studies in Honour of Jarmila Panevová, Karolinum, pp. 106–132. Charles University Press, Prague, Czech Republic (1998)
16. Hajič, J., et al.: Prague Dependency Treebank 2.0. CDROM. Cat. No. LDC2006T01. Linguistic Data Consortium, Philadelphia, PA (2006), http://ufal.mff.cuni.cz/pdt2.0 ISBN: 1-58563-370-4
17. Hajič, J., Panevová, J., Urešová, Z., Bémová, A., Kolářová, V., Pajas, P.: PDT-VALLEX: Creating a Large-Coverage Valency Lexicon for Treebank Annotation. In: Proceedings of the 2nd Treebanks and Linguistic Theories Workshop. Växjö, Sweden, November 14-15, pp. 57–68 (2003)
18. Hajičová, E.: Old linguists never die, they only get obligatorily deleted. Computational Linguistics 32(4), 457–469 (2006)
19. Jelinek, F.: Continuous Speech Recognition by Statistical Methods. Proceedings of the IEEE 64(4), 532–536 (1976)
20. Jelinek, F.: Statistical Methods For Speech Recognition. The MIT Press, Cambridge, MA (1998)
21. Jelinek, F.: Some of My Best Friends Are Linguists. In: Invited talk at the occasion of the Antonio Zampolli Award presented to Frederick Jelinek at the LREC 2004 conference, Lisbon, Portugal (2004)
22. Jelinek, F., Bahl, L.R., Mercer, R.L.: Design of a Linguistic Statistical Decoder for the Recognition of Continuous Speech Recognition by Statistical Methods. IEEE Transactions on IT 21(3), 250–256 (1975)
23. Jurafsky, D., Martin, J.H.: Speech and Language Processing. Prentice-Hall, Englewood Cliffs (2000)
24. Manning, C.D., Schütze, H.: Foundations of Statistical Natural Language Processing. The MIT Press, Cambridge, MA (2000)
25. Marcus, M.P., Santorini, B., Marcinkiewicz, M.A.: Building a Large Annotated Corpus of English: The Penn Treebank. Computational Linguisitics 19(2), 313–330 (1993)
26. Meyers, A., Reeves, R., Macleod, C., Szekely, R., Zielinska, V., Young, B., Grishman, R.: The NomBank Project: An Interim Report. In: HLT-NAACL Workshop: Frontiers in Corpus Annotation. Boston, Massachusetts, USA, pp. 24–31 (2004)
27. http://www.coli.uni-saarland.de/projects/sfb378/NEGRA-en.html
28. Och, F.J.: Large-scale Machine Translation: Challenges and Opportunities. Invited talk at NAACL/HLT 2007, Rochester, NY, USA (April 22-27, 2007)

29. Palmer, M.S., Gildea, D., Kingsbury, P.: The Proposition Bank: An Annotated Corpus of Semantic Roles. Computational Lingusitics 31(1), 71–105 (2005)
30. Ribarov, K., Bémová, A., Vidová Hladká, B.: When a statistically oriented parser was more efficient than a linguist: a case of treebank conversion. The Prague Bulletin of Mathematical Linguistics 86, 21–38 (2006)
31. Robinson, J.J.: Case, category and configuration. Journal of Linguistics 6, 57–80 (1969)
32. Robinson, J.J.: Depenency structures and transformational rules. Language 46, 259–285 (1970)
33. Sgall, P.: Zur Frage der Ebenen im Sprachsystem. Travaux linguistiques de Prague 1, 95–106 (1964)
34. Sgall, P.: Generative Bschreibung und die Ebenen des Sprachsystems. In: presented at the Second International Symposium in Magdeburg, Germany. Zeichen und System der Sprache III 1966, Berlin, pp. 225–239 (1964)
35. Sgall, P., Hajičová, E., Panevová, J.: The Meaning of the Sentence in its Semantic and Pragmatic Aspects. Reidel - Academia, Dordrecht - Prague (1986)
36. http://www.ims.uni-stuttgart.de/projekte/TIGER/TIGERCorpus
37. Vidová Hladká, B.: The Czech Academic Corpus version 1.0 has been released. The Prague Bulletin of Mathematical Lingustics 86, 57–58 (2006)
38. Weaver, W.: Translation. Memorandum. Reprinted. In: Locke, W.N., Booth, A.D. (eds.) Machine Translation of Languages: Fourteen Essays, pp. 15–23. MIT Press, Cambridge (1949)
39. Žabokrtský, Z., Lopatková, M.: Valency Frames of Czech Verbs in VALLEX 1.0. In: HLT-NAACL Workshop: Frontiers in Corpus Annotation. Boston, Massachusetts, USA, pp. 70–77 (2004)

# Some Special Problems of Speech Communication

Heinrich Niemann

Institute of Pattern Recognition,
Martensstr. 3, 91058 Erlangen, Germany
niemann@informatik.uni-erlangen.de

**Abstract.** We start with a brief overview of our work in speech recognition and understanding which led from monomodal (speech only) human-machine dialog to multimodal human-machine interaction and assistance. Our work in speech communication initially had the goal to develop a complete system for question answering by spoken dialog [7,15]. This goal was achieved in various projects funded by the German Research Foundation [14] and the German Federal Ministry of Education and Research [16]. Problems of multilingual communication were considered in projects supported by the European Union [2,4,10]. In the Verbmobil project the speech-to-speech translation problem was investigated and it turned out that *prosody* and the recognition of *emotion* was important and extremely useful – if not indispensible – to disambiguate utterances and to influence the dialog strategy [3,17]. Multimodal and multimedia aspects of human-machine communication became a topic in the follow-up projects Embassi [11], SmartKom [1], FORSIP [12], and SmartWeb [9].

The SmartWeb project [19], which involves 17 partners from companies, research institutes, and universities, has the general goal to provide the foundations for multimodal human-machine communication with distributed semantic web services using different mobile devices, hand-held, mounted in a car or to a motor cycle. It uses speech and video signals as well as signals from other sensors, e.g. ECG or skin resistance. A special problem in human-machine interaction and assistance is the question whether the user speaks to the machine or not, that is, the distinction of on- and off-talk. It is shown how on-/off-talk can be classified by the combination of prosodic and image features. Using additional sensors the user state in general is estimated to give further cues to the dialog control. This may be used, for example, to avoid input from the dialog system in a situation where a driver is under stress.

In other projects the special problem of children's speech processing was considered [20]. Among others it was investigated whether a manual correction of automatically computed fundamental frequency $F_0$ and word boundaries might have a positive effect on the automatic classification of the 4 classes anger, motherese, emphatic, and neutral; this was not the case, leading to the conclusion that presently there is no need for improved $F_0$ algorithms in emotion recognition. The word accuracy (WA) of native and non-native English speaking children was investigated; it was shown that non-native speakers (age 10 – 15) achieve about the same WA as children aged 6 – 7 using a speech recognizer trained with native children speech. The recognizer also was used to develop an automatic scoring of the pronunciation quality of children learning English.

A special problem are impairments of speech which may be congenital (e.g. the cleft lip and palate) or acquired by disease (e.g. cancer of the larynx). Impairments are, among others, treated with speech training by speech therapists. They

V. Matoušek and P. Mautner (Eds.): TSD 2007, LNAI 4629, pp. 11–13, 2007.
© Springer-Verlag Berlin Heidelberg 2007

score the speech quality subjectively according to various criteria. The idea is that the WA of an automatic speech recognizer should be highly correlated with the human rating. Using speech samples from laryngectomees it is shown that the machine rating is about as good as the rating of five human experts and can also be done via telephone. This opens the possibility of an objective and standardized rating of speech quality.

## Acknowledgment

The above presentation is based on the work of members of the Speech Processing and Understanding Group (headed by E. Nöth) at the Chair for Pattern Recognition, Erlangen, in particular A. Batliner [3], C. Hacker [5], T. Haderlein [6], F. Hönig [8], A. Horndasch [9], A. Maier [13], E. Nöth [18], S. Steidl [21].
    The references provide further material.

## References

1. Adelhardt, J., Shi, R., Frank, C., Zeißler, V., Batliner, A., Nöth, E., Niemann, H.: Multimodal User State Recognition in a Modern Dialogue System. In: Günter, A., Kruse, R., Neumann, B. (eds.) KI 2003. LNCS (LNAI), vol. 2821, pp. 591–605. Springer, Heidelberg (2003)
2. Aretoulaki, M., Harbeck, S., Gallwitz, F., Nöth, E., Niemann, H., Ivanecki, J., Ipsic, I., Pavešić, N., Matoušek, V.: SQEL: A Multilingual and Multifunctional Dialog System. In: Mannell, R.H., Robert-Ribes, J. (eds.) Proc. International Conference on Spoken Language Processing (ICSLP), Sydney, Australia, pp. 855–858 (1998)
3. Batliner, A., Huber, R., Niemann, H., Nöth, E., Spilker, J., Fischer, K.: The Recognition of Emotion. In: Wahlster, W. (ed.) Verbmobil: Foundations of Speech-to-Speech Translation, pp. 122–130. Springer, Heidelberg (2000)
4. Gallwitz, F., Nöth, E., Niemann, H.: Recognition of Out-of-Vocabulary Words and their Semantic Category. In: Matoušek, V., Niemann, H. (eds.) Proc. of the 2nd SQEL Workshop on Multi-Lingual Information Retrieval Dialogs, University of West Bohemia, Pilsen, pp. 114–121 (1997)
5. Hacker, C., Cincarek, T., Maier, A., Heßler, A., Nöth, E.: Boosting of Prosodic and Pronunciation Features to Detect Mispronunciations of Non-Native Children. In: Proc. Int. Conf. Acoustics, Speech and Signal Processing (ICASSP). Honolulu, Hawaii, vol. 4, pp. 197–200 (2007)
6. Haderlein, T., Nöth, E., Schuster, M., Eysholdt, U., Rosanowski, F.: Evaluation of Tracheoesophageal Substitute Voices Using Prosodic Features. In: Hoffmann, R., Mixdorff, H. (eds.) Proc. Speech Prosody, 3rd International Conference, Dresden, pp. 701–704 (2006)
7. Hein, H.W., Niemann, H.: Expert knowledge for automatic understanding of continuous speech. In: Proc. First European Signal Processing Conference (EUSIPCO), Lausanne, Switzerland, pp. 647–651. North Holland Publ. Comp., Amsterdam (1980)
8. Hönig, F., Batliner, A., Nöth, E.: Fast Recursive Data-Driven Multi-Resolution Feature Extraction for Physiological Signal Classification. In: 3rd Russian-Bavarian Conference on Biomedical Engineering, Erlangen (to appear, 2007)
9. Horndasch, A., Nöth, E., Batliner, A., Warnke, V.: Phoneme-to-Grapheme Mapping for Spoken Inquiries to the Semantic Web. In: Proc. Interspeech, 2006 ICSLP, 10th International Conference on Spoken Language Processing. Pittsburgh, pp. 13–16 (2006)

10. Kuhn, T., Niemann, H., Schukat-Talamazzini, E.G., Eckert, W., Rieck, S.: Context-Dependent Modeling in a two Stage HMM Word Recognizer for Continuous Speech. In: Vandewalle, J., Boite, R., Moonen, M., Oosterlinck, A. (eds.) Signal Processing IV, Theories and Applications, Proc. EUSIPCO-92, pp. 439–442. Elsevier, Amsterdam (1992)

11. Ludwig, B., Niemann, H., Klarner, M., Görz, G.: Content and Context in Dialogue Systems. In: Wilks, Y. (ed.) Proc. of the 3rd Bellagio Workshop on Human-Computer Conversation, Sheffield, pp. 105–111 (2000)

12. Ludwig, B., Klarner, M., Reiß, P., Görz, G., Niemann, H.: A Pragmatics First Approach to Analysis and Generation of Discourse Relations. In: Buchberger, E. (ed.) 7. Konferenz zur Verarbeitung natürlicher Sprache (KONVENS), Vienna, pp. 117–124 (2004)

13. Maier, A., Nöth, E., Nkenke, E., Schuster, S.: Automatic Assessment of Children's Speech with Cleft Lip and Palate. In: Fifth Slovenian and First International Language Technologies Conference. Ljubljana, pp. 31–35 (2006)

14. Mast, M., Kummert, F., Ehrlich, U., Fink, G., Kuhn, T., Niemann, H., Sagerer, G.: A Speech Understanding and Dialog System with a Homogeneous Linguistic Knowledge Base. IEEE Trans. on Pattern Analysis and Machine Intelligence 16, 179–194 (1994)

15. Niemann, H., Brietzmann, A., Mühlfeld, R., Regel, P., Schukat, G.: The Speech Understanding and Dialog System EVAR. In: De Mori, R., Suen, C.Y. (eds.) New Systems and Architectures for Automatic Speech Recognition and Synthesis. NATO ASI Series F16, pp. 271–302. Springer, Heidelberg (1985)

16. Niemann, H., Sagerer, G., Ehrlich, U., Schukat-Talamazzini, E.G., Kummert, F.: The Interaction of Word Recognition and Linguistic Processing in Speech Understanding. In: Laface, P., De Mori, R. (eds.) Speech Recognition and Understanding, Recent Advances, Trends and Applications. NATO ASI Series F75, pp. 425–453. Springer, Heidelberg (1992)

17. Nöth, E., Batliner, A., Kießling, A., Kompe, R., Niemann, H.: VERBMOBIL: The Use of Prosody in the Linguistic Components of a Speech Understanding System. IEEE Trans. on Speech and Audio Processing 8, 519–532 (2000)

18. Nöth, E., Hacker, C., Batliner, A.: Does Multimodality Really Help? The Classification of Emotion and of On/Off-Focus in Multimodal Dialogues – Two Case Studies. In: Proc. 49th International Symposium ELMAR-2007, Zadar, Croatia (to appear, 2007)

19. Reithinger, N., Herzog, G., Blocher, A.: SmartWeb – Mobile Broadband Access to the Semantic Web. Künstliche Intelligenz, 2, 30–33 (2007)

20. Steidl, S., Stemmer, G., Hacker, C., Nöth, E., Niemann, H.: Improving Children's Speech Recognition by HMM Interpolation with an Adult's Speech Recognizer. In: Michaelis, B., Krell, G. (eds.) Pattern Recognition. LNCS, vol. 2781, pp. 600–607. Springer, Heidelberg (2003)

21. Steidl, S., Levit, M., Batliner, A., Nöth, E., Niemann, H.: Of All Things the Measure is Man. – Classification of Emotions and Inter-Labeler Consistency. In: Proceedings of ICASSP 2005 - International Conference on Acoustics, Speech, and Signal Processing. Philadelphia, pp. 317–320 (2005)

# Recent Advances in Spoken Language Understanding

Renato De Mori[1,2]

[1] School of Computer Science, Mc Gill University, Canada
[2] Laboratoire d'Informatique, Université d'Avignon, France

**Abstract.** This presentation will review the state of the art in spoken language understanding. After a brief introduction on conceptual structures, early approaches to spoken language understanding (SLU) followed in the seventies are described. They are based on augmented grammars and non stochastic parsers for interpretation.

In the late eighties, the Air Travel Information System (ATIS) project made evident problems peculiar to SLU, namely, frequent use of ungrammatical sentences, hesitations, corrections and errors due to Automatic Speech Recognition (ASR) systems. Solutions involving statistical models, limited syntactic analysis, shallow parsing, were introduced.

Automatic learning of interpretation models, use of finite state models and classifiers were also proposed. Interesting results were found in such areas as concept tags detection for filling slots in frame systems, conceptual language models, semantic syntax--directed translation, stochastic grammars and parsers for interpretation, dialog event tagging.

More recent approaches combine parsers and classifiers and reconsider the use of probabilistic logics. Others propose connectionist models and latent semantic analysis. As interpretation is affected by various degrees of imprecision, decision about actions should depend on information states characterized by the possibility of having competing hypotheses scored by confidence indicators. Proposed confidence measures at the acoustic, linguistic and semantic level will be briefly reviewed.

V. Matoušek and P. Mautner (Eds.): TSD 2007, LNAI 4629, p. 14, 2007.

# Transformation-Based Tectogrammatical
# Dependency Analysis of English

Václav Klimeš

Institute of Formal and Applied Linguistics, Faculty of Mathematics and Physics,
Charles University, Prague, Czech Republic*
klimes@ufal.mff.cuni.cz

**Abstract.** We present experiments with automatic annotation of English texts,
taken from the Penn Treebank, at the dependency-based tectogrammatical layer,
as it is defined in the Prague Dependency Treebank. The proposed analyzer,
which is based on machine-learning techniques, outperforms a tool based on
hand-written rules, which is used for partial tectogrammatical annotation of English now, in the most important characteristics of tectogrammatical annotation.
Moreover, both tools were combined and their combination gives the best results.

## 1 Introduction

Wall Street Journal collection (WSJ) is the largest subpart of the Penn Treebank[1]. It
consists of texts from the Wall Street Journal, its volume is one million words and it is
syntactically annotated using constituent syntax.

The Prague Dependency Treebank (PDT), now in version 2.0 ([1]), is a long-term research project aimed at a complex, linguistically motivated manual annotation of Czech
texts. It is being annotated at three layers: morphological, analytical, and tectogrammatical. The Functional Generative Description theory ([2]) is the main guidance for
principles and rules of annotation of PDT.

In Section 1, we briefly describe the layers and ways of annotation of PDT and WSJ
and the machine-learning toolkit used. The evaluation method used for reporting results
is given in Sect. 2. The algorithm used by our analyzer is described in Sect. 3 and we
propose our results and closing remarks in Sect. 4.

### 1.1 Layers of Annotation of the Prague Dependency Treebank

At the *morphological* layer, shortly *m-layer*, the morphological lexical entry (represented by a *lemma*) and values of morphological categories (a *morphological tag*,
shortly *m-tag*, i. e. the combination of person, number, tense, gender, verbal voice, etc.)
are assigned to each word.

---

* This research was supported by the grant of the Grant Agency of the Czech Republic
No. 405/03/0913 and the Information Society project of the Grant Agency of the Academy
of Sciences of the Czech Republic No. 1ET101470416.

[1] http://www.cis.upenn.edu/~treebank/, LDC Catalog No. LDC99T42.

V. Matoušek and P. Mautner (Eds.): TSD 2007, LNAI 4629, pp. 15–22, 2007.

At the second and third (*analytical and tectogrammatical*) layers of annotation, the sentence is represented as a rooted tree. Edges usually represent the relation of dependency between two nodes: the governor and the dependent. As some edges are of rather technical character (e. g. those adjacent to nodes representing punctuation marks at the analytical layer or those capturing coordination and apposition constructions), we denote the pair of adjacent nodes with general terms "parent" and "child".

Each token (word or punctuation) from the original text becomes a node at the *analytical* (surface-syntactic) layer, shortly *a-layer*, and a label (attribute) called *analytical function*, shortly *afun*, is assigned to each node, describing the type of surface dependency relation of the node to its parent. The original word order position of the corresponding token is also kept as a separate attribute.

The *tectogrammatical* layer ([3]), shortly *t-layer*, captures the deep (underlying) structure of the sentence. Nodes represent only autosemantic words (i. e. words with their own meanings); synsemantic (auxiliary) words and punctuation marks can only affect values of attributes of the autosemantic words they belong to. Nodes may be created for several reasons, e. g. when rules of valency "dictate" to fill ellipses. For the sake of filling ellipses, nodes can also be copied. Relative position of nodes at the t-layer can differ from the position of their counterparts at the a-layer, if they exist. At the t-layer, as many as 39 attributes can be assigned to nodes. In the data, mainly the following attributes are filled:

- *(Deep) functor* captures the tectogrammatical function of a node relative to its parent, i. e. the type of the modification. Special functors are used for child-parent relations that are of a technical nature.
- The t_lemma attribute means "tectogrammatical lemma" and it arises from the (morphological) lemma of the corresponding token or it has a special value. When a node does not originate from the a-layer, the attribute has a special value indicating that it represents a general participant, or an "empty" governing verb predicate and so on.
- For connecting the t-layer with the lower layers, a link to corresponding autosemantic word at the a-layer (if any) is stored in the a/lex.rf attribute, and links to corresponding synsemantic words (if any) are stored in the a/aux.rf attribute.
- In the val_frame.rf attribute, the identifier of the corresponding valency frame, which is kept in a separate valency lexicon, is stored.
- The is_member attribute states whether the node is a member of a coordination.
- The is_generated attribute expresses whether the node is new at the t-layer. When set, it does not necessarily mean that the node has no counterpart at the a-layer, since it can be a "copy" of another t-layer node and more nodes can refer to one a-layer node through their a/lex.rf attribute.

## 1.2 Dependency Annotation of the Penn Treebank

An analyzer ([4]) has been developed for the annotation of PDT-like annotated Czech texts at the t-layer, given the annotation at the m-layer and the a-layer. In order to be able to use it for English constituency-annotated data, we created the clone of the analyzer and we converted and annotated the data by respective sequence of components of *TectoMT* (the used version is from March 2007), an unpublished automatic system developed by Zdeněk Žabokrtský (his early experiments in this field are described in [5]). The system creates annotation layers as similar to the Czech ones as possible; for English, guidelines exist for the tectogrammatical annotation only and are presented in [6].

Each word from an input text gets its (morphological) lemma and tag. The text is processed by a parser, which appoints its phrase structure and the head of each phrase. The heads are then repaired by hand-written rules so that they correspond better to principles used when annotating Czech texts. Using this information, the annotation at the so called *phrase layer*, shortly *p-layer*, is created: it corresponds to the m-layer enriched with phrase structure.

Technical conversion of p-layer phrase structures into the a-layer is performed using the information about phrase heads (roughly speaking, the head of a phrase becomes the parent of the other nodes in the phrase). This a-layer is the same as in PDT, but the values of afuns are not filled. Since our analyzer relies on afuns heavily, we had to approximate them. The best results were obtained with the following approach: if the terminal node is a head of some phrases, the tag of the smallest phrase is used as its afun[2]; otherwise, the m-tag of the terminal is used. Data in this state of annotation are the input of our analyzer.

Partial tectogrammatical annotation is created by—and our analyzer is compared to—another set of components of TectoMT, which applies hand-written rules using information from lower layers: it deletes synsemantic nodes and copies some information from them into the governing autosemantic nodes; it assigns functors on the basis of lemmas, m-tags and function tags; and it adds new nodes in a few cases. Human annotators correct and complete only the tectogrammatical annotation, not the annotation at the lower layers.

## 1.3  fnTBL

For the machine learning part of our tool, we have chosen, mainly for its speed and relatively low memory consumption, the *fnTBL* toolkit ([7]), a fast implementation of the transformation-based learning mechanism. However, the toolkit is aimed at classification tasks only and is not capable of processing tree structures; how we overcame these drawbacks is described at appropriate places.

The rules which the toolkit learns are specified by *rule templates* which have to be designed manually before the learning can start. A rule template is a subset of possible names of features together with the name of a feature that bears information about the class the sample belongs to (since it is possible to perform more classification tasks at once). A rule is an instance of a template: particular values are assigned to all the features. The rule is interpreted in the following way: a sample belongs to the given class if the given features have the given values.

---

[2] If function tags are filled, which is the case only when human-annotated data are used (see Sect. 4), the function tag of the phrase used is appended to the phrase tag.

## 2   Evaluation

Since two tectogrammatical trees constructed over the same sentence do not necessarily contain the same number of nodes, the first step of evaluation must be the *alignment* of nodes: a node from one tree is paired with a node from the other tree and each node is a part of at most one such pair. Only after that step, attributes of the paired nodes can be compared. The evaluation is based on our own alignment procedure, but its description is beyond the scope of this paper.

We define *precision*, $P$, for any attribute assignment to be the number of pairs where both nodes have a matching value of the attribute divided by the total number of nodes in the test annotation; and we define *recall*, $R$, as the number of pairs with the correct value of the attribute divided by the total number of nodes in the correct, human-revised annotation. We also report *F-measure*, $F$: it is, as usual, the equally weighted harmonic mean of precision and recall.

When we want to compare the structure, we have to slightly modify this approach. We define a node to be correctly placed if the node and its parent are aligned and the counterpart of the parent of the node in question is the parent of the counterpart of the node in question.[3] Then the numerator of the fractional counts from the paragraph above is the number of correctly placed nodes.

The a/aux.rf attribute is a set attribute holding unordered links to several nodes of the a-layer and thus it requires special treatment. In order not to complicate the evaluation, we evaluate it *en bloc* and consider it to be correct only if the whole set of links matches its counterpart from the correct annotation.

## 3   The Algorithm

The tectogrammatical attributes set by our tool are divided into two groups. The first group consists of those with values assigned based on training: structure, functor, t_lemma, and val_frame.rf. The remaining attributes— a/aux.rf, a/lex.rf, is_generated, and is_member — are, mainly due to their rather technical than linguistic nature, being set by hand-coded procedures of the tool.

When we want to process the tree structures by the fnTBL toolkit, we have to convert features of a set of nodes (e. g. of children of a node) to fixed-size pieces of information. The biggest disadvantage of this conversion is that the toolkit cannot employ the tree structure in its current state—it remains the same during the whole process of training or classification and cannot be modified until the process is finished.

Not only for this reason, we split the processes of training and classification into several phases. After each phase of processing, a set of trees being processed is modified according to the rules used by this phase and the modified trees then serve as the input for the next phase. After analysing the data carefully, we split the process into four phases; their description and intended aim follows.

1. Deletion of nodes (synsemantic words or majority of punctuation) and assignment of functors to the remaining nodes;

---

[3] Informally, the "same" node has to depend on the "same" parent in both trees.

2. relocation of nodes (mainly rhematizers and phrases having different structure at the a-layer and the t-layer), copying of nodes (filling ellipses), and creation (insertion) of "inner" nodes (again, filling ellipses); assignment of functors of copied or newly created nodes;
3. creation of leaf nodes (missing valency modifications) and assignment of their functors; assignment of the val_frame.rf attribute;
4. assignment of the t_lemma attribute.

### 3.1   Phase 1: Deletion of Nodes and Assignment of Functors

Assignment of functors to nodes is an elementary classification task. It can easily involve the task of deletion of nodes by assigning a special functor value meaning "deleted" as well. However, when a node to be deleted has children, one of them should take its place and become the new parent of its siblings. We call this node *successor*. That is why we enriched the rule templates with another type: deleting the parent of a node together with appointing this node the successor of its parent. Both types of rule templates have to exist, since the second type cannot cause the deletion of leaf nodes.

A node is deleted if a rule of any type states so. If the node has children and its successor is not appointed, it is deleted only if it has the only child (which is not to be deleted as well)—and this child becomes its successor. Otherwise, the node is retained.

For both types of rules we used lemmas, afuns, and m-tags of the node in question, of its parent and of its grandparent. Rule templates can contain four features at maximum.

### 3.2   Phase 2: Relocation, Copying and Creation of Nodes

The transformations this phase should perform are rather complex; therefore, we describe each of them with a single formula as a subtree-to-subtree transformation. A transformation is the smallest possible, i. e. besides nodes being created, deleted, copied or moved, only nodes required in order for the original and resultant structures to be trees are involved in it.

We explain the mechanism on two (real) transformations. The first one is A(B,C,D)->D(B,C)+ADVP-TMP+VP|D. Each node is denoted with a letter and its children follow it in parentheses. Siblings on the left side of a transcription follow their surface order. The whole transformation is recorded at its node B. Since these pieces of information may not suffice to fully characterize C and following nodes, their afuns, serving as their identification, are attached (after the last parenthesis). In the example, the nodes C and D have afuns ADVP-TMP and VP. When a node is deleted,[4] its successor node is given as well (after the | sign). In the example, the successor of the node A is the node D. With the second transformation A(B)->*(A,B):LOC, another phenomenon is illustrated: a new node is denoted with an asterisk and its functor is attached (after the colon). A lowercase letter (not present in the examples) denotes a copy of the node with the uppercase variant of the same letter; the functor of such node can have a special value meaning "the same functor as the original node has".

---

[4] This is given implicitly: it occurs on the left side, but not on the right side.

Although it would be the best if fnTBL had access to features of all the nodes involved in a transformation, variable and possibly high number of the nodes needs a compromise solution. After exploring the most frequent transformations occurring in the data, we decided to use features of five groups of nodes: the node in question (B), its parent (A), its children, its left siblings and its right siblings. Attributes from all the layers, namely lemma, m-tag, afun, and functor, are chosen as features; and where there are more nodes in a group, the respective values are merged into one string and considered to be atomic. This way we got 20 feature types. Each template contains at most five features and each group of features is represented by at most two of its features in a template.

### 3.3   Phase 3: Creation of Leaf Nodes and `val_frame.rf` Assignment

Although the `val_frame.rf` attribute contains a link into the valency lexicon to a description of the valency frame of a word, we do not use this attribute for the assignment of valency modifications. The most important reason for not doing so is that the information in the valency lexicon is not complete: there are some very complex rules for transforming valency frames depending on certain properties of the word in question or of its valency members (e. g. passivization), and since they are the same for all valency frames, they are not expressed in the valency lexicon.

If we could determine the valency frame of a word at its occurrence, adding missing obligatory valency slots would be simple. However, we can learn valency frames from a corpus directly only with difficulties, since certain valency member which is sometimes present and sometimes not may indicate, on one hand, the only valency frame with the member being optional, or, on the other hand, two frames: the first one with the member being obligatory, and the second one without it at all. That is why we make decisions about the obligatoriness of a valency member one by one: each functor corresponds to one feature which states whether (or, in case of free modifications, how many times) a member with the functor should occur at an occurrence of a word. From the 65 functors, only 18 were chosen: those occurring at least twice in the training data at nodes that are new at the t-layer and that have no children, i. e. at potential valency members.

The set of features used in templates consists of a lemma (since valency is the property of a word) and m-tag (since part of speech is encoded there, which can partially characterize valency, e. g. a verb usually has its actor; besides, it contains verbal voice, which also affects the valency frame). Every rule template consists of either lemma, or m-tag, or both of them.

If a wrong functor was assigned to a valency member in Phase 1, an extra error can occur by creating a superfluous node with the correct functor. In order to avoid it, we created rule templates for correcting the functors as well. Corrections are recorded at the node being altered and the following information is used in rules: the lemma of the word which valency frame we are interested in of, and the functor, the morphemic realization[5], and the lemma (hopefully useful for the identification of phrases) of the valency member. Each rule template contains either the functor, or the morphemic realization of a valency member; it can also contain the lemma of the valency member, or of its parent, or of both.

---

[5] This is also called "subcategorization information".

The *morphemic realization* contains roughly the same information as in the traditional description of valency frames: lemmas of all nouns, prepositions, subordinating conjunctions and possessive endings being auxiliary words; and part of speech of the synsemantic word (distinguishing only verbs, nouns, pronouns, adjectives, adverbs, and numerals).

When applying rules, functors are repaired first, then the number of valency members is determined, those missing in a tree are created and their functors are set. When there is an extra node, it is not deleted.

### 3.4   Phase 4: Assignment of the `t_lemma` Attribute

For the assignment of the `t_lemma` attribute, different pieces of information are important depending on whether a node originates from the a-layer or not. In order to be able to design separate sets of templates for both cases, we divided the task into two classifications; only one of them is performed for each node.

For a node originating from the a-layer, there are three templates: the first one considers only its lemma. The second one considers lemmas of synsemantic words belonging to the node in question as well (this is useful in case of verbs with particles). The third one considers the m-tag of the node only.

For a node not originating from the a-layer, there are two templates. Both of them contain the functor of the node; one of them contains the m-tag of the parent of the node as well.

## 4   Results and Closing Remarks

The table below summarizes the F-measure of results of our experiments. We present results for both machine-annotated data and for data adopted from WSJ, i. e. those with human-annotated m-tags, phrase structure and function tags; for these data, the respective steps of annotation described in Subsect. 1.2 were left out. In all cases, the same test data were used and they consisted of 7,605 tokens in 318 sentences. The set used for training of our tool consisted of 50,362 tokens in 2,013 sentences.

First, figures for baseline annotations are presented. The baselines were created from lower layers of an annotation in the following way. Structure, `t_lemma` (identical to morphological lemma), and `is_member` were adopted from the a-layer. Links to the a-layer and the `is_generated` attribute were then set in an obvious way. All functors were set to RSTR, their most common value, and the `val_frame.rf` attribute remained empty.

Then, we present results for TectoMT and our analyzer. Moreover, these tools were combined in the following way: the t-layer annotation created by TectoMT was the input of our tool instead of the a-layer annotation. The rule templates in Phase 1 were altered to take advantage of the pre-annotation: the functors of used nodes and morphemic realization of the node in question were added.

Our tool clearly outperforms TectoMT in structure and functors, which are the most important characteristics of tectogrammatical annotation. It turned out that the tool originally aimed at Czech can analyze English after relatively small changes as well.

| attribute | machine-annotated | | | | human-annotated | | | |
|---|---|---|---|---|---|---|---|---|
| | base | TMT | our | cmbn | base | TMT | our | cmbn |
| structure | 53.0% | 78.6% | **80.3%** | 80.1% | 55.0% | 82.1% | 84.1% | **85.0%** |
| functor | 18.3% | 57.0% | 65.6% | **70.3%** | 17.8% | 57.7% | 71.6% | **71.9%** |
| val_frame.rf | 66.8% | 78.7% | **84.6%** | 84.5% | 66.1% | 80.0% | 85.2% | **86.1%** |
| t_lemma | 65.4% | 91.2% | **90.3%** | **90.3%** | 64.0% | **93.6%** | 92.2% | 93.4% |
| a/lex.rf | 74.2% | 92.8% | **93.3%** | 93.1% | 71.8% | 94.0% | 94.6% | **95.2%** |
| a/aux.rf | 47.0% | 66.0% | **68.2%** | **68.2%** | 45.2% | **93.0%** | 82.2% | 92.7% |
| is_member | 73.3% | 87.1% | **88.8%** | 88.6% | 73.0% | 88.0% | 90.5% | **91.0%** |
| is_generated | 74.3% | 92.6% | 92.1% | **92.7%** | 71.9% | 93.8% | 93.3% | **95.1%** |

**Fig. 1.** The F-measure of results of baseline annotation (*base*), TectoMT (*TMT*), our analyzer (*our*) and the combination of both tools (*cmbn*); the best results in each category are typed bold

Moreover, we can expect an improvement of the results of our analyzer when more training data are available. However, the combination of both tools is what gives the best results.

# References

1. Hajič, J., et al.: Prague Dependency Treebank 2.0, Linguistic Data Consortium, Philadelphia LDC Catalog No. LDC2006T01 (2006), http://ufal.mff.cuni.cz/pdt2.0/
2. Sgall, P., Hajičová, E., Panevová, J.: The Meaning of a Sentence in Its Semantic and Pragmatic Aspects. Academia – Kluwer, Praha – Amsterdam (1986)
3. Hajičová, E., Panevová, J., Sgall, P.: A Manual for Tectogrammatic Tagging of the Prague Dependency Treebank. ÚFAL/CKL Technical Report TR-2000-09. Charles University, Prague (2000)
4. Klimeš, V.: Transformation-Based Tectogrammatical Analysis of Czech. In: Sojka, P., Kopeček, I., Pala, K. (eds.) Proceedings of Text, Speech and Dialogue 2006, Springer, Heidelberg (2006)
5. Kučerová, I., Žabokrtský, Z.: Transforming Penn Treebank Phrase Trees into (Praguian) Tectogrammatical Dependency Trees, Prague Bulletin of Mathematical Linguistic, Prague, vol. 78, pp. 77–94 (2002)
6. Cinková, S., et al.: Annotation of English on the Tectogrammatical Level. Technical Report No. TR-2006-35. ÚFAL MFF, Charles University, Prague (2006)
7. Ngai, G., Florian, R.: Transformation-Based Learning in the Fast Lane. In: Proceedings of NAACL 2001, Pittsburgh, PA, pp. 40–47 (2001)

# Multilingual Name Disambiguation
# with Semantic Information*

Zornitsa Kozareva, Sonia Vázquez, and Andrés Montoyo

Departamento de Lenguajes y Sistemas Informáticos
Universidad de Alicante
{zkozareva,svazquez,montoyo}@dlsi.ua.es

**Abstract.** This paper studies the problem of name ambiguity which concerns the discovery of the different underlying meanings behind a name. We have developed a semantic approach on the basis of which a graph-based clustering algorithm determines the sets of the semantically related sentences that talk about the same name. Our approach is evaluated with the Bulgarian, Romanian, Spanish and English languages for various couples of city, country, person and organization names. The yielded results significantly outperform a majority based classifier and are compared to a bigram co-occurrence approach.

## 1 Introduction and Related Work

The phenomenon of building multiple alternative linguistic structures for a single input is called ambiguity [1]. This problem forms part of many Natural Language Processing (NLP) tasks such as part-of-speech (POS) tagging where one and the same word "water" can refer to the noun or the verb form, or word sense disambiguation where the meaning of the word "plant" depends on the context in which it is utilized, or syntactic analysis where a sentence can have more than one parsing tree.

Presently, researchers have developed various approaches for the resolution of these NLP tasks, however there are few approximations for the resolution of name ambiguity. This problem forms part of search engines, information retrieval and question answering systems. The difficulty of the task depends on the number of underlying names behind an ambiguous name. For instance, a search about "Michael Hammond" leads to several different individuals who are all professors at the University of Arizona, Warwick, Toronto or Southampton. The question is which one of these referents we are actually looking for and interested in. The resolution of name ambiguity can help us identify the relevant name and also improve the performance of the search engine.

Early work in the field of name disambiguation is that of [2] who proposed a cross-document coreference resolution algorithm which uses vector space model to resolve the ambiguities between people sharing the same name. The approach is evaluated on 35 different mentions of John Smith and reaches 85% f-score. Another approach for the discovery of the underlying meanings of a name is that of [3]. They collected information from the web such as age, date of birth, family relationships, associations

---

* This research has been funded by QALLME number FP6 IST-033860 and TEX-MESS number TIN2006-15265-C06-01.

V. Matoušek and P. Mautner (Eds.): TSD 2007, LNAI 4629, pp. 23–30, 2007.
© Springer-Verlag Berlin Heidelberg 2007

with other entities in order to create feature vectors characterizing the names. Then this information is used by an agglomerative clustering algorithm to separate the different individuals according to their category.

[4] collect sentences with polysemous words, mingle them together and later identify the name which refers to the sentence. They developed a bigram co-occurrence approach that captures indirect relations such as *workplace* and *ergonomics* from the *ergonomics science* and *workplace science* co-occurrences. The second order vectors are used to cluster the sentences and thus disambiguate the conflated names. Their approach is initially evaluated in English, but later on [5] proved to obtain similar performance with the Bulgarian, Romanian and Spanish languages. Apart from the static corpus, [6] explored the Web as corpus for person name disambiguation. They observed the effect of the training data feature selection versus the data in which the ambiguous names occurred, as well as the impact of different association measures for the identification of lexical features. The performance of the method ranges from 66 to 95% depending on the ambiguous name pairs.

Considering the pros and cons of the previously developed name disambiguation approaches, we decided to use semantic information and to group together the sentences that have strong semantic similarity. Our hypothesis is that sentences talking about the same fact or event are highly probable to refer to the same name. To our knowledge, this is the first approach which takes advantage of exploiting Latent Semantic Analysis for the resolution of the name disambiguation problem. An advantage of our approach is that it does not consider any manually annotated data, dictionary or biographic information, but rather a set of text snippets containing the ambiguous names. Moreover, given the ample presence of the freely available texts, it is easy to evaluate the effectiveness of our approach and to prove its multilingual issue by conducting series of exhaustive evaluations with the Bulgarian, Spanish, Romanian and English languages.

## 2   Name Disambiguation with Semantic Information

In the core of our approach lays Latent Semantic Analysis (LSA)[1] [7] which is a knowledge induction and representation technique, where the similarity of the meaning of the words or the passages is established on the basis of the analysis of large text corpora. LSA estimates the similarity on the basis of mathematical analysis from where the deeper "latent" semantic relations between the words are inferred. Such method is considered better than the simple frequency, co-occurrence counts or correlations, because it can relate sentences that express the same meaning expressed by different words such as synonyms, antonyms among others.

The first step in LSA is to represent explicitly terms and documents in a rich, high dimensional space, allowing the underlying "latent", semantic relationships between terms and documents to be exploited. LSA relies on the constituent terms of a document to suggest the document's semantic content. However, the LSA model views the terms in a document as somewhat unreliable indicators of the concepts contained in the document. It assumes that the variability of word choice partially obscures the semantic structure of the document.

---

[1] http://infomap-nlp.sourceforge.net/

The second step in LSA is to reduce the dimension of the term-by-document space, so that the underlying, semantic relationships between the documents are revealed, and much of the "noise" (differences in word usage, terms that do not help distinguish documents, etc.) is eliminated. LSA statistically analyzes the patterns of word usage across the entire document collection, placing documents with similar word usage patterns near to each other in the term-by-document space, and allowing semantically-related documents to be closer even though they may not share the same terms.

Taking into consideration these properties of LSA, we thought that instead of constructing the traditional term-by-document matrix, we can construct a term-by-sentence matrix with which we can find a set of sentences that are semantically related and talk about the same name. The rows of the term-by-sentence matrix stand for a unique word from the sentence in which the name has to be disambiguated (the target sentence), while the columns stand for the rest of the sentences. A cell in the term-by-sentence matrix represents the number of times a given word of its row from the target sentence co-occurs in a given sentence denoted by its column. When two columns of the term-by-sentence matrix are similar, this means that the two sentences contain similar words and are therefore very likely to be semantically related. When two rows are similar, then the corresponding words occur in most of the same sentences and are likely to be semantically related.

The cells in the matrix are weighted by a function which expresses the importance of the word in the sentence and the informativeness of the word in the sentence. After the term weighting is performed, Singular Value Decomposition (SVD) is applied in order to find correlations among the rows and columns. SVD decomposes the original term-by-sentence matrix $T$ into the product of three other matrices $T=U\Sigma_k V^T$, where $\Sigma_k$ is the diagonal $k \times k$ matrix containing the $k$ singular values of $T$, $\sigma_1 \geq \sigma_2 \ldots \sigma_k$, and $U$ and $V$ are the column-orthogonal matrices. When the three matrices are multiplied together, the original term-by-sentence matrix is re-composed. We choose $k' \ll k$ obtaining the approximation $T \simeq U\Sigma_{k'} V^T$. More specifically, in the experiments of this paper, we used the matrix $T' = U\Sigma_{k'}$, whose rows represent the term vectors in the reduced space using only 300 dimensions (i.e. $k' = 300$). According to the study of [8] this is the best dimension reduction.

The sentence similarity in the resulting vector space can be measured in various ways. In our approximation, the semantic similarity values among the target sentences and the rest of the sentences are used for the construction of a new similarity-based rectangular matrix $S$. This matrix is processed by a Pole-Based Overlapping Clustering (PoBOC) [9] algorithm where each object which in our case is a sentence with ambiguous name, is being separated into a meaningful cluster. PoBOC finds so called *poles* in which the *monosemous* names are placed and thus the name disambiguation process terminates. The performance of the name disambiguation approach is evaluated in terms of precision, recall, f-score and accuracy.

## 3   Data Description

The main focus of our study is to disambiguate person, organization and location names discovering the underlying meanings behind a name. In order to facilitate the evaluation of our approach, we adapted the experimental setting of [5], where the name ambiguity in the data is caused by the conflation of largely unambiguous names. For instance, [5] take all of the sentences that talk about the president George Bush and the president Bill Clinton, and convert them into an ambiguous pair by replacing the name occurrences with the George Bush-Bill Clinton label. According to the study of [10], the pseudo-word pairs which are created from individually unambiguous words are still related in some way. Another motivation for the creation of the pseudo-name pairs is related to the evaluation and the lack of hand-annotated name disambiguation data, so the pseudo-names are ambiguous to our method, but during the evaluation we already know their true identity.

The name disambiguation is performed with the freely available corpus of [5][2]. We choose to work with this corpus, because it can provide us with a relevant feedback and we can conduct a comparative study to already developed approaches. Moreover, the data is created for four different languages: Bulgarian, Romanian, English and Spanish which is ideal for our multilingual study. The NE examples are extracted from large news corpora where different unambiguous names are conflated. The selected names and their distribution are shown in Table 1.

## 4    Experimental Evaluation and Comparative Study

### 4.1    Experimental Setup

We have performed two experiments for our multilingual name disambiguation approach in order to observe whether the information provided by a group of semantically related sentences provides more relevant judgement for the name disambiguation process compared to the evidence of the most semantically similar sentence.

In the first experiment, the matrix for the clustering algorithm is built from the semantic similarity values given by LSA where the cells represent the semantic relatedness scores of the target sentence and the rest of the sentences. This experiment is denoted as a set of semantically similar sentences (SSS).

In the second experiment, the matrix is built only from the highest semantic similarity value which LSA returns for the most similar sentence and the target one. The rest of the sentences obtain the value 0, indicating no similarity at all. This experiment is denoted as a set of one sentence (SOS).

### 4.2    Results and Discussion

Table 1 shows the obtained results in terms of accuracy for the two experiments we have carried out. These results are compared to the bigram co-occurrence approach of [5] and to a baseline classifier which corresponds to the majority sense of the conflated name pair. For instance, one of the conflated name pairs for the Bulgarian language is France, Germany and Russia. The number of examples for these names are 1726, 2095 and 2645 respectively. According to this distribution, Russia determines the majority sense for the Fr-Ge-Ru conflated pair. The baseline for this ambiguous pair is obtained

---

[2] http://www.d.umn.edu/ tpederse/Data/cicling2006-data.zip

**Table 1.** Results for the multilingual name disambiguation

| Name | Distribution | Maj | SSS | SOS | Diff | Ted | Diff | Comp |
|---|---|---|---|---|---|---|---|---|
| **Bulgarian** | | | | | | | | |
| PS-IK-GP | 318+524+811=1653 | 49.06% | **44.40%** | 41.92% | -4.66 | 58.68% | +9.62 | -14.28 |
| NM-NV-SS | 645+849+976=2470 | 39.51% | **40.04%** | 40.00% | **+0.53** | 59.39% | +19.88 | -19.35 |
| BSP-SDS | 2921+4680=7601 | 61.57% | **67.55%** | 64.72% | **+5.98** | 57.31% | -4.26 | +10.24 |
| Fr-Ge-Ru | 1726+2095+2645=6466 | 40.91% | **38.40%** | 37.58% | -2.51 | 41.60% | +0.69 | -3.2 |
| Va-Bu | 1240+1261=2501 | 50.42% | 56.93% | **57.82%** | **+7.4** | 66.09% | +15.71 | -8.27 |
| **English** | | | | | | | | |
| BC-TB | 1900+1900=3800 | 50.00% | **78.79%** | 73.39% | **+28.79** | 80.95% | +30.95 | -2.16 |
| BC-TB-EB | 1900+1900+1900=5700 | 33.33% | **49.21%** | 48.31% | **+15.88** | 47.93% | +14.60 | +1.28 |
| IBM-Mi | 2406+3401=5807 | 58.57% | 82.86% | **86.21%** | **+27.64** | 63.70% | +5.13 | +22.51 |
| Me-Ug | 1256+1256=2512 | 50.00% | **59.00%** | 58.76% | **+9** | 59.16% | +9.16 | -0.16 |
| Me-In-Ca-Pe | 1500+1500+1500+1500=6000 | 25.00% | **41.57%** | 41.26% | **+16.57** | 28.78% | +3.78 | +12.79 |
| **Romanian** | | | | | | | | |
| TB-AN | 1804+1932=3736 | 51.34% | **62.37%** | 61.40% | **+11.03** | 51.34% | +0.00 | +11.03 |
| TB-II-AN | 1948+1966+2301=6215 | 37.02% | 45.50% | **47.64%** | **+10.62** | 39.31% | +2.29 | +8.33 |
| PD-PSD | 3264+2037=5301 | 61.57% | **57.10%** | 56.21% | -4.47 | 77.70% | +16.13 | -20.6 |
| Br-Bu | 2559+2310=4869 | 52.56% | **66.33%** | 66.19% | **+13.77** | 63.67% | +11.11 | +2.66 |
| Fr-SUA-Ro | 3890+1370+2396=7656 | 50.81% | **53.16%** | 51.08% | **+2.35** | 52.66% | +1.85 | +0.5 |
| **Spanish** | | | | | | | | |
| YA-BC | 1004+2340=3344 | 69.98% | **85.68%** | 84.03% | **+15.7** | 77.72% | +7.74 | +7.96 |
| JP-BY | 1447+1450=2897 | 50.05% | **93.82%** | 92.30% | **+43.77** | 87.75% | +37.70 | +6.07 |
| OTAN-EZLN | 1093+1093=2186 | 50.00% | **93.37%** | 91.72% | **+43.37** | 69.81% | +19.81 | +23.56 |
| NY-WA | 1517+2418=3935 | 61.45% | 52.58% | **53.01%** | -8.44 | 54.69% | -6.76 | -1.68 |
| NY-Br-Wa | 1517+1748+2418=5683 | 42.55% | **35.02%** | 34.81% | -7.53 | 42.88% | +0.33 | -7.86 |

by the normalization of the number of occurrences of Russia over the total number of examples in the conflated pair. The 40.91% baseline indicates the disambiguation power of a system whose answer is always Russia.

According to the yielded results, the two clustering methods, outperform the majority baseline classifier from 0.5 to 43%. In general the SSS clustering performs better than the SOS. This demonstrates that more contextual evidence for the conflated names brings better representation and more precise judgement while discovering the underlying meanings of a name. However, it is interesting to note that the performance of the conflated names whose sentences appear in very similar contexts such as the two Bulgarian cities Varna and Burgas (Va-Bu), the IBM and Microsoft (IBM-Mi) companies for English, the persons Trian Basescu, Ion Iliescu and Adrian Nastase (TB-II-AN) for Romanian and the cities New York and Washington (NY-WA) for Spanish, obtained better results with the SOS method. This is explainable, because when the sentences referring to different names appear in similar contexts, their feature vectors after the SVD process lie on the same line or are very close to each other. For this reason the similarity density around the sentences confuses the clustering algorithm. Therefore, when only the most semantically similar sentence is considered, it eliminates the semantic dispersion introduced by the rest of the sentences.

During the error analysis we found out that for the Bulgarian language, the person names PS-IK-GP and NM-NV-SS are not classified very well. Among all languages, Bulgarian obtains the lowest performance. This is due to the lack of a POS tagger, which we apply a-priory to the texts that LSA has to classify. In addition, we use the POS information not only to remove stop words, but also to encode the syntactic categories of the words. It is known that LSA treats the words as tokens and it is not aware of their grammatical categories. However, we found a way to encode the syntactic categories

**Table 2.** Results for the individual name disambiguation

| Bulg | Prec | Rec | Acc | F | Rom | Prec | Rec | Acc | F |
|------|------|-----|-----|---|-----|------|-----|-----|---|
| PS | 24.27 | 7.86 | 12.88 | 11.87 | TB | 49.08 | 44.25 | 28.46 | 46.54 |
| IK | 36.22 | 35.87 | 42.89 | 36.04 | II | 43.25 | 46.13 | 31.63 | 44.64 |
| GP | 51.43 | 64.24 | 42.89 | *57.12* | AN | 45.15 | 46.02 | 31.63 | 45.58 |
| NM | 37.74 | 30.07 | 22.79 | 33.47 | Fr | 38.83 | 8.75 | 41.66 | 14.29 |
| NV | 39.33 | 43.46 | 22.79 | 41.29 | SUA | 48.08 | 36.72 | 13.06 | 41.64 |
| SS | 43.03 | 43.64 | 25.10 | *43.33* | Ro | 56.12 | 78.92 | 41.66 | 65.59 |
| BSP | 59.51 | 48.75 | 67.55 | 53.59 | TB | 61.53 | 60.47 | 62.36 | *61.00* |
| SDS | 71.25 | 79.29 | 67.55 | *75.06* | AN | 63.47 | 64.13 | 62.36 | *63.80* |
| Va | 56.12 | 60.96 | 56.93 | *58.44* | PD | 49.01 | 35.49 | 57.10 | 41.17 |
| Bu | 58.13 | 52.97 | 56.93 | *55.43* | PSD | 63.91 | 70.58 | 57.10 | 67.08 |
| Fr | 30.75 | 23.00 | 28.13 | 26.31 | Br | 66.47 | 58.61 | 66.33 | 62.29 |
| Ge | 35.56 | 31.69 | 32.26 | 33.51 | Bu | 66.85 | 73.30 | 66.33 | 69.93 |
| Ru | 43.26 | 53.76 | 28.13 | *47.94* | | | | | |
| Eng | Prec | Rec | Acc | F | Spa | Prec | Rec | Acc | F |
| BC | 43.46 | 43.42 | 31.89 | 43.44 | NY | 26.60 | 24.52 | 15.78 | 25.52 |
| TB | 49.65 | 49.26 | 33.84 | 49.45 | Br | 31.76 | 30.03 | 15.78 | 30.87 |
| EB | 51.88 | 52.26 | 31.89 | 52.07 | Wa | 43.09 | 45.20 | 28.47 | *44.12* |
| BC | 72.28 | 75.89 | 73.39 | *74.04* | YA | 84.58 | 63.94 | 85.67 | 72.83 |
| TB | 74.66 | 70.89 | 73.39 | *72.73* | BC | 85.99 | 95.00 | 85.67 | *90.27* |
| IBM | 82.35 | 74.64 | 82.86 | 78.30 | JP | 96.14 | 91.29 | 93.82 | *93.65* |
| Mi | 83.17 | 88.67 | 82.86 | *85.84* | BY | 91.72 | 96.34 | 93.82 | *93.97* |
| Me | 59.30 | 57.08 | 58.75 | *58.17* | OTAN | 93.56 | 93.13 | 93.36 | *93.35* |
| Ug | 60.04 | 60.42 | 58.75 | *60.23* | EZLN | 93.16 | 93.59 | 93.36 | *93.38* |
| Me | 34.06 | 38.06 | 19.80 | 35.95 | NY | 40.12 | 36.98 | 53.01 | 38.49 |
| In | 43.25 | 43.20 | 21.08 | 43.22 | WA | 62.06 | 63.06 | 53.01 | *62.56* |
| Ca | 37.58 | 42.66 | 20.95 | 39.96 | | | | | |
| Pe | 55.73 | 41.13 | 19.80 | 47.33 | | | | | |

of the words through corpus transformation. For instance, the sentence "pass me the water because I want to water the flowers" is transformed into "pass#v water#n I#pron want#v water#v flowers#n" and LSA has two distinct instances of water: the noun and the verb. Obviously this transformation changes LSA's space and this is proven with the low performance obtained for the Bulgarian language.

In addition to these observation, we noticed that the cities Varna and Burgas form part of weak named entities such as the University of Varna, the University of Burgas, the Major house of Varna or the Sport Hall of Burgas. Although the strong entities are embedded into the weak ones, practically Varna and Burgas change their semantic category from city to university or major house. This creates additional ambiguity in our already ambiguous names. In order to improve the performance of our method, we need a better data generation process where the mixture of weak and strong entities will be avoided.

The best performance of our name disambiguation approach is obtained for the person and organization names in the Spanish language. The performance reaches 93% which is with 43% better than the majority baseline and with 10% better than the bigram co-occurrence approach of [5]. Our method has an increment of 6% for the John Pail and Boris Yeltsin pseudo-name pair, and 20% increment for the OTAN-EZLN. According to the $z'$ statistics with confidence level of 0.975, the obtained differences between our approach and the rest of the approaches are statistically significant.

It is interesting to note that our approach achieves good results not only for the disambiguation of two names but also for three or four names. Our observation is supported by the 41% correct name disambiguation for the Mexico, India, California and Peru pair, which is with 17% better than the baseline and with 13% better than the co-occurrence approach.

Apart from the performance of the general name disambiguation, we show in Table 2 the individual performances of the names. The general observation is that a name which has many examples, performs significantly better compared to the rest of the names. This is related to the semantic evidence which LSA gathers for the conflated names.

This event is observed in four of the five conflated name pairs for the Bulgarian language and in three of the five conflated name pairs for Spanish. When the number of examples in the conflated pair are balanced, the names are equally disambiguated. We denote these results in Table 2 in bold and italic, as in the case of Me-Ug, where Mexico reaches 58% and Uruguay reaches 60%. The only language where we observe some variation is Romanian, however we are not familiar with the structure of this language and its properties, therefore we cannot provide a logical explanation about the obtained results.

## 5   Conclusions

This paper presents a multilingual name disambiguation approach which uses text semantic information. We have conducted two experimental setups for the clustering algorithm in order to study whether a set of similarity sentences or the most similar sentences performs better during the name disambiguation process. According to the obtained results for the four languages, SSS achieved better results and provides more information during the disambiguation. However, when two sentences have similar contexts and belong to different names, SOS disambiguates the names much better. According to the obtained results, our approach yields higher results than the majority baseline and bigram co-occurrence classifiers.

During the exhaustive evaluation study, we found out that the presence of a POS tagger is very important for the better functioning of LSA. In addition, the general classification errors are due to the low quality data and the double ambiguity created from the embedding of the strong entities into the weak ones.

In the future, we want to expand our approach to cross-lingual name disambiguation, to incorporate information from the web [6] as well as to establish relations among names [8].

## References

1. Jurafski, D., Martin, J.: Speech and Language Processing: An Introduction to Natural Language Processing, Speech Recognition, and Computational Linguistics. Prentice-Hal, Englewood Cliffs (2000)
2. Bagga, A., Baldwin, B.: Entity-based cross-document coreferencing using the vector space model. In: Proceedings of the Thirty-Sixth Annual Meeting of the ACL and Seventeenth International Conference on Computational Linguistics, pp. 79–85 (1998)

3. Mann, G.S., Yarowsky, D.: Unsupervised personal name disambiguation. In: Proceedings of the seventh conference on Natural language learning at HLT-NAACL 2003, pp. 33–40 (2003)
4. Kulkarni, A.: Unsupervised discrimination and labeling of ambiguous names. In: Proceedings of 43rd Annual Meeting of the Association for Computational Linguistics (2005)
5. Pedersen, T., Kulkarni, A., Angheluta, R., Kozareva, Z., Solorio, T.: An unsupervised language independent method of name discrimination using second order co-occurrence features. In: Gelbukh, A. (ed.) CICLing 2006. LNCS, vol. 3878, pp. 208–222. Springer, Heidelberg (2006)
6. Pedersen, T., Kulkarni, A.: Unsupervised discrimination of person names in web contexts. In: Proceedings of the Eighth International Conference on Intelligent Text Processing and Computational Linguistics (2007)
7. Foltz, P.W.: Using latent semantic indexing for information filtering. In: Proceedings of the ACM SIGOIS and IEEE CS TC-OA conference on Office information systems, pp. 40–47 (1990)
8. Turney, P.D.: Human-level performance on word analogy questions by latent relational analysis. Technical report, Institute for Information Technology, National Research Council of Canada (2004)
9. Cleuziou, G., Martin, L., Vrain, C.: Poboc: An overlapping clustering algorithm, application to rule-based classification and textual data. In: ECAI, pp. 440–444 (2004)
10. Nakov, P., Hearst, M.: Category-based pseudowords. In: NAACL '03: Proceedings of the 2003 Conference of the North American Chapter of the Association for Computational Linguistics on Human Language Technology, pp. 67–69 (2003)

# Inducing Classes of Terms from Text[*]

Pablo Gamallo, Gabriel P. Lopes, and Alexandre Agustini

[1] Faculdade de Filologia, Universidade de Santiago de Compostela, Spain
pablogam@usc.es
[2] Faculdade de Ciências e Tecnologia, Universidade Nova de Lisboa, Portgual
gpl@di.fct.unl.pt
[3] Departamento de Informática, PUCRS, Brazil
agustini@inf.pucrs.br

**Abstract.** This paper describes a clustering method for organizing in semantic classes a list of terms. The experiments were made using a *POS* annotated corpus, the ACL Anthology, which consists of technical articles in the field of Computational Linguistics. The method, mainly based on some assumptions of *Formal Concept Analysis*, consists in building bi-dimensional clusters of both terms and their lexico-syntactic contexts. Each generated cluster is defined as a semantic class with a set of terms describing the extension of the class and a set of contexts perceived as the intensional attributes (or properties) valid for all the terms in the extension. The clustering process relies on two restrictive operations: *abstraction* and *specification*. The result is a concept lattice that describes a domain-specific ontology of terms.

## 1 Introduction

This paper describes a method for clustering terms into semantic classes using as input a domain-specific corpus and a preliminary list of terms in the same domain. The corpus is the ACL Anthology, which consists of technical articles published by journals and conferences on Computational Linguistics. The method consists in building bi-dimensional clusters of both terms and their properties. Each cluster is the result of either merging or unified their constituents (i.e., terms and properties). The properties of a cluster/class are represented by those lexico-syntactic contexts co-occurring in the corpus with all terms of the class.

The basic intuition underlying our approach is that similar classes of terms can be aggregated to generate either more specific or more generic classes, without inducing odd associations between terms and their properties/contexts. A new *specific* class is generated when the properties of the constituent classes are merged (intension expansion), while the terms are intersected (intension reduction). A new *generic* class is generated when the properties are intersected (intension reduction), while the constituent terms are merged (extension expansion). Intersecting either terms or properties allows us to generate tight clusters with representative and prototypical constituents. These tight clusters can be perceived as centroids to classify both new terms and properties. The theoretical background our work is mainly based on is *Formal Concept Analysis*

---

[*] This work has been supported by the Spanish Government, within the project GaricoTer.

V. Matoušek and P. Mautner (Eds.): TSD 2007, LNAI 4629, pp. 31–38, 2007.

(FCA) [1,2]. The clusters we acquired have all the features of "formal concepts" in FCA. Figure 1 shows a class consisting of a set of terms and a set of properties learnt by our system. The cluster represents a formal concept with a term extension ("Natural Language Processing", "Speech Processing", etc) and a descriptive intension ("research in", "area of", etc). The clustering algorithm only selects those properties that can co-occur with all terms in the extensional set. Each crossing line in the figure represents the binary relation "co-occurs with" between a property (or local context) and a term of the class.

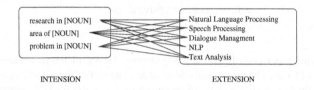

INTENSION                                          EXTENSION

**Fig. 1.** An example of bi-dimensional cluster generated by our method

Let's note that our algorithm learnt the main areas in NLP are Text, Speech, and Dialogue. This is in accordance with the *TSD* conference name.

This article is organised as follows. Section 2 introduces some related work. Then sections 3 and 4 describes two complementary clustering methods: by abstraction and by specification. Finally, in section 5, experiments, results, and an evaluation protocol are discussed.


## 2   Related Work

In order to induce semantically homogeneous clusters of words (tokens, types or lemmas), some approaches compare the semantic similarity between $< word, context >$ pairs and sets of those pairs. These sets are perceived as semantic classes, also called "selection types" [3,4]. Given two vocabularies, $W$ and $LC$, which represent respectively the set of words and the set of local contexts, a semantic class is defined as a pair $< LC', W' >$, where $LC' \subseteq LC$ and $W' \subseteq W$. In this model, the same word or context can in principle belong to more than one class. So, the positive side of these approaches is that they try to take into account polysemy. Some difficulties arise, however, in the process of class generation. Those approaches propose a clustering algorithm in which each class is represented by the centroid distributions of all of its members. This is in conflict with the fact that many words and local contexts can significatively involve more than one semantic dimension. As a result, the clustering method appears to be too greedy since it overgenerates many wrong associations between words and local contexts.

To avoid this problem, a more recent approach tried to limit the information contained in the centroids by introducing a process of "clustering by committee" [5]. The centroid of a cluster is constructed by taking into account only a subset of the cluster members. This subset, called "committee", contains the more representative members (prototypes) of a class. So, the main and more difficult task of such an approach is to first identify a list of committees, i.e., a list of semantically homogeneous clusters. Committees represent basic semantic classes of similar words and are useful for word classification.

Other approaches also try to identify homogenous clusters representing basic semantic classes. The main difference with regard to the former methods is that each basic cluster is constituted, not by similar words, but by a set of similar local contexts [6,7,8,9]. The method is focused on computing the semantic similarity between local contexts. Words are no more seen as objects to be clustered but as features of contexts. These are taken as the objects of the clustering process. As local contexts turn out to be less polysemic than words, it is assumed that searching for classes of homogeneous contexts is an easier task than to find tight classes of semantically related words. The main drawback, however, is that local contexts are less frequent than words and, then, they are more sparse.

The method proposed in this paper considers that clusters are bi-dimensional objects consisting of both words (or terms) and contexts (or properties). Our main contribution is the use of very restrictive operations (*specification* and *abstraction*) in the process of building tight clusters. Using these operations, we aim at solving the overgeneration problem. In the next section, we will describe a clustering algorithm that makes use of the abstraction operation.

## 3   Clustering by Abstraction

It consists in building generic classes from very specific clusters.

### 3.1   The Input: Specific Classes

The input of this clustering process is a list of very *specific* classes of terms, i.e., each class consists of a small set of terms (small extension) and a large set of properties/contexts (big intension). To build these specific classes, three tasks are required: first, the training corpus is annotated to extract co-occurrences between word (or multiword) lemmas and their lexico-syntactic contexts. Second, a list of terms which are relevant in the domain is selected. And third, we compute similarity between the terms of this list and all word (or multi-word) lemmas occurring in the corpus.

**Corpus Processing.**  The corpus is first *POS* tagged and then binary dependencies are extracted using pattern matching techniques (articles and pronouns are previously removed). From each binary dependency, two complementary lexico-syntactic contexts are selected. For instance, given a binary dependency: "entry-in-thesaurus", two contexts/properties are selected: < entry in [NOUN] > and < [NOUN] in thesaurus >. We follow the notion of *co-requirement* introduced in [9].

Finally, each lexico-syntactic context is associated to its co-occurring word lemmas to build a collocation database. Each word lemma can be viewed as a vector and each lexico-syntactic context correspond to a feature. Before starting the clustering process, sparse contexts are removed from the vectors. A context is sparse if it has high word dispersion. Dispersion is defined as the number of different word lemmas occurring with a lexico-syntactic context divided by the total number of different word lemmas in the training corpus. So, the vector space is only constituted by those lexico-syntactic contexts whose word dispersion is lower than an empirically set threshold.

**List of Terms.** The startpoint to build the specific classes is a list of terms. This list can be selected by manual intervention using pre-existing glossaries, or by unsupervised analysis of domain-specific documents (terminology extraction). In our experiments, we used as startpoint an existing glossary of terms.

**Similarity.** To build the input classes, we compute the Dice similarity between each term and the rest of word and multi-word lemmas. Similarity between a term, $t$, and a word or multi-word lemma, $w$, which is not in the starting list of term samples, is computed as follows:

$$Dice(t, w) = \frac{2 * \sum_i min(f(t, c_i), f(w, c_i))}{f(t) + f(w)}$$

where $f(t, c_i)$ represents the number of times $t$ co-occurs with the context $c_i$. Likewise, $f(w, c_i)$ represents the number of times $w$ co-occurs with the context $c_i$. For each term, its top-$k$ most similar lemmas, where $k = 5$, are selected. For instance, in our experiments on a corpus consisting of articles published in conferences and journals on Computational Linguistics, the 5 most similar words to the term "thesaurus" are "bilingual lexicon", "bilingual dictionary", "taxonomy", "lexical resource", and "ontology". As it was expected, words that are similar to a given term in a specific domain are also terms in that domain.

Now, a set of specific classes can be generated. Given the top-5 most similar lemmas to a given term, we build 5 very specific classes by aggregating the term to each similar word in the list. The intension of each class consists of those contexts that are shared by the two similar terms. Table 1 shows the five basic classes associated to "thesaurus".

The specific classes built from the previously selected terms are the input of the clustering process.

### 3.2    Generating Intermediate Classes

The first step of the clustering process is to build a set of "intermediate classes" (neither very specific nor very abstract). For this purpose, we use a clustering algorithm inspired by [10]. To explain this algorithm, let's take the specific classes in Table 1. The first basic class, 0110, is considered as the starting centroid. Then, we search for other centroids among the other 4 classes. A class is a centroid if it is not similar to the previously identified centroids. Here, we consider that two objects are similar if they share more than 50% of the properties/contexts. In our example, there is only one centroid more: class 0112. Finally, each one of the remaining classes is aggregated to a centroid if they are similar, i.e. if thy share more than 50% of their contexts. In our example, both 0111 and 0113 are aggregated to 0110, while 0114 is joined to 0112. All aggregations are made using the operator of abstraction since each generated cluster is obtained

**Table 1.** 5 specific classes built from the term "thesaurus"

| class | extension | intension |
|-------|-----------|-----------|
| 0110 | {thesaurus, bilingual_lexicon} | {<access to [N]>, <[N] construction>, <entry in [N]>, <machine-readable [N]>, <compilation of [N]>, <headword of [N]>, <online [N]>, <consult [N]>, ... } |
| 0111 | {thesaurus, bilingual_dictionary} | {<access to [N]>, <term in [N]>, <entry in [N]>, <machine-readable [N]>, <compilation of [N]>, <[N] entry>, <headword of [N]>, <online [N]>, <consult [N]>, <define in [N]>, ...} |
| 0112 | {thesaurus, taxonomy} | {<[N] relation>, <[N] construction>, <concept of [N]>, <relation in [N]>, <node of [N]>, <node in [N]>, <[N] generation>, <[N] concept>, <[N] from dictionary>, <monolingual [N]>, <[N] of domain>, ...} |
| 0113 | {thesaurus, lexical_resource} | {<access to [N]>, <word from [N]>, <entry in [N]>, <machine-readable [N]>, <online [N]>, <consult [N]>, <define in [N]>, <computerized [N]> ...} |
| 0114 | {thesaurus, ontology} | {<[N] relation>, <[N] construction>, <concept of [N]>, <relation in [N]>, <node of [N]>, <class in [N]>, <[N] generation>, <[N] concept>, <[N] from dictionary>, <monolingual [N]>, <[N] of domain>, ...} |

**Table 2.** Intermediate classes built from "thesaurus"

| class | extension | intension |
|-------|-----------|-----------|
| $CL\_015$ | {thesaurus, bilingual lexicon, bilingual dictionary, lexical resource} | {<access to [N]>, <[N] construction>, <entry in [N]>, <machine-readable [N]>, <headword of [N]>, <online [N]>, <consult [N]>, ... } |
| $CL\_123$ | {thesaurus, taxonomy, ontology} | {<[N] relation>, <[N] construction>, <concept of [N]>, <relation in [N]>, <node of [N]>, <[N] generation>, <[N] concept>, <[N] from dictionary>, <[N] of domain>, ...} |

by intersecting the two constituent intensions. Following this algorithm, we obtain two intermediate classes (see Table 2).

Let's note that this clustering algorithm allows us to put the same term in different classes (soft clustering). Terms, even if they are well-defined technical expressions, can be used in a corpus with a high degree of ambiguity. For instance, "thesaurus" is considered in our training corpus either as a repository of entries (lexical resource) or as a concept structure (ontology). Our algorithm found other polysemous terms. For instance, "thematic role" was aggregated with "case slot" into a class characterised by the property <fill [N]>, whereas it was put into another cluster with terms such as "grammatical function" or "semantic relation", where they share the property <assignment of [N]>. In sum, a thematic role can be viewed either as a recipient (slot) to be filled by an entity or as a linguistic function the entity is assigned to.

Furthermore, the generated clusters contain word lemmas that were not in the starting list of terms (e.g., "hierarchy" and "bilingual lexicon"). Indeed, lemmas appearing with terms in a cluster must also be considered as terms. This clustering process is repeated for the remaining input classes associated to the other terms of the list. Intermediate classes are the input of the following clustering step.

### 3.3   Generating Generic Classes by Hierarchical Clustering

A standard hierarchical clustering takes as input the intermediate classes to generate more generic ones. For this purpose, we make use of an open source software: Cluster 3.0[1]. In this step, the clustering conditions are the same: the similarity threshold is still 50%, and classes are aggregated with the operation of abstraction. Table 3 illustrates two generic classes containing "thesaurus" as a member. They are the result of putting together the intermediate classes depicted above in Table 2 with other similar intermediate classes.

**Table 3.** Generic classes built from "thesaurus"

| class | extension | intension |
|---|---|---|
| $NODE\_09$ | {thesaurus, automatic thesaurus, bilingual terminology, terminology, bilingual lexicon, bilingual dictionary, lexical resource} | {<[N] construction>, <entry in [N]> } |
| $NODE\_21$ | {thesaurus, hierarchical structure, hierarchy, taxonomy, ontology, type hierarchy} | {<concept of [N]>, <node of [N]>, <[N] of domain>} |

### 3.4   Classifying Unknown Words

So far, the generated clusters have been loosing relevant information step by step, since they were aggregated using intersecting operations. Besides that, the intersecting aggregations did not allow us to infer context-word associations that were not attested in the training corpus. As has been mentioned above, our objective was to design a very restrictive clustering strategy so as to avoid overgeneralisations.

In order to both reintroduce lost information and learn new context-word associations, the last step aims at assigning unknown words to the generic classes generated by the clustering algorithm. An unknown word is assigned to one or more classes if it satisfies two conditions: 1) it co-occurs with more than 50% of the contexts constituting the class intension, and 2) it is one of the top-10 most similar words to, at least, one of the terms in the class extension. We assume that those words being correctly classified should been considered as terms. So, classification can also be perceived as a kind of automatic term extraction.

---

[1] http://bonsai.ims.u-tokyo.ac.jp/~mdehoon/software/cluster/software.htm

## 4   Clustering by Specification

The only difference with regard to the former strategy is that the collocation database is viewed now as a collection of word vectors. Each unique context corresponds to a vector and each word lemma corresponds to a feature. The input classes are then generic concepts whose intension consists of two similar context/properties (instead of two similar terms). The extension is the set of terms and word lemmas shared by those contexts. For instance, <entry in [N]> was considered to be the most similar context to <[N] entry>. Both contexts represent the intension of a generic class whose extension consists of terms such as "bilingual dictionary", "bilingual lexicon", etc. Further clustering steps will make this generic class more specific (by intersecting the extension and unifying the intension with other similar classes). As a result, a very specific class is generated. The main problem of this algorithm is that lexico-syntactic contexts are more sparse than word lemmas. Classification of unknown contexts was not performed. This task is part of our current research.

## 5   Experiments and Evaluation

Experiments have been carried out over a large corpus of 25 million word tokens: the ACL Anthology[2]. It consists of technical articles published in conferences and journals in the domain of Computational Linguistics. The main drawback is that the corpus is very noisy because of the .pdf to .txt conversion process. *POS* tagging was made with freely available Tree-Tagger[3].

The starting glossary of terms contains 175 entries. It is a manual selection from the glossary appearing in the appendix of the "Oxford HandBook of Computational Linguistics", edited by Ruslan Mitkov[4]. We learnt 201 generic classes using the abstraction algorithm. These classes contain 803 different terms. So, as well as being organised terms in classes, we extracted more than 600 new terms. On the other hand, the specification algorithm gave rise to 803 specific classes with 297 different terms.

Measuring the correctness of the acquired classes of terms is not an easy task. We are not provided with a gold standard against to which results can be compared. So, we evaluated the capacity of the algorithm to classify unknown terms into existing generic clusters and the quality of the clusters themselves. The test data consisted of 160 randomly selected classifications. Then they were given to 3 human judges for evaluation. The evaluation protocol was inspired by [11]. Judges scored between 1 and 5 each test classification. Score 1 means that the cluster is non-sensical (and so term classification). Score 2 means that the term was oddly classified in a correct cluster. 3 means the cluster is correct but the evaluator is undecided about the correctness of classification. 4 means the term fits with the general sense of the cluster (which is correct). Finally, 5 means the term fits well with the cluster (which must be correct).

---

[2] http://wing.comp.nus.edu.sg/ min/dAnth/acl/

[3] http://www.ims.uni-stuttgart.de/projekte/corplex/TreeTagger/DecisionTreeTagger.html

[4] An electronic version of part of this glossary can be found in
http://turing.iimas.unam.mx/črodriguezp/gloss/index.htm

Table 4 illustrates the results of evaluation. Note that most clusters are meaningful since only about 5% of classifications are non-sensical (Classif. 1).

**Table 4.** Evaluation of Word Classification

|           | Judge_1 | Judge_2 | Judge_3 |
|-----------|---------|---------|---------|
| Classif. 1 | 3.94%   | 3.82%   | 8.12%   |
| Classif. 2 | 7.89%   | 5.73%   | 11.25%  |
| Classif. 3 | 12.50%  | 12.10%  | 17.50%  |
| Classif. 4 | 21.05%  | 26.11%  | 10.62%  |
| Classif. 5 | 54.60%  | 52.22%  | 52.50%  |
| **Number of Tests** | 160 | | |
| **Average Score** | 4.10 | | |
| **Average Difference** | 0.15 | | |

In further research, we intend to develop a process of context/property classification using the classes learnt by means of the specification operation. In this process, each specific class will be assigned unknown lexico-syntactic contexts that were not involved in the previous clustering steps.

# References

1. Hereth, J., Stumme, G., Wille, R., Wille, U.: Conceptual knowledge discovery - a human-centered approach. Journal of Applied Artificial Intelligence 17(3), 288–301 (2003)
2. Priss, U.: Formal concept analysis in information science. Information Science and Technology 40, 521–543 (2006)
3. Pereira, F., Tishby, N., Lee, L.: Distributional clustering of english words. In: ACL'93, Columbos, Ohio, pp. 183–190 (1993)
4. Roth, M.: Two-dimensional clusters in grammatical relations. In: AAAI-95 (1995)
5. Pantel, P., Lin, D.: Discovering word senses from text. In: ACM SIGKDD, Edmonton, Canada, pp. 613–619 (2002)
6. Faure, D., Nédellec, C.: Asium: Learning subcategorization frames and restrictions of selection. In: Nédellec, C., Rouveirol, C. (eds.) Machine Learning: ECML-98. LNCS, vol. 1398, Springer, Heidelberg (1998)
7. Allegrini, P., Montemagni, S., Pirrelli, V.: Example-based automatic induction of semantic classes through entropic scores. Linguistica Computazionale, 1–45 (2003)
8. Reinberger, M.L., Daelemans, W.: Is shallow parsing useful for unsupervised learning of semantic clusters? In: Gelbukh, A. (ed.) CICLing 2003. LNCS, vol. 2588, pp. 304–312. Springer, Heidelberg (2003)
9. Gamallo, P., Agustini, A., Lopes, G.: Clustering syntactic positions with similar semantic requirements. Computational Linguistics 31(1), 107–146 (2005)
10. Allegrini, P., Montemagni, S., Pirrelli, V.: Learning word clusters from data types. In: Coling-2000, pp. 8–14 (2000)
11. Lin, D., Pantel, P.: Induction of semantic classes from natural language text. In: SIGKDD-01, San Francisco (2001)

# Accurate Unlexicalized Parsing
# for Modern Hebrew

Reut Tsarfaty and Khalil Sima'an

Institute for Logic, Language and Computation, University of Amsterdam
Plantage Muidergracht 24, 1018TV Amsterdam, The Netherlands

**Abstract.** Many state-of-the-art statistical parsers for English can be viewed
as Probabilistic Context-Free Grammars (PCFGs) acquired from treebanks consisting of phrase-structure trees enriched with a variety of contextual, derivational
(e.g., markovization) and lexical information. In this paper we empirically investigate the applicability and adequacy of the unlexicalized variety of such parsing models to Modern Hebrew, a Semitic language that differs in structure and
characteristics from English. We show that contrary to experience with parsing
the WSJ, the markovized, head-driven unlexicalized variety does not necessarily
outperform plain PCFGs for Semitic languages. We demonstrate that enriching
unlexicalized PCFGs with morphologically marked agreement features percolated up the parse tree (e.g., definiteness) outperforms plain PCFGs as well as a
simple head-driven variation on the MH treebank. We further show that an (unlexicalized) head-driven variety enriched with the same features achieves even
better performance. We conclude that morphologically rich languages introduce
an additional dimension of parametrization that is orthogonal to the horizon-
tal/vertical dimensions proposed before [1] and its contribution is essential and
complementary.

Parsing Modern Hebrew (MH) as a field of study is in its infancy. Although a syntactically annotated corpus has been available for quite some time [2] we know of only two
studies attempting to parse MH using supervised methods.[1] The reason state-of-the-art
parsing models are not immediately applicable to MH is not only that their adaptation
to the MH data and annotation scheme is not trivial, but also that they do not guarantee
to yield comparable results. The MH treebank is small, the internal phrase- and clause-
structures are relatively flat and variable, multiple annotated dependencies complicate
the selection of a single syntactic head, and a plentiful of disambiguating morphological
features are not exploited by current state-of-the-art models for parsing, e.g., English.
This paper provides a theoretical overview of the MH data and an empirical evaluation of different dimensions of parameters for learning treebank grammars which break
independence assumptions irrelevant for Semitic languages. We illustrate the utility of
a three-dimensional parametrization space for parsing MH and obtain accuracy results

---

[1] The studies we know of are [2] which uses a DOP tree-gram model and 500 training sentences,
and [3] which uses a treebank PCFG in an integrated system for morphological and syntactic
disambiguation. Both achieved around 60-70% accuracy.

V. Matoušek and P. Mautner (Eds.): TSD 2007, LNAI 4629, pp. 39–47, 2007.
© Springer-Verlag Berlin Heidelberg 2007

that are comparable to those obtained for Modern Standard Arabic (75%) using a lexicalized parser [4] and a much larger treebank.

# 1  Dimensions of Unlexicalized Parsing

The factor that sets apart vanilla treebank Probabilistic Context-Free Grammars (PCFGs) [5] from unlexicalized extensions as proposed by, e.g., [6,1], is the choice of statistical parametrization that embodies weaker independence assumptions. Recent studies on accurate unlexicalized parsing models outline two dimensions of parametrization. The first, proposed by [6], is the annotation of parent categories, effectively conditioning on aspects of a node's generation history, and the second encodes a head-outward generation process [7] in which the head is generated followed by outward Markovian sister generation processes. Klein and Manning [1] systematize the distinction between these two forms of parametrization by drawing them on a *horizontal-vertical* grid: parent-ancestor encoding is *vertical* ($v$) (external to the rule) whereas head-outward generation is *horizontal* ($h$) (internal to the rule). By varying the value of the parameters along the grid they tune their treebank grammar to achieve better performance. This two-dimensional parametrization[2] was shown to improve parsing accuracy for English [7,4] as well as other languages, e.g., German [8] Czech [9] and Chinese [10]. However, results for languages different than English still lag behind.[3]

We claim that for various languages including the Semitic family, e.g. Modern Hebrew (MH) and Modern Standard Arabic (MSA), the horizontal and vertical dimensions of parameters are insufficient for encoding linguistic information relevant for breaking false independence assumptions. In Semitic languages, arguments may move around rather freely and the phrase-structure of clause level categories is often shallow. For such languages agreement features play a role in disambiguation at least as important as vertical and horizontal histories. Here we propose to add a third dimension of parametrization that encodes morphological features orthogonal to syntactic categories, such as those realizing syntactic agreement. These features are percolated from surface forms in a bottom-up fashion and they express information that is orthogonal to the previous two. We refer to this dimension as *depth* ($d$) as it can be visualized as a dimension along which parallel tree structures labeled with syntactic categories encode an increasing number of morphological features at all levels of constituency. These structures lie in a three-dimensional coordinate-system we refer to as $(v, h, d)$.

This work focuses on MH and explores the empirical contribution of the three dimensions of parameters to analyzing different syntactic categories. We present extensive experiments that lead to improved performance as we increase the number of dimensions which are exploited across all levels of constituency. In the next section we review characterizing aspects of MH (and other Semitic languages) highlighting the special role of morphology and the kind of dependencies witnessed by morphosyntactic processes. In section 3 we describe the method and procedure for the empirical evaluation of

---

[2] Typically accompanied with various category-splits and lexicalization.

[3] The learning curves over increasing training data (e.g., for German [8]) show that treebank size cannot be the sole factor to account for the inferior performance.

unlexicalized parsing models for MH. In section 4 we report and analyze our results, and in section 5 we conclude.

## 2    Dimensions of Modern Hebrew Grammar

### 2.1    Modern Hebrew Structure

Phrases and sentences in MH, as well as Arabic and other Semitic languages, have a relatively flexible phrase structure. Subjects, verbs and objects can be inverted and prepositional phrases, adjuncts and verbal modifiers can move around rather freely. The factors that affect word-order in the language are not necessarily syntactic and have to do with rhetorical and pragmatic factors as well. To illustrate, figure 1 shows two syntactic structures that express the same grammatical relations yet vary in their order of constituents. The level of freedom in the order of internal constituents also varies between categories, and figure 1 further illustrates that within noun-phrase categories determiners always precede nouns.[4]

Within the flexible phrase structure it is typically morphological information that provides cues for the grammatical relations between surface forms. In figure 1, for example, it is agreement on gender and number that reveals the subject-predicate dependency. Agreement features also help to reveal the relations between higher levels of constituents, as shown in figure 2.

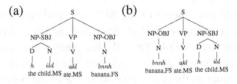

**Fig. 1.** Word Order in MH Phrases (marking the agreement features M(asculine), F(eminine), S(ingular))

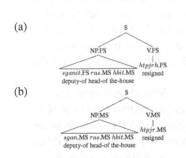

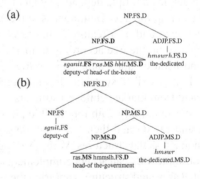

**Fig. 2.** Phrase-Level Agreement Features (marking M(asculine), F(eminine), S(ingular))

**Fig. 3.** Definiteness as Phrase-Level Agreement (marking M(asculine), F(eminine), S(ingular), D(efiniteness))

Figure 2(a) further shows that selecting the child constituents that contribute the agreement features is not a trivial matter. Consider, for instance, definiteness in MH,

---

[4] See [11] and [12] for formal and statistical accounts of noun phrases in MH.

which is morphologically marked (as a prefix to the stem) and behaves as a syntactic property [13]. Definite nouns exhibit agreement with other modifying phrases as shown in figure 3. The agreement on definiteness helps to determine the level of attachment in, e.g., the complex structure of an NP construct-state (*smixut*) or attachment to predicates in verbless sentences.[5] Figure 3(a) further illustrates that definiteness may be percolated from a different form (*hbit*.MS.**D** the-house) than the one determining the gender of the phrase (*sganit*.**FS** deputy-of).

Agreement features are thus helpful in disambiguating syntactic structures and they operate not only at the lexical level but also manifest relations between higher levels of constituents. For MH, features percolated from multiple surface forms manifest multiple kinds of dependencies and jointly determine the features of higher level constituents. Determining such features requires bi-dimensional percolation which does not coincide with head or parent dependencies, and we propose to view it as taking place along an orthogonal dimension we call *depth*.

## 2.2   The Modern Hebrew Treebank Scheme

The annotation scheme of the MH treebank[6] aims to capture the morphological and syntactic properties of MH we described, and differs from, e.g., the WSJ Penn treebank annotation scheme [14]. The MH Treebank is built over word segments, and the yields of the syntactic trees do not correspond to space delimited words but rather to morphological segments that carry distinct syntactic roles (i.e., each corresponding to a single POS tag). The POS categories assigned to segmented words are decorated with features such as gender, number, person and tense, and these features are percolated higher up the tree according to pre-defined syntactic dependencies [15]. Since agreement features of non-terminal constituents may be contributed by multiple children, the annotation scheme defines multiple dependency labels that guide the percolation of different features higher up the tree. Definiteness in the MH treebank is treated as a segment at the POS-tag level and as a feature at the level of non-terminals. As any other feature, it is percolated up the tree according to marked dependency labels. Table 1 lists the features and feature-values annotated on top of syntactic categories in the MH treebank, and table 2 describes syntactic dependencies which define the features that are to be percolated from marked child constituents.

In order to comply with the flexible phrase structure in MH, clausal categories (S, SBAR and FRAG and their corresponding interrogatives SQ, SQBAR and FRAGQ) are annotated as flat structures. Verbs (VB tags) always attach to a VP mother (however only non-finite VBs can accept complements under the same VP parent). NP and PP are annotated as nested structures capturing the recursive structure of construct-state nouns, numerical expressions and possession and an additional category PREDP is added to account for sentences in MH that lack a copular element. The scheme also features null

---

[5] Present tense predicative sentences in MH lack a copular element.

[6] Version 2.0 of the MH treebank was made available to us in January 2007 and is currently publicly available at http://mila.cs.technion.ac.il/english/index.html along with a complete annotation guide and illustrative examples.

**Table 1.** Morphological Features in the MH Treebank Annotation Scheme

**Table 2.** Dependency Labels in the MH Treebank Annotation Scheme

| Feature | Value | Value Encoded |
|---|---|---|
| gender | Z | masculine |
| gender | N | feminine |
| number | Y | singular |
| number | R | plural |
| definiteness | H | definite |
| definiteness | U | underspecified |

| Dependency Type | Features Percolated |
|---|---|
| DEP_HEAD | all |
| DEP_MAJOR | gender |
| DEP_NUMBER | number |
| DEP_DEFINITE | definiteness |
| DEP_ACCUSATIVE | case |
| DEP_MULTIPLE | all (e.g., conjunction) |

elements that mark traces and functional elements that mark, e.g. SBJ, OBJ, which we strip off and ignore throughout this study.

### 2.3 Treebank Grammars for Modern Hebrew

In MH there are various aspects that provide indication for the expansion possibilities of a node. Firstly, the variability in the order and number of an expansion of a non-terminal node depends on its label (e.g., while NP structures may involve nested recursive derivations, S level constituents are usually flat). Additional indication comes from the node's syntactic context. S nodes appearing under SBAR, for instance, are less shallow than those under TOP as they often involve non-finite VPs under which more modifiers can be attached. Further, although the generation of child nodes in a phrase-structure revolves, as in English, around a syntactic head, the order in which they are generated may not be as strict. Finally, morphological features indicating agreement between surface forms percolate up the tree indicating multiple dependencies. We propose to take such complementary information into account. The practice of having morphological features orthogonal to a constituency structure is familiar from theories of syntax (e.g., HPSG, LFG), however here we propose to frame it as an additional dimension for statistical estimation, a proposal which, to the best of our knowledge, has not been empirically explored before.

## 3 Experimental Setup

In this work we set out to empirically investigate a refined space of parameters for learning treebank grammars for MH. The models we implement use the *vertical* ($v$, parental history), *horizontal* ($h$, markovized child generation) and *depth* ($d$, orthogonal morphology) dimensions, and we instantiate $d$ with the definitensess feature as it has the least amount of overlap with features determining the head.

We use version 2.0 of the MH treebank [2] which consists of 6501 sentences from the daily newspaper 'ha'aretz' and employ the syntactic categories, POS categories and morphological features annotated therein. The data set is split into 13 sections consisting of 500 sentences each. We use the first section (section 0) as our development set and the last section (section 12) as our test set. The remaining sentences (sections 1–11) are used for training. After cleaning the data set we remain with a *devset* of

**Table 3.** Transforms over the MH Treebank

| |
| --- |
| **Lexicalize** select and percolate lexical heads and their categories for markovization |
| **Linearize** linearize RHS of CFG productions (using [16]) |
| **Decorate** annotate contextual/morphological features on top of syntactic categories |

**Table 4.** Implementing Different Parametrization Options using Transforms

| Name | Params | Description | Transforms used |
| --- | --- | --- | --- |
| DIST | $h = 0$ | 0-order Markov process | **Lexicalize(category), Linearize(distance)** |
| MRK | $h = 1$ | 1-order Markov process | **Lexicalize(category), Linearize(distance, neighbor)** |
| PA | $v = 1$ | Parent Annotation | **Decorate(parent)** |
| DEF | $d = 1$ | Definiteness Percolation | **Decorate(definiteness)** |

483 sentences (average length in word segments 48), a *trainset* of 5241 sentences (53) and a *testset* of 496 sentences (58).[7]

Our methodology is similar to the one used by [6] and [1]. We transform our training set according to certain parametrization decisions and learn different treebank grammars according to different instantiations of one, two, and three dimensions of parameters (tables 3 and 4 show the transforms we use to instantiate different parameters).

The input to our parser is a sequence of word segments (each corresponding to a single POS-tag). This setup assumes partial morphological disambiguation (e.g., segmentation) but we do *not* provide the parser with POS tags information.[8] We train a treebank PCFG on the resulting treebank using relative frequency estimates, and we use BitPar, an efficient general-purpose PCFG parser [17], to parse unseen sentences.[9]

We evaluate our models using EVALB focusing on bare syntactic categories. We report the average F-measure for sentences of length up to 40 and for all sentences ($F_{\leq 40}$ and $F_{All}$ respectively), once including punctuation marks (WP) and once excluding them (WOP). For selected models we show a break-down of the average $F_{All}$ (WOP) measure for different categories.

## 4   Results and Analysis

In a series of experiments we evaluated models that instantiate one, two or three dimensions in a coordinate-system defined by the parameters $(v, h, d)$. We set our baseline model at the $(0, 0, 0)$ point of the coordinate-system and compared its performance to a simple treebank PCFG and to different combinations of parameters. Table 5 shows

---

[7] Since this work is only the first step towards the development of a broad-coverage statistical parser for MH (and other Semitic languages) we use only the development set and leave our test set untouched.

[8] This setup makes our results comparable to parallel studies in other languages.

[9] We smooth pre-terminal rules by providing the parser with statistics on "rare words" distribution. The frequency defining "rare words" is tuned empirically and set to 1.

**Table 5.** Multi-Dimensional Parametrization of Treebank Grammars (Head-Driven Models are Marked $h \neq \infty$): $F_{\leq 40}$, $F_{ALL}$ Accuracy Results on Section 0

| Name | Params $(v, h, d)$ | $F_{ALL}$ WP | $F_{\leq 40}$ WP | $F_{ALL}$ WOP | $F_{\leq 40}$ WOP |
|------|------|------|------|------|------|
| BASE | $(0, 0, 0)$ | 66.56 | 68.20 | 67.59 | 69.24 |
| PCFG | $(0, \infty, 0)$ | 65.17 | 66.63 | 66.11 | 67.7 |
| PCFG+DEF | $(0, \infty, 1)$ | 67.53 | 68.78 | 68.7 | **70.37** |
| PA | $(1, 0, 0)$ | 68.87 | 70.48 | 69.64 | 70.91 |
| MRK | $(0, 1, 0)$ | 66.69 | 68.14 | 67.93 | 69.37 |
| DEF | $(0, 0, 1)$ | 68.85 | 69.92 | 70.42 | **71.45** |
| PA+MRK | $(1, 1, 0)$ | 69.97 | 71.48 | 70.69 | 71.98 |
| MRK+DEF | $(0, 1, 1)$ | 69.46 | 70.79 | 71.05 | 72.37 |
| PA+DEF | $(1, 0, 1)$ | 71.15 | 72.34 | 71.98 | **72.91** |
| PA+MRK+DEF | $(1, 1, 1)$ | 72.34 | 73.63 | 73.27 | **74.41** |

**Table 6.** The Contribution of the *horizontal* and *depth* Dimensions ($v > 0$ Marks Parent Annotation, $h > 0$ Marks 1-Order Markov Process): $F_{All}$ (WOP) per Syntactic Category on Section 0

| $(v, h, d)$ | $(0, 0, 1)$ $v = 0$ | $(0, 1, 0)$ | $(1, 0, 1)$ $v > 0$ | $(1, 1, 0)$ |
|------|------|------|------|------|
| ADJP | 76.42 | **76.62** | **81.34** | 80.12 |
| ADVP | 72.65 | **74.77** | **79.66** | 78.19 |
| NP | **75.28** | 74.74 | **79.29** | 77.66 |
| VP | 71.10 | **71.80** | **75.69** | 73.89 |
| S | 74.41 | **78.08** | 76.04 | **79.49** |
| SBAR | 56.65 | **63.62** | 59.59 | **65.65** |
| SQ | 50.00 | **54.55** | 44.44 | 40.00 |
| FRAG | **56.00** | 53.85 | **61.02** | 58.62 |

the accuracy results for parsing section 0 for all models. The first outcome of our experiments is that our head-driven baseline performs slightly better than a vanilla treebank PCFG. Because of the variable phrase-structure a simple PCFG does not capture relevant generalization about sentences' structure in the language. However, enriching a vanilla PCFG with orthogonal morphological information (definiteness in our case) already performs better than our baseline unlexicalized model. In comparing the contribution of three one-dimensional models we observe that the depth dimension contributes the most to parsing accuracy. These results demonstrate that incorporating dependency information marked by morphology is important to analyzing syntactic structures at least in as much as the main head-dependency is. The results for two-dimensional models re-iterate this conclusion by demonstrating that selecting the depth dimension is better than not doing so. Notably, the configuration most commonly used by current state-of-the-art parsers for English (i.e., $(v, h, 0)$, cf. [1]) performs slightly worse than the ones incorporating a depth feature. A three-dimensional annotation strategy achieves the best accuracy results among all models we tested.[10] The error reduction rate from a plain PCFG is more than 20%, providing us with a new, much stronger, lower bound on the performance of unlexicalized treebank grammars in parsing MH.

The general trend observed in our results is that higher dimensionality is better. Different dimensions provide different sorts of information which are complementary. As further illustrated in table 6 the internal structure of different syntactic constituents may benefit to a different extent from information provided by different dimensions. Table 6 shows the breakdown of the $F_{All}$(WOP) accuracy results for the main syntactic categories in our treebank. In the lack of parental context ($v = 0$) the Markovian head-outward process ($h = 1$) encodes information relevant for disambiguating the flat variable phrase-structures. The morphological dimension ($d = 1$) helps to determine

---

[10] The addition of an orthogonal *depth* dimension to the *horizontal-vertical* space goes beyond mere "state-splits" (cf. [1]) as it does not only encode refined syntactic categories but also signals linguistically motivated co-occurrences between them.

the correct labels and attachment via the agreement with modifiers within NP structures. In the presence of a vertical history ($v = 1$) that provides cues for the expansion possibilities of nodes, the contribution of an orthogonal morphological feature ($d = 1$) is even more substantial. Accuracy results for phrase-level categories (ADJP, ADVP NP and VP) are better for the $v/d$ combination than for the $v/h$ one. Yet, high-level clausal categories (S and SBAR) benefit from head-outward markovization processes ($h = 1$) which encode additional rhetoric, pragmatic, and perhaps extra linguistic knowledge that govern order-preferences in the genre.

## 5   Conclusion

Tuning the dimensions and values of the parameters in a treebank grammar is largely an empirical matter, but our results point out that the selection of parameters for statistical estimation should be in tune with our linguistic knowledge of the factors licensing grammatical structures in the language. Morphologically rich languages introduce an additional dimension into the expansion possibilities of a node which is orthogonal to the vertical [6] and horizontal [7] dimensions systematized by [1]. Via a theoretical and empirical consideration of syntactic structures and morphological definiteness in MH we show that a combination of multiple orthogonal dimensions of parameters is invaluable for boosting the performance of unlexicalized parsing models. Our best model provides us with a new, strong, baseline for the performance of treebank grammars for MH.

## References

1. Klein, D., Manning, C.: Accurate Unlexicalized Parsing. In: Dignum, F.P.M. (ed.) ACL 2003. LNCS (LNAI), vol. 2922, pp. 423–430. Springer, Heidelberg (2004)
2. Sima'an, K., Itai, A., Winter, Y., Altman, A., Nativ, N.: Building a Tree-Bank of Modern Hebrew Text. In: Traitement Automatique des Langues (2001)
3. Tsarfaty, R.: Integrated Morphological and Syntactic Disambiguation for Modern Hebrew. In: Proceedings of SRW COLING-ACL (2006)
4. Bikel, D.: Intricacies of Collins' Parsing Model. Computational Linguistics 30(4) (2004)
5. Charniak, E.: Tree-Bank Grammars. In: AAAI/IAAI, vol. 2, pp. 1031–1036 (1996)
6. Johnson, M.: PCFG Models of Linguistic Tree Representations. Computational Linguistics 24(4), 613–632 (1998)
7. Collins, M.: Head-Driven Statistical Models for Natural Language Parsing. Computational Linguistics (2003)
8. Dubey, A., Keller, F.: Probabilistic Parsing for German using Sister-Head Dependencies. In: Dignum, F.P.M. (ed.) ACL 2003. LNCS (LNAI), vol. 2922, Springer, Heidelberg (2004)
9. Collins, M., Hajic, J., Ramshaw, L., Tillmann, C.: A Statistical Parser for Czech. In: Proceedings of ACL, College Park, Maryland (1999)
10. Bikel, D., Chiang, D.: Two Statistical Parsing Models Applied to the Chinese Treebank. In: Second Chinese Language Processing Workshop, Hong Kong (2000)
11. Wintner, S.: Definiteness in the Hebrew Noun Phrase. Journal of Linguistics 36, 319–363 (2000)
12. Goldberg, Y., Adler, M., Elhadad, M.: Noun Phrase Chunking in Hebrew: Influence of Lexical and Morphological Features. In: Proceedings of COLING-ACL (2006)

13. Danon, G.: Syntactic Definiteness in the Grammar of Modern Hebrew. Linguistics 39(6), 1071–1116 (2001)
14. Marcus, M., Kim, G., Marcinkiewicz, M., MacIntyre, R., Bies, A., Ferguson, M., Katz, K., Schasberger, B.: The Penn Treebank: Annotating Predicate-Argument Structure (1994)
15. Milea, A.: Treebank Annotation Guide. MILA, Knowledge Center for Hebrew Processing (2007)
16. Hageloh, F.: Parsing using Transforms over Treebanks. Master's thesis, University of Amsterdam (2007)
17. Schmid, H.: Efficient Parsing of Highly Ambiguous Context-Free Grammars with Bit Vectors. In: Proceedings of ACL (2004)

# Disambiguation of the Neuter Pronoun and Its Effect on Pronominal Coreference Resolution

Véronique Hoste[1], Iris Hendrickx[2], and Walter Daelemans[2]

[1] LT3 - Language and Translation Technology Team,
University College Ghent, Groot-Brittaniëlaan 45, Ghent,
Belgium
veronique.hoste@hogent.be
[2] CNTS - Language Technology Group,
University of Antwerp, Universiteitsplein 1, Antwerp,
Belgium
iris.hendrickx@ua.ac.be, walter.daelemans@ua.ac.be

**Abstract.** Coreference resolution, determining the appropriate discourse referent for an anaphoric expression, is an essential but difficult task in natural language processing. It has been observed that an important source of errors in machine-learning based approaches to this task, is the wrong disambiguation of the third person singular neuter pronoun as either referential or non-referential. In this paper, we investigate whether a machine learning based approach can be successfully applied to the disambiguation of the neuter pronoun in Dutch and show a modest potential effect of this disambiguation on the results of a machine learning based coreference resolution system for Dutch.

## 1 Introduction

Coreference resolution, the task of determining the appropriate discourse referent for a given anaphoric expression, has gained increasing popularity in natural language processing research and has become a key component in applications such as information extraction, question answering, automatic summarization, etc. in which text understanding is of major importance.

In this paper we focus on pronominal coreference resolution, and more specifically on the improvement of a machine learning system for automatic pronominal coreference resolution through the automatic disambiguation of "het" (Eng.: "it") as either referential or non-referential. The focus is on the classification of the Dutch neuter "het", in contrast to most of the related work which is mainly oriented towards English. In order to classify the third person singular neuter pronoun, two different types of approaches have been proposed for English: rule-based strategies ([1], [2]) and machine learning approaches ([3], [4]).

Although the existing approaches to the automatic identification of the different uses of the third person singular neuter pronouns are always motivated by the task of pronominal coreference resolution, this effect of the automatic classification of "it" on resolution performance has to our knowledge not yet been investigated, except by [5] who performed a more global comparison of resolution perfomance with and without

V. Matoušek and P. Mautner (Eds.): TSD 2007, LNAI 4629, pp. 48–55, 2007.

the detection of non-anaphoric constituents. The goal of this paper is twofold. In a first step, we investigate whether the distribution of the different uses of the Dutch neuter pronoun is similar to the one reported for English. Furthermore, we investigate whether the machine learning approach, successfully applied for the English "it", can be easily ported to the automatic classification of the Dutch "het". In a second step, we evaluate the effect of this classification on a learning approach for Dutch pronominal coreference resolution as described in [6]. Since the coreference resolution system is designed to detect coreferential chains between nominal constituents, we are mainly interested in the detection of the pronouns referring to antecedent NPs.

The remainder of this paper is organized as follows. Section 2 introduces the data used for the experiments and describes the annotation and the distribution of phenomena of interest compared to English data used for the same task. The experimental set-up and results are described in Section 3. The effect of the separate disambiguation component on overall anaphora resolution is discussed in Section 4, and Section 5 summarizes the main findings of the paper.

## 2   Data Sets

For the experiments, the focus was on Dutch coreference resolution. Two corpora were annotated with information on the third person neuter pronoun: KNACK, á corpus of news magazine texts (106,011 tokens) and SPECTRUM, a corpus with medical encyclopedia texts (133,887 tokens). Two linguists annotated the corpora in parallel in accordance with the annotation guidelines described below, which are based on the general Dutch grammar (ANS)[1]. As input, the annotators received free text in which all occurrences of "het" were marked, the majority of which involved "het" as definite article. For the annotation of the personal pronoun "het", the annotators had to differentiate between the non-referential use of the pronoun as in example (1) and its referential use. In example (1), "het" is part of an idiomatic expression and not referential.

(1)   Leopold haalt scherp uit naar onder meer Hubert Pierlot, de eeuwige zondebok met wie hij **het** niet kon vinden.
English: Leopold sharply attacks among others Pierlot, the eternal scapegoat with whom he can't get on.

We distinguished between the following four types of referential use: (i) reference to preceding "het" words as in example (2), (ii) reference to a preceding clause as in example (3), (iii) "het" as anticipatory subject (4) and finally, "het" as subject of a nominal predicate (5).

(2)   Weet je waar **mijn boek** is? Ik heb **het** niet gezien.
English: Do you know where **my book** is? I haven't seen **it**.

(3)   **Leopold III kwam aan de macht nadat zijn vader in 1934 was verongelukt in Marche-les-Dames.** Volgens historicus Jan Verwelkenhuyzen ging toen het gerucht dat de Duitsers **het** zo hadden gewild.

---

[1] http://oase.uci.ru.nl/*sim*ans/

English: **Leopold III came to power after his father died in an accident in 1934 in Marche-les-Dames.** According to historian Jan Verwelkenhuyzen, there was a rumour that the Germans had wanted **it** that way.

(4)  **Het** lijkt er namelijk op **dat de bevolking van Zimbabwe haar huisbakken dictator meer dan beu is**.
English: **It** seems **the population of Zimbabwe has had it with its home-grown dictator**.

(5)  **Het** zijn, voorlopig althans, **slechts schuchtere signalen**.
English: **It** is, for now, **only a weak signal**.

On the Knack data, a kappa agreement score was obtained of 0.74; on the Spectrum data, the kappa score was 0.81. After this first annotation round, both annotators re-annotated the texts jointly in order to reach a consensus annotation. In total, 6560 occurrences of "het" were annotated, of which 844 are pronominal. Table 1 gives an overview of the distribution of the different uses of the neuter pronoun in both annotated Dutch corpora. For English, we also provided the number of times the "it" refers to a preceding noun phrase. The table reveals that the English corpora which have been previously used for the automatic classification of "it" all show a large number (>67%) of occurrences of "it" in which the pronoun refers to a preceding noun phrase. For Dutch, however, this percentage drops to around 20% for the newspaper texts, whereas for the medical texts nearly half of the "het" occurrences refer to an NP. Taking the Dutch corpora as a whole, three categories show a similar distribution: the non-referential use (30.1%), the reference to a preceding noun phrase (32.6%) and the neuter pronoun as anticipatory subject (23.3%).

**Table 1.** Distribution of the pronominal "het" in the different data sets

|  | DUTCH | | | ENGLISH |
|  | Knack | Spectrum | Total |  |
|---|---|---|---|---|
| Pronominal use | 507/2890 | 337/3670 | 884/6560 |  |
| Non-referential | 39.0% | 17.7% | 30.1% |  |
| Ref - preceding clause | 5.7% | 0.3% | 3.5% |  |
| **Ref - noun phrase** | **21.3%** | **49.5%** | **32.6%** | MUC-6 **74.4%** |
|  |  |  |  | MUC-7 **80.7%** |
|  |  |  |  | [4]    **67.9%** |
|  |  |  |  | [3]    **69.6%** |
| Ref - anticipatory subject | 19.9% | 28.5% | 23.3% |  |
| Ref - anticipatory object | 5.1% | 1.2% | 3.5% |  |
| Ref - nominal predicate | 8.9% | 3.9% | 6.9% |  |

# 3  Experimental Setup

For the construction of the machine learning data sets, the following preprocessing steps were taken. Lemmatization was performed using a memory-based lemmatizer trained on a lexicon derived from the Spoken Dutch Corpus (CGN)[2], a 10-million word corpus of spoken Dutch. Part-of-speech tagging and text chunking were performed by the memory-based tagger MBT[7], which was also trained on the CGN corpus. For all occurrences of "het", a feature vector was built consisting of 38 features. These features include positional information (sentence number and position in sentence), information on the focus word itself (wordform, part-of-speech and chunk information), furthermore information on the wordform, lemma and part-of-speech of five words before and after the focus word, and finally information on the use of a preposition before the focus word [1]. Based on the assumption that verbs which occur more often with "het" indicate the non-anaphoric use of the pronoun, we included a last feature for which the association strength was calculated between "het" as a subject and its accompanying verb. This association strength was represented by mutual information scores and was based on the Dutch Twente News Corpus (500 million words). A minimal cut-off frequency of 1000 was chosen.

For the classification of the different uses of "het", we used a memory-based learning algorithm, as was also previously applied to this task by [4] and [3]. Memory-based learning (a $k$-nearest neighbor approach) is a lazy learning approach that stores all training data in memory. At classification time, the algorithm classifies new instances by searching for the nearest neighbors to the new instance using a similarity metric, and extrapolating from their class. In our experiments we use the TIMBL [7] software package[3] that implements a version of the $k$-nn algorithm optimised for working with linguistic datasets and that provides several similarity metrics and variations of the basic algorithm. Since these different parameters, individually and in combination, can strongly affect the functioning of the algorithm, we performed joint feature selection and parameter optimization by means of a generational genetic algorithm as described in [6] and as shown in Figure 1. Given the modest size of the data sets, leave-one-out was used for validation. The following parameters were varied: the number of nearest neighbours, expressed by $k$, the distance metric and the model to extrapolate from the nearest neighbours. For the three data sets, viz. Knack, Spectrum and the concatenation of the two, optimization led to a selection of a high $k$ value (9 for Spectrum and the concatenated data; 16 for Knack) and to the selection of exponential decay distance weighted voting and of gain ratio (a normalised version of information gain) as distance metric for the three data sets. Feature selection led to an omission of the feature informing on the position of the word in the sentence and to a selection in the local context features. For the Knack data, the association strength feature was also filtered out.

Tables 2 gives an overview of the overall and 5-ary classification results of the optimized memory-based classifier. It shows a 30% improvement over the most frequent sense accuracy for the three data sets. The results also show that for some subtypes of referential use, there is too little evidence in the training data to train an accurate classifier on. The non-referential use of "het", on the other hand, can be detected with a reliability of >70%. For the "het" which refers to a preceding noun phrase, divergent F-scores are obtained: 39.1% for Knack, as opposed to 83.4% for Spectrum.

---

[2] http://lands.let.ru.nl/cgn
[3] http://ilk.uvt.nl

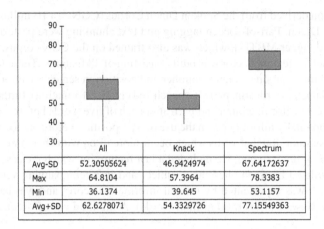

**Fig. 1.** Optimization results for the three data sets. The graphs show the difference between the best and the worst parameter and feature subset combination per data set. The boxes in the graphs represent averages and deviations.

**Table 2.** Performance in terms of accuracy, precision, recall and F-score of TIMBL on the three data sets. Both the overall and 5-ary classification results of the optimized memory-based classifier are given. As baseline score, the most frequent class, i.e. reference to a preceding NP, was taken and kept constant over all data sets.

| | Knack | | | Spectrum | | | Total | | |
|---|---|---|---|---|---|---|---|---|---|
| Baseline | 21.3 | | | 49.5 | | | 32.6 | | |
| Accuracy | 57.40 | | | 78.34 | | | 64.81 | | |
| | P | R | F | P | R | F | P | R | F |
| Non-referential | 60.20 | 89.39 | 71.95 | 85.11 | 71.43 | 77.67 | 67.49 | 75.20 | 71.14 |
| Ref - preceding clause | 0.00 | 0.00 | 0.00 | 0.00 | 0.00 | 0.00 | 0.00 | 0.00 | 0.00 |
| Ref - noun phrase | 49.30 | 32.41 | 39.11 | 78.72 | 88.62 | 83.38 | 63.16 | 69.82 | 66.32 |
| Ref - anticipatory subject | 52.34 | 55.45 | 53.85 | 74.51 | 79.17 | 76.77 | 62.77 | 73.60 | 67.76 |
| Ref - anticipatory object | 100.00 | 42.31 | 59.46 | 0.00 | 0.00 | 0.00 | 100.00 | 23.33 | 37.84 |
| Ref - nominal predicate | 54.55 | 26.67 | 35.82 | 0.00 | 0.00 | 0.00 | 66.67 | 20.69 | 31.58 |

## 4    Effect on Pronominal Coreference Resolution

The automatic disambiguation of the singular neuter pronoun finds it motivation in the difficulty to handle these cases in automatic coreference resolution. Our focus is on a classification based approach to coreference resolution, as for example described by [8], [9], [10], [6] and others, in which information about potentially coreferring pairs of NPs is represented as a set of feature vectors which are then classified by a machine learning algorithm as being coreferential or not. Instances are created between every NP and all of its preceding NPs. Sometimes, the search scope is limited through the application of distance restrictions or linguistically motivated filters (see for example [11,12,13]). Applied to the case of the Dutch pronominal "het", this implies that for each occurrence of the pronoun, an instance is created. The automatic detection of the non-referential uses of the pronoun could lead to the creation of instances solely for the occurrences of "het" which do refer to a preceding noun phrase.

In a postprocessing phase, a complete coreference chain is built between the pairs of NPs that were classified as being coreferential. If we consider the task of pronominal coreference resolution, two types of errors can occur on the coreference chain level, namely precision and recall errors. In a coreferential chain, all discourse entities (mostly noun phrases) referring to each other are gathered in one single chain. The recall errors are caused by classifying positive instances as being negative. These false negatives cause missing links in the coreferential chains, as exemplified in (6) and (7), in which the pronoun was classified as being not coreferential with any of the preceding NP's.

(6) The company will work with Sega Enterprises of Japan, SegaSoft and Time Warner Interactive to test the software. **It** will be sold starting this summer. (MUC-7)

(7) Maar voorzitter Spiritus-Dassesse gelooft niet in het nieuwe plan. **Het** lijkt te veel op het vorige. (KNACK)
English: But chairwoman Spiritus-Dassesse does not have faith in the new plan. **It** resembles the previous one too much.

The precision errors on the other hand are caused by classifying negative instances as being positive and create spurious links in the coreference chains, as shown in (8), in which the pronoun is erroneously linked to "the US government" and in (9), in which an antecedent is sought for the non-referential "het".

(8) **Hughes Electronics Corp.** has paid *the U.S. government* $4 million to settle a 1990 lawsuit filed by two former employees who accused **it** of lying to the Pentagon. (MUC-7)

(9) Een god van *het vuur*. Als vice-minster van Defensie heeft Paul Wolfowitz eigenlijk een bescheiden job in de Amerikaanse regering. Hoe komt **het** dan dat hij zoveel invloed heeft in het Witte Huis? (KNACK)
English: A god of *the fire*. As a vice minister of Defense, Paul Wolfowitz has a rather insignificant job in the American government. How is **it** possible that he has such an influence in the White House?

In order to evaluate the effect of this classification of "het" on pronominal coreference resolution for Dutch, we performed a 10-fold cross-validation experiment using TIMBL on the 242 annotated Knack documents, which were also annotated with coreferential chain information. The search scope for instance construction was reduced to 3 sentences and the instances consist of a set of features encoding morphological-lexical, syntactic, semantic, string matching and positional information sources [6].

The following experiments were conducted. In a first experiment, the output of the experiments described in Section 3 was used as the basis for filtering (Predicted). This implies that only the instances of "het" which were classified as referring to a preceding NP, were taken into account for coreference resolution. However, given the low F-score on this category in the Knack corpus, we performed a second experiment in order to assess the upper bound of potential performance increase. For this experiment we used the annotated corpus as an oracle to filter out all NPs not referring to a preceding noun phrase (Oracle). Table 3 shows the classification results before and after filtering on the instances in which "het" occurs as a potential anaphor. It reveals that filtering leads to a large reduction of the instances, but to a decrease in F-score. Furthermore, the expected potential performance increase is low (2%).

**Table 3.** Classification performance on the instances in which "het" occurs as potential anaphor. These are the results before and after filtering.

|  | #number | accuracy | precision | recall | F-score |
|---|---|---|---|---|---|
| Default | 9322 | 97.71 | 11.58 | 7.86 | 9.36 |
| Oracle | 1719 | 90.98 | 19.23 | 8.13 | 11.43 |
| Predicted | 2604 | 95.20 | 7.69 | 4.94 | 6.02 |

The low classification results show that filtering is insufficient to detect the correct antecedent for a given anaphor. In addition to filtering, more effort should be put in new features on top of the current 39 morphological-lexical, syntactic, semantic, string matching and positional features in order to detect the appropriate referent for an anaphoric "het". Consider for example the two instances in (10), which contain too little evidence to decide on a positive or negative classification.

(10)   (het ) (aids ) 1 5 heeft , aangezien WW(pv,tgw,met-t) LET() VG(onder) een
klasse van LID(onbep,stan,agr) N(soort,ev,basis,zijd,stan) VZ(init) dist_lt_two
appo_no jpron_yes 0 0 0 def_yes num_yes 0 0 0 0 0 0 0 0 I-OBJ 0 0 object 0 0 0
0 0 POS
(het) (Het ogenschijnlijke doel) 19 155 heeft, aangezien WW(pv,tgw,met-t)
LET() VG(onder) een klasse van LID(onbep,stan,agr) N(soort,ev,basis,zijd,stan)
VZ(init)
dist_gt_two appo_no jpron_yes 0 0 0 def_yes num_yes 0 0 0 0 0 0 0 0 I-OBJ I-SU
0 object GEN_NEUT 0 0 0 0 NEG

## 5   Concluding Remarks

We have shown that in a classification-based machine learning approach to coreference resolution for Dutch, the accurate disambiguation of "het" (it) as being referential or not can lead to modest improved performance on the resolution of pronominal coreference. We developed a machine learning based system for the disambiguation of referential or non-referential use of "het" using memory-based learning and genetic algorithm based joined optimization of feature selection and algorithm parameter selection. Results show that filtering is a first step towards an improved pronominal resolution and that the selection of the appropriate referent for an anaphoric "het" remains problematic. In addition to filtering, more effort should be invested in discriminating features capturing the relationship between an anaphoric "het" and its referent.

# References

1. Paice, C., Husk, G.: Towards an automatic recognition of anaphoric features in english text: the impersonal pronoun 'it'. Computer Speech and Language 2, 109–132 (1987)
2. Lappin, S., Leass, H.: An algorithm for pronominal anaphora resolution. Computational Linguistics 20(4), 535–561 (1994)
3. Boyd, A., Gegg-Harrison, W., Byron, D.: Identifiying non-referential it: a machine learning approach incorporating linguistically motivated patterns. In: Proceedings of the ACL Workshop on Feature Engineering for Machine Learning in NLP, pp. 40–47 (2005)
4. Evans, R.: Applying machine learning toward an automatic classification of it. Literary and Linguistic Computing 16(1), 45–57 (2001)
5. Ng, V., Cardie, C.: Identifying anaphoric and non-anaphoric noun phrases to improve coreference resolution. In: Proceedings of the 19th International Conference on Computational Linguistics (COLING-2002) (2002)
6. Hoste, V.: Optimization Issues in Machine Learning of Coreference Resolution. PhD thesis, Antwerp University (2005)
7. Daelemans, W., van den Bosch, A.: Memory-based Language Processing. Cambridge University Press, Cambridge (2005)
8. McCarthy, J.: A Trainable Approach to Coreference Resolution for Information Extraction. PhD thesis, Department of Computer Science, University of Massachusetts, Amherst MA (1996)
9. Soon, W., Ng, H., Lim, D.: A machine learning approach to coreference resolution of noun phrases. Computational Linguistics 27(4), 521–544 (2001)
10. Ng, V., Cardie, C.: Combining sample selection and error-driven pruning for machine learning of coreference rules. In: Proceedings of the 2002, Conference on Empirical Methods in Natural Language Processing (EMNLP-2002), pp. 55–62 (2002)
11. Yang, X., Zhou, G., Su, J., Tan, C.L: Coreference resolution using competition learning approach. In: Proceedings of the 41st Annual Meeting of the Association for Compuatiational Linguistics (ACL03), pp. 176–183. Sapporo, Japan (2003)
12. Uryupina, O.: Linguistically motivated sample selection for coreference resolution. In: Proceedings of DAARC-2004 (2004)
13. Hendrickx, I., Hoste, V., Daelemans, W.: Evaluating hybrid versus data-driven coreference resolution. In: Anaphora: Analysis, Algorithms and Applications (LNAI 4410) (2007)

# Constructing a Large Scale Text Corpus Based on the Grid and Trustworthiness

Peifeng Li[1,2], Qiaoming Zhu[1], Peide Qian[1], and Geoffrey C. Fox[2]

[1] School of Computer Science & Technology, Soochow University, Suzhou, 215006, China
[2] Community Grids Lab, Indiana University, Bloomington, IN 47404
{pfli, qmzhu, pdqian}@suda.edu.cn, gcf@indiana.edu

**Abstract.** The construction of a large scale corpus is a hard task. A novel approach is designed to automatically build a large scale text corpus with low cost and short building period based on the trustworthiness. It mainly solves two problems: how to automatically build a large scale text corpus on the Web and how to correct mistakes in the corpus. As Grid provides the infrastructure for processing large scale data, our approach uses Grid to collect and process language materials on the Web in the first stage. Then it picks out untrustworthy language materials in the corpus according to their trustworthiness, and checks them manually by users. After the check finishes, our approach computes the trustworthiness of each checked result and selects those ones with the highest trustworthiness as the correct results.

## 1 Introduction and Related Work

Text corpora are used for modeling language in many language technology applications including information retrieval, text classification, speech recognition, etc. Researches on text corpora started at 1970s. Recently there are a lot of corpora in use, such as Brown, LOB, COBUILD, LONGMAN, BNC, ICE, TreeBank, etc [1, 2].

The growing needs in the fields of Corpus Linguistics and Natural Language Processing (NLP) have led to an increasing demand for text corpora [3]. The automation of corpora development has therefore become an important and active field of research.

The Web is obviously a great source of data for corpora development. It is either considered as a corpus by itself [4] or as a huge databank to look for specific texts to be selected and similar to a specific corpus [5]. In a word, automatically constructing corpora on the Web is a rapid and economical approach to build a large scale corpus.

Current researches on constructing text corpora on the Web mainly focus on how to collect texts from the Web and how to process them to build a corpus. The Linguistic Data Consortium (LDC) has corpora for twenty languages [6] while using Web search engines to collect texts from the Web. Jones and Ghani [7, 8] propose an approach of automatically generating Web-search queries and collecting documents from the Web to build a language-specific corpus. LE [9] uses Web resources to collect and make efficient textual corpora and then propose a set of filtering tools for fast language model construction in minority languages. Rayson [10] presents a proposal to facilitate the use

V. Matoušek and P. Mautner (Eds.): TSD 2007, LNAI 4629, pp. 56–65, 2007.

of an annotated web as corpus by alleviating the annotation bottleneck for corpus data drawn from the web. He described a framework for large scale distributed corpus annotation using P2P technology to meet this need. Resnik [11] uses the STRAND system for mining parallel texts on the Web and then automatically constructed parallel corpora for English and French. Zhang [12] describes a system that overcomes the limitation – it automatically collects high quality parallel bilingual corpora from the web by using multiple features to identify parallel texts via a $k$-nearest-neighbor classifier.

However, constructing a corpus is really a complex task. Above researches only focus on how to collect texts from the Web. Besides, the high-performance NLP needs large scale corpora [13]. Constructing a corpus with lower cost, shorter building period, high efficiency and satisfied quality is still a major problem in corpora. Unfortunately, there are rarely any researches in this area. So our purpose is to provide a basic and general framework for building a corpus automatically. In this paper, we propose an approach to automatically build a large scale text corpus based on the trustworthiness, which not only could provide a distributed and large scale computational environment Grid to process the corpus, but also present a method to select the untrustworthy language materials by using the trustworthiness formula to be checked by users. We note that in the following sections, the "language materials" are shortened as LMTs while one item of LMTs is shorted as LMT.

## 2 Methodology

The Web is immense, open and diverse. It contains hundreds of billions of words of text that can be used for building a corpus. To collect and process such enormous data and construct a large scale corpus timely and rapidly, a large scale computational environment is required. On the other hand, the process of constructing a large scale corpus consists of many stages, including selecting text sources, crawling texts, cleaning texts, standardizing LMTs, etc. So it could be regarded as a workflow. Obviously a distributed or parallel computing environment could meet its requirement for efficiency in each stage.

Grid [14] consists of a group of distributed computers that can provide the large scale computation and mass storage to build a corpus. Original LMTs come from the web, where various Web pages are conveniently available to feed into Grid for rapid crawling, Grid can process original LMTs while store them into the corpus by using many NLP techniques. As an economic computing environment, the cost of using Grid resources is lower than a cluster computer.

Another issue of building a corpus automatically is that due to too many mistakes in the corpus, the quality of the corpus is lower than the manual corpus. To improve the quality of a corpus is a hard and necessary task for corpus developers. It's impossible to check all of LMTs manually, as it would cause long building period and high cost. Therefore, we design a model to compute the trustworthiness of each LMT in the corpus and then pick out the untrustworthy ones to be checked by people. Large part of LMTs would be corrected by manual work. As manual check also exist mistakes, our approach allows two or more people to check the same LMT and then chooses the result with the highest trustworthiness as the answer and stores it into the corpus.

## 3  Constructing the Corpus Based on Grid

### 3.1  The Framework to Construct a Corpus

We designed a framework to construct a corpus and provide the services to users based on Information Grid. The framework is showed as figure 1. This framework not only can provide the computational ability to collect and process LMTs rapidly and efficiently, but can also provide the Web Services and interface to end users.

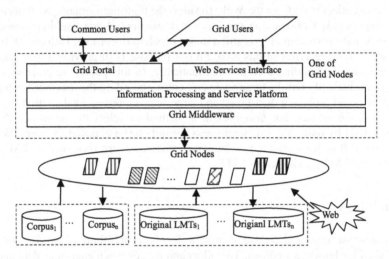

**Fig. 1.** The Framework to construct the corpus and issue the services based on Information Grid

In Figure 1, the Grid is built on Globus [15] and consists of many computing nodes which are used to process LMTs. The functions of each node are as follows:

1) To crawl on the Internet and then download Web pages from Websites;
2) To transfer a Web page to pure text and store it as the original LMT on Grid;
3) To process and transfer the original LMT into a processed LMT and then store it into the corpus;
4) To correct the processed LMT by people and keep its trustworthiness under the threshold;
5) To use database techniques to organize the corpus and distribute it over the Grid for better performance;
6) To provide interface and services for users to access the corpus.

Grid Middleware is the kernel of Grid computing and it provides functions such as remote process management, resource allocation, storage access and authorization etc. Globus Toolkit 4 [16] is used by our framework as the middleware.

The IPSP (Information Processing and Service Platform) [17] provides following functions: resource management, jobs and services scheduling, implementation of services, and corpus deploy, etc.

Portal typically provides personalized capabilities to its users by a pathway to other components. Grid portal is an interface for users to access Grid resources. In our framework, it provides a Website for users to access services of IPSP and the interface for Web applications.

Users are divided into two types: Grid users and common users. Grid users can access the corpus on a Grid node and it's easy and direct to access the corpus through the Grid environment; common users can only access the corpus via the Grid portal on a non-Grid node.

## 3.2  Collecting LMTs Automatically

In our framework there are a group of grid nodes available, so a distributed computing model is designed to collect and process original LMTs from Websites.

There are many solutions mentioned above for collecting LMTs from the Web, email servers, messaging system, etc. We choose the robot method to collect texts. Firstly, we build a parallel web-robot (or web-spider) [18] to crawl on special URLs and download web pages simultaneously. Then we build a cleaning tool [19] to preprocess original web pages and to extract useful contents from HTML files or emails. Finally, we use NLP tools to build a corpus.

## 3.3  Checking LMTs

To improve the quality of automatically building corpus, manual correction is necessary. But it's impossible for people to check all LMTs in a large scale corpus due to its enormous size. Therefore, we put forward an approach that provides a set of formulae to calculate the trustworthiness of each LMT, and then picks out untrustworthy ones to be checked by people. In this way, only a small part of LMTs need to be checked. It's very efficient and could improve the quality of the corpus.

In our experiments, we found out that mistakes in the corpus mainly came from two aspects: the ambiguity of the LMT itself and the shortcoming of algorithms. We set up the following two rules to pick out untrustworthy LMTs to be checked:

1) If the LMT itself is ambiguous for one algorithm, then it is tagged as an unchecked LMT.
2) According to a special algorithm, a threshold $\theta$ is set and all LMTs with the trustworthiness $T$ which is less than $\theta$ should also be tagged as unchecked LMTs.

After above two steps, a checking tool is provided for people to correct those mistakes. To insure the quality of checked LMTs, checking rules are as follows:

1) To provide a unified and convenient tool;
2) Each unchecked LMT should be checked by multiple people;
3) For each LMT, if the result is still not agreed by all people, then the trustworthiness should be calculated and then pick out the result with the highest trustworthiness;
4) For each user, to adjust the contents of unchecked LMTs to meet his ability according to the result of checked LMTs which checked by such user. A good rule allows customization of selecting appropriate unchecked LMTs for a special user.

## 4  The Trustworthiness of LMTs

In this section, firstly we define the formulae to compute the trustworthiness of each LMT in the corpus. Secondly, we assign a threshold value to split LMTs into two parts: trustworthy LMTs set and untrustworthy LMTs set. Thirdly, the untrustworthy LMTs are checked by users. At last, the trustworthiness of all checked LMTs are computed and then trustworthy results are updated into the corpus.

### 4.1  The Definition of the Trustworthiness

**Definition 1.** Trustworthiness, denoted by $T$, shows the degree of something to be trusted. The value of the trustworthiness is between 0 and 1, while 0 means the lowest trustworthiness and 1 means the highest trustworthiness.

The trustworthiness is decided by three factors: Competency, Sincerity and Difficulty. The Competency indicates the knowledge and techniques to accomplish a task. The Sincerity indicates the average quality of all finished task. The Difficulty estimates the difficult level to accomplish a task. The Competency has a nearly invariable value while the other two vary frequently.

We assume that $C$ expresses the Competency, $S$ expresses the Sincerity and $D$ expresses the Difficulty. The relation between the trustworthiness and three factors is defined as:

**Definition 2.** The trustworthiness $T$:

$$T = S @ C @ (1-D) \quad \text{or} \quad t = \sigma(c, s, d)$$

where @ is an operator, $\sigma$ is a function, $c \in C$, $s \in S$, $d \in D$, $t \in T$, and $c, s, d, t \in (0, 1)$.

For a user/program $x$ and a task $y$, and the trustworthiness of task $y$ processed by $x$ can be expressed as $t_x(y$

$$t_x(y) = (c\ (y), s\ (y), d\ (y)) \tag{1}$$

where $\forall y \subset z, t_x(y) \leq t_x(z)$.

**Definition 3.** Operator @ definition:

$$\begin{aligned}
\text{Multiple:} \quad & \sigma_{mul}(c, s, d) = s \times c \times (1-d) \\
\text{Minimum:} \quad & \sigma_{min}(c, s, d) = min(s, c, (1-d)) \\
\text{Combination:} \quad & \sigma_{com}(c, s, d) = 1-(1-s) \times (1-c) \times d
\end{aligned}$$

where $\sigma_{mul}(c, s, d) \leq \sigma_{min}(c, s, d) \leq \sigma_{com}(c, s, d)$.

The application could select one of above operators to calculate the trustworthiness. Which one should be chosen depends on the application itself. Otherwise, a threshold needs to be set for a decision – which value is viewed trustworthy and which one is untrustworthy. The threshold of the trustworthiness is defined as follows:

**Definition 4.** The threshold of the trustworthiness, denoted by $\theta$, is the lowest value which could be trusted. The LMT with the trustworthiness $T$ are less than $\theta$ is untrustworthy while the LMT with the trustworthiness $T$ is higher than $\theta$ is trustworthy.

There are two types trustworthiness of LMTs: *TLMP* and *TLMU*. *TLMP* represents trustworthiness of LMTs produced by a program; *TLMU* represents trustworthiness of LMTs produced by a group users.

## 4.2 TLMP and Its Algorithm

*TLMP* represents trustworthiness of LMTs which is produced by a program. Sincerity of the application is undoubted, so it's set to 1. Competency could be regarded as the precision of the application's algorithm while Difficulty is based on the processing contents. Therefore, *TLMP* is defined as follow:

**Definition 5.** For an application $x$ and LMT $y$, *TLMP* is defined as:

$$TLMP_x(y) = \sigma_{mul}(c_x(y),1,d_x(y)) = c_x(y)\times(1-d_x(y))$$

In the following section, we take the email classification as an example to illustrate how to use *TLMP* to pick out untrustworthy LMTs based on the trustworthiness. The basic principle of email classification is to calculate the similarity values between the unclassified mail and classified mails, and then decide the categorization of the unclassified mail by comparing its similarity.

We assume that the number of categories is $k$, so each category is denoted as $Catalog_i$ ($1 \le i \le k$). For each email $m$, the probability of $m$ belonged to $Catalog_i$ is denoted as $P_m(Catalog_i)$.

All $P_m(Catalog_i)$ can be calculated by a specified algorithm, such as ME, SVM, KNN and Bayes etc. In email categorization we assume that $m$ should be classified to category $j$ if the $P_m(Catalog_j)$ is the highest probability.

We find the reason why mails are classified to irrelevant category they are mainly generated from limitations of algorithms. Currently all algorithms applied to the email categorization are not perfect. The precisions of all algorithms are between 80% and 95%. If we are able to know which email is categorized incorrectly, the precision would be improved significantly. But it's impossible! If we can find such a solution, the problem of text/email categorization would be solved completely and the precision of such algorithm would reach 100%. Therefore, we explore another approach to approximate that effect. If we can pick out a set of emails that are mainly incorrect-categorized emails, the precision may get improved. From the experiments with the algorithm ME, SVM and KNN, we find that most incorrect-categorized emails have following characteristics:

There are multiple Probabilities with high values and it's hard for the algorithm to distinguish them. For example, for an email $m$, $P_m(Catalog_3)$=0.2345, $P_m(Catalog_6)$=0.2343 and $P_m(Catalog_{10})$=0.2348. Actually the algorithm always chose the category with the highest probability (0.2348) as the category for $m$, but it may always be a wrong decision since these three probabilities are so close.

However, it's possible to compare the probabilities and to pick out the untrustworthy email when some of its high probabilities are almost equal.

For an application $x$ and an email $y$, the trustworthiness in the mail classification is defined as $TLMP_x(y)=c_x(y)\times d_x(y)$ while $c_x(y)$ is equal to the precision of the algorithm, and $d_x(y)$ could be computed as follows:

1) For $i=1$ to $k$, to compute the probability $P_y(Catalog_i)$;
2) To pick out the $P_y(Catalog_i)$ with the highest value, noted it as $Max(P_y(Catalog_i))$;
3) To compute the $\Delta_i$ between $Max(P_y(Catalog_i))$ and other $P_y(Catalog_j)$:

$$\Delta_i=Max(P_y(Catalog_i)) -P_y(Catalog_j) \quad (1\leq j\leq k \text{ and } j\neq i) \tag{2}$$

4) To set up a threshold $\theta$, and then count the number $m$ while $\Delta_i \leq \theta$ for $i=1$ to $k$;
5) So $d_x(y)$ can be calculated as follows:

$$d_x(y)=Max(P_y(Catalog_i))\times Min(\Delta_i)/m \tag{3}$$

while $Min(\Delta_i)$ is the minimum one of $\Delta_i$. So the trustworthiness of email classification is:

$$TA_x(y)=c_x(y)\times Max(P_y(Catalog_i))\times Min(\Delta_i)/m \tag{4}$$

We have tested above algorithm on our mail corpus which includes 58872 category-defined mails with 9 categories. We choose 1/3 emails randomly for each category as test set, and the rest as training set. We use ME [20] to classify the test set and the result is showed as in Table 1.

**Table 1.** The result of ME categorizer

| Total number | Correct number | Incorrect number | micro- p |
|---|---|---|---|
| 19624 | 18243 | 1381 | 0.930 |

From our many experiments, $c_x(y)$ is set to 0.93 and the threshold $\theta$ is set to 0.35. For each mail in the test set, $TA_x(y)$ is calculated and the result is showed in Table 2.

**Table 2.** The result of trustworthiness

| $TA_x(y)$ | Categories | Number |
|---|---|---|
| $\geq\theta$ | Correct-categorized mails | 17045 |
| | Incorrect-categorized mails | 323 |
| $<\theta$ | Correct-categorized mails | 1198 |
| | Incorrect-categorized mails | 1059 |

Table 2 shows that 1198 correct-categorized emails were untrustworthy while 1059 incorrect-categorized emails also were untrustworthy. So the total number of emails that need to be checked is 2257 (11.25% of all emails). The lower number of correct-categorized emails that are regarded as untrustworthy and the more incorrect-categorized emails that are regarded as untrustworthy, the better the algorithm is. Similarly the lower number of emails that need to be checked is, the better the algorithm is. Even though it's impossible to pick out all the incorrect-categorized emails by our algorithm, but if all the untrustworthy emails are classified correctly, the micro-p would be increased to 98.4% and it's an acceptable score.

## 4.3 TLMU

*TLMU* represents trustworthiness of LMTs which produced by users. Sincerity is related to the quality of a user's finished work with same type of LMTs. Competency is based on a user's type. And Difficulty is also based on the contents of LMTs.

We assume all LMTs are divided into $n$ types and each type is denoted as $y_i$ $(1 \leq i \leq n)$. And $y_{i,j}$ means a special user who checks LMTs with the type $y_i$ in $j^{th}$ time. *TLMU* is defined as follows:

**Definition 5.** The trustworthiness of user $x$ checking LMTs with type $y_i$ in his $j^{th}$ time is defined as:

$$TLMU_x(y_{i,j}) = \sigma_{com}(c_x(y_{i,j}), s_x(y_{i,j}), d_x(y_{i,j})) = 1 - (1 - c_x(y_{i,j}))(1 - s_x(y_{i,j}))d_x(y_{i,j}) \tag{5}$$

$c_x(y_{i,j})$ means the competency of user $x$ when he is checking LMTs with the type $y_i$ in his *jth* time. All users are divided into four types: experts, staffs, registered users and unregistered users. Their initialized values $c_x(y_{i,0})$ are set to 0.9, 0.7, 0.5 and 0.3 respectively. $c_x(y_{i,j})$ is a dynamic variable and it will be adjusted after a user's each checking work on the LMTs with type $y_i$. We use formula (6) to adjust $c_x(y_{i,j})$ after user $x$ have checked LMTs with type $y_i$ in his $(j-1)^{th}$ time:

$$c_x(y_{i,j}) = \begin{cases} c_x(y_{i,j-1}) + TLMU_x(y_{i,j-1}) \times c_x(y_{i,j-1}) & TLMU_x(y_{i,j}) \geq \theta \\ c_x(y_{i,j-1}) - (1 - TLMU_x(y_{i,j-1})) \times c_x(y_{i,j-1})/2 & TLMU_x(y_{i,j}) < \theta \end{cases} \tag{6}$$

All initialized values $s_x(y_{i,0})$ are set to 0.5 and it will also be adjusted after a user's each checking work on the LMTs with type $y_i$. We use formula (7) to adjust $s_x(y_{i,j})$:

$$s_x(y_{i,j}) = \begin{cases} s_x(y_{i,j-1}) + \dfrac{\sum_{z \in U} s_z(y_{i,j-1})}{n} \times s_x(y_{i,j-1}) & TLMU_x(y_{i,j}) \geq \theta \\ s_x(y_{i,j-1}) - \dfrac{\sum_{z \in U} s_z(y_{i,j-1})}{n} \times s_x(y_{i,j-1}) & TLMU_x(y_{i,j}) < \theta \end{cases} \tag{7}$$

where $U$ means a set of users who are assigned to check LMTs with type $y_i$ in his $(j-1)^{th}$ time, and $n$ is the total number of users in the set $U$.

$d_x(y_{i,j})$ is defined as formula (8):

$$d_x(y_{i,j}) = 1 - \frac{1}{n \times l} \tag{8}$$

where $n$ is the total number of users to check LMT $y_{i,j}$, and $l$ is the total number of the different results.

We have evaluated *TMLU* on above email corpus. The test set is 2257 untrustworthy mails. The number of users is 82, including 2 experts, 10 staffs, 50 registered users and 20 unregistered users. The unregistered users are simulated by computers, so the results are random. In our experiment, each email is checked by 10 users and each user checks 100 emails. The experimental results are showed in Table 3.

The precision of the test set is 97.4% and the Micro-p for entire email corpus is increased to 98.1%. So the micro-p score is 5.1% higher than before. This result also shows that there are 58 incorrect-categorized emails, but it's acceptable because the number of incorrect-categorized mails is very small for entire corpus and building a zero-error corpus is impossible, especially for a large scale corpus. Besides, the manual correction also can't avoid mistakes as well.

**Table 3.** Test results after checking work

| Total number | Correct number | Incorrect number | Precision | Micro-p |
|---|---|---|---|---|
| 2257 | 2139 | 58 | 0.974 | 0.981 |

## 5  Conclusion

This paper proposes a new method to build a large scale text corpus and provides an algorithm to correct mistakes in the corpus. Although the precision of our corpus is lower than the corpus built by experts, but our approach has many advantages that manual corpus can't achieve, such as low cost, short building period, large scale and high efficiency.

Currently, we have built more than 10 corpora, including classification corpora, parallel corpora, POS corpora, email corpora and phrase corpora, etc. According to our previous work, we consider that the framework based on Grid is a really good environment for building a large corpus. It can be used to build any kind of corpora. However, our approach to checking the corpus has some limitations. We have only applied it to some kinds of corpora, such as classification corpora, parallel corpora and POS corpora. We think it's very hard to check some corpus by our approach, especially for N-gram corpus. Our future work is to optimize the trustworthiness formula and extend the test scope to reduce the size of language materials that need to be checked and then improve the precision of corpus. We also plan to apply our approach to build more different corpora.

**Acknowledgments.** The authors would like to thank three anonymous reviewers for their comments on this paper. This research was supported by the National Natural Science Foundation of China under Grant No. 60673041 and Nation 863 Project of China under Grant No. 2006AA01Z147.

## References

1. Kennedy, G.: An Introduction to Corpus Linguistics. Longman, London (1998)
2. Biber, D., Conrad, S., Reppen, D.: Corpus Linguistics: Investigating Language Structure and Use. Cambridge University Press, Cambridge (1998)
3. Fairon, C.: Corporator: A Tool for Creating RSS-based Specialized Corpora. In: Proc. of the 2nd Int. Workshop on Web as Corpus, pp. 43–49 (2006)
4. Renouf, A.: WebCorp: Providing a Renewable Energy Source for Corpus Linguistics. Language and Computers 48(1), 39–58 (2003)
5. Kilgariff, A., Grefenstette, G.: Introduction to the Special Issue on the Web as Corpus. Computational Linguistics 29(3), 333–347 (2003)
6. Liberman, M., Cieri, C.: The Creation, Distribution and Use of Linguistic Data. In: Proc. of the 1st Int. Conf. on Language Resources and Evaluation (1998)

7. Jones, R., Ghani, R.: Automatically Building a Corpus for a Minority Language from the Web. In: 38th Meeting of the ACL, Proc. of the Student Research Workshop. Hong Kong, pp. 29–36 (2000)
8. Ghani, R., Jones, R., Mladenic, D.: Building Minority Language Corpora by Learning to Generate Web Search Queries. Knowledge and Information Systems 7(1), 56–83 (2005)
9. Le, B.V., Bigi, B., Besacier, L., Castelli E.: Using the Web for Fast Language Model Construction in Minority Languages. In: Eurospeech, pp. 3117–3120 (2003)
10. Rayson, P., Walkerdine, J., Fletcher, H.W., Kilgarriff, A.: Annotated Web as Corpus. In: Proc. of the 2nd Int. Workshop on Web as Corpus, pp. 27–33 (2006)
11. Resnik, P., Smith, A.N.: The Web as a Parallel Corpus. Computational Linguistics 29(3), 349–380 (2003)
12. Zhang, Y., Wu, K., Gao, J., Vines, P.: Automatic Acquisition of Chinese-English Parallel Corpus from the Web. In: Proc. of ECIR-06, 28th European Conf. on Information Retrieval (2006)
13. Banko, M., Brill, E.: Scaling to Very Very Large Corpora for Natural Language Disambiguation. In: Proc. of the 39th Annual Meeting on Association for Computational Linguistics, pp. 26–33 (2001)
14. Foster, I., Kesselman, C., Nick, J., et al.: Computer Grid Services for Distributed System Integration. IEEE Computer 35(6), 37–46 (2002)
15. Globus project: Globus homepage (2006), http://www.globus.org/
16. Globus project: Globus Toolkit 4.0 Release Manuals (2005), http://www.globus.org/toolkit/docs/4.0/
17. Li, P., Zhu, Q., Zhi, L.: The Design of a Grid Resource Management System Oriented to Information Service. Computer Engineering (2007)
18. Gong, Z., Zhu, Q., Li, P.: Implementation of Web Information Extraction System Based on Similar Pages. Computer Application 26(08), 1983–1986 (2006)
19. Zhu, Q., Gong, Z., Li, P., et al.: An Unsupervised Framework for Robust Web-based Information Extraction. Journal of Chinese Language and Computing 16(3), 157–168 (2006)
20. Li, P., Zhu, Q., Li, J.: A ME Model Based on Feature Template for Chinese Text Categorization. In: Proc. of the 2006 Int. Conf. on Machine Learning, Model, Technologies and Applications, pp. 242–248 (2006)

# Disambiguating Hypernym Relations
# for *Roget's* Thesaurus

Alistair Kennedy[1] and Stan Szpakowicz[1,2]

[1] School of Information Technology and Engineering, University of Ottawa
Ottawa, Ontario, Canada
akennedy@site.uottawa.ca, szpak@site.uottawa.ca
[2] Institute of Computer Science, Polish Academy of Sciences
Warsaw, Poland

**Abstract.** *Roget's* Thesaurus is a lexical resource which groups terms by semantic relatedness. It is *Roget's* shortcoming that the relations are ambiguous, in
that it does not *name* them; it only shows that there *is* a relation between terms.
Our work focuses on disambiguating hypernym relations within *Roget's* Thesaurus. Several techniques of identifying hypernym relations are compared and
contrasted in this paper, and a total of over 50,000 hypernym relations have been
disambiguated within *Roget's*. Human judges have evaluated the quality of our
disambiguation techniques, and we have demonstrated on several applications
the usefulness of the disambiguated relations.

## 1 Introduction

*Roget's* Thesaurus has proven useful in several applications, including determining semantic similarity between terms [1]. *Roget's* is a good resource for Natural Language
Processing, not the least because it contains many terms and phrases not found in other
lexical resources. One factor limits the usefulness of *Roget's*: unlike in *WordNet* [2],
the relations between terms are not named. Instead, *Roget's* clusters terms according
to certain kinds of implicit semantic relatedness. Although it is usually clear to people
that words in the Thesaurus are related, it is not always clear in what way. In this paper,
we describe methods of disambiguating hypernym relations in *Roget's* Thesaurus. To
demonstrate that this is useful, we show how these relations can improve *Roget's* capacity for solving problems of semantic similarity, synonym identification and analogy
identification. We work with the 1987 version of *Penguin's Roget's Thesaurus* [3].

### 1.1 Semantic Distances in *Roget's* Thesaurus

*Roget's* Thesaurus has been implemented in Java as an Electronic Lexical Knowledge
Base (ELKB) [4]. An 8-level hierarchy for grouping words and phrases in the Thesaurus induces a measure of semantic distance between words/phrases [1]. A distance
is calculated as the length of the shortest path through the hierarchy between two given
terms. A score reflects the level at which both words/phrases appear. The Semicolon
Group contains the most closely related terms, while the Class is the broadest category:

V. Matoušek and P. Mautner (Eds.): TSD 2007, LNAI 4629, pp. 66–75, 2007.
© Springer-Verlag Berlin Heidelberg 2007

- distance 0 – the same *Semicolon Group*
- distance 2 – the same *Paragraph*
- distance 4 – the same *Part of Speech*
- distance 6 – the same *Head*
- distance 8 – the same *Head Group*
- distance 10 – the same *Sub-Section*
- distance 12 – the same *Section*
- distance 14 – the same *Class*
- distance 16 – different *Classes*, or a word or phrase not found

The Part of Speech group found in *Roget's* Thesaurus does not contain all terms/phrases within a particular part of speech, only those terms of a given POS related to a particular subject (Head). There can also be cross references between Heads in the Thesaurus. An example of a paragraph appears in Figure 1. Each line is a semicolon group.

> support, underpinning, (703 aid);
> leg to stand on, point d'appui, footing, ground, terra firma;
> hold, foothold, handhold, toe-hold, (778 retention);
> life jacket, lifebelt, (662 safeguard);
> life-support machine or system;

**Fig. 1.** The first paragraph from Head 218

### 1.2 Related Work on Discovering Hypernyms

It is a time-consuming task to construct a large lexical resource that would be as trustworthy as *WordNet*: much work must be done manually. In recent years there has been research on ways to construct such lexical resources automatically from a corpus, in particular by creating hypernym hierarchies. Often people apply patterns similar to those proposed by Hearst [5], with modifications to improve precision and recall [6,7,8]. People have also considered Machine Learning in the identification of hypernyms in text [9], and mined dictionaries for relations [10], including relations other than hypernyms. In recent years some systems, such as Espresso [11], have been designed to identify a variety of different semantic relations from text. Similar research has been done on labeling semantic classes using *is a* relations [12].

## 2   Potential Relations in *Roget's* Thesaurus

We need to know where in *Roget's* hierarchy we can generally encounter hypernymy. To find out, we took relations from *WordNet* and counted how many of them mapped to *Roget's* Thesaurus at various levels of granularity. We decided that relations would have to be between terms/phrases in the same Semicolon Group, Paragraph or Part of Speech. This eliminates the need for word sense disambiguation (into *Roget's* word senses), since the same word in two different senses rarely appears in the same Part of Speech. We found a total of 57,478 relations in the same Part of Speech, 45,481 in the same Paragraph and 15,106 in the same Semicolon Group. We found relatively few relations in the same Semicolon Group, compared to the Paragraph and Part of Speech. Since about 80% of all relations found at the Part of Speech level also appeared in the same Paragraph, we chose to focus on disambiguating relations in the Paragraph.

## 3   Identifying Relations

To identify hypernym relations in *Roget's* Thesaurus, we look at a variety of resources, using a variety of techniques. For each of the identified hypernym relations we attempted to find all the places where this relation appears in *Roget's*. If both terms in the relation are found to be in the same *Roget's* paragraph, the relation in *Roget's* is disambiguated and labeled as a hypernym relation. To accomplish this effectively, the ELKB is used to generate all morphological forms of the terms. This is necessary since many words in *Roget's* are not in their base form.

*Roget's* Thesaurus is a large resource, with over 50,000 unique nouns and noun phrases, many of them absent from other lexical resources or corpora. We applied three different methods of identifying hypernym/hyponym relations: including lexical resources – *WordNet* and *OpenCyc*; search in dictionaries – Longman Dictionary of Contemporary English (LDOCE) and Wiktionary; and examine large corpora using patterns proposed by Hearst [5] – the British National Corpus (BNC) and the Waterloo MultiText (WMT) corpus. It is our overall research plan to identify hypernymy in as many ways as possible, to allow a multi-faceted disambiguation of hypernym relations in *Roget's* Thesaurus.

### 3.1   Identifying Hypernyms in Existing Ontologies

Existing lexical resources are an obvious source of lexical relations. We worked both with *WordNet* [2] and *OpenCyc* [13]. The relation between a pair of words in *Roget's* is labeled as a hypernym if the two words have a hypernym/hyponym relationship in *WordNet*. The hypernyms in *WordNet* can be any distance from each other in the hypernym tree. The only requirement is that both hypernym and hyponym appear in the same Paragraph in *Roget's*. We have identified 53,404 relations using *WordNet*.

*OpenCyc* is a freely distributed version of *Cyc*, a large general knowledge base. Although not intended as a lexical ontology, it contains a hierarchy of classes and subclasses, called "genls". Phrases are also included in *Cyc*, generally rendered as a single word; for example "PlatonicIdea" stands for "platonic idea". *OpenCyc* contains only a fraction of the relations from the full version of *Cyc*, but we still identified 1,608 relations.

### 3.2   Identifying Hypernyms in Dictionaries

A second source of hypernym/hyponym pairs are machine-readable dictionaries, among them LDOCE [14], often used in the past to find relations in text. We identify hypernym relations in LDOCE using patterns similar to the two presented by Nakamura and Nagao [10].

Nakamura and Nagao [10] have shown these patterns to work well for LDOCE. We also tried to apply them to Wiktionary [15]. This is somewhat more difficult.

Wiktionary, unlike LDOCE, is not built by professionals (not systematically), so patterns frequent in LDOCE may not appear as frequently in Wiktionary. In the end, we found 5,153 hypernyms in LDOCE and 4,483 in Wiktionary that appear in *Roget's*.

### 3.3  Identifying Hypernyms in a Large Corpus

We identify hypernym relations from text, using the six patterns proposed by Hearst [5] on two different resources: the BNC [16] and the Waterloo MultiText System [17]. The BNC already labels each word/phrase with a part-of-speech tag, which is convenient for implementing Hearst's patterns. We used them across all the BNC and discovered almost 30,000 relations, but only 1332 relations appeared in the same *Roget's* paragraph.

The WMT corpus [17] contains half a terabyte of queryable Web data. We ran queries for specific terms in conjunction with Hearst's patterns, for example "such *NP* as *football*" or "Protestant and other *NP*". First we compiled a list of terms that had no hypernyms assigned by any other method we describe in this paper[1]. The list contained 26,430 unique terms. Of the 26,430 unique words searched for, 15,443 had at least one phrase retrieved using this method. Once the phrases have been extracted, they were tagged using Brill's tagger. Since the WMT corpus does not count punctuation in its patterns, many of the extracted sentences could not match Hearst's patterns due to incorrect or irregular punctuation. For 11,392 relations both terms appear in the same *Roget's* Paragraph.

### 3.4  Labeling a Hypernym Network in *Roget's*

The methods we have presented identified 68,717 unique hypernyms, appearing 92,675 times in *Roget's* Thesaurus. The difference is due to the fact that some hypernyms appear in more than one paragraph. Once this has been done, we removed all cycles and redundant hypernym links. A cycle is a series of hypernym links where a term can eventually become its own hypernym, of the form "A *is a* B *is a* ... *is a* C *is an* A". We fix cycles by removing the link that is least likely to be correct. In Section 4.1 we discuss how we determine the accuracy of the hypernyms based on scores assigned by human evaluators. We found 3,756 cycles; the average cycle length was 3.8 links.

Redundant hypernym links appear when there is a series of relations "A *is a* B *is a* ... *is a* C" and also a link "A *is a* C". The relation "A *is a* C" is not incorrect, but it is redundant. We dropped 30,068 redundant hypernym links. After these two fixes, 58,851 non-unique hypernym relations remained.

## 4  Evaluation

### 4.1  Manual Evaluation of Hypernyms

We asked five evaluators, fluent in English, to evaluate the automatically acquired relations as true or false hypernymy. We sampled 200 pairs from each of the six resources.

---

[1] Two other methods of inferring hypernyms from synonyms were attempted, with poor results. They are not included.

The evaluators did not know from which resource the samples came. One evaluation was incomplete. Table 1 shows the scores from each evaluator (R1..R5) as well as the average score Av and Fleiss' Kappa K [18].

**Table 1.** Raters R1-R5: kappa, precision, recall, F-measure

| | R1 | R2 | R3 | R4 | R5 | Av | K | Total | P | R | F |
|---|---|---|---|---|---|---|---|---|---|---|---|
| **BNC** | .68 | .61 | .75 | .67 | .62 | .66 | .44 | 1,332 | .663 | .017 | .034 |
| **CYC** | .95 | .78 | .86 | .86 | .89 | .87 | .38 | 1,608 | .865 | .021 | .041 |
| **LDOCE** | .85 | .59 | .82 | .73 | .93 | .78 | .27 | 5,153 | .782 | .067 | .123 |
| **WMT** | .72 | .39 | .51 | .51 | .57 | .54 | .37 | 11,392 | .536 | .147 | .231 |
| **Wiki** | .73 | .56 | .75 | - | .87 | .73 | .17 | 4,483 | .726 | .057 | .107 |
| **WN** | .86 | .52 | .80 | - | .77 | .74 | .11 | 53,404 | .735 | .690 | .712 |

WMT was the least accurate resource. This is likely due to our using WMT as a last resort: to find relations for terms/phrases not found by any other method. Such terms may be less frequent or may represent concepts harder to identify.

The kappa scores – see Table 1, column K – were not high, particularly for the hypernyms identified in Wiktionary and *WordNet*. These kappa results are somewhat lower than the score of 0.51 shown in Rydin [7] on a similar problem. The low kappa scores and the fact that *WordNet* scored relatively poorly suggests that people are not always good judges of hypernymy. *WordNet's* low scores may be because some hypernyms links appear to be closer to synonymy than to actual hypernymy or because it has many infrequent words senses, of which evaluators may not have been aware.

It is possible to evaluate each resource using precision and recall – see Table 1. Precision (P) is the accuracy of the resource and recall (R) is the proportion of relations found in that resource. Also shown are the total number of relations found in each resource (Total) and the F-measure (F).

## 4.2   Combining the Hypernyms from the Resources

The sets of hypernyms we have produced had to be combined in a way that promises high accuracy. Table 1 shows average accuracy for each resource. These results can be used to determine new accuracies for hypernyms that come from two or more resources. The counts of co-occurring hypernyms for 1-6 resources are 61581, 5839, 1102, 171, 21 and 3.

We used the accuracy assigned to each hypernym pair to break cycles, as discussed in Section 3.4. Let the probability of a false hypernym classified as true in resource A be $P(A)$ (it is 1 - Accuracy from Table 1). When a hypernym pair $x$ appears in just one resource, the probability of error is $P(x) = P(A)$. If $x$ is found in more than one resource, $P(x)$ is calculated as $P(x)=P(A) * P(B) *... * P(Z)$. Once we have determined the probability of error for each hypernym pair, we can determine the average error for the entire set of hypernyms. The total average accuracy is 73.1% over all 68,717 hypernym pairs. Due to the low kappa scores (Table 1), this accuracy may not be entirely reliable. With this method, some disambiguated hypernyms have extremely high accuracy. The hypernyms "drill *is a* tool", "crow *is a* bird" and "cactus *is a* plant" were found in all 6 resources and had probabilities close to 1.0.

### 4.3   Evaluation Through Applications

The last method of evaluating the disambiguated hypernyms in *Roget's* Thesaurus is to test the enhanced Thesaurus on the same applications on which the original (unenhanced) *Roget's* system was tested. The chosen applications make use of a new semantic similarity function that accounts for hypernyms. We start off with a function presented in Jarmasz and Szpakowicz [1] (as seen in Section 1.1), and adjust it for hypernymy. If the two terms are direct hypernyms/hyponyms of each other, we increase the score by 4. We add 3 if there are two hypernym/hyponym links between the terms, 2 for three links, 1 for four links. A penalty of -1 applies to both words if they have no hypernym/hyponym links; this is done because sometimes the relations between a word and the other words in its Head, Paragraph and even Semicolon Group are not clear. If no hypernym for that term exists in its Head, then it becomes more likely that the Paragraph that contains the word does not really represent its true sense. We chose these values because they add reward/penalty to the original similarity function without completely overwhelming it. All this gives a range of scores -2..20, which we shift up to 0..22 to get only non-negative values.

**Semantic Distance and Correlation with Human Annotators.** We tested the new and old semantic similarity measures on three data sets: Miller & Charles [19], Rubenstein & Goodenough [20] and Finkelstein et al. [21][2]. We measured the Pearson product-moment correlation coefficient between the numbers given by human judges and those achieved by the two systems. The results appear in Table 2 where we compare the original and enhanced semantic distance function. We considered only nouns for this task. We found that the improvement on Rubenstein & Goodenough [20] and Finkelstein et al. [21] was statistically significant with a P-value $p < 0.05$ using a Paired Student t-test.

**Table 2.** Results for semantic distances and choosing the correct synonym

| Data Set | Orig | Enh | | | ESL | | TOEFL | | RDWP | |
|---|---|---|---|---|---|---|---|---|---|---|
| | | | | | Orig | Enh | Orig | Enh | Orig | Enh |
| Miller & Charles | 0.773 | 0.836 | | | | | | | | |
| Rubenstein & Goodenough | 0.781 | 0.838 | Right | 38 | 42 | 58 | 58 | 201 | 205 |
| Finkelstein et al. | 0.411 | 0.435 | Wrong | 12 | 8 | 22 | 22 | 99 | 95 |
| | | | Ties | 3 | 0 | 5 | 5 | 23 | 13 |

**Synonym Identification Problems.** The same semantic similarity function can also work for the problem of identifying a correct synonym of a word from a group of

---

[2] The WordSimilarity-353 Test Collection is available at: http://www.cs.technion.ac.il/~gabr/resources/data/wordsim353/wordsim353.html

candidates. We tried a method similar to that found in Jarmasz and Szpakowicz [1]. We used three data sets for this application: Test Of English as a Foreign Language (TOEFL) [22], English as a Second Language (ESL) [23] and Reader's Digest Word Power Game (RDWP) [24]. See Table 2 for the results of the original and the enhanced system. We show the number of correct, incorrect and tied answers; ties are *also* counted as incorrect.

**Analogy Identification Problems.** In an analogy identification problem we get words $W_1, W_2$ and we choose among several other pairs the pair linked by the same relation as $W_1, W_2$. In this way, it is a relation disambiguation problem since the relations between the words is not known. We worked with 374 SAT analogy questions [25] where the correct analogy is selected among five possibilities. The focus of our work is on a subset of the 374 SAT analogy problem where we can identify hypernym relations using the enhanced *Roget's* Thesaurus. For the sake of completeness we do tests on the entire data set, but – since only a fraction of the relations are hypernyms – we cannot expect any improvements to be very large. Let the words in the original pair be $A$ and $B$ and in the candidate pair $C$ and $D$. The distance formula is as follows[3]:

$$dist = |semDist(A, B) - semDist(C, D)| + 1/ \\ (semDist(A, C) + semDist(B, D) + 1)$$

Each word in the data set had previously been labeled with its part of speech. The candidate pair with the lowest distance score is chosen as the correct analogy. We can also modify the function by checking for hypernym analogies. If both the original word pair and one of the analogy candidates are linked by hypernymy, we can prefer that candidate. In such "hypernymy matching", we take into account the number of hypernym links between two terms in the original pair and the potential analogy pair; *dist* is altered by this formula:

$$distAlt = dist - (k - |hypernymDist(A, B) - hypernymDist(C, D)|)$$

Here, $k$ is a constant. Ideally $k$ should be a suitably high number, so that pairs of words that are both related by hypernymy are favoured above pairs that are not both related by hypernymy. In our case, the selected analogy pair rarely had a *dist* score greater than 8, so we chose $k = 8$. We tested four different variations on this algorithm – see Table 3 – on hypernym questions, and all questions. There are 24 cases where the original word pair can be matched by hypernyms. Of these 24 pairs, six more were found to be correct using this new system. All three enhanced systems show considerable improvement over simply using the original semantic distance function without any sort of hypernym matching. All three enhanced methods were found to be statistically significant with a P-value $p < 0.05$ using a Paired Student t-test on the 24-case subset, though not on the full SAT analogy dataset.

---

[3] The formula comes from Jarmasz, M., Nastase, V., Szpakowicz, S.: Roget's Thesaurus as an Electronic Lexical Knowledge Base for Natural Language Processing (submitted to *Language Resources and Evaluation*).

**Table 3.** Results for choosing the correct analogy from a set of candidates

| System | Right | Wrong | Ties | Omit |
|---|---|---|---|---|
| Hypernyms Only | | | | |
| Original | 7 | 15 | 2 | |
| Original with Hypernym Matching | 13 | 9 | 2 | |
| Enhanced | 13 | 10 | 1 | |
| Enhanced with Hypernym Matching | 14 | 9 | 1 | |
| All Data | | | | |
| Original | 124 | 226 | 14 | 10 |
| Original with Hypernym Matching | 130 | 220 | 14 | 10 |
| Enhanced | 129 | 231 | 4 | 10 |
| Enhanced with Hypernym Matching | 130 | 230 | 4 | 10 |

# 5   Conclusion

It was difficult to get strong agreement between raters. With kappa scores ranging between .11 and .44, the rater agreement was not high at all. When we average the accuracy for each of the hypernyms identified with each resource (as determined by averaging the results from the human annotators), the hypernyms disambiguated in *Roget's* Thesaurus are 73% accurate. The accuracy of the hypernyms identified ranges from nearly 100% to as low as 53%.

The enhanced *Roget's* Thesaurus worked better than the original ELKB for most of the data sets on which it was tested. We found statistically significant improvements for the data in Rubenstein & Goodenough [20] and Finkelstein et al. [21]. Improvements on these two data sets as well as on the Miller & Charles data [19] are fairly substantial given the already high scores obtained using the unenhanced *Roget's* Thesaurus. We also found small improvements for the ESL [23] and RDWP [24] data sets. The TOEFL set [22] did not show any improvement, but it did no worse either.

We also found some improvement in solving the SAT Analogy questions [25]. Using the improved semantic distance function did improve the results for answering SAT questions, but the best improvements came from matching hypernym analogies to hypernym solutions. This system could be more effective if more hypernym relations as well as other relations were disambiguated in *Roget's*. The problem of solving analogy questions is not an easy one for people or machines. The most successful system that we are aware of is 56% accurate on the same SAT data set [26], while the average college-bound high-school student gets about 57% accuracy.

# Acknowledgment

Our research is supported by the Natural Sciences and Engineering Research Council of Canada (NSERC) and the University of Ottawa. We would also like to thank Dr. Diana Inkpen, Anna Kazantseva, Darren Kipp and Dr. Vivi Nastase for reading this paper and providing many useful comments.

# References

1. Jarmasz, M., Szpakowicz, S.: Roget's thesaurus and semantic similarity. In: Proc. Conference on Recent Advances in Natural Language Processing (RANLP 2003), pp. 212–219 (2003)
2. Fellbaum, C. (ed.): WordNet – An electronic lexical database. MIT Press, Cambridge, Massachusetts, London, and England (1998)
3. Kirkpatrick, B. (ed.): Roget's Thesaurus of English Words and Phrases. Penguin, Harmondsworth, Middlesex, England (1987)
4. Jarmasz, M., Szpakowicz, S.: The design and implementation of an electronic lexical knowledge base. In: Proc. 14th Biennial Conference of the Canadian Society for Computational Studies of Intelligence (AI 2001), pp. 325–334 (2001)
5. Hearst, M.A.: Automatic acquisition of hyponyms from large text corpora. In: Proc. 14th Conference on Computational linguistics, pp. 539–545 (1992)
6. Caraballo, S.A., Charniak, E.: Determining the specificity of nouns from text. In: Proceedings the Joint SIGDAT Conference on Empirical Methods in Natural Language Processing (EMNLP) and Very Large Corpora (VLC), pp. 63–70 (1999)
7. Rydin, S.: Building a hyponymy lexicon with hierarchical structure. In: Proc. SIGLEX Workshop on Unsupervised Lexical Acquisition, ACL'02, pp. 26–33 (2002)
8. Cederberg, S., Widdows, D.: Using LSA and noun coordination information to improve the precision and recall of automatic hyponymy extraction. In: Proc. Seventh Conference on Natural Language Learning at HLT-NAACL 2003, pp. 111–118 (2003)
9. Snow, R., Jurafsky, D., Ng, A.Y.: Learning syntactic patterns for automatic hypernym discovery. In: Saul, L.K., Weiss, Y., Bottou, L. (eds.) Advances in Neural Information Processing Systems, vol. 17, pp. 1297–1304. MIT Press, Cambridge, MA (2005)
10. Nakamura, J., Nagao, M.: Extraction of semantic information from an ordinary english dictionary and its evaluation. In: Proc 12th Conference on Computational linguistics, Morristown, NJ, USA, Association for Computational Linguistics, pp. 459–464 (1988)
11. Pantel, P., Pennacchiotti, M.: Espresso: Leveraging generic patterns for automatically harvesting semantic relations. In: Proc. 21st International Conference on Computational Linguistics and 44th Annual Meeting of the Association for Computational Linguistics, Sydney, Australia, Association for Computational Linguistics (July 2006), pp. 113–120 (2006)
12. Pantel, P., Ravichandran, D.: Automatically labeling semantic classes. In: Proc. 2004 Human Language Technology Conference (HLT-NAACL-04), pp. 321–328 (2004)
13. Lenat, D.B.: Cyc: A large-scale investment in knowledge infrastructure. Communications of the ACM 38(11) (November 1995)
14. Procter, P.: Longman Dictionary of Contemporary English. Longman Group Ltd. (1978)
15. Wiktionary: Main page - wiktionary (2006),
    http://en.wiktionary.org/wiki/Main_Page/
16. Burnard, L.: Reference guide for the british national corpus (world edition) (2000)
17. Clarke, C.L.A., Terra, E.L.: Passage retrieval vs. document retrieval for factoid question answering. In: SIGIR '03: Proc. 26th Annual International ACM SIGIR Conference on Research and Development in Information Retrieval, pp. 427–428. ACM Press, New York (2003)
18. Fleiss, J.L.: Statistical Methods for Rates and Proportions, 2nd edn. John Wiley & Sons, New York (1981)
19. Miller, G.A., Charles, W.G.: Contextual correlates of semantic similarity. Language and Cognitive Process 6(1), 1–28 (1991)
20. Rubenstein, H., Goodenough, J.B.: Contextual correlates of synonymy. Communication of the ACM 8(10), 627–633 (1965)

21. Finkelstein, L., Gabrilovich, E., Matias, Y., Rivlin, E., Solan, Z., Wolfman, G., Ruppin, E.: Placing search in context: the concept revisited. In: WWW '01: Proc. 10th International Conference on World Wide Web, pp. 406–414. ACM Press, New York (2001)
22. Landauer, T., Dumais, S.: A solution to Plato's problem: The latent semantic analysis theory of acquisition, induction, and representation of knowledge. Psychological Review 104, 211–240 (1997)
23. Turney, P.: Mining the web for synonyms: Pmi-ir versus lsa on toefl. In: Flach, P.A., De Raedt, L. (eds.) ECML 2001. LNCS (LNAI), vol. 2167, pp. 491–502. Springer, Heidelberg (2001)
24. Lewis, M. (ed.): Readers Digest, 158(932, 934, 935, 936, 937, 938, 939, 940), 159(944, 948). Readers Digest Magazines Canada Limited (2000-2001)
25. Turney, P., Littman, M., Bigham, J., Shnayder, V.: Combining independent modules to solve multiple-choice synonym and analogy problems. In: Proceedings International Conference on Recent Advances in Natural Language Processing (RANLP-03), pp. 482–489 (2003)
26. Turney, P.: Similarity of semantic relations. Computational Linguistics 32(3), 379–416 (2006)

# Dependency and Phrasal Parsers
# of the Czech Language: A Comparison*

Aleš Horák[1], Tomáš Holan[2], Vladimír Kadlec[1], and Vojtěch Kovář[1]

[1] Faculty of Informatics
Masaryk University
Botanická 68a, 602 00 Brno, Czech Republic
{hales,xkadlec,xkovar3}@fi.muni.cz
[2] Faculty of Mathematics and Physics
Charles University
Malostranské nám. 25, CZ-11800 Prague, Czech Republic
Tomas.Holan@mff.cuni.cz

**Abstract.** In the paper, we present the results of an experiment with comparing the effectiveness of real text parsers of Czech language based on completely
different approaches – stochastic parsers that provide dependency trees as their
outputs and a meta-grammar parser that generates a resulting chart structure representing a packed forest of phrasal derivation trees.

We describe and formulate main questions and problems accompanying such
experiment, try to offer answers to these questions and finally display also factual
results of the tests measured on 10 thousand Czech sentences.

## 1  Introduction

During last ten years a number of syntax parsers of the Czech language have been implemented with the concentration to real parsing of real texts (in contrast to theoretical
and demonstration parsers created in 80s and 90s of the last century).

Some of those "real text parsers" came into existence in the team around the Prague
Dependency Treebank [1], we will call them as the Prague parsers although the best
ones of them are variants of parsers of British or American authors.

The other set of compared parsers are variants of the parser designed and implemented in the team of NLP laboratory at Masaryk university in Brno (the synt parser
[2]), thus we call it the Brno parser in the context of this paper.

Although all these parsers are tested and used for several years already, their implementations are running more or less independently and no rigorous comparison of their
effectiveness has been done yet.

This paper tries to formulate all problems that have hindered such comparison so far,
then offers a solution of them and finally present the results of the actual comparison.
The Prague parsers have already been compared and rated all together, so the novelty

---

* This work has been partly supported by the Academy of Sciences of Czech Republic under
the projects T100300414, T100300419 and 1ET100300517 and by the Ministry of Education
of CR within the Center of basic research LC536 and by the Czech Science Foundation under
the project 201/05/2781.

V. Matoušek and P. Mautner (Eds.): TSD 2007, LNAI 4629, pp. 76–84, 2007.
© Springer-Verlag Berlin Heidelberg 2007

in this comparison is the Brno parser `synt` that is based on completely different approaches than the Prague parsers.

## 2   The Compared Parsers

In this section, we will shortly describe the parsers used in the prepared measuring and comparison.

### 2.1   The Prague Parsers – Basic Characteristics

The set of dependency parsers selected and denoted as the Prague parsers contains the following representatives:

**McD.** McDonnald's maximum spanning tree parser [3],
**COL.** Collins's parser adapted for PDT [4],
**ZZ.** Žabokrtský's rule-based dependency parser [5],
**AN.** Holan's parser ANALOG – it has no training phase and in the parsing phase it searches in the training data for the most similar local tree configuration [6],
**L2R, R2L, L2R3, R2L3.** Holan's pushdown parsers [7],
**CP.** Holan's and Žabokrtský 's combining parser [5],

The selection of Prague parsers was limited to the parsers contained in CP, which is currently the parser with the best known results on PDT including also other parsers like, e.g., Hall and Novák's corrective modeling parser [8] or Nilsson, Nivre and Hall's graph transformation parser [9]. These parsers were not included in the comparison, since currently we do not have their results for all sentences of the testing data set.

The pushdown parsers, during their training phase, create a set of premise-action rules, and apply it during the parsing phase. In the training phase, the parser determines the sequence of actions which leads to the correct tree for each sentence (in case of ambiguity, a pre-specified preference ordering of the actions is used). During the parsing phase, in each situation the parser chooses the premise-action pair with the highest score. In the tests, we have measured four versions of the pushdown parser: L2R – the basic pushdown parser (left to right), R2L – the parser processing the sentences in reverse order, L23 and R23 – the parsers using 3-letter suffices of the word forms instead of the morphological tags.

### 2.2   The Brno Parser – Basic Characteristics

In contrast to the Prague parsers, the Brno parser `synt` is based only on its meta-grammar, the parser does not have any training phase used to learn the context dependencies of the input texts. All rules that guide the analysis process are developed by linguistic and computer experts with all the drawbacks it can bring (see the Section 3.5 for a description of some of them). The advantage of this process is a better adaptation to yet undescribed language phenomena.

The current meta-grammar contains about 250 meta-rules that allow to describe in a human-maintainable way all possible rules used as the actual input for the chart parsing algorithm formed by 2800 generated rules plus feature agreement tests and other contextual actions used for pruning the resulting chart. This meta-grammar describes more than 90 % of sentences from the PDTB-1.0 corpus (the predecessor of PDT-2.0).

The involved chart parsing algorithm uses a modified *Head-driven chart parser* [10], which provides a very fast parsing of real-text sentences with an average time of 0.07 sec/sentence.

## 3   The Principal Differences of the Parsers

The most principal difference between the parsers is, of course, the underlying formalism and methodology of the parsing process. This is however not the sort of difference that would cause problems in the parser comparison. In this section, we will concentrate on the problems arising with different input and output data structures, different morphological and syntactical tagging and different presuppositions on the input text that all need to be resolved before we can start with the real comparison.

### 3.1   Q1: The Input Format

The input of the Brno parser is either a tagged text (from corpus or from other tagged source) with morphological tags compatible with the tagset of the Czech morphological analyser called Ajka [11] or a plain text (divided into sentences), which is then processed with Ajka. Since Ajka does not resolve ambiguities on the morphological level,[1] the Brno parser generally counts with the possibility of ambiguous surface level tokens.

The Prague parsers use as their input also text split into individual sentences, but with unambiguous morphological tags obtained from Hajič's morphological analyser and tagger [12].

Both morphological analysers (and thus both parser groups) use different morphological tagging systems, which are not 1:1 translatable to each other. However, the differences do not affect the most important morphological features from the point of view of the syntactic analysis, so we have used an automatic conversion with some information stripping.

### 3.2   Q2: Dependency Trees vs. Phrasal Trees

The output of Prague parsers is formed by dependency trees or graphs, whereas the output of the Brno parser is basically formed by packed shared forest of phrasal trees. The Brno parser includes the possibility of sorting the trees of the shared forest and output $N$ trees with the highest *tree rank* (a value obtained as a combination of several "figures of merit," see [13]).

This difference in the output format plus the fact that the Brno team does not yet have a large testing tree bank of phrasal trees for measurements[2] was the cause of the biggest problems in the comparison. Since the measurements had to be done on

---

[1] Ajka provides all possible combinations of morphological features of the input words.

[2] Such tree bank of about 5 thousand phrasal trees is being prepared during this year.

several thousands of sentences, we have decided to use the PDT-2.0 tree bank[3] [14]. Since this tree bank provides only the dependency trees for more than 80 thousand Czech sentences, we have decided to convert them to phrasal trees using the Collin's conversion tool [15] and then measure the differences between the Brno parser output and this "phrasal PDT-2.0" using the *PARSEVAL* and the *Leaf-ancestor assessment* techniques (for more details see the Section 4).

### 3.3  Q3: One Resulting Tree vs. (Shared) Forest

The output of the Brno parser is formed by the resulting *chart* structure, which encompasses a whole forest of derivation trees (all of them, however, have the same root nonterminal that represents the successful analysis).

In order to be able to provide a comparison of this forest with the one tree obtained from PDT 2.0 conversion procedure, we have for each sentence extracted first 100 (or less) trees sorted according to the *tree rank*. Each of these trees was then compared to the one from PDT and the results are displayed with the following 3 numbers: a) *best trees* – one tree from the set that is most similar to the desired tree is selected and compared; b) *first tree* – the tree with the highest tree rank is selected and compared; and c) *average* – the average of all trees is presented.

### 3.4  Q4: Projective vs. Non-projective Trees

The output of the Brno parser is always in the form of projective trees, but a non-projective phrase can, in some cases, be analysed with the mechanism of different rule levels, that allow to handle special kinds of phrases. Nevertheless, the Brno parser is not suitable for analysing non-projective sentences at the moment. In the future, we will have to provide techniques like corrections for non-projective parses described in [8].

On the other hand, the output of the Prague parsers, as a set of dependency edges between words, can cross the word surface order without problems. Thus it can represent projective as well as non-projective sentences.

According to the Prague Dependency Treebank statistics, PDT contains approximately 20 % of non-projective sentences. The sentences selected for comparison are thus not limited to only projective sentences, but the results are counted separately for projective and non-projective sentences.

### 3.5  Q5: The Testing Data Set

For the measuring and comparison of parser effectiveness, we definitely need syntactically annotated data. Such data are available for the dependency trees in PDT. The tree bank has three parts – the training part (train), the testing part for development (d-test) and the testing part for evaluation (e-test).

Since the Prague parsers use the first two sets for development and because there is no such similar tree bank available for the phrasal trees from the Brno parser, we have

---

[3] The Prague Dependency Treebank, version 2.0, was created by the Institute of Formal and Applied Linguistics, http://ufal.mff.cuni.cz

decided to use the PDT e-test part (approx. 10 thousand sentences) for the comparison and we will try to overcome the differences between the parser outputs.

One important difference regarding the testing data set is the fact that the Brno parser does not have any training or learning phase – it is purely grammar based parser. The drawback of this fact is that the Brno parser cannot automatically adapt to kinds of texts that were not intended for analysis. The parser is designed to analyse only sentences of the usual structure. Since the Czech language is a representative of free-word-order languages, the parser allows an analysis of many possible word combinations that can form even very "wild" Czech sentences, however, it refuses to analyse texts like PDT sentences e-test#00017: "*10 - 3 %*" or e-test#00554: "*Dítě 4 - 10 let : 1640* (Child 4–10 years:1640)." The Prague parsers, thanks to their stochastic nature, do not have any problems in analysing such kinds of sentences.

## 4   The Results

As we have described in the Section 3, we have decided to use the PDT-2.0 e-test part, where the morphological tags were automatically converted from the Prague tags to the Ajka tags without ambiguities. The e-test set contains approximately 10 thousand syntactically annotated dependency trees. To get trees comparable to Brno parser output, we needed to convert these dependency trees to phrasal trees.

The conversion proceeded in two steps: first, the PDT-2.0 dependency trees in PML format (the default format in PDT-2.0) were converted into the CSTS format (earlier format of PDT) with PDT tool btred. Then, the Collin's conversion tool [15] was used to obtain PDT-2.0 phrasal trees similar to the output of the Brno parser. The statistical features of the e-test set are:

- 10148 sentences (173586 words)
- 7732 projective sentences
- 2416 non-projective sentences
- 87.7 % Brno parser coverage

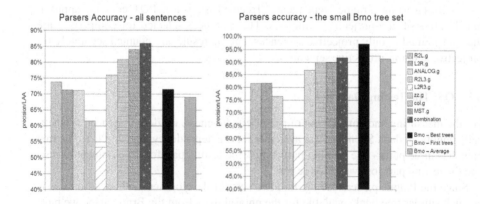

**Fig. 1.** The difference of the results with measuring on the converted PDT trees and on the small Brno tree set

**Table 1.** The results of the Prague parsers (precision = recall)

| Parser | all sentences | non-projective | projective |
|--------|--------------|----------------|-----------|
| R2L | 73.845 % | 69.823 % | 75.735 % |
| L2R | 71.315 % | 67.297 % | 73.204 % |
| ANALOG | 71.077 % | 66.625 % | 73.169 % |
| R2L3 | 61.648 % | 58.276 % | 63.233 % |
| L2R3 | 53.276 % | 49.672 % | 64.912 % |
| zz | 75.931 % | 74.177 % | 76.755 % |
| col | 80.905 % | 75.634 % | 83.383 % |
| MST | 83.984 % | 82.230 % | 84.809 % |
| CP | 85.85 % | 83.434 % | 86.979 % |

**Table 2.** The results of the Brno parser on the e-test set

| | cross-brackets | precision | recall | LAA |
|---|----------------|-----------|--------|-----|
| all sentences | | | | |
| Best trees | 4.473 | 60.228 % | 60.645 % | 71.5 % |
| First trees | 6.229 | 47.306 % | 50.778 % | 69.1 % |
| Average | 5.799 | 45.627 % | 46.584 % | 69.0 % |
| projective sentences | | | | |
| Best trees | 3.619 | 66.718 % | 68.663 % | 73.1 % |
| First trees | 5.289 | 53.028 % | 57.630 % | 70.6 % |
| Average | 4.942 | 50.859 % | 52.552 % | 70.5 % |
| non-projective sentences | | | | |
| Best trees | 7.251 | 39.615 % | 35.727 % | 65.6 % |
| First trees | 9.325 | 29.275 % | 29.699 % | 63.5 % |
| Average | 8.625 | 29.112 % | 28.097 % | 63.3 % |

Since the Brno parser does not provide output for all sentences in the e-test set (see the discussion in the Section 3.5), the actual comparison was run only on those sentences from e-test, that were accepted by the Brno parser.

## 4.1   Measuring Techniques

The methodology for measuring the results of dependency parsing is usually defined as computation of the precision and recall of the particular dependency edges in the resulting graph/tree. These quantities are measured for each lexical item and the result is then computed as an average precision and average recall throughout the whole set.

In the case of phrasal trees we use the two following measures, PARSEVAL and leaf-ancestor assessment (LAA).

The PARSEVAL scheme utilizes only the bracketing information from the parser output to compute three values:

- *crossing bracket* – the number of brackets in the tested analyzer's parse that cross the tree bank parse.

**Table 3.** The results of the Brno and Prague parsers on the small Brno tree set

|              | cross-brackets | precision | recall | LAA |
|--------------|---------------|-----------|--------|-----|
| Best trees   | 0.792 | 89.519 % | 92.274 % | 97.2 % |
| First trees  | 2.132 | 70.849 % | 74.358 % | 92.6 % |
| Average      | 2.311 | 63.330 % | 64.453 % | 91.4 % |
| R2L          |       |          | 81.472 % |     |
| L2R          |       |          | 81.634 % |     |
| ANALOG       |       |          | 76.537 % |     |
| R2L3         |       |          | 63.754 % |     |
| L2R3         |       |          | 57.201 % |     |
| zz           |       |          | 86.650 % |     |
| col          |       |          | 90.129 % |     |
| MST          |       |          | 89.889 % |     |
| CP           |       |          | 91.912 % |     |

- *recall* – a ratio of the number of correct brackets in the analyzer's parse to the total number of brackets in the tree bank parse.
- *precision*– a ratio of the number of correct brackets in the analyzer's parse to the total number of brackets in the parse.

There are several known limitations [16] of the PARSEVAL technique. It is not clear whether this metric can be used for comparing parsers with different degrees of structural fineness since the score on this metric is tightly related to the degree of the structural detail.

The leaf-ancestor assessment [17,18] measure is more complicated than PARSE-VAL. It considers a lineage for each word in the sentence, that is, the sequence of node-labels found on the path between leaf and root nodes in the respective trees. The lineages are compared by their edit distance, each of them having the score between 0 and 1. The score of the whole sentence is then defined as the mean similarity of the lineage-pairs for its respective leaves.

Since it considers not only boundaries between the phrases, the LAA measure is supposed to be more objective than the PARSEVAL, even at non-projective sentences. In this comparison we used the Geoffrey Sampson's LAA implementation, available at http://www.grsampson.net/Resources.html.

### 4.2   Problems and Discussion

Overall results of the Prague parsers testing are presented in the Table 1 in the form of percentage of correct dependendences for the whole set of sentences, for non-projecive and for projective only. The results of the Brno parser on the whole testing set (with manual tagging from PDT-2.0), e-test is displayed in the Table 2.

The experiment of comparing the results of parsers with dependency and phrasal outputs has opened several problems that we have tried to cope with. One of the main causes of these problems were the incompatibilities between the "phrasal PDT" trees and phrasal trees from the Brno parser. This was also the main source of low precision and recall of the parser. In order to prove this thesis, we have (manually) prepared a small set of phrasal trees[4] in the form of the Brno parser trees and repeated the measurements for this subset. The improvement of the results of the Brno parser on this small subset may be seen in the Table 3 and in the Figure 1.

## 5  Conclusions and Future Directions

In the paper, we have described a thorough comparison of the techniques and outputs of the two groups of parsers of the Czech language – the stochastic dependency Prague parsers and the meta-grammar phrasal Brno parser. We have summarized and discussed all the problems of a comparison of such different approaches and we have presented the measured results of the experiment. The results show that the Prague stochastic parser are better for general textual data, which do not have to follow (Czech) grammatical structures. However, it is not easy to give such conclusion for proper sentences.

In the future development, we would like to repeat this tests on another set of input data, namely on the prepared Brno phrasal tree bank. The question is whether this different testing set will shuffle the table of results significantly or it will stay more or less the same.

## References

1. Hajič, J.: Building a syntactically annotated corpus: The Prague Dependency Treebank. In: Issues of Valency and Meaning, Prague, Karolinum, pp. 106–132 (1998)
2. Horák, A., Kadlec, V.: New meta-grammar constructs in Czech language parser synt. In: Matoušek, V., Mautner, P., Pavelka, T. (eds.) TSD 2005. LNCS (LNAI), vol. 3658, pp. 85–92. Springer, Heidelberg (2005)
3. McDonald, R.: Discriminative learning and spanning tree algorithms for dependency parsing. PhD thesis, University of Pennsylvania (2006)
4. Hajič, J., Collins, M., Ramshaw, L., Tillmann, C.: A Statistical Parser for Czech. In: Proceedings ACL'99, Maryland, USA (1999)
5. Holan, T., Žabokrtský, Z.: Combining Czech Dependency Parsers. In: Sojka, P., Kopeček, I., Pala, K. (eds.) TSD 2006. LNCS (LNAI), vol. 4188, pp. 95–102. Springer, Heidelberg (2006)
6. Holan, T.: Genetické učení závislostních analyzátorů. In: Sborník semináře ITAT 2005. UPJŠ, Košice (2005)
7. Holan, T.: Tvorba závislostního syntaktického analyzátoru. In: Wiil, U.K. (ed.) MIS 2004. LNCS, vol. 3511, Springer, Heidelberg (2005)
8. Hall, K., Novák, V.: Corrective modeling for non-projective dependency parsing, 42–51 (2005)
9. Nilsson, J., Nivre, J., Hall, J.: Graph transformations in data-driven dependency parsing. In: Proceedings of the 21st Conference on Computational Linguistics and 44th Annual Meeting of the ACL, Sydney, pp. 257–264 (2006)
10. Horák, A., Kadlec, V., Smrž, P.: Enhancing best analysis selection and parser comparison. In: Sojka, P., Kopeček, I., Pala, K. (eds.) TSD 2002. LNCS (LNAI), vol. 2448, pp. 461–467. Springer, Heidelberg (2002)

---

[4] For 100 sentences randomly chosen from the e-test projective sentences.

11. Sedláček, R.: Morphemic Analyser for Czech. PhD thesis, Masaryk University (2005)
12. Hajič, J.: Disambiguation of Rich Inflection (Computational Morphology of Czech). Karolinum, Charles University Press, Prague, Czech Republic (2004)
13. Horák, A., Smrž, P.: Best analysis selection in inflectional languages. In: Proceedings of the 19th international conference on Computational linguistics, Taipei, Taiwan, Association for Computational Linguistics, pp. 363–368 (2002)
14. Hajič, J.: Complex Corpus Annotation: The Prague Dependency Treebank, Bratislava, Slovakia, Jazykovedný ústav Ľ. Štúra, SAV (2004)
15. Collins, M.: dep2phr – conversion between dependency and phrase structures (1998), http://ufal.mff.cuni.cz/pdt/Utilities/dep2phr/
16. Bangalore, S., Sarkar, A., Doran, C., Hockey, B.A.: Grammar & parser evaluation in the XTAG project (1998),
http://www.cs.sfu.ca/~anoop/papers/pdf/eval-final.pdf
17. Sampson, G.: A Proposal for Improving the Measurement of Parse Accuracy. International Journal of Corpus Linguistics 5(01), 53–68 (2000)
18. Sampson, G., Babarczy, A.: A test of the leaf-ancestor metric for parse accuracy. Natural Language Engineering 9(04), 365–380 (2003)

# Automatic Word Clustering in Russian Texts

Olga Mitrofanova, Anton Mukhin, Polina Panicheva, and Vyacheslav Savitsky

Department of Mathematical Linguistics
Faculty of Philology, St.-Petersburg State University
Universitetskaya emb. 11, 199034 St.-Petersburg, Russia
alkonost-om@yandex.ru, antonmuhin@gmail.com,
ppolin@yandex.ru, v_savitsky@yahoo.com

**Abstract.** The paper deals with development and application of automatic word clustering (AWC) tool aimed at processing Russian texts of various types, which should satisfy the requirements of flexibility and compatibility with other linguistic resources. The construction of AWC tool requires computer implementation of latent semantic analysis (LSA) combined with clustering algorithms. To meet the need, Python-based software has been developed. Major procedures performed by AWC tool are segmentation of input texts and context analysis, co-occurrence matrix construction, agglomerative and $K$-means clustering. Special attention is drawn to experimental results on clustering words in raw texts with changing parameters.

## 1 Introduction

Recent advances in development of linguistic research tools encourage solution of the problems dealing with semantic data extraction from text corpora. One of the most relevant issues is automatic word clustering (AWC) – a procedure which provides data on hierarchical structure of the lexicon which are indispensable in construction of NLP-oriented lexicographic modules (dictionaries, thesauri and ontologies), word sense disambiguation, automatic text indexing, document clustering, information retrieval, etc. AWC procedures are widely used in various NLP systems, e.g.:

- COALS (http://dlt4.mit.edu/~dr/COALS/),
- InfoMap (http://infomap.stanford.edu),
- Google-Sets (http://labs.google.com/sets),
- DSM (http://clg.wlv.ac.uk/demos/similarity/),
- SenseClusters (http://senseclusters.sourceforge.net/), etc.

However, only few AWC modules aimed at Russian corpora processing have been developed for research purposes (e.g.: [1], [2], [3]); thereby, the necessity of constructing such devices seems quite evident.

The aim of the discussed project is elaboration and application of AWC tool for Russian. It is claimed that the proposed tool should allow processing texts of various types and size (raw and morphologically tagged texts, monolingual and multilingual parallel texts, texts of various genres, small and large corpora, etc.), it should be flexible and compatible with other linguistic resources.

V. Matoušek and P. Mautner (Eds.): TSD 2007, LNAI 4629, pp. 85–91, 2007.
© Springer-Verlag Berlin Heidelberg 2007

The project is implemented in stages: first, the environment for processing raw texts is constructed; second, functions which enable operating on morphologically tagged texts are introduced. The paper presents results of AWC tool development achieved so far with regard to computer implementation, as well as experimental data on AWC procedures performed on raw monolingual and multilingual parallel texts.

## 2  AWC Techniques

It is implied that AWC may be successfully performed on the basis of co-occurrence data obtained from corpora. Thus, AWC procedure requires realization of latent semantic analysis (LSA) (e.g.: [4], [5], [6]) and clustering algorithms (e.g.: [7], [8], [9]).

From a linguistic point of view, LSA is based on the possibility of detecting semantic similarity of words by comparing their syntagmatic properties (co-occurrence or distribution analysis). From a technical standpoint, LSA involves construction of vector-space models for processed texts; it means that the sets of contexts for each word are represented as distribution vectors in $N$-dimensional space.

It is possible to evaluate semantic similarity of words by measuring distances between their distribution representations. Numerous metrics are used for the given purpose, e.g. Euclidean measure, Hamming measure, Chebyshev measure, cosine measure, etc. The selection of metrics often depends on qualitative parameters of processed texts. In our case, Euclidean measure was chosen as a basic metric. Results of measuring semantic distances are applied in clustering: words having similar distribution representations as a rule reveal similarity of meaning and should be included into the same cluster.

General approaches to clustering are exposed in hierarchical (agglomerative, divisive), partitioning ($K$-means, $K$-medoid, etc.), hybrid algorithms. Certain linguistic tasks require application of special clustering techniques, e.g. CBC [10], MajorClust [11]. The choice of a particular algorithm is determined by experimental conditions (corpora size, required speed of clustering, constraints for the number of resulting clusters, etc.). At the first stage of the project preference was given to basic clustering algorithms (agglomerative and $K$-means).

Data extracted from texts through AWC procedure admit cogent linguistic interpretation.

## 3  Computer Implementation of AWC

Python-based AWC software developed and adjusted within the project framework maintains a set of conjoined modules performing text preprocessing, agglomerative clustering, $K$-means clustering. Such parameters as names of input files (processed texts and sets of words subjected to clustering), context window size [$\pm s$], weight assignment for context items (*yes/no*), distance metric, clustering technique, ultimate number of clusters ($C$), etc. are determined by users.

The first module provides text preprocessing. Context segmentation is carried out in accordance with a particular context window size. Automatic weight assignment may be done for lexical items taking into account their positions in contexts. The given module is responsible for such operations as forming distribution representations of words, measuring semantic distances, building co-occurrence matrix. The second and the third modules support agglomerative and $K$-means clustering respectively. An output file contains co-occurrence data and clustering results.

## 4   Experimental Results on AWC with Various Parameters

In order to determine research potential of AWC tool and to evaluate its effectiveness, a series of experiments on clustering words in raw texts was fulfilled, namely:

- automatic clustering of Russian most frequent and polysemous verbs in the experimental corpus of verbal contexts;
- automatic clustering of descriptors in scientific texts included into the corpus on Corpus Linguistics;
- automatic clustering of words in parallel texts: the original English text of the fairy story «Animal Farm» by G. Orwell and its translation into Russian.

### 4.1   Automatic Clustering of Verbs in Experimental Corpus

AWC proves to be quite productive with respect to distinguishing verbs of different semantic classes. Data on contextual neighbours of verbs and on verbal valency frames extracted from corpora play a crucial role in multilevel text analysis, therefore AWC tool was employed in trial processing of the experimental corpus of verbal contexts. The given corpus containing over 100 000 tokens was formed on the basis of a large Russian corpus *Bokr'onok* built in St.-Petersburg State University [1], [12]. Both raw and morphologically tagged versions of the experimental corpus are accessible.

Trial AWC procedures gave promising results. Major clustering parameters are as follows: context window size $s = \pm 5$, weight assignment for context items – *yes*, distance metric – Euclidean measure, clustering technique – agglomerative.

Experimental procedure was carried out for a set of the most frequent and polysemous verbs belonging to different semantic classes:

- verbs of intellectual activity: *dumat'* (*think*), *ponimat'* (*understand*), etc.;
- verbs of perception: *videt'* (*see*), *smotret'* (*look*), etc.;
- verbs of transmission: *brat'* (*take*), *dat'* (*give*), etc.;
- verbs of functioning: *delat'* (*do*), *rabotat'* (*work*), etc.;
- verbs of management: *deržat'* (*hold*), *brosat'* (*throw*), etc.;
- verbs of movement: *idti* (*walk*), *jehat'* (*drive*), etc.;
- verbs of location: *stojat'* (*stand*), *ležat'* (*lay*), etc.

Clustering of the given verbs was performed successfully, e.g.:

[*idti* (*walk*), *jehat'* (*drive*) [*videt'* (*see*), *smotret'* (*look*)]];
[*idti* (*walk*), *jehat'* (*drive*) [*delat'* (*do*), *rabotat'* (*work*)]];
[*brat'* (*take*), *dat'* (*give*) [*videt'* (*see*), *smotret'* (*look*)]];
[*deržat'* (*hold*), *brosat'* (*throw*) [*dumat'* (*think*), *ponimat'* (*understand*)]];
[*stojat'* (*stand*), *ležat'* (*lay*) [*dumat'* (*think*), *ponimat'* (*understand*)]].

Data on semantic distances for the given verbs seem to be reliable. Fluctuation of distance values indicates that diffrences in verb distribution are more or less significant.

The verbs belonging to the same class show similarity of distributions (their distance values are low), e.g.:

$D$ (*dumat'* (*think*), *ponimat'* (*understand*)) = 0,107;
$D$ (*delat'* (*do*), *rabotat'* (*work*)) = 0,117.

The verbs belonging to different classes reveal diverse distributions (their distance values are high), e.g.:

$D$ (*ponimat'* (*understand*), *ležat'* (*lay*)) = 0,152;
$D$ (*videt'* (*see*), *idti* (*walk*)) = 0,151.

Difference in distribution is also registered for verbs belonging to the same semantic class but being in contrastive relations, e.g.:

$D$ (*brat'* (*take*), *dat'* (*give*)) = 0,131.

However, such verbs are clustered correctly, so that the difference in their distribution seems to be less significant than the difference in distribution for verbs representing separate semantic classes.

### 4.2 Automatic Clustering of Descriptors in Scientific Texts

AWC shows considerable promise in processing terminological items and domain-restricted texts. In such cases clustering data contribute much to adequate domain modelling and ensure development of lexicographic and ontological systems. Linguistic resource involved in the experiment is the corpus on Corpus Linguistics being developed in St.-Petersburg State University and Institute of Linguistic Studies, RAS. The corpus contains texts of research papers in Russian [13], [14], [15], [16]. Each text of the corpus is supplied with metadata which include bibliographic passport and a set of 10 relevant descriptors (key words) indicating the topic of the paper. E.g., text № 2002_72_79 gets such a set of descriptors: [*arhiv* (*archive*), *bank* (*bank*), *dannyje* (*data*), *korpus* (*corpus*), *massiv* (*array*), *poisk* (*retrieval*), *razmetka* (*annotation*), *tekst* (*text*), *format* (*format*), *češskij* (*Czech*)], etc.

Major clustering parameters are as follows: context window size $s = \pm5$, weight assignment for context items – *yes*, distance metric – Euclidean measure, clustering techniques – agglomerative and $K$-means, ultimate number of clusters $C = 3, 5, 7, 9$.

It is worth noting that the results furnished by agglomerative and $K$-means clustering differ distinctly with regard to central cluster size and filling: e.g. text № 2002_72_79, $C = 5$:

agglomerative clustering
[*arhiv* (*archive*), *bank* (*bank*), *massiv* (*array*), *format* (*format*) [*razmetka* (*annotation*) [*češskij* (*Czech*) [*poisk* (*retrieval*) [[*tekst* (*text*), *korpus* (*corpus*)] *dannyje* (*data*)]]]]];

$K$-means clustering
[[*arhiv* (*archive*)] [*bank* (*bank*)] [*razmetka* (*annotation*)] [*dannyje* (*data*), *korpus* (*corpus*), *poisk* (*retrieval*), *tekst* (*text*), *format* (*format*), *češskij* (*Czech*)] (*massiv* (*array*))]].

Alongside revealing semantic relations in the sets of terminological items, trial AWC procedure allows exposure of nuclear descriptors characteristic of Corpus Linguistics domain: *korpus* (*corpus*), *tekst* (*text*), *razmetka* (*annotation*), *poisk* (*retrieval*), *dannyje* (*data*), etc.

Clustering results may be involved in the comparison of documents with partly coinciding sets of descriptors, e.g. text № 2002_72_79 and text № 2002_55_64 contain similar descriptors, their sets being structured uniformly:

[*massiv* (*array*) [*dannyje* (*data*) [*korpus* (*corpus*), *tekst* (*text*)]]];

at the same time, text № 2002_72_79 and text № 2006_16_24 are equally characterized by descriptors *korpus* (*corpus*), *tekst* (*text*), *format* (*format*), *razmetka* (*annotation*), *poisk* (*retrieval*), although they are ordered in different ways:

text № 2002_72_79
[*format* (*format*) [*razmetka* (*annotation*) [*poisk* (*retrieval*) [*tekst* (*text*), *korpus* (*corpus*)]]]];

text № 2006_16_24
[*razmetka* (*annotation*) [[[*korpus* (*corpus*), *tekst* (*text*)] *format* (*format*)] [*poisk* (*retrieval*)]]].

Given the texts which share common descriptors, similar clustering results prove adherence of the texts to the same topic, and conversely, differences in their clustering provide evidence on divergence of corresponding texts in regards to their subject matter.

## 4.3  Automatic Clustering of Words in Parallel Texts

Processing multilingual parallel texts with the help of AWC tool enables us to compare relations between lexical items within semantic classes in contrasting languages and thus to verify adequacy of translation.

The original English text of the fairy-story «Animal Farm» by G. Orwell and its translation into Russian were involved in trial processing, Russian text size of about 24 000 tokens, English text size of about 30 000 tokens.

Clustering within a group of words denoting human beings, animals and birds (over 50 words occurring in both texts) was carried out with the following clustering parameters: context window size $s = \pm 5$, weight assignment for context items – *yes*, distance metric – Euclidean measure, clustering technique – agglomerative.

AWC tool proved to be quite efficient in distinguishing nouns within microgroups, e.g. in differentiating generic and specific names, in proper assignment of generic names, in revealing kinship hierarchy, e.g.:

[*voron* [*ovca, životnoje*]],          [*raven* [*sheep, animal*]];
[*cypl'onok* [*koška, životnoje*]],     [*chicken* [*cat, animal*]];
[*os'ol* [*utka, ptica*]],              [*donkey* [*duck, bird*]];
[*koza* [*ut'ata, ptica*]],             [*goat* [*ducklings, bird*]];
[*ptica* [*utka, golub'*]],             [*bird* [*duck, pigeon*]];
[*cypl'onok* [*kurica, petuh*]],        [*chicken* [*hen, cockerel*]].

In most cases clustering results obtained for Russian and English items coincided, although some peculiar examples of divergent clustering output were found as well, e.g.:

[*kobyla (mare)* [*žereb'onok (foal), lošad'( horse)*]],
[*foal* [*horse, mare*]];

[*ptica (bird)* [*celovek (man), životnoje (animal)*] *borov (boar)*],
[*bird* [*boar* [*man, animal*]]].

The difference in clustering indicates that relations within pairs "original text item vs. translation equivalent" are asymmetrical, possible explanations being different frequency of terms in the original and in translation as well as particular properties of the plot of the analyzed text.

## 5  Conclusions and Work in Progress

AWC tool developed for Russian ensures effective clustering of lexical items in raw texts. Trial procedures involving specialized software confirm reliability of implemented research techniques (LSA and clustering algorithms) forming actual basis of AWC tool. Experimental data show a wide range of possible applications of AWC in linguistic analysis and NLP systems.

Work in progress includes

- software elaboration (user interface perfection, introduction of cluster visualization mode, implementation of additional distance metrics and clustering algorithms, e.g. MajorClust);
- linguistic experiments on AWC with regard to morphologically tagged texts of different size and various genres, mono- and multilingual texts (preliminary results of clustering Russian verbs in synsets taking into account POS tag distributions seem to be rather encouraging).

Further development of the project allows embedding of AWC tool into multilevel linguistic research environment equipped with a corpus manager and subsidiary modules.

## Acknowledgments

The project is supported by the RF Presidential Grant № MK-9701.2006.6. The authors are grateful to Prof. Mikhail Alexandrov, Prof. Irina Azarova, Evguenia Malaia, Anna Marina and Prof. Viktor Zakharov for valuable advices and inspiring discussions concerning the project in question. The authors would like to thank anonymous reviewers for their helpful comments.

## References

1. Azarova, I.V., Marina, A.S.: Avtomatizirovannaja klassifikacija kontekstov pri podgotovke dannyh dl'a kompjuternogo tezaurusa RussNet. In: Kompjuternaja lingvistika i intellektual'nyje tehnologii: Trudy meždunarodnoj konferencii Dialog–2006, Moscow, pp. 13–17 (2006)
2. Baglej, S.G., Antonov, A.V., Meškov, V.S., Suhanov, A.V.: Klasterizacija dokumentov s ispol'zovanijem metainformacii. In: Kompjuternaja lingvistika i intellektual'nyje tehnologii: Trudy meždunarodnoj konferencii Dialog–2006, Moscow, pp. 38–45 (2006)

3. Križanovskij, A.A.: Avtomatizirovannoje postrojenije spiskov semantičeski blizkih slov na osnove rejtinga tekstov v korpuse s giperssylkami i kategorijami. In: Kompjuternaja lingvistika i intellektual'nyje tehnologii: Trudy meždunarodnoj konferencii Dialog–2006, Moscow, pp. 297–302 (2006)
4. Landauer, Th., Foltz, P.W., Laham, D.: Introduction to Latent Semantic Analysis. Discourse Processes 25, 259–284 (1998)
5. Smrž, P., Rychlý, P.: Finding Semantically Related Words in Large Corpora. In: Matoušek, V., Mautner, P., Mouček, R., Tauser, K. (eds.) TSD 2001. LNCS (LNAI), vol. 2166, pp. 108–115. Springer, Heidelberg (2001)
6. Pekar, V.: Linguistic Preprocessing for Distributional Classification of Words. In: Proceedings of the COLING–04 Workshop on Enhancing and Using Electronic Dictionaries. Geneva, pp. 15–21 (2004)
7. Stein, B., Niggemann, O.: On the Nature of Structure and its Identification. In: Widmayer, P., Neyer, G., Eidenbenz, S. (eds.) WG 1999. LNCS, vol. 1665, pp. 122–134. Springer, Heidelberg (1999)
8. Lin, D., Pantel, P.: Induction of Semantic Classes from Natural Language Text. In: Proceedings of ACM Conference on Knowledge Discovery and Data Mining KDD–01. San Francisco, CA, pp. 317–322 (2001)
9. Shin, S.-I., Choi, K.-S.: Automatic Word Sense Clustering Using Collocation for Sense Adaptation. In: Proceedings of the Second International WordNet Conference GWC–2004, Brno, Czech Republic, pp. 320–325 (2004)
10. Pantel, P.: Clustering by Committee. Ph.D. Dissertation, Department of Computing Science, University of Alberta (2003), http://www.isi.edu/ pantel/Content/publications.htm
11. Stein, B., Meyer zu Eissen, S.: Document Categorization with MajorClust. In: Proceedings of the 12th Workshop on Information Technology and Systems WITS–02. Barcelona, Spain, pp. 91–96 (2002)
12. Azarova, I.V., Sinopal'nikova, A.A.: Ispol'zovanije statistiko-kombinatornyh svojstv korpusa sovremennyh tekstov dl'a formirovanija struktury kompjuternogo tezaurusa RussNet. In: Trudy meždunarodnoj konferencii Korpusnaja lingvistika – 2004, St.-Petersburg, pp. 5–15 (2004)
13. Doklady naučnoj konferencii Korpusnaja lingvistika i lingvističeskije bazy dannyh – 2002. St.-Petersburg (2002)
14. Trudy meždunarodnoj konferencii Korpusnaja lingvistika – 2004. St.-Petersburg (2004)
15. Trudy meždunarodnoj konferencii MegaLing–2005: Prikladnaja lingvistika v poiske novyh putej. St.-Petersburg (2005)
16. Trudy meždunarodnoj konferencii Korpusnaja lingvistika – 2006. St.-Petersburg (2006)

# Feature Engineering in Maximum Spanning Tree Dependency Parser

Václav Novák and Zdeněk Žabokrtský

Institute of Formal and Applied Linguistics, Charles University
Malostranské nám. 25, CZ-11800 Prague, Czech Republic
{novak,zabokrtsky}@ufal.mff.cuni.cz

**Abstract.** In this paper we present the results of our experiments with modifications of the feature set used in the Czech mutation of the Maximum Spanning Tree parser. First we show how new feature templates improve the parsing accuracy and second we decrease the dimensionality of the feature space to make the parsing process more effective without sacrificing accuracy.

## 1 Introduction

Dependency parsing has become an increasingly popular discipline in the field of Natural Language Processing. There are numerous systems engaged in competitions such as CoNLL Shared Task 2006 [1]. The most successful parsers use syntactically annotated corpora for supervised training. While most of the attention is naturally drawn to the algorithms employed by the parsers, in our paper we focus on the problem of feature extraction, which is to some extent orthogonal to the problem of finding the appropriate statistical models and algorithms.

Our exploration is limited to the *a-layer* (analytical layer, layer of surface syntax) data of the Prague Dependence Treebank (PDT 2.0),[1] a corpus of Czech texts annotated with syntactical information consisting mainly of dependency relationships and labels of these relationships [2]. At this layer, PDT 2.0 contains 1,500,000 syntactically annotated tokens (around 80,000 sentences) divided into 80% for training, 10% as the development test set and 10% as the evaluation test set. Examples of a-layer sentence representations is given Figures in Figure 1 and in Figure 2. The choice of using only one treebank and one language enables us to experiment with linguistically motivated features and features taking into account the particular annotation scheme.

Our feature functions are evaluated using Ryan McDonald's implementation of Maximum Spanning Tree (MST) parser with Margin Infused Relaxed Algorithm (MIRA) for estimating the optimal feature weights [3], a modification of which was the best performing system for PDT 2.0 at the CoNLL Shared Task 2006. Other parsers with high accuracy reported on PDT 2.0 use feature extraction as well [4,5].

Experiments described in [6] show that there is still a space for accuracy improvement of automatic parsing: the combination of several parsers can reduce errors by 10%. At least a part of this improvement may be caused by the fact that the parsers use their own specific feature functions which detect some relationships hidden to the

---

[1] http://ufal.mff.cuni.cz/pdt2.0/

V. Matoušek and P. Mautner (Eds.): TSD 2007, LNAI 4629, pp. 92–98, 2007.

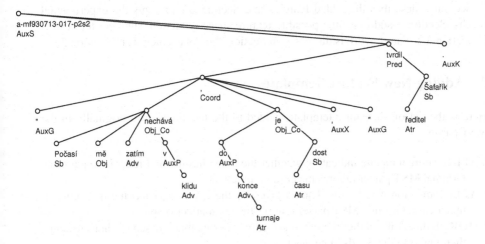

**Fig. 1.** Sample a-layer tree: PDT 2.0 surface-syntactic representation of the sentence *"Počasí mě zatím nechává v klidu, do konce turnaje je času dost," tvrdil ředitel Šafařík.* ("The weather leaves me calm now, there is enough time till the end of the tournament" director Šafařík said.)

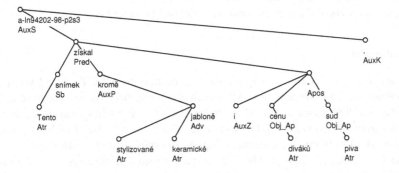

**Fig. 2.** Sample a-layer tree: PDT 2.0 surface-syntactic representation of the sentence *Tento snímek získal kromě stylizované keramické jabloně i cenu diváků – sud piva.* (Besides the stylized ceramic apple tree, this movie obtained also the audience prize—a barrel of beer).

others. This shows that by combining additional features we should be able, in principle, to increase the accuracy of any of these parsers. In practice this is hindered by speed and memory requirements of the algorithms. However, MIRA turns out to be able to effectively estimate a sufficiently large number of features and thus enables us to freely experiment with various feature templates.

Using millions of features requires the user to use at least a 64bit machine and the model is still being loaded for about 9 minutes. To decrease the hardware requirements we explored the possibility to exclude some of the features in the final model and show the impact on both parsing speed and accuracy.

Section 2 describes the added features and Section 2.1 presents the experimental results. Section 3 addresses the possible feature space reduction, showing the results in Section 3.2. Future work is discussed in Section 4 and we conclude by Section 5.

## 2   Adding New Feature Templates

Here is the list of all feature templates added to the templates used originally in the MST parser:

**CAPI.** Feature template indicating whether the token has the first letter in upper case. Original MST parser ignores the case of all letters.

**CAPL.** Indication whether the lemma given by the lemmatizer has the first letter in upper case. Original MST parser ignores the lemmatizer output.

**COOR.** Indication whether the two tokens are "coordinable" (based on their distance, their tags and the words in between)

**DECT.** Individual positions of the 15-letter tags as individual features. Original MST parser uses only 2-letter tags proposed by [5].

**FULT.** Full 15-letter morphological tags of candidate words as new features

**LEME.** Lemmata technical suffices given by the lemmatizer. As described in [7], these suffices distinguish proper names, locations, etc.

**5TAG.** First 5 letters of the tags as opposed to the full 15-letter tags.

### 2.1   Evaluation

Preliminary tests on a portion of train data showed that the **DECT** and **5TAG** feature templates decrease the accuracy. After excluding them from the set, we were able to train the modified parser on the whole training data. The modified parser achieved 84.69% accuracy, which is 2.9% relative error reduction from the baseline MST parser accuracy of 84.24%. If any of the feature templates is omitted, the parsing accuracy decreases as summarized in Table 1. The feature space dimension increased from original 15,486,593 to 21,817,386 in the best model.

**Table 1.** Parsing accuracy with various feature templates included. The ✓ marks indicate the included templates.

| CAPI | CAPL | COOR | FULT | LEME | Accuracy | Feature Space Dimension |
|------|------|------|------|------|----------|-------------------------|
| ✓ | ✓ | ✓ | ✓ | ✓ | **84.69%** | 21,817,386 |
|   | ✓ | ✓ | ✓ | ✓ | 84.61% | 21,817,362 |
| ✓ |   | ✓ | ✓ | ✓ | 84.66% | 21,817,362 |
| ✓ | ✓ |   | ✓ | ✓ | 84.69% | 21,816,918 |
| ✓ | ✓ | ✓ |   | ✓ | 84.58% | 16,989,434 |
| ✓ | ✓ | ✓ | ✓ |   | 84.65% | 19,908,745 |

## 3   Feature Space Reduction

The original feature space dimension (15,486,593) for the MST parser trained on PDT 2.0 train data causes significant difficulties for practical parsing purposes. In this section we explore the effects of reducing the dimension by removing features with the lowest weights. This approach differs from the typical frequency cutoff [8] in that it's applied after the training phase.

### 3.1   Selection of the Least Important Features

The absolute weights ogive in Figure 3 shows that about one third of the features has exactly zero weight after the MIRA training. Removing these features from the feature space doesn't change the parsing outcome at all.

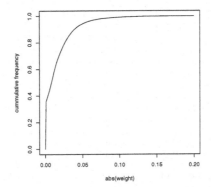

**Fig. 3.** Ogive of absolute values of feature weights

The next step is to remove features with the lowest absolute values of weights. After removing these features, the model is retrained to ensure that all the remaining features receive the correct weights.

### 3.2   Evaluation

The experimental results are presented in Figure 4. The $x$-axis of all the graphs is the threshold for weight absolute value. The experiments show that introducing the threshold significantly reduces the size of the model file as well as the initial loading time. It has also a positive impact on the parsing speed. Most importantly, the parsing accuracy deterioration is much slower than the decrease in the number of features. Setting the threshold to less than 0.02 even doesn't affect the accuracy at all.

– Figure 4(a) shows the loading time of the model in seconds. The loading time is the initialization time of the parser.

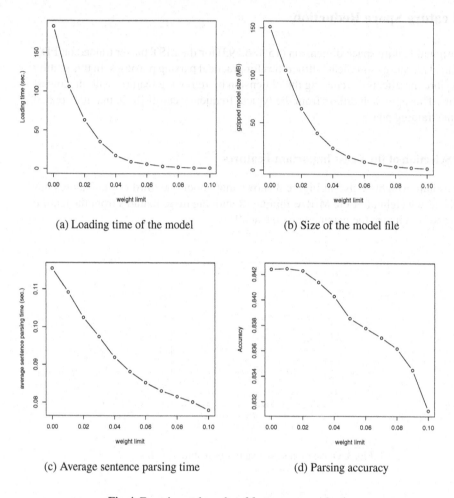

(a) Loading time of the model          (b) Size of the model file

(c) Average sentence parsing time          (d) Parsing accuracy

**Fig. 4.** Experimental results of feature space reduction

- Figure 4(b) shows the size of the model file loaded by the parser. The loading time is linear with the model size.
- Figure 4(c) shows the average sentence parsing time with the initialization time ignored.
- Figure 4(d) shows the unlabeled parsing accuracy, the number of correctly attached nodes.

The experiments show that after reducing the feature space dimension to 0.07% of the original size, loading time as well as size of the model decreases drastically – from 7 minutes to 1 sec., resp. from 231 MB to 1.9 MB – while the accuracy drops only by 1.11 percent points (from 84.24% to 83.13%). Values for other thresholds can be derived by combining Table 2 with the graphs in Figure 4.

**Table 2.** Feature space dimension with various weight absolute value thresholds

| Weight Threshold | Feature Space Dimension |
|---|---|
| 0.00 | 9,856,377 |
| 0.01 | 6,833,618 |
| 0.02 | 4,121,610 |
| 0.03 | 2,438,019 |
| 0.04 | 1,429,533 |
| 0.05 | 858,809 |
| 0.06 | 540,734 |
| 0.07 | 353,412 |
| 0.08 | 237,704 |
| 0.09 | 163,321 |
| 0.10 | 114,621 |

## 4   Future Work

The improvement caused by introducing the coordination feature template should be increased by introducing a more precise rules for coordination ability of toke pairs. Most of the useful added feature templates are not PDT 2.0 dependent and we should test them on other parsers and treebanks as well.

## 5   Conclusion

We have shown that by adding more feature templates we can improve the state of the art dependency parser accuracy for PDT 2.0 Czech texts. We have described the feature templates and we have shown that when we are able to train over a large feature space, the addition of full 15-letters morphological tags for Czech outperforms the 2-letters tags commonly used since [5].

We have also shown that, for practical purposes, the MST parser model size can be decreased to a fraction of the original size without a large loss of the accuracy. Moderately pruned models with the threshold less than 0.02 can even maintain the same accuracy while saving over 70% of resources.

## Acknowledgment

This work was supported by Czech Academy of Science grants 1ET201120505 and 1ET101120503, and by Czech Ministry of Education, Youth and Sports projects LC536 and MSM0021620838. The views expressed are not necessarily endorsed by the sponsors.

We also thank Ryan McDonald for providing us with his parser implementation.

## References

1. Buchholz, S., Marsi, E.: CoNLL-X Shared Task on Multilingual Dependency Parsing. In: Màrquez, L., Klein, D. (eds.) Proceedings of CoNLL-X, New York City, USA (2006)
2. Hajič, J. et al.: Prague Dependency Treebank 2.0. CD-ROM, Linguistic Data Consortium, LDC Catalog No.: LDC2006T01, Philadelphia (2006)

3. McDonald, R., Pereira, F., Ribarov, K., Hajič, J.: Non-projective dependency parsing using spanning tree algorithms. In: HLT '05: Proceedings of the conference on Human Language Technology and Empirical Methods in Natural Language Processing, Vancouver, British Columbia, Canada, Association for Computational Linguistics, pp. 523–530 (2005)
4. Nivre, J., Hall, J., Nilsson, J.: Memory-Based Dependency Parsing. In: Ng, H.T., Riloff, E. (eds.) Proceedings of the Eighth Conference on Computational Natural Language Learning (CoNLL), Boston, Massachusetts, USA, pp. 49–56 (2004)
5. Collins, M., Hajič, J., Ramshaw, L., Tillmann, C.: A statistical parser for Czech. In: Proceedings of the 37th Annual Meeting of the ACL, College Park, MD, USA, Association for Computational Linguistics (1999)
6. Holan, T., Žabokrtský, Z.: Combining Czech Dependency Parsers. In: Sojka, P., Kopeček, I., Pala, K. (eds.) TSD 2006. LNCS (LNAI), vol. 4188, pp. 95–102. Springer, Heidelberg (2006)
7. Hana, J., Zeman, D., Hajič, J., Hanová, H., Hladká, B., Jeřá, E.: Manual for Morphological Annotation, Revision for the Prague Dependency Treebank 2.0. Technical Report TR-2005-27, ÚFAL MFF UK, Prague, Czech Rep. (2005)
8. Malouf, R., van Noord, G.: Wide coverage parsing with stochastic attribute value grammars. In: Su, K.-Y., Tsujii, J., Lee, J.-H., Kwong, O.Y. (eds.) IJCNLP 2004. LNCS (LNAI), vol. 3248, Springer, Heidelberg (2005)

# Automatic Selection of Heterogeneous Syntactic Features in Semantic Similarity of Polish Nouns

Maciej Piasecki[1], Stanisław Szpakowicz[2,3], and Bartosz Broda[1]

[1] Institute of Applied Informatics, Wrocław University of Technology, Poland
maciej.piasecki@pwr.wroc.pl
[2] School of Information Technology and Engineering, University of Ottawa
szpak@site.uottawa.ca
[3] Institute of Computer Science, Polish Academy of Sciences

**Abstract.** We present experiments with a variety of corpus-based measures applied to the problem of constructing semantic similarity functions for Polish nouns. Rich inflection in Polish allows us to acquire useful syntactic features without parsing; morphosyntactic restrictions checked in a large enough window provide sufficiently useful data. A novel feature selection method gives the accuracy of 86% on the WordNet-based synonymy test, an improvement of 5% over the previous results.

## 1 Introduction

The versatility of WordNet [5] as a resource for Natural Language Processing (NLP) must be the envy of researchers who work in languages other than English. Short of translating WordNet as is, the construction of a comparable resource is highly labour-intensive. It is a wide-open question how much of a wordnet can be created automatically. Steps already attempted include automated acquisition of semantic relation, e.g., [7,14]. It begins with a *semantic similarity function* (SSF): $L \times L \rightarrow R$, which maps pairs of *lexical units* (LUs) into real numbers. A LU is a word type or lexeme, which in inflected languages groups forms with different values of morphological categories such as number, gender etc.

We discuss the design of a SSF for Polish noun LUs based on a corpus of Polish. We assume relatively simple language tools (for example, no parsing), very limited manual intervention in the design process and fully automated application of the resulting similarity-measuring tool. The larger context is a computer-assisted development of a Polish WordNet [4,18] (Polish WN), in which a support system suggests synonymy, hypernymy and possibly other semantic relations to the lexicographer. We have decided first to work with nouns, for which we can effectively evaluate the accuracy of our measure using the preliminary version of Polish WN; its noun hierarchy has already been reasonably well developed. We will consider only noun unigrams, and will assume that the corpus includes initially no other information than sequences of words.

In contrast with the extraction of semantic relations based on lexico-syntactic patterns, e.g., [7,14], we begin by assigning a similarity score to any pair of nouns. We treat this as a good basis for a broad-coverage method that combines pattern and statistics, building on the classic idea of distributional similarity [9]: the more often two LUs

V. Matoušek and P. Mautner (Eds.): TSD 2007, LNAI 4629, pp. 99–106, 2007.

occur in similar contexts, the more similar their meanings are. The context can be a document or its part, e.g., a fixed-size text window moving across a document. In [8,12,22] the description of LU occurrences in contexts was enriched with attributes expressing syntactic relation to some other words. In *Latent Semantic Analysis* (LSA) [10], a context is the whole document but the created occurrence matrix is next transformed by *Singular Value Decomposition* (SVD) [1] to a matrix of reduced dimensions. LUs are compared using the cosine measure on the reduced matrix. SSFs applied to vectors describing the contextual distribution include Jaccard [8] and other probabilistic measures [3,6]. An experimental comparative analysis appears in [22].

SSFs are often evaluated in relation to an existing lexical resource, e.g., [8,12]. In LSA the comparison with the *Test of English as a Foreign Language* (TOEFL) gave 64.4% of hits. In [20] 73.75% hits in TOEFL were achieved, recently increased to 97.5% [21], practically solving the problem. [6] proposed a *WordNet-based Synonymy Test* (WBST); WordNet is used to generate "a large set of questions identical in format to those in the TOEFL". The best reported result for nouns is 75.8% [6].

Research on the acquisition of semantic relations has focussed on English. Polish introduces its own challenges: rich inflection (e.g., up to 14 nominal and 119 verbal forms) and virtually free word order (this makes simple positional patterns and even shallow parsing problematic). LSA has been applied to a Polish corpus of short news [15]. [16] discuss a SSF based on noun-adjective interaction, with 81.15% accuracy in WBST (section 4), a very promising result even though only one of the possible syntactic constraints was applied and the values of SSF were not directly comparable among noun pairs differing in both nouns. Here we extend that work [16] by exploring heterogeneous syntactic constraints and making SSF values comparable among different tested noun pairs. We work with KIPI, the largest corpus of Polish [19]. The only NLP tool is TaKIPI [17], a general-purpose morphosyntactic tagger of Polish. Its accuracy (93.44% overall, 86.3% for ambiguous words) is lower than that of a typical English tagger (about 97%), but [16] found no significant negative influence of such accuracy on SSF.

## 2   Checking Restrictions

We represent lexemes with morphological base forms. From documents tagged by TaKIPI, an $N \times C$ matrix $\mathbf{M}$ is created; $C$ is the number of lexico-syntactic features used, $N$ – the number of noun base forms, $\mathbf{M}[n, c]$ – the number of occurrences of the $n$-th noun base form with the $c$-th feature. We worked with nouns in the preliminary version of Polish WN [4,18] ($N_{WN}$) – the 4611 most frequent nouns from KIPI. This helped evaluation – section 4.

Neither predicate-argument structures nor dependency structures [12,22] are readily available for Polish. Following [16], we identified potential syntactic indicators of the semantic properties of nouns, which it appears feasible to check using the constraint language JOSKIPI [17]: modification of a noun by an adjective; coordination of two nouns; modification by a noun in genitive; the presence of a verb for which a given noun in a given case can be an argument.

We show one example: half of a simplified version of a constraint that tests the presence in a window of an particular adjective which agrees on the appropriate attributes with a noun in the window centre. The other half is analogous.

```
or(llook(-1,-5,$A, and(inter(pos[$A],  {adj}),
                       inter(base[$A], {"adjectiveBaseForm"}),
                       agrpp(0, $A, {nmb, gnd, cas}, 3))),
   rlook(1,5,$A,and(...))))
```

The operator `llook` seeks a condition-satisfying word among 5 positions to the left. The part of speech must be `adj`, the base form must match exactly and number/gender/case agreement with the noun must hold. $M[n, c_a]$ is incremented when a condition with $a$ holds for a window with $n$. For each matrix column a constraint including a different base form is assigned. The final form of the condition (developed by repeating manual tests on KIPI and redesign) is this:

– search for a specific adjective or adjectival participle with number/gender/case agreement,
– test for acceptable modifiers (e.g., adverbs or numerals) between the adjective/participle and the noun,
– test for the absence of several other words or phrases.

The $\pm 5$-word window[1] works well [16]: the linguistic constraints eliminate false associations with adjectives further from the noun. Tests for the range reduced to 2 gave no significant change.We designed similar constraints for the other potential indicators, with the lowest accuracy when testing the presence of the nearest verb for a noun in a given case (free word order means many ways in which the noun is not an argument of a verb in the window). We based the constraints on 15768 adjectives, 45651 nouns and 15414 verbs from KIPI.

In the matrix construction phase, we iterate on corpus documents and nouns $n \in N_{WN}$ in the window centre with $\pm 5$ words around $n$ but not crossing sentence boundaries. Occurrences of adjectives and nouns are identified in the window; verbs are sought in the whole sentence, as the expected distance of the association can be larger. If the lexico-syntactic constraint $c$ holds for $n$, $M[n, c_a]$ is incremented. The resulting matrix can have up to 200 000 columns, each accounting for heterogeneous features.

## 3    Feature Selection and Similarity Measures

The nearly 200 000 features in $M$ are computationally manageable but it is unlikely that all of them are needed to calculate the similarity for each noun pair. We follow a fairly typical blueprint for SSF computation.

1. Global selection of features on the basis of global statistical evaluation and perhaps a heuristic assessment.

---

[1] The ranges 2–6 plus 10 had also been tested for adjective constraints, with higher results for the range about 5.

2. Transformation of the matrix cells (or rows), possibly depending on the compared nouns (or rows).
3. Local selection of features for the comparison of two nouns.
4. Similarity calculation for a pair of row vectors.

The novelty is the distinction between global and local feature selection. *Global selection* excludes from further processing features that are not good discriminators for matrix rows. It also improves the efficiency of SSF calculation. We tested these methods: a heuristic elimination of adjectives with uncertain selectivity, e.g., function words classified as adjectives in KIPI (determiners, quantifiers) and ordinal numbers (we already eliminated precisely these adjective in the construction phase), entropy of a column as the measure of noise it introduces, and threshold $tf_c$ for the minimal number of occurrences of feature $c$. Any known method of cell transformation will do: LSA-like [10], *t-test* between a noun and a feature, *Mutual Information* and other functions surveyed in [22].

*Local feature selection* – aiming for nearly optimal comparison of nouns – is implicitly present in co-occurrence retrieval models [22] (CRM). The feature sequences selected from the rows for both nouns may have to be padded (usually with zeroes) if the similarity measure requires equal-size vectors.

Our blueprint seems to encompass many, if not all, methods of SSF construction. For instance, to reimplement CRM, one needs the identity function for global selection, some weight function analysed in CRM for transformation, local selection by the condition $M[n, c_a] > 0$ (applied after transformation) and the CRM F-score as the similarity measure. SSF construction not encompassed by CRM, e.g., Lin's method [12], can also be handily reimplemented.

The results in [16] and preliminary experiments with several different methods point to serious problem in normalising the values generated by SSF in relation to the frequencies of nouns and features in corpus: SSFs generate values in different ranges for different target nouns. It is therefore impossible to compare the strength of similarity between different pairs of nouns, since the same value, produced by the same SSF, can mean high similarity in one case and quite poor similarity in another case. We note that feature values in the matrix depend on frequencies, that no corpus is perfectly balanced, and that weighting functions alone do not solve the problem. We need generalisation. Applying SVD to very sparse matrices does not help [16]. We assume that similarity of two types of objects depends more on what significant features characterise them than on those features' "strength". So, we compare the relative ranking of importance of noun features before comparing their level.

With these assumptions in mind, we propose a *Rank Weight Function*, whose construction has been inspired by the *neighbour set comparison technique* introduced in [11] and modified in [22].

1. Weighted values of the cells are recalculated using a weight function $f_w$:
   $\forall_c M[n_i, c] = f_w(M[n_i, c])$.
2. Features in a row vector $M[n_i, \bullet]$ are sorted in the ascending order on the weighted values.
3. The $k$ highest-ranking features are selected; e.g. $k = 1000$ works well.
4. For each selected feature $c_j$: $M[n_i, c_j] = k - rank(c_j)$.

To compare two nouns, we align the subvectors of ranks, with padding with zeroes where necessary. Now we can apply any similarity measure to vectors of ranks (the cosine measure is an a example). Section 5 discusses those we experimented with. Such transformation of features to rank values introduces a form of generalisation: the relative importance of features is preserved, but the exact values are mapped to a common level.

## 4 Evaluation

We have run the WBST test [6] based of the current contents of Polish WN. For each $n \in N_{WN}$ we randomly chose its synonym from Polish WN, and created all ordered question/answer pairs for every two words in the same synset. Polish WN is fine-grained [4], so a synset often includes only one LU (its meaning is defined by other lexical-semantic relations). We had to extend the basic WBST procedure by considering singleton synsets and their direct hypernyms. To a synonym pair we add three other words not in the same (possibly extended) synset. The evaluation task is to solve the synonymy problem: use the SSF to choose one of four possible answers. The accuracy is calculated as the number of correct answers, with 25% for random selection. An example of a synonymy problem: dział (*department, section*) – fundacja(*foundation, charity*), instytut (*institute*), **komórka**(*division, cell*), ranga (*rank*).

As an additional validation of our results, we tested 24 native speakers of Polish on 2 sets of randomly generated synonymy problem. A set included 79 problems. The average score was 89.29%, and interjudge agreement within one set, measured by Cohen's kappa [2], ranged between 0.19 and 0.47.

## 5 Experiments

We worked with the set $N_{WN}$ of 4611 most frequent nouns from KIPI. Unless noted otherwise, we tested all SSFs on WBST with 4780 problems generated for nouns in KIPI at least 1000 times (the threshold used in [6,16]). In all tests, vectors were padded with zeroes after local selection. Table 1 shows the results.

**Table 1.** WBST experiments: S – adj. stop-list, N – noise threshold, mCom – minimal number of common features, TSC – t-score threshold, mFF – minimal feature frequency

| | Sim. measure | Features | Global Selection | Local Selection | Noise | $WBST$ |
|---|---|---|---|---|---|---|
| 1 | $cos$ | all | S | | 0% | 60% |
| 2 | $cos$ | adj | S | | 0% | 81.15% |
| 3 | $cos$ | noun | S | | 0% | 77.01% |
| 4 | $cos$ | adj+noun | S | | 0% | 82.72% |
| 5 | $cos$ | adj+noun | S,N,mFF | mCom,TSC | 10% | 83.66% |
| 6 | $CRM$ | adj+noun | S,mFF | $D_{mi}$ | 0% | 85.59% |
| 7 | $cos$ | adj+noun | S,N,mFF | TSC,RWF(t-score) | 1% | 86.09% |

**Exp. 1:** a naive join of all features (adjective modification, noun coordination, cased noun's co-occurrence with a verb), 138354 values, no global or local selection, only LSA logarithmic scaling and row entropy normalisation [16] (LongEnt), plus the cosine measure. The accuracy of some verb matrices alone was only about 30%. **Exp. 2, Exp. 3:** identifiy the two best matrices, adjective modification and noun co-ordination, respectively, LongEnt transformation; in the joined matrices of 61419 columns there were only 0.5% non-zero cells. **Exp. 4:** combine the two; these features describe nouns much better than verb constraints, and are to some extent complementary.

**Exp. 5:** as **4** but with *global selection*: minimal frequency of a feature $mFF > 5$ plus 5% features with the highest column noise eliminated, and *local selection* by $tscore(n, f) \geq 2.567$ [13]:

$$tscore(n, f) = \frac{M[n,f] - \frac{TF_n TF_f}{W}}{\sqrt{\frac{TF_n TF_f}{W}}},$$

where $TF_n$, $TF_f$ are the total frequencies of nouns/feature words, and $W$ is number of words processed, plus a threshold for the number of features common to both nouns $mCom > 1\%$ as a constraint for the similarity to be $> 0$ (otherwise similarity is set to 0).

In **Exp. 5** t-score test was used to select only strongly associated features for a noun. The role of local selection based on t-score appeared to dominate: changes in the noise threshold 0%–95% were not significant (but the maximum was for 5%), also the increase of the minimal frequency to 20 did not cause any significant change. The increase of $mCom$ to 5% decreased the accuracy to 70.67% (if all answers have similarity 0, the first one is chosen). A manual inspection of the generated lists of the 20 nouns most similar to the target noun (*pseudo-synsets*) revealed that $mCom = 1\%$ results in pseudo-synsets easier to understand.

**Exp. 6:** as in an instance of CRM [22] with *global selection*, $mFF > 5$, and Mutual Information as a *weighting function*

$$D_{mi}(\mathbf{M}[n, f]) = log(\frac{P(n,f)}{P(f)P(n)}), \text{ where } P(n, f) = \frac{M[n,f]}{N}, N = \sum_i TF_{n_i},$$

$$P(f) = \frac{tf_f}{\sum_i tf_{f_i}}, P(n) = \frac{TF_n}{N}, TF_n \text{ is the frequency of the noun } n \text{ in IPIC,}$$

$N$ — total number of noun occurrences, $tf_f$ — frequency of feature $f$ (sum across the column of $\mathbf{M}$), and *local selection* incorporated in *CRM similarity measure* $sim_{CRM}(n_1, n_2) = \gamma m_h(P(n_1, n_2), R(n_1, n_2)) + (1-\gamma) m_a(P(n_1, n_2), R(n_1, n_2))$, where $\beta, \gamma \in [0, 1]$ are parameters, *precision*

$$P(n_1, n_2) = \frac{\sum_{F(n_1) \cap F(n_2)} D(n_1, f)}{\sum_{F(n_1)} D(n_1, f)} \text{ recall } R(n_1, n_2) = \frac{\sum_{F(n_1) \cap F(n_2)} D(n_2, f)}{\sum_{F(n_2)} D(n_2, f)}, \text{ and}$$
$$F(n_i) = f : D_{mi}(\mathbf{M}[n_i, f]) > 0.$$

We took $\beta = 0.2$ i $\gamma = 0.25$ as for the experiments presented in [22]. Introducing noise threshold changed the result insignificantly. Despite of the good WBST result (85.59%), a manual analysis of pseudo-synsets revealed many associations hard to explain. Such SSF is not useful in semi-automatic work on a wordnet.

**Exp. 7:** the proposed *Rank Weight Function* (RWF) was applied for weighting before local selection; *global selection*: as in **6**, *local selection*: $R$ best features (50–1500 tested) selected using RWF applied to the results of t-score, and the *cosine measure*. The best result of 86.09% – when selecting the best $R = 350$ features in RWF. $R = 1500$ worsened the results slightly. A manual analysis of pseudo-synsets showed better quality than all other methods and than [16]. We also tested the use of $D_{mi}$ instead t-score in RWF, with a much worse result.

## 6  Conclusions

Our SSF construction method is general enough to allow re-implementation of different known methods or construct new versions via flexible combination of elements. Automatic selection of features – on criteria derived from statistics (t-score) and information theory (noise) – works well when we combine features of similar quality, e.g., adjective modification and noun coordination. The techniques do not separate adequately low-accuracy features (e.g. cased nouns co-occurring with verbs – simulated recognition of predicate-argument relations hurts accuracy). The distinction between global and local selection has a technical character but it helps to optimise the set of parameters and their values.

We proposed *Rank Weight Function* (RWF) that compares the relative importance of features instead of their exact values. An SSF built on its basis with *t-score* ranking (SSF$_{RWF(t-score)}$) achieved the accuracy of 86.09% for the nouns occurring $> 1000$ times in KIPI. This result is better than achieved by any other type of SSF and surpasses the result in [16] (and in [6], but clearly those experiments are not directly comparable). SSF$_{RWF(t-score)}$ also produces pseudo-synsets that people understand and that read better than pseudo-synsets produced by other methods, e.g., re-implementation of CRM$_{MI}$ in spite of the fact that CRM$_{MI}$ achieved a similar result 85.59% in WBST. Moreover, SSF$_{RWF(t-score)}$ produces more comparable results for different pair of nouns: similarity above 0.3 usually means very good association, while similarity below 0.2 is rather suspect. The difference in the quality of pseudo-synsets produced by CRM$_{MI}$ and SSF$_{RWF(t-score)}$ raises doubts about WBST. We will investigate this issue further.

We find it promising that our method successfully exploits morpho-syntactic associations in an inflected language with rich morphology and free word order, and requires relatively simple tools and a general corpus. We expect it to be effective for other inflected languages, among them other Slavic languages.

## Acknowledgement

Work financed by the Polish Ministry of Education and Science, project No. 3 T11C 018 29.

## References

1. Berry, M.: Large scale singular value computations. International J. of Supercomputer Applications 6(1), 13–49 (1992)
2. Cohen, J.: A coefficient of agreement for nominal scales. Educational and Psychological Measurement 20, 3–46

3. Dagan, I., Lee, L., Pereira, F.: Similarity-based method for word sense disambiguation. In: Proc. 35th Annual Meeting of the ACL, pp. 56–63 (1997)
4. Derwojedowa, M., Piasecki, M., Szpakowicz, S., Zawisławska, M.: Polish WordNet on a Shoestring. In: Proc. Biannual Conf. of the Society for Computational Linguistics and Language Technology, Universität Tübingen, pp. 169–178 (2007)
5. Fellbaum, C. (ed.): WordNet — An Electronic Lexical Database. MIT, Cambridge (1998)
6. Freitag, D., Blume, M., Byrnes, J., Chow, E., Kapadia, S., Rohwer, R., Wang, Z.: New experiments in distributional representations of synonymy. In: Proc. 9th ACL Conf. on Computational Natural Language Learning, pp. 25–32 (2005)
7. Girju, R., Badulescu, A., Moldovan, D.: Automatic discovery of part-whole relations. Computational Linguistics 32(1), 83–135 (2006)
8. Grefenstette, G.: Evaluation techniques for automatic semantic extraction: Comparing syntactic and window based approaches. In: Proc. ACL Workshop on Acquisition of Lexical Knowledge from Text, Columbus, SIGLEX'93, pp. 143–153 (1993)
9. Harris, Z.S.: Mathematical Structures of Language. Interscience Publishers (1968)
10. Landauer, T., Dumais, S.: A solution to Plato's problem: The latent semantic analysis theory of acquisition. Psychological Review 104(2), 211–240 (1997)
11. Lin, D.: Using syntactic dependency as local context to resolve word sense ambiguity. In: Proc. 35th Annual Meeting of the Association for Computational Linguistics and 8th Conf. of EACL, Madrid, pp. 64–71 (1997)
12. Lin, D.: Automatic retrieval and clustering of similar words. In: Proc. COLING, pp. 768–774 (1998)
13. Manning, C.D., Schütze, H.: Foundations of Statistical Natural Language Processing. The MIT Press, Cambridge (2001)
14. Pantel, P., Pennacchiotti, M.: Esspresso: Leveraging generic patterns for automatically harvesting semantic relations. In: Proc. ACL, pp. 113–120 (2006)
15. Piasecki, M.: LSA based extraction of semantic similarity for Polish. In: Proc. Multimedia and Network Information Systems 2007. Wrocław University of Technology, pp. 99–107 (2006)
16. Piasecki, M., Broda, B.: Semantic similarity measure of Polish nouns based on linguistic features. In: Abramowicz, W. (ed.) BIS 2007, vol. 4439, Springer, Heidelberg (2007)
17. Piasecki, M., Godlewski, G.: Effective Architecture of the Polish Tagger. In: Sojka, P., Kopeček, I., Pala, K. (eds.) TSD 2006. LNCS (LNAI), vol. 4188, Springer, Heidelberg (2006)
18. plWordNet: The homepage of the Polish Wordnet project (2007),
    http://plwordnet.pwr.wroc.pl/main/?lang=en
19. Przepiórkowski, A.: The IPI PAN Corpus, Preliminary Version. Institute of Computer Science PAS (2004)
20. Turney, P.: Mining the web for synonyms: PMI-IR versus LSA on TOEFL. In: Proc. Twelfth European Conf. on Machine Learning, pp. 491–502. Springer, Berlin (2001)
21. Turney, P., Littman, M., Bigham, J., Shnayder, V.: Combining independent modules to solve multiple-choice synonym and analogy problems. In: Proc. International Conf. on Recent Advances in NLP (2003)
22. Weeds, J., Weir, D.: Co-occurrence retrieval: A flexible framework for lexical distributional similarity. Computational Linguistics 31(4), 439–475 (2005)

# Bilingual News Clustering Using Named Entities and Fuzzy Similarity

Soto Montalvo[1], Raquel Martínez[2], Arantza Casillas[3], and Víctor Fresno[2]

[1] GAVAB Group, URJC
soto.montalvo@urjc.es
[2] NLP&IR Group, UNED
{raquel,vfresno}@lsi.uned.es
[3] Dpt. Electricidad y Electrónica, UPV-EHU
arantza.casillas@ehu.es

**Abstract.** This paper is focused on discovering bilingual news clusters in a comparable corpus. Particularly, we deal with the news representation and with the calculation of the similarity between documents. We use as representative features of the news the cognate named entities they contain. One of our main goals consists of proving whether the use of only named entities is a good source of knowledge for multilingual news clustering. In the vectorial news representation we take into account the category of the named entities. In order to determine the similarity between two documents, we propose a new approach based on a fuzzy system, with a knowledge base that tries to incorporate the human knowledge about the importance of the named entities category in the news. We have compared our approach with a traditional one obtaining better results in a comparable corpus with news in Spanish and English.

## 1 Introduction

The huge amount of documents written in different languages that are available electronically, leads to develop tools to manage such amount of information for filtering, retrieving and grouping purposes. Grouping or clustering automatically related multilingual documents can be a useful task for the processing and management of the multilingual information.

Multilingual Document Clustering (MDC) involves dividing a set of $n$ documents, written in different languages, into a specified number $k$ of clusters, so that the documents most similar to other documents will be in the same cluster. Meanwhile a multilingual cluster is composed of documents written in different languages, a monolingual cluster is composed of documents written in one language. MDC has many applications in task such as Cross-Lingual Information Retrieval, the training of parameters in statistics based machine translation, or the alignment of parallel and non parallel corpora, among others. MDC is normally applied with parallel [19] or comparable corpus [14], [4], [8], [13], [18], [21]. In the case of the comparable corpora, the documents usually are news articles.

MDC systems have developed different solutions to group related documents. The strategies employed can be classified in two main groups: (1) the ones which use

V. Matoušek and P. Mautner (Eds.): TSD 2007, LNAI 4629, pp. 107–114, 2007.

translation technologies, and (2) the ones that transform the document into a language-independent representation.

Regarding the first strategy, some authors use machine translation systems ([13], [2]); whereas others translate the document word by word consulting a bilingual dictionary ([14], [18], [4]). One of the crucial issues regarding the methods based on document or feature translation is the correctness of the proper translation. Bilingual resources usually suggest more than one sense for a source word and it is not a trivial task to select the appropriate one. Although word-sense disambiguation methods can be applied, these are not free of errors.

The strategy that uses language-independent representation tries to normalize or standardize the document contents in a language-neutral way; for example: by mapping text contents to an independent knowledge representation, such as thesaurus ([21], [20], [17]), or by recognizing language-independent text features inside the documents ([19], [7], [3]). Both strategies can be either used isolated or combined. Methods based on language-independent representation also have limitations. For instance, those based on thesaurus depend on the thesaurus scope. Numbers or dates identification can be appropriate for some types of clustering and documents; however, for other types of documents or clustering it could not be so relevant and even it could be a source of noise. None of the revised works use as unique knowledge for clustering the cognate named entities shared between both sides of the comparable corpora.

In this work we study two of the crucial issues of MDC: document representation and document similarity calculation. We represent the documents only by means of cognate Named Entities (NE), considering them like independent language features. We propose a document representation by means of different vectors, one per each NE category taken into account. Regarding the similarity, we propose a similarity measure based on a fuzzy rule system, that combines information about the shared NE category. We have tested this approach with a comparable corpus of news written in English and Spanish, obtaining better results than with a traditional approach. We also wanted to confirm preliminary results indicating that the use of only NE is a good source of knowledge for multilingual news clustering since some authors ([5], [6]) suggest that it is not.

In the following section we present our approach for MDC for both, document representation and document similarity calculation phases. Section 3 describes the corpora, as well as the clustering algorithm used, the experiments and results. Finally, Section 4 summarizes the conclusions and the future work.

## 2 Multilingual News Clustering: Our Approach

Our approach is made up of the combination of two proposals for two outstanding aspects in MDC: document representation and document similarity calculation. In this work, we did not study the clustering algorithms, but we used a partitional one instead. In a previous work [15] we obtained preliminary encouraging MDC results by using only the cognate named entities as news features; therefore, in this one we have gone more deeply into the study of the representation of news by means of NE. In that work we proposed a clustering algorithm that grouped news according to the number and category of the shared NE among news. That algorithm is highly dependent on the thresholds and the corpora. However, in this work we tried to propose a more robust approach for document representation and comparison, that could be used with any clustering algorithm. The two proposals are described in detail in the following subsections.

## 2.1 Document Representation: Cognate Named Entities Selection

It is well known NE play an important role in news documents. We want to exploit this characteristic by means of considering them like the only distinguishing features of the documents. In addition, we take into account its specific category as well.

We just consider the following NE categories: PERSON, ORGANIZATION and LOCATION. Other categories, such as DATE, TIME or NUMBER are not taken into account in this work because we think they can lead to group documents with few content in common. However, PERSON, ORGANIZATION or LOCATION NE can be suitable to find common content in documents in a multilingual news corpus.

In most of the MDC approaches the document representation is based on the vector-space model. In this model each document is considered to be a vector, where each component represents the weight of a feature in the document. Usually each document is represented with only one vector that contains all the features. Nevertheless, we have generated several feature vectors for each document. In fact, three different partial feature vectors represent each document, one per NE category taken into account: one vector represents the PERSON NE, other vector the LOCATION NE, and the third one the ORGANIZATION NE. Other work where more than one feature vector per document is used is [5]. They use two vectors, one representing the NE and the other one the nouns.

Our proposal allows to use information about the category of the cognate NE that the news can have in common in order to determine the documents similarity. To represent the comparable corpus we only consider the NE that appear in all the languages involved. This decision considerably reduces the number of features taken into account. The cognate NE identification between languages, as well as the PERSON NE coreference resolution is based on the use of the Levenshtein edit-Distance function (LD), described in detail in [15].

Once the vocabulary (set of the cognate NE shared) has been defined, each element of the feature vectors must be weighted using a weighting function. We use the TF-IDF weighting function, which combines *Term Frequency* (TF) and *Inverse Document Frequency* (IDF) to weight terms.

## 2.2 Document Similarity Using a Fuzzy System

Most of the clustering algorithms determine the grouping according to the similarity (or the distance) among the documents. In this case, the similarity has to be calculated from the three partial feature vectors which represent the documents content by means of the cognate NE.

Other works that represent the content of each document with more than one vector such as [5], fix coefficients for each vector and use different functions in order to carry out linear combinations. In [17] the authors calculate the similarity between clusters in different languages using three vectors with a relative weighted impact of 70%, 20% and 10%, respectively.

In this work, we propose the use of a fuzzy logic system to combine the partial similarities obtained from the three partial vectors. A fuzzy reasoning system sets suitably model the uncertainty inherent to human reasoning processes, by embodying his knowledge and expertise in a set of linguistic expressions that manages words instead of numerical values [11] [9]. Our main aim is to incorporate the human knowledge about the importance of the category of the NE of the news.

Three linguistic variables are defined: PERSON, LOCATION, and ORGANIZA-TION similarity. Each variable represents the similarity between two documents according to the content of the respective vectors. These partial similarities are considered as the inputs of a fuzzy rule system. The linguistic variable values are inputs in the antecedent of each IF-THEN rule in the knowledge base. The rule consequent has a unique variable: the global DOCUMENT similarity. In Figure 1 we present the input and output of the linguistic variables.

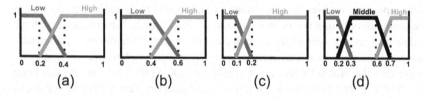

**Fig. 1.** Linguistic variables: (a) PERSON similarity, (b) LOCATION similarity, (c) ORGANIZA-TION similarity, (d) DOCUMENT similarity

The fuzzy set values in the linguistic variables (see values in Figure 1) are defined taken into account the following assumption: we have considered that organization NE are more distinguishing in news, following by person NE, and finally, location NE. For example, is more likely that the same location NE appears in several news of different topic; less probable that a person NE does, and finally, even less that an organization NE does.

The three partial similarities of two documents, corresponding to the input to the three linguistic variables, are obtained with the cosine distance function ($cosine(d_1, d_2)$) normalized to the largest vocabulary dimension: $sim_{part}(d_1, d_2) = cosine(d_1, d_2) \times \frac{dim}{dim_h}$, where $dim$ is the dimension of the partial vectors, and $dim_h$ is the dimension of the largest partial vector.

The knowledge base of the fuzzy system is expressed like a set of IF-THEN rules. These rules try to register the reasoning, or even the common sense that the humans would employ in order to calculate the global similarity between two news from their partial similarities. In this case, we defined the rules based on the same assumption used to determine the fuzzy set values. With these criteria we defined the following 8 rules:

```
if PS is Low and LS is Low and OS is Low then S is Low
if PS is low and LS is Low and OS is High then S is Middle
if PS is Low and LS is High and OS is Low then S is Middle
if PS is Low and LS is High and OS is High then S is Middle
if PS is High and LS is Low and OS is Low then S is Middle
if PS is High and LS is Low and OS is High then S is Middle
if PS is High and LS is High and OS is Low then S is Middle
if PS is High and LS is High and OS is High then S is High
```

where PS, LS and OS represent the similarity between PERSON NE, LOCATION NE, and ORGANIZATION NE respectively; and S represents the resultant document similarity.

## 3  Evaluation

In this Section, first the corpus is described; next, the clustering algorithm used; and finally, the experiments and the results are presented.

### 3.1  Corpus

A Comparable Corpus is a collection of similar texts in different languages or in different varieties of a language. In this work we compiled a collection of news written in Spanish and English belonging to the same period of time. The news are categorized and come from the news agency EFE compiled by HERMES project (http://nlp.uned.es/hermes/index.html). That collection can be considered like a comparable corpus. The articles belong to a variety of IPTC categories [10], including "politics", "crime law & justice", "disasters & accidents", "sports", "lifestyle & leisure", "social issues", "health", "environmental issues", "science & technology" and "unrest conflicts & war", but without subcategories.

First, we performed a linguistic analysis of each document by means of *Freeling* tool [1]. Specifically we carried out: morpho-syntactic analysis, lemmatization, and recognition and classification of NE; the *NEClassifier* Software [16] is used to detect and classify NEs in the English documents.

We have used five subsets of that collection to evaluate the experiments carried out: $S1$, $S2$, $S3$, $S4$ and $S5$. Subset $S1$ consists of 192 news, 100 in Spanish and 92 in English; subset $S2$ consists of 179 news, 93 in Spanish and 86 in English; subset $S3$ consists of 150 news, 79 in Spanish and 71 in English; subset $S4$ consists of 137 news, 71 in Spanish and 66 in English; and finally, subset $S5$ consists of 63 news, 35 in Spanish and 28 in English. Some articles belong to more than one IPTC category according to the automatic categorization. We were interested in a MDC which goes beyond the high level IPTC categories, making clusters of smaller granularity. So, in order to test the MDC results we carried out a manual clustering with each subset. Three persons read the documents and grouped them considering the content of each one. They judged independently and only the identical resultant clusters were selected. The human clustering solution of subset $S1$ is composed of 33 multilingual clusters and 2 monolingual clusters; subset $S2$ has 33 multilingual clusters; 26 multilingual clusters has the human solution of subset $S3$; subset $S4$ has 24 multilingual clusters and 2 monolingual; and the solution of subset $S5$ has 8 multilingual clusters and 2 monolingual.

## 3.2   Clustering Algorithm

Since our objective was not to propose a clustering algorithm, we selected one from the well known CLUTO library [12], the "Direct" algorithm. The input to the "Direct" clustering algorithm is the similarity matrix generated by the fuzzy system, as well as the number of clusters.

## 3.3   Experiments and Results

The quality of the results are determined by means of an external evaluation measure, the F-measure [22]. This measure compares the human solution with the system one. The F-measure combines the precision and recall measures:

$$F(i,j) = \frac{2 \times Recall(i,j) \times Precision(i,j)}{Precision(i,j) + Recall(i,j)}, \tag{1}$$

where $Recall(i,j) = \frac{n_{ij}}{n_i}$, $Precision(i,j) = \frac{n_{ij}}{n_j}$, $n_{ij}$ is the number of members of cluster human solution $i$ in cluster $j$, $n_j$ is the number of members of cluster $j$ and $n_i$ is the number of members of cluster human solution $i$. For all the clusters:

$$F = \sum_i \frac{n_i}{n} max\{F(i)\} \tag{2}$$

The closer to 1 the F-measure value the better.

In order to compare our approach to one traditional, we also represented the documents by means of a vector with all the NE, and we used a well known similarity measure to compute the similarity of two documents: the cosine distance.

The results of the experiments are shown in Table 1. The first column shows the F-measure values for the different subsets of corpus used with the fuzzy approach. The second column shows the F-measure values with the traditional approach, and the third one represent the total number of cognate NE and the number per NE categories used in the corpus representation.

**Table 1.** Clustering results

|    | Fuzzy App. | Traditional. App. | Number of NE |
|----|-----------|-------------------|--------------|
| $S1$ | 0.81 | 0.72 | 815 (288 PER, 348 LOC, 179 ORG) |
| $S2$ | 0.87 | 0.72 | 783 (280 PER, 331 LOC, 172 ORG) |
| $S3$ | 0.85 | 0.76 | 627 (217 PER, 267 LOC, 143 ORG) |
| $S4$ | 0.77 | 0.74 | 587 (218 PER, 235 LOC, 134 ORG) |
| $S5$ | 0.92 | 0.77 | 259 (94 PER, 115 LOC, 50 ORG) |

The best results of the F-measure are obtained with the fuzzy approach, so represent the documents by means of different vectors per NE category and combine them with

**Table 2.** Average percentage of cognate NE identified

|              | Named Entities Cognates |
| ------------ | :---------------------: |
| Person       | 66%                     |
| Location     | 89%                     |
| Organization | 53%                     |

a fuzzy system seems to be a suitable approach for bilingual news clustering. Moreover, as we can see in the Table 2 the average percentage of cognate NE identified is not very high, therefore by improving the cognate NE identification maybe could be possible to obtain better results.

## 4 Conclusions and Future Work

In this paper we have presented an approach for bilingual news clustering. It is made up of two proposals for two outstanding aspects in MDC: feature selection and document similarity calculation. We represent the documents, the news, only by means of cognate Named Entities (NE). Regarding the similarity, we propose a similarity measure based on a fuzzy rule system. These rules try to incorporate the human knowledge about the importance of the category of the named entities of the news.

The experiments were carried out with five different comparable corpus of news written in English and Spanish, by comparing our approach with a traditional one. The clustering algorithm used belongs to the CLUTO library. Our approach obtained better results with all the corpora, so it seems to be appropriate for bilingual news clustering. The main advantage of using only cognate NE is that no translation resources are needed. On the other hand, the cognate identification approach needs the languages involved in the corpora to be of the same alphabet and linguistic family.

Future work will include the compilation of more news corpora in order to confirm these results and conclusions. We will also explore the application of our approach to other type of documents, such as web pages. We think that the improvement of the cognate identification will also increase the accuracy of the MDC results.

## Acknowledgements

We wish to thank the anonymous reviewers for their helpful and instructive comments. This work has been partially supported by MCyT TIN2006-15265-C06-02 and by CAM CCG06-URJC/TIC-0603.

## References

1. Atserias, J., Casas, B., Comelles, E., González, M., Padró, L., Padró, M.: FreeLing 1.3. Syntactic and semantic services in an open-source NLP library. In: Proceedings of the LREC'06. Genoa, Italy (2006), http://garraf.epsevg.upc.es/freeling/

2. Braschler, M., Ripplinger, B., Schuble, P.: Experiments with the Eurospider Retrieval System for CLEF 2001. In: Peters, C., Braschler, M., Gonzalo, J., Kluck, M. (eds.) CLEF 2001. LNCS, vol. 2406, pp. 102–110. Springer, Heidelberg (2002)
3. Chau, R., Yeh, C., Smith, K.A.: A Neural Network Model for Hierarchical Multilingual Text Categorization. In: Wang, J., Liao, X.-F., Yi, Z. (eds.) ISNN 2005. LNCS, vol. 3497, Springer, Heidelberg (2005)
4. Chen, H., Lin, C.: A Multilingual News Summarizer. In: Proceedings of 18th International Conference on Computational Linguistics, pp. 159–165 (2000)
5. Friburger, N., Maurel, D.: Textual Similarity Based on Proper Names. Mathematical Formal Information Retrieval (MFIR'02), 155–167 (2002)
6. Gang, W.: Named Entity Recognition and An Apply on Document Clustering. MCSc Thesis. Dalhousie University, Faculty of Computer Science, Canada (2004)
7. García-Vega, M., Martínez-Santiago, F., Urea-López, L.A., Martín-Valdivia, M.T.: Generación de un tesauro de similitud multilinge a partir de un corpus comparable aplicado a CLIR. Procesamiento del Lenguaje Natural, vol. 28 (2002)
8. Gliozzo, A., Strapparava, C.: Cross language Text Categorization by acquiring Multilingual Domain Models from Comparable Corpora. In: Proceedings of the ACL Workshop on Building and Using Parallel Texts, pp. 9–16 (2005)
9. Hansen, B.K.: Analog forecasting of ceiling and visibility using fuzzy sets. In: Proceedings of the AMS2000 (2000)
10. IPTC - NAA Information Interchange Model Version 4, http://www.iptc.org/std/IIM/4.1/specification/IIMV4.1.pdf
11. Isermann, R.: On Fuzzy Logic Applications for Automatic Control Supervision and Fault Diagnosis. IEEE Trans.Syst. Man and Cybern. 28, 221–235 (1998)
12. Karypis, G.: CLUTO: A Clustering Toolkit. Technical Report: 02-017. University of Minnesota, Department of Computer Science, Minneapolis, MN 55455 (2002)
13. Lawrence, J.L.: Newsblaster Russian-English Clustering Performance Analysis. Columbia computer science Technical Reports, http://www1.cs.columbia.edu/library/2003.html
14. Mathieu, B., Besancon, R., Fluhr, C.: Multilingual document clusters discovery. RIAO'2004, pp. 1–10 (2004)
15. Montalvo, S., Martínez, R., Casillas, A., Fresno, V.: Multilingual Document Clustering: an Heuristic Approach Based on Cognate Named Entities. In: Proceedings of COLING-ACL 2006, pp. 1145–1152 (2006)
16. NEClassifier (2004), http://l2r.cs.uiuc.edu/cogcomp/software.php
17. Pouliquen, B., Steinberger, R., Ignat, C., Ksper, E., Temikova, I.: Multilingual and cross-lingual news topic tracking. In: Proc. of the CoLing'2004, pp. 23–27 (2004)
18. Rauber, A., Dittenbach, M., Merkl, D.: Towards Automatic Content-Based Organization of Multilingual Digital Libraries: An English, French, and German View of the Russian Information Agency Novosti News. In: Proceedings of RCDL01 (2001)
19. Silva, J., Mexia, J., Coelho, C., Lopes, G.: A Statistical Approach for Multilingual Document Clustering and Topic Extraction form Clusters. Pliska Studia Mathematica Bulgarica 16, 207–228 (2004)
20. Steinberger, R., Pouliquen, B., Ignat, C.: Exploting multilingual nomenclatures and language-independent text features as an interlingua for cross-lingual text analysis applications. In: SILTC (2004)
21. Steinberger, R., Pouliquen, B., Scheer, J.: Cross-Lingual Document Similarity Calculation Using the Multilingual Thesaurus EUROVOC. In: Gelbukh, A. (ed.) CICLing 2002. LNCS, vol. 2276, Springer, Heidelberg (2002)
22. van Rijsbergen, C.J.: Foundations of evaluation. Journal of Documentation 30, 365–373 (1974)

# Extractive Summarization of Broadcast News: Comparing Strategies for European Portuguese

Ricardo Ribeiro and David Martins de Matos

L²F/INESC ID Lisboa
Rua Alves Redol, 9, 1000-029 Lisboa, Portugal
{rdmr, david}@l2f.inesc-id.pt
http://www.l2f.inesc-id.pt

**Abstract.** This paper presents the comparison between three methods for extractive summarization of Portuguese broadcast news: feature-based, Maximal Marginal Relevance, and Latent Semantic Analysis. The main goal is to understand the level of agreement among the automatic summaries and how they compare to summaries produced by non-professional human summarizers. Results were evaluated using the ROUGE-L metric. Maximal Marginal Relevance performed close to human summarizers. Both feature-based and Latent Semantic Analysis automatic summarizers performed close to each other and worse than Maximal Marginal Relevance, when compared to the summaries done by the human summarizers.

## 1 Introduction

According to Furui [1], speech summarization poses new challenges when compared to text summarization: issues like recognition errors and disfluencies must be taken into account. Both Furui [1] and Murray, Renals and Carletta [2] agree that although there is a lot of work in text summarization, when it comes to spoken language things are not quite the same.

Most of the work concentrates on sentence extraction, using methods like Latent Semantic Analysis (LSA) [3], Maximal Marginal Relevance (MMR) [4] or feature-based approaches [1]: Maskey and Hirschberg [5] compare several kinds of features for speech summarization of broadcast news, to conclude that the combination of lexical, acoustic/prosodic, structural, and discourse features improves the scoring of sentences to be included in a summary; Murray, Renals and Carletta [2] compare LSA, MMR, and feature-based approaches to summarization of meeting recordings and investigate how the results are influenced by the errors of the automatic speech recognition system, to conclude that LSA has a better performance than the other approaches (although MMR has a comparable performance), and that it had a minimal deterioration in the presence of recognition errors.

SSNT [6,7] is a system which aims to selectively disseminate multimedia contents, mainly TV broadcast news. The system is based on an automatic speech recognition module, that generates the transcriptions used by the topic segmentation, topic indexing, and title&summarization modules. User profiles enable the system to deliver e-mails containing relevant news stories. These messages contain the name of the news

V. Matoušek and P. Mautner (Eds.): TSD 2007, LNAI 4629, pp. 115–122, 2007.
© Springer-Verlag Berlin Heidelberg 2007

service, a generated title, a summary, a link to the corresponding video segment, and a classification according to a thesaurus used by the broadcasting company.

In this document, we present the results of a task that had as its main goal the improvement of the quality of the summaries provided by the multimedia dissemination system, by comparing different strategies for extractive summarization of Portuguese broadcast news: feature-based, MMR, and LSA.

This document is organized as follows: section 2 introduces the broadcast news processing system; section 3 details the summarization task, presenting the experiments done, and the results of the comparison; final remarks close the document.

## 2   Selective Dissemination of Multimedia Contents

SSNT is a system for selective dissemination of multimedia contents, working primarily with Portuguese broadcast news services. It is composed by three main blocks: a capture block, a processing block, and a service block. For the present work, the most important are the capture and processing blocks, depicted in figure 1 (the service block is responsible for user interaction).

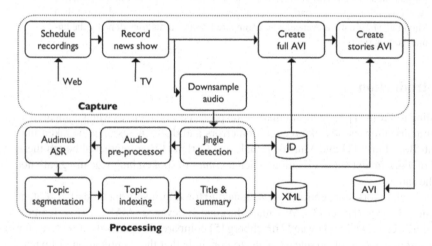

**Fig. 1.** Capture and processing blocks of the SSNT system

The capture block comprehends the processes responsible for scheduling recordings, recording TV shows, and producing two independent streams: a video stream and an uncompressed audio stream, which feeds the processing block.

The processing block is responsible for the automatic speech recognition, topic segmentation, topic indexing, title and summary generation, and for assembling the information needed to compose the final result: currently, a summary is composed by the first $n$ transcribed segments of each news story and a title consisting on the first transcribed segment.

The automatic speech recognition module, based on a hybrid speech recognition system that combines Hidden Markov Models with Multilayer Perceptrons, with an average word error rate of 24.2% [8], greatly influences the performance of the subsequent modules.

The topic segmentation and topic indexing modules were developed by Amaral and Trancoso [9]. Topic segmentation is based on clustering and groups transcribed segments into stories. The algorithm relies on a heuristic derived from the structure of the news services: each story starts with a segment spoken by the anchor. This module achieved an F-measure of 61.6% [8]. The main identified problem was boundary deletion: a problem which impacts the summarization task. Topic indexing is based on a hierarchically organized thematic thesaurus used by the broadcasting company. The hierarchy has 22 thematic areas on the first level, for which the module achieved a correctness of 91.4% [8].

At the time of the experiment, sentence segmentation was not being done. So, we developed a simple sentence boundary detection module using MegaM[1] maximum entropy model optimization package to be applied after the topic segmentation and indexing module. Using the corpora developed in the ALERT European project, the module achieved a precision of 49.3%, a recall of 47.5%, and an F-measure of 48.4% in the test corpus.

## 3    Extractive Summarization

To summarize is to obtain, from a given information source, the most relevant content and delivering it according to a specified context [10]. Since the result is expected to be in conformance with a specific context, the summarization process must have several parameters that can be manipulated to adjust the output. In the case of extractive summarization, the *relation to source* is that all that is in the summary is in the information source.

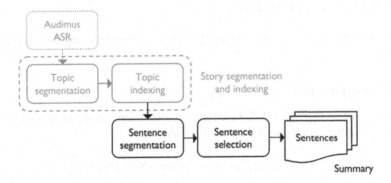

**Fig. 2.** Global idea of the summarization process

Considering the described background, and taking as ultimate input the output of the automatic speech recognition module, the idea is to take each previously segmented

---

[1] http://www.cs.utah.edu/~hal/megam/

and indexed story, divide it into sentences, and select the most relevant sentences to constitute the summary. The process as whole can be seen in figure 2.

There are several strategies for sentence selection when building extract-based summaries. From the initial work of Edmundson [11], where sentence selection is done through the linear combination of features, to more recent approaches like MMR, Maximum Entropy [12], or LSA, it is not completely clear which perform best [2].

Another issue in sentence-based extractive summarization is the specification of a *compression rate*: longer sentences have more information and tend to have higher scores, which means that is easy for them to appear in a summary. The problem is that reducing the content of an input source to 10% of its sentences is different of extracting only 10% of the words or characters (and usually, it means a summary with more than 10% of the words – or characters – of the content of the input source).

### 3.1 Feature-Based Summarization

The following features were used in a linear combination for sentence scoring:

*TF-ISF [13].* TF-ISF means *term frequency-inverse sentence frequency* and is a measure similar to TF-IDF.

*Sentence position.* The position of the sentence in the story. Usually, news stories introduce the most relevant information in the beginning.

*Sentence length.* Too short sentences and too long sentences are penalized. The first ones may be only remarks, speaker changes, greetings, etc., while the second ones can compromise the idea of summarizing by selecting too much information.

*Number of stop words.* Sentences with a large number of stop words (in proportion) have little information and are penalized.

*Number of keywords.* Sentences that contain the words used for indexing the story are more relevant.

### 3.2 MMR Summarization

We used the Lemur Toolkit[2] for MMR-based sentence selection. For the query required by MMR, we used the keywords selected in story indexing.

$$\arg\max_{S_i} \left[ \lambda(Sim_1(S_i, Q)) - (1 - \lambda)(\max_{S_j} Sim_2(S_i, S_j)) \right]$$

Where $Sim_1$ and $Sim_2$ are similarity metrics that do not have to be different; $S_i$ are the yet unselected sentences and $S_j$ are the previously selected ones; $Q$ is the query; and, $\lambda$ is a parameter that allows to configure the result to be from a standard relevance-ranked list ($\lambda = 1$) to a maximal diversity ranking ($\lambda = 0$).

---

[2] http://www.lemurproject.org/

### 3.3 LSA Summarization

For LSA summarization, we implemented a module following the ideas of Gong and Liu [14] and using, for matrix operations, the GNU Scientific Library[3].

LSA is based on the singular vector decomposition (SVD) of the term-sentence frequency $m \times n$ matrix, $M$. $U$ is an $m \times n$ matrix of left singular vectors; $\Sigma$ is the $n \times n$ diagonal matrix of singular values; and, $V$ is the $n \times n$ matrix of right singular vectors (this decomposition is only possible if $m \geq n$):

$$M = U\Sigma V^T$$

### 3.4 Results

For this experiment, we asked two groups of people – one working in speech and language processing with information about the context of the evaluation, and other completely unrelated to the previous context – to produce summaries based on the output of the automatic speech recognition module (story- and sentence-segmented). Evaluation data consisted of a Portuguese news program that the topic segmentation and indexing module divided into 8 stories. The size of the stories ranged from 10 to 100 sentences. The summaries were created using a compression rate of 10% of the original size. The evaluation metric used was ROUGE-L [15] (based on the longest common subsequence).

As shown in figure 3, MMR clearly outperformed both feature-based and LSA summarizers (which had a comparable performance). Although working with broadcast news instead of meeting recordings, this result is not completely in accordance to the one presented by Murray, Renals and Carletta [2], where the best summarizer was LSA-based, and both MMR and LSA outperformed feature-based summarization. Nevertheless, it is observed by those authors that feature selection can be a non-trivial problem: this can explain the worse results of the feature-based summarizer. LSA summarization appears to suffer from the errors produced by the previous modules, although this conclusion merits further study.

Given that Portuguese is a highly inflective language [16] and the nature of the used methods, a lemmatized version of the evaluation was also performed: we lemmatized the input stories and the human summaries; generated the automatic summaries from the lemmatized stories and compared them to the lemmatized human summaries using the ROUGE-L metric.

Surprisingly enough, the feature-based and MMR summarizers had worse results, while LSA had slightly better results. Nevertheless, although in terms of recall the feature-based summarizer performed best, MMR still had the best results. One possible reason for these results is that LSA is based on a term-sentence frequency matrix that tries to capture the relation between the terms and the sentences, through means of the relevance of the terms in the context of all sentences, which gains from the lemmatization process. Even though this reason could also imply a better performance from the feature-based summarizer, since it uses TS-ISF as a feature, that did not happen due to the weight of this feature.

---

[3] http://www.gnu.org/software/gsl/

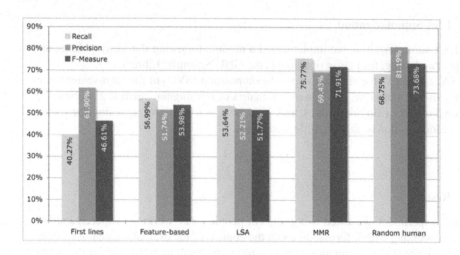

**Fig. 3.** ROUGE-L scores

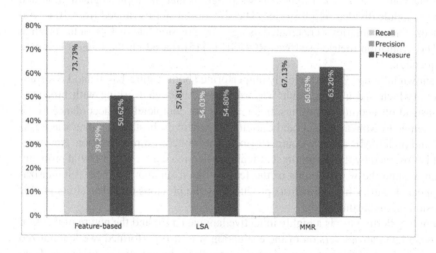

**Fig. 4.** ROUGE-L scores for lemmatized stories

Regarding human summarization, there is considerable agreement between the human summarizers in a small number of sentences, in spite of the very noisy nature of the input, and some dispersion on the rest of each summary (see the performance of the *Random human* summarizer in figure 3). An interesting aspect was that human summarizers tend to ignore sentence boundaries, joining sentences that were only relevant if considered together. In the presence of story segmentation problems, having, for example, two or more stories in the same segment, human summarizers tend to aggregate in the same summary several sub-summaries, each corresponding to a different story. Another issue is that sentences selected by humans usually have a small amount of word recognition errors and are mostly well-structured. One should take this into account by including the confidence level of the recognition in the sentence selection process. For instance, the speech recognition module has specific language models adapted to each anchor, what means that summaries including sentences spoken by the anchor will have less recognition errors (and more relevant information, since in broadcast news most relevant information is usually introduced by the anchors).

## 4 Final Remarks

We compared the performance of three extractive summarization approaches when given as input the result of a processing chain consisting of automatic speech recognition, story segmentation and indexing, and sentence boundary detection. The accumulation of errors in all these phases made this task especially hard. This aspect was also a source of complaints from several human summarizers.

Results were obtained using the ROUGE-L metric. MMR performed close to human summarizers. Both, feature-based and LSA automatic summarizers, performed close to each other and worse than MMR, when compared to the summaries done by the human summarizers. All these approaches produced better summaries than the approach of selecting the first $n$ lines of a story.

These results, although using a language other than English, are comparable to state-of-the-art experiments. Yet, as pointed out by Furui [1], more work in the steps that precede the summarization is needed. Automatic speech recognition and sentence segmentation must be improved to allow the production of better summaries.

## Acknowledgments

This work was partially supported by FCT Project NLE-GRID (POSC/PLP/60663/2004).

## References

1. Furui, S.: Recent Advances in Automatic Speech Summarization. In: Proc. of the IEEE/ACL Workshop on Spoken Language Technology, IEEE, Los Alamitos (2006)
2. Murray, G., Renals, S., Carletta, J.: Extractive Summarization of Meeting Records. In: Proc. of the 9th EUROSPEECH - INTERSPEECH 2005 (2005)
3. Landauer, T.K., Foltz, P.W., Laham, D.: An Introduction to Latent Semantic Analysis. Discourse Processes 25 (1998)
4. Carbonell, J., Goldstein, J.: The Use of MMR, Diversity-Based Reranking for Reordering Documents and Producing Summaries. In: Proc. of the ACM SIGIR 1998 (1998)
5. Maskey, S., Hirschberg, J.: Comparing Lexical, Acoustic/Prosodic, Strucural and Discourse Features for Speech Summarization. In: Proc. of the 9th EUROSPEECH - INTERSPEECH 2005 (2005)
6. Neto, J.P., Meinedo, H., Amaral, R., Trancoso, I.: A System for Selective Dissemination of Multimedia Information. In: Proc. of the ISCA Workshop on Multilingual Spoken Document Retrieval 2003 (2003)

7. Trancoso, I., Neto, J.P., Meinedo, H., Amaral, R.: Evaluation of an alert system for selective dissemination of broadcast news. In: Proc. of the 8$^{th}$ EUROSPEECH - INTERSPEECH 2003 (2003)
8. Amaral, R., Meinedo, H., Caseiro, D., Trancoso, I., Neto, J.P.: Automatic vs. Manual Topic Segmentation and Indexation in Broadcast News. In: Proc. of the IV Jornadas en Tecnologia del Habla (2006)
9. Amaral, R., Trancoso, I.: Improving the topic indexation and segmentation modules of a media watch system. In: Proc. of the 8$^{th}$ International Conference on Spoken Language Processing (INTERSPEECH 2004 - ICSLP) (2004)
10. Mani, I.: Automatic Summarization. John Benjamins Publishing Company, Amsterdam (2001)
11. Edmundson, H.P.: New methods in automatic abstracting. Journal of the Association for Computing Machinery 16(2), 264–285 (1969)
12. Osborne, M.: Using Maximum Entropy for Sentence Selection. In: Proc. of the ACL Workshop on Automatic Summarization (2002)
13. Neto, J.L., Santos, A.D., Kaestner, C.A.A., Freitas, A.A.: Document Clustering and Text Summarization. In: Proc. of the 4$^{th}$ International Conference on Practical Applications of Knowledge Discovery and Data Mining (2000)
14. Gong, Y., Liu, X.: Generic Text Summarization Using Relevance Measure and Latent Semantic Analysis. In: Proc. of the ACM SIGIR 2001 (2001)
15. Lin, C.Y.: Looking for a Few Good Metrics: ROUGE and its Evaluation. In: Proc. Working Notes of NTCIR-4. vol. (supl. 2) (2004)
16. Laporte, E.: Resolução de Ambiguidades. In: Tratamento das Línguas por Computador – Uma introdução à Linguística Computacional e as suas aplicações. Editorial Caminho (2001)

# On the Evaluation of Korean WordNet

Altangerel Chagnaa[1], Ho-Seop Choe[2], Cheol-Young Ock[1], and Hwa-Mook Yoon[2]

[1] School of Computer Engineering and Information Technology,
University of Ulsan, South Korea
{goldenl, okcy}@mail.ulsan.ac.kr
[2] Information System Development Team, Korean Institute of Science and Technology
Information, Daejeon, South Korea
{hschoe, hmyoon}@kisti.re.kr

**Abstract.** WordNet has become an important and useful resource for the natural language processing field. Recently, many countries have been developing their own WordNet. In this paper we show an evaluation of the Korean WordNet (U-WIN). The purpose of the work is to study how well the manually created lexical taxonomy U-WIN is built. Evaluation is done level by level, and the reason for selecting words for each level is that we want to compare each level and to find relations between them. As a result the words at a certain level (level 6) give the best score, for which we can make a conclusion that the words at this level are better organized than those at other levels. The score decreases as the level goes up or down from this particular level.

## 1 Introduction

WordNet [6] has become an important and useful resource for the natural language processing field. Recently, many countries have been developing their own WordNet. Korean WordNet U-WIN has been developing since 2002. Manually building such a lexical resource is very hard and time consuming work.

In this paper we show an evaluation of the Korean WordNet (U-WIN). The purpose of the work is to study how well the manually created lexical taxonomy U-WIN is built. Evaluation is done level by level, and the reason for selecting words for each level is that we want to compare each level and to find relations between them. By analyzing the evaluation we can obtain very useful information about manually built lexical taxonomies. Further, we can use this kind of information for automatically constructing such knowledge resources. We used the traditional K-Means algorithm in this work.

The remainder of this paper is organized as follows. In the next subsection, related works are mentioned and in section 2, there is a brief introduction of the U-WIN and the data sources used in the experiment. Experiments and results are discussed in section 3. Finally we end the paper with the conclusion and future works.

### 1.1 Related Works

There has been no work which directly evaluates WordNet semantics until now, to the best of our knowledge. But there are some works related to word sense discovery [1],

V. Matoušek and P. Mautner (Eds.): TSD 2007, LNAI 4629, pp. 123–130, 2007.

word clustering [3], learning taxonomies [2] etc. They all used WordNet as an answer set and mostly its synset information. There are also some works for analyzing WordNet from the conceptual point of view such as [4]. They have analyzed the WordNet's top-level synset taxonomy and proposed a new top-level taxonomy which is more conceptually rigorous. However, our work here is expressed as s semantic evaluation of the word senses in the taxonomy.

## 2 Data Resources

In this section we introduce the Korean WordNet and monolingual dictionary, Kum Sung, from which noun syntactic information is extracted. We have used a manually constructed mapping table between noun senses in each of these two resources. The semantic tag set used in the Kum Sung dictionary is different from the semantic tag used in U-WIN (because these are two different dictionaries).

### 2.1 Korean WordNet, U-WIN

Here we give a brief introduction to Korean WordNet, namely User-Word Intelligent Network (U-WIN) [5] which is being developed at the University of Ulsan and started in 2002. It is aimed to be a large scale lexical knowledge base which is useful for various fields such as linguistics, Korean information processing, information retrieval, machine translation and semantic web etc. Recently this has been used in many technologies as language education systems, automatic vocabulary learning systems, automatic construction of compound nouns, and explanation construction, automatic construction of concept system in particular professional fields, ontology-based semantic annotation, word sense disambiguation, semantic tagging and expanding query in information retrieval etc.

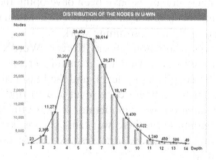

**Fig. 1.** Distribution of nodes in U-WIN

The base knowledge used for constructing UWIN is the Korean Standard Dictionary. The Korean Standard Dictionary contains very detailed information about Korean words and their senses. U-WIN has many kinds of semantic relations (such as Subclass_of, Csynonym_of, Psynonym_of, Part_of, Antonym_of and Related_to etc.)

and conceptual relations. Here we have used only Subclass_of relations (IS_A and Kind_of relations) in the taxonomy. There are 23 top-level nodes. Every node in the taxonomy corresponds to a certain word sense and has its unique ID. Thus every polysemous word has appropriate IDs (entries) for each of its senses. Taxonomy goes down to the depth of 14 levels and the distribution of the nodes in each depth is shown in the figure above. Currently it has a vocabulary of about 300,000 words.

## 2.2 Korean Monolingual Dictionary, Kum Sung

Kum Sung is one of the Korean monolingual dictionaries, which has about 149,644 entries of head words and about 172,020 senses of these headwords. We extracted noun syntactic information from the text in the explanations. This dictionary is PoS and semantically tagged. There are a total 24,932 distinct nouns and 8,172 distinct verbs (totally 296,442 pairs of verb-noun) in the explanations.

In the current experiment we used only one type of syntactic relation which is an object-verb relation. From those syntactic relations, 38,724 distinct verb-object noun pairs (3,820 distinct verbs and 11,904 distinct nouns) are extracted. From 11,904 nouns, we have mapped 5048 words to U-WIN senses.

Also, note that this dictionary has actually tagged in homonymous level, in other words homonymous words have same tags. That means one word in this dictionary can be mapped to many senses of the nouns in U-WIN. We have mapped each word to only one sense in U-WIN, thus one word belongs to only one position in the taxonomy.

## 3 Experiment and Results

### 3.1 Experiment

We carried out experiments for the word senses in each level in the taxonomy (except level 8). In the following texts word means a particular word sense. The procedure of the experiment is shown in the following list. In each depth of the taxonomy:

1. Select the noun senses in U-WIN from the nouns used in the definition of dictionary.
2. Create answer set by cutting beneath the parent level of the nouns.
3. Extract noun and verb syntactic information from the definitions of dictionary.
4. Calculate the relationship between nouns and verbs.
5. Cluster nouns by K-Means (selected number of clusters from step 2)
6. Compare the result with the answer set.

In this experiment we only used object-verb relation as noun and verb syntactic relation. Thereafter, mutual information (MI) [7] and t-score [8] are selected as the measure of relationship between a noun and a verb (lexical association measure), thus each noun is expressed as a vector of mutual information/t-score. In the case of the answer set, it is created by just cutting from the position above the selected words (parent node of the words) i.e. words which have the same parent are in the same group of the answer set. After cutting, single word groups are removed from the answer set and remaining words are used in the further experiment. Table 1 shows the statistics after creation of the answer set and the column Key Cluster shows the number of groups in the answer set.

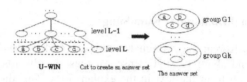

**Fig. 2.** Creating answer set for the words at level L in U-WIN

The words in 4 top-level nodes (THING, ACTION, LIVING THING and SHAPE) of the U-WIN are selected in the experiment. Most of the word senses in the syntactic relations found from the dictionary explanation are in these top-level nodes. Therefore a total of 2,155 word senses are used in this experiment.

For the comparison with the answer set we used F-measure. We set the alpha coefficient as 1, thus it is simplified as shown in the following equation. P and R are the precision and Recall of the resulting cluster vs. a group in the answer set. For further simplification, X is the number of words in a cluster, Y is the number of words in a group and Z is the number of the common nouns in both the cluster and group.

$$\text{F-Measure} = (1+\alpha)*P*R/(\alpha*P+R)=2*P*R/(P+R)=2*Z/(X+Y) \qquad (1)$$

For each cluster in the result set of K-Means algorithm, we selected one group in the answer set which have the highest F-measure score. And the overall clustering result is the average of the F-Measures of the clusters. For a more detailed example, see the following figure.

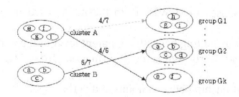

**Fig. 3.** An example of selecting a corresponding group for the cluster using F-Measure. *Cluster A* has the highest score to *Group Gk* among the other groups, so the score for this cluster is 4/6=0.667.

Note that in the tables and figures of the following sections, the experiment for level 8 includes the words in level 8 plus all the words under this level. The reason is that the number of words at this level is very small.

**Table 1.** The statistics of the nouns and verbs used in the experiment. The NV pairs column indicates the number of object verb relations found from the corpus.

| Top-level (number of words) | Depth | Nouns | NV pairs | Features | Key Cluster |
|---|---|---|---|---|---|
| THING (1070) | 4 | 193 | 1215 | 574 | 34 |
| | 5 | 278 | 1093 | 532 | 57 |
| | 6 | 223 | 750 | 406 | 60 |
| | 7 | 194 | 699 | 351 | 33 |
| | 8 | 182 | 533 | 315 | 31 |
| ACTION (488) | 4 | 132 | 505 | 268 | 25 |
| | 5 | 113 | 358 | 208 | 37 |
| | 6 | 128 | 384 | 225 | 43 |
| | 7 | 73 | 241 | 162 | 18 |
| | 8 | 42 | 115 | 84 | 11 |
| LIVING THING (323) | 4 | 62 | 360 | 259 | 5 |
| | 5 | 99 | 378 | 252 | 25 |
| | 6 | 54 | 140 | 103 | 18 |
| | 7 | 55 | 160 | 119 | 18 |
| | 8 | 53 | 131 | 98 | 14 |
| SHAPE (274) | 4 | 100 | 485 | 289 | 26 |
| | 5 | 83 | 278 | 181 | 26 |
| | 6 | 39 | 178 | 127 | 9 |
| | 7 | 29 | 85 | 75 | 7 |
| | 8 | 23 | 42 | 31 | 6 |

In the figure below, the average number of features per word (NV pairs divided by number of words) is shown for each level in top level nodes.

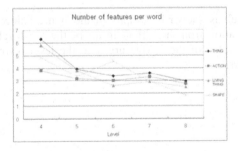

**Fig. 4.** Average number of features/predicates per word for words extracted from dictionary

The reason for selecting words for each level is that we want to compare each level and to find relations between them. From the table and figure above, we can observe that roughly, the number of features is decreasing as the level goes down to a deeper one. Naturally the number of features will decrease as the number of words decreases, but we can also see that of the average number of feature per word tends to decrease.

## 3.2 Results and Discussion

The following table shows the result of the experiment in each level (or depth) of the U-WIN and for each association measure. For convenient comparison, the columns in the previous table are repeated. The average F-Measure of the THING set is relatively lower than others. This is related to the number of words in the set. As shown in the table, the number of words in this set is bigger than in the other sets.

**Table 2.** Experimental result. Column *Key Cluster* shows the number of clusters.

| Top-level | Depth | Nouns | Key Cluster | Fmeasure (MI) | Fmeasure (t-score) |
|-----------|-------|-------|-------------|---------------|--------------------|
| THING | 4 | 193 | 34 | 0.310 | 0.309 |
| | 5 | 278 | 57 | 0.313 | 0.308 |
| | 6 | 223 | 60 | 0.364 | 0.389 |
| | 7 | 194 | 33 | 0.273 | 0.311 |
| | 8 | 182 | 31 | 0.293 | 0.348 |
| ACTION | 4 | 132 | 25 | 0.311 | 0.343 |
| | 5 | 113 | 37 | 0.421 | 0.418 |
| | 6 | 128 | 43 | 0.452 | 0.478 |
| | 7 | 73 | 18 | 0.424 | 0.450 |
| | 8 | 42 | 11 | 0.429 | 0.381 |
| LIVING THING | 4 | 62 | 5 | 0.333 | 0.331 |
| | 5 | 99 | 25 | 0.325 | 0.372 |
| | 6 | 54 | 18 | 0.480 | 0.448 |
| | 7 | 55 | 18 | 0.406 | 0.447 |
| | 8 | 53 | 14 | 0.434 | 0.362 |
| SHAPE | 4 | 100 | 26 | 0.426 | 0.420 |
| | 5 | 83 | 26 | 0.401 | 0.390 |
| | 6 | 39 | 9 | 0.386 | 0.360 |
| | 7 | 29 | 7 | 0.393 | 0.426 |
| | 8 | 23 | 6 | 0.445 | 0.528 |

The experimental result is shown by graphs in the figure below. We can draw a very interesting observation from here that, in all of the top level nodes (except SHAPE), the maximum value of the F-Measure is at the level 6 (regardless of the association measures used).

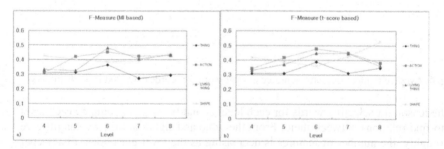

**Fig. 5.** F-measure versus taxonomy depth graph for top-level nodes *THING, ACTION, LIVING THING* and *SHAPE*: a) Result for MI based evaluation, b) Result for t-score based evaluation

In the case of the SHAPE set, the number of words is very small compared with the others, and drastically decreases as it goes down to deeper levels. Naturally, clustering performance increases as the number of elements to be clustered, decreases.

The average number of features for the SHAPE set at level 6 is the highest among those of all the other sets at this level (over 4 features per word, see figure 2 for more detail) but it does not give a good result.

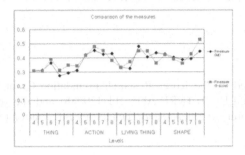

**Fig. 6.** Comparison of association measures

For the comparison of the association measures used in the experiment, t-score measure shows slightly better performance than that of MI. The reason for the difference in the performance is that MI gives much weight to rare pairs of nouns and verbs. However, the overall shape of the experiment is the same regardless of the measures.

The experimental result shows that the words of U-WIN, at the level 6, gives the best score for the automatic clustering method. The score decreases as the level goes up or down from this level. We can draw a conclusion that the words at this level are better organized than those at other levels. Further, as it goes up from this level, words have more abstract senses and as it goes down from this level, words have more concrete senses. As we see from the distribution of the words in U-WIN (see figure 1), most of the words are in levels 5 and 6.

## 4   Conclusion and Future Works

The paper has presented the evaluation of the manually constructed Korean lexical taxonomy U-WIN. Words in the taxonomy are clustered and compared to the answer set, at each level. Traditional method K-Means, is used for clustering and the result is evaluated with F-Measure. Information about nouns is extracted from Korean PoS and semantically tagged monolingual dictionary Kum Sung.

The experiment gives an interesting result. The words of U-WIN, at the level 6, gives the best score, for which we can make a conclusion that the words at this level are better organized than those at other levels. Further, as it goes up from this level, words have more abstract senses and as it goes down from this level, words have more concrete senses. Also, as the level in the taxonomy goes down to the bottom, the average features per word in that level decreases. We think that such manually built lexical networks have similar properties; at a certain level words are better organized than others.

In the case of association measures used for the noun and verb relationship, the t-score measure gives a slightly better performance than the mutual information measure. However, the overall shape of the experiment is the same regardless of the measures.

For the future work, we will extend our research to use the same feature set in each level (common features of the levels in top-level node), and also to include more syntactic relations.

## Acknowledgments

This work was supported by the Korea Research Foundation Grant Funded by the Korea Government (MOEHRD) (KRF-2004-211-420088) and the research project consigned by Korea Institute of Science and Technology Information (KISTI).

## References

1. Pantel, P., Lin, D.: Discovering Word Senses from Text. In: Proceedings of ACM SIGKDD Conference on Knowledge Discovery and Data Mining, pp. 613–619 (2002)
2. Cimiano, P., Hotho, A., Staab, S.: Learning Concept Hierarchies from Text Corpora using Formal Concept Analysis. Journal of Artificial Intelligence 24, 305–339 (2005)
3. Cicurel, L., Bloehdorn, S., Cimiano, P.: Clustering of Polysemic Words. In: Advances in Data Analysis: Proceedings of the 30th Annual Conference of the German Classification Society (GfKl), Springer, Heidelberg (2006)
4. Gangemi, A., Guarino, N., Oltramari, A.: Conceptual Analysis of Lexical Taxonomies: The Case of WordNet Top-Level. In: Welty, C., Barry, S. (eds.) Formal Ontology in Information Systems. Proceedings of FOIS2001, pp. 285–296. ACM Press, New York (2001)
5. Ho-Seop, Ch., Jee-Hui, I., Cheol-young, O.: Large Scale Korean Intelligent Lexical Network. Hangeul 273, 125–151 (2006)
6. Miller, George, A.: Christiane Fellbaum, Katherine J. Miller.: Five papers on WordNet, ftp://ftp.cogsci.princeton.edu/pub/wordnet/5papers.ps
7. Church, K., Hanks, P.: Word Association Norms, mutual Information and Lexigography. Computational Linguistics 16, 22–29 (1990)
8. Manning, C., Schutze, H.: Foundations of Statistical Natural Language Processing. MIT Press, Cambridge, MA (1999)

# An Adaptive Keyboard with Personalized Language-Based Features

Siska Fitrianie and Leon J.M. Rothkrantz

Man-Machine Interaction, Delft University of Technology,
Mekelweg 4, 2628CD Delft, the Netherlands
{s.fitrianie,l.j.m.rothkrantz}@ewi.tudelft.nl

**Abstract.** Our research is about an adaptive keyboard, which autonomously adjusts its predictive features and key displays to current user input. We used personalized word prediction to improve the performance of such a system. Prediction using common English dictionary (represented by the British National Corpus) is compared with prediction using personal data, such as personal documents, chat logs, and personal emails. A user study was also conducted to gather requirements for a new keyboard design. Based on these studies, we developed a personalized and adaptive on-screen keyboard for both single-handed and zero-handed users. It combines tapping-based and motion-based text input with language-based acceleration techniques, including personalized and adaptive task-based dictionary, frequent character prompting, word completion, and grammar checker with suffix completion.

## 1 Introduction

With embedded technology and connectivity, mobile devices and wearable computers are progressively smaller and more powerful. Such devices offer users freeing one or both hands for mobile activity demands. Alternative input devices have been developed to support operating the devices, such as single-handed (e.g. joystick, pen and touchscreen, trackball, and mouse) and zero-handed input devices (e.g. head-mouse or gaze-tracker). These input devices are also used to assist disabled people for interacting with computers [17]. This type of users may have lost the use of one or both hands. Some of them rely on computers to bridge communication with others.

Despite of these developments, text input is still a bottle-neck [15]. Improvement in the input method performance is still highly desired. While speech input [12] and handwriting recognition technology [3] continue to improve, pointing-based character entry is still the most popular to use. With pointing-based (on-screen) keyboards, inserting characters is strictly sequential. The distance to travel from one key to the next [14] and time for distinguishing an individual character from the group [5] have major effects on the text entry performance. Familiarity with the location of characters [10] and visual cues to draw attention to the next most probable character(s) in a currently typed word [11][22] can facilitate the performance. The predictive feature can also suggest word completion beginning with the characters that have been inputted so far. The user can select the suggested word or continue to input until the desired word appears.

V. Matoušek and P. Mautner (Eds.): TSD 2007, LNAI 4629, pp. 131–138, 2007.

In this paper, we present our studies on a comparison of common English and personalized dictionary for improving the word prediction of an adaptive keyboard. We also gathered user requirements for developing such a system. The results are used to develop a new on-screen keyboard that can collect knowledge about user linguistics compositions and use the knowledge to alter its future interaction. It has an n-gram based word-level prediction based on the user's personal way of formulating language, the user's task and the English syntax.

The paper is structured as follows. In section 2, we present related work. We continue with presenting our studies in section 3 and 4, respectively. Then, our keyboard model and its word prediction are described in section 5. Finally, we conclude the work in section 6.

## 2   Related Work

Some alternative keyboard layouts (other than QWERTY) with movement minimizing were developed recently, such as (1) tapping-based (clicking-based) entries, e.g. Fitaly [16], (2) motion-based (gesture-based) entries, e.g. Cirrin [13], and (3) hybrid-based entries, e.g. ATOMIC [22]. Cirrin (Fig. 2(a)) arranges the characters inside the perimeter of an annulus. The most commonly used digrams are nearest to each other, therefore distances traveled between characters are shorter than QWERTY. However, since there is not any predictive feature, a user must attend to the interface when entering text.

Some adaptive input techniques have been developed with predictive features. Dasher uses prediction by partial matching, in which a set of previous symbols in the uncompressed symbol stream is used to predict the next symbol in the stream [19]. It employs continuous input by dynamically arranging characters in multiple columns positioning the next most likely character near the user's cursor pointer in boxes sized according to their relative probabilities. An icon-based keyboard developed by Fitrianie et al. [7] rearranges most relevant icons to the user's input context (on or) around the center with different icon sizes according to their relative probabilities.

Word prediction/completion can improve entry performance but searching through its word list is considered as tedious and disruptive [2]. Moreover, since statistical models are considered weak in capturing long-distance co-occurrence relations between words, a small amount of improvement on word prediction can be achieved by using syntactic information in the prediction, such as part-of-speech n-gram information [8]. In contrast, Windmill uses a parsing algorithm for excluding implausible or ungrammatical words from its word prediction's input [20]. Most of these grammar checkers employ a part-of-speech tagger and a set of pattern matching rules [9].

## 3   Experiment: Common or Personalized Dictionary

The purpose of this experiment was to find out: (1) can a word prediction be improved by using personalized dictionary? and (2) which and how personal data should be used? We collected four datasets from: (1) common English from British National Corpus (BNC - 166261 words) [1], (2) 4.4 MB personal documents in the multimodal communication field (19121 words), such as documents, spreadsheets, and schedulers, (3)

7.2 MB corporate e-mails (13046 words) from the Enron Co., an energy company in Texas [4], and (4) 4.2 MB chat-logs (15432 words) that contain discussions about life aftertime, science and aliens [21].

As a first step, we compared the coverage of the BNC to words and bigrams in personal datasets. 5500 most frequent words (at least 20 times) of the personal datasets were selected. Table 1 shows that in average 87% of words in personal datasets and about 74% of the union of all personal datasets are covered by the BNC. Most words that are not covered by the BNC are names and specific terms, e.g. "xtag", "wordnet", and "website" in the personal documents, "teleconference" and "unsubscribe" in the e-mails, and "lol" (laugh out loud), "yup" (OK), and emoticons in the chat-logs. We selected bigrams containing words that were covered by the BNC. Table 1 shows that although all words are covered, their combinations may not, which are terminologies in a specific domain. For example, (1) in the personal documents: "input fusion", "modality conversion" and "multimodal dialogue" in the field of multimodal system and "shallow parsing", "pattern matching" and "speech recognition" in the field of NLP - they are considered as high frequent bigrams (at least 29 times), (2) in the e-mails: "employee meeting", "management report" and "retirement plans" in corporate domain and "intended recipient", "conference call" and "video connection" in communication field, and (3) in the chat-logs: "immune system", "orbital path", "aftertime life" and "underground shelters".

**Table 1.** The coverage of BNC toward the personal datasets

| | #Words | BNC Cov. (166261 words) | $A \cup B \cup C$ Words Cov. | #Bigrams | BNC Cov. (726000 bigrams) | $A \cup B \cup C$ Bigrams Cov. |
|---|---|---|---|---|---|---|
| A:Personal Docs | 5500 | 4982 (90%) | 49% | 54829 | 33994 (62%) | 56% |
| B:E-mails | 5500 | 4740 (86%) | 49% | 10505 | 7016 (83%) | 11% |
| C:Chat Logs | 5500 | 4754 (86%) | 49% | 36801 | 29809 (81%) | 37% |
| $A \cap B \cap C$ | 1685 | 1674 (99%) | 15% | 2426 | 2348 (96%) | 2.4% |
| $A \cup B \cup C$ | 11168 | 9579 (85%) | | 89275 | 68742 (77%) | |

These experimental results show that user personal word usage has a strong correlation with the user's task context. The coverage of the BNC to the intersection of the personal datasets is quite high. However, among the personal datasets share only a small amount. The reason could be that the datasets were from a specific context and/or not from the same source.

In a second step, we simulated word completion without any statistical model using hash-tables. Fig. 1(a) shows user character entries to serve as a prefix before a completion of a word. Different columns show that some characters are necessary for completing the word, e.g. for "thermometer" needs "t", "h", "e", "r", "m", and "o" to distinguish it from "thermal". Fig. 1(b) shows that on average 3.6% of the cases, a user is able to select an intended word in just one entry. Almost similar coverage in all datasets occurs

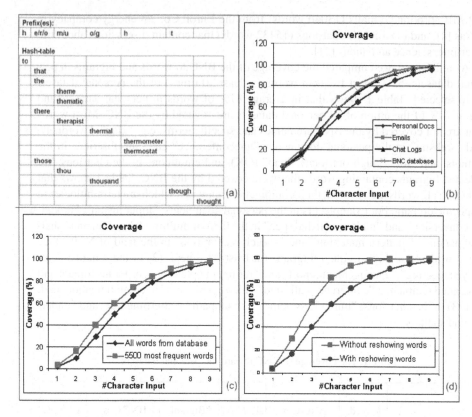

**Fig. 1.** (a) A part of a hierarchical hash-table for the first character "t" (schematic view - read from left to right), (b) coverage of 5500 most frequent words, (c) average coverage of all words and 5500 most frequent words, and (d) average coverage of the 5500 most frequent words with reshowing words and without reshowing words

for every prefix. Fig. 1(c) shows the degradation of the performance of the completion if a complete set of the datasets is used due to the inclusion of lower frequency words. Fig. 1(d) shows if the completion is not reshowing the same word completions once these words have been shown for a given word being entered. For example, when "ther" is written, "thermal" is one possible completion. If "m" is inputted next, a better option is to show a different word completion, e.g. "thermometer".

## 4   User Study: Requirements

Dasher is an adaptive on-screen keyboard system for both single and zero-handed users. As reported in [19], although the creators claimed that Dasher needs short training time, its users' text entry rate is less than QWERTY layout's. Moreover, typical writing errors were spelling and syntax errors, which were reduced after some training. There was not any report about its user's satisfaction.

We conducted an interview with a Dasher user, to gather more requirements for developing a new adaptive on-screen keyboard. Our participant is a computer science student. He suffers from cerebral palsy, which impairs physical movement and limits speech. To enable him to communicate he uses a computer device. He has used Dasher for two years with a head-tracker device. The only reason is because Dasher is a motion-based text entry. Although our participant claimed that Dasher is easy to use, but he needed some time to learn it. He always uses Dasher's word prediction. The boxes sizes and color contrasts are very important for him as visual cues for next character selections. However, because the character arrangement constantly changes, Dasher demands its user's visual attention to dynamically react to the changing layout. This makes him dizzy after some time. Moreover, it is not always easy to correct errors, since Dasher's interface does not provide fast error recovery button/menu. The current implementation helps him in writing text and documents, but is less suitable for writing in specific context like daily talks, e-mailing, chatting, emergency noting and programming. It is desirable to have such a text entry device that works in specific domains with a personalized vocabulary.

## 5   A Personalized-Adaptive On-Screen Keyboard

Fig. 2(b) shows our developed on-screen keyboard. We adopt the design of the Cirrin ([13] - Fig. 2(a)) by displaying all characters in a circular way. Therefore, the interface can gives visual cues, such as different key sizes and color contrasts, for frequently used characters according to their relative probabilities without changing the character layout. Besides this cue, our developed keyboard offers a fast input, less visually demanding and fast error recovery by four ways: (1) the most likely completions of the partially typed word (both user's input and its completion shown in the middle of circle), (2) combining both tapping- and motion-based input (tapping is easier for novice users - [22]), (3) adding space and backspace into the circle for fast error recovery, and (4) each suggested word is shown once after it is rejected by selecting the next character for better language coverage. An additional matrix 6 x 5 is placed on the right side of the circle for numbers, shift, return, control, period, punctuations and comma.

When entering a word, a user may begin with the tapping mode and continue with the motion mode, or vice versa, or only one of them. New selections will be appended to the previous selections. In the motion mode, dragging starts and ends in the middle of the circle. When the user stops dragging, a space will be added at the end of the word. When a space is selected, the input will be flushed to the user's text area. Selecting a backspace on a space will return the previous inputted word back to the middle of the circle. The user can select a word completion with a single tap (in tapping mode) or a left-to-right line motion (in motion mode) in the middle of the circle.

Our developed word prediction consists of several components (Fig. 2(c)). It has two main *dictionaries*, such as (1) a common dictionary(from the BNC) and (2) a user-personal dictionary, which consists of sub-dictionaries for every user's task context (i.e. writing documents, e-mailing, and chatting). They consist of unigram, bigram and trigram list, which include part-of-speech tags and frequency. The *learning component* updates both dictionaries by two ways: (1) extracting inputs during interaction and (2)

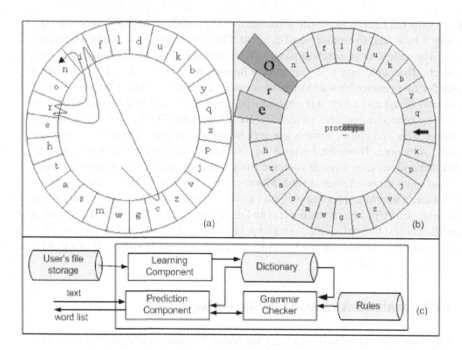

**Fig. 2.** (a) Classic Cirrin, (b) personalized-adaptive Cirrin, and (c) schematic view of our developed word prediction

extracting the user's file storage (scheduled). The *prediction component* generates three lists of suggestions after the first character is inputted, such as (a) from the common dictionary, (b) from the personal dictionaries, and (c) based on the context of user's task. The probability of a sentence is estimated with the use of Bayes rule, where $h_i$ is the relevant history when predicting a word $w_i$:

$$P(w_1, w_2, ..., w_n) = \prod_1^n P(w_i \mid w_1, ..., w_{i-1}) = \prod_1^n P(w_i \mid h_i) \qquad (1)$$

The *grammar checker* excludes syntactically implausible words from the suggestion lists and includes suffix completions, in five steps. First, using Qtag POS tagger [18], it parses the user's input and results the highest probability part-of-speech of each word. Second, this component splits a POS-tagged input into chunks of noun phrase, verb phrase and preposition phrase for detecting noun pluralism and verb tense. Third, it creates all forms for each word in the three suggestion lists from the prediction component. Currently we use thirteen suffixes, such as: "s", "ed", "er", "est", "ly", "able", "full", "less", "ing", "ion", "ive", "ment", and "nest". Using WordNet [6], each new form is verified. Since a word form may be ambiguous and adhere to more forms, all word forms are added to the suggestion lists with the same probability. Four, the grammar checker uses a rule-based approach to check each suggestion whether it is confirmed by grammatical, ungrammatical or out of scope of the grammar. The ungrammatical ones are discarded from the lists. Finally, this component will choose the highest probability word from the context-based dictionary preceded the personal and common dictionary. The personal dictionary will be chosen preceded the common dictionary, if the context-based suggestion list is empty.

# 6 Conclusion and Discussion

In our study we found that the word completion shows better performance using a relatively small dictionary containing the most frequent words. This may indicate that a personalized task based dictionary can offer a more efficient word completion than a large common dictionary. We believe that this can also imply to the accuracy of the word prediction if syntactically implausible words are also excluded from its prediction space. In this way, besides saving time and energy in inputting, a text entry system can also assist the users in the composition of well-formed text. In addition, the number of user inputs for a desired word can be reduced if the system takes an assumption that a suggested word is rejected after the user selects the next character. Therefore, the user can have a better language coverage since each suggestion word is shown only once.

An adaptive single- and zero-handed Cirrin-based on-screen keyboard with personalized language-based techniques acceleration, which include personalized and adaptive task-based dictionary, frequent character prompting, word completion, and grammar checker with suffix completion, has been developed. It allows both tapping and motion-based input. The system's predictive features enable it to display a syntactically plausible word completion and characters in different key sizes and color contrasts according to their relative probabilities.

## Acknowledgments

The research reported here is part of the Interactive Collaborative Information Systems (ICIS) project, supported by the Dutch Ministry of Economic Affairs, grant nr: BSIK03024.

## References

1. A British National Corpus: Unigrams and Bigrams (2007), Retrieved on January 5, 2007, from http://natcorp.ox.ac.uk
2. Anson, D.K., Moist, P., Przywara, M., Wells, H., Saylor, H., Maxime, H.: The Effects of Word Completion and Word Prediction on Typing Rates using On-Screen Keyboards. In: Proc. of RESNA, Arlington, RESNA Press, Virginia (2005)
3. Biem, A.: Minimum Classification Error Training for Online Handwriting Recognition. In: IEEE Trans. on Pattern Analysis and Machine Intelligence, vol. 7(28), pp. 1041–1051. IEEE Computer Society, CA (2006)
4. Corrada-Emmanuel, A.: (n.d.). Enron E-mail Dataset Research. Retrieved in January 5, 2007, from http://ciir.cs.umass.edu/corrada/enron/
5. Eriksen, B.A., Eriksen, C.W.: Effects of Noise Letters Upon the Identification of a Target Letter in a Non-Search Task. Perception and Psychophysics 16, 143–149 (1974)

6. Fellbaum, C.: WordNet - An Electronic Lexical Database. The MIT Press, Cambridge (1998)
7. Fitrianie, S., Datcu, D., Rothkrantz, L.J.M.: Human Communication based on Icons in Crisis Environments, HCII, China. LNCS. Springer, Heidelberg (to appear, 2007)
8. Garay-Vitoria, N., Abascal, J.: A Comparison of Prediction Techniques to Enhance the Communication Rate. In: Stary, C., Stephanidis, C. (eds.) ERCIM Workshop 2004. LNCS, vol. 3196, Springer, Heidelberg (2004)
9. Heidorn, G.: Intelligent Writing Assistant. In: Dale, R., Moisl, H., Somers, H. (eds.) A Handbook of Natural Language Processing: Techniques and Applications for the Processing of Language as Text. Marcel Dekker (2000)
10. MacKenzie, I.S., Zhang, S.X., Soukoreff, R.W.: Text Entry using Soft Keyboards. Behaviour and Information Technology 18, 235–244 (1999)
11. Magnien, L., Bouraoui, J.L., Vigouroux, N.: Mobile Text Input with Soft Keyboards - Optimization by Means of Visual Clues, Mobile HCI, pp. 337–341. Springer, Heidelberg (2004)
12. Maier, A., Haderlein, T., Noth, E.: Environmental Adaptation with a Small Data Set of the Target Domain. In: Sojka, P., Kopeček, I., Pala, K. (eds.) TSD 2006. LNCS (LNAI), vol. 4188, pp. 431–437. Springer, Heidelberg (2006)
13. Mankoff, J., Abowd, G.D.: Cirrin: A Word-Level Unistroke Keyboard for Pen Input. ACM UIST'98, pp. 213–214 (1998)
14. Isokoski, P.: Manual Text Entry - Experiments, Models, and Systems (Ph.D. thesis), Dept. of Comp. Sciences, University of Tampere, Finland (2004)
15. Karlson, A., Bederson, B., Contreras-Vidal, J.: Understanding Single Handed Use of Handheld Devices. In: Jo, L. (ed.) Handbook of Research on User Interface Design and Evaluation for Mobile Technology (in press, 2006)
16. Langendorf, D.J.: Textware Solution's Fitaly Keyboard V1.0 Easing the Burden of Keyboard Input. WinCELair Review (1988)
17. Sussman, V.: Opening Doors to an Inaccessible World. U.S. News and World Report, 85 (September 1994)
18. Tufis, D., Mason, O.: Tagging Romanian Texts - a Case Study for QTAG, a Language Independent Probabilistic Tagger. In: Proc. of LREC. Spain, pp. 589–596 (1998)
19. Ward, D.A., Blackwell, A., MacKay, D.: Dasher - a Data Entry Interface Using Continuous Gesture and Language Models. ACM UIST, pp. 129–136. ACM, NY (2000)
20. Wood, M.E.J., Lewis, E.: Windmill The Use of a Parsing Algorithm to Produce Predictions for Disabled Persons. In: Proc. of Autumn Conference on Speech and Hearing, vol. 18(9) pp. 315–322 (1996)
21. ZetaTalk: Chat Logs - ZetaTalk Live December 2001-May 2003, Retrieved in January 5, 2007, from http://www.zetatalk3.com/index/zetalogs.htm
22. Zhai, S., Kristensson, P.-O.: Shorthand Writing on Stylus Keyboard, ACM CHI, pp. 97–104 (2003)

# An All-Path Parsing Algorithm for Constraint-Based Dependency Grammars of CF-Power

Tomasz Obrębski

Adam Mickiewicz University
Umultowska 87, 61-614 Poznań, Poland
obrebski@amu.edu.pl

**Abstract.** An all-path parsing algorithm for a constraint-based dependency grammar of context-free power is presented. The grammar speci-fies possible dependencies between words together with a number of constraints. The algorithm builds a packed representation of ambiguous syntactic structure in the form of a dependency graph. For certain types of ambiguities the graph grows slower than the chart or parse forest.

## 1   Introduction

The problem of parsing context-free languages has been investigated in depth for nearly half a century. The computational complexity of standard algorithms (e.g. Earley's, CYK) is $O(n^3|G|^2)$ where $n$ is the length of the sentence and $|G|$ is the size of the grammar. The factor $|G|^2$ proved to be an important limitation in real-word applications. The main problem is the growth of the data structure representing alternative syntactic interpretations assigned to fragments of the sentence (typically chart [3] or parse forest [6]). For this reason, parsers of context-free power are considered too slow for some applications, in particular for those relating to text corpora processing.

In this paper we are going to present a parsing algorithm whose time complexity is less influenced by the size of the grammar. It builds an ambiguous dependency structure in which certain types of ambiguities are packed more concisely as compared to substring-based representations, such as charts or parse forests. The algorithm subsumes a constraint-based formulation of a grammar in dependency paradigm. Only projective structures are considered.

## 2   The Grammatical Formalism

Constraint-based grammar formulation stands in opposition to generative one. Within the latter approach, a grammar specifies which basic grammatical constructs are possible and how they may be combined. Within the former approach, grammatical description is formulated by stating what are the necessary conditions (constraints) the syntactic structure must meet to be considered correct.

Below, we will introduce a grammatical formalism, called Constraint–based Dependency Grammar, which is a possible formulation of a minimal constraint-based system of CF power. 'Minimal' means that it implements all and only those elements which are necessary to achieve context-free power[1]. CbDG is in fact a hybrid grammatical system, composed of a generative backbone (connectivity rules) and a constraint extension.

It includes the following components: $\Sigma$ — a finite set of word-forms, $C$ — a finite set of syntactic categories, $C_{root} \subseteq C$ — a subset of root categories, $T$ — a finite set of dependency types, $\mathcal{L} \subseteq \Sigma \times C$ — a lexicon, $\mathcal{R} \subseteq C \times C \times T$ — connectivity rules expressing possible connections between words, $T_{left}, T_{right} \subseteq T$ — subsets of dependency types with surface direction restricted to left and right, respectively, $T_{sgl} \subseteq T$ — a subset of dependency types which may appear at most once for each head, $T_{obl} : C \longrightarrow 2^T$ — a function assigning to each category the set of dependency types obligatory for that category.

## 3   The Core Algorithm

We start with a simplified system reduced to the generative backbone of CbDG $(\Sigma, C, C_{root}, \mathcal{L}, \mathcal{R})$ to introduce the main idea of the algorithm. The input to the algorithm is a word–graph with a separate node for each possible assignment of syntactic category to a word-form. The set of nodes of this graph will be denoted by $W$. The syntactic category of $w$ will be denoted by $c_w$. The output is a dependency graph: a packed representation of possible dependency trees. Its set of nodes is the same as that of the word graph. Arcs represent possible syntactic dependencies: they are triples $(u, \tau, w)$, written as $u \xrightarrow{\tau} w$, where $u, w \in W$, $u$ is the head, $w$ is the dependent, $\tau$ is the dependency type.

> while there are unprocessed elements in $W$
>   $w \leftarrow$ a minimal unprocessed element from $W$
>   $lh_w^* \leftarrow \{w\}, \quad lv_w \leftarrow \bigcup_{u \in ln_w} lh_u^*$
>   while there are unchecked candidates in $lv_w$
>     $u \leftarrow$ a maximal unchecked candidate from $lv_w$
>     forall $\tau \in T$
>       if $\langle c_w, c_u, \tau \rangle \in \mathcal{R}$
>         add arc $w \xrightarrow{\tau} u$, extend $lv_w$ with $lv_u$
>       if $\langle c_u, c_w, \tau \rangle \in \mathcal{R}$
>         add arc $u \xrightarrow{\tau} w$, extend $lh_w^*$ with $lh_u^*$

**Fig. 1.** Dependency graph construction — without constraints

We will introduce several auxiliary relations defined on nodes of the dependency graph: $ln$ — left neighbour (in the surface ordering), $lh$ — left head (a head preceding its dependent in surface order), $ld$ — left dependent, $lh^*$ — transitive closure of $lh$, $ld^*$ — transitive closure of $ld$, $lv = lh^* \cdot ln \cdot ld^*$, called *visibility relation*, expressing the fact of being a transitive left head of a left neighbour of a transitive left dependent of

---

[1] As shown by Obrębski and Graliński [4], these are: (1) lexical ambiguity, (2) a distinguished subset of root categories, (3) at least binary branching, (4) the ability to restrict the surface position of the dependent wrt the head, (5) the ability to express that a specific dependent have to be present, (6) the ability to express that a specific dependent must not be repeated.

compute $roots = \{w \mid bos \in lv_w \wedge w \in lv_{eos} \wedge c_w \in C_{root}\}$
choose $r$ from $roots$ and build $\mathbf{V}(bos, r)$ and $\mathbf{V}(r, eos)$

build $\mathbf{V}(u, w)$
    if $u \in ln_w$ then done
    else  choose arc $w \xrightarrow{\tau} v$ such that $u \in lv_v$ and build $\mathbf{V}(u, v)$ and $\mathbf{V}(v, w)$  /Fig. 3 a)/
      or
      choose node $v$ such that $u \in lh_v^* \wedge v \in ln_w$ and build $\mathbf{H}(u, v)$       /Fig. 3 a')/

build $\mathbf{H}(u, w)$
    if $u = w$ then done
    else choose arc $v \xrightarrow{\tau} w$ such that $u \in lh_v^*$ and build $\mathbf{H}(u, v)$ and $\mathbf{V}(v, w)$  /Fig. 3 b)/

**Fig. 2.** Tree generation algorithm — without constraints

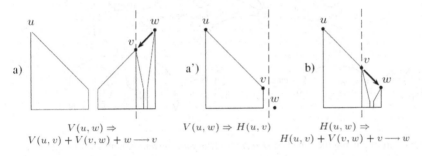

**Fig. 3.** Tree construction steps

a given node. If the pair of nodes $(u, w) \in lv$ we will say that $u$ is visible from $w$ (on the left). Visibility is the necessary condition for two nodes be connectable.

The algorithm will compute the set of arcs and for each node $w$ the set of nodes which are in the relations $lh^*$ and $lv$ with $w$, noted $lh_w^*$ and $lv_w$.

The algorithm constructing the dependency graph is given in Fig. 1. The qualifiers 'minimal' and 'maximal' refer to the surface ordering.

The algorithm generating trees from the graph is shown in Fig. 2. It recursively builds subtrees delimited by pairs of nodes related by $lh^*$ relation (subtree of type $\mathbf{H}$) and $lv$ relation (subtree of type $\mathbf{V}$). In order to present the main idea in a clear way, the indeterministic version is presented. Auxiliary nodes $bos$ and $eos$ ($bos$ precedes all the initial nodes, $eos$ follows all the final nodes) are introduced to simplify the specification. The tree construction steps are illustrated in Fig. 3. It is important to note that all choices made in the algorithm are successful, there are no blind paths. The transformation of the algorithm into a deterministic version iterating through all solutions is fairly straightforward.

## 4   Example

Let us consider the following Polish sentence *Dzieci wzięły psy na spacer.*
(Kids$_{\text{nom or acc}}$ took dogs$_{\text{nom or acc}}$ for a walk$_{\text{acc}}$) and the grammar:

$\Sigma$ = {dzieci, wzięły, psy, na, spacer}
$C$ = {$N_{nom}$, $N_{acc}$, V, P}
$C_{root}$ = {V}
$T$ = {subj, cmpl, mod, pcmpl}
$\mathcal{L}$ = {(dzieci,$N_{nom}$),(dzieci,$N_{acc}$),(wzięły,V),(psy,$N_{nom}$),(psy,$N_{acc}$),(na,P),(spacer,$N_{acc}$)}
$\mathcal{R}$ = {(V,subj,$N_{nom}$), (V,cmpl,$N_{acc}$), (V,mod,P), (N,mod,P), (P,pcmpl,$N_{acc}$)}

The dependency graph produced by the algorithm from Fig. 1 is shown in Fig. 4. Two examples of dependency trees generated by the tree construction algorithm are also presented: one is correct and the other — highly defective due to the lack of constraints.

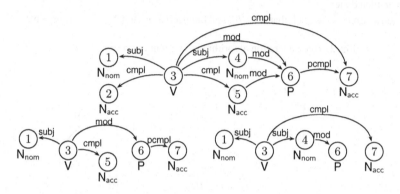

**Fig. 4.** The result of parsing the sentence *Dzieci wzięły psy na spacer.* without constraints: dependecy graph (above) and two of possible dependency trees (below).

## 5   Introducing Constraints

The introduction of direction constraints ($T_{left}$, $T_{right}$) is trivial: no change in the algorithm is needed, except that the direction of arcs has to be checked.

In order to handle singleness and obligatoriness constraints, for each node additional information will be stored: $req_w$ — the set of dependency types still required by the node $w$ and $excl_w$ — the set of dependency types forbidden for the node $w$, containing those of dependency types declared as single which were already attached to $w$. The pair ($req_w$, $excl_w$) will be called *node properties* and will be denoted by $\pi_w$. A node $w$ will be called *saturated* if $req_w = \emptyset$, otherwise *unsaturated*.

The attachment of an arc may change the properties of the head–node. As all arcs in the graph are only possible connections and we have to keep track of all the alternatives, the original node (without the connection and with old properties) must also be retained. The solution is to make a duplicate of the node each time its properties change and attach the arc to the duplicate. A node introduced as the result of the duplication operation will be called the *clone*, the original node — the *ancestor*.

The information on the clone–ancestor relationship will be stored in the arc which caused the duplication. An arc will now be a 4–tuple ($h$, $h_{anc}$, $\tau$, $d$), written $h/h_{anc} \xrightarrow{\tau} d$, where the additional element $h_{anc}$, is the head's ancestor. Arcs which did not cause the head–node duplication will have $h_{anc} = h$. This information is necessary for proper tree construction: when an arc $u/u' \stackrel{\tau}{\longrightarrow} w$ will be selected for the tree, in further steps the node $u'$ will be used instead of $u$.

By properly setting the contents of $lh^\star$ and $lv$ sets we can ensure that 1) the inclusion of a clone in a tree will imply the inclusion of the arc which caused its creation, 2) all nodes in the tree will be saturated.

The complete algorithm is given in Fig. 5. The obvious improvement of the algorithm

---

while there are unprocessed nodes
  $w \leftarrow$ a minimal unprocessed node
  if $w$ is a base–node
    $lh^\star_w \leftarrow \{w\}$
    $lv_w \leftarrow \bigcup_{\text{starurated } u \in ln_w} lh^\star_u \ \cup\ \bigcup_{\text{unstarurated } u \in ln_w} \{u\}$
    $\pi_w \leftarrow (obl(c_w), \emptyset)$
  while there are unchecked candidates in $lv_w$
    $u \leftarrow$ a maximal unchecked candidate form $lv_w$
    forall $\tau \in T$
      if $\langle c_w, c_u, \tau \rangle \in \mathcal{R}$ and $constr(w, u, \tau)$ and $u$ saturated
        if $\pi_w \oplus \tau = \pi_w$
          add arc $w/w \stackrel{\tau}{\longrightarrow} u$,   add $lv_u$ to $lv_w$
        else
          add clone $w'$ with $\pi_{w'} = lv_{w'} = \emptyset$   $lh^\star_{w'} = lh^\star_w \setminus \{w\} \cup \{w'\}$   $\pi_w \oplus \tau$
          add arc $w'/w \stackrel{\tau}{\longrightarrow} u$,   add $lv_u$ to $lv_{w'}$
      if $\langle c_u, c_w, \tau \rangle \in \mathcal{R}$ and $constr(u, w, \tau)$
        if $\pi_u \oplus \tau = \pi_u$
          add arc $u/u \stackrel{\tau}{\longrightarrow} w$, extend $lh^\star_w$ with $lh^\star_u$ if $u$ saturated, with $\{u\}$ otherwise
        else
          add clone $u'$ with $lv_{u'} = lv_u$   $lh^\star_{u'} = lh^\star_u \setminus \{u\} \cup \{u'\}$   $\pi_{u'} = \pi_u \oplus \tau$
          add arc $u'/u \stackrel{\tau}{\longrightarrow} w$, extend $lh^\star_w$ with $lh^\star_{u'}$ if $u'$ saturated, with $\{u'\}$ otherwise

where
$$constr(h, d, \tau) \stackrel{\text{def}}{=} (\tau \in T_{left} \Rightarrow d < h) \wedge (\tau \in T_{right} \Rightarrow h < d) \wedge \tau \notin excl_h$$

$$\pi_w \oplus \tau \stackrel{\text{def}}{=} (rq_w \setminus \{\tau\}, ex_w \cup \{\tau\} \cap T_{sgl})$$

**Fig. 5.** Dependency graph construction — with constraints

---

results from the fact that clones with equal properties and equal $lv$ and $lh^\star$ sets are indistinguishable and may be merged. We did not include this feature in the algorithm specification, as this would complicate it even more.

The algorithm for generating trees remains almost unchanged: the main difference is that while selecting an arc the information on the heads's ancestor must be taken into account and computations must continue with the ancestor instead of the original head.

## 6 Example — Continued

Let us extend the grammar from Sec. 4 with constraints:

$T_{left} = \emptyset,\ T_{right} = \{pcmpl\}$,
$T_{sgl} = \{subj, cmpl, pcmpl\},\ T_{obl}(P) = \{pcmpl\},\ T_{obl}(c) = \emptyset$ for all $c \neq P$

In Fig. 6 the extended dependency graph for the example sentence is presented, created with the algorithm from Fig. 5. Clones are drawn with dotted lines. Numbers below arcs indicate the arc head's ancestors. If the number is not given, the ancestor is the same node. From this graph only four trees may be generated, differing in the choice of the subject and complement and in PP attachment.

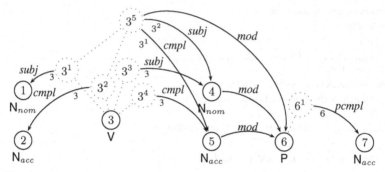

**Fig. 6.** The dependency graph created for the sentence *Dzieci wzięły psy na spacer.* — with constraints.

## 7 Computational Complexity

The time complexity of the graph construction algorithm without constraints (Fig. 1) is $O(n^3|T|)$, where $n$ is the number of nodes in the word graph, which is proportional to the length of the input. We assumed the linear complexity of the set union computation. As this operation may be efficiently implemented as bit–vector sum, the implementational complexity may be reduced to nearly $O(n^2|T|)$ (precisely $O(n^2\frac{n}{machine\text{-}word\text{-}size}|T|)$). Thus the core–algorithm complexity does not depend on the size of the grammar (to be precise: on these components of the grammar which may be large in real applications, such as the set of categories or the set of connection rules). The time complexity of the full algorithm is the same with respect to the size of the graph, but the size of the graph is no more proportional to the length of the input. It also depends on the number of clones created during analysis, which in turn depends on the number of singleness and obligatoriness constraints.

## 8 Comparison to Related Works

The basic idea behind the core algorithm is similar to the well-known simple single-path algorithms for parsing with binary dependency rules, described, among others, by Covington [2]. By removing the unique head requirement and keeping separate list of head/dependent candidates for each word, we obtained an all-path version. The additional information attached to the nodes of the resulting graph structure ($lv$ and $lh^\star$ sets) is necessary for retrieving parse trees.

The idea of using a dependency graph to represent ambiguous dependency structures was investigated also by Barbero and Lombardo [1]. However, their approach is much different from ours. They use standard packing techniques of subtree sharing and local packing. A node in Barbero and Lombardo's graph corresponds to a subtree with fixed leftmost position. In our approach a node is not related to any specific subtree, instead the node stores the information on possible left contexts in which it may occur ($lv$ and $lh^*$ sets may be interpreted in this way). After creating a link the set of possible left contexts is updated all at once using efficient set operations.

The fact that the elements of the structure used to represent alternatives (graph nodes in our case) are not tied to specific substrings of the input distinguishes our approach also from other commonly used techniques: charts [3], parse forests [6]. The interesting consequence of this feature is that certain types of ambiguities may be represented in a more compact way.

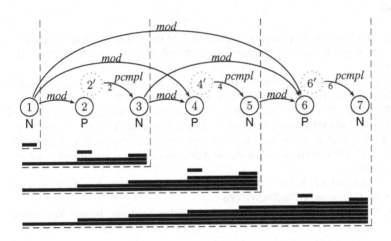

**Fig. 7.** The growth of the dependency graph compared to the growth of the number of well-formed substrings

We will compare the growth of a dependency graph representation to the growth of a number of well-formed substrings on two examples. In Fig. 7 we present a comparison for a sequence of prepositional phrases. In the upper part of the figure the dependency graph is shown, below the different well-formed substrings are enumerated (horizontal bars). The number of well-formed substrings grows proportionally to the square of the number of PP's in the sequence, while the growth of the graph (measured in number of nodes, as this factor is important for the time complexity) is linear. This is because connections which are not subject to branching–level constraints (mod) does not result in introducing any new nodes. In contrast, in the case of a sequence of coordinative constructions, i.e. (N), (N,CONJ,N), (N,CONJ,N,CONJ,N),... where the connection N → CONJ is single and the connection CONJ → N is obligatory for CONJ, the growth of both representations is quadratic.

# 9  Conclusion

We have presented an algorithm for a constraint-based dependency grammar of CF power. It uses the dependency graph to represent ambiguous syntactic structure. Both the algorithm and the graph have certain properties interesting from the point of view of real-word applications: parsing time does not depend on the number of categories and the number of connection rules; for certain types of ambiguities graph representation is more compact than that based on well-formed substrings. The algorithm was implemented in a parser of Polish (dgp, a new component of UAM Text Tools [5]). The processing speed of several thousand words per second (graph construction only, sentences of 10-30 words) makes it potentially interesting for some tasks related to corpora processing.

# References

1. Barbero, C., Lombardo, V.: Dependency graphs in natural language processing. In: Gori, M., Soda, G. (eds.) Topics in Artificial Intelligence. LNCS, vol. 992, pp. 115–126. Springer, Heidelberg (1995)
2. Covington, M.: A dependency parser for variable–word–order languages. Technical Report AI-1990-01, The University of Georgia, Athens, Georgia 30602 (1990)
3. Kay, M.: Algorithm schemata and data structures in syntactic processing. In: Grosz, B., et al. (ed.) Readings in Natural Language Processing, Morgan Kaufmann, Los Altos, CA (2000)
4. Obrębski, T., Graliński, F.: Some notes on generative capacity of dependency grammars. In: Proceedings of COLING 2004 Workshop Recent Advances in Dependency Grammar, Geneve, pp. 65–71 (2004)
5. Obrębski, T., Stolarski, M.: UAM Text Tools — a flexible NLP architecture. In: Proceedings of LREC 2006, Genoa, pp. 2259–2262 (2006)
6. Tomita, M.: Efficient Parsing for Natural Language. Kluwer Academic, Boston, MA (1986)

# Word Distribution Based Methods for Minimizing Segment Overlaps

Joe Vasak and Fei Song

Department of Computing and Information Science
University of Guelph
Guelph, Ontario, Canada N1G 2W1
{jvasak, fsong}@uoguelph.ca

**Abstract.** Dividing coherent text into a sequence of coherent segments is a challenging task since different topics/subtopics are often related to a common theme(s). Based on lexical cohesion, we can keep track of words and their repetitions and break text into segments at points where the lexical chains are weak. However, there exist words that are more or less evenly distributed across a document (called document-dependent or distributional stopwords), making it difficult to separate one segment from another. To minimize the overlaps between segments, we propose two new measures for removing distributional stopwords based on word distribution. Our experimental results show that the new measures are both efficient to compute and effective for improving the segmentation performance of expository text and transcribed lecture text.

## 1  Introduction

Text segmentation is to divide a document into a set of segments that are related to topic/subtopics, where a topics are about the main ideas and subtopics provide supporting ideas for the topics. Text segmentation forms the foundation for further text processing such as topic-based information retrieval [5], text summarization [11], question and answering, and information visualization based on topic structures [2].

Most text segmentation methods rely on lexical cohesion to capture cohesive segments [4]. Lexical cohesion involves the selection and tracking of words and their repetitions. Reyner [9] states that word repetition produces overlaps within segments and across segment boundaries, where word connections tend to be strong within individual segments, but weak between different segments; thus, lexical cohesion provides us a basis to break a text into a sequence of cohesive segments.

There are two kinds of documents that have been widely tested for text segmentation systems: news feed and expository text. We call the former "stream" documents, and the latter, "coherent" documents. News feed typically contains a set of unrelated topics, while the expository text is made up of topics related to a common theme(s). As stated in [7], topic transitions and changes are relatively obvious in stream documents, while subtopic are usually subtler and more difficult to detect in coherent documents. In terms of segment boundaries, there are few overlaps in stream documents since topics are often unrelated, but there may be a more overlaps in coherent documents since the subtopics are related to a common theme(s).

V. Matoušek and P. Mautner (Eds.): TSD 2007, LNAI 4629, pp. 147–154, 2007.

This paper focuses on minimizing the overlaps between segments in coherent documents. Building on the previous work of [7], we propose two new methods for removing words that are more or less uniformly distributed across a document (called distributional stopwords). Although our methods are also based on the distributions of words across a document, we differ from [7] in that we model word distribution by the positions of the sentences in which a word appears, called sentence distribution of a word, to formulate our measures for word removal: distribution significance and distribution difference.

## 2   Previous Work on Text Segmentation

Researchers have focused on three relevant concepts for text segmentation: semantic density, text characteristics, and word distribution.

Semantic density deals with schemes that resemble Youmans' observation that a new topic is often introduced by the heavy use of new word types within a short distance and the number of word types in a given range measures the semantic density [12]. Thus, the dense regions of word types correspond to the segments. Major schemes based on this concept include: adjacency comparison [5], language model [11], and topology ([9] [2]; [7]).

Another observation by Youmans [13] states that fixed intervals of grouped vocabulary are apparent within text. Here, the intervals indicate static patterns in documents that could be exploited and implemented for text segmentation. Major schemes based on this concept are: preferred document length [6], average segment length [11], and average number of segments [9], [2].

Finally, Youmans [12] observed that words of the same type can span over long distances, making it difficult to partition text into segments, since such words may appear across the segment boundaries. Schemes for dealing with such kinds of words include: tracking the dependencies of words over long distances [8] and the removal of document-dependent stopwords [7].

## 3   Word Distribution

Skorochod'ko [10] describes the semantic structures of text in terms of graphs where the nodes correspond to sentences, and the arcs, the semantic relations between the sentences. Such structures capture both information and semantic loads of the text, since each sentence describes a new situation or a new aspect of an object, and the links between the sentences represent semantic relations between them.

Although there can be many kinds of semantic relations between sentences, the simplest form of such relations is lexical chaining, where sequences of semantically related words are used to represent lexical cohesion in text. We use lexical chaining to model word distribution through word repetition at the sentence level, where the distribution is all the sentence occurrences of a word in the text. The occurrences of a word contribute to the information load which is the frequency count for a fragment of text, and the sentence positions contribute to the semantic load which is some measure given sentence positions. Thus, word distribution helps approximate both information and semantic loads of coherent documents.

Using word distribution, we can identify words that are more or less uniformly distributed, called distributional stopwords, regardless of their word frequencies in the text. Such words behave like stopwords in that they spread apart evenly in many segments and do not help distinguishing between different segments. Furthermore, unlike general stopwords that usually have high frequencies in text (such as "the", "of", etc.), distributional stopwords are document-dependent and can occur at all frequency levels.

Figure 1 shows the distributions of selected words from Stargazer text [5]. We can see that words like "form" and "scientist" can be classified as distributional stopwords, since they are distributed evenly over the text and are covered by many segments. On the other hand, words like "star" and "species" are highly concentrated in certain regions and are good indicators for text segmentation.

```
Sentence:   05   10  15   20   25   30   35   40  45  50   55   60   65  70   75   80  85   90   95
------------------------------------------------------------------------------------------------------
14    form   1       111 1    1                         1 1     1  1       1     1       1     1
8 scientist                11              1    1        1              1      1  1
5     space 11       1   1                                                        1
25     star  1                1                     11 22  111112  1 1   1   11 1111       1
5    binary                                         11 1              1                         1
4   trinary                                         1  1              1                         1
8 astronomer 1               1                      1 1              1  1      1  1
7     orbit  1                    1                           12   1 1
6      pull                     2  1 1                        1  1
16    planet  1   1        11              1              1       21 11111              1       1
7    galaxy  1                              1                       1 11   1            1
4     lunar        1  1    1         1
19     life 1 1  1                          1      11 1  11 1  1               1 1     1 111  1 1
27     moon       13 1111  1 1 22 21  21       21       11 1
3      move                               1  1  1
7  continent                           2 1 1 2 1
3  shoreline                                 12
6      time        1                     1  1  1     1                                          1
3     water                         11       1
6       say                         1 1       1      11                      1
3    species                        1  1  1
------------------------------------------------------------------------------------------------------
Sentence:   05   10  15   20   25   30   35   40  45  50   55   60   65  70   75   80  85   90   95
```

**Fig. 1.** Distribution of selected terms from the Stargazer text, with a single digit frequency per sentence number (vertical lines indicate segment boundaries and blanks indicate a frequency of zero)

In fact, distributional stopwords not only offer no help for distinguishing between segments; they actually interfere with the placement of segment boundaries. Since the distributional stopwords are spread apart more or less evenly, they are likely to appear across the segment boundaries, adding noise in the overlaps between segments. As shown in Fig. 1, word 'form' will interfere within segments 6-9, because its occurrences in that region is somewhat spread apart.

Therefore, it is desirable to remove distributional stopwords when we segment each particular document, and by doing so, we can minimize the word overlaps between segments improve the segmentation accuracy.

## 4   New Measures for Distributional Stopword Removal

The recent work by Ji and Zha [7] uses word frequency over multiple partition levels to remove document-dependent stopwords. A document is first partitioned evenly into multiple levels: two segments, three segments, four segments, and so on. At each level, word frequency is used to calculate the variance of a word over the average word frequency. After that, the average variance of the word is computed across all the partition levels. Words with the average variance below a threshold are then identified as document-dependent stopwords and removed for text segmentation.

We believe that this method has two potential problems: computing word variances over multiple partition levels is computationally expensive, and averaging the word variances may reduce the distinctions between different words, since at the coarse/low partition levels, the variances tend to be close to the average word frequencies. For this reason, we propose two new measures for removing distributional stopwords: Distribution Difference, and Distribution Significance.

### 4.1   Distribution Difference

This method tries to emphasize the distribution difference by identifying an appropriate partition level for computing the variance of a specific word. Let m be the number of sentences in a given document; and $|w_i|$, the number of sentences containing word $w_i$. Then, the appropriate block size $B_i$ should be inversely propositional to the sentence frequency $|w_i|$: the higher the sentence frequency $|w_i|$, the smaller the suitable block size $B_i$. We formally define $B_i = \alpha \times \frac{m}{|w_i|}$, where $\alpha$ is a tuning parameter for further adjusting the block size $B_i$.

Based on the appropriate block size $B_i$, we can form a partition that contains $|B_i| = \lfloor \frac{m}{B_i} \rfloor$ number of blocks. In addition, we can compute the average word frequency $\overline{m_i} = B_i \times \frac{|w_i|}{m}$. Then, we can use the following formula to define the distribution difference measure:

$$[\sum_{j=1}^{|B_i|} |(m_{ij} - \overline{m_i})|] \times \frac{1}{|B_i|} \qquad (1)$$

where $m_{ij}$ is the number of sentences containing word $w_i$ in the $j^{th}$ block of the appropriate partition.

Essentially, we model a uniform distribution of a word $w_i$ by its average word frequency $\overline{m_i}$, and identify an appropriate partition that is proportional to the uniform distribution. Then, based on the actual word distribution, we compute the difference of the word $w_i$ block by block in the appropriate partition. If a word is more or less uniformly distributed, the distribution difference measure will be low, making it a good candidate for word removal.

A key difference of this measure from Ji and Zha's method is that we only use one partition level to compute the word difference and the word distribution is directly measured against the average word frequency of the word. As a result, our measure is more efficient to compute and potentially more discriminative.

## 4.2   Distribution Significance

Alternatively, we can capture the relevance of a word over the uniform distribution by a significance measure similar to word distribution density in [3]. For all sentence occurrences of the word $w_i$, we measure the distances between the adjacency pairs of sentences in terms of the number of sentences between them, denoted as $dist(j, j + 1)$ for $j = 1, 2, \ldots, |w_i|$. The significance between each adjacency pair is simply the inversion of the distance between the adjacency pair. Thus, we can use the following formula to define the distribution significance measure:

$$[ \sum_{j=1}^{|w_i|-1} \frac{1}{dist(j, j+1)} ] \times \frac{1}{|w_i|} \tag{2}$$

where $|w_i|$ the total number of sentences containing the word $w_i$.

The intuition behind this method is that dense words have multiple occurrences close to each other and thus are more relevant for text segmentation. We try to capture the word distribution of a word by scaling the distance between adjacent words, which focus on local content such that it is given a higher value. Dividing with respect to the content need to define the distribution $|w_i|$, we capture the relevance of the word distribution. Thus, defining both localized and uniformly distributed words, where localized words receive a high value and uniform words receive a low value.

Our approach is different from Ji and Zha's method in that we incorporate word frequency with sentence position to define a global adjacency distance weighing measure. As a result, we can capture the significance of the word relative to text segments given some representation of distance.

## 5   Experiments

The following experiments are conducted on transcribed lecture data, and written expository data. We obtained the AI Lectures data set from Malioutov and Barzily [8]. The corpus contains 23 sets of lectures, which differ in subject matter, style, segmentation granularity, and have on average twelve segments per lecture where a typical segment is about a half page.

The Mars data set consists of three selected sections (Chap. 1 Sec. 1 and 3, Chap. 2 Sec. 2) from the book "Mars" by Percival Lowell published in 1895 [7]. Six editors were involved in marking the Mars corpus, and they were asked to mark paragraph boundaries at which the topics are changed. We further reviewed the judgement results and followed [5] to set 'true' boundaries at the threshold of three or more judgements, since there are often disagreements among different judgements.

### 5.1   Experimental Procedure

We take a list of tokenized sentences as input ([2], [11]). Then, a maximin entropy sentence boundary detection program converts the text into a sequence of sentences. Next, punctuation marks and general stopwords are removed from each sentence using a stopword list. Finally, we stem the remaining words by applying a Porter stemming algorithm.

The removal of distributional words is done by applying a threshold. The resulting documents with the distributional stopwords removed are then tested on three baseline text segmentation systems. For all distributional stopword removal methods, we start with a threshold value of 1 and decrease the value by 0.05 each time until it becomes 0. The runs with the lowest $P_k$ values will be recorded along with the optimal threshold *OpTh* values.

The segmentation results are evaluated with the $P_k$ measure [1], which is a probabilistic error metric that accounts for near misses in aligning the computed segmentation structure with the reference structure. The probabilistic error metric is 0 if we have a perfect segmentation match; otherwise, its value is in the range between 0 and 100. The value for $k$ is set to half the average of the reference segments.

## 5.2 Experiment 1 - Baseline Results

The baseline results are obtained by testing the two corpora on the following algorithms: TextTiling [5], C99 [2], and U00 [11]. Table 1 shows the segmentation results without distributional stopword removal.

**Table 1.** Baseline results on AI Lectures and Mars corpora with $P_k$ values

| Corpora | TextTiling | C99 | U00 |
|---------|-----------|-------|-------|
| Lectures | 55.97 | 51.00 | 35.68 |
| Mars | 50.90 | 42.63 | 32.04 |

## 5.3 Experiment 2 - AI Lectures Data Set

On the AI Lectures data set, we compare our two new measures with Ji and Zha's method for distributional stopwords. The results in Table 2 show that both new measures produced better results on TextTiling and C99 by a factor of 1.85 for distribution significance and 0.72 for distribution difference than Ji and Zha's measure. However, Ji and Zha's measure outperformed on U00 by a factor of 1.11 over the others.

**Table 2.** Results on AI Lectures with $P_k$ and $OpTh$ values for different word distribution methods

| Measure | OpTh | TextTiling | OpTh | C99 | OpTh | U00 |
|---------|------|-----------|------|-------|------|-------|
| Ji and Zha | 0.8 | 53.41 | 0.4 | 49.37 | 0.05 | **35.49** |
| difference | 0.8 | 52.18 | 0.9 | 49.17 | 0.6 | 37.03 |
| significance | 0.2 | **52.10** | 0.5 | **46.98** | 0.05 | 36.17 |

## 5.4 Experiment 3 - Mars Data Set

We also compare our two new measures with Ji and Zha's method for distributional stopword removal on the Mars data set. Experimental results in Table 3 show that the distribution significance measure produced better results on all three algorithms

**Table 3.** Results on Mars with $P_k$ and $OpTh$ values for different word distribution methods

| Measure | OpTh | TextTiling | OpTh | C99 | OpTh | U00 |
|---|---|---|---|---|---|---|
| Ji and Zha | 0.1 | 49.11 | 0.9 | 34.97 | 0.05 | 31.32 |
| difference | 0.75 | 51.44 | 1.0 | 32.97 | 0.1 | **29.59** |
| significance | 0.05 | **47.99** | 0.1 | **31.77** | 0.03 | 30.27 |

by a factor of 1.79, while the distribution difference measure is better on C99 and U00 by a factor of 1.86, then Ji and Zha's measure.

Compared with the baseline, our results showed that on the Transcribed Lecture data and the Expository data, the distribution measures outperformed TextTiling, C99, and U00 by a factor of 1.46/3.39 for Ji and Zha, 1.42/3.86 for distribution difference, and 2.47/5.8 for distribution significance respectfully.

## 6    Conclusions and Future Work

We proposed two new measures for removing distributional stopwords based on word distribution, which provide good approximations for capturing both information and semantic loads. Our measures are aimed at minimizing the segment overlaps by removal distributional stopwords and should be particularly useful for segmenting coherent text.

The results show that our measures are both efficient to compute and effective for improving segmentation performance. In particular, distribution significance outperforms the other measures for both corpora and distribution difference performs better than Ji and Zha's method in many cases.

Further experiments are required on a larger Mars corpus and different expository data sets. We also intend to apply our methods to web pages with rich contents. Such pages are difficult to classify, since they often contain multiple topics. However, by segmenting them into individual topics, we can potentially do classification at the topic level and then merge the results for more accurate classification.

## Acknowledgements

The authors acknowledge the financial support of CITO, a division of Ontario Centres of Excellence (OCE). We would also like to thank Netsweeper Inc. for marking the segmentation structures of the Mars corpus, Masao Utiyama for his text segmentation system, and Freddy Choi for his publicly available toolkit.

## References

1. Beeferman, D., Berger, A., Lafferty, J.D.: Statistical Models for Text Segmentation. Machine Learning 34(1-3), 177–210 (1999)
2. Choi, F.Y.Y.: Advances in Domain Independent Linear Text Segmentation. In: Proceedings of the NAACL 00, pp. 26–33 (2000)

3. Dais, G., Alves, E.: Discovering Topic Boundaries for Text Summarization Based on Word Co-occurrence. In: Proceedings of the RANLP (2005)
4. Michael, A.K., Halliday, M.A.K., Hasan, R.: Cohesion in English. Longman, New York (1976)
5. Hearst, M.: Multi-Paragraph Segmentation of Expository Text. In: Proceedings of the ACL, pp. 9–16 (1994)
6. Heinonen, O.: Optimal Multi-Paragraph Text Segmentation by Dynamic Programming. In: proceedings of the COLING-ACL (1998)
7. Ji, X., Zha, A.: Domain-independent Text Segmentation Using Anisotropic Diffusion and Dynamic Programming. In: Proceedings of the ACM SIGIR, pp. 322–329 (2003)
8. Malioutov, I., Barzily, R.: Minimum Cut Model for Spoken Lecture Segmentation (2006)
9. Jeffery, C., Reynar, J.C.: Topic Segmentation: Algorithms and Application. Ph.D. Thesis, University of Pennsylvania (1998)
10. Skorochod'ko, E.F.: Adaptive method of automatic abstracting and indexing. In: Proceedings of the IFIP, vol. 71, pp. 1179–1182 (1972)
11. Utiyama, M., Isahara, H.: A Statistical Model for Domain-Independent Text Segmentation. In: Proceeedings of the ACL, pp. 491–498 (2001)
12. Youmans, G.: Measuring Lexical Style and Competence: The Type-Token Vocabulary Curve. Style 24, 584–599 (1990)
13. Youmans, G.: A new Tool for Discourse analysis: The Vocabulary-Management Profile. Language 67(4), 763–789 (1991)

# On the Relative Hardness of Clustering Corpora*

David Pinto[1,2] and Paolo Rosso[1]

[1] Department of Information Systems and Computation,
Polytechnic University of Valencia, Spain
Faculty of Computer Science
[2] B. Autonomous University of Puebla, Mexico
{dpinto, prosso}@dsic.upv.es

**Abstract.** Clustering is often considered the most important unsupervised learning problem and several clustering algorithms have been proposed over the years. Many of these algorithms have been tested on classical clustering corpora such as Reuters and 20 Newsgroups in order to determine their quality. However, up to now the relative hardness of those corpora has not been determined. The relative clustering hardness of a given corpus may be of high interest, since it would help to determine whether the usual corpora used to benchmark the clustering algorithms are hard enough. Moreover, if it is possible to find a set of features involved in the hardness of the clustering task itself, specific clustering techniques may be used instead of general ones in order to improve the quality of the obtained clusters. In this paper, we are presenting a study of the specific feature of the vocabulary overlapping among documents of a given corpus. Our preliminary experiments were carried out on three different corpora: the train and test version of the R8 subset of the Reuters collection and a reduced version of the 20 Newsgroups (Mini20Newsgroups). We figured out that a possible relation between the vocabulary overlapping and the F-Measure may be introduced.

## 1 Introduction

Clustering deals with finding a structure in a collection of unlabeled data [2]. When dealing with raw text corpora, the discovering of the most appropiate features can help on the selection of methods and techniques for determining the possible intrinsic grouping in those sets of unlabeled data. Therefore, this study would be of high benefit. As far as we know, research works in this field nearly have not been carried out in literature. We found just one attempt for determining the relative hardness of the Reuters-21578[1] clustering collection [1], but this research work neither derived formulae for determining the hardness of these corpora nor the possible set of features that are involved in the clustering hardness. A related work which could be considered in order to observe the hardness of a given corpus (with respect to a specific clustering algorithm) is partially

---

* The term 'hardness' is employed like in [1] where this term was introduced to analyse the relative hardness of the Reuters corpora.
[1] http://www.daviddlewis.com/resources/testcollections/reuters21578/

V. Matoušek and P. Mautner (Eds.): TSD 2007, LNAI 4629, pp. 155–161, 2007.

presented in [3] and [4]. In these research works, the author discusses internal clustering quality measures, such as the one from the Dunn Index family, which showed to perform well in the experiments presented by Bezdek et al. in [5,6], among others.

Reuters-21578 (now Reuters RCV1 and RCV2) and 20 Newsgroups[2] are well-known collections which have been used for benchmarking clustering algorithms. However, the fact that several clustering methods may obtain bad results over those corpora does not necessarily imply that they are difficult to be clustered. Further investigation needs to be done in order to determine whether the current clustering corpora are easy clustering instances or not.

We are interested in investigating two aspects: a set of possible features hypothetically related with the hardness of the clustering task, as well as the definition of a formula for the easy evaluation of the relative hardness of a given clustering corpus. We empirically know that at least three components are involved: (i) the size of the clustering texts, (ii) the broadness of the corpora domain and, (iii) whether the documents are single or multi categorized. In the our preliminary experiments, we have investigated the possible relationship between the vocabulary overlapping of a given text corpus with its F-Measure obtained with the MajorClust clustering algorithm [7].

The rest of this paper is structured as follows. In Section 2 we briefly describe the main characteristics of the corpora used in our preliminary experiments. In Section 3 we introduce the used formula and the employed approach to split the corpus in order to calculate the relative hardness for all the possible combinations of two or more categories. Section 5 shows the experimental results we obtained. Finally, conclusions are drawn and the necessary further work to be done is discussed.

## 2  Datasets

The preliminary experiments were carried out by using three different corpora: the R8 version of the Reuters collection (train and test) and, partially, a reduced version of the 20 Newsgroups named "Mini20Newsgroups". We have pre-processed each corpus eliminating punctuation symbols, stopwords and, thereafter, applying the Porter stemmer. The characteristics of each corpus after the pre-processing are given in Table 1.

Table 1. Characteristics of Reuters-R8 and Mini20Newsgroups

|  | R8-Train | R8-Test | Mini20Newsgroups |
|---|---|---|---|
| **Size** | ≈2,500 KBytes | ≈900 KBytes | ≈1,900 KBytes |
| **Documents** | 5,839 | 2,319 | 2,000 |
| **Categories** | 8 | 8 | 20 |

## 3  Calculating the Relative Hardness of a Corpus

In order to determine the Relative Hardness (RH) of a given corpus, we have considered the vocabulary overlapping among the texts of the corpus. In our experiments, we

---

[2] http://people.csail.mit.edu/jrennie/20Newsgroups/

have used the well-known Jaccard coefficient for calculating the overlapping. We considered all the possible combinations of more than two categories from the corpus and for each of them we calculated its RH. For instance, for a given corpus of $n$ categories, $2^n - (n + 1)$ possible subcorpora will be obtained: e.g. for the R8 (eight categories) we obtained 247 subsets.

Thereafter, we calculated their RHs as follows: given a corpus $C_i$ made up of $n$ categories (CAT), the RH of $C_i = \{CAT_1, CAT_2, ..., CAT_n\}$ is:

$$RH(C_i) = \frac{1}{n(n-1)/2} \times \sum_{j,k=1;j<k}^{n} Similarity(CAT_j, CAT_k), \tag{1}$$

where the similarity among categories is obtained by using the Jaccard coefficient in order to determine their overlapping (see Eq. (2)). However, more sophisticated measures also could be used, such as the one presented in [8] in the plagiarism degree calculation framework.

$$Similarity(CAT_j, CAT_k) = \frac{|CAT_j \bigcap CAT_k|}{|CAT_j \bigcup CAT_k|} \tag{2}$$

In the above formula we have considered each category $j$ as the "document" obtained by concatenating all the documents belonging to the category $j$.

## 4   Clustering the Datasets

In order to evaluate the relative hardness formula used in the experiments, we have carried out an unsupervised clustering of all the documents of each subcorpus obtained for each dataset. We have chosen the MajorClust clustering algorithm [7] due to its peculiarity of taking into account both, the inside and outside similarities among the clusters obtained during its execution. In order to keep independent the validation with respect to RH, we have used the tf-idf formula for calculating the input similarity matrix for MajorClust. Each evaluation was performed with the F-Measure formula which is calculated as follows: given a set of clusters $\{G_1, \ldots, G_m\}$ and a set of classes $\{C_1, \ldots, C_n\}$, the $F$-measure between a cluster $i$ and a class $j$ is given by the following formula.

$$F_{ij} = \frac{2 \cdot P_{ij} \cdot R_{ij}}{P_{ij} + R_{ij}}, \tag{3}$$

where $1 \leq i \leq m, 1 \leq j \leq n$. $P_{ij}$ and $R_{ij}$ are defined as follows:

$$P_{ij} = \frac{\text{Number of texts from cluster } i \text{ in class } j}{\text{Number of texts from cluster } i}, \tag{4}$$

and

$$R_{ij} = \frac{\text{Number of texts from cluster } i \text{ in class } j}{\text{Number of texts in class } j}. \tag{5}$$

The global performance of a clustering method is calculated by using the values of $F_{ij}$, the cardinality of the set of clusters obtained, and normalising by the total number of documents in the collection ($|D|$). The obtained measure is named $F$-measure and it is shown in Equation (6).

$$F = \sum_{1 \leq i \leq m} \frac{|G_i|}{|D|} \max_{1 \leq j \leq n} F_{ij}. \tag{6}$$

## 5   Correlation Between Relative Hardness and F-Measure

Our preliminary experiments were carried out on the train and test version of the Reuters R8 collection and, partially, also on a reduced version of the 20 Newsgroups. In Figure 1 we may see the possible correlation between the relative hardness of the (i) train and (ii) test versions of the R8 collection with respect to the F-Measure obtained by using the MajorClust clustering algorithm. The smaller is the value of RH (x-axis) the higher is the obtained F-Measure (y-axis) and viceversa for both corpora. The relative hardness vs. $F$-measure correlation was calculated for all possible corpora variants of R8 (247).

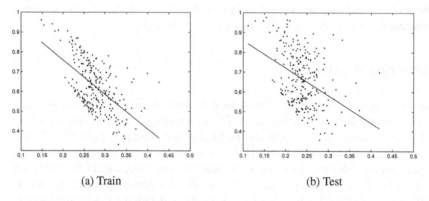

(a) Train                                          (b) Test

**Fig. 1.** Evaluation of all R8 subcorpora (more than two categories per corpus)

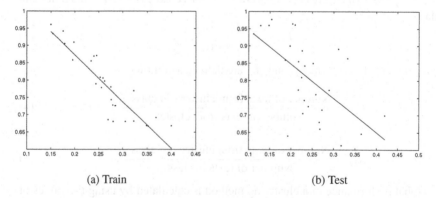

(a) Train                                          (b) Test

**Fig. 2.** Evaluation of single pairs of the R8 categories

**Table 2.** The most related categories of the R8 collection

<table>
<tr><td colspan="3">(a) Train</td><td colspan="3">(b) Test</td></tr>
<tr><td>**RH value**</td><td>**Category**</td><td>**Category**</td><td>**RH value**</td><td>**Category**</td><td>**Category**</td></tr>
<tr><td>0.426</td><td>trade</td><td>monex-fx</td><td>0.419</td><td>monex-fx</td><td>interest</td></tr>
<tr><td>0.399</td><td>monex-fx</td><td>interest</td><td>0.364</td><td>trade</td><td>monex-fx</td></tr>
<tr><td>0.367</td><td>trade</td><td>crude</td><td>0.332</td><td>trade</td><td>interest</td></tr>
<tr><td>0.362</td><td>monex-fx</td><td>crude</td><td>0.317</td><td>trade</td><td>crude</td></tr>
<tr><td>0.352</td><td>trade</td><td>interest</td><td>0.311</td><td>monex-fx</td><td>crude</td></tr>
</table>

**Table 3.** The least related categories of the R8 collection

<table>
<tr><td colspan="3">(a) Train</td><td colspan="3">(b) Test</td></tr>
<tr><td>**RH value**</td><td>**Category**</td><td>**Category**</td><td>**RH value**</td><td>**Category**</td><td>**Category**</td></tr>
<tr><td>0.188</td><td>interest</td><td>earn</td><td>0.186</td><td>interest</td><td>acq</td></tr>
<tr><td>0.180</td><td>acq</td><td>ship</td><td>0.154</td><td>ship</td><td>earn</td></tr>
<tr><td>0.173</td><td>ship</td><td>earn</td><td>0.147</td><td>acq</td><td>ship</td></tr>
<tr><td>0.153</td><td>grain</td><td>acq</td><td>0.128</td><td>grain</td><td>earn</td></tr>
<tr><td>0.147</td><td>grain</td><td>earn</td><td>0.111</td><td>grain</td><td>acq</td></tr>
</table>

**Table 4.** The most related categories of the Mini20Newsgroups collection

| **RH value** | **Category** | **Category** |
|---|---|---|
| 0.3412 | talk politics guns | talk politics misc |
| 0.3170 | alt atheism | talk religion misc |
| 0.3092 | talk politics guns | talk religion misc |
| 0.3052 | talk politics misc | talk religion misc |
| 0.3041 | soc religion christian | talk religion misc |
| 0.2988 | sci crypt | talk politics guns |
| 0.2985 | soc religion christian | talk politics misc |
| 0.2958 | soc religion christian | talk politics guns |
| 0.2932 | talk politics mideast | talk politics misc |
| 0.2905 | sci electronics | sci space |
| 0.2868 | comp sys ibm pc hardware | comp sys mac hardware |

In order to easily visualise the correlation between RH and F-Measure, we have plotted the polynomial approximation of degree one.

In Figure 2 we may see the possible correlation between the relative hardness of each pair of categories of the R8 collection and the F-Measure obtained again by using the MajorClust clustering algorithm. The same conclusion is obtained: the smaller is the value of RH (x-axis) the higher is the obtained F-Measure (y-axis) and viceversa.

In order to fully appreciate the RH formula, the most and least related pairs of categories for the R8 dataset are presented in Tables 2 and 3, respectively. The RH value associated with each pair was calculated with the same formula presented in Section 3. Some preliminary experiments were carried out also with the Mini20Newsgroups dataset and the most and least related pairs of categories are shown in Tables 4 and 5, respectively.

**Table 5.** The least related categories of the Mini20Newsgroups collection

| RH value | Category | Category |
|----------|----------|----------|
| 0.1814 | comp os mswindows misc | rec sport hockey |
| 0.1807 | misc forsale | talk politics misc |
| 0.1804 | misc forsale | talk religion misc |
| 0.1803 | comp sys ibm pc hardware | talk politics mideast |
| 0.1798 | comp os mswindows misc | talk religion misc |
| 0.1789 | alt atheism | comp os mswindows misc |
| 0.1767 | alt atheism | misc forsale |
| 0.1751 | misc forsale | soc religion christian |
| 0.1737 | comp os mswindows misc | soc religion christian |
| 0.1697 | misc forsale | talk politics mideast |
| 0.1670 | comp os mswindows misc | talk politics mideast |

## 6 Conclusions

We have observed that it is possible to introduce a measure to determine the relative hardness of clustering corpora based on the vocabulary overlapping. The obtained results show that there exists a correlation between the $F$-measure and the RH formula. With respect to the analysis carried out in [1], the introduced formula in our research work relies only on the vocabulary overlapping and it does not use any classifier. In fact, we use the MajorClust clustering algorithm only to evaluate the quality of the proposed formula by employing the $F$-measure. Therefore, the introduced RH formula may be used in an unsupervised manner in order to determine the relative hardness of clustering corpora.

## 7 Further Work

As future work, we need to investigate the correlation between the relative hardness and the F-Measure also on the Mini20Newsgroups dataset. Moreover, we are interested in evaluate both, the vocabulary overlapping and the term frequencies. This will allow us to further investigate whether the use of the tf-idf formula in the same context improves the current results or not. Besides, we would like to investigate the possible relationship the RH-Measure could have with cluster validity measures, such as the Density Expected Measure (DEM) which quantifies the similarity within clusters [?]. Moreover, we plan to determine the correlation between RH-Measure and the F-Measure through rank correlation coefficients such as Spearman's and Kendall's ones [4]. The final aim of this research work is to determine the level of hardness of a narrow-domain corpus, such as hep-ex [9], from a clustering task perspective.

## Acknowledgements

We would like to thank Dr. Mikhail Alexandrov from the Autonomous University of Barcelona for his invaluable feedback during the development of this research. This work has been partially supported by the MCyT TIN2006-15265-C06-04 research project, as well as by the BUAP-701 PROMEP/103.5/05/1536 grant.

# References

1. Debole, F., Sebastiani, F.: An analysis of the relative hardness of reuters-21578 subsets. Journal of the American Society for Information Science and Technology 56(6), 584–596 (2005)
2. Zaïane, O.R.: Principles of knowledge discovery in databases - Ch. 8: Data clustering (1999), online-textbook  http://www.cs.ualberta.ca/zaiane/courses/cmput690/slides/Chapter8/
3. Meyer zu Eissen, S., Stein, B.: Analysis of clustering algorithms for web-based search. In: Karagiannis, D., Reimer, U. (eds.) PAKM 2002. LNCS (LNAI), vol. 2569, pp. 168–178. Springer, Heidelberg (2002)
4. Meyer zu Eissen, S.: On Information Need and Categorizing Search. Dissertation, University of Paderborn (2007)
5. Bezdek, J.C., Pal, N.R.: Cluster validation with generalized dunn's indices. In: 2nd International two-stream conference on ANNES, pp. 190–193 (1995)
6. Bezdek, J.C., Li, W.Q., Attikiouzel, Y., Windham, M.: Geometric approach to cluster validity for normal mixtures. Soft Computing 1(4), 166–179 (1997)
7. Stein, B., Nigemman, O.: On the nature of structure and its identification. In: Widmayer, P., Neyer, G., Eidenbenz, S. (eds.) WG 1999. LNCS, vol. 1665, pp. 122–134. Springer, Heidelberg (1999)
8. Kang, N.O., Gelbukh, A., Han, S.Y.: Ppchecker: Plagiarism pattern checker in document copy detection. In: Sojka, P., Kopeček, I., Pala, K. (eds.) TSD 2006. LNCS (LNAI), vol. 4188, pp. 661–667. Springer, Heidelberg (2006)
9. Pinto, D., Jiménez-Salazar, H., Rosso, P.: Clustering abstracts of scientific texts using the transition point technique. In: Gelbukh, A. (ed.) CICLing 2006. LNCS, vol. 3878, pp. 536–546. Springer, Heidelberg (2006)

# Indexing and Retrieval Scheme
# for Content-Based Multimedia Applications

Martynov Dmitry and Eugenij Bovbel

Belarusian State University, Minsk, Belarus
dimamrt@tut.by

**Abstract.** Rapid increase in the amount of the digital audio collections demands a generic framework for robust and efficient indexing and retrieval based on the aural content. In this paper we focus our efforts on developing a generic and robust audio-based multimedia indexing and retrieval framework. First an overview for the audio indexing and retrieval schemes with the major limitations and drawbacks are presented. Then the basic innovative properties of the proposed method are justified accordingly. Finally the experimental results and conclusive remarks about the proposed scheme are reported.

## 1 Introduction

Rapid increase in the amount of the digital audio collections presenting various formats, types, durations, and other parameters that the digital multimedia world refers, demands a generic framework for robust and efficient indexing and retrieval based on the aural content. From the content-based multimedia retrieval point of view the audio information is mostly unique and significantly stable within the entire duration of the content and therefore, the audio can be a promising part for the content-based management for those multimedia collections accompanied with an audio track. Traditional key-word based search engines usually cannot provide successful audio retrievals since they usually require manual annotations that are obviously unpractical for large multimedia collections.

The usual approach for indexing is to map database primitives into some high dimensional vector space that is so called feature domain. Among many variations, careful selection of the feature sets allows capturing the semantics of the database items. The number of features extracted from the raw data is often kept large due to the naive expectation that it helps to capture the semantics better. Content-based similarity between two database items can then be assumed to correspond to the (dis-) similarity distance of their feature vectors. Henceforth, the retrieval of a similar database items with respect to a given query (item) can be transformed into the problem of finding such database items that gives feature vectors, which are close to the query feature vector. This is QBE. QBE is costly and CPU intensive especially for large-scale multimedia databases since the number of similarity distance calculations is proportional with the database size. This fact brought a need for indexing techniques, which will organize the database structure in such a way that the query time and I/O access amount could be reduced [1, 2].

V. Matoušek and P. Mautner (Eds.): TSD 2007, LNAI 4629, pp. 162–169, 2007.
© Springer-Verlag Berlin Heidelberg 2007

There are some techniques to speed up QBE but all they may not provide efficient retrieval scheme from the user's point of view due to their strict parameter dependency. All QBE alternatives have some common drawbacks. First of all, the user has to wait until all of the similarity distances are calculated and the searched database items are ranked accordingly. This might take a significant time if the database size is large. In order to speed up the query process, it is a common application design procedure to hold all features of database items into the system memory first and then perform the calculations. Therefore, the growth in the size of the database and the set of features will not only increase the query time but it might also increase the minimum system memory requirements such as memory capacity and CPU power [3].

All the systems based on these techniques achieved a certain performance; however present some more limitations and drawbacks. All techniques are designed to work in pre-fixed audio parameters. It is a fact that the aural content is totally independent from such parameters. And, they are mostly designed either for short sound files bearing a unique content or manually selected (short) sections. However, in a multimedia database, each clip can contain multiple content types, which are temporally (and also spatially) mixed with indefinite durations. Even the same content type (i.e. speech or music) may be produced by different sources (people, instruments, etc.) and should therefore, be analyzed accordingly.

For the past three decades, researchers proposed several indexing techniques that are formed mostly in a hierarchical tree structure that is used to cluster (or partition) the feature space. Initial attempts such as KD-Trees [4] and R-tree [5] are the first examples of Spatial Access Methods (SAMs). But, especially for content-based indexing and retrieval in large-scale multimedia databases, SAMs have several drawbacks and significant limitations. By definition an SAM-based indexing scheme partitions and works over a single feature space. However, a multimedia database can have several feature types, each of which might also have multiple feature subsets. In order to provide a more general approach to similarity indexing for multimedia databases, several efficient Metric Access Methods (MAMs) are proposed. The generality of MAMs comes from the fact that any MAM employs the indexing process by assuming only the availability of a similarity distance function, which satisfies three trivial rules: symmetry, non-negativity and triangular inequality. Therefore, a multimedia database might have several feature types along with various numbers of feature subsets all of which are in different multi-dimensional feature spaces. The MAMs so far addressed present several shortcomings. Contrary to SAMs, these metric trees are designed only to reduce the number of similarity distance computations, paying no attention to I/O costs (disk page accesses). They are also intrinsically static methods in the sense that the tree structure is built once and new insertions are not supported. Furthermore, all of them build the indexing structure from top to bottom and hence the resulting tree is not guaranteed to be balanced.

As a summary, the indexing structures so far addressed are all designed to speed up any QBE process by using some multidimensional index structure. However, all of them have significant drawbacks for the indexing of large-scale multimedia databases.

## 2 New Approaches

### 2.1 Indexing Scheme: Cellular Tree

To improve the retrieval feasibility and efficiency databases need to be indexed in some way and traditional methods are no longer adequate. It is clear that the nature of the search mechanism is influenced heavily by the underlying architecture and indexing system employed by the database. Therefore, we present a novel indexing technique, *Cellular Tree (CT)*, which is designed to bring an effective solution especially for indexing multimedia databases.

*CT* is a dynamic, cell–based and hierarchically structured indexing method, which is purposefully designed for query operations and advanced browsing capabilities within large-scale multimedia databases. It is mainly a hierarchical clustering method where the items are partitioned depending on their relative distances and stored within cells on the basis of their similarity proximity. The similarity distance function implementation is a *black-box* for the *CT*. Furthermore, *CT* is a self-organized tree, which is implemented by genetic programming principles. This basically means that the operations are not externally controlled; instead each operation such as item insertion, removal, mitosis, etc. are carried out according to some internal rules within a certain level and their outcomes may uncontrollably initiate some other operations on the other levels. Yet all such "reactions" are bound to end up in a limited time, that is, for any action (i.e. an item insertion), its consequent reactions cannot last indefinitely due to the fact that each of them can occur only in a higher level and any *CT* body has naturally limited number of levels.

#### 2.1.1 Cell Structure

A cell is the basic container structure in which similar database items are stored. Ground level cells contain the entire database items. Each cell further carries a Minimum Spanning Tree (MST) where the items are spanned via its nodes. This internal MST stores the minimum (dis-) similarity distance of each individual item to the rest of the items in the cell. All cell items are used as vantage points for any (other) cell item. These item-cell distance statistics are mainly used to extract the cell compactness feature. In this way we can have a better idea about the similarity proximity of any item instead of comparing it only with a single item (i.e. the cell nucleus) and hence a better compactness feature. The compactness feature calculation is also a black-box implementation and we use a regularization function obtained from the statistical analysis using the MST and some cell data. This dynamic feature can then be used to decide whether or not to perform mitosis within a cell at any instant. If permission for mitosis is granted, the MST is again used to decide where the partition should occur and the longest branch is a natural choice. Thus an optimum decision can be made to enhance the overall compactness of the cell with no additional computational cost. Furthermore, the MST is used to find out an optimum nucleus item after any operation is completed within the cell. In *CT*, the cell size is kept flexible, which means there is no fixed cell size that cannot be exceeded. However, there is a maturity concept for the cells in order to prevent a mitosis operation before the cell reaches a certain level of maturity. Therefore, using a similar argument for the organic cells, a maturity cell size (e.g. 6) is set for all the cells in *CT* body (level independent).

### 2.1.2 Level Structure

*HCT* body is hierarchically partitioned among one or more levels, as one sample example shown in Figure 1. Apart from the top level, each level contains various numbers of cells that are created by mitosis operations occurring at that level. The top level contains a single cell and when this cell splits, and then a new cell is created at the level above. The nucleus item of each cell in a particular level is represented on the higher level.

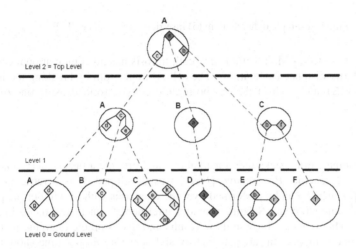

**Fig. 1.** A Sample 3-level CT body

Each level dynamically tries to maximize the compactness of their cells although this is not a straightforward process to do since the incoming items may not show a similarity to the items present in the cells and therefore, such dissimilar item insertions will cause a temporary degradation on the overall (average) compactness of the level. So each level, while analyzing the effects of the (recent) incoming items on the overall level compactness, should employ necessary management steps towards improving compactness in due time (i.e. with future insertions). Within a period of time (i.e. during a number of insertions or after some number of mitosis), each level updates its compactness threshold according to the compactness feature statistics of mature cells, into which an item was inserted.

### 2.1.3 CT Operations

There are mainly three *CT* operations: cell mitosis, item insertion and removal. Cell mitosis can only happen after any of the other two *CT* operations occurs. Both item insertion and removal are generic *HCT* operations that are identical for any level. Insertions should be performed one item at a time. However, item removals can be performed on a cell-based, i.e., any number of items in the same cell can be removed simultaneously.

By means of the proposed dynamic insertion technique, the MST is initially used and updated only whenever necessary. A sample dynamic insertion operation is illustrated in Figure 2.

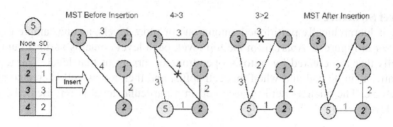

**Fig. 2.** A sample dynamic item (5) insertion into a 4-node MST

Due to the presence of MST within each cell, mitosis has no computational cost in terms of similarity distance calculations. The cell is simply split by breaking the longest branch in MST and each of the newborn child cells is formed using one of the MST partitions.

## 2.2 Query

In order to eliminate drawbacks mentioned above and provide a faster query scheme, a novel retrieval scheme, *Progressive Query*, was developed. It is a retrieval (via QBE) technique, which can be performed over the databases with or without the presence of an indexing structure. Scheme provides intermediate query results during the query process. The user may browse these results and may stop the ongoing query in case the results obtained so far are satisfactory and hence no further time should unnecessarily be wasted, so it may perform the overall query process faster (within a shorter total query time).

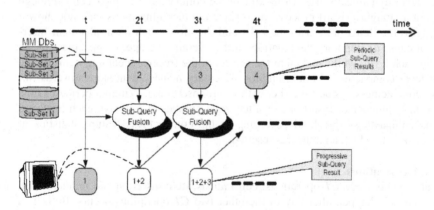

**Fig. 3.** *Progressive Query* Overview

The principal idea behind the new design is to partition the database items into some subsets within which individual (sub-)queries can be performed. Therefore, a sub-query is a fractional query process that is performed over any sub-set of database items. Once a sub-query is completed over a particular sub-set, the incremental retrieval results (belonging only to that sub-set) should be fused (merged) with the last overall retrieval result to obtain a new overall retrieval result, which belongs to the items where query operation so far covers from the beginning of the operation. The order of the database items processed is a matter of the indexing structure of the database. If the database is not indexed at all, simply a sequential or random order can be chosen. In case the database has an indexing structure, a *query path* can be formed in order to retrieve the most relevant items at the beginning during a query operation.

Obviously, *query path* is nothing but a special sequence of the database items, and when the database lacks an indexing structure, it can be formed in any convenient way such as sequentially or randomly. Otherwise, the most advantageous way to perform query is to use the indexing information so that the most relevant items can be retrieved in earlier sub-query steps.

Query operation over *CT* is executed synchronously over two parallel processes: *CT tracer* and a generic process for sub-query formation using the latest *query path* segment. *CT tracer* is a recursive algorithm, which traces among the *CT* levels in order to form a *query path* (segment) for the next *sub-query* update.

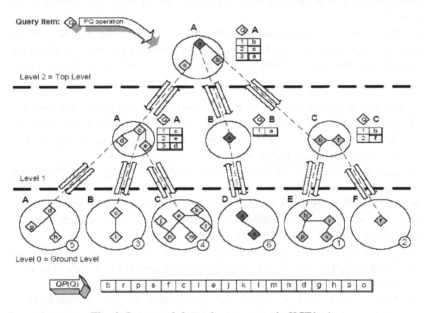

**Fig. 4.** Q*uery path* formation on a sample *HCT* body

## 3  Experiments

Like it was mentioned before the similarity distance function, all decision rules are "black boxes" for our indexing scheme. So it is possible to work with any type of multimedia data. In our work we made an indexing scheme for dataset that contains audio records with speech of different speakers.

## 3.1 Feature Extraction and Distance Function

Feature vectors were formed on the basis of cepstral parameters, as these parameters are optimal for speech signal processing. Cepstral parameters form an orthogonal set, thus making it possible to calculate Euclidian distance between characteristic vectors. It was shown, that Mel-frequency cepstral coefficients in particular possess good resolution and reasonable noise tolerance.

## 3.2 Experimental Results

The main point of our experimental researches was to compare new Progressive Query with normal Query by Example (NQ) when first one is working over Cellular Tree indexing scheme. All experiments are carried out on an Athlon64 2800+computer with 1024 MB memory.

The first difference is that if *NQ* is chosen, then the user has to wait till the whole process is completed but if *PQ* is chosen then the retrieval results will be updated periodically (with the user-defined period value) each time a new sub-query is accomplished.

*PQ* and *NQ* eventually converge to the same retrieval result at the end. Also in the abovementioned scenarios they are both designed to perform exhaustive search over the entire database. However *PQ* has several advantages over *NQ* in the following aspects:

*System Memory Requirement:* The memory requirement is proportional to the database size and the number of features present in a *NQ* operation. Due to the partitioning of the database into sub-sets, *PQ* will reduce the memory requirement by the number of *sub-query* operations performed.

*"Earlier and Better" Retrieval Results:* Along with the ongoing process *PQ* allows intermediate query results (*PSQ* steps), which might sometimes show equal or 'even better' performance than the final (overall) retrieval result. This is obviously an advantage for *PQ* since it proceeds within sub-queries performed in (smaller) sub-sets whereas *NQ* always has to proceed through the entire database.

*Query Accessibility:* This is the major advantage that *PQ* provides. Stopping an ongoing query operation is an important capability in the user point of view. The user can stop it any time (i.e. when the results are so far satisfactory).

*Overall Retrieval Time (Query Speed):* The overall query time is the time elapsed from the beginning of the query to the end of the operation. For *PQ,* since the retrieval is a continuous process with *PSQ* series, the *overall retrieval* means that *PQ* proceeds over the entire database and its process is finally completed. As mentioned earlier, at this point both *PQ* and *NQ* should generate identical retrieval results for a particular queried item. If *PQ* is completed with only one sub-query, then it basically performs a *NQ* operation. As experimentally verified, *PQ's* overall retrieval time is 0-20% faster than *NQ* retrievals (depending on the number of sub-query series) if *NQ* memory requirement does not exceed the system memory.

# 4 Conclusion

In this paper we focused our efforts on developing a generic and robust audio-based multimedia indexing and retrieval scheme. Indexing structure – Cellular Tree – is a dynamic, parameter independent and flexible cell (node) sized indexing structure, which is optimized to achieve as many focused cells as possible using aural descriptors with limited discrimination factors. Retrieval scheme – Progressive Query – an efficient retrieval technique (via QBE), which works with both indexed and non-indexed databases, it is the unique query method which may provide "faster" retrievals and provides "Browsing" capability between instances (sub-queries) of the ongoing query. *CT* is particularly designed to work with *PQ* in order to provide the earliest possible retrievals of relevant items. Since the ultimate measure of any system performance is the satisfaction of the system user, the most important property achieved is therefore its continuous user interaction, which provides a solid control and enhanced relevance feedback mechanism along with the ongoing query operation. The experiments demonstrate the superior performance achieved by *PQ* over *CT* in terms of speed, minimum system requirements, user interaction and possibility of better retrievals as compared with the traditional query scheme.

# References

1. Wold, E., Blum, T., Keislar, D., Wheaton, J.: Content-based Classification, Search, and Retrieval of Audio, IEEE Multimedia Magazine, Fall, pp. 27–36 (1996)
2. Foote, J.T.: Content-Based Retrieval of Music and Audio. In: Proc. of SPIE, vol. 3229, pp. 138–147 (1997)
3. Cheikh, F.A., Cramariuc, B., Gabbouj, M.: Relevance feedback for shape query refinement. In: Proc. of IEEE International Conference on Image Processing, ICIP 2003, Barcelona, Spain (September 14-17, 2003)
4. Bentley, J.L.: Multidimensional binary search trees used for associative searching. In: Proc. of Communications of the ACM, vol. 18(9), pp. 509–517 (September 1975)
5. Guttman, A.: R-trees: a dynamic index structure for spatial searching. In: Proc. Of ACM SIGMOD, pp. 47–57 (1984)
6. Graham, R.L., Hell, O.: On the history of the minimum spanning tree problem. Annual Hist. Comput. 7, 43–57 (1985)
7. Uhlmann, J.K.: Satisfying General Proximity/Similarity Queries with Metric Trees. Information Processing Letters 40, 175–179 (1991)
8. Zhang, T., Kuo, C.–C.J.: Hierarchical Classification of Audio Data for Archiving and Retrieving. In: Proc. of IEEE Int. Conf. on Acoustics, Speech, Signal Proc. Phoenix, pp. 3001–3004 (March 1999)

# Automatic Diacritic Restoration
# for Resource-Scarce Languages

Guy De Pauw[1], Peter W. Wagacha[2], and Gilles-Maurice de Schryver[3,4]

[1] CNTS - Language Technology Group, University of Antwerp, Belgium
guy.depauw@ua.ac.be
[2] School of Computing and Informatics, University of Nairobi, Kenya
waiganjo@uonbi.ac.ke
[3] African Languages and Cultures, Ghent University, Belgium
[4] Xhosa Department, University of the Western Cape, South Africa
gillesmaurice.deschryver@ugent.be

**Abstract.** The orthography of many resource-scarce languages includes diacritically marked characters. Falling outside the scope of the standard Latin encoding, these characters are often represented in digital language resources as their unmarked equivalents. This renders corpus compilation more difficult, as these languages typically do not have the benefit of large electronic dictionaries to perform diacritic restoration. This paper describes experiments with a machine learning approach that is able to automatically restore diacritics on the basis of local graphemic context. We apply the method to the African languages of Cilubà, Gĩkũyũ, Kĩkamba, Maa, Sesotho sa Leboa, Tshivenda and Yoruba and contrast it with experiments on Czech, Dutch, French, German and Romanian, as well as Vietnamese and Chinese Pinyin.

## 1 Introduction

Language corpus compilation for resource-scarce languages is often done by web crawling the (limited) available content on the Internet [1] or by scanning and "OCRing" hard copy resources [2]. This poses a problem for languages that have diacritically marked characters in their orthography. Despite an increasing awareness of encoding issues, OCR research on orthographically rich languages [3], and the development of specialized computer keyboards [4], many of the digital and digitized language resources use the standard Latin alphabet, with accented characters represented by their unmarked equivalents. While language users can perform real-time disambiguation of unmarked text while reading, a lot of phonological, morphological and lexical information is lost this way, that could be useful in the context of language technology.

Typical diacritic restoration methods employ large lexicons to translate words without diacritics into the properly annotated format. This type of information source is however not digitally available for most resource-scarce languages, many of which make extensive use of diacritically marked characters. In this paper we describe experiments with a machine learning approach that tries to predict the placement of diacritics on the basis of local graphemic context, thereby circumventing the need for a digital dictionary.

V. Matoušek and P. Mautner (Eds.): TSD 2007, LNAI 4629, pp. 170–179, 2007.
© Springer-Verlag Berlin Heidelberg 2007

The focus in this paper will be on seven African languages: Cilubà (Congo, Central Africa), Gĩkũyũ, Kĩkamba and Maa (Kenya, Eastern Africa), Sesotho sa Leboa and Tshivenda (South Africa) and Yoruba (Nigeria, Western Africa). We contrast the results with those obtained on better resourced languages: Czech, Dutch, French, German, Romanian and Vietnamese. To isolate its performance on predicting tonal diacritics, we also investigate the technique on Chinese Pinyin data.

We first look at previous work on diacritic restoration in Section 2, highlighting the grapheme-based approach to diacritic restoration. Section 3 discusses the languages and data sets used in the experiments. We then outline the experimental results in Section 4 and conclude with some pointers to future work in Section 5.

## 2   Grapheme-Based Diacritic Restoration

Most of the automatic diacritic restoration methods [5,6,7] tackle both the actual task of retrieving diacritics of unmarked text and the related tasks of part-of-speech tagging and word-sense disambiguation. Although complete diacritic restoration ideally involves a large amount of syntactic and semantic disambiguation, this type of analysis can typically not be done for resource-scarce languages. Moreover, these methods rely heavily on lookup procedures in large lexicons, which are usually not available for such languages.

Mihalcea (2002) describes an alternative diacritics restoration method that uses a machine learning technique operating on the level of the grapheme [8,9]. By backing off the problem from the word level to the grapheme level, it opens up the possibility of diacritic restoration for languages that have no electronic word lists available. Applied to Romanian, Czech, Hungarian and Polish, the technique achieves very high accuracy scores of up to 99% on the grapheme level [9]. Similar work on Gĩkũyũ [10] has likewise yielded encouraging results.

The general idea of the approach coined in [8,9,10] is that local graphemic context encodes enough information to solve this disambiguation problem. It projects diacritic restoration as a standard classification problem, that can be solved by a machine learning algorithm.

| Left | Left | Left | Left | Left | Focus | Right | Right | Right | Right | Right | Class |
|------|------|------|------|------|-------|-------|-------|-------|-------|-------|-------|
| - | - | - | - | - | **m** | b | u | r | i | - | m |
| - | - | - | - | m | **b** | u | r | i | - | - | b |
| - | - | - | m | b | **u** | r | i | - | - | - | ũ |
| - | - | m | b | u | **r** | i | - | - | - | - | r |
| - | m | b | u | r | **i** | - | - | - | - | - | i |

**Fig. 1.** Training Instances for the Gĩkũyũ word "mbũri" (goat)

To this end, training instances in the form of fixed feature vectors are extracted for the graphemes of the words in the corpus. We illustrate this in Figure 1, using an example from one of the target languages under investigation in this paper, i.e. the Gĩkũyũ word "mbũri" (goat). Using a sliding window, the instance describes eleven features for each grapheme: an ambiguous focus letter, e.g. the Latin character "u", the left context of the focus grapheme and its right context. These features are associated with a class, in this case the diacritically marked character "ũ". The instances can then be used to train a machine learning algorithm which can consequently classify new instances.

Touted as language independent, the scalability of this technique to small data sets and its applicability to non Indo-European data sets, has so far not extensively been investigated. Furthermore, the experimental results presented in [8,9] do not provide an appropriate task-oriented evaluation of the approach. In this paper, we wish to address these issues by adjusting the experimental setup of the technique and re-evaluating it on a more varied array of languages and data sets.

## 3   The Data Sets

In this section we will outline the available data sets for the languages under investigation. While a detailed overview of the orthography of all these languages would fall beyond the scope of this paper, we will attempt to quantify the disambiguation challenges that our diacritic restoration method faces on the respective languages.

Table 1 provides some quantitative information for the data sets. For Dutch, German and Maa we used the readily available word lists. For each of the other languages, we extracted a word list of unique word forms (column **Types**) from a language corpus, consistently discarding English word forms often found in web crawled corpora. Table 1 further describes the number of non-Latin characters (column **n**) found in the word list and the percentage of words with at least one diacritic (column **T(d)**).

The most informative quantification of the diacritic disambiguation problem is the "lexical diffusion" metric (**LexDif**). To arrive at this value, we first convert all types to latinized word forms, whereby sometimes multiple types converge to the same Latin form. The **LexDif** value is then calculated by dividing the number of types by the number of latinized word forms. It thus expresses the average number of orthographic alternatives per Latin form. Since our grapheme-based technique can only predict one single possible alternative for a given latinized word form, this column describes the degree of resolvability of our approach: the higher the lexical diffusion value, the more inherently unsolvable the diacritic restoration problem.

*Cilubà.* The manually compiled corpus [11] for this Congolese Bantu language includes almost twenty non-Latin characters. Tonal marking in the orthography causes high values for the **T(d)** and **LexDif** metrics, indicating a significant disambiguation challenge.

*Gĩkũyũ and Kĩkamba.* These closely related Kenyan Bantu languages have manually compiled corpora available to them [2]. Both have two frequently used diacritically marked characters. The languages are tonal, but tone is not marked in the orthography. Previous diacritic restoration work on Gĩkũyũ [10] showed the grapheme-based approach to be effective for this language, despite the extensive use of diacritics in the orthography.

**Table 1.** Information on data sets used in the experiments: number of **tokens** and **types** in the corpus; number of diacritically marked characters (**n**); percentage of types with one or more diacritics (**T(d)**); average number of possible orthographic instantiations of the same Latin form (**LexDif**)

| Language | Tokens | Types | n | T(d) | LexDif |
|---|---|---|---|---|---|
| Cilubà | 144.7k | 20.0k | 17 | 71.8 | 1.17 |
| Gĩkũyũ | 14.8k | 9.1k | 2 | 64.9 | 1.03 |
| Kĩkamba | 38.3k | 9.7k | 2 | 65.7 | 1.07 |
| Maa | 22.2k | 22.2k | 11 | 46.9 | 1.05 |
| Sesotho sa Leboa | 6.9M | 157.8k | 1 | 23.3 | 1.04 |
| Tshivenda | 249.0k | 9.6k | 5 | 18.2 | 1.03 |
| Yoruba | 65.6k | 4.2k | 21 | 61.3 | 1.26 |
| Czech | 123.9k | 105.8k | 15 | 66.3 | 1.05 |
| Romanian | 3.3M | 146.9k | 5 | 39.9 | 1.05 |
| French | 23.2M | 258.6k | 19 | 21.0 | 1.04 |
| Dutch | 301.9k | 301.9k | 18 | 1.5 | 1.00 |
| German | 365.6k | 365.6k | 4 | 23.9 | 1.03 |
| Vietnamese | 2.6M | 50.9k | 26 | 61.3 | 1.21 |
| Chinese Pinyin | 73.5k | 12.0k | 25 | 97.1 | 1.12 |

*Maa.* For this Kenyan Nilotic language, spoken by the Maasai, we used the online Maa dictionary[1] as our data set. We restricted the disambiguation problem to eleven characters (representing phonemes) and discarded tonal markings. The complete tonally marked orthography includes more than 40 characters and can not be handled with a data set of this size.

*Sesotho sa Leboa.* As one of the eleven official languages of South Africa, this Bantu language has a considerable corpus [12]. With only one diacritically marked character and no tonal markings, the **LexDif** column nevertheless indicates a surprisingly hard disambiguation problem.

*Tshivenda.* As one of the smaller official Bantu languages of South Africa, a more modest corpus was manually assembled for the purposes of this paper. The orthography contains quite a few non-Latin characters, but has no tonal marking.

*Yoruba.* The **LexDif** value for this Nigerian Defoid language indicates a similar challenge as for Cilubà, also counting a considerable number of special characters and tonal markings. The corpus material was compiled from sources supplied by Paa Kwesi Imbeah (kasahorow.org) and Kevin Scannell (web crawler "An Crúbádán").

*Indo-European languages.* For the experiments on Czech we used a word list extracted from the DESAM corpus [13]. The Romanian data set is the same used for the experiments in [9]. The word list for French was extracted from a corpus of French newspaper text (Le Monde). For Dutch and German, we used the readily available lexical databases of CELEX [14].

---

[1] http://darkwing.uoregon.edu/~dlpayne/Maa%20Lexicon/lexicon/main.htm

*Vietnamese.* The data set for this Mon-Khmer language was compiled by Le An Ha [15]. The orthography employed in this corpus makes heavy use of diacritics, marking both phonemic variants and tonal characteristics. The high **LexDif** value and the large number of diacritically marked characters predict a complicated disambiguation problem, similar to Yoruba.

*Chinese Pinyin.* This data set[2] contains a latinized version of the Mandarin Chinese orthography. The diacritics only mark tone, no phonemic variations. Experiments on this data set will allow us to isolate the performance of the technique on predicting tonal diacritics.

## 4   Experiments

### 4.1   Experimental Setup

Given that the grapheme-based diacritic restoration approach can principally predict only one single alternative, it simulates a (unigram) lexicon lookup approach. In a practical context, one would therefore be expected to combine the lexicon lookup approach for known words and use the grapheme-based approach for out-of-vocabulary words. This consequently means it should be evaluated primarily on the basis of its performance on unknown words.

In the experiments described in [8,9], instances for graphemes are extracted from a corpus of plain text. The individual instances are then divided into a training set and test set. Making this division on the grapheme level, rather than the word level, means that there will be a significant amount of instances in the test set that have an exact match in the training set. While the experimental results reported in [8,9] are solid, we believe that this methodology does not constitute an appropriate evaluation of the diacritic restoration problem, since the performance on unknown words cannot be established in this manner.

We therefore opt for a significantly different experimental setup, that will allow for a more task-oriented evaluation. Rather than first processing the corpus and dividing the individual instances into a training and test set, we randomly divide the lexicon of unique word forms into ten parts. For each experiment during the 10-fold cross validation, we extract instances from nine partitions, used to train the machine learning algorithm, and evaluate it on the instances extracted from the test set, consisting of unknown words (Section 4.3). In a final experiment (Section 4.4) we also measure performance on plain text data.

### 4.2   Memory-Based Learning

The instances extracted from the training set are used to train a TiMBL classifier [16], an implementation of the machine learning technique of memory-based learning. The scope of the experiments prevented a thorough exploration of parameter and feature settings. The experimental results were obtained by using the standard settings, except for an increased k-value of 3.

---

[2] Compiled from http://www.inference.phy.cam.ac.uk/dasher

Interestingly, while other machine learning algorithms like maximum entropy learning and support vector machines are typically able to outperform memory-based learning on many NLP tasks, these algorithms were not able to improve on TiMBL's performance for these experiments, often significantly underperforming. Furthermore, previous experiments using trigram-based processing [10] showed a significant accuracy increase for this task on the Gĩkũyũ data set. After rigid pre-processing of the lexicons, the trigram approach, typically providing more noise-robust output, was no longer observed to yield significant increases in accuracy.

### 4.3 Experimental Results: Unknown Words

Following up on the new experimental setup described in Section 4.1, we also provide a different, more task-oriented evaluation. Whereas [8,9] provide accuracy scores on the grapheme level, we opt to primarily evaluate the technique on the word level, i.e. the percentage of words in the test that have been predicted completely correctly. Table 2 nevertheless also provides the average accuracy with which latinized graphemes have been disambiguated.

The baseline model identifies candidate graphemes for diacritic marking and chooses the most frequent solution observed in the training set. For French and Dutch for instance these invariably equal to the unmarked characters. This trivial baseline already achieves a very high accuracy for Dutch and Tshivenda (Table 2) because of the limited use of diacritics in these languages. While the disambiguation problem in Sesotho sa Leboa seems limited with only one diacritically marked character, the baseline results confirm the difficulty of the problem.

**Table 2.** Word level and grapheme level accuracy scores on unknown words (**Ci**: Cilubà, **Gĩ**: Gĩkũyũ, **Kĩ**: Kĩkamba, **Ma**: Maa, **Se**: Sesotho sa Leboa, **Ts**: Tshivenda, **Yo**: Yoruba, **Cz**: Czech, **Ro**: Romanian, **Fr**: French, **Du**: Dutch, **Ge**: German, **Vi**: Vietnamese, **Ch**: Chinese Pinyin)

| Word | Ci | Gĩ | Kĩ | Ma | Se | Ts | Yo | Cz | Ro | Fr | Du | Ge | Vi | Ch |
|---|---|---|---|---|---|---|---|---|---|---|---|---|---|---|
| Baseline | 28.2 | 48.7 | 58.4 | 53.1 | 76.2 | 81.8 | 35.4 | 33.7 | 60.6 | 75.2 | 98.5 | 78.3 | 29.4 | 6.7 |
| MBL | 36.6 | 74.9 | 73.5 | 58.6 | 90.1 | 89.3 | 40.6 | 74.4 | 83.2 | 88.2 | 99.6 | 92.7 | 63.1 | 31.5 |

| Grapheme | Ci | Gĩ | Kĩ | Ma | Se | Ts | Yo | Cz | Ro | Fr | Du | Ge | Vi | Ch |
|---|---|---|---|---|---|---|---|---|---|---|---|---|---|---|
| Baseline | 69.8 | 58.9 | 66.7 | 76.8 | 50.6 | 87.2 | 54.0 | 83.2 | 92.5 | 93.8 | 99.7 | 83.1 | 65.8 | 40.4 |
| MBL | 77.4 | 83.1 | 80.4 | 85.4 | 80.9 | 92.9 | 68.2 | 95.2 | 97.3 | 97.2 | 99.9 | 94.3 | 82.7 | 69.0 |

The grapheme-based memory-based learning approach (**MBL** in Table 2) is able to improve both word level and grapheme level accuracy scores for all data sets, with a particularly encouraging increase in accuracy for Gĩkũyũ, Kĩkamba, Sesotho sa Leboa, Czech, Romanian and Vietnamese. Note how for Czech and Romanian a modest increase of accuracy on the grapheme level has a major impact on the accuracy on the word level. Interestingly, the grapheme accuracy scores for Czech and Romanian are well below those reported in [8,9]. Since we use the same machine learning algorithm and same data, we hypothesize that the difference is due to evaluating the task on unseen words, rather than evaluating it on graphemes, extracted from a combination of known and unknown words.

While the results for Cilubà and Yoruba have improved significantly, the diacritic restoration problem is still far from solved for these languages. The trailing results compared to the other African languages, are caused by the tonal markings present in these languages. Tonal diacritics can simply not be solved on the level of the grapheme. Particularly the problem of floating tones needs to be resolved on the sentence level. The increase in accuracy reported on these languages is mainly due to the restoration of diacritics that indicate phonemic alternatives.

This hypothesis is further corroborated by the results on Chinese Pinyin. Diacritics in this data set solely mark tone. While there is a significant increase using the machine learning approach, the results are still severely lacking. Note that the **LexDif** metric (Table 1) was able to predict the trailing results for Cilubà, Yoruba and Chinese Pinyin.

A special case is the language pair Gĩkũyũ and Kĩkamba. Closely related with a very similar orthography, we conducted some combination experiments. In the first experiment, we isolated a Kĩkamba test set and added the Gĩkũyũ data set to the Kĩkamba training set. Word-level accuracy decreased 5.4% compared to a plain Kĩkamba training set (67.1% vs 72.5%). A reverse experiment with a Gĩkũyũ test set yielded a decrease of 6.1% (67.4% vs 73.5%). In a second set of experiments, we solely used Gĩkũyũ training data to classify the Kĩkamba test set and vice versa. Word-level accuracy on the Gĩkũyũ test set was 55.8%, and 52.3% on the Kĩkamba test set. Since these results indicate the orthography of the languages is to some extent similar, re-using the data may bootstrap a basic diacritic restoration method for other closely related languages such as Kĩembu or Kĩmerũ.

### 4.4   Experimental Results: Plain Text

For the languages for which we had a plain text corpus available (all except Maa), we conducted some experiments measuring the effectiveness of our technique on a text containing both known and unknown words. Table 3 displays the results for these experiments. The baseline model for this experiment implements the lexicon lookup method (**LLU**). In this approach, the training set lexicon is used to translate the unmarked words in the test set into the associated diacritically marked words using a unigram model. Particularly for languages with a large training lexicon, this is the baseline to beat. The second method is the grapheme-based memory-based learning approach (**MBL**). The third method combines the two, using lexicon lookup for known words, and MBL for unknown words (**LLU+MBL**).

The results show that for Dutch and German, the lexicon lookup model scores quite well. For the former, this is almost a solved problem. Not surprisingly, the smaller lexicon for French yields a more modest score for the plain text test set. Using the MBL method, there is only a small decrease for French, Dutch and German compared to the lexicon lookup approach. These results are encouraging, since they give an indication of the relative accuracy of the grapheme-based approach, compared to the standard lexicon lookup approach.

For languages with a larger corpus, like Sesotho sa Leboa, Czech and Romanian, the combined approach outperforms all other alternatives, but rather surprisingly, despite

**Table 3.** Word level accuracy scores on plain text

| Word | Ci | Gĩ | Kĩ | Se | Ts | Yo | Cz | Ro | Fr | Du | Ge | Vi | Ch |
|---|---|---|---|---|---|---|---|---|---|---|---|---|---|
| LLU | 77.0 | 77.3 | 79.4 | 97.6 | 97.7 | 67.8 | 61.8 | 94.0 | 89.1 | 99.9 | 96.2 | 74.5 | 78.5 |
| MBL | 85.3 | 92.4 | 91.6 | 99.2 | **99.4** | **76.8** | 89.2 | 96.5 | 88.3 | 99.8 | 95.3 | 73.5 | **83.9** |
| LLU+MBL | 79.6 | 91.5 | 90.4 | **99.4** | 99.2 | 68.5 | **90.1** | **96.6** | **89.3** | **99.9** | **96.8** | **75.5** | 80.3 |

the considerable size of the training lexicon, MBL still significantly outperforms the lexicon lookup method.

As expected, the score for the lexicon lookup approach is quite low for the resource-scarce languages of Cilubà, Gĩkũyũ, Kĩkamba, Tshivenda and Yoruba. For each of these, the grapheme-based approach also outperforms the combined approach by a significant margin. This means that a typical training set for these resource-scarce languages does not yet contain enough lexical information to enable accurate lexicon lookup approaches. This projects the grapheme-based approach as the more robust diacritic restoration method for resource-scarce languages.

Also note that the word level accuracy scores on plain text are a lot higher than those for unknown words. This is particularly true for the Chinese Pinyin data set. We hypothesize that the artificially inflated scores are the effect of using small domain-specific corpora, with typically a restricted lexicon. This provides further support to the claim that the diacritic restoration task is preferably to be evaluated on unknown words, to truly measure its effectiveness in a practical context.

## 5  Conclusion and Future Work

In this paper we have presented experiments with a grapheme-based machine learning approach for diacritic restoration. We described a new experimental approach to this task, that enables a more task-oriented evaluation of this particular disambiguation problem. The difference in results between disambiguating unknown words and known words provides some indication that previously reported results were overstated. We also introduced the metric "lexical diffusion" that is able to predict the difficulty of the diacritic restoration problem for a given language.

Focusing on resource-scarce African languages, we showed that the machine learning approach is indeed to a great extent language independent. But while the method is able to predict diacritics for phonemic variants of the same Latin character with a high degree of accuracy, there are considerable issues when dealing with languages that mark tonality in the orthography. Future research will extend the technique to predict multiple variants of the same latinized word form, combined with contextual sentence models to trigger the correct tonal pattern of a word.

Since for most African languages there is an almost one-to-one mapping between phoneme and grapheme, an effective diacritic restoration method for African languages is almost tantamount to grapheme-to-phoneme conversion. Particularly given the more than encouraging results on processing plain text, the machine learning approach presented in this paper warrants further investigation on a larger array of African languages.

In the meantime, we are confident that the proposed diacritic restoration method can significantly speed up corpus development for the resource-scarce languages under investigation in this paper, as it provides an effective tool to process and enhance unmarked digital language resources.

## Acknowledgments and Demo

The research presented in this paper was made possible through the support of the VLIR-IUC-UON programme. The first author is funded as a Postdoctoral Fellow of the Research Foundation - Flanders (FWO). We would like to thank Martine Coene, Paa Kwesi Imbeah (kasahorow.org), Rada Mihalcea, Kevin Scannell, Le An Ha, Mercy Nevhulaudzi, M.J. Mafela, Pauline Githinji and Ruth Wambua for their co-operation.

Demonstration systems of the diacritic restoration method presented in this paper, are available at http://aflat.org.

## References

1. de Schryver, G.M.: Web for/as corpus: A perspective for the African languages. Nordic Journal of African Studies 11/2, 266–282 (2002)
2. Wagacha, P., De Pauw, G., Getao, K.: Development of a corpus for Gĩkũyũ using machine learning techniques. In: Proceedings of LREC workshop - Networking the development of language resources for African languages, Genoa, Italy, ELRA, pp. 27–30 (2006)
3. Hussain, F., Cowell, J.: Amharic character recognition using a fast signature based algorithm. In: Proceedings of the IEEE conference on Image Visualisation 2003, London, UK, pp. 384–389. IEEE Computer Society Press, Los Alamitos (2003)
4. Bailey, D.: Creating a South African keyboard. In: Afrilex 2006, the user perspective in lexicography, programme and abstracts, Pretoria, South Africa (SF)[2] Press, pp. 17–18 (2006)
5. Yarowsky, D.: A comparison of corpus-based techniques for restoring accents in Spanish and French text. In: Proceedings of the Second Annual Workshop on Very Large Corpora, Kyoto, Japan, pp. 19–32 (1994)
6. Tufiş, D., Chiţu, A.: Automatic diacritics insertion in Romanian texts. In: Proceedings of the International Conference on Computational Lexicography, Pecs, Hungary, pp. 185–194 (1999)
7. Simard, M.: Automatic insertion of accents in French text. In: Proceedings of the Third Conference on Empirical Methods in Natural Language Processing, Granada, Spain, pp. 27–35 (1998)
8. Mihalcea, R.F.: Diacritics restoration: Learning from letters versus learning from words. In: Gelbukh, A. (ed.) CICLing 2002. LNCS, vol. 2276, pp. 339–348. Springer, Heidelberg (2002)
9. Mihalcea, R.F., Nastase, V.: Letter level learning for language independent diacritics restoration. In: Proceedings of CoNLL-2002, Taipei, Taiwan, pp. 105–111 (2002)
10. Wagacha, P., De Pauw, G., Githinji, P.: A grapheme-based approach for accent restoration in Gĩkũyũ. In: Proceedings of the Fifth International Conference on Language Resources and Evaluation, Genoa, Italy, ELRA, pp. 1937–1940 (2006)
11. de Schryver, G.M.: Bantu Lexicography and the Concept of Simultaneous Feedback (MA dissertation). Ghent University, Ghent, Belgium (1999)

12. de Schryver, G.M.: Corpus-based statements of meaning versus descriptions of actual language use in dictionaries. Culture, Language and Identity (CLIDE) Seminar, University of the Western Cape, Bellville, South Africa (2007)
13. Pala, K., Rychlý, P., Smrž, P.: DESAM - annotated corpus for Czech. In: Jeffery, K.G. (ed.) SOFSEM 1997. LNCS, vol. 1338, pp. 523–530. Springer, Heidelberg (1997)
14. Baayen, R.H., Piepenbrock, R., van Rijn, H.: The CELEX lexical data base on CD-ROM. Linguistic Data Consortium, Philadelphia, PA (1993)
15. Ha, L.: A method for word segmentation in Vietnamese. In: Proceedings of the Corpus Linguistics 2003 Conference, pp. 282–287 (2003)
16. Daelemans, W., Zavrel, J., van den Bosch, A., van der Sloot, K.: TiMBL: Tilburg Memory Based Learner, version 5.1, Reference Guide. ILK Technical Report 04-02, Tilburg University (2004)

# Lexical and Perceptual Grounding of a Sound Ontology

Anna Lobanova, Jennifer Spenader, and Bea Valkenier

Artificial Intelligence Department,
University of Groningen, The Netherlands
{a.lobanova,j.spenader,b.valkenier}@ai.rug.nl

**Abstract.** Sound ontologies need to incorporate source unidentifiable sounds in an adequate and consistent manner. Computational lexical resources like Word-Net have either inserted these descriptions into conceptual categories, or make no attempt to organize the terms for these sounds. This work attempts to add structure to linguistic terms for source unidentifiable sounds. Through an analysis of WordNet and a psycho-acoustic experiment we make some preliminary proposal about which features are highly salient for sound classification. This work is essential for interfacing between source unidentifiable sounds and linguistic descriptions of those sounds in computational applications, such as the Semantic Web and robotics.

## 1 Sounds Without Identifiable Sources

Bumps, rattles and rumbles: languages are filled with expressions to name sounds that we cannot identify according to their origin, the most common way to describe a sound. The ability to describe and distinguish between sounds is an essential cognitive skill, as sounds are one of our major sources of information about our environment.

In computational applications, ontologies are valuable resources for relating different concepts, often for the purpose of inference. For example, it is important to know that a *melody* is a part of a *song* which in turn is a kind of musical piece. In particular with the Semantic Web, having cognitively grounded ontologies available to serve as the backbone of search engines is more necessary than ever before. For source identifiable sounds, existing semantic ontologies are often already sufficient, e.g. for describing something as *the sound of a car engine* or *the sound of running water* the hierarchies for the source concepts *car engine* and *running water* are already present. Additionally, for source identifiable sound names like *a scream, a bark* or *a whinny*, the sounds can be integrated into classifications already present, e.g. dogs or horses.

The challenge is dealing with source unidentifiable sounds such as *click, clank, plop, thud, screech* and *rattle*. These sounds cannot be categorized by linking them to a source concept.

Our aim is to discover what features of source unidentifiable sounds are perceived as relevant to their classification, and to use them to develop an ontological structure for source unidentifiable sounds that captures the way in which listeners perceive them. Further, we are interested in if and how linguistic patterns might support an ontological structure. The lexical means available to describe sounds may offer clues to the features most salient to their classification.

V. Matoušek and P. Mautner (Eds.): TSD 2007, LNAI 4629, pp. 180–187, 2007.

In section 2 we discuss a number of examples of source unidentifiable sounds that seem to fall into different feature groups, in section 3 we discuss previous ontological attempts, first looking at some work on source identifiable sounds, and finally focusing on WordNet [1]. We show that WordNet does not organize source unidentifiable sounds in a consistent and sufficient way for computational applications. In section 4 we present a psycho-acoustic experiment we conducted in order to examine which features humans use when classifying sounds and whether some features are more salient than others. Based on these results in section 5 we propose that some features are more salient than others and when identified correctly they can be used to structure sounds in an ontology.

## 2   Features of Source Unidentifiable Sounds

There are sounds with a clear source, and sounds where the source is not clear or not known at all. The source typically functions as its description. For example, bells toll, horns toot and knocking can be an effect of fingers touching the surface of a door. While sound source identification in the examples above is relatively easy, the sound of *swish* is not, since it can be produced by fallen leaves and the wind (nature), curtains (material), or by a gramophone record (a plastic object). Further, the sound of *whack* can be a result of almost anything from someone's hand (body part) to wings of birds (animal part). And what about sounds like *a thunk, a whiz* or *a throb*?

At least three different perspectives can describe sounds. For source unidentifiable sounds, source based descriptions are obviously not possible. Sounds can also be described according to their acoustic properties. This has the advantage of being entirely objective, but has the disadvantage of potentially being completely incompatible with the way in which humans perceive sounds.[1] Since our aim is to make a classification that will allow humans to categorize and relate sounds they perceive to other sounds, the third perspective, descriptions based on perceptually relevant aspects of sounds, seems most promising.

But what are the perceptually relevant aspects of source unidentifiable sounds? We began by listening to a large number of source unidentifiable sounds and identifying salient features that seemed to help characterize the sounds, finally identifying five features:

1. **Repetitiveness:** A drum is repetitive, a sigh is not.
2. **Continuousness:** If the sound is interrupted by silences it is not continuous. Drumming is (−)continuous while a sigh is (+)continuous. Repetitive sounds can be continuous, e.g. the toll of a bell, or not continuous, e.g. tapping.
3. **Duration:** If the sound is produced by an ongoing process it exhibits a durational aspect. A sigh, for example, is produced by the ongoing flow of air, while a click is not.
4. **Harmonicity:** Has to do with how pleasant a sound is. The toll of a bell is much more harmonic than a sigh or the beat of a drum.

---

[1] This is then analogous to the correct biological classification of a tomato as a berry, while most people would consider it to be a vegetable, and expect to find it among the vegetables in the grocery store.

5. **Pitch:** Has to do with whether or not a sound can show pitch variation. For example, a sigh or swish doesn't, but a screech or ring does.

We consider these features to be highly salient but it is not clear which features are most relevant for classifying linguistic terms for source unidentifiable sounds. This can be analogous to an initial classification of animals where it would be determined that whether or not an animal could fly is a less salient feature than whether or not they give birth to leave offspring, since the latter distinguishes mammals from birds, while the former only distinguishes birds like penguins from e.g. robins. In order to reliably determine which features are more salient than others we will need to do some psycho-acoustic experiments. But first let's see what classification attempts have already been made.

## 3   Proposed Ontologies for Sounds

### 3.1   Previous Work on Sound Ontologies

Most of the work on sound classification is done in the area of sound and speech recognition. There is no consistency as to which criteria should be used when distinguishing different sounds. [4], for example, divided sounds into 3 classes: speech, music and sound texture. [4] does not give a clear definition of sound textures but some examples include the sounds of a copy machine, fish tank babbling, waterfall, applause, and so on, in other words, source identifiable sounds that are not music or speech. In a psycho-acoustic experiment [4] asked the participants to cluster sound textures in order to find out which features people find salient. Participants differed radically in their classifications. One possible explanation is that some participants used the *source* of the sound as the main feature, while others used such perceptual features as *periodicity* and *smoothness* leading to different classifications.

[3] proposed to make a sound ontology where sounds were grouped according to their acoustic features into such sound classes as music and speech. All individual sounds in these groups could be listed together with their attributes (that is acoustic features) like *frequency*, *timber* or *rhythm*, and connected with each other by the ontological relationships *part-of* and *isa*. The main aim of such an ontology was to provide enough information about the features and to enable sound segregation from an input sound mixture. Because of its specific purpose, the attributes, or features listed in the ontology are acoustic in nature (e.g. AM/FM modulation, power spectrum, formant) and no lexical information is given. Similarly, [5] looked at 13 acoustic features in their real-time computer models in order to see whether these features can help to distinguish between music and speech sounds. They report that the best model used only 3 out of 13 features (namely, 4 Hz Modulation Energy, Var Spectral Flux and Pulse Metric). These results suggest that some features are more salient than others. The formal properties of these salient features might have some overlap with the basic perceptual features we used in our psycho-acoustic experiment. For example, modulation energy is related to the *loudness* and *repetitiveness*, spectral "Flux" might have to do with the *continuity* and pulse metric might overlap with our feature *repetitiveness*. However, it is possible that humans use very different features, and it is not clear how well these features carry over to source unidentifiable sounds.

## 3.2 WordNet's Sound Classification

The WordNet ontology includes source identifiable as well as source unidentifiable sounds. Unlike traditional dictionaries where all words and their meanings are enumerated in the alphabetical order, WordNet was originally based on psycholinguistic principles trying to capture the way words and meanings are represented in humans. We concentrate on WordNet because it is currently the most widely used lexical resource in computational linguistics. All words are organized in the so-called synsets, or sets of synonymous words, hierarchically organized via such semantic relations as hyponymy and hypernymy. Each lexical entry has a definition and often an example of use. WordNet contains four categories: nouns, verbs, adjectives and adverbs but our main focus is on the nouns describing sounds. Nouns belong to one of nine hierarchies, each associated with a top level concept called a unique beginner. Since the same lexical string can have more than one meaning and belong to different synsets, it can occur in several different hierarchies.

Sounds are organized in the WordNet according to their senses and not their features. The string *sound* has 8 senses but only 6 are relevant: - $sound_1$: the particular auditory effect produced by a given cause; - $sound_2$: auditory sensation: the subjective sensation of hearing something; - $sound_3$: mechanical vibrations transmitted by an elastic medium; - $sound_4$: the sudden occurrence of an audible event; - $sound_5$: the audible part of a transmitted signal; - $sound_6$: (phonetics) an individual sound unit of speech without concern as to whether or not it is a phoneme of some language;

$Sound_1$, $sound_5$ and $sounds_6$ occur in the hierarchy with the unique beginner *Abstraction*, $sound_2$ is in the hierarchy *Psychological Feature*, $sound_3$ is in the hierarchy *Phenomenon* and $sound_4$ is in the hierarchy *Event*. It is important to point out that WordNet does not distinguish between source identifiable and unidentifiable sounds per se because its main point is to represent sounds as to the main concepts they imply, for example, whether it is an instance of such basic cognitive process as sensation (such as *music*) or whether it is an occurrence of an audible event (such as *drumbeat*). This kind of approach seems to be insufficient because of the inconsistencies it causes.

For example, one of the problems with sound representation in WordNet is that when two sounds are on the same low level of a hierarchy, the difference between them can only be elicited from their definitions. Consider sounds *throbbing* and *knocking*, two terminal sister leaves of the *Event* hierarchy with $sound_4$. According to the WordNet, *throbbing* is a sound "with a strong rhythmic beat", and *knocking* is a sound of "knocking as on a door or in an engine or bearing". Only these definitions provide information about the quality of the sound *throbbing* (namely, that this sound is strong and rhythmic), and about the source of the sound *knocking* (namely, a door or an engine).

However, in some cases the way of distinction between sister terms is not possible since some of them (especially sounds that belong to the same synset) share the same definition. For example, both *click* and *clink* belong to the same synset, hence, share the same hierarchy and base type. They are also described by the same definition of "a short light metallic sound" and no example of use is given. Likewise, a "plop-and-a-plunk" problem is rather an evident example of the inconsistent representation of sounds at the lower levels of sound hierarchies. Sounds *plop* and *plunk* occur in the same *Event* hierarchy, however, while *plunk* is a direct hyponym of $sound_4$, *plop* is linked to sound$_4$ indirectly via *noise*. There is no clear criterion to consider *plop* (defined as "the noise of a rounded object dropping into liquid without splash") as a hyponym of *noise* (defined as "sound of any kind (especially unintelligible or dissonant sound)") while *plunk* (defined as "a hollow twanging sound") as its sister.

In summary, although the idea of organizing sounds as to the possible mental representation of sounds is very appealing, the current state of affairs in WordNet proves to be inconsistent and insufficient.

One of the plausible ways to proceed is to look at the definitions of sounds more closely. As has been mentioned above, some definitions provide enough information for distinguishing one sound from another. Namely, *throbbing* is rhythmic, *clicks* and *clinks* are short and light and *plunk* is hollow and twanging. These descriptive words seem to be very good indicators of how people perceive and describe sounds (as in "I heard a short click") because they represent the perceptual features of sounds. These descriptive words are what we use as the basis for our psycho-acoustic experiment we present in the next section.

## 4   Experiment

Are the features identified by the experimenters in section 2 perceived by listeners as relevant to classifying sounds? Participants were presented with three sounds and asked to choose the sound that differs most from the other two.

### 4.1   Method and Materials

We used 26 questions consisting of three sounds each as stimuli. All of the sounds are real life sounds taken from the Auvidis sound library [2]. We were careful to choose sounds for which the source was difficult, or not possible to determine. All files were cropped to a uniform duration of 80 milliseconds. After initial selection, we decided which features characterize each sound sample. The stimuli can be subdivided into three types. Type Simple (16/26) questions consisted of two sounds similar on all features and one sound that differed on one or more of these features from the other two. In the example below sound S2 differs on feature F2:

S1: F1(+)F2(+)...   S2: F1(+)F2(-)...   S3: F1(+)F2(+)...

This type of questions will be used to test whether participants actually perceive the features. If this is the case, the sound that differs on one feature will be determined as the least similar. In questions from Type Complex (7/26) a set of three sounds consists of two pairs, where one sound belongs to both pairs. Within each pair the sounds share features. Since this set-up creates a conflict of several features, participant's choice will show which feature is more dominant. For example,

S1: F1(+)F2(-)...   S2: F1(+)F2(+)...   S3:F1(-)F2(+)...

The sounds S1 and S2 share feature F1 and the sounds S2 and S3 share feature F2. This type of question will be used to test whether one feature is more salient than another. If, for example, feature F2 is more salient than feature F1, sound S1 will be experienced as being more different, because it does not share this feature. Finally Type Control consisted of three questions (3/26) where two out of three sounds were exactly the same. These control questions tested whether participants were paying attention. Additionally five times during the experiment participants were asked to explain their choice in comments.

The experiment was done online and results were stored in a database. Thirty one adult native Dutch speakers took part in the experiment. Two of the participants were excluded for reporting hearing problems and one participant was excluded for giving a wrong answer on one of the control questions.

## 4.2   Results and Discussion

Questions of type Simple were meant to test whether people were sensitive to the features identified by the experimenters. From the 16 type Simple questions ten questions were answered as expected, the stimulus that differed on the feature dimension identified by the experimenters was chosen significantly more often than chance ($\chi^2$, p-value 0.001). However, in four cases participants consistently chose a stimulus different from the stimulus predicted by the experimenters ($\chi^2$, p-value 0.001). The participants agreed on which sound was different but this was not what the experimenters predicted from the features identified. Since all five features identified were presented in the type Simple stimuli answered as predicted as well as in the type Simple stimuli not answered as predicted the results are difficult to interpret. What is, however, striking is the high degree of agreement among participants as to which sound was different (cf. [4]'s results), suggesting that if the correct salient features could be identified, what sound participants will judge as different should be predictable.

Among questions of type Complex, five out of seven answers were significantly different from a uniform distribution. However, these results do not indicate that one feature was consistently considered to be more dominant than the other feature, so not much can be concluded about which features might be more dominant than the others. In four cases participants chose the sound exhibiting the most features ($\chi^2$, p-value 0.001). These sounds may be considered more complex, and *sound complexity* might also be a salient feature.

Participants' explanations about their choices were not always easy to interpret. For example, participants did not report that a sound was chosen because it exhibited "a different tone color". Instead they reported that the sound they chose was "more sharp" or the sound was "more dull". All these descriptions were interpreted and labeled. In most of the cases the reports were consistent with the chosen sound. For example, the participants who chose the third sound for a given question gave another description than the participants that chose the first sound. The features that were determined by the experimenters were all mentioned at least once. Furthermore participants referred to *tone color*, *changing through time* and *on-/offset* characteristics.

There are three possible explanations why we didn't obtain the clear results we had hoped for. First, it could be that many of the stimuli were too complex, making it hard to compare. We removed a number of questions because we thought they were too simple but that may have been a mistake. Second, it could be that additional features play key roles. Features mentioned by the participants might be a good starting point to look for other salient characteristics in future work. Third, it's possible that the features are hierarchically ordered but in such a way that some of our stimuli sets made it difficult to compare, or led to comparisons in a way we did not expect.

But because people were quite consistent in their evaluation of sounds, and because this to a certain degree was similar to our expectations we are quite optimistic that further experiments with more stimuli will help us determine the actual hierarchical characteristics of the features.

## 5   Conclusions

How can we use our observations about the shortcomings of sound classification in WordNet and the experimental results to propose a classification for source unidentifiable sounds?

If we examine the features we have studied again, *pitch, duration, harmonicity, continuity* and *repetitiveness*, what characteristics do these have compared to features we chose not to focus on? One important characteristic is that the values of these features are consistent across all tokens of a given type of sound type. Taking each sound term, such as *clink, rattle* or *plonk* as a type, the feature *absolute pitch*, which we did not choose to study, conspicuously does not have these characteristics. A *clink* could have a high pitch or a low pitch, and both would still be instances of *clinks*. The same goes for *rattle*: high-pitched rattle or a low-pitched rattle are both possible rattle tokens. Based on this, even though absolute pitch might be a salient feature for classification of some sounds in the experiment, its ability to vary among tokens of the same type make it an inappropriate choice for classification.

The feature *pitch* we used has to do with having or not having pitch, so e.g. *swish, thump* and *gurgle* are all examples that are (−)*pitch*, and *pitch* seems to remain consistent for all tokens of each of these types. The experimental results also suggest this is a salient feature, and could be used to split sound types into two sets.

Examining a number of sound types, long *durations* seem to be consistent across e.g. *rattle* or *hum* tokens, while short *durations* are also characteristics of *clinks* and *plonks*. Further *repetitiveness* and *continuous* seem to be associated with *a long duration*. Thus it seems that these might be lower branches. Both these features also seem to be consistent among tokens of the same type.

But which feature would make a better initial split: *pitch* or *duration*? Unfortunately, the results from the experiment were not clear enough to allow us to make this decision, and more tests are needed.

The function of the feature *harmonicity* is also not clear. It might be a feature allowing us to split the set of sounds with short duration into +*harmonic* (e.g. *pling* or *booing*) from those that seem to be (−)*harmonic* (e.g. *click* or *plonk*), but it might be necessary to do a classification experiment to see if subjects would agree with this division.

As for the other features pointed out by experiment participants, such as e.g. *tone color*, we will have to do more research. Our results certainly suggest that the choice of salient features in a sound ontology has to be empirically grounded. It seems possible to make principled decisions to structure source unidentifiable sounds in an ontology, a result that should be useful for many applications, both those under development such as for searching media on the Semantic Web, and applications in the future, e.g. robots that can describe sounds they heard as humans would.

# References

1. Fellbaum, C.: WordNet: An Electronic Lexical Database. MIT Press, Cambridge, MA (1998)
2. Mercier, D.: Sound Library. AUVIDIS (1989)
3. Nakatani, T., Okuno, H.G.: Sound Ontology for Computational Auditory Scene Analysis. In: AAAI/IAAI, pp. 1004–1010 (1998)
4. Saint-Arnaud, N.: Classification of Sound Textures. M.S. Thesis in Media Arts and Sciences, Massachusetts institute of Technology (1995)
5. Scheirer, E., Slaney, M.: Construction and Evaluation of a Robust Multifeature Speech/music Discriminator. In: Proc. ICASSP-97, Munich (1997)

# Named Entities in Czech:
# Annotating Data and Developing NE Tagger[*]

Magda Ševčíková, Zdeněk Žabokrtský, and Oldřich Krůza

Faculty of Mathematics and Physics, Charles University
Malostranské nám. 25, CZ-11800 Prague, Czech Republic
{sevcikova,zabokrtsky,kruza}@ufal.mff.cuni.cz

**Abstract.** This paper deals with the treatment of Named Entities (NEs) in Czech. We introduce a two-level NE classification. We have used this classification for manual annotation of two thousand sentences, gaining more than 11,000 NE instances. Employing the annotated data and Machine-Learning techniques (namely the top-down induction of decision trees), we have developed and evaluated a software system aimed at automatic detection and classification of NEs in Czech texts.

## 1   Introduction

After the series of Message Understanding Conferences (MUC; [1]), processing of NE became a well established discipline within the NLP domain (see [2] for a survey of NE related research), usually motivated by the needs of Information Extraction, Question Answering, or Machine Translation. For English, one can find literature about attempts at rule-based solutions for the NE task as well as machine-learning approaches, be they dependent on the existence of labeled data (such as CoNLL-2003 shared task data), unsupervised (using redundancy in NE expressions and their contexts, see e.g. [3]) or a combination of both (such as [4], in which labeled data are used as a source of seed for an unsupervised procedure exploiting huge unlabeled data).

For Czech, the situation is different. To our best knowledge, until the presented work there have been no data with explicitly annotated NE instances available for Czech. Although there are several other types of available resources potentially usable for recognition and classification of NE (e.g. gazetteers, or technical lemma suffixes used at the morphological layer of PDT [5]), we have not found any published attempt concerning development of NE taggers for Czech[1].

This paper is structured as follows: in Section 2 we introduce our classification of NE, which we have used for annotating sample sentences as described in Section 3. Section 4 presents our NE tagger trained on the annotated data. The summary is given in Section 5.

---

[*] The research reported on in this paper was supported by the projects 1ET101120503, MSM0021620838, MSMT CR LC536, GD201/05/H014, and GA UK 643/2007.

[1] Even if some approaches developed for English are claimed to be language independent, it is obvious that they cannot be straightforwardly applied to Czech because of its rich inflection.

V. Matoušek and P. Mautner (Eds.): TSD 2007, LNAI 4629, pp. 188–195, 2007.
© Springer-Verlag Berlin Heidelberg 2007

## 2 Proposed Two-Level NE Classification

There is no generally accepted typology of Named Entities. One can see two trends: from the viewpoint of unsupervised learning, it is advantageous to have just a few coarse-grained categories (cf. the NE classification developed for MUC conferences or the classification proposed in [3], where only persons, locations, and organizations were distinguished), whereas those interested in semantically oriented applications prefer more informative (finer-grained) categories (e.g. [6] with eight types of person labels, or Sekine's Extended NE Hierarchy, cf. [7]).

Therefore we have proposed a two-level NE classification, as depicted in Figure 1. The first level corresponds to rough categories (called *NE supertypes*) such as person names, geographical names etc., whereas the second level provides a more detailed classification: e.g. within the supertype of geographical names, the *NE types* of names of cities/towns, names of states, names of rivers/seas/lakes etc. are distinguished. If more robust processing is necessary, only the first level (NE supertypes) can be used, whereas the second level (NE types) comes into play when more subtle information is needed. Each NE type is encoded by a unique two-character tag (e.g., gu for names of cities/towns, gc for names of states; a special tag, such as g_, makes it possible to leave the NE type underspecified).

Besides the terms of NE type and supertype, we use also the term *NE instance*, which stands for a continuous subsequence of tokens expressing the entity in a given text. In the simple plain-text format, which we use for manual annotations, the NE instances are marked as follows: the word or the span of words belonging to the NE is delimited by symbols < and >, with the former one immediately followed by the NE type tag (e.g. <pf *John*> loves <pf *Mary*>).

The annotation scheme allows for the embedding of NE instances. There are two types of embedding. In the first case, the NE of a certain type can be embedded in another NE (e.g., the river name can be part of a name of a city as in <gu *Ústí nad* <gh *Labem*>>). In the second case, two or more NEs are parts of a (so-called) *container NE* (e.g., two NEs, a first name and a surname, form together a person name container NE such as in <P<pf *Paul*> <ps *Newman*>>). The container NEs are marked with a capital one-letter tag: P for (complex) person names, T for temporal expressions, A for addresses, and C for bibliographic items. A more detailed description of the NE classification can be found in [8].

## 3 Annotating Data

We have created the data with labeled NE instances by the following procedure:

1. We have randomly selected 2000 sentences from the Czech National Corpus[2] from the result of the query ([word=".*[a-z0-9]"] [word="[A-Z].*"]) (this query makes the relative frequency of NEs in the selection higher than the corpus average, which makes the subsequent manual annotation much more effective, even if it may slightly bias the distribution of NE types).

---

[2] http://ucnk.ff.cuni.cz

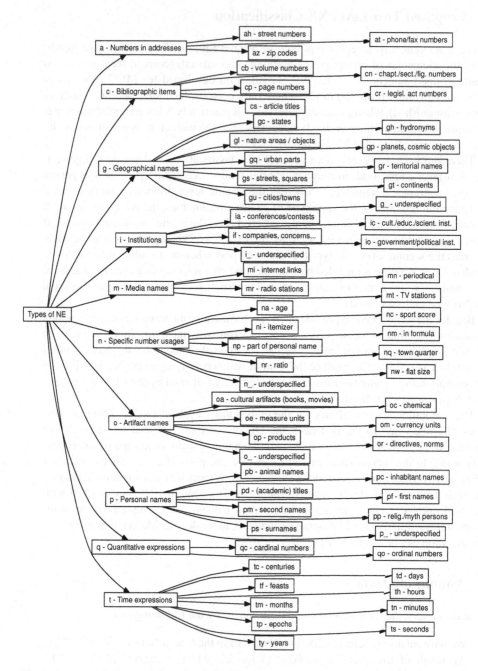

**Fig. 1.** Proposed two-level classification of NEs in Czech. Note that the (detailed) NE types are divided into two columns just because of the space reasons here.

2. The data (simple line-oriented plain-text files, editable in any text editor) have been manually annotated (i.e., enriched with starting and ending symbols and tags for NE types or NE containers) by two annotators in parallel. Differently annotated instances have been checked and decided by a third person. 11,644 NE instances have been detected in the sample.

3. The sentences have been enriched with morphological tags and lemmas using Jan Hajič's tagger shipped with Prague Dependency Treebank 2.0 ([5]).

4. The data have been divided into training, development test and evaluation test parts (8:1:1), and converted into pairs of XML files (source sentences enriched with morphological attributes are stored in an XML file with the same format as m-files in PDT 2.0, whereas the NE instances are represented in a separated XML file in the form of triples (i) reference to the first token of the given NE instance, (ii) reference to the last token of the instance, (iii) type of the NE instance according to the two-level classification; see [8] for more detail).

After the last step, the data have been 'frozen' and prepared for experiments with NE taggers described in Section 4 below.[3]

Sample of annotated text (after the second step):

```
Britský mediální umělec <p_ Sting>, vlastním jménem <P<pf Gor-
don> <ps Sumner>>, který má vystoupit <T<td 14.> <tm června>>
v pražské <ic Sportovní hale> , bude s největší pravděpodobno-
stí bydlet se svým devětadvacetičlenným týmem pod krycím jmé-
nem v některém z pražských hotelů .
<P<pd Ing .> <pf Karel> <ps Hennhofer> , <pd PhD>> , zastupu-
jící
ostravskou divizi <ic Technické inspekce " DOM-ZO <n_ 13> ">
popsal nové pojetí systému managementu jakosti podle normy
<or ČSN EN ISO <nr 9001:2001>> .
```

# 4 Development of NE Tagger

## 4.1 Task Definition

Our goal was to create a program for automatic processing of NEs in Czech texts enriched with morphological annotation. It has been subdivided into two 'subgoals': (i) the word or the span of words belonging to the NE should be delimited, (ii) a type/container tag chosen from a given set (cf. Section 2) should be assigned to the detected NE instance.

A NE instance is correctly recognized if and only if both the span was correctly detected and the correct tag was assigned. However, we also present results based on less strict rules (only the supertype is taken into consideration) to provide the reader with more insight into the principles of the NE tagger.

---

[3] The data are available also for other researchers by request from the authors.

## 4.2 Decomposition of the Task

The two proposed subgoals are inherently different: the first one (i.e., detecting the span of a NE instance) involves searching the input text for tokens or sequences of tokens forming NE instances (such a sequence can be of an arbitrary length, limited only by the length of the sentence) whereas the second one (i.e., assigning the tag) is the assignment of a class to the detected NE instance.

Concerning the first subgoal, a considerable simplification can be reached by limiting the length of a NE instance we attempt to recognize. With such an approach, we can tackle the NE instance detection as a classification task as well. The price is that we will not be able to detect NE instances longer than the given upper bound any more. For our system, a bound of two words has been set[4].

These two subgoals have been solved separately for single-word NE instances (see subtask 1) and for two-word NE instances (subtask 2). As for multi-word NE instances, only one type of such NE instances has been detected, namely multi-word names of Czech towns (subtask 3; i.e., only one tag can be assigned). Thus, there are three subtasks to be solved:

1. Detection and classification of single-word NE instances.
2. Detection and classification of two-word NE instances.
3. Detection and classification of one type of multi-word NE instances.

The first two subtasks have been approached as a feature-based classification. The third one has been solved by specially crafted algorithms. The feature sets are different for each subtask.

## 4.3 Implementation

The system has been implemented in Perl, it uses the *Rulequest c5.0* classifier. Firstly, the training of the system will be described, then the analysis will be briefly sketched.

**Training.** Is divided into (i) the preparation of the data for the classifier and (ii) the use of the classifier to construct a prediction method. The data sets for each subtask are prepared in the same way, except for the different feature sets.

We distinguish categorial and boolean features. Features used for detection – and subsequently also for classification – of single-word NE instances are the following:

– Categorial: How many times the lemma occurs in training data.
– Boolean: The word form is capitalized.
– Categorial: The NE type denoted by the technical lemma suffix as used at the morphological layer of PDT 2.0.
– Boolean: The token is the only capitalized word (first word excluded) or the only number in the sentence.
– Boolean: The form is a single capital letter.
– Boolean: The lemma denotes a month.

---

[4] NE instances of length up to two words cover more than 87 % of all NE instances in the training data.

- Boolean: The form is a number adjacent to another number.
- Boolean: The form matches a simple time expression pattern (e.g., *16:30*).
- Categorial: The lemma; OTHER value for non-frequent lemmas.
- Boolean: The lemma is included in the list of Czech town names.
- Tag-based features: In the positions of the morphological tag used in PDT 2.0, the part-of-speech information, information concerning the gender, number etc. is encoded. Values on these tag positions are treated as categorial features.
- Contextual features: Features representing presence or absence of trigger words in the immediate neighborhood; the list of around 600 trigger words (words that signal the beginning or the end of a NE instance, such as *president*) was semimanually extracted from the training data.

Features used for detection of two-word NE instances as well as for their classification are the following:

- Categorial: Part-of-speech pattern:
  Two-letter symbol formed by concatenation of the parts of speech of the two words in the scope (again, OTHER is used for infrequent combinations).
- Categorial: Single-word prediction pattern:
  The pair of NE types predicted by the system for the individual words of the bigram. In case of rare pairs, OTHER is used instead.
- Categorial: Capitalization pattern:
  Each word of the bigram is classified as (i) being the first word in the sentence, (ii) being capitalized or (iii) neither of the above. The couple of these categories is then used as the feature value.
- Boolean: The words agree in number, gender and case.
- Categorial: How many times the lemmas occur in the training data next to each other.

The **analysis** (i.e., detection and classification of NE instances in the unseen data) is performed for each sentence in three phases, as mentioned in Subsection 4.2 above.

For each word, features for single-word NE instance detection are evaluated. Based on these features, the prediction method decided whether the word is a NE instance. If so, then features used for classification of single-word NE instances are evaluated. The prediction method assigns a tag to the NE instance. For each bigram, the same procedure is executed as for each word, using feature sets for two-word NE instances. The last phase involves searching the sentence for an occurrence of a multi-word Czech town name (as the only one multi-word NE instance to be detected). In case such NE instance is found, the words are marked as a NE instance of type gu.

## 4.4   Results and Evaluation

We evaluate the results using *precision*, *recall* and *f-measure*. We present results based on three definitions of a correctly detected (and classified) NE instance:

- The span of the NE instance was detected correctly.
- The span of the NE instance was detected correctly and a correct supertype tag (i.e., the first character of the NE type tag) was assigned.

**Table 1.** Results of the capitalization-based baseline classifier

|           | Correct type | Correct supertype | Correct span |
|-----------|--------------|-------------------|--------------|
| Precision | 0.16         | 0.29              | 0.68         |
| Recall    | 0.16         | 0.29              | 0.68         |
| F-measure | 0.16         | 0.29              | 0.68         |

**Table 2.** Results of the baseline classifier based on the in-data occurrence (the precision and recall results seem to be identical only because of rounding)

|           | Correct type | Correct supertype | Correct span |
|-----------|--------------|-------------------|--------------|
| Precision | 0.54         | 0.57              | 0.59         |
| Recall    | 0.33         | 0.34              | 0.36         |
| F-measure | 0.40         | 0.43              | 0.45         |

**Table 3.** Final results of the developed system (precision/recall/F-measure)

|                   | All NE inst.      | One-word NE inst. | Multi-word NE inst. |
|-------------------|-------------------|-------------------|---------------------|
| Correct type      | 0.74 / 0.54 / 0.62 | 0.72 / 0.69 / 0.70 | 0.93 / 0.22 / 0.35  |
| Correct supertype | 0.81 / 0.59 / 0.68 | 0.79 / 0.76 / 0.78 | 0.95 / 0.22 / 0.36  |
| Correct span      | 0.88 / 0.64 / 0.75 | 0.87 / 0.84 / 0.86 | 0.98 / 0.23 / 0.37  |

– Both the span of the NE instance was detected correctly and a correct NE type tag was assigned.

Two baselines have been suggested. The first one recognized every capitalized word (excluding sentence-first words) as a NE instance of the most frequent type (ps, i.e. a surname; see Table 1). The second baseline (see Table 2) checked each word and bigram for presence in the training data (effectively evaluating the corresponding feature described above) and marked it as a NE instance if an occurrence has been found. The type of the NE instance was denoted as the type of one of the NE instances found in training data.

The final results are shown in Table 3. The F-measure equal to 0.62 seems to be a rather low number, but it is necessary to take into account the very high number of employed NE types (and thus very low baselines). Restricting the task in any dimension (i.e. the kind of entities sought, the number of types, etc.) improves the performance considerably. The weakest part is the recall in recognizing multi-word NE instances, which is no surprise as such entities are frequent in real-world data and we have no satisfactory method to deal with them yet. The results are comparable with those of HAREM competitors ([9]) and of Sassano and Utsuro ([10]).

## 5   Conclusion

We believe that the contributions of our work are the following: (i) we have introduced a detailed two-level NE classification verified on authentic corpus data, (ii) we have manually annotated a substantive sample of Czech sentences with the proposed NE tags, and (iii) we have developed and evaluated a NE tagger for Czech. To our knowledge, the presented work is novel for Czech in all three aspects.

As for future work, we plan to use also unlabeled data for the development of NE taggers, and to study the status of NE instances at more abstract layers of linguistic representation, especially at the tectogrammatical layer as implemented in PDT 2.0.

# References

1. Grishman, R., Sundheim, B.: Message Understanding Conference - 6: A Brief History. In: Proceedings of the 16th International Conference on Computational Linguistics (COLING), vol. I, pp. 466–471 (1996)
2. Sekine, S.: Named Entity: History and Future (2004), http://www.cs.nyu.edu/~sekine/papers/NEsurvey200402.pdf
3. Collins, M., Singer, Y.: Unsupervised Models for Named Entity Classification. In: Proceedings of the Conference on Empirical Methods in Natural Language Processing and Very Large Corpora (EMNLP/VLC), pp. 189–196 (1999)
4. Talukdar, P.P., Brants, T., Liberman, M., Pereira, F.: A Context Pattern Induction Method for Named Entity Extraction. In: Proceedings of the 10th Conference on Computational Natural Language Learning (CoNLL-X), pp. 141–148 (2006)
5. Hajič, J., Panevová, J., Hajičová, E., Sgall, P., Pajas, P., Štěpánek, J., Havelka, J., Mikulová, M., Žabokrtský, Z., Ševčíková, M.: Prague Dependency Treebank 2.0 (2006)
6. Fleischman, M., Hovy, E.: Fine Grained Classification of Named Entities . In: Proceedings of the 19th International Conference on Computational Linguistics (COLING), vol. I, pp. 267–273 (2002)
7. Sekine, S.: Sekine's Extended Named Entity Hierarchy (2003), http://nlp.cs.nyu.edu/ene/
8. Ševčíková, M., Žabokrtský, Z., Krůza, O.: Zpracování pojmenovaných entit v českých textech. ÚFAL MFF UK, Praha (2007)
9. Santos, D., Seco, N., Cardoso, N., Vilela, R.: HAREM: An Advanced NER Evaluation Contest for Portuguese. In: Proceedings of the 5th International Conference on Language Resources and Evaluation (LREC), pp. 1986–1991 (2006)
10. Sassano, M., Utsuro, T.: Named Entity Chunking Techniques in Supervised Learning for Japanese Named Entity Recognition. In: Proceedings of the 18th International Conference on Computational Linguistics (COLING), vol. II, pp. 705–711 (2000)

# Identifying Expressions of Emotion in Text

Saima Aman[1] and Stan Szpakowicz[1,2]

[1] School of Information Technology and Engineering, University of Ottawa, Ottawa, Canada
[2] Institute of Computer Science, Polish Academy of Sciences, Warszawa, Poland
{saman071, szpak}@site.uottawa.ca

**Abstract.** Finding emotions in text is an area of research with wide-ranging applications. We describe an emotion annotation task of identifying emotion category, emotion intensity and the words/phrases that indicate emotion in text. We introduce the annotation scheme and present results of an annotation agreement study on a corpus of blog posts. The average inter-annotator agreement on labeling a sentence as emotion or non-emotion was 0.76. The agreement on emotion categories was in the range 0.6 to 0.79; for emotion indicators, it was 0.66. Preliminary results of emotion classification experiments show the accuracy of 73.89%, significantly above the baseline.

## 1 Introduction

Analysis of sentiment in text can help determine the opinions and affective intent of writers, as well as their attitudes, evaluations and inclinations with respect to various topics. Previous work in sentiment analysis has been done on a variety of text genres, including product and movie reviews [9, 18], news stories, editorials and opinion articles [20], and more recently, blogs [7].

Work on sentiment analysis has typically focused on recognizing valence – positive or negative orientation. Among the less explored sentiment areas is the recognition of types of emotions and their strength or intensity. In this work, we address the task of identifying expressions of emotion in text. Emotion research has recently attracted increased attention of the NLP community – it is one of the tasks at Semeval-2007[1]; a workshop on emotional corpora was also held at LREC-2006[2].

We discuss the methodology and results of an emotion annotation task. Our goal is to investigate the expression of emotion in language through a corpus annotation study and to prepare (and place in the public domain) an annotated corpus for use in automatic emotion analysis experiments. We also explore computational techniques for emotion classification. In our experiments, we use a knowledge-based approach for automatically classifying emotional and non-emotional sentences. The results of the initial experiments show an improved performance over baseline accuracy.

The data in our experiments come from blogs. We wanted emotion-rich data, so that there would be ample examples of emotion use for analysis. Such data is

---

[1] http://nlp.cs.swarthmore.edu/semeval/tasks/task14/summary.shtml
[2] http://www.lrec-conf.org/lrec2006/IMG/pdf/programWSemotion-LREC2006-last1.pdf

V. Matoušek and P. Mautner (Eds.): TSD 2007, LNAI 4629, pp. 196–205, 2007.
© Springer-Verlag Berlin Heidelberg 2007

expected in personal texts, such as diaries, email, blogs and transcribed speech, and in narrative texts such as fiction. Another consideration in selecting blog text was that such text does not conform to the style of any particular genre *per se*, thus offering a variety in writing styles, choice and arrangement of words, and topics.

## 2 Related Work

Some researchers have studied emotion in a wider framework of *private states* [12]. Wiebe et al. [20] worked on the manual annotation of private states including emotions, opinions, and sentiment in a 10,000-sentence corpus (the MPQA corpus) of news articles. Expressions of emotions in text have also been studied within the *Appraisal Framework* [5], a functional theory of the language used for conveying attitudes, judgments and emotions [15, 19]. Neither of these frameworks deals exclusively with emotion, the focus of this paper.

In a work focused on learning specific emotions from text, Alm et al. [1] have explored automatic classification of sentences in children's fairy tales according to the basic emotions identified by Ekman [3]. The data used in their experiments was manually annotated with emotion information, and is targeted for use in a text-to-speech synthesis system for expressive rendering of stories. Read [14] has used a corpus of short stories, manually annotated with sentiment tags, in automatic emotion-based classification of sentences. These projects focus on the genre of fiction, with only sentence-level emotion annotations; they do not identify emotion indicators within a sentence, as we do in our work.

In other related work, Liu et al. [4] have utilized real-world knowledge about affect drawn from a common-sense knowledge base. They aim to understand the semantics of text to identify emotions at the sentence level. They begin with extracting from the knowledge base those sentences that contain some affective information. This information is utilized in building affective models of text, which are used to label each sentence with a six-tuple that corresponds to Ekman's six basic emotions [3]. Neviarouskaya et al. [8] have also used a rule-based method for determining Ekman's basic emotions in the sentences in blog posts.

Mihalcea and Liu [6] have focused in their work on two particular emotions – *happiness* and *sadness*. They work on blog posts which are self-annotated by the blog writers with *happy* and *sad* mood labels. Our work differs in the aim and scope from those projects: we have prepared a corpus annotated with rich emotion information that can be further used in a variety of automatic emotion analysis experiments.

## 3 The Emotion Annotation Task

We worked with blog posts we collected directly from the Web. First, we prepared a list of seed words for six basic emotion categories proposed by Ekman [3]. These categories represent the distinctly identifiable facial expressions of emotion – *happiness, sadness, anger, disgust, surprise* and *fear*. We took words commonly used in the context of a particular emotion. Thus, we chose "happy", "enjoy", "pleased" as seed words for the *happiness* category, "afraid", "scared", "panic" for the *fear* category, and so on. Next, using the seed words for each category, we retrieved blog posts containing one or more of those words. Table 1 gives the details of the datasets thus collected. Sample examples of annotated text appear in Table 2.

**Table 1.** The details of the datasets

| Dataset | # posts | # sentences | Collected using seed words for |
|---------|---------|-------------|-------------------------------|
| Ec-hp | 34 | 848 | *Happiness* |
| Ec-sd | 30 | 884 | *Sadness* |
| Ec-ag | 26 | 883 | *Anger* |
| Ec-dg | 21 | 882 | *Disgust* |
| Ec-sp | 31 | 847 | *Surprise* |
| Ec-fr | 31 | 861 | *Fear* |
| Total | 173 | 5205 | |

**Table 2.** Sample examples from the annotated text

| |
|---|
| I have to look at life in her perspective, and it would <u>break anyone's heart</u>. (*sadness, high*) |
| We stayed in a tiny mountain village called Droushia, and these people brought hospitality to <u>incredible</u> new heights. (*surprise, medium*) |
| But the rest of it came across as a <u>really angry</u>, <u>drunken rant</u>. (*anger, high*) |
| And I <u>reallllly want</u> to go to Germany – <u>dang</u> terrorists are making flying overseas <u>all scary</u> and <u>annoying</u> and expensive though!! (*mixed emotion, high*) |
| I <u>hate</u> it when certain people always seem to be better at me in everything they do. (*disgust, low*) |
| Which, to be honest, was making Brad <u>slightly nervous</u>. (*fear, low*) |

Emotion labeling is reliable if there is more than one judgment for each label. Four judges manually annotated the corpus; each sentence was subject to two judgments. The first author of this paper produced one set of annotations, while the second set was shared by the three other judges. The annotators received no training, though they were given samples of annotated sentences to illustrate the kind of annotations required. The annotated data was prepared over a period of three months.

The annotators were required to label each sentence with the appropriate emotion category, which describes its affective content. To Ekman's six emotions [3], we added *mixed emotion* and *no emotion*, resulting in eight categories to which a sentence could be assigned. While sentiment analysis usually focuses on documents, this work's focus is on the sentence-level analysis. The main consideration behind this decision is that there is often a dynamic progression of emotions in the narrative texts found in fiction, as well as in the conversation texts and blogs.

The initial annotation effort suggested that in many instances a sentence was found to exhibit more than one emotion – consider (1), for example, marked for both *happiness* and *surprise*. Similarly, (2) shows how more than one type of emotion can be present in a sentence that refers to the emotional states of more than one person.

(1) Everything from trying to order a baguette in the morning to asking directions or talking to cabbies, we were always <u>pleasantly surprised</u> at how open and <u>welcoming</u> they were.

(2) I <u>felt bored</u> and wanted to leave at intermission, but my wife was <u>really enjoying</u> it, so we stayed.

We also found that the emotion conveyed in some sentences could not be attributed to any basic category, for example in (3). We decided to have an additional category called *mixed emotion* to account for all such instances. All sentences that had no emotion content were to be assigned to the *no emotion* category.

(3) It's like everything everywhere is going crazy, so we don't go out any more.

In the final annotated corpus, the *no emotion* category was the most frequent. It is important to have *no emotion* sentences in the corpus, as both *positive* and *negative* examples are required to train any automatic analysis system. It should also be noted that in both sets of annotations a significant number of sentences were assigned to the *mixed emotion* category, justifying its addition in the first place.

The second kind of annotations involved assigning emotion intensity (*high, medium,* or *low*) to all emotion sentences in the corpus, irrespective the emotion category assigned to them. No intensity label was assigned to the *no emotion* sentences. A study of emotion intensity can help recognize the linguistic choices writers make to modify the strength of their expressions of emotion. The knowledge of emotion intensity can also help locate highly emotional snippets of text, which can be further analyzed to identify emotional topics. Intensity values can also help distinguish borderline cases from clear cases [20], as the latter will generally have higher intensity.

Besides labeling the emotion category and intensity, the secondary objective of the annotation task was to identify spans of text (individual words or strings of consecutive words) that convey emotional content in a sentence. We call them emotion indicators. Knowing them could help identify a broad range of affect-bearing lexical tokens and possibly, syntactic phrases. The annotators were permitted to mark in a sentence any number of emotion indicators of any length.

We considered several annotation schemes for emotion indicators. First we thought to identify only individual words for this purpose. That would simplify calculating the agreement between annotation sets. We soon realized, however, that individual words may not be sufficient. Emotion is often conveyed by longer units of text or by phrases, for example, the expressions "can't believe" and "blissfully unaware" in (4). It would also allow the study of the various linguistic features that serve to emphasize or modify emotion, as the use of word "blissfully" in (4) and "little" in (5).

(4) I <u>can't believe</u> this went on for so long, and we were <u>blissfully unaware</u> of it.

(5) The news brought them <u>little happiness</u>.

## 4 Measuring Annotation Agreement

The interpretation of sentiment information in text is highly subjective, which leads to disparity in the annotations by different judges. Difference in skills and focus of the judges, and ambiguity in the annotation guidelines and in the annotation task itself also contribute to disagreement between the judges [11]. We seek to find how much the judges agree in assigning a particular annotation by using metrics that quantify these agreements.

First we measure how much the annotators agree on classifying a sentence as an emotion sentence. Cohen's kappa [2] is popularly used to compare the extent of consensus between judges in classifying items into known mutually exclusive categories. Table 3 shows the pair-wise agreement between the annotators on emotion/non-emotion labeling of the sentences in the corpus. We report agreement values for pairs of annotators who worked on the same portion of the corpus.

**Table 3.** Pair-wise agreement in emotion/non-emotion labeling

|       | a↔b  | a↔c  | a↔d  | average |
|-------|------|------|------|---------|
| Kappa | 0.73 | 0.84 | 0.71 | 0.76    |

**Table 4.** Pair-wise agreement in emotion categories

| Category      | a↔b  | a↔c  | a↔d  | average |
|---------------|------|------|------|---------|
| happiness     | 0.76 | 0.84 | 0.71 | 0.77    |
| sadness       | 0.68 | 0.79 | 0.56 | 0.68    |
| anger         | 0.62 | 0.76 | 0.59 | 0.66    |
| disgust       | 0.64 | 0.62 | 0.74 | 0.67    |
| surprise      | 0.61 | 0.72 | 0.48 | 0.60    |
| fear          | 0.78 | 0.80 | 0.78 | 0.79    |
| mixed emotion | 0.24 | 0.61 | 0.44 | 0.43    |

Within the emotion sentences, there are seven possible categories of emotion to which a sentence can be assigned. Table 4 shows the value of kappa for each of these emotion categories for each annotator pair. The agreement was found to be highest for *fear* and *happiness*. From this, we can surmise that writers express these emotions in more explicit and unambiguous terms, which makes them easy to identify. The *mixed emotion* category showed least agreement which was expected, given the fact that this category was added to account for the sentences which had more than one emotions, or which would not fit into any of the six basic emotion categories.

Agreement on emotion intensities can also be measured using kappa, as there are distinct categories – *high, medium,* and *low.* Table 5 shows the values of inter-annotator agreement in terms of kappa for each emotion intensity. The judges agreed more when the emotion intensity was high; agreement declined with decrease in the intensity of emotion. It is a major factor in disagreement that where one judge perceives a low-intensity, another judge may find no emotion.

**Table 5.** Pair-wise agreement in emotion intensities

| Intensity | a↔b | a↔c | a↔d | average |
|-----------|-----|-----|-----|---------|
| High | 0.69 | 0.82 | 0.65 | 0.72 |
| Medium | 0.39 | 0.61 | 0.38 | 0.46 |
| Low | 0.31 | 0.50 | 0.29 | 0.37 |

Emotion indicators are words or strings of words selected by annotators as marking emotion in a sentence. Since there are no predefined categories in this case, we cannot use kappa to calculate the agreement between judges. Here we need to find agreement between the sets of text spans selected by the two judges for each sentence.

Several methods of measuring agreement between sets have been proposed. For our task, we chose the measure of agreement on set-valued items (MASI), previously used for measuring agreement on co-reference annotation [10] and in the evaluation of automatic summarization [11]. MASI is a distance between sets whose value is 1 for identical sets, and 0 for disjoint sets. For sets A and B it is defined as:

MASI = J * M, where the Jaccard metric is

$$J = |A \cap B| / |A \cup B|$$

and monotonicity is

$$M = \begin{cases} 1, & \text{if } A = B \\ 2/3, & \text{if } A \subset B \text{ or } B \subset A \\ 1/3, & \text{if } A \cap B \neq \phi, A - B \neq \phi, \text{ and } B - A \neq \phi \\ 0, & \text{if } A \cap B = \phi \end{cases}$$

If one set is monotonic with respect to another, one set's elements always match those of the other set – for instance, in annotation sets {crappy} and {crappy, best} for (6). However, in non-monotonic sets, as in {crappy, relationship} and {crappy, best}, there are elements not contained in one or the other set, indicating a greater degree of disagreement. The presence of monotonicity factor in MASI therefore ensures that the latter cases are penalized more heavily than the former.

While looking for emotion indicators in a sentence, often it is likely that the judges may identify the same expression but differ in marking text span boundaries. For example in sentence (6) the emotion indicator identified by two annotators are "crappy" and "crappy relationship", which essentially refer to the same item, but disagree on the placement of the span boundary. This leads to strings of varying lengths. To simplify the agreement measurement, we split all strings into words to ensure that members of the set are all individual words. MASI was calculated for each pair of annotations for all sentences in the corpus (see Table 6).

(6) We've both had our share of crappy relationship, and are now trying to be the best we can for each other.

We adopted yet another method of measuring agreement between emotion indicators. It is a variant of the IOB encoding [13] used in text chunking and named entity recognition tasks. We use IO encoding, in which each word in the sentence is labeled as being either In or Outside an emotion indicator text span, as shown in (7).

(7) Sorry/I  for/O  the/O  ranting/I  post/O,  but/O  I/O  am/O  just/O  really/I annoyed/I.

Binary IO labeling of each word in essence reduces the task to that of word-level classification into non-emotion and emotion indicator categories. It follows that kappa can now be used for measuring agreement; pair-wise kappa values using this method are shown in Table 6. The average kappa value of 0.66 is lower than that observed at sentence level classification. This is in line with the common observation that agreement on lower levels of granularity is generally found to be lower.

**Table 6.** Pair-wise agreement in emotion indicators

| Metric | a↔b | a↔c | a↔d | average |
|--------|-----|-----|-----|---------|
| MASI | 0.59 | 0.66 | 0.59 | 0.61 |
| Kappa | 0.61 | 0.73 | 0.65 | 0.66 |

## 5  Automatic Emotion Classification

Our long-term research goal is fine-grained automatic classification of sentences on the basis of emotion categories. The initial focus is on recognizing emotional sentences in text, regardless of their emotion category. For this experiment, we extracted all those sentences from the corpus for which there was consensus among the judges on their emotion category. This was done to form a gold standard of emotion-labeled sentences for training and evaluation of classifiers. Next, we assigned all emotion category sentences to the class "EM", while all no emotion sentences were assigned to the class "NE". The resulting dataset had 1466 sentences belonging to the EM class and 2800 sentences belonging to the NE class.

### 5.1  Feature Set

In defining the feature set for automatic classification of emotional sentences, we were looking for features which distinctly characterize emotional expressions, but are not likely to be found in the non-emotional ones. The most appropriate features that distinguish emotional and non-emotional expressions are obvious emotion words present in the sentence. To recognize such words, we used two publicly available lexical resources – the General Inquirer [16] and WordNet-Affect [17].

The General Inquirer (GI) is a useful resource for content analysis of text. It consists of words drawn from several dictionaries and grouped into various semantic categories. It lists different senses of a term and for each sense it provides several tags indicating the different semantic categories it belongs to. We were interested in the tags representing emotion-related semantic categories. The tags we found relevant are *EMOT* (emotion) – used with obvious emotion words; *Pos/Pstv* (positive) and *Neg/Ngtv* (negative) – used to indicate the valence of emotion-related words; *Intrj* (interjections); and *Pleasure* and *Pain*.

WordNet-Affect (WNA) assigns a variety of affect labels to a subset of synsets in WordNet. We utilized the publicly available lists[3] extracted from WNA, consisting of emotion-related words. There are six lists corresponding to the six basic emotion categories identified by Ekman [3].

Beyond emotion-related lexical features, we note that the emotion information in text is also expressed through the use of symbols such as emoticons and punctuation (such as "!"). We, therefore, introduced two more features to account for such symbols. All features are summarized in Table 7 (the feature vector represented counts for all features).

**Table 7.** Features Used in emotion classification

| GI Features | WN-Affect Features | Other Features |
|---|---|---|
| Emotion words | Happiness words | Emoticons |
| Positive words | Sadness words | Exclamation ("!") and |
| Negative words | Anger words | question ("?") marks |
| Interjection words | Disgust words | |
| Pleasure words | Surprise words | |
| Pain words | Fear words | |

## 5.2 Experiments and Results

For our binary classification experiments, we used Naïve Bayes, and Support Vector Machines (SVM), which have been popularly used in sentiment classification tasks [6, 9]. All experiments were performed using stratified ten-fold cross validation. The naïve baseline for our experiments was 65.6%, which represents the accuracy achieved by assigning the label of the most frequent class (which in our case is NE) to all the instances in the dataset. Each sentence was represented by a 14-value vector, representing the number of occurrences of each feature type in the sentence. Table 9 shows the classification accuracy obtained with the Naïve Bayes and SVM text classifiers. The highest accuracy achieved was 73.89% using SVM, which is higher than the baseline. The improvement is statistically significant (we used the paired t-test, $p=0.05$).

To explore the contribution of different feature groups to the classification performance, we conducted experiments using (1) features from GI only, (2) features from WordNet-Affect only, (3) combined features from GI and WordNet-Affect, and (4) all features (including the non-lexical features). We achieved the best results when

**Table 8.** Emotion classification accuracy

| Features | Naïve Bayes | SVM |
|---|---|---|
| GI | 71.45% | 71.33% |
| WN-Affect | 70.16% | 70.58% |
| GI+WN-Affect | 71.7% | 73.89% |
| **ALL** | **72.08%** | **73.89%** |

[3] http://www.cse.unt.edu/~rada/affectivetext/data/WordNetAffectEmotionLists.tar.gz

all the features were combined. While the use of non-lexical features does not seem to affect results of SVM, it did increase the accuracy of the Naïve Bayes classifier. This suggests that a combination of features is needed to improve emotion classification results.

The results of the automatic emotion classification experiments show how external knowledge resources can be leveraged in identifying emotion-related words in text. We note, however, that lexical coverage of these resources may be limited, given the informal nature of online discourse. For instance, one of the most frequent words used for *happiness* in the corpus is the acronym "lol", which does not appear in any of these resources. In future experiments, we plan to augment the word lists obtained from GI and WordNet-Affect with such words. Furthermore, in our experiments, we have not addressed the case of typographical errors and orthographic features (for e.g. "soo sweeet") that express or emphasize emotion in text.

We also note that the use of emotion-related words is not the sole means of expressing emotion. Often a sentence, which otherwise may not have an emotional word, may become emotion-bearing depending on the context or underlying semantic meaning. Consider (8), for instance, which implicitly expresses *fear* without the use of any emotion bearing word.

(8)     What if nothing goes as planned?

Therefore to be able to accurately classify emotion, we need to do contextual and semantic analysis as well.

# 6   Conclusion and Future Work

We address the problem of identifying expressions of emotion in text. We describe the task of annotating sentences in a blog corpus with information about emotion category and intensity, as well as emotion indicators. An annotation agreement study shows variation in agreement among judges for different emotion categories and intensity. We found the annotators to agree most in identifying instances of fear and happiness. We found that agreement on sentences with high emotion intensity surpassed that on the sentences with medium and low intensity. Finding emotion indicators in a sentence was found to be a hard task, with judges disagreeing in identifying precisely the spans of text that indicate emotion in a sentence.

We also present the results of automatic emotion classification experiments, which utilized knowledge resources in identifying emotion-bearing words in sentences. The accuracy is 73.89%, significantly higher than our baseline accuracy.

This paper described the first part of an ongoing work on the computational analysis of expressions of emotions in text. In our future work, we will use the annotated data for fine-grained classification of sentences on the basis of emotion categories and intensity. As discussed before, we plan to incorporate methods for addressing the special needs of the kind of language used in online communication. We also plan on using a corpus-driven approach in building a lexicon of emotion words. In this direction, we intend to start with the set of emotion indicators identified during the annotation process, and further extend that using similarity measures.

# References

1. Alm, C.O., Roth, D., Sproat, R.: Emotions from text: machine learning for text-based emotion prediction. In: Proc. of the Joint Conf. on Human Language Technology/Empirical Methods in Natural Language Processing (HLT/EMNLP), pp. 579–586 (2005)
2. Cohen, J.: A coefficient of agreement for nominal scales. Educational and Psychological Measurement 20, 37–46 (1960)
3. Ekman, P.: An Argument for Basic Emotions. Cognition and Emotion. 6, 169–200 (1992)
4. Liu, H., Lieberman, H., Selker, T.: A Model of Textual Affect Sensing using Real-World Knowledge. In: Proc. of the Int'l Conf. on Intelligent User Interfaces (2003)
5. Martin, J.R., White, P.R.R.: The Language of Evaluation: Appraisal in English, Palgrave, London (2005), http://grammatics.com/appraisal/
6. Mihalcea, R., Liu, H.: A corpus-based approach to finding happiness. In: The AAAI Spring Symposium on Computational Approaches to Weblogs, Stanford, CA (2006)
7. Mishne, G., Glance, N.: Predicting Movie Sales from Blogger Sentiment. In: AAAI 2006 Spring Symposium on Computational Approaches to Analysing Weblogs (2006)
8. Neviarouskaya, A., Prendinger, H., Ishizuka, M.: Analysis of affect expressed through the evolving language of online communication. In: Proc. of the 12th Int'l Conf. on Intelligent User Interfaces (IUI-07), Honolulu, Hawaii, pp. 278–281 (2007)
9. Pang, B., Lee, L., Vaithyanathan, S.: Thumbs up? Sentiment Classification using Machine Learning Techniques. In: Proc. Conf. on EMNLP (2002)
10. Passonneau, R.: Computing reliability for coreference annotation. In: Proc. International Conf. on Language Resources and Evaluation, Lisbon (2004)
11. Passonneau, R.J.: Measuring agreement on set-valued items (MASI) for semantic and pragmatic annotation. In: Proc. 5th Int'l Conf. on Language Resources and Evaluation (2006)
12. Quirk, R., Greenbaum, S., Leech, G., Svartvik, J.: A Comprehensive Grammar of the English Language. Longman, New York (1985)
13. Ramshaw, L.A., Marcus, M.P.: Text chunking using transformation-based learning. In: Proc. Third ACL Workshop on Very Large Corpora (1995)
14. Read, J.: Recognising affect in text using pointwise mutual information. Master's thesis, University of Sussex (2004)
15. Read, J., Hope, D., Carroll, J.: Annotating expressions of Appraisal in English. In: The Proc. of the ACL Linguistic Annotation,Workshop, Prague (2007)
16. Stone, P.J., Dunphy, D.C., Smith, M.S., Ogilvie, D.M., et al.: The General Inquirer: A Computer Approach to Content Analysis. The MIT Press, Cambridge (1966)
17. Strapparava, C., Valitutti, A.: WordNet-Affect: an affective extension of WordNet. In: Proc. 4th International Conf. on Language Resources and Evaluation, Lisbon (2004)
18. Turney, P.D.: Thumbs Up or Thumbs Down? Semantic Orientation Applied to Unsupervised Classification of Reviews. In: Proc. the 40th Annual Meeting of the ACL, Philadelphia (2002)
19. Whitelaw, C., Garg, N., Argamon, S.: Using Appraisal Taxonomies for Sentiment Analysis. In: Proc. of the 2nd Midwest Comp., Linguistic Colloquium, Columbus (2005)
20. Wiebe, J., Wilson, T., Cardie, C.: Annotating expressions of opinions and emotions in language. Language Resources and Evaluation 39(2-3), 165–210 (2005)

# ECAF: Authoring Language
# for Embodied Conversational Agents

Ladislav Kunc[1] and Jan Kleindienst[2]

[1] Department of Computer Science and Engineering, FEE, CTU in Prague
Technická 2, Praha 6, 166 27, Czech Republic
[2] IBM Česká Republika
V Parku 2294/4, Praha 4, 148 00, Czech Republic

**Abstract.** Embodied Conversational Agent (ECA) is the user interface metaphor that allows to naturally communicate information during human-computer interaction in synergic modality dimensions, including voice, gesture, emotion, text, etc. Due to its anthropological representation and the ability to express human-like behavior, ECAs are becoming popular interface front-ends for dialog and conversational applications. One important prerequisite for efficient authoring of such ECA-based applications is the existence of a suitable programming language that exploits the expressive possibilities of multimodally blended messages conveyed to the user. In this paper, we present an architecture and interaction language ECAF, which we used for authoring several ECA-based applications. We also provide the feedback from usability testing we carried for user acceptance of several multimodal blending strategies.

## 1 Introduction

Natural interactions involve both verbal and non-verbal communication acts. In this paper, we introduce the ECA authoring language, called ECAF (Embodied Conversational Agent Facade), which focuses on the needs of developers that build multimodal applications with ECA-based interfaces. We followed the user-centered design approach when designing the ECAF language, and many features that exist in the language had been added based on the direct developers feedback. The developer requirements were thus important guidance leading the language expressive capabilities and the ability to operate the avatar at real-time, rather then relying on the pre-computed animated behavior.

To make the 3D avatar believable it is important for it to express certain human quality such as personality, affectiveness, sympathy, naturalness, etc. Some of these qualities can be expressed using straightforward algorithmic means, e.g. adding Perlin noise to the head movement in the idle state as described later. As user-centered design was the main driving point of the project, the paper also summarizes the developer and user feedback on the applications developed in ECAF.

### 1.1 Background

The authoring environment for avatar-based applications is fragmented, and the researchers tend to develop their own languages. The existing languages tend to fall to

V. Matoušek and P. Mautner (Eds.): TSD 2007, LNAI 4629, pp. 206–213, 2007.

different categories of abstractions, accenting the respective design priorities that the authors were following in support of their projects. One class of languages supports the modeling of human behavior at the very high level, such as Human Markup Language [4]. HML is backed-up by the Internet repository system and a set of tools, that tags various verbal and non-verbal communication cues used in human-to-human interactions. Its complex design makes it difficult for authors to get down to the level of plain animation when needed.

What seems to be the prevailing concept in ECA language design is the notion of independent communication channels. These channels, such as head, speech, gestures, body, expressions, etc. are mixed and matched to a multimodal communication act that the avatar as the "anthropological" output device delivers to the user. Examples of languages supporting the channel mixing concept include VHML [11], SMILE-AGENT [7] and RRL [9].

The Web application domain has brought its own set of XML-based languages that help Web page designers enhance human-machine interaction experience. Multimodal Presentation Mark-up Language (MPML) [6] builds on the body of the Microsoft Agent to create predefined animation sequences. Behavior Expression Animation Toolkit (BEAT) [1] processes the XML input description in the form of a tree containing both verbal and non-verbal signals, to produce the synchronized animated sequence on the output.

Most of these languages and toolkits tend to split the application authoring task into (a) off-line step for pre-processing and (b) the real-time step of running the preprocessed animations. This comes with the implicit disadvantage, that the animations are realized on the closed set of pre-computed behaviors. The ECAF language and toolkit we propose in this paper is designed for direct runtime processing and is thus able to react to the dynamically evolving applications context that involves combination of behaviors not foreseen during the animation design.

## 2  The ECAF Architecture

The basic concept of the architecture design of the ECAF framework is the client-server communication paradigm (see Fig. 1). The server listens on specific network port and receives commands through a bi-directional communication link, mainly exercising two ECAF commands – ACT and SPEAK, which we describe in more detail in Section 3. The client controls the behavior of the avatar by connecting to the server and sending a stream of ACT and SPEAK commands performed in real-time. In our case we use ChiliX [2] library for transporting commands through TCP/IP network.

At the heart of the talking head server is the open-source Expression toolkit [3], which we have significantly modified to match it up with the ECAF language capabilities. This toolkit renders and controls the 3D head model. It displays the head model with aid of OpenGL system [10]. Movements of the face are simulated by the model of human muscles. These muscles are controlled by parameters in the Expression toolkit.

Since the ECAF avatar supports lip-synchronization, we incorporated the support for IBM eViaVoice Text-To-Speech engine [5]. The synthesizer outputs various speech parameters during synthesis, from which we use the phoneme sequences that we translate to visemes used for lip synchronization.

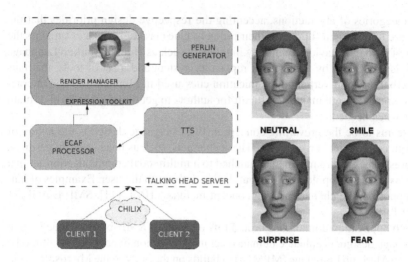

**Fig. 1.** System architecture and examples of talking head expressions

**Random Movements of the Head.** To achieve realistic dynamic appearance of the avatar, we implemented random movements of the head. These movements are generated by 1-dimensional Perlin noise generator [8]. Perlin generator is a function that adds controllable pseudo-random noise to the avatar head's movement. Based on user feedback, we tuned the Perlin noise generator to obtain pretty realistic appearance of the head. The level of Perlin noise is a configurable parameter that can be set by the application designer.

## 3   ECAF Language

In this section, we present the ECAF Language, scripting and controlling language for authoring applications based on synthetic avatars (Talking Heads). The ECAF is designed with the user-center design approach, where most of the features are responding to direct needs of developers. We intended to design markup based language with fast learning curve, real-time performance, and built-in extensibility to support future avatar functionalities.

**Realtime Behavior Control.** As already indicated, the talking head could be controlled through two basic control elements – the ACT and SPEAK. With the aid of the ACT command, we immediately change the appearance and behavior of the Talking Head. For example, we can send command to turn head 20 degrees right, which results in real-time response showing the animation of turning head.

The SPEAK command is closely connected to the speech synthesizer. It prescribes the utterance which is sent to the synthesizer and the head shows individual visemes[1] as this utterance is synthesized and played through a sound card. By these two

---

[1] The approximation of lips and face shape that corresponds to the phoneme.

**Table 1.** Communication channels supported by ECAF

| Metachannels | Communication channel | Affected by |
|---|---|---|
| speech | voice | `<speak>` |
| visual | head turning | `<gesture head_angle="">` |
| | eyes pointing | `<gesture eye_horiz="" eye_vert="">` |
| | face expression | `<gesture expr="" expr_scale="">` |
| | background picture | `<gesture background="">` |
| | text window | `<text>` and `<gesture text="">` |
| | head size | `<gesture zoom="">` |
| | head position | `<gesture head_x="" head_y="">` |
| | body posture | `<gesture posture="">` |

high-level commands we are able to cover many of the communication channels supported by ECAF as depicted in Table 1.

**The ACT command.** The ACT command is realized by the XML `<act>` element. The ACT command has only *one* XML element child. Action carried by this command is specified by one of these elements: `<text>`, `<stop>` `<start_capture>`, `<stop_capture>`, `<gesture>`.

`<text>` element. As the talking head is a multimodal application, we need support for showing textual output. Such a plain text is only child of a `<text>` element. The toolkit displays a rectangle window near the head and writes the text into this window. New line is marked by the $ symbol. For example, displaying the text: *"Traffic jam ↩ Nuselsky most"*, is realized by issuing the command: `<act><text>Traffic jam$Nuselsky most</text></act>`

`<stop>` element. When this command is received by the talking head, it tries to immediately stop the synthesis[2]. The head will also return into the neutral position and display the default expression.

`<start_capture>` and `<stop_capture>` elements are used for capturing video and audio track of the real-time animations for debugging and archiving purposes.

`<gesture>` element is the most complex element in our language. It controls several talking head channels including movements and expressions. These channels are affected by attributes and attribute's value of this element. As there are nine independent communication channels supported by the ECAF talking head, the gesture element can contain these attributes:

1. *head_angle* – Integer attribute value which represents the angle of a head turning in the horizontal plane.
2. *eye_horiz* – Integer attribute value which drives the talking head eyes turning in the horizontal plane.
3. *eye_vert* – This attribute which has the same meaning as parameter *eye_vert*, but in the vertical plane.

---

[2] Immediately means in the case of IBM eViaVoice synthesizer after next word.

4. *expr* – A value of the *expr* parameter is string from the table of supported face expressions. The head will show up this expression. Expression strings are for example: neutral, smile, anger, ....

5. *expr_scale* – A floating point number that depicts a scale of an expression. Number 1.0 represents full expression, while 0.0 displays no expression. The parameter *expr_scale* must be used only in combination with the *expr* parameter.

6. *idle* – A string from the table of the idle movements that turns on the idle movement of the head. For example, *FlyOut* means that the talking head will start an animation of the head disappearing in the perspective.

7. *background* – The attribute contains the name of the image file to be loaded as the window background.

8. *text* – Text child has the same meaning as the `<text>` element.

9. *zoom* – This attribute takes in a floating point number that denotes the zoom of the head. This allows to make the head bigger or smaller depending on the application scenario.

10. *head_x* and *head_y* – These parameters control the position of the head in the window. They allow to move the head over the background image onto the desired coordinates depending on the application scenario.

```
<ecaf>
  <speak src="helloworld.wav">
    <phoneme time="0.611">E</phoneme>
    <phoneme time="0.787">l</phoneme>
    <phoneme time="0.845">o</phoneme>
    <gesture head_angle="15">
      <phoneme time="0.925">w</phoneme>
      <phoneme time="1.06">r</phoneme>
      <phoneme time="1.170">l</phoneme>
      <phoneme time="1.252">d</phoneme>
    </gesture>
  </speak>
</ecaf>
```

**Example 1.** Speech recorded in a sound file scenario example

**The SPEAK command.** The SPEAK command is encoded by the elements `<speak>` and `</speak>`. Its content is the text to be spoken. This text is send to the speech synthesizer which produced the artificial speech delivered in realtime. The movements of avatar lips are synchronized with this speech. The SPEAK command can be combined with the `<gesture>` element (and all it's options) to modify the appearance of the head during the speech.

There is an alternate way how the developers can utilize the SPEAK channel. In case they need to lip-synchronize on the pre-recorded sound (such as .wav file), the `<speak>` element will contain the name of a file as the attribute. The utterance will be represented by phoneme elements with a value of the phoneme including the time stamp of the beginning of this phoneme. You can see small illustration in Example 1.

**Scenario Script.** So far we have shown how the client application sends the SPEAK and ACT commands to the talking head server according to the application state. For testing purposes, and also for creation of static demo applications, it is possible to write the sequence of ECAF commands into an XML script file. Such a script is rooted within the `<ecaf>` and `</ecaf>` elements, which contains the sequence of `<speak>`, `<act>` and `<sleep>` elements. The `<sleep>` element is the execution command parameterized by the floating point number. The client program that reads scenario script file will stop sending next command for the time of seconds specified by the `<sleep>` element. The demonstration of the scenario files is presented in Examples 1 and 2.

```
<ecaf>
  <speak>Hello everybody!</speak>
  <speak>How are you?<gesture expr="smile" expr_scale="0.8">
    I am fine.</gesture>Now I will turn eyes and head
    <gesture eye_vert="20"> I see the
    <gesture head_angle="-25" persistent="true" />top.</gesture>
  </speak>
  <speak><gesture head_angle="0" persistent="true" />I will turn
    head and <gesture expr="smile">
    give you smile. <gesture head_angle="25">Look at me now!
    I am flexible.</gesture></gesture>
  </speak>
  <speak>I can show you, how I can <gesture expr="smile">mix
    expressions. <gesture expr="left blink">
    It's hard but it works</gesture></gesture>
  </speak>
  <act><gesture idle="LookAndSleep" persistent="true" /></act>
  <act><text>I'm sleeping$just now</text></act>
</ecaf>
```

**Example 2.** Text-To-Speech scenario example

# 4 Case Studies

While developing the current ECAF toolkit, several student projects were using ECAF for authoring in parallel to provide direct developer feedback. We outline several sample applications in this section. Each application has undergone the usability testing, which results we briefly summarize in conclusion.

**Talking Head as Virtual Secretary.** In this application, the talking head as a virtual secretary welcomes the user in the morning, highlights important information about a coming day, cautions about incoming calls and visits, about duties and meetings, about lunch time, etc. Using the ECAF expression means, the developers implemented two secretary personas – one very temperament with expressive behavior and one that was strict and conservative.

**Weather Forecast.** Here the avatar moved over a forecast map to highlight the respective weather patterns. The data for weather forecasting was retrieved from the RSS weather feeds. Through the usability testing the goal was to find a proper and entertaining mix of speech, expressions, text and background images.

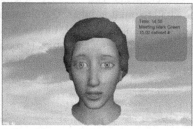

**Fig. 2.** Weather forecast and text window scenario

**Recipe Reader.** The task was to model the avatar as the recipe reader and the kitchen wizard. The talking head presented recipes, e.g. the recipe of a garlic soup and advised the cook on the respective steps during the soup preparation. The usability testing was to answer whether the spoken information is sufficient or if the avatar should also display the text of the recipe.

## 5    Conclusion

The proliferation of animated characters can be boosted by the existence of effective programming languages and architectures supporting real-time generation of behavior based on the higher-level expressive concepts. In this paper we have introduced such a system called ECAF, which development was propelled by the needs of avatar application developers.

The ECAF language builds on the concept of blended communication channels, that comprise the multimodal communication acts rendered to the user at real-time. We accent effectiveness, practicability, and extensibility as the key requisites of the ECAF language design.

Many applications have been already designed using ECAF, out of which we sketched a few. It turns out that it is good to keep the modal blending possibilities wide (ECAF supports nine channels), as the application authors are to decide the proper modality mix, given the target set of users, the environment, and the occasion. This is especially important, as our usability testing confirms the basic existence of two distinct user personas: one group tends to enjoy emotional and very lively avatars, while the other group requests presentation of facts without emotions and with the conservative behavior. The toolkits must be able to effectively support both.

## Acknowledgements

We like to thank the students developing the applications mentioned above for the very useful feedback on the development of the ECAF toolkit. This includes P. Muzikova, L. Dolejsky, V. Koutnik, L. Eimut, T. Fidler and others.

## References

1. Cassell, J., Vilhjálmsson, H., Bickmore, T.: BEAT: the Behavior Expression Animation Toolkit. In: ACM SIGGRAPH. Computer Graphics Proceedings, Annual Conference Series, ACM Press, New York (2001) (Reprinted in this Volume)
2. Fleury, P., Cuřín, J., Kleindienst, J.: CHiLiX – Connecting Computers. CHIL Technology Catalogue, http://chil.server.de/servlet/is/8923/
3. Gedalia, P.: The Expression Toolkit An Open-Source Procedural Facial Animation System, http://expression.sf.net
4. HumanML. Human markup language, http://www.humanmarkup.org
5. IBM: Embedded ViaVoice Multiplatform Software Development Kit (2005)
6. Ishizuka, M., Tsutsui, T., Saeyor, S., Dohi, H., Zong, Y., Prendinger, H.: MPML: A multimodal presentation markup language with character agent control functions. In: Achieving Human-like Behavior in Interactive Animated Agents, Proceedings of the AA'00, pp. 50–54. Workshop, Barcelona (2000)
7. Not, E., Balci, K., Pianesi, F., Zancanaro, M.: Synthetic Characters as Multichannel Interfaces. In: Proceedings of the ICMI'05, Trento, Italy, pp. 200–207 (2005)
8. Perlin, K.: Image Synthesizer. Computer Graphics Journal 3, 287–296 (1985)
9. Piwek, P., Krenn, B., Schröder, M., Grice, M., Baumann, S., Pirker, H.: RRL: a rich representation language for the description of agents behaviour in NECA. In: Falcone, R., Barber, S., Korba, L., Singh, M.P. (eds.) AAMAS workshop 2002. LNCS (LNAI), vol. 2631, Springer, Heidelberg (2003)
10. Segal, M., Akeley, K.: The OpenGL Graphics System: A Specification (1993)
11. VHML. Virtual Human Markup Language, http://www.vhml.org

# Dynamic Adaptation of Language Models in Speech Driven Information Retrieval*

César González-Ferreras and Valentín Cardeñoso-Payo

Departamento de Informática, Universidad de Valladolid, 47011 Valladolid, Spain
{cesargf,valen}@infor.uva.es

**Abstract.** This paper reports on the evaluation of a system that allows the use of spoken queries to retrieve information from a textual document collection. First, a large vocabulary continuous speech recognizer transcribes the spoken query into text. Then, an information retrieval engine retrieves the documents relevant to that query. The system works for Spanish language. In order to increase performance, we proposed a two-pass approach based on dynamic adaptation of language models. The system was evaluated using a standard IR test suite from CLEF. Spoken queries were recorded by 10 different speakers. Results showed that the proposed approach outperforms the baseline system: a relative gain in retrieval precision of 5.74%, with a language model of 60,000 words.

## 1 Introduction

Using the web to access information is becoming mainstream. People are used to access the world wide web using a personal computer. Moreover, when users access the web, searching is usually the starting point. As more information becomes available, better information retrieval technology is required.

There is also a proliferation of mobile devices that allow access to the web anytime and everywhere. However, their user interface is limited by small displays and input devices (keypad or stylus). Speech can be used to overcome those limitations and provide a more usable interaction. Furthermore, using speech as the input to an information retrieval engine is a natural and effective way of searching information in a mobile environment.

This paper reports on the evaluation of a system that allows users to search information using spoken queries. The front end is a large vocabulary continuous speech recognizer (LVCSR) which transcribes the query from speech to text and puts it through an information retrieval (IR) engine to retrieve the set of documents relevant to that query. A two-pass approach is proposed in order to increase performance over the baseline system. In the first pass a set of documents relevant to the query are retrieved and used to dynamically adapt the language model (LM). In the second pass the adapted language model is used and the list of documents is presented to the user. The system is designed for Spanish language. The performance of the system was evaluated using a test suite from CLEF, which is an evaluation forum similar to TREC. We recorded 10 speakers reading the queries. Results of different experiments showed that the proposed approach

---

* This work has been partially supported by *Consejería de Educación de la Junta de Castilla y León* under project number VA053A05.

V. Matoušek and P. Mautner (Eds.): TSD 2007, LNAI 4629, pp. 214–221, 2007.

outperforms the baseline system. We report a relative reduction in OOV word rate of 15.21%, a relative reduction in WER of 6.52% and a relative gain in retrieval precision of 5.74%, with a LM of 60,000 words.

The structure of the paper is as follows: Section 2 presents some related work; Section 3 explains the system in detail; Section 4 describes the experiments and the analysis of results; in Section 5 we discuss about factors that affect system performance; conclusions and future work are presented in Section 6.

## 2  Related Work

This is the first work, to our knowledge, on speech driven information retrieval for Spanish language. Experiments for other languages have been reported in the bibliography. All the experiments employed a similar methodology: a standard IR test suite (designed to evaluate IR systems using text queries) was used; some speakers reading the queries were recorded; finally, system performance was evaluated and compared with the results obtained using text queries.

First experiments in speech driven information retrieval were described in [1], for English language. Results showed that increasing WER reduces precision and that long spoken queries are more robust to speech recognition errors than short ones. A system designed for mobile devices was presented in [2], and several experiments in Chinese were reported. Retrieval performance on mobile devices with high-quality microphones (for example PDA) was satisfactory, although the performance over cellular phone was significantly worse. Some experiments in Japanese were presented in [3]. The performance of the system was improved using the target document collection for language modeling and a bigger vocabulary size. More experiments with the same test collection were described in [4]. Techniques for combining outputs of multiple LVCSRs were evaluated and improvement in speech recognition and retrieval accuracies was achieved.

## 3  System Description

The objective of the system is to retrieve all the documents relevant to a given spoken query. The system is based on a two-pass strategy, as shown in Fig. 1. In the first pass, the spoken query made by the user is transcribed into text by the speech recognizer, using a general LM. Next, a list of documents relevant to that query are retrieved by the information retrieval engine. Then, using those documents, a dynamic adaptation of the LM is carried out. In the second pass, the speech recognizer uses the adapted LM instead of the general LM. Finally, the list of documents relevant to the query is obtained and presented to the user. In the following sections we describe in detail each component of the system.

### 3.1  Speech Recognition

We used SONIC, a large vocabulary continuous speech recognizer from the University of Colorado [5]. It is based on continuous density hidden Markov model (HMM) technology and implements a two-pass search strategy using token-passing based recognition. We trained acoustic and language models for Spanish language.

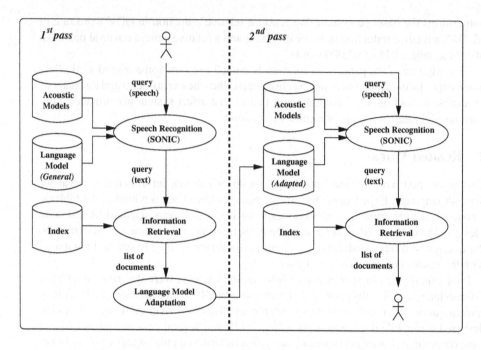

**Fig. 1.** Architecture of the system, based on a two-pass strategy

Acoustic models were triphone HMMs with associated gamma probability density functions to model state durations. Standard feature extraction was used: 12 Mel Frequency Cepstral Coefficients (MFCC) and normalized log energy, along with the first and second order derivatives. We used Albayzin corpus to train gender independent acoustic models [6] (13,600 sentences read by 304 speakers).

Word based trigram language models were created, with three different vocabulary sizes: 20,000, 40,000 and 60,000 words. To train the **general LM** (used in the first pass), the target document collection was used because this can result in an adaptation of the LVCSR to the given task and provides better system performance [3]. EFE94 document collection is composed of one year of newswire news (511 Mb) and has 406,762 different words. The vocabulary was created selecting the most frequent words found in the documents. We used SRILM statistical language modeling toolkit [7], with Witten Bell discounting.

### 3.2   Information Retrieval

We used a modified version of an information retrieval engine developed for Spanish language [8]. It is based on the vector space model and on term frequency-inverse document frequency (TF-IDF) weighting scheme. We also used a stop word list to remove function words and a stemming algorithm[1] to reduce the dimensionality of the space.

---

[1] Snowball stemmer: http://snowball.tartarus.org

The similarity of a query $q$ with each document $d_i$ in the document collection is calculated as follows:

$$sim(d_i, q) = \sum_{t_r \in q} w_{r,i} \times w_{r,q} \tag{1}$$

$$w_{r,i} = (1 + log(tf_{r,i})) \times log\left(\frac{N}{df_r}\right) \tag{2}$$

$$w_{r,q} = log\left(\frac{N}{df_r}\right) \tag{3}$$

where $w_{r,i}$ is the weight of the term $t_r$ in the document $d_i$; $w_{r,q}$ denotes the weight of the term $t_r$ in the query $q$; $tf_{r,i}$ represents the frequency of the term $t_r$ in the document $d_i$; $df_r$ denotes the number of documents in the collection that contain the term $t_r$; $N$ is the total number of documents in the collection.

### 3.3  Language Model Adaptation

The system dynamically adapts the LM to the query made by the user. A two-pass strategy is applied: in the first pass, a general LM is used for speech recognition and the documents retrieved are used to train an adapted LM; in the second pass, the adapted LM is used to obtain the final list of documents. We compared two different approaches to create the adapted language model:

- **Topic LM:** a LM trained using the 1000 documents obtained in the first pass. The vocabulary was built selecting the most frequent words found in that documents.
- **Interpolated LM:** a combination of the *general LM* and the *topic LM*. Because of the limited amount of data available to train the topic LM, we decided to merge it with the general LM. Linear interpolation was employed to combine both models. The interpolation coefficient was computed using the EM algorithm (training data was divided into two portions: one was used to train the topic LM and the other was used to estimate the interpolation coefficient). The vocabulary was built as follows: first, words from the adaptation data were selected based on frequency; second, if the desired vocabulary size was not reached, words from the general corpus were added, based also on frequency.

## 4  Experiments

We measured the performance of the system using CLEF 2001, a standard IR test suite. CLEF organizes evaluation campaigns in a similar way to TREC. Its aim is to develop an infrastructure for the testing and evaluation of information retrieval systems operating on European languages, under standard and comparable conditions [9].

In the following sections we describe the experimental set-up and present the results of the different experiments.

## 4.1  Experimental Set-Up

We used CLEF 2001 Spanish monolingual IR test suite, which includes a document collection, a set of topics and relevance judgments. The document collection has 215,738 documents of the year 1994 from EFE newswire agency (511 Mb). Topics simulate user information needs and are used to build the queries. There are 49 topics and each of them has three parts: a brief title statement, a one-sentence description and a more complex narrative. Relevance judgments determine the set of relevant documents for each topic, and were created using pooling techniques.

We expanded CLEF 2001 test suite to include spoken queries. We used the description field of each topic as query (mean length of 16 words, ranging from 5 to 33). We recorded 10 different speakers (5 male and 5 female) reading the queries. Headset microphone was used under office conditions, at 16 bit resolution and 16 kHz sampling frequency.

## 4.2  Results

Spoken queries were processed by the system and the 1000 most relevant documents (sorted by relevance) were retrieved for each query. Mean average precision (MAP) was calculated using relevance judgments. The same methodology of CLEF was used to evaluate the results [9]. Results for different configurations of the system are shown in Table 1:

- **Text:** results using text queries (for comparison purposes).
- **General LM:** results obtained in the first pass, using only the *general LM* (baseline system).
- **Topic LM:** results obtained in the second pass, using *topic LM* as the adapted LM (see Sect. 3.3).
- **Interpolated LM:** results obtained in the second pass, using *interpolated LM* as the adapted LM (see Sect. 3.3).

Three different vocabulary sizes were used: 20k, 40k and 60k words. For each different configuration we report the out of vocabulary word rate (OOV), the word error rate (WER) and the mean average precision (MAP).

## 4.3  Analysis of Results

The results of the baseline system were improved by the use of dynamically adapted LM. There was a reduction in OOV and WER, and the reduction was larger using interpolated LM: a relative reduction in OOV word rate of 15.21% and a relative reduction in WER of 6.52%, for a vocabulary of 60k words. There was an increase in MAP, but in contrast with OOV and WER, the use of topic LM yielded to slightly better results: a relative gain in MAP of 5.74%, for a vocabulary of 60k words.

As a possible explanation, we argue that both adapted LMs provided better estimates than the general LM, because they were trained using documents with a semantic relatedness with the current query. Both adapted LMs obtained better estimates for content

**Table 1.** System performance for different system configurations, using a vocabulary of 20k, 40k and 60k words (OOV: out of vocabulary word rate; WER: word error rate; MAP: mean average precision)

|  | OOV | WER | MAP |
|---|---|---|---|
| Text | – | – | 0.4475 |
| 20k general LM | 6.77% | 24.2% | 0.3013 |
| 20k topic LM | 3.27% | 20.3% | 0.3412 |
| 20k interpolated LM | 3.27% | 19.6% | 0.3393 |
| 40k general LM | 2.81% | 19.0% | 0.3267 |
| 40k topic LM | 2.38% | 18.9% | 0.3557 |
| 40k interpolated LM | 2.08% | 17.8% | 0.3534 |
| 60k general LM | 2.17% | 18.4% | 0.3412 |
| 60k topic LM | 2.36% | 18.5% | 0.3608 |
| 60k interpolated LM | 1.84% | 17.2% | 0.3591 |

words, which affected retrieval performance. However, interpolated LM also provided better estimates for function words (the reason for the reduction in WER), which had no effect in retrieval performance.

Experiments with different vocabulary sizes showed that adapted LM improved performance in all cases. Better absolute results were obtained with a vocabulary of 60k words and higher relative increase with a vocabulary of 20k words. As an interesting result, the performance using 20k topic LM (two-pass strategy) was equivalent to the performance of 60k general LM (one-pass).

We also analyzed the results of each individual query. Most of the queries had small loss of precision while some queries had high loss of precision. It means that in general queries did well, but there were some that did badly.

**Table 2.** Comparison between systems in the state of the art and our system (MAP-T: mean average precision using text queries; MAP-S: mean average precision using spoken queries; Ploss: loss in MAP of spoken queries compared with text queries)

|  | Test Suite | MAP-T | MAP-S | Ploss |
|---|---|---|---|---|
| Barnett | TIPSTER | 0.3465 | 0.3020 | 12.84% |
| Chang | TREC-5 | 0.3580 | 0.2570 | 28.21% |
|  | TREC-6 | 0.4890 | 0.4630 | 5.32% |
| Fujii | NTCIR-3 | 0.1257 | 0.0766 | 39.06% |
| Matsushita | NTCIR-3 | 0.1181 | 0.0820 | 30.57% |
| Our system | CLEF01 | 0.4475 | 0.3608 | 19.37% |

In Table 2 we compare our results with other systems. For each system, the loss in MAP of spoken queries compared with text queries is calculated. We claim that our system has a performance comparable with the systems in the state of the art, although the comparison is not conclusive, since there are many factors that affect system performance: language, IR test suite, length of the queries, vocabulary size of the recognizer and retrieval model of the IR engine. Moreover, there are some important differences between our experiments and the experiments reported by other researchers. Barnett et al. [1] used long queries (50-60 words) for the experiment, which are more robust to speech recognition errors than shorter ones. Chang et al. [2] used a gender dependent LVCSR, with speaker and channel adaptation. They also reported a performance for TREC-6 queries significantly better compared to TREC-5 queries for similar settings of the system, but no explanation for this was reported. Fujii et al. [3] and Matsushita et al. [4] used a larger document collection (100 Gb, about 10 million documents) which makes the task more difficult. Overall, the loss in MAP of our system is well inside the margins of those contributions.

## 5   Discussion

Speech driven information retrieval is a task with an open vocabulary, and better results are obtained with larger vocabulary language models, because of better vocabulary coverage. In our experiments, when 20k LM was used, the majority of the errors were due to OOV words. As we increased the size of the vocabulary, OOV word errors decreased and most of the errors were regular speech recognition errors. However, increasing vocabulary size indefinitely is impractical. We have presented an approach based on dynamic adaptation of language models that obtains a reduction of OOV word rate, and allows the system to improve performance without increasing the vocabulary size.

Each speech recognition error has big impact on retrieval accuracy: keywords with important semantic content may be missing, and some relevant documents are not retrieved. Moreover, words not related with the query may be introduced, making the system retrieve documents not related with the query. Surprisingly, there are some queries that improve when speech recognition errors occur.

There is a mismatch between speech recognition and information retrieval: speech recognition favors frequent words, because they are more likely to be said (higher probability in the LM); whereas information retrieval favors infrequent words, because they usually carry more semantic content (using TF-IDF weighting scheme). The worst case happens with proper nouns: they are effective query terms, but because of their low frequency, they have low probability in the LM or even are not included in the vocabulary of the speech recognizer.

## 6   Conclusions

In this paper we describe a system which allows the use of spoken queries to retrieve information from a textual document collection. We used a standard IR test suite to evaluate the performance of the system. The results showed that dynamic adaptation of language models provided better results than the baseline system. We reported a relative reduction in OOV word rate of 15.21%, a relative reduction in WER of 6.52% and a relative gain in retrieval precision of 5.74% for a LM of 60,000 words.

Overall, these results are encouraging and show the feasibility of building speech driven information retrieval systems. Although the performance was not as good as using text input, the system can help in overcome the limitations of mobile devices, and can also be useful in situations where speech is the only possible modality (for example, while driving a car).

As future work, we plan to extend the system with spoken dialog capabilities, because user interaction can provide valuable feedback to improve the retrieval process.

# References

1. Barnett, J., Anderson, S., Broglio, J., Singh, M., Hudson, R., Kuo, S.W.: Experiments in Spoken Queries for Document Retrieval. In: Eurospeech (1997)
2. Chang, E., Seide, F., Meng, H.M., Chen, Z., Shi, Y., Li, Y.: A System for Spoken Query Information Retrieval on Mobile Devices. IEEE Transactions on Speech and Audio Processing 10(8), 531–541 (2002)
3. Fujii, A., Itou, K.: Building a Test Collection for Speech-Driven Web Retrieval. In: Eurospeech (2003)
4. Matsushita, M., Nishizaki, H., Nakagawa, S., Utsuro, T.: Keyword Recognition and Extraction by Multiple-LVCSRs with 60,000 Words in Speech-driven WEB Retrieval Task. In: ICSLP (2004)
5. Pellom, B., Hacioglu, K.: Recent Improvements in the CU SONIC ASR System for Noisy Speech: The SPINE Task. In: ICASSP (2003)
6. Moreno, A., Poch, D., Bonafonte, A., Lleida, E., Llisterri, J., Mariño, J.B., Nadeu, C.: AL-BAYZIN Speech Database: Design of the Phonetic Corpus. In: Eurospeech (1993)
7. Stolcke, A.: SRILM – an Extensible Language Modeling Toolkit. In: ICSLP (2002)
8. Adiego, J., Fuente, P., Vegas, J., Villarroel, M.A.: System for Compressing and Retrieving Structured Documents. UPGRADE 3(3), 62–69 (2002)
9. Braschler, M., Peters, C.: CLEF Methodology and Metrics. In: Peters, C., Braschler, M., Gonzalo, J., Kluck, M. (eds.) CLEF 2001. LNCS, vol. 2406, Springer, Heidelberg (2002)

# Whitening-Based Feature Space Transformations in a Speech Impediment Therapy System

András Kocsor[1,2], Róber Busa-Fekete[1], and András Bánhalmi[1]

[1] MTA-SZTE Research Group on Artificial Intelligence
H-6720 Szeged, Aradi vértanúk tere 1., Hungary
{kocsor, busarobi, banhalmi}@inf.u-szeged.hu
[2] Applied Intelligence Laboratory Ltd., Petőfi S. Sgt. 43., H-6725 Szeged, Hungary

**Abstract.** It is quite common to use feature extraction methods prior to classification. Here we deal with three algorithms defining uncorrelated features. The first one is the so-called whitening method, which transforms the data so that the covariance matrix becomes an identity matrix. The second method, the well-known Fast Independent Component Analysis (FastICA) searches for orthogonal directions along which the value of the non-Gaussianity measure is large in the whitened data space. The third one, the Whitening-based Springy Discriminant Analysis (WSDA) is a novel method combination, which provides orthogonal directions for better class separation. We compare the effects of the above methods on a real-time vowel classification task. Based on the results we conclude that the WSDA transformation is especially suitable for this task.

## 1 Introduction

The primary goal of this paper is twofold. First we would like to deal with a unique group of feature extraction methods, namely with the uncorrelated ones. The uncorrelation can be carried out by using the well-known whitening method. After whitening among the linear transformations precisely the orthogonal ones preserve the property that the data covariance matrix remains the identity matrix. Thus following the whitening process we can apply any feature extraction method, which resulted in orthogonal feature directions. This kind of method composition in every case leads to uncorrelated features. Among the possibilities we selected two methods from the orthogonal family. The first one is the Fast Independent Component Analysis proposed by Hyvärinen and Oja [8], while the second one, recently introduced, is the Springy discriminant Analysis [9]. In this paper we investigate a version of this method combined with the whitening process. Our second aim here is to compare the effects of the above methods on a speech recognition task. We try to apply them on a real-time vowel classification task, which is one of the basic building blocks of our speech impediment therapy system [10].

Now without loss of generality we shall assume that, as a realization of multivariate random variables, there are $n$-dimensional real attribute vectors in a compact set $\mathcal{X}$ over $\mathbb{R}^n$ describing objects in a certain domain, and that we have a finite $n \times k$ sample matrix $X = (\mathbf{x}_1, \ldots, \mathbf{x}_k)$ containing $k$ random observations. Actually, $\mathcal{X}$ constitutes the initial

V. Matoušek and P. Mautner (Eds.): TSD 2007, LNAI 4629, pp. 222–229, 2007.

feature space and $X$ is the input data for the linear feature extraction algorithms which defines a linear mapping

$$h : \mathcal{X} \rightarrow \mathbb{R}^m$$
$$\mathbf{z} \rightarrow V\mathbf{z} \tag{1}$$

for the extraction of a new feature vector. The $m \times n$ $(m \leq n)$ matrix of the linear mapping – which may inherently include a dimension reduction – is denoted by $V$, and for any $\mathbf{z} \in \mathcal{X}$ we will refer to the result $h(\mathbf{z}) = V\mathbf{z}$ of the mapping as $\mathbf{z}^*$. With the linear feature extraction methods we search for an optimal matrix $V$, where the precise definition of optimality can vary from method to method. Now we will decompose $V$ in a factorized form, i.e. we assume that $V = WQ$, where $W, Q$ are orthogonal matrices and $Q$ transforms the covariance matrix into the identity matrix. We will obtain $Q$ by the whitening process, which can easily be solved by an eigendecomposition of the data covariance matrix (cf. Section 2). For the $W$ matrix, which further transforms the data, we can apply various objective functions. Here we will find each particular direction of the optimal $W$ transformations *one-by-one*, employing a $\tau : \mathbb{R}^n \rightarrow \mathbb{R}$ objective function for each direction (i.e. row vectors of $W$) separately. We will describe the Fast Independent Component Analyses (FastICA), and the Whitening-based Springy Discriminant Analysis (WSDA) via defining different $\tau$ functions.

The structure of the paper is as follows. In Section 2 we introduce the well-known whitening process, which is followed in Section 3 and 4 by the description of Independent Component Analysis and Springy Discriminant Analysis, respectively. Section 5 deals with the experiments, than in Section 6 we round off the paper with some concluding remarks.

## 2   The Whitening Process

Whitening is a traditional statistical method for turning the data covariance matrix into an identity matrix. It has two steps. First, we shift the original sample set $\mathbf{x}_1, \ldots, \mathbf{x}_k$ with its mean $E\{\mathbf{x}\}$, to obtain data

$$\mathbf{x}_1' = \mathbf{x}_1 - E\{\mathbf{x}\}, \ldots, \mathbf{x}_k' = \mathbf{x}_k - E\{\mathbf{x}\}, \tag{2}$$

with a mean of $\mathbf{0}$. The goal of the next step is to transform the centered samples $\mathbf{x}_1', \ldots, \mathbf{x}_k'$ via an orthogonal transformation $Q$ into vectors $\mathbf{z}_1 = Q\mathbf{x}_1', \ldots, \mathbf{z}_k = Q\mathbf{x}_k'$, where the covariance matrix $E\{\mathbf{z}\mathbf{z}^\top\}$ is the unit matrix. If we assume that the eigenpairs of $E\{\mathbf{x}'\mathbf{x}'^\top\}$ are $(\mathbf{c}_1, \lambda_1), \ldots, (\mathbf{c}_n, \lambda_n)$ and $\lambda_1 \geq \ldots \geq \lambda_n$, the transformation matrix $Q$ will take the form $[\mathbf{c}_1 \lambda_1^{-1/2}, \ldots, \mathbf{c}_t \lambda_t^{-1/2}]^\top$. If $t$ is less than $n$ a dimensionality reduction is employed.

*Whitening transformation of arbitrary vectors.* For an arbitrary vector $\mathbf{z} \in \mathcal{X}$ the whitening transformation can be performed using $\mathbf{z}^* = Q(\mathbf{z} - E\{x\})$.

*Basic properties of the whitening process.* $i$) for every normalized $\mathbf{v}$ the mean of $\mathbf{v}^\top \mathbf{z}_1, \ldots, \mathbf{v}^\top \mathbf{z}_k$ is set to zero, and its variance is set to one; $ii$) for any matrix $W$ the covariance matrix of the transformed, whitened data $W\mathbf{z}_1, \ldots, W\mathbf{z}_k$ will remain a unit matrix if and only if $W$ is orthogonal.

## 3 Independent Component Analysis

Independent Component Analysis [8] is a general purpose statistical method that originally arose from the study of blind source separation (BSS). An application of ICA is
unsupervised feature extraction, where the aim is to linearly transform the input data
into uncorrelated components, along which the distribution of the sample set is the least
Gaussian. The reason for this is that along these directions the data is supposedly easier
to classify.

For optimal selection of the independent directions, several objective functions were
defined using approximately equivalent approaches. Here we follow the way proposed
by A. Hyvärinen et al. [8]. Generally speaking, we expect these functions to be non-
negative and have a zero value for the Gaussian distribution. Negentropy is a useful
measure having just this property, which is used for assessing non-Gaussianity (i.e. the
least Gaussianity). The negentropy of a variable $\eta$ with zero mean and unit variance is
estimated by using the formula

$$J_G(\eta) \approx \left(E\{G(\eta)\} - E\{G(\nu)\}\right)^2, \tag{3}$$

where $G : \mathbb{R} \to \mathbb{R}$ is an appropriate non-quadratic function, $E$ again denotes the
expectation value and $\nu$ is a standardized Gaussian variable. The following three choices
of $G(\eta)$ are conventionally used: $\eta^4$, $\log(\cosh(\eta))$ and $-\exp(-\eta^2/2)$. It should be
mentioned that in Eq. (3) the expectation value of $G(\nu)$ is a constant, its value only
depending on the selected $G$ function.

In Hyvärinen's FastICA algorithm for the selection of a new direction $\mathbf{w}$ the following $\tau$ objective function is used:

$$\tau_G(\mathbf{w}) = \left(E\{G(\mathbf{w}^\top \mathbf{z})\} - E\{G(\nu)\}\right)^2, \tag{4}$$

which can be obtained by replacing $\eta$ in the negentropy approximant Eq. (3) with $\mathbf{w}^\top \mathbf{z}$,
the dot product of the direction $\mathbf{w}$ and sample $\mathbf{z}$. FastICA is an approximate Newton
iteration procedure for the local optimization of the function $\tau_G(\mathbf{w})$. Before running the
optimization procedure, however, the raw input data $X$ must first be preprocessed – by
whitening it.

Actually property $i)$ of the whitening process (cf. Section 2) is essential since Eq.
(3) requires that $\eta$ should have a zero mean and variance of one hence, with the substi-
tution $\eta = \mathbf{w}^\top \mathbf{z}$, the projected data $\mathbf{w}^\top \mathbf{z}$ must also have this property. Moreover, after
whitening based on property $ii)$ it is sufficient to look for a new orthogonal base $W$ for
the preprocessed data, where the values of the non-Gaussianity measure $\tau_G$ for the base
vectors are large. Note that since the data remains whitened after an orthogonal transformation, ICA can be considered an extension of PCA. The optimization procedure of
the FastICA algorithm can be found in Hyvärinen's work [8].

*Transformation of test vectors.* For an arbitrary test vector $\mathbf{z} \in \mathcal{X}$ the ICA transformation can be performed using $\mathbf{z}^* = WQ(\mathbf{z} - E\{\mathbf{x}\})$. Here $W$ denotes the orthogonal
transformation matrix we obtained as the output from FastICA, while $Q$ is the matrix
obtained from whitening.

# 4    Whitening-Based Springy Discriminant Analysis

Springy discriminant analysis (SDA) is a method similar to Linear Discriminant Analysis (LDA), which is a traditional supervised feature extraction method [4,9]. Because SDA belongs to the supervised feature extraction family, let us assume that we have $r$ classes and an indicator function $\mathcal{L} : \{1, \ldots, k\} \rightarrow \{1, \ldots, r\}$, where $\mathcal{L}(i)$ gives the class label of the sample $\mathbf{x}_i$. Let us further assume that we have preprocessed the data using the whitening method, the new data being denoted by $\mathbf{z}_1, \ldots, \mathbf{z}_k$.

The name Springy Discriminant Analysis stems from the utilization of a spring & antispring model, which involves searching for directions with optimal potential energy using attractive and repulsive forces. In our case sample pairs in each class are connected by springs, while those of different classes are connected by antisprings. New features can be easily extracted by taking the projection of a new point in those directions where a small spread in each class is obtained, while different classes are spaced out as much as possible. Now let $\delta(\mathbf{w})$, the potential of the spring model along the direction $\mathbf{w}$, be defined by

$$\delta(\mathbf{w}) = \sum_{i,j=1}^{k} \left( (\mathbf{z}_i - \mathbf{z}_j)^\top \mathbf{w} \right)^2 [M]_{ij}, \tag{5}$$

where

$$[M]_{ij} = \begin{cases} -1, \text{ if } \mathcal{L}(i) = \mathcal{L}(j) \\ 1, \text{ otherwise} \end{cases} \quad i,j = 1, \ldots, k. \tag{6}$$

Naturally, the elements of matrix $M$ can be initialized with values different from $\pm 1$ as well. The elements can be considered as a kind of force constant and can be set to a different value for any pair of data points.

It is easy to see that the value of $\delta$ is largest when those components of the elements of the same class that fall in the given direction $\mathbf{w}$ ($\mathbf{w} \in \mathbb{R}^n$) are close, and the components of the elements of different classes are far at the same time.

Now with the introduction of the matrix

$$D = \sum_{i,j=1}^{k} (\mathbf{z}_i - \mathbf{z}_j) (\mathbf{z}_i - \mathbf{z}_j)^\top [M]_{ij} \tag{7}$$

we immediately obtain the result that $\delta(\mathbf{w}) = \mathbf{w}^\top D \mathbf{w}$. Based on this, the objective function $\tau$ for selecting relevant features can be defined as the Rayleigh quotient $\tau(\mathbf{w}) = \delta(\mathbf{w}) / \mathbf{w}^\top \mathbf{w}$. It is straightforward to see that the optimization of $\tau$ leads to the eigenvalue decomposition of $D$. Because $D$ is symmetric, its eigenvalues are real and its eigenvectors are orthogonal. The matrix $W$ of the SDA transformation is defined using those eigenvectors corresponding to the $m$ dominant eigenvalues of $D$.

*Transformation of test vectors.* For an arbitrary vector $\mathbf{z} \in \mathcal{X}$ the Whitening-Based SDA transformation can be performed using $\mathbf{z}^* = WQ(\mathbf{z} - E\{\mathbf{x}\})$.

# 5   Experiments and Results

In the previous sections three linear feature space transformation algorithms were presented. *Whitening* concentrates on those uncorrelated directions with the largest variances. *FastICA* besides keeping the directions uncorrelated, chooses directions along which the non-Gaussianity is large. *WSDA* creates attractive forces between the samples belonging to the same class and repulsive forces between samples of different classes. Then it chooses those uncorrelated directions along which the potential energy of the system is maximal. In this section we discuss these methods on the real-time vowel recognition tests. The motivation for doing this is to improve the recognition accuracy of our speech impediment therapy system, the 'SpeechMaster'. Besides reviewing 'SpeechMaster' here we will talk about the extraction of the acoustic features, the way the transformations were applied, the learners we employed and, finally, about the setup and evaluation of the real-time vowel recognition experiments.

**The 'SpeechMaster'.** An important clue to the process of learning to read for alphabetical languages is the ability to separate and identify consecutive sounds that make words and to associate these sounds with its corresponding written form. To learn to read in a fruitful way young learners must, of course, also be aware of the vowels and be able to manipulate them. Many children with learning disabilities have problems in their ability to process phonological information. Furthermore, phonological awareness teaching has also great importance for the speech and hearing handicapped, along with improving the corresponding articulatory strategies of tongue movement.

The 'SpeechMaster' software developed by our team seeks to apply speech recognition technology to speech therapy and the teaching of reading. Both applications require a real-time response from the system in the form of an easily comprehensible visual feedback. With the simplest display setting, feedback is given by means of flickering letters, their identity and brightness being adjusted to the speech recognizer's output [10]. In speech therapy it is intended to supplement the missing auditive feedback of the hearing impaired, while in teaching reading it is to reinforce the correct association between the phoneme-grapheme pairs. With the aid of a computer, children can practice without the need for the continuous presence of the teacher. This is very important because the therapy of the hearing impaired requires a long and tedious fixation phase. Experience shows that most children prefer computer exercises to conventional drills. In the 'SpeechMaster' system the real-time vowel recognition module has a great importance, this is why we chose this task for testing the uncorrelated feature extraction methods.

**Evaluation Domain.** For training and testing purposes we recorded samples from 160 normal children aged between 6 and 8. The ratio of girls and boys was 50% - 50%.The speech signals were recorded and stored at a sampling rate of 22050Hz in 16-bit quality. Each speaker uttered all the 12 isolated Hungarian vowels, one after the other, separated by a short pause. The recordings were divided into a train and a test set in a ratio of 50% - 50%.

**Acoustic Features.** There are numerous methods for obtaining representative feature vectors from speech data, but their common property is that they are all extracted from 20-30 ms chunks or "frames" of the signal in 5-10 ms time steps. The simplest possible feature set consists of the so-called bark-scaled filterbank log-energies (FBLE). This means that the signal is decomposed with a special filterbank and the energies in these filters are used to parameterize speech on a frame-by-frame basis. In our tests the filters were approximated via Fourier analysis with a triangular weighting, as described in [6].

It is known from phonetics that the spectral peaks (called formants) code the identity of vowels [11]. To estimate the formants, we implemented a simple algorithm that calculates the gravity centers and the variance of the mass in certain frequency bands [1]. The frequency bands are chosen so that they cover the possible place of the first, second and third formants. This resulted in 6 new features altogether.

A more sophisticated option for the analysis of the spectral shape would be to apply some kind of auditory model. We experimented with the In-Synchrony-Bands-Spectrum of Ghitza [5], because it is computationally simple and attempts to model the dominance relations of the spectral components. The SBS model analysis the signal using a filterbank that is approximated by weighting the output of a FFT - quite similar to the FBLE analysis. In this case, however, the output is not the total energy of the filter, but the frequency of the component that has the maximal energy.

**Feature Space Transformation.** When applying the uncorrelated feature extraction methods (see Section 2, 3 and 4) we invariably kept only 8 of the new features. We performed this severe dimension reduction in order to show that, when combined with the transformations, the classifiers can yield the same scores in spite of the reduced feature set. Naturally, when we applied a certain transformation on the training set before learning, we applied the same transformation on the test data during testing.

**Classifiers.** Describing the mathematical background of the learning algorithms applied is beyond the scope of this article; in the following we specify only the parameters applied.

*Gaussian Mixture Modeling (GMM).* In the GMM experiments, three Gaussian components were used and the expectation-maximization (EM) algorithm was initialized by $k$-means clustering [4]. To find a good starting parameter set we ran it 15 times and used the one with the highest log-likelihood. In every case the covariance matrices were forced to be diagonal.

*Artificial Neural Networks (ANN).* In the ANN experiments we used the most common feed-forward multilayer perceptron network with the backpropagation learning rule [2]. The number of neurons in the hidden layer was set to 18 in each experiment (this value was chosen empirically, based on preliminary experiments). Training was stopped based on the cross-validation of 15% of the training data.

*Projection Pursuit Learning (PPL).* Projection pursuit learning is a relatively little-known modelling technique [7]. It can be viewed as a neural net where the rigid sigmoid function is replaced by an interpolating polynomial. In each experiment, a model with 8 projections and a 5th-order polynomial was applied.

*Support Vector Machines (SVM).* Support vector machines is a classifier algorithm that is based on the ubiquitous kernel idea [12]. In all the experiments with SVM the radial basis kernel function was applied.

**Experiments.** In the experiments 5 feature sets were constructed from the initial acoustic features described above. *Set1* contained the 24 FBLE features. In *Set2* we combined

**Table 1.** Recognition errors for each feature set as a function of the transformation and classification applied

| feature set | classifier | none(all) | Whitening(8) | FastICA(8) | WSDA(8) |
|---|---|---|---|---|---|
| | GMM | 16.38 | 14.21 | 16.45 | 14.32 |
| | ANN | 10.34 | 9.85 | 9.93 | 9.42 |
| Set1 (24) | PPL | 11.04 | 10.46 | 10.69 | 10.02 |
| | SVM | 9.93 | 10.12 | 8.95 | **8.05** |
| | GMM | 13.33 | 11.21 | 13.33 | 12.33 |
| | ANN | 7.43 | 7.35 | 7.36 | **5.25** |
| Set2 (30) | PPL | 9.37 | 8.41 | 6.54 | 6.23 |
| | SVM | 8.33 | 6.85 | 6.66 | 5.43 |
| | GMM | 25.90 | 22.34 | 25.90 | 23.67 |
| | ANN | 20.00 | **18.41** | 19.58 | 19.65 |
| Set3 (24) | PPL | 20.48 | 19.43 | 19.58 | 19.33 |
| | SVM | 19.65 | 20.08 | 18.88 | 19.48 |
| | GMM | 13.95 | 12.21 | 15.90 | 13.67 |
| | ANN | 10.27 | 9.79 | 8.05 | 8.48 |
| Set4 (48) | PPL | 10.48 | 8.80 | 9.37 | 9.31 |
| | SVM | 9.09 | 9.46 | 8.26 | **7.41** |
| | GMM | 15.48 | 12.46 | 13.33 | 12.72 |
| | ANN | 8.68 | 7.31 | 6.45 | 7.41 |
| Set5 (54) | PPL | 8.26 | 9.05 | 7.36 | 7.09 |
| | SVM | 9.37 | 9.11 | 5.76 | **5.64** |

*Set1* with the gravity center features, so *Set2* contained 30 measurements. *Set3* was composed of the 24 SBS features, while in *Set4* we combined the FBLE and SBS sets. Lastly, in *Set5* we added all the FBLE, SBS and gravity center features, thus obtaining a set of 54 values.

In the classification experiments every transformation was combined with every classifier on every feature set. The results are shown in Table 1. In the header Whitening, FastICA, WSDA stand for the linear uncorrelated feature space transformation methods. The numbers shown are the recognition errors on the test data. The number in parentheses denotes the number of features preserved after a transformation. The best scores of each set are given in bold.

**Results and Discussion.** Upon inspecting the results the first thing one notices is that the SBS feature set (*Set3*) did about twice as badly as the other sets, no matter what transformation or classifier was tried. When combined with the FBLE features (*Set1*) both the graity center and the SBS features brought some improvement, but this improvement is quite small and varies from method to method.

When focusing on the performance of the classifiers, we see that ANN, PPL and SVM yielded very similar results. They, however, consistently outperformed GMM, which is still the method most commonly used in speech technology today. This can be attributed to the fact that the functions that a GMM (with diagonal covariances) is able to represent are more restricted in shape than those of ANN or PPL.

As regards the transformations, an important observation is that after the transformations the classification scores did not get worse compared to the classifications when no transformation was applied. This is so in spite of the dimension reduction, which shows that some features must be highly redundant. Removing some of this redundancy by means of a transformation can make the classification more robust and, of course, faster. Comparing the methods, we may notice that WSDA brought significant improvement on the recognition accuracy. Maybe this is due to the supervised nature of the method.

# 6   Conclusions

In this paper three linear uncorrelated feature extraction algorithms (Whitening, FastICA and WSDA) were presented, and applied to real-time vowel classification. After inspecting the test results we can confidently say that it is worth experimenting with these methods in order to obtain better classification results. The Whitening-based Springy Discriminant Analysis brought a notable increase in the recognition accuracy despite applying a severe dimension reduction. This transformation could greatly improve our phonological awareness teaching system by offering a robust and reliable real-time vowel classification, which is a key part of the system.

# Acknowledgments

A. Kocsor was supported by the János Bolyai fellowship of the Hungarian Academy of Sciences.

# References

1. Albesano, D., Mori, R.D., Gemello, R., Mana, F.: A study on the effect of adding new dimensions to trajectories in the acoustic space. In: Proc. of Eurospeech'99, pp. 1503–1506 (1999)
2. Bishop, C.M.: Neural Networks for Pattern Recognition. Oxford Univerisity Press Inc., New York (1996)
3. Duda, R.O., Hart, P.E., Stork, D.G.: Pattern Classification. John Wiley & Sons, New York (2001)
4. Fukunaga, K.: Statistical Pattern Recognition. Academic Press, New York (1989)
5. Ghitza, O.: Auditory Nerve Representation Criteria for Speech Analysis/Synthesis. IEEE Transaction on ASSP 35, 736–740 (1987)
6. Huang, X., Acero, A., Hon, H.W.: Spoken Language Processing. Prentice Hall, Englewood Cliffs (2001)
7. Hwang, J.N., Lay, S.R., Maechler, M., Martin, R.D., Schimert, J.: Regression Modeling in Back-Propagation and Projection Pursuit Learning, IEEE Trans. on Neural Networks 5, 342–353 (1994)
8. Hyvärinen, J., Oja, E.: A fast fixed-point algorithm for independent component analysis. Neural Comp. 9, 1483–1492 (1997)
9. Kocsor, A., Tóth, L.: Application of Kernel-Based Feature Space Transformations and Learning Methods to Phoneme Classification. Appl. Intelligence 21, 129–142 (2004)
10. Kocsor, A., Paczolay, D.: Speech Technologies in a Computer-Aided Speech Therapy System. In: Miesenberger, K., Klaus, J., Zagler, W., Karshmer, A.I. (eds.) ICCHP 2006. LNCS, vol. 4061, pp. 615–622. Springer, Heidelberg (2006)
11. Moore, B.C.J.: An Introduction to the Psychology of Hearing, Acad. Pr. (1997)
12. Vapnik, V.N.: Statistical Learning Theory. John Wiley & Sons Inc., NY (1998)

# Spanish-Basque Parallel Corpus Structure: Linguistic Annotations and Translation Units

A. Casillas[1], A. Díaz de Illarraza[2], J. Igartua[2], R. Martínez[3],
K. Sarasola[2], and A. Sologaistoa[2]

[1] Dpt. Electricidad y Electrónica, UPV-EHU
arantza.casillas@ehu.es
[2] IXA Taldea
{jipdisaa,webigigj,kepa.sarasola,jibsofra}@ehu.es
[3] NLP&IR Group, UNED
raquel@lsi.uned.es

**Abstract.** In this paper we propose a corpus structure which represents and manages an aligned parallel corpus. The corpus structure is based on a stand-off annotation model, which is composed of several XML documents. A bilingual parallel corpus represented in the proposed structure will contain: (1) the entire corpus together with its corresponding linguistic information, (2) translation units and alignment relations between units of the two languages: paragraphs, sentences and named entities. The proposed structure permits to work with the corpus both as an annotated corpus with linguistic information, and as a translation memory.

## 1 Introduction

A compiled parallel corpus with additional linguistic information can be a helpful resource for different purposes such as training datasets for inductive programs, learning models for machine translation, cross-lingual information retrieval, automatic descriptor assignment, document classification, cross-lingual document similarity or other linguistic applications.

Nowadays we can find several bilingual compiled corpora with extra-information where majority languages are involved. However, it is difficult to find this type of corpus where one of the involved languages is a minority language.

There are two official languages in the Spanish side of the Basque Country: Basque and Spanish and most of the main public institutions such as the Basque Government or universities publish their official documentation in both official languages. Therefore, we thought that a linguistically tagged Spanish-Basque corpus would be a very valuable resource for the research community because nowadays, this type of Spanish-Basque corpus is rare, so there are not enough reference corpora to consult.

In this paper, we propose a bilingual parallel corpus structure that contains two types of information: (1) aligned translation units with their corresponding linguistic information, and (2) the whole documents with their linguistic information. We propose such a rich structure because we want our corpus resources to be general and useful for different tasks in language technology research. Once the linguistic and alignment

V. Matoušek and P. Mautner (Eds.): TSD 2007, LNAI 4629, pp. 230–237, 2007.

information is added to the corpus, it is possible to extract different language resources from it such as translation memories or gazetteers.

Next section presents some related work on bilingual corpus representation. In Section 3 the structure of the bilingual corpus, which includes aligned translation units and linguistic information, is described. Section 4 explains the characteristics of the bilingual corpus collected and the steps carried out to add linguistic and alignment information. We present, in Section 5, a web interface for managing the parallel corpus and its additional information. Finally, conclusions and future work are added.

## 2    Related Work

Similar work representing bilingual corpora, in which at least one of the involved languages is a minority language has been carried out by: [14], [7] and [13].

In [14] are presented procedures and formats used in building a newspaper bilingual corpus for Croatian-English. The author compares the two different ways to encode parallel corpus using XML: alignment by storing pointers in separate documents and translation memory (TMX) inspired encoding. In one of his papers, he concludes that it is better to use the former due to the DTD's simplicity, because the original document keeps more unchanged, and because even with the stand-off way there is no problem to keep aligned sentences together in the same element while retaining upper levels of text encoding.

In [7], the authors present a bilingual Slovene-English corpus. They also use a stand-off model to represent the bilingual corpus, so that the linguistic information is stored in separate files.

The authors of [13] present a freely available parallel corpus containing European Union documents of mostly legal nature. This corpus is available in TEI-compliant XML stand-off format, and includes marked-up texts and bilingual alignment information for all language pairs involved. Particularly they include alignment information at paragraph and sentence level.

## 3    Bilingual Parallel Corpus Structure

The two main features that characterize our corpus structure are: the richness of the linguistic information represented and the inclusion of relationships between units of the two languages which have the same meaning. In a previous work [5] we presented a preliminary bilingual parallel corpus structure. In this work, we have improved the access to the supplementary information added to the bilingual corpus. The corpus structure proposed is based on the data model presented in [3], which represents and manages monolingual corpora with linguistic annotations based on a stand-off annotation and a typed feature structure. This representation may be seen as composed of several XML documents. Figure 1 shows a graphical representation of the currently implemented document model for the bilingual parallel corpus. This representation includes: linguistic information, translation units (paragraphs, sentences and entities) and alignment relations. Next in this section, we will present the content of the XML documents that constitute the proposed data model.

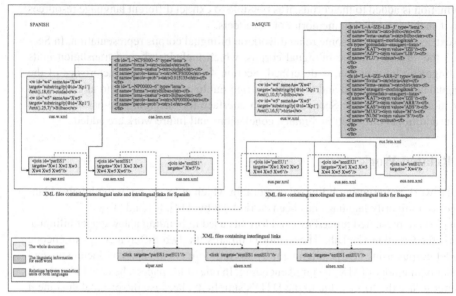

Figure 1: Corpus Structure: XML documents and their contents

**Fig. 1.** Corpus Structure: XML documents and their content

We have carried out two different processes with this corpus: (1) Detecting and aligning translation units, (2) Adding linguistic information to each monolingual subcorpus. Within the proposed corpus structure, we have merged the output information of both processes. The final structure of the corpus is composed of the manuscript texts and of several files to define stand-off annotations; these annotations contain the linguistic information and the delimitation of the units detected and aligned. The information to be exchanged among the different tools to manage this corpus is complex and diverse. Because of this complexity, we decided to use Feature Structures (FSs) to represent this information [3]. Feature structures are coded following the TEIs DTD for FSs [12], and Feature Structure Definition descriptions (FSD) have been thoroughly defined for each document created. The documents created as input and output of the different tools are coded in XML. The use of XML for encoding the information flowing between programs forces us to describe each document in a formal way, which offers advantages to keep coherence, reliability and maintenance. This structure avoids unnecessary redundancies in the representation of linguistic features of repeated units. The annotations that contain the linguistic information are saved into four XML documents:

- *eus.w.xml* and *cas.w.xml*, which contain single-word tokens in Basque and Spanish respectively.
- *eus.lem.xml* and *cas.lem.xml*, which keep for each single-word token of the two languages: its lemma, its syntactic function and some significant features of the morphological analysis. Words can be ambiguous and correspond to more than one lemma or syntactic function.

In order to represent the annotations that delimit translation units we have created six XML documents:

- *eus.par.xml* and *cas.par.xml*, which are used to delimit the paragraphs detected in the bitext. Paragraphs are delimited with references to their first single-word token and their last single-word token.
- *eus.sen.xml* and *cas.sen.xml*, which contain the sentences of the parallel corpus by means of references to their first and last single-word token.
- *eus.nen.xml* and *cas.nen.xml*, which keep the name entities.

We have also created XML documents that relate units of the two languages with the same meaning, in order to parallelize the two monolingual corpora.

- *alpar.xml*, this document is used to relate the paragraphs delimited in the files *cas.par.xml* and *eus.par.xml*. Each paragraph in one language is related to its corresponding paragraph (or paragraphs) in the other language, using the paragraph identifiers.
- *alsen.xml*, the relations between corresponding sentences from both languages are saved in this document. It is possible to set up 1-1 or N-M alignments.
- *alnen.xml*: name entities are aligned by means of this document. Relations of 1-1 and N-M are contemplated.

While translation memories take translation units as their primary "corpus", the corpus structure proposed contains the whole documents and the translation units detected and aligned. In the case of pure translation memories, only the units are saved, that is, the source text, the context from which the units come from, does not exist. In addition, our corpus structure does not contain redundant information, for example one word can appear many times but its linguistic information is only stored once. Wherever a word appears one link is inserted to get its linguistic representation.

## 4    Adding Information to the Bilingual Parallel Corpus

We have compiled a bilingual parallel corpus of 3 million words. This corpus is composed of two types of documents: official (about 2 million words) and non-official documents (about 1 million words). The official documents are from local governments and from the University of the Basque Country. Mainly, they are edicts, bulletins, letters or announcements. We have also collected some books and non-official documents that have been translated into Basque by this public university. They are about various subjects: fossils, music, education, etc.

Starting from the original plain text we are successively enriching the information contained in this corpus. The process consists of the following steps:

1. Obtaining the texts: we have downloaded the government official publications from [6], and collected the available documents from the university. Actually, we continue with this collecting work and we download official publications from different websites every day. We have also got in touch with the editors of the public university to get more publications of this type.

2. Normalization of the texts into a common format: is this second phase, we have processed manually all the official publications because the documents were incomplete or there were some mistakes. On the contrary, there was no need to pre-process the books. In both cases, we have converted and saved all the documents into ASCII format.

3. Tokenization: this involves linguistic analysis for the isolation of words.

4. Segmentation: to determine the boundaries of different types of units such as paragraphs, sentences and entities (person, location, organization). Due to the morphosyntactic differences between Spanish and Basque it was necessary to execute particular algorithms for each language in the detection process.

5. Alignment: the units detected in both languages are aligned in this phase. With the alignment process we have related the Spanish and Basque units of the same type that have the same meaning. Nevertheless, the alignment algorithms are independent of the language pair. The algorithms that we have executed to detect and align the different units are explained in more detail in [10] and [11].

6. Lematization and morphosyntactic analysis: in this phase the lemma, number, gender and case of each word are recognized. FreeLing package [9] has been used for generating Spanish linguistic information. In the case of Basque, we have used a set of different linguistic processing tools. The parsing process starts with the outcome of the morphosyntactic analyzer MORFEUS [2]. It deals with all the lexical units of a text, both simple words and multiword units, using the lexical database for Basque EDBL [1]. From the obtained results, grammatical categories and lemmas are disambiguated. The disambiguation process is carried out by means of linguistic rules (CG grammar) and stochastic rules based on Markovian models [8] with the aim of reducing the set of parsing tags for each word taking into account its context. Once morphosyntactic disambiguation has been performed, we have morphosyntactically fully disambiguated text.

## 5   Environment for Consulting Translation Units and Their Linguistic Annotations

We have extended the graphical web interface EULIA [4] in order to browse the aligned parallel corpora. EULIA is an environment to coordinate NLP tools and exploits the data generated by these tools. The interface shows the translation units detected in both sides of the corpus and the relations between units of the two languages. The main functions that we have incorporated are two:

- Consulting and browsing the translation units.
- Consulting and browsing the linguistic annotations attached to translation units. Figure 2 shows the linguistic annotations for the Basque version of a named entity.

We have tagged and aligned a sample bilingual Spanish-Basque corpus of 100.000 words in the XML parallel corpus structure explained in Section 3. The bilingual sample corpus is distributed in 160 paired documents and each monolingual document is stored in a file. The original Spanish corpus size is 340,2 KB (57.000 words) and the original Basque corpus size is 331 KB (44.000 words). The compiled corpus size is 78,3 MB;

**Table 1.** Number of elements detected in the bilingual corpus

|  | Spanish | Basque | Alignment |
|---|---|---|---|
| Num. Paragraphs | 2533 | 2535 | 2533 |
| Num. Sentences | 2549 | 2596 | 2549 |
| Num. Named entities | 2104 | 1999 | 1008 |

26,91 MB are used to store the Spanish linguistic information, 29,19 MB for the Basque linguistic information and 22,2 MB for the alignment information. Table 1 shows the number of paragraphs, sentences and named entities detected in each language and the number of alignments detected for each element.

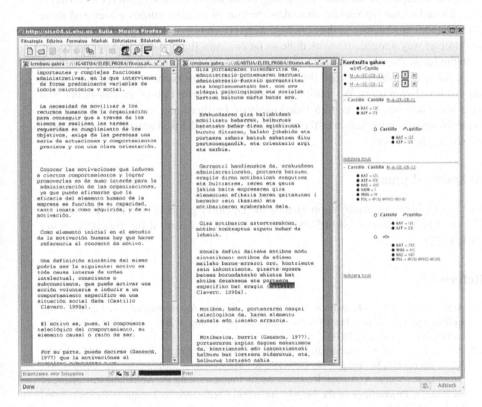

**Fig. 2.** The web environment for consulting the linguistic information of a named entity

## 6   Conclusion and Future Work

In this paper we have proposed a corpus structure which represents and manages an aligned parallel corpus. The proposed structure supports linguistic information of the texts, as well as information of the alignment of the translation units detected. The corpus structure is based on a stand-off annotation model and may be seen as composed of several XML documents. The resultant XML files contain the following information: (1) the whole document, (2) the linguistic information for each word, and (3) relations between translation units of both languages. This means that we have obtained mainly two resources: a translation memory and a morphosyntactically tagged parallel corpus. The main disadvantage of our proposal is that it needs more space than a translation memory or than a tagged corpus. Nevertheless, we think this representation will ensure the use of this "small" corpus in different tasks in language technology research. A bilingual parallel corpus represented in the proposed structure can be used as a translation memory for the automatic translation process or can be employed as a tagged parallel corpus for research in corpora based machine translation, machine learning, document clustering, cross-lingual information retrieval and other language applications.

Instead of repeating the same processing of the texts once and again for so different research tasks, our representation makes the use of parallel corpus easier and more efficient, adding structure to the corpus to keep coherence, reliability and maintenance in a simpler way. Indeed, the work done so far confirms the scalability of our approach.

In the future, we are planning to include a new level of alignment at phrase or chunk level. Also, we want to extend the web environment for editing the relations of the aligned translation units. We think that this is a very useful option for translators who want to create a personalized translation memory. Furthermore, we are thinking of storing the linguistic information of the parallel corpus in a XML database so as to reduce the size of the corpora.

## Acknowledgments

This work has been partially supported by the OpenMT project (TIN2006-15307-C03-01), the Ehiztari project (SPE05UN3206) and the University of the Basque Country (GIU05/52).

## References

1. Aldezabal, I., Ansa, O., Arrieta, B., Artola, X., Ezeiza, A., Hernández, G., Lersundi, M.: EDBL: a general lexical basis for the automatic processing of Basque. In: IRCS Workshop on linguistic databases (2001)
2. Aduriz, I., Agirre, E., Aldezabal, I., Alegria, I., Ansa, O., Arregi, X., Arriola, J.M., Artola, X., de Ilarraza, A.D., Ezeiza, N., Gojenola K., Maritxalar, A., Maritxalar, M., Oronoz, M., Sarasola, K., Soroa, A., Urizar, R., Urkia, M.: A Framework for the Automatic Processing of Basque. In: Proceedings of the First International Conference on Language Resources and Evaluation (1998)
3. Artola, X., de Illarraza, A.D., Ezeiza, N., Gojenola, K., Labaka, G., Salogaistoa, A., Soroa, A.: A framework for representing and managing linguistic annotations based on typed feature structures. In: RANLP (2005)
4. Artola, X., de Ilarraza, A.D., Ezeiza, N., Gojenola, K., Sologaistoa, A., Soroa, A.: EULIA: a graphical web interface for creating, browsing and editing linguistically annotated corpora. In: LREC (2004)
5. Casillas, A., de Illarraza, A.D., Igartua, J., Martínez, R., Sarasola, K.: Compilation and Structuring of a Spanish-Basque Parallel Corpus. In: 5th SALTMIL Workshop on Minority Languages

6. Euskal Herriko Agintaritzaren Ofiziala (EHAA), http://www.euskadi.net
7. Erjavec, T.: Compiling and using the IJS-ELAN Parallel Corpus. Informatica 26, 299–307 (2002)
8. Ezeiza, N., Aduriz, I., Alegria, I., Arriola, J.M., Urizar, R.: Combining Stochastic and Rule-Based Methods for Disambiguation in Agglutinative Languages. In: Proceedings of COLING-ACL'98 (1998)
9. FreeLing 1.5 An Open Source Suite of Language Analyzers, http://garraf.epsevg.upc.es/freeling/
10. Martínez, R., Abaitua, J., Casillas, A.: Bitext Correspondences through Rich Mark-up. In: Proceedings of the 17th International Conference on Computational Linguistics (COLING'98) and 36th Annual Meeting of the Association for Computational Linguistics (ACL'98), pp. 812–818 (1997)
11. Martínez, R., Abaitua, J., Casillas, A.: Aligning tagged bitext. In: Proceedings of the Sixth Workshop on Very Large Corpora, pp. 102–109 (1998)
12. MarSperberg-McQueen, C.M., Burnard, L.: Guidelines for Electronic Text Encoding and Interchange. TEI P3 Text Encoding Initiative (1994)
13. Steinberger, R., Pouliquen, B., Widiger, A., Ignat, C., Erjavec, T., Tufis, D., Varga, D.: The JRC-Acquis: A Multilingual Aligned Parallel Corpus with 20+ Languages. In: LREC, pp. 2142–2147 (2006)
14. Marko, T.: Building the Croatian-English Parallel Corpus. In: LREC (2000)

# An Automatic Version of the Post-Laryngectomy Telephone Test

Tino Haderlein[1,2], Korbinian Riedhammer[1], Andreas Maier[1,2], Elmar Nöth[1],
Hikmet Toy[2], and Frank Rosanowski[2]

[1] Universität Erlangen-Nürnberg, Lehrstuhl für Mustererkennung (Informatik 5)
Martensstraße 3, 91058 Erlangen, Germany
Tino.Haderlein@informatik.uni-erlangen.de
http://www5.informatik.uni-erlangen.de
[2] Universität Erlangen-Nürnberg, Abteilung für Phoniatrie und Pädaudiologie
Bohlenplatz 21, 91054 Erlangen, Germany

**Abstract.** Tracheoesophageal (TE) speech is a possibility to restore the ability
to speak after total laryngectomy, i.e. the removal of the larynx. The quality of
the substitute voice has to be evaluated during therapy. For the intelligibility evaluation of German speakers over telephone, the Post-Laryngectomy Telephone
Test (PLTT) was defined. Each patient reads out 20 of 400 different monosyllabic words and 5 out of 100 sentences. A human listener writes down the words
and sentences understood and computes an overall score. This paper presents
a means of objective and automatic evaluation that can replace the subjective
method. The scores of 11 naïve raters for a set of 31 test speakers were compared
to the word recognition rate of speech recognizers. Correlation values of about
0.9 were reached.

## 1 Introduction

In 20 to 40 percent of all cases of laryngeal cancer, total laryngectomy has to be performed, i.e. the removal of the entire larynx [1]. For the patient, this means the loss of
the natural voice and thus the loss of the main means of communication. One possibility of voice restoration is the tracheoesophageal (TE) substitute voice. In TE speech,
the upper esophagus, the pharyngo-esophageal (PE) segment, serves as a sound generator (see Fig. 1). The air stream from the lungs is deviated into the esophagus during
expiration via a shunt between the trachea and the esophagus. Tissue vibrations of the
PE segment modulate the streaming air and generate a substitute voice signal. In order to
force the air to take its way through the shunt into the esophagus and allow voicing, the
patient usually closes the tracheostoma with a finger. In comparison to normal voices,
the quality of substitute voices is "low". Inter-cycle frequency perturbations result in a
hoarse voice [2]. Furthermore, the change of pitch and volume is limited which causes
monotone voice. Acoustic studies of TE voices can be found for instance in [3,4]. The
reduced sound quality and problems such as the reduced ability of intonation or voiced-
voiceless distinction [5,6] lead to worse intelligibility. For the patients, this means a
deterioration of quality of life as they cannot communicate properly.

V. Matoušek and P. Mautner (Eds.): TSD 2007, LNAI 4629, pp. 238–245, 2007.

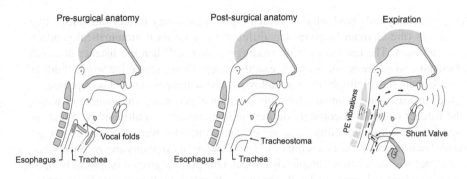

**Fig. 1.** Anatomy of a person with intact larynx *(left)*, anatomy after total laryngectomy *(middle)*, and the substitute voice *(right)* caused by vibration of the pharyngoesophageal segment (pictures from [10])

In speech therapy and rehabilitation, a patient's voice has to be evaluated by the therapist. An automatically computed, objective measure would be a very helpful support for this task. In current automatic evaluation, usually sustained vowels are analyzed and the voice quality is rated. However, for criteria like intelligibility not just a voice sample but a speech sample is needed. Moerman et al. [7] investigated recordings of a short text that contained 18 words. Correlations to human ratings were only given for the "overall impression" of the substitute voice ($r = 0.49$), so no direct comparisons to our study are possible. In previous work, we showed that an automatic speech recognition (ASR) system can be used to rate the intelligibility in close-talking speech of post-laryngectomy speakers [8,9]. The telephone is a crucial part of the patients' social life, and it is necessary for them to have a means of communication that does not require them to leave their home. Therefore, intelligibility on a telephone reflects an everyday communication situation which is important for the patient. In this paper, we will present an automatic version of an introduced standard test for intelligibility over the telephone.

This paper is organized as follows: In Sect. 2, the Post-Laryngectomy Telephone Test will be explained. The test data will be introduced in Sect. 3. Section 4 will give some information about the speech recognition system. Section 5 contains the results, and Sect. 6 will give a short outlook on future work.

## 2 The Post-Laryngectomy Telephone Test

The Post-Laryngectomy Telephone Test (PLTT, [11]) was developed in order to represent the communication situation outside the patient's usual environment (the family) by taking into account both voice and language. The patient calls a naïve rater over a standard landline telephone. The rater should not know about the text material of the test and may not have any hearing impairment.

The PLTT vocabulary consists of 400 monosyllabic words and 100 sentences, each of them written on an individual card. For one session, 22 words and 6 sentences are randomly chosen. The first two words and the first sentence are not taken into account for evaluation. Instead, they are supposed to allow the listener to adapt to the speaker.

The speaker may only read what is written on the cards. Any further utterances, like e.g. articles (the German language has different ones for each grammatical gender), are not allowed. The test begins with reading the words. When the listener does not understand a word, he or she may say exactly once: "Please repeat the word." Further feedback about the intelligibility is not allowed. The sentences may not be repeated.

Three measures are computed from the listening experiment. The number of words $w$ the listener understood correctly during the first attempt is multiplied by 5 and represents the word intelligibility $i_{word}$ in percent. Words that were repeated do not get a point. Each sentence $s$ gets a score $c_s$ of 0 to 2 points. Two points are assigned when the sentence was understood completely correct. One point is given if one word is missing or not understood correctly. In all other cases, the reader gets no point. The sentence intelligibility $i_{sent}$ in percent is the resulting sum of points multiplied with 10. The total intelligibility $i_{total}$ is then given by

$$i_{total} = \frac{i_{word} + i_{sent}}{2} = \frac{1}{2}\left(5w + 10\sum_{s=1}^{5} c_s\right). \tag{1}$$

The test was shown to be valid, reliable and objective [11], and it was also applied to laryngectomized persons before: Patients with shunt valves reached an average PLTT result between 70 and 80 [12]. A reason why the test should be done via telephone was also given: A quiet room does not reflect a real-world communication situation as noise is present almost everywhere. In a noise-free environment, the voice rehabilitation progress would be overestimated. The telephone situation is easy to maintain and thus suitable for practical use. But like each evaluation that involves human raters, this test is subjective for many reasons, like the listener's hearing abilities or experience with TE voices. Other persons might not be able to understand or reproduce the results. For these reasons, an objective and automatic version of the PLTT using automatic speech recognition was desired.

## 3  Test Data

A test set of PLTT recordings (*pltt_8kHz*) from 31 laryngectomees was available where each recording contained all words and sentences the respective speaker read out. The speakers were 25 men and 6 women (63.4±8.7 years old) with tracheoesophageal substitute speech. They were provided with a shunt valve of the Provox® type [13]. The data were recorded with a dialogue system provided by Sympalog Voice Solutions[1]. The audio files were also segmented by hand so that each word and sentence was stored in a separate file. This was done in order to explore whether the automatic evaluation is influenced by noise or non-verbals between the words in the full recordings. This database is denoted as *pltt_seg_8kHz*.

The human listeners were 8 male and 3 female students (average age: 22.5±1.2 years). None of them had experience with voice and speech analysis. For recording the PLTT, each patient got a unique sheet of paper with instructions and 22 words

[1] http://www.sympalog.com

and 6 sentences that were randomly chosen. The first two words and the first sentence were neither used for human nor for automatic evaluation. The raters listened to the *pltt_seg_8kHz* data set. They could pause the play-back to write down the understood utterance.

## 4   The Speech Recognition System

The speech recognition system used for the experiments was developed at the Chair of Pattern Recognition in Erlangen. It can handle spontaneous speech with mid-sized vocabularies up to 10,000 words. The latest version is described in detail in [14]. The system is based on semi-continuous Hidden Markov Models (HMM). It can model phones in a context as large as statistically useful and thus forms the so-called polyphones, a generalization of the well-known bi- or triphones. The HMMs for each polyphone have three to four states; for the PLTT experiments, the codebook had 500 classes with full covariance matrices. The short-time analysis applies a Hamming window with a length of 16 ms, the frame rate is 10 ms. The filterbank for the Mel-spectrum consists of 25 triangle filters. For each frame, a 24-dimensional feature vector is computed. It contains short-time energy, 11 Mel-frequency cepstral coefficients, and the first-order derivatives of these 12 static features. The derivatives are approximated by the slope of a linear regression line over 5 consecutive frames (56 ms). A zerogram language model was used so that the results are only dependent on the acoustic models.

The baseline system for the experiments in this paper was trained with German dialogues from the VERBMOBIL project [15]. The topic in these recordings is appointment scheduling. The data were recorded with a close-talking microphone at a sampling frequency of 16 kHz and quantized with 16 bit. The speakers were from all over Germany and covered most regions of dialect. They were, however, asked to speak standard German. About 80% of the 578 training speakers (304 male, 274 female) were between 20 and 29 years old, less than 10% were over 40. This is important in view of the test data because the fact that the average age of our test speakers is more than 60 years may influence the recognition results. 11,714 utterances (257,810 words) of the VERBMOBIL-German data (12,030 utterances, 263,633 words, 27.7 hours of speech) were used for the training and 48 (1042 words) for the validation set, i.e. the corpus partitions were the same as in [14].

A speech recognition system can only recognize the words stored in its vocabulary list. This list had to be created from the words and sentences occurring in the PLTT. This, however, is not enough to simulate the human listener. A human being knows more words than those occurring in the test which might cause misperceptions. In order to simulate this in the automatic test, the vocabulary list of the recognizer had to be extended by words phonetically similar to those of the actual vocabulary. This was done by the definition of a modified Levenshtein distance for phonetic transcriptions. It involved a weighting function which assigns phoneme pairs that sound similar (e.g. /s/ and /z/) a low weight and thus finds the desired words [16]. In this way, the basic PLTT vocabulary that consisted of 738 words (*PLTT-small*) was expanded to 1017 words (*PLTT-large*). The additional words and their transliterations were taken from the CELEX dictionary [17]. The VERBMOBIL baseline training set was

**Table 1.** Human evaluation results $i_{word}$, $i_{sent}$ and $i_{total}$; Pearson's correlation $r$ and Spearman's correlation $\rho$ are calculated between the respective rater and the average of the remaining 10 raters

| rater | $i_{word}$ | | | | $i_{sent}$ | | | | $i_{total}$ | | | |
|---|---|---|---|---|---|---|---|---|---|---|---|---|
| | $\mu$ | $\sigma$ | $r$ | $\rho$ | $\mu$ | $\sigma$ | $r$ | $\rho$ | $\mu$ | $\sigma$ | $r$ | $\rho$ |
| BM | 31.4 | 21.5 | 0.91 | 0.90 | 48.1 | 30.4 | 0.88 | 0.88 | 39.8 | 22.6 | 0.92 | 0.90 |
| BT | 29.2 | 20.7 | 0.83 | 0.80 | 52.2 | 29.5 | 0.88 | 0.87 | 40.7 | 21.8 | 0.90 | 0.90 |
| CV | 43.9 | 20.1 | 0.90 | 0.90 | 45.9 | 28.6 | 0.83 | 0.81 | 44.9 | 22.5 | 0.89 | 0.89 |
| GM | 39.4 | 21.4 | 0.87 | 0.83 | 48.1 | 28.9 | 0.88 | 0.86 | 43.8 | 23.3 | 0.93 | 0.93 |
| HT | 43.0 | 21.3 | 0.89 | 0.84 | 57.2 | 28.6 | 0.77 | 0.76 | 50.1 | 22.4 | 0.86 | 0.86 |
| KC | 41.7 | 20.8 | 0.93 | 0.91 | 46.9 | 28.8 | 0.65 | 0.60 | 44.3 | 20.8 | 0.85 | 0.83 |
| MM | 47.2 | 23.5 | 0.91 | 0.90 | 50.0 | 28.7 | 0.79 | 0.78 | 48.6 | 23.9 | 0.92 | 0.92 |
| PC | 34.1 | 21.9 | 0.87 | 0.83 | 48.4 | 29.5 | 0.87 | 0.85 | 41.3 | 22.4 | 0.92 | 0.92 |
| SM | 51.1 | 21.3 | 0.82 | 0.82 | 59.4 | 24.0 | 0.82 | 0.75 | 55.2 | 19.2 | 0.90 | 0.86 |
| ST | 53.6 | 23.0 | 0.92 | 0.92 | 67.5 | 29.6 | 0.72 | 0.64 | 60.6 | 23.4 | 0.86 | 0.85 |
| WW | 40.3 | 19.2 | 0.89 | 0.90 | 56.6 | 25.2 | 0.87 | 0.89 | 48.4 | 19.7 | 0.94 | 0.94 |

downsampled to 8 kHz sampling frequency, a VERBMOBIL recognizer was trained and the vocabulary changed to the *PLTT-small* or *PLTT-large* word list, respectively. For both cases, a polyphone-based and a monophone-based recognizer version were created. Monophones were supposed to be more robust for recognition of highly pathologic speech because each of them is trained with more data than a polyphone model.

## 5   Results

Table 1 shows the PLTT results of the single raters. Although they had never heard TE voices before, the inter-rater correlation for the total intelligibility $i_{total}$ is greater than 0.8 for all persons. However, perceptive results vary strongly among the raters. The difference in the average of $i_{total}$ for the "best" and the "worst" rater is more than 20 points which shows how strongly the test depends on the particular listener. The standard deviation is very similar for all raters, however. The recognition results and the PLTT measures both for recognizers and human raters are assembled in Table 2. Since the first part of a PLTT session consists of single words, not only the word accuracy (WA) but also the word recognition rate (WR) was computed. It is based on the word accuracy, but the number of words wrongly inserted by the recognizer is not counted. In comparison to the human WA which reached 55%, the automatic recognition rates are much lower due to the following reasons: First of all, the recognizers were trained with normal speech. This simulates a naïve listener who has not heard TE voices before, i.e. the kind of listener that is required for the PLTT. The average WA for close-talking recordings of laryngectomees was determined at approx. 30% [8,9]. Here, the results are even lower: The speakers had read another text right before the PLTT and were therefore exhausted. The bad signal quality of the telephone transmission and the fact that the training data of the recognizers were just downsampled and not real telephone speech had also negative influence. No sentence was recognized completely correct according to the PLTT rules. For this reason, $i_{sent}$ was 0 for all recognizers. WA and

**Table 2.** Average word accuracy (WA), word recognition rate (WR), and the PLTT measures $i_{word}$, $i_{sent}$ and $i_{total}$ for speech recognizers and human raters

| data set | pltt_8kHz | | | | pltt_seg_8kHz | | | | |
|---|---|---|---|---|---|---|---|---|---|
| vocabulary | PLTT-small | | PLTT-large | | PLTT-small | | PLTT-large | | raters |
| recog. units | mono | poly | mono | poly | mono | poly | mono | poly | |
| $\mu(WA)$ | 10.0 | 1.8 | 8.0 | −0.1 | 9.2 | 0.3 | 7.4 | −1.5 | 55.1 |
| $\sigma(WA)$ | 14.8 | 20.4 | 13.5 | 19.9 | 14.7 | 21.4 | 12.9 | 20.2 | 21.4 |
| $\mu(WR)$ | 17.3 | 16.6 | 14.4 | 13.7 | 16.4 | 15.6 | 14.2 | 13.2 | 55.3 |
| $\sigma(WR)$ | 13.2 | 12.6 | 9.3 | 11.2 | 9.9 | 10.8 | 8.7 | 10.3 | 21.4 |
| $\mu(i_{word})$ | 17.8 | 13.1 | 14.5 | 10.9 | 14.1 | 11.1 | 12.0 | 9.4 | 41.4 |
| $\sigma(i_{word})$ | 15.1 | 13.0 | 12.8 | 10.9 | 13.8 | 11.6 | 12.7 | 11.1 | 21.3 |
| $\mu(i_{sent})$ | 0.0 | 0.0 | 0.0 | 0.0 | 0.0 | 0.0 | 0.0 | 0.0 | 52.8 |
| $\sigma(i_{sent})$ | 0.0 | 0.0 | 0.0 | 0.0 | 0.0 | 0.0 | 0.0 | 0.0 | 28.3 |
| $\mu(i_{total})$ | 8.9 | 6.6 | 7.3 | 5.5 | 7.0 | 5.5 | 6.0 | 4.7 | 47.1 |
| $\sigma(i_{total})$ | 7.5 | 6.5 | 6.4 | 5.8 | 6.9 | 5.8 | 6.4 | 5.6 | 22.0 |

**Table 3.** Pearson's correlation $r$ and Spearman's correlation $\rho$ between the speech recognizers' results ("rec") and the human raters' average values ("hum")

| data set | pltt_8kHz | | | | pltt_seg_8kHz | | | |
|---|---|---|---|---|---|---|---|---|
| vocabulary | PLTT-small | | PLTT-large | | PLTT-small | | PLTT-large | |
| recognition units | mono | poly | mono | poly | mono | poly | mono | poly |
| $r(WA_{rec}, WA_{hum})$ | 0.72 | 0.71 | 0.73 | 0.70 | 0.73 | 0.67 | 0.71 | 0.69 |
| $\rho(WA_{rec}, WA_{hum})$ | 0.85 | 0.82 | 0.83 | 0.82 | 0.83 | 0.79 | 0.80 | 0.80 |
| $r(WR_{rec}, WA_{hum})$ | 0.82 | 0.86 | 0.81 | 0.83 | 0.87 | 0.88 | 0.86 | 0.87 |
| $\rho(WR_{rec}, WA_{hum})$ | 0.88 | 0.94 | 0.89 | 0.92 | 0.91 | 0.91 | 0.89 | 0.91 |
| $r(WA_{rec}, i_{total,hum})$ | 0.71 | 0.72 | 0.72 | 0.71 | 0.72 | 0.67 | 0.71 | 0.70 |
| $\rho(WA_{rec}, i_{total,hum})$ | 0.84 | 0.81 | 0.83 | 0.80 | 0.81 | 0.76 | 0.79 | 0.79 |
| $r(WR_{rec}, i_{total,hum})$ | 0.81 | 0.88 | 0.82 | 0.85 | 0.85 | 0.89 | 0.86 | 0.89 |
| $\rho(WR_{rec}, i_{total,hum})$ | 0.86 | 0.93 | 0.87 | 0.92 | 0.88 | 0.90 | 0.90 | 0.90 |

WR for the human raters were computed from the raters' written transliteration of the audio files.

Although the automatic recognition yielded so bad results, the correlation to the human ratings was high (see Table 3). The reason is that the crucial measure is not the average of the recognition rate but its range or standard deviation, respectively. It was not the task of the experiments to optimize the mean recognition rate. For this reason, voices with low quality often receive negative values of WA. Nevertheless, the distribution of these values corresponds well to the measures obtained by the human listeners. The best correlation between an automatic measure and the overall PLTT result $i_{total}$ was reached for WR on the polyphone-based recognizers. Both Pearson's correlation $r$ and Spearman's correlation $\rho$ are about 0.9.

The outcome of these experiments is that the PLTT can be replaced by an objective, automatic approach. The question whether monophone-based or polyphone-based recognizers are better for the task could not be answered. When the word accuracy WA was compared to $i_{\text{total}}$, monophones were advantageous; when the word recognition rate WR was used instead, the polyphone-based recognizers were closer to the human rating. There are also some cases in which the correlation is slightly better when each word and sentence is processed separately, but in general the long *pltt_8kHz* recordings which contain the entire test can be used without prior segmentation.

## 6  Conclusion and Outlook

In this paper, an approach for the automation of the Post-Laryngectomy Telephone Test (PLTT) was presented. The correlation between the overall intelligibility score that is usually computed by a human listener and the word recognition rate of a speech recognizer was about 0.9. A difference between the human and the machine evaluation was that the automatic version does not process words again it did not "understand" on first attempt. This is not necessary since the result would be the same. Furthermore, a word that was not understood by the listener on first attempt does not get a point anyway, so it is not necessary to consider word repetition in the automatic version at all.

Adaptation of the speech recognizers to the signal quality might enhance the recognition results. Experiments in order to find out whether also the correlation to the human results will get better will be part of future work. Another aspect that will be taken into consideration are reading errors by the patient that have to be identified before the intelligibility measure is computed.

## Acknowledgments

This work was partially funded by the German Cancer Aid (Deutsche Krebshilfe) under grant 106266. The responsibility for the contents of this study lies with the authors.

## References

1. van der Torn, M., Mahieu, H., Festen, J.: Aero-acoustics of silicone rubber lip reeds for alternative voice production in laryngectomees. J. Acoust. Soc. Am. 110(5 Pt 1), 2548–2559 (2001)
2. Schutte, H., Nieboer, G.: Aerodynamics of esophageal voice production with and without a Groningen voice prosthesis. Folia Phoniatr. Logop 54(1), 8–18 (2002)
3. Robbins, J., Fisher, H., Blom, E., Singer, M.: A Comparative Acoustic Study of Normal, Esophageal, and Tracheoesophageal Speech Production. J. Speech Hear. Disord. 49(2), 202–210 (1984)
4. Bellandese, M., Lerman, J., Gilbert, H.: An Acoustic Analysis of Excellent Female Esophageal, Tracheoesophageal, and Laryngeal Speakers. J. Speech Lang. Hear. Res. 44(6), 1315–1320 (2001)
5. Gandour, J., Weinberg, B.: Perception of Intonational Contrasts in Alaryngeal Speech. J. Speech Hear. Res. 26(1), 142–148 (1983)
6. Searl, J., Carpenter, M.: Acoustic Cues to the Voicing Feature in Tracheoesophageal Speech. J. Speech Lang. Hear. Res. 45(2), 282–294 (2002)

7. Moerman, M., Pieters, G., Martens, J., van der Borgt, M., Dejonckere, P.: Objective evaluation of the quality of substitution voices. Eur. Arch. Otorhinolaryngol. 261(10), 541–547 (2004)
8. Schuster, M., Nöth, E., Haderlein, T., Steidl, S., Batliner, A., Rosanowski, F.: Can You Understand Him? Let's Look at His Word Accuracy – Automatic Evaluation of Tracheoesophageal Speech. In: Proc. IEEE Int. Conf. on Acoustics, Speech and Signal Processing (ICASSP). Philadelphia, PA (USA), vol. I, pp. 61–64 (2005)
9. Schuster, M., Haderlein, T., Nöth, E., Lohscheller, J., Eysholdt, U., Rosanowski, F.: Intelligibility of laryngectomees' substitute speech: automatic speech recognition and subjective rating. Eur. Arch. Otorhinolaryngol. 263(2), 188–193 (2006)
10. Lohscheller, J.: Dynamics of the Laryngectomee Substitute Voice Production. Berichte aus Phoniatrie und Pädaudiologie. Shaker Verlag, Aachen (Germany), vol. 14 (2003)
11. Zenner, H.: The postlaryngectomy telephone intelligibility test (PLTT). In: Herrmann, I. (ed.) Speech Restoration via Voice Prosthesis, pp. 148–152. Springer, Heidelberg (1986)
12. de Maddalena, H., Zenner, H.: Evaluation of speech intelligibility after prosthetic voice restoration by a standardized telephone test. In: Algaba, J. (ed.) Proc. 6th International Congress on Surgical and Prosthetic Voice Restoration After Total Laryngectomy, San Sebastian (Spain), pp. 183–187. Elsevier Science, Amsterdam (1996)
13. Hilgers, F., Balm, A.: Long-term results of vocal rehabilitation after total laryngectomy with the low-resistance, indwelling Provox voice prosthesis system. Clin. Otolaryngol. 18(6), 517–523 (1993)
14. Stemmer, G.: Modeling Variability in Speech Recognition. Studien zur Mustererkennung. Logos Verlag, Berlin, vol. 19 (2005)
15. Wahlster, W. (ed.): Verbmobil: Foundations of Speech-to-Speech Translation. Springer, Berlin (2000)
16. Riedhammer, K.: An Automatic Intelligibility Test Based on the Post-Laryngectomy Telephone Test. Student's thesis, Lehrstuhl für Mustererkennung (Chair for Pattern Recognition), Universität Erlangen–Nürnberg, Erlangen (2007)
17. Baayen, R.H., Piepenbrock, R., Gulikers, L.: The CELEX Lexical Database (Release 2). Linguistic Data Consortium, Philadelphia, PA (USA) (1996)

# Speaker Normalization Via Springy Discriminant Analysis and Pitch Estimation

Dénes Paczolay[1], András Bánhalmi[1], and András Kocsor[2]

[1] Research Group on Artificial Intelligence,
Hungarian Academy of Sciences and University of Szeged
H-6720 Szeged, Aradi vértanúk tere 1., Hungary
{pdenes, bandras}@inf.u-szeged.hu
[2] Applied Intell. Lab. Ltd., Petőfi S. Sgt. 43., H-6725 Szeged, Hungary
kocsor@ail.hu

**Abstract.** Speaker normalization techniques are widely used to improve the accuracy of speaker independent speech recognition. One of the most popular group of such methods is Vocal Tract Length Normalization (VTLN). These methods try to reduce the inter-speaker variability by transforming the input feature vectors into a more compact domain, to achieve better separations between the phonetic classes. Among others, two algorithms are commonly applied: the Maximum Likelihood criterion-based, and the Linear Discriminant criterion-based normalization algorithms. Here we propose the use of the Springy Discriminant criterion for the normalization task. In addition we propose a method for the VTLN parameter determination that is based on pitch estimation. In the experiments this proves to be an efficient and swift way to initialize the normalization parameters for training, and to estimate them for the voice samples of new test speakers.

## 1 Motivation

Speaker normalization is a very useful tool for decreasing the inter-speaker variability of speech samples, which can in turn significantly improve the performance of speaker-independent continuous speech recognizers. The most commonly used methodology performs Vocal Tract Length Normalization (VTLN), which is realized by a non-linear warping of the frequency axis. Most VTLN algorithms lead to a multidimensional optimization problem that can take a long time to solve. The most popular technique that is based on the maximum likelihood (ML) criterion requires a lot of slow iteration steps (basically, runs of the speech recognizer). Another family of methods that is built on the technique of linear discriminant (LD) analysis involves matrix operations with quite large matrices. In this paper we examine an alternative to the LD-based technique: the springy discriminant (SD) criterion-based method, which circumvents several numerical problems of LD. We also propose a speed-up technique that can be applied for the fast computation of both the LD and SD algorithms. Finally, we examine the relationship between the optimal warping factor and the pitch of a speaker, and based on our findings we propose a method that can be used for estimating the warping factor from the pitch directly, or for the initialization of the LD and SD algorithms.

V. Matoušek and P. Mautner (Eds.): TSD 2007, LNAI 4629, pp. 246–253, 2007.
© Springer-Verlag Berlin Heidelberg 2007

## 2   Vocal Tract Length Normalization

One of the major physiological sources of inter-speaker variation is the variation in the vocal tract length of speakers. This can be compensated for by warping the frequency axis, and several sophisticated warping functions have been proposed for this in the literature [1,2]. One of these which will be used here is the bilinear warping function. The goal of all the VTLN methods is to assign to each speaker a warping factor in such a way that the transformed feature vectors should be optimal in some sense. A common feature of them is that they transform the feature vectors of the phone classes – obtained from the speech samples of a variety of speakers – into a smaller domain, without significantly reducing the average inter-class distances. The only real difference is the way they assign a warping factor to each particular speaker. In the following we first discuss the two frequently used optimization criteria, and then suggest an alternative one.

### 2.1   VTLN with Maximum Likelihood (ML) Criterion

In the case of ML the criterion for choosing the optimal warping factor $\alpha$ is based on the probability of the speaker's utterance calculated using an acoustic probability model. Here this probability value is obtained via a Continuous Speech Recognizer (CSR). The optimal $\alpha$ value is chosen like so. The feature vectors of a particular speaker's voice sample are transformed using all the possible $\alpha$ values. The CSR is given all the transformed feature vector series as input data, and it computes the likelihood values as output data. *The $\alpha$ value corresponding to the transformation with the maximal likelihood value should be chosen.*

Algorithmically, the criterion above is used in the following way. The starting-point is a CSR model trained on non-warped data. Then, for each speaker a warping factor value is found using the ML criterion. The CSR model is next trained on the warped feature vectors. The warping factor determination and the model retraining are iterated until there is no change in the $\alpha$ values.

A few problems were reported in the literature when applying ML-based VTLN. Firstly, the previously described iterative method will converge to a local optima, which is incidentally a common feature of all iterative methods (ML, SD). Furthermore, the quality of the local optima found is closely related to the initialization of the warping factors. Usually, the initial values are set to correspond to the identity transformation. Here we propose a more sophisticated parameter initialization method instead.

And secondly, experience tells us that the average value of the warping factors gets bigger after each iteration, so after some iterations most of the warping factor values will be set to the highest constant, and at the same time the speech recognition performance will worsen. Acero [3] suggested an idea for tackling this problem; namely that the average of the warping factors should be set to a constant value after each iteration.

### 2.2   VTLN with Linear Discriminant (LD) Criterion

Linear Discriminant Analysis (LDA) is a technique which finds particular directions in the space of the feature vectors. These directions are optimal in the sense that the feature vectors in these directions have the highest inter-class covariance, while at the same time the average within-class covariance is the lowest. LDA finds these directions by computing the eigenvectors of an eigenvalue problem, and the eigenvalues give a measure of the optimality of these eigenvectors [4]. Thus the eigenvalues can be used for estimating the optimality of an input space transformation, which here is the VTLN one [5].

More precisely, when adapting this technique to the VTLN problem, the iterative algorithm is similar to the one described for ML. By optimizing the LD criterion we *choose a warping factor which results in the largest maximal eigenvalues* (obtained using different warping factors). Here the classes are the different vowels, and the product of the 5 maximal eigenvalues were maximized to obtain the optimal $\alpha$. The optimization method iterates the following steps until there is no change in the warping factor values, or the number of iterations reaches a predefined limit:

- Form the $X$ dataset by transforming the feature vectors of all the speakers using the actual warping factor values (the warping factors are initially set to correspond to the identity transformation).
- Assign the optimal warping factor to each speaker by selecting the warping factor that is the best according to the LD criterion. For this, replace the rows in the $X$ dataset corresponding to the particular speaker being tested.

This iterative method has the same problem as ML: in most cases only a local optimum is achieved, hence choosing better initial values for the warping factors can be important for getting better results.

The time-complexity of the optimization method can be greatly reduced using the following update formulas for the mean vectors and covariance matrices. Let $\mu$ and $\Sigma$ be the mean and covariance of the $X$ dataset. Let $X_P$ stand for the data to be added to $X$, and $X_N$ the data to be removed from $X$. Then the update formulas are ($X_{new} = X \cup X_P \backslash X_N$):

$$\mu_{new} = \frac{|X| \cdot \mu + \sum_{x \in X_P} x - \sum_{x \in X_N} x}{|X| + |X_P| - |X_N|}. \tag{1}$$

$$\Sigma_{new} = \frac{(\Sigma + \mu \cdot \mu^T)|X| + \sum_{x \in X_P} x \cdot x^T - \sum_{x \in X_N} x \cdot x^T}{|X| + |X_P| - |X_N|} - \mu_{new} \cdot \mu_{new}^T. \tag{2}$$

Because the iterative algorithm always replaces the same set of examples, $\sum x \cdot x^T$ and $\sum x$ can be computed in advance for all the classes, all the speakers, and all the warping factors.

## 3   Springy Discriminant Analysis

LDA is not the only possible transformation which attempts to find those directions in the input space that are optimal in some sense with respect to class separation. Earlier, we suggested a new formulation for the objective function, the so-called "Springy Discriminant Analysis" (SDA), which yields similar results to LDA, but the transformation matrix is orthogonal and it has no numerical problems like that of matrix inversion. Although it was originally proposed in a non-linear form [6], the authors later published the corresponding linear version as well [7].

Let $\varphi(\mathbf{v})$, the potential of the spring model along the direction $\mathbf{v}$, be defined by

$$\varphi(\mathbf{v}) = \sum_{i,j=1}^{k} M_{ij} \left( (\mathbf{x}_i - \mathbf{x}_j)^\top \mathbf{v} \right)^2, \tag{3}$$

where

$$M_{ij} = K_{\mathcal{L}(i)\mathcal{L}(j)} \cdot \delta(i,j), \quad \delta(i,j) = \begin{cases} -1, & \text{if } \mathcal{L}(i) = \mathcal{L}(j) \\ 1, & \text{otherwise} \end{cases} \tag{4}$$

$$K_{\mathcal{L}(i)\mathcal{L}(j)} = K_{\mathcal{L}(j)\mathcal{L}(i)}, \quad K_{\mathcal{L}(i)\mathcal{L}(j)} > 0, \quad i,j = 1,\ldots,k.$$

Here $\mathcal{L}(i)$ means the class of the $i$th point. The initialization of the elements of the matrix $K_{\mathcal{L}(i)\mathcal{L}(j)}$ will be analyzed later. The elements can be considered as a kind of force constant between the points of different and same classes.

It is easy to see that the value of $\varphi$ is larger when those components of the elements of the same class that fall in the given direction $\mathbf{v}$ ($\mathbf{v} \in \mathbb{R}^n$) are closer, and the components of the elements of different classes are farther at the same time.

Now with the introduction of the matrix

$$D = \sum_{i,j=1}^{k} M_{ij} (\mathbf{x}_i - \mathbf{x}_j)(\mathbf{x}_i - \mathbf{x}_j)^\top \tag{5}$$

we immediately obtain the result that $\varphi(\mathbf{v}) = \mathbf{v}^\top D \mathbf{v}$. Based on this, the objective function $\tau$ can be defined as the Rayleigh quotient

$$\tau(\mathbf{v}) = \frac{\varphi(\mathbf{v})}{\mathbf{v}^\top \mathbf{v}} = \frac{\mathbf{v}^\top D \mathbf{v}}{\mathbf{v}^\top \mathbf{v}}. \tag{6}$$

Obviously, the optimization of $\tau$ leads to the eigenvalue decomposition of $D$. Because $D$ is symmetric, its eigenvalues are real and its eigenvectors are orthogonal. The matrix $V$ of the SDA transformation is defined using those eigenvectors corresponding to the $m$ dominant eigenvalues of $D$.

The form of the $D$ matrix will now be rewritten. Let us break up the sum into parts. For each variable, let take each $x_i$ from a particular class, and do the same with $x_j$.

$$D = \sum_{i,j=1}^{k} M_{ij}(x_i - x_j)(x_i - x_j)^T = \sum_{l,k=1}^{|C|} \sum_{\substack{i:x_i \in C(l), \\ j:x_j \in C(k)}} M_{ij}(x_i - x_j)(x_i - x_j)^T. \tag{7}$$

Here $C$ denotes the set of the classes. The classes are defined as $C(l) = \{x_i : \mathcal{L}(i) = l\}$. Now consider one pair of classes, and let us subtract from each $x_i$ the mean of the class ($\mu_l$) $x_i$ belongs to. That is,

$$lM_{ij}(x_i - x_j)(x_i - x_j)^T = M_{ij}((x_i - \mu_l) - (x_j - \mu_k) + (\mu_l - \mu_k))$$
$$\times ((x_i - \mu_l) - (x_j - \mu_k) + (\mu_l - \mu_k))^T \qquad (8)$$

After resolving the factors, summing the expression and dropping the zero components we get:

$$\sum_{\substack{i:x_i \in C(l), \\ j:x_j \in C(k)}} \{M_{ij}(x_i - \mu_l)(x_i - \mu_l)^T + (x_j - \mu_k)(x_j - \mu_k)^T +$$

$$+ (\mu_l - \mu_k)(\mu_l - \mu_k)^T\}. \qquad (9)$$

then using the definition of the $M$ matrix, we have

$$K_{l,k} \cdot |C(k)| \cdot |C(l)| \cdot \{\text{mdif}[C(l), C(k)] + \text{cov}[C(l)] + \text{cov}[C(k)]\} \cdot \delta(l, k), \quad (10)$$

where

$$\text{cov}[C(l)] = \frac{1}{|C(l)|} \sum_{i:x_i \in C(l)} (x_i - \mu_l)(x_i - \mu_l)^T \qquad (11)$$

$$\text{mdif}[C(l), C(k)] = (\mu_l - \mu_k)(\mu_l - \mu_k)^T. \qquad (12)$$

So summing over all the pairs of classes:

$$lD = \sum_{k,l=1}^{|C|} K_{l,k} \cdot |C(l)||C(k)|\text{mdif}[C(l), C(k)] +$$

$$+ \sum_{\substack{k,l=1, \\ (k \neq l)}}^{|C|} K_{l,k} \cdot |C(l)||C(k)|(\text{cov}[C(l)] + \text{cov}[C(k)]) - \qquad (13)$$

$$- 2 \cdot \sum_{l=1}^{|C|} K_{l,l} \cdot |C(l)|^2 \text{cov}[C(l)].$$

This form of the $D$ matrix shows that the optimality of the within- and inter-class variances can be achieved only if the $K$ matrix contains suitable values. The first sum tells us that we should be interested in those directions where the projected means of the classes are as far apart as possible. The within-class covariances are handled explicitly by the two other sums. Looking at the second and third components, the within-class covariance will be minimized if, for all $l$,

$$K_{l,l} - \sum_{k \neq l} K_{l,k} > 0, \text{ so } \frac{K_{l,l}}{\sum_{k \neq l} K_{l,k}} > 1. \qquad (14)$$

Thus $K$ is a positive, symmetric, diagonal dominant matrix. When finding the optimal directions the value of the quotient above affects the weight of the within-class covariance and the weight of the variance of the class means.

### 3.1  VTLN with Springy Discriminant (SD) Criterion

The SD criterion can be applied for VTLN using the same iterative algorithm as that described for LD (see Section 2.2). The SD criterion-based method tries to find the warping factor corresponding to the maximum of the largest eigenvalues. Here the products of the first 2 largest eigenvalues were used to choose the best $\alpha$ value.

The iterative method may again, of course, find just a local optimum. Hence, as before, choosing better initial values for the warping factors can be important for getting better results. The computations can be greatly speeded up using the update formulas for the mean vector and the covariance matrix defined above (Section 2.2).

## 4  Relation Between Pitch and Optimal Warping Factor

In our experiments we observed that there was a definite correlation between the pitch and the optimal warping factor for different speakers (see Fig. 1). This connection can be exploited in estimating the warping parameter values. First, using this connection for the initialization of the warping factors in the iterative methods discussed earlier can help overcome the problem of convergence to a local optimum. Another important point is that the warping parameters for a new speaker should be determined by using as little speech data as possible. It is especially important in LVCSR systems where a new speaker's voice needs to be adjusted to an already normalized acoustic model.

Several pitch estimation methods have been proposed that give different performances. Here we will apply the one that was suggested in [8], and which has proven to be superior to the usual methods. This method essentially runs five basic methods in parallel, and chooses the best value of the five via a voting procedure. For more details about this method, see [8].

Figure 1 shows the distribution of the optimal warping factors plotted against the pitch values. The close connection between these features suggests a way of estimating the warping factor using the pitch value of a particular speaker. For this purpose we used a quadratic function, its coefficients being determined by applying quadratic regression on the warping factor values and pitch values obtained from the training database.

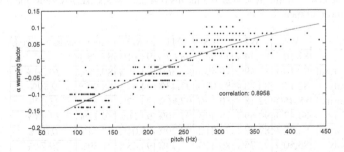

**Fig. 1.** The density of the warping factor values plotted against the pitch values. The solid curve shows the quadratic regression on these two variables. In this experiment the value of the correlation between these two parameters was 0.89.

## 5    Experimental Results

The continuous speech recognizer used for testing purposes and for the ML-criterion was a HMM based Viterbi-decoding system with an $n$-gram language model. It was developed within the framework of a medical dictation project [9].

For training purposes we used samples recorded from 100 speakers at a 22050 Hz sampling rate and 16-bit quality. Each speaker uttered 6 long sentences (16 words per sentence on average) and 12 phonetically rich words. The database contained a total of about 10,800 words (85,300 phonemes) in about 100 minutes of recorded speech samples. For the test phase, speech samples of 5 speakers (3 women: W1, W2, W3 and 2 men: M1, M2) were taken. Each speaker uttered 20-20 different paragraphs, and the total length of the recorded speech was about 15 minutes for each particular speaker. The Hidden Markov Models of the CSR system were trained on the conventional mel-frequency cepstral coefficient (MFCC) based features using a 25 ms frame window at a frame rate of 100 frames per sec. In addition, the phone models for 44 phone classes had the usual 3-state left-to-right structure. For frequency warping the bilinear function was used . The LD and SD criteria were then computed using a dataset containing 32 dimensional normalized Mel Filter Bank vectors. Here the number of the vowel classes was 10 and, on average, 16 sample frames were selected per speaker per phone. In our tests pitch estimation was performed on 400 vowel frames on average, and the warping factor was estimated using the quadratic function shown in Fig. 1. Table 1 summarizes the results we obtained for the speech recognition accuracy using the above methods.

The results show that in most cases the application of the normalization methods brought an increase in recognition accuracy. However, for the various methods there are some "outlier" speakers (W1, M1) where the accuracy fell, and this caused a reduction

**Table 1.** Recognition accuracies and word error rate reductions when using VTLN with 3 criteria (ML, LD, SD). The "poly" case means that HMM training was performed using $\alpha$ values computed via the corresponding quadratic function. The "2-i" extension means that the model was trained after two iteration steps, where the normalization algorithm was re-initialized using the quadratic regression obtained after the first iteration. The last column shows the average accuracy score for the 5 speakers. Here the W1, W2, W3 means that the test speakers were women, while M1 and M2 sign those of men.

|  |  | W1 | W2 | W3 | M1 | M2 | avg. | avg. wer. red. |
|---|---|---|---|---|---|---|---|---|
| Baseline | | 87.10% | 85.74% | 73.94% | 98.07% | 93.91% | 87.75% | - |
| ML | normal | 90.10% | 87.59% | 79.49% | 94.36% | 93.48% | 89.00% | *10.20%* |
| | poly | 82.15% | 89.28% | 81.07% | 97.43% | 94.57% | 88.90% | *9.39%* |
| LD | normal | 88.45% | 92.18% | 80.59% | 97.86% | 93.77% | 90.57% | *23.02%* |
| | poly | 84.47% | 90.89% | 79.49% | 97.50% | 93.26% | 89.12% | *11.18%* |
| SD | normal | 84.32% | 87.27% | 78.81% | 93.29% | 95.01% | 87.74% | *0.00%* |
| | poly | 86.50% | 92.75% | 86.42% | 97.79% | 93.48% | 91.39% | *29.71%* |
| LD 2-i | normal | 81.55% | 90.49% | 75.45% | 96.84% | 91.42% | 87.11% | *-5.22%* |
| | poly | 88.22% | 90.41% | 76.41% | 96.50% | 95.01% | 89.31% | *12.73%* |
| SD 2-i | normal | 89.80% | 95.97% | 86.08% | 98.43% | 96.11% | 93.28% | *45.14%* |
| | poly | 84.32% | 94.28% | 83.20% | 97.36% | 94.06% | 90.64% | *23.59%* |

in the average accuracy as well. When the HMMs were trained using the $\alpha$ values got via quadratic regression the average word error rate reduction values became much more stable, but not always better than those got from the models trained using the optimal $\alpha$ values. Applying quadratic regression to initialize the warping factors can greatly increase the accuracy, but it can cause a reduction too. On average, the best method was the VTLN with the SD criterion and initialization based on the pitch values, which combination attained a word error rate reduction of 45%.

# 6   Future Work

In this paper we examined a springy discriminant (SD)-based algorithm for speaker normalization, and the possibility of exploiting pitch estimation to find the optimal warping factor value. The average word error rate improvements we obtained look promising, but our scores show a high variance among the five speakers we chose to test our system with. Thus in the near future we plan to repeat the test on a larger training and test database, from which the trends will hopefully become clearer. We also intend to test the LD and SD methods by using their non-linearized (kernel-based) versions [6] instead of the linear ones. Lastly, we found that the performance of our system can improve quite a lot when applying speaker adaptation. Hence we intend to investigate whether the combined application of speaker normalization and speaker adaptation can bring about further improvements to our speech recognition system.

# References

1. Eide, E., Gish, H.: A parametric approach to vocal tract length normalization. In: Proc. of ICASSP, Munich, pp. 1039–1042 (1997)
2. Wegmann, S., McAllaster, D., Orloff, J., Peskin, B.: Speaker normalization on coversational telephone speech. In: Proc. of ICASSP, pp. 339–341 (1996)
3. Acero, A.: Acoustical and environmental robustness in automatic speech recognition. Kluwer Academic Publishers, Dordrecht (1993)
4. Saon, G., Padmanabhan, M., Gopinath, R.A.: Eliminating inter-speaker variability prior to linear discriminant transforms. In: Proc. of ASRU (2001)
5. Westphal, M., Schultz, T., Waibel, A.: Linear discriminant – a new criterion for speaker normalization. In: Proc. ICSLP '98, Morgan Kaufmann Publishers, San Francisco (1998)
6. Kocsor, A., Kovács, K.: Kernel springy discriminant analysis and its application to a phonological awareness teaching system. In: Sojka, P., Kopeček, I., Pala, K. (eds.) TSD 2002. LNCS (LNAI), vol. 2448, pp. 325–328. Springer, Heidelberg (2002)
7. Kocsor, A., Tóth, L.: Application of kernel-based feature space transformations and learning methods to phoneme classification. Applied Intelligence 21, 129–142 (2004)
8. Bánhalmi, A., Kocsor, A., Kovács, K., Tóth, L.: Fundamental frequency estimation by combinations of various methods. In: Proc. of 7th Nordic Signal Processing Symposium (NORSIG) (2006)
9. Bánhalmi, A., Paczolay, D., Tóth, L., Kocsor, A.: First results of a Hungarian medical dictation project. In: Proc. of IS-LTC, pp. 23–26. Morgan Kaufmann Publishers, San Francisco (2006)

# A Study on Speech with Manifest Emotions

Horia-Nicolai Teodorescu[1,2] and Silvia Monica Feraru[2]

[1] Institute for Computer Science, Romanian Academy, Bd. Carol I nr 8, Iasi, Romania
[2] Technical University of Iasi, Iasi, Romania
{hteodor, mferaru}@etc.tuiasi.ro

**Abstract.** We present a study of the prosody – seen in a broader sense – that supports the theory of the interrelationship function of speech. "Pure emotions" are meant to show a relationship of the speaker with the general context. The analysis goes beyond the basic prosody, as related to pitch trajectory; namely, the analysis also aims to determine the change in higher formants. The refinement in the analysis asks for improved tools. Methodological aspects are discussed, including a discussion of the limitations of the currently available tools. Some conclusions are drawn.

## 1 Introduction

The study of spoken languages has significantly progressed during the last decades, due to the dramatic increase of interest fuelled by applications like virtual reality, video-games, human-computer speech interaction, security, and medical applications. Speech is a subtle and rich communication; it transfers not only the linguistic information, but also information about the personality and the emotional state of the speaker. The emotion is a motivation-related answer adapted to the social environment. The prosody is a communication means which includes the attitude and the emotions [7]; it also contains information about the speaker and about the environment. In Section 2, we review the state of the art in emotional speech analysis. In Section 3, we present the essentials of our research methodology. In Section 4, we summarize the results. The final section includes a brief discussion and conclusions.

## 2 Existing Approaches

For comparison with our approach, we briefly present in this section several researches on emotional voice and related databases, for the languages Greek [3], German [6], Danish [5], and Spanish [4]. Numerous other databases, for other languages, exist, but we limit our presentation to the ones described below because they are quite different and illustrate well the variety of approaches in the literature.

There is no general agreement on the classification of emotions. According to [3], emotions are classified in "basic" emotions, with different intensity levels, and in "non-basic" emotions (the "mixed" emotions). Without entering details, we chose four "basic" emotions for our analysis, happiness, anger, sadness, and neutral tone. At

V. Matoušek and P. Mautner (Eds.): TSD 2007, LNAI 4629, pp. 254–261, 2007.
© Springer-Verlag Berlin Heidelberg 2007

least some of these emotions (furry/anger and joy/happiness) are known to be produced at the level of the limbic system; this means that they are "elementary" emotions, not states of the mind produced together by several mind processes occurring in various sites of the brain.

The creation of an emotional database requires a number of speakers who simulate the emotions in different contexts [2]. A different set of subjects (the evaluators) listen to the recordings and seek to identify the emotion that the speaker has tried to simulate. The experimental analysis of Buluti, Narayanan and Syrdal [1] showed that the emotion's recognition is not perfect for emotions like sorrow, sadness, joy, and the neutral tone. The recognition rate was 92% for the neutral tone, 89% for sorrow, 89.7% for sadness, and 67.3% for happiness. The recordings have been made by actors, in most cases, or by persons with professional voices [12].

The recordings for the Greek database reported in [3] were made by actors in a professional studio. The goal of that research was to improve the naturalness of synthesized voice. The recordings were made in three different contexts, namely i) in order to reflect the reaction of the speaker to a concrete stimulus (authentic emotion); ii) preparing the environment in order to help psychologically the speaker to simulate the indicated emotion; iii) simulating the emotions only by imagining a context. The study was oriented towards the evaluation of the simulated emotional states by free answers (86.9%) and false answers (89.6%).

The German emotional database [6] contains six basic emotions: anger, happiness, fear, sadness, disgust, boredom and neutral tone. The recordings have been realized by professional actors. The validation commission recognized 80% of the simulated emotional states. The database contains files with sentences and words, the results of the perception tests, and the results of measuring the fundamental frequency, the energy, the duration, the intensity, and the rhythm.

The Danish database [5] contains recordings of two words, nine sentences and two fragments of fluent speech, simulating happiness, surprise, sadness, anger and neutral tone, spelled by professional actors. The emotional states were recorded in a theatre room with an excellent acoustic. The emotions were correctly recognized in 67% of the cases. The happiness state was mostly confused with surprise; the sadness state was confused with the neutral tone; 75% of the people listening to the recordings said that it was difficult to identify the recorded emotions. Each voice recording has attached video information. The database also contains information about the profile of the speakers. To increase the consistency of the assessment of the emotion in the speech, the researchers used a questionnaire for the assessing persons; questions such as "how the emotions identification seems like", "what are the factors which bring to the correct identification of the emotions", etc. help fine-tuning the assessment.

The Spanish database [4] contains recordings with seven emotions: happiness, desire, fear, fury, surprise, sadness, and disgust; eight actors have pronounced the sentences. Every speaker recorded the every sentence for three times, with various levels of intensity of the emotions. The validation of the recorded emotions was made by a test based on the questions: "mark the emotion which was recognized in each recording", "mark the credibility level of the speaker", and "specify if the emotional state was recognized and at what level". The goal of the study was to describe a useful methodology in the validation of the simulated emotional states. The quoted researchers derived a set of rules describing the behavior of the main parameters of the emotional speech, in view of synthesizing emotional speech. The analyzed parameters are the fundamental frequency trajectory, time, and rhythm. They obtained the following characteristics of the emotional modulation [4]: i) for the joy state: "increase of the average tone, increase of the variability of the tone, quick modulations of the tone, [...] stable intensity, decrease [of] the silence time; [...]"; ii) for fury state: "variation of the emotional intonation structure, short number of pauses, increase of the intensity from beginning till the end, variation of timber, increase of the energy; [...]"; iii) sadness state: "decrease of the average tone, decrease of the variability of the tone, no inflexions of the intonation, decrease of the average intensity."

The above-quoted analysis leaves many unanswered questions on the variation of objective parameters, like formants, from one emotion to another. In the research reported here, we specifically address the characterization of emotions using the objective parameters for the states reported in [4]. We also contrast the characteristics of the voice for the above emotions with the normal (i.e., no emotion) speech. The comparison of our results with the results reported in [4] may help identify inter-language variations for the emotional speech.

## 3 Methodology, Database, and Analysis Tools

We place a high emphasis on the methodology of acquiring and analyzing emotional speech; this justifies the length of this section. In the first place, while we value the use of dramatic actors to produce emotional speech, we argue that speech by "normal" people should be the primary focus of a sound research in the field, if it were to obtain results for everyday applications. The database contains short sentences or phrases fragments, with different emotional states. We recall that the emotions investigated are sadness, joy, fury and neutral tone. The files are classified in class A (feminine voice) and class B (masculine voice). The speakers are persons aged between 25-35 years, born and educated (higher education) in the middle area of Moldova (Romania), without manifested pathologies. The recordings use a sampling frequency of 22050 Hz, 24 bits. Every speaker pronounced the sentence for three times, following the recording protocol. The persons have been previously informed about the objective of the project and they signed an informed consent in accordance with to the Protection of Human Subjects Protocol of the U.S. Food and Drug Administration and with Ethical Principles of the Acoustical Society of America.

The database contains two types of protocols, namely the recording technical protocol and the recording documentation one. The recording protocol contains information about the noise, the microphone used, the soundboard etc. The documentation protocol contains information on the speaker's profile – linguistic, ethnic, medical, educational, and professional information; for details, see [10].

The sentences are: 1. *Vine mama.* (Mother is coming) 2. *Cine a facut asta.* (Who did that?) 3. *Ai venit iar la mine.* (You came back to me) 4. *Aseară.* (Yesterday evening). The consistency of the emotional content in the speech recordings has been verified by the evaluators; the emotion confusion matrix has proved that all emotions are identified with a rate of more than 67% by the evaluators.

Today, there is no standard model for the emotional annotation process [13]. The sentences have been annotated using the Praat™ software (www.praat.org) at several levels: phoneme, syllable, word, and sentence. In this paper, the analysis refers only to the sentences "Aseară" and "Vine mama", as pronounced by eight persons, three times each, i.e., a total of 192 recordings. The values of the formants were determined using four tools: Praat™, Klatt analyzer™ (www.speech.cs.cmu.edu/comp.speech/ Section5/Synth/klatt.kpe80.html), GoldWave™ (www.goldwave.com), and Wasp™ (www.wasp.dk). Every tool produces a value for each formant. We compared the obtained values for the emotional states. In case where three out of four values are increased, the conclusion is that the values of the formants increase. Where there are clear discrepancies between subjects or between results obtained with different tools, we cannot draw any conclusion and we say that the values of the formants are fluctuant. There are cases when the tools (e.g. Praat) cannot determine the value of the formant (see the table 1).

**Table 1.** The values of F0 [Hz] obtained with several tools, for the one-word utterance „Aseară", for the states *happiness* and *sadness* (person # 77777m)

| Tools | Happiness | | |
|---|---|---|---|
| | F0 / a | F0 / ea | F0 / ă |
| GoldWave™ | 100-150 | 100-200 | 400-500 |
| Wasp™ | 117 | 166 | 467-490 |
| Klatt analyzer™ | 106 | 176 | 476 |
| Praat™ | 112 | 165 | 481 |

| Tools | Sadness | | |
|---|---|---|---|
| | F0 / a | F0 / ea | F0 / ă |
| GoldWave™ | 80-150 | 100-150 | 100-150 |
| Wasp™ | 78 | 77 | 74 |
| Klatt analyzer™ | 88 | 123 | 106 |
| Praat™ | undefined | 81 | 479 |

We have been confronted with several problems in the determination of the formants, namely with large disagreements between values provided by different applications. According to Klatt Analyzer™, the F1 formant for vowel *i* is "missing". Notice that, sometimes, it is difficult to determine visually the formants in the spectrograms using the GoldWave™. The difficulties are largely due to the imprecision of the definitions of the pitch and of the formants, especially for non-stationary signals. The nonlinear behavior of the phonatory organ, which is well documented in the medical literature as well as in the recent info-linguistic literature, [8], [9], determines a lack of significance of the parameters defined in the frame of the linear theory of speech analysis.

The differences in the results obtained with various tools reflect the theoretical limits of the formant parameters, as well as the capabilities of the various approximation methods used in the tools. These inconsistencies are one reason why the results we report should be considered preliminary, although we made every effort to obtain the results according to the best present knowledge.

With respect to other voice databases, we are insisting on some methodological aspects, namely the using of "natural voices", i.e. non-artist speakers which show "everyday emotions", and the use of inter-validated tools to determinate the formants and intra-validated (per speaker) emotional utterance.

## 4  Results of the Analysis of Speech with Manifest Emotions

The main results obtained in the analysis are listed in the Tables 2-5. The main rules we obtained, based on the results on the analyzed eight subjects, are listed at the end of this Section. In the tables, "-" means a decrease of the obtained values in first emotion, compared to the second emotional state; these couples of states can be happiness compared with sadness, fury compared with sadness, happiness with fury, and any emotion compared to normal tone. Also, "+" means an increase, while "±" means fluctuant, i.e. no conclusion can be derived. The "a", "ea", and "ă" represent the first vowel, the diphthong, respectively the last vowel in the word "aseară".

As a general conclusion, the states happiness and sadness, on one side, and fury from sadness, on the other side can be easily distinguished in all cases. It is more difficult to distinguish between happiness and fury. The utterance analyzed in the tables 2, 3, and 4 is "Aseară"; in table 5, the utterance is "Vine mama".

**Table 2.** The tendency for the F0, F1, F2 formants for the eight persons (- =increase, + =decrease, ± =fluctuant) for the states *happiness* and *sadness* versus *happiness* and *fury*

| Subject | F0 | | | F1 | | | F2 | | |
|---------|------|------|------|------|------|------|------|------|------|
|         | a | ea | ă | a | ea | ă | a | ea | Ă |
| 20048f | -/± | +/+ | +/± | ±/± | ±/+ | ±/- | +/+ | +/+ | +/+ |
| 01312f | ±/+ | +/+ | +/± | +/± | +/± | +/± | +/- | +/+ | +/+ |
| 55555f | ±/- | +/± | -/- | +/- | +/- | -/- | +/± | +/± | ±/± |
| 123456f | -/- | +/+ | +/+ | +/± | +/+ | +/+ | +/+ | +/+ | +/+ |
| 77777m | +/± | +/± | +/+ | +/+ | +/- | +/+ | ±/+ | +/+ | +/+ |
| 263315m | +/± | +/± | ±/± | +/- | +/± | -/- | +/+ | +/- | +/± |
| 14411f | ±/± | -/- | -/± | -/- | -/- | +/- | -/- | ±/± | -/- |
| 26653m | ±/± | -/± | ±/± | -/± | -/± | -/- | -/- | -/- | ±/- |
| General | ±/ ± | +/ ± | partly+/ ± | partly+/ ± | partly+/ ± | ±/ - | partly+/ partly+ | +/ partly+ | partly+/ partly + |

For the utterance "Aseară", the obtained values for the F0, F1, F2 formants of the diphthong "ea" for all the persons increase in the happiness compared with sadness state; the values for the F1, F2 formants of the first vowel "a" and for the F2 formant of the last vowel "ă" in the word "aseară" increase too (see table 2).

For the utterance "Aseară", the obtained values for the F0 formant of the diphthong "ea" for all the persons increase in fury state compared with neutral tone; the values for the F1 formant of the first vowel "a", of the diphthong "ea" in the word "aseară" increase too (table 3). Notice that, from table 2, no significant conclusions can be derived related to (joy; furry), except that the values for the formants are fluctuant (table 2); thus, the states (joy; furry) can not be reliably distinguished.

**Table 3.** The tendency for the F0, F1, F2 formants for the seven persons, for *fury* and *sadness*

| Subject | F0 | | | F1 | | | F2 | | |
|---|---|---|---|---|---|---|---|---|---|
| | a | ea | ă | a | ea | ă | a | ea | Ă |
| 20048f | - | ± | ± | ± | - | + | - | + | ± |
| 01312f | - | + | + | + | + | + | + | + | + |
| 55555f | + | + | - | + | + | ± | + | + | ± |
| 123456f | ± | ± | + | + | + | ± | - | + | + |
| 77777m | + | + | + | + | + | ± | ± | + | ± |
| 263315m | + | + | ± | + | + | ± | + | + | + |
| 14411f | + | - | ± | - | ± | + | - | ± | - |
| General | + | partly+ | ± | partly+ | partly+ | ± | ± | + | ± |

For the sentence "Aseară", the obtained values (table 4) for the F0 and F2 formants of the diphthong "ea" and, partly, of the last vowel "ă", for all the persons, increase in the couple (happiness; neutral tone); the values for the F1 and, partly, F2 formants of the first vowel "a" respectively of the diphthong "ea" increase too. Table 5 shows, for the sentence "Vine mama", the same problems as table 2, with regard to the fluctuations of the formants values.

**Table 4.** The tendencies for the F0, F1, F2 formants for the seven persons, for *happiness* and neutral tone

| Subject | F0 | | | F1 | | | F2 | | |
|---|---|---|---|---|---|---|---|---|---|
| | a | ea | ă | a | ea | ă | a | ea | ă |
| 20048f | - | + | + | + | ± | ± | + | + | + |
| 01312f | ± | + | ± | ± | + | + | - | + | + |
| 55555f | - | + | ± | + | + | ± | ± | ± | + |
| 123456f | - | + | + | + | + | ± | + | + | + |
| 77777m | + | + | + | + | ± | + | + | + | + |
| 263315m | ± | + | - | ± | ± | - | + | ± | + |
| 14411f | ± | - | ± | - | - | - | - | ± | - |
| General | ± | + | ± | + | ± | ± | + | + | + |

Notice in table 5, that the obtained values for the F0 formant of the vowels "e" in the word "vine", the first "a" (a1) and the last "a" (a2) in the word "mama", increase in the happiness state compared with sadness state, for all the persons.

Comparing the results obtained on both phrases, the additional rules are derived.

- The obtained value for the F0, F1, and F2 formants in the couple (happiness; sadness state) increase in both situations, but the increasing tendency is more evident in the case of the sentence "Vine mama". The values for the F0 formant increase for all subjects, while for the sentence "Aseară" there is a subject not obeying the rule (see tables 2 and 5).
- In the couple (happiness; neutral tone), the fluctuations of the values for the formants are similar for the "Aseară" and "Vine mama" sentences.
- In the couple (fury; sadness), the obtained results in the case of "Vine mama" are more fluctuant compared with the results for the sentence "Aseară".

- In the couple (fury; neutral tone), the obtained results in the case of "Aseară" are more fluctuant compared with the results for the sentence "Vine mama".

**Table 5.** The tendencies for the F0, F1, F2 formants for the six persons, for *happiness* and *fury* versus for *happiness* and *sadness*

| Subject | F0 | | | F1 | | | F2 | | |
|---|---|---|---|---|---|---|---|---|---|
| | e | a1 | a2 | e | a1 | a2 | e | a1 | a2 |
| 20048f | +/+ | +/+ | +/+ | -/+ | -/+ | +/+ | -/+ | -/- | +/+ |
| 01312f | +/+ | ±/+ | +/+ | -/- | +/+ | ±/+ | ±/± | +/+ | +/+ |
| 55555f | -/+ | ±/+ | ±/+ | -/± | ±/- | -/+ | -/± | ±/+ | ±/+ |
| 123456f | +/+ | +/+ | +/+ | +/+ | +/+ | ±/+ | +/+ | -/+ | ±/± |
| 77777m | ±/+ | ±/+ | +/+ | ±/+ | ±/+ | ±/+ | +/+ | +/+ | +/+ |
| 263315m | ±/+ | +/+ | ±/+ | ±/+ | +/+ | -/+ | +/+ | +/+ | +/+ |
| General | ±/+ | ±/+ | partly+/+ | ±/partly+ | ±/partly+ | ±/partly+ | ±/partly+ | ±/partly+ | partly+/partly+ |

The recognition systems of emotional states must be trained by speaker in order to distinguish the fury. The emotional intra-speaker states can be clearly distinguished, but we cannot specify the emotional inter-speaker states.

Comparing the prosody for happiness with that for fury, in the Romanian language, we noticed amazing similarities with the other European language. Even more remarkable, the other two languages where the same findings where reported on ambiguity between happiness and fury are of different roots than Romanian: while the Romanian is a Latin language with Slavic influence, the other two languages are German and Greek. We hypothesize that all Indo-European languages have a similar representation of emotion and the same resemblance between happiness and fury. This general conclusion on similarity of emotion representation in European languages, disregarding their particular roots, is preserved for all emotions.

## 5  Discussion and Conclusions

The reported research had the general but somewhat diffuse aim of determining whether there are prosodic features that support the interrelationship theory of language. The choice of the paralinguistic features in prosody, selected for the analysis, has been motivated by the analysis of manifest, intentional emotions.

For sentences uttered with manifest emotional load in the Romanian language, we found that most informative regarding the emotions is the change of the pitch. This conclusion is compatible with some findings reported for other languages. In contrast, we found that the accented vowels do not carry significantly more emotional information than the non-accented vowels; rather, the opposite is true. This conclusion is a departure from findings by other authors, for different languages. We need to further analyze this issue to determine its validity for a larger number of sentences and subjects. We also found that some higher formants, F1 and F2, in both accented and non-accented vowels, are also essential in conveying emotional information, at least from the perspective of the voice personality of some speakers.

Regarding the available speech analysis tools, we conclude that no tool provides irrefutable results. While we used four tools and compared the results, no one is significantly better than the others are. We have indicated a methodology to choose a stable section of the vowels for the analysis, to improve consistency in measurements, but even using this methodology, the lack of good formant extractors restricts today possibilities of obtaining high confidence in the results.

## Acknowledgments

Research partly performed in relation to the Romanian Academy "priority research" theme "Cognitive Systems" and to a CEEX grant.

## References

1. Buluti, M., Narayanan, S.S., Syrdal, A.K.: Expressive speech synthesis using a concatenative synthesizer (accessed May 9, 2006),
   http://sail.usc.edu/publications/BulutNarayananSyrdal.pdf
2. Douglas-Cowie, E., Campbell, N., Cowie, R., Roach, P.: Towards a new generation of databases. Speech Communication 40, 33–60 (2003)
3. https://nats-www.informatik.uni-hamburg.de/intern/proceedings/2004/LREC/pdf/41.pdf (accessed May 9, 2006)
4. Iriondo, I., et al.: Validation of an acoustical modelling of emotional expression in Spanish using speech synthesis techniques (accessed July 01, 2006),
   http://serpens.salleurl.edu/intranet/pdf/239.pdf
5. Engberg, I.S., Hansen, A.V.: Documentation of the Danish Emotional Speech Database (accessed July 01, 2006), http://kom.aau.dk/~tb/speech/Emotions/des.pdf
6. http://pascal.kgw.tu-berlin.de/emodb/ (accessed July 01, 2006)
7. Teodorescu, H.N.: A proposed theory in prosody generation and perception: the multi-dimensional contextual integration principle of prosody. In: Burileanu, C. (Coordinator), Trends in Speech Technology, pp. 109–118. Romanian Academy Publishing House, Bucharest, Romania (2005)
8. Loscos, A., Bonada, J.: Emulating rough and growl voice in spectral domain. In: Proc. of the 7th Int. Conference on Digital Audio Effects (DAFX-04), Naples, Italy, October 5-8, 2004 (accessed November 12, 2006),
   http://www.iua.upf.es/mtg/publications/DAFX04-aloscos.pdf
9. Sun, X.: Pitch determination and voice quality analysis using subharmonic-to-harmonic ratio (accessed November 12, 2006),
   http://www.ling.northwestern.edu/~jbp/sun/sun02pitch.pdf
10. http://www.etc.tuiasi.ro/sibm/romanian_spoken_language/index.htm
11. Teodorescu, H.N., Feraru, M., Trandabat, D.: Studies on the Prosody of the Romanian Language: The Emotional Prosody and the Prosody of Double-Subject Sentences. In: Burileanu, C., Teodorescu, H.N. (eds.) Advances in Spoken Language Technology, Editura Academiei Române, pp. 171–182 (2007) ISBN 978-973-27-1516-1
12. Teodorescu, H.N., Feraru, M.: Trandabat, Nonlinear assessment of the professional voice pleasantness, Biosignal, June 28-30, 2006, Brno, Czech Republic, pp. 63–66 (2006)
13. Ordelman, R., Poel, M., Heylen, D.: Emotion annotation in the AMI Project. In: Extended Abstract–Humaine Workshop, Paris, March 2005 (accessed May 9, 2006),
   http://wwwhome.cs.utwente.nl/~heylen/Publicaties/emo-anno-REV.pdf

# Speech Recognition Supported by Prosodic Information for Fixed Stress Languages

György Szaszák and Klára Vicsi

Budapest University of Technology and Economics
Dept. of Telecommunications and Media Informatics
Budapest, Hungary
{szaszak, vicsi}@tmit.bme.hu
http://alpha.ttt.bme.hu/speech/

**Abstract.** In our paper we examine the usage prosodic features in speech recognition, with a special attention payed to agglutinating and fixed stress languages. The used prosodic features, acoustic-prosodic pre-processing, and segmentation in terms of prosodic units are presented in details. We use the expression "prosodic unit" in order to make a difference from prosodic phrases, which are longer. We trained a HMM-based prosodic segmenter reliing on fundamental frequency and intensity of speech. The output of the prosodic segmenter is used for N-best lattice rescoring in parallel with a simplified bigram language model in a continuous speech recognizer, in order to improve speech recognition performance. Experiments for Hungarian language show a WER reduction of about 4% using a simple lattice rescoring.

## 1 Introduction

Prosodic features are an integral part of every spoken language utterance. They provide cues for the listener for the decoding of syntactic and semantic structure of the message. Moreover, they also contribute to the expression of speaker's emotions and communicative intent. On different linguistic levels, accents, breaks, rhythm, speaking rate, etc. play an important role in syntactic classification of each element of a message. In addition to this, intonation carries information also about sentence modality, which might be very important in several speech applications (like information retreival systems for example, where it is crucial in terms of correct speech understanding to be able to differentiate questions from statements). This information contained in prosody can and should be exploited in automatic speech recognition in order to improve speech recognition and understanding performance by decoding information that is only carried by prosody and also by ensuring a redundant parameter sequence for the whole speech decoding process itself.

Hungarian, as a language of the Finno-Ugrian family, is a highly agglutinating language (a noun might even have more than 1000 different forms) characterized by a relatively free word order. Syntactic relations are expressed by case endings (the number of possible case endings is also very high, Hungarian distinguishes about 18 cases). The agglutinating property also results in a relatively higher average word length. In

V. Matoušek and P. Mautner (Eds.): TSD 2007, LNAI 4629, pp. 262–269, 2007.

addition to this, Hungarian is a fixed stress language, stress falls always on the first syllable of a word. Due to this "atypical" organization of such languages, standard methods developed mainly for English speech recognition are not directly applicable for highly agglutinating languages. Well-known problems are the larger size of vocabulary and the radically increased complexity of statistical language models - characteristics of all such languages. We beleive that prosody might help to overcome a part of these difficulties.

Using prosodic features in automatic speech recognition is not a trivial task, however, several attempts prooved to be succesful in this domain. Veilleux and Ostendorf [1] presented an N-best rescoring algorithm based on prosodic features and they have significantly improved hypothesis ranking performance. Kompe et al. presented a similar work for German language in [2]. Gallwitz et al. described a method for integrated word and prosodic phrase boundary recognition [3].

This paper presents how prosodic features can be used to improve speech recognition of agglutinating and fixed stress languages. Finally, we summarize the results of such an experiment in which a prosodic segmenter was integrated into a Hungarian speech recognizer.

## 2   Exploiting Prosodic Information

### 2.1   Acoustic Prosodic Features

For representation of prosody, fundamental frequency ($F_0$), energy level and time course were measured. $F_0$ and energy, measured at the middle of vowels of the utterance, and the duration of the vowels in the syllable (nucleus duration) are presented in a Hungarian sentence in Fig 1. In this example, the peaks of energy and fundamental frequency clearly represent the first syllables of the words. Of course, a strict word-level segmentation based on prosody is not always feasible, since usually not all words have accent within a sentence. (Such a word level segmentation algorithm was described in [4].) Fig. 1 illustrates that syllable prominence in Hungarian is governed mainly by intensity and $F_0$. Syllable length, it was found, is not greatly influenced by the stress [4].

### 2.2   Acoustic Prosodic Pre-processing

The extraction of prosodic information is performed using the Snack package of KTH [5]. A window of 25 $ms$ is used for the extraction of both $F_0$ (AMDF method) and intensity. The frame rate was 10 $ms$. The $F_0$ contour is firstly filtered with our anti-octave jump tool to eliminate frequency halving or doubling. This is followed by a smoothing with a 5 point mean filter (5 point means 50 ms) and then the log values of $F_0$ are taken which are linearly interpolated. The interpolation does not affect pauses in $F_0$ longer than 250 $ms$. Interpolation is omitted if the initial value of $F_0$ after the $F_0$-gap is higher than a threshold value. This threshold value depends on the last measured $F_0$ values and equals the 110 % of the average $F_0$ value of the three last voiced frames before the gap (unvoiced period). The intensity contour is simply mean filtered, using again a 5 point filter. Hereafter, delta and acceleration coefficients are appended to both $F_0$ and intensity streams. These coefficients are computed with a regression-based formula. The

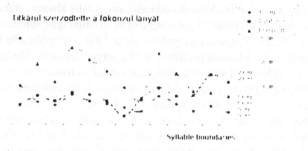

Titkárul szerződtette a főkonzul lányát

Syllable boundaries.

**Fig. 1.** $F_0$ and intensity levels measured at the middle of vowels and duration of the vowels in the syllables in a Hungarian sentence 'Titkárul szerződtette a főkonzul lányát.' The syllable sequence is presented on the X axis.

regression is performed in 3 different steps with increasing regression window length: firstly with a window of $\pm 10$ frames, secondly with a window of $\pm 25$ frames and finally, a window of $\pm 50$ frames is used. This means that the final feature vector consists of 14 elements (original $F_0$ and intensity data + 3-3 delta + 3-3 acceleration components for both of them).

### 2.3    Speech Material

The speech material was chosen from the BABEL [6] continuous read speech database-and consisted of 1600 sentencesfrom 22 speakers. The database was segmented in terms of prosodic features by an expert, reliing on waveform, $F_0$, intensity and subjective judgement after listening to the utterance. During the segmentation, prosodically well marked units were looked for, which were found to be tipically short speech segments consisting of one or some more word(s). Forward in the article, these units will be called prosodic units in order to differentiate them from prosodic phrases, especially because prosodic units are considerably shorter than prosodic phrases. Prosodic units in fixed stress, agglutinating Hungarian that has free word order usually consist of an initial stressed word followed by other, non or slightly stressed words. Labeling was based on the following considerations: the typical $F_0$ - and often also the intensity - contour of such units is a less or more strong accent, followed by a slowly or suddenly falling $F_0$ (intensity). Of course, higher lingustic levels and their associated prosodic features also influence the prosody as in case of some questions and phrase endings that do not co-occur with sentence endings a slowly rising contour is the most likely. In this latter case, intonation often suppresses normal (high) accents which translates into an inversed (low) accent. This means that in this case the first syllable of a word has the lowest $F_0$ / intensity value. However, all labels were placed in a manner that they coincide with word boundaries. The used 7 labels are presented in Table 1.

### 2.4    Automatic Prosodic Segmentation

In order to carry out prosodic segmentation, a small HMM set was trained for all the 7 prosodic unit types in Table 1. The training methodology of the prosodic segmenter is

**Table 1.** Prosodic units used for segmentation

| Prosodic label | Description |
|:---:|:---:|
| me | Sentence onset unit |
| fe | Strongly stressed unit |
| fs | Stressed unit |
| mv | Low sentence ending |
| fv | High sentence or phrase ending |
| s | Slowly falling $F_0$ / intensity (after fv) |
| sil | Silence |

identical with the one used in [4], the acoustic pre-processing is done as shown in section 2.2. Here, the recognition of prosodic units is based on Viterbi decoding, similar to the case of speech recognition. However, as prosodic features are processed as vectors of 14 elements and a strict prosodic grammar is used, the prosodic decoding process is very fast. During prosodic recognition, a sophisticated sentence model is used as a constraint. This supposes that each sentence is built from one or more prosodic phrase and constituting prosodic phrases are composed from prosodic units. The sentence model requires that each sentence begin by a sentence onset unit (*me*) and end by either a low sentence ending unit (*mv*) or a phrase ending unit (*fv*). Stressed units (*fe* and *fs*) are allowed to appear within a sentence. The *fe* symbol refers to a stronger accent typically found at the beginning of a new phrase within a sentence. The slowly falling unit (*s*) is allowed optionally, but only immediately after a phrase ending (*fv*). Between all sentences, a silence (*sil*) is supposed.

Since prosodic features (often called also supra-segmental feautures) contain information on higher level than basic recognition units (ie. phonemes), it is worth blurring the boundaries of the prosodic units. This means that the prosodic unit boundaries predicted by the prosodic module are transformed to a probability density function that predicts rather the likelihood of prosodic unit boundaries than their exact location in time. To implement this in an easy way, a cosine function in the $[-\pi, \pi]$ interval is matched against the predicted prosodic unit boundary $\pm \Delta T$ interval. Thus, within the interval of $[t_B - \Delta T, t_B + \Delta T]$ the $L_B$ log likelihood of prosodic unit boundary is defined as:

$$L_B(t) = C * cos(\frac{\pi}{2\Delta T}t) + C \qquad (1)$$

where $t_B$ is the location of the predicted prosodic unit boundary on the time axis, $t \epsilon [t_B - \Delta T, t_B + \Delta T]$, $C$ is a constant. Otherwise the value of $L_B$ is regarded to be 0. During the experiments presented in section 4, $\Delta T$ was equal to 10 frames (100 ms).

## 3    Speech Recognition Using Prosody

In our speech recognizer with integrated prosodic segmenter module, speech recognition is done in two subsequent stages. In the first stage, an N-best lattice is generated. The output of the prosodic segmenter is then combined with this lattice using a rescoring algorithm presented in section 3.2. The rescored lattice is then used

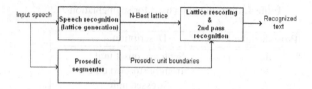

**Fig. 2.** Speech decoding process using prosodic features

in the second pass recognition that yields the final recognition results. This process is illustrated in Fig. 2.

### 3.1 First Pass Recognition

In the first pass of speech decoding Viterbi algorithm-based standard lattice generation and prosodic segmentation is done in parallel. Currently, the speech recognizer (lattice generator) and the prosodic segmenter modules work independently, hence no information about word or phone alignment hypotheses are transferred from the speech recognizer to the prosodic segmenter module.

In this stage, a special bigram language model is used for recognition. In order to investigate the effect of prosodic information, bigram weights can even be set to an equal value. In this case, the bigram model will contain only binary information whether a word sequence is allowed or not. In this way, a part of the information contained in a statistical bigram is supposed to be replaced by prosodic information. This approach also allows to partly overcome diffculties in bigram language model generation and usage in case of highly agglutinating languages, hence for large vocabulary applications for such languages, generating an N-gram language model is often blocked by data sparsity problem.

### 3.2 Lattice Rescoring

The rescoring of the N-best lattice created in the first pass speech recognition process is based on the output of the prosodic segmenter. The principle of rescoring is to remunerate word or word-chain candidates whose boundaries match the prosodic segmentation and to punish those which contain a prosodically predicted unit boundary within themselves. To do this, all arcs are collected from the word hypotheses graph (arcs correspond to distinct words in such graphs). Then, for each arc, two additional scores are calculated in addition to the $Sc_{original}$ original score in the lattice given by the speech recognizer. All scores used are interpreted as logaritmic scores.

The renumerating score ($Sc_{renum}$) is computed based on the degree of co-occurence of prosodic unit boundaries and the initial and terminal nodes of the arc:

$$Sc_{renum} = w_a L_B(t_{start}) + w_b L_B(t_{end}) \tag{2}$$

where $L_B$ is computed from (1), $t_{start}$ and $t_{end}$ refer to the start ans end nodes (times) of the arc, $w_a$ and $w_b$ are constant weights. For simplicity, $w_a = w_b = 0.5$ in (2). A similar punishment ($Sc_{punish}$) score is computed for in-word frames of each arc:

$$Sc_{punish} = \sum_{i=k+1}^{N-k-1} L_B(i) \tag{3}$$

where $L_B$ is got from (1), $N$ is the number of frames in the arc, $k$ is an integer $2k < N, k = \Delta T$. The reason for the skip in (3) defined by $k$ is to allow prosodic unit boundaries to be a bit shifted. The highest scores will be added to arcs which fit the best the prosodic segmentation. The new score ($Sc_{rescored}$) in the rescored lattice is given by:

$$Sc_{rescored} = w_o Sc_{original} + w_c(Sc_{renum} - Sc_{punish}) \tag{4}$$

where $w_o$ and $w_c$ are the corresponding weights. In experiments shown in section 4, the weights used were $w_o = 1$ and $w_c = 2.5$.

### 3.3  Second Pass Recognition

At this final step, the rescored lattice is used as a grammar for speech recognition. Recognition at this stage is very fast, because possible hypotheses are evaluated by simply parsing the word graph defined by the lattice. The final recognition result is output at the end of the second pass.

## 4  Experimental Tests

For experimental tests, a medical application for Hungarian language [7] was selected which ensures the transcription of medical reports in radiology (abdominal ultrasonography). This task has a relatively small vocabulary (5000 words). The speech recognizer itself was a state of the art speech recognizer trained by the HTK toolkit [8]. This recognizer uses phoneme models (37 models for Hungarian), speech is pre-processed in the "standard" way using 39 coefficients (12 MFCC + energy + 1st and 2nd order deltas) and 10 $ms$ frame rate, the output distribution is described by 32 Gaussians. The recognizer was trained with acoustic data from the Hungarian Reference Database [7] using approx. 8 hours of speech. In addition to this baseline recognizer, the system was extended by the prosodic segmenter unit described in 2.4, as presented in section 3. During experiments, only a simplified bigram language model was used. This means that the bigram model contained only binary information whether a word sequence is allowed or not. (For reasons explained in section 3.1). The word insertion log probability was set to -25, the prosodic score scale factor was 2.5 (see equation (4)). These settings were obtained after empirical optimization.

The performance (ratio of correctly recognized words and WER reduction) of the extended system was compared to the baseline recognizer. Results for 6 selected test sequences are presented in Table 2. The selection of these presented utterances was representative. As it can be seen in Table 2, rescoring of the N-best lattice based on the output of the prosodic segmenter usually improves the performance of the recognition system. However, in some cases (for example ID 16 in Table 2), performance is worse after rescoring (WER reduction is less than 0). On the other hand, a sometimes very

**Table 2.** Correctly recognized words in case of the baseline speech recognizer ("Baseline") and of the recognizer extended by the prosodic module ("Prosodic") and WER reduction for 6 representatively selected medical reports

| Speaker ID | Words correct [%] | | WER reduction [%] |
|---|---|---|---|
| | Baseline | Prosodic | |
| 03 | 71.2 | 78.9 | 10.9 |
| 07 | 78.8 | 80.6 | 3.6 |
| 08 | 84.6 | 84.6 | 0.0 |
| 10 | 70.8 | 72.2 | 2.0 |
| 16 | 68.3 | 66.7 | -2.4 |
| 19 | 83.8 | 90.5 | 8.1 |
| All 20 | 75.99 | 78.89 | 3.82 |

high WER reduction was reached (for example ID 03). In this case we usually found utterances of proper and careful - but of course natural - prononciation, also in terms of prosodic constituents. The overall results for all 20 test medical reports show total WER reduction of 3.82%.

Test utterances which show a slightly higher WER after prosodic rescoring of the word hypotheses graph were further analysed in order to reveal the reason for the increased WER values. It was found, that the higher WER was caused by the errors of prosodic segmentation, in which case prosodic unit boundaries were placed in the middle of words. On the other hand, omission of a prosodic unit boundary by the prosodic segmenter did not influence system performance. If the correctness of prosodic segmentation is defined as the ratio of the number of correctly placed prosodic unit boundaries and of the number of all placed prosodic unit boundaries, this ratio was found to be 79.7% in case of medical reports. A prosodic unit boundary was regarded as correct if it did not deviate more than $\Delta T = 100ms$ from real prosodic unit boundary placed by human expert. In our opinion, the reason for a considerable part of prosodic segmentation errors was the supra-segmental nature of prosodic features (since the typical duration of a phoneme is about 80-100 ms in Hungarian, and we had to use a severe lattice rescoring in terms of the allowed prosodic unit boundary deviation defined by equation (1)). By allowing communication between prosodic segmenter and first pass speech recognizer modules (ie. by using phoneme - and hence syllable - alignment information in the prosodic segmenter), this error can be further minimized.

## 5   Conclusions

In our paper we presented the use of a prosodic segmenter in speech recognition providing a more detailed prosodic segmentation than detection of phrase boundaries. The output of the prosodic segmenter was used to rescore N-best lattices in parallel with a simplified language model. Obtained results show a WER reduction of 3.8% for a Hungarian medical speech recognition task. The detailed prosodic segmentation also can provide a syntactic analysis of spoken utterances in natural language processing, however, this issue has not been investigated yet in the paper. The approach is adaptable to all fixed stress languages or means an alternative or additional information source for language modeling in case of agglutinating languages, where N-gram language model generation and usage reveal several difficulties.

# References

1. Veilleux, N.M., Ostendorf, M.: Prosody/parse scoring and its application in ATIS. In: Human Language and Technology. In: Proceedings of the ARPA workshop, Plainsboro, pp. 335–340 (1993)
2. Kompe, R., Kiessling, A., Niemann, H., Nöth, H., Schukat-Talamazzini, E.G., Zottman, A., Batliner, A.: Prosodic scoring of word hypothesis graphs. In: Procs. of the European Conference on Speech Communication and Technology. Madrid, pp. 1333–1336 (1995)
3. Gallwitz, F., Niemann, H., Nöth, E., Warnke, V.: Integrated recognition of words and prosodic phrase boundaries. In: Speech Communication, vol. 36, pp. 81–95 (2002)
4. Vicsi, K., Szaszák, G.Y.: Automatic Segmentation of Continuous Speech on Word Level Based on Supra-segmental Features. International Journal of Speech Technology 8(4), 363–370 (2005)
5. Sjölander, K., Beskow, J.: Wavesurfer - an open source speech tool. In: Procs. ICSLP-2000. Beijing, China, vol. 4. pp. 464–467 (2000)
6. Roach, P.: BABEL: An Eastern European multi-language database. In: Procs. of ICSLP-1996. Philadelphia (1996)
7. Hungarian Reference Database:
   http://alpha.tmit.bme.hu/speech/hdbMRBA.php
8. Young, S.J.: The HTK Hidden Markov Model Toolkit: Design and Philosophy. Technical Report TR.153, Department of Engineering. Cambridge University, UK (1993)

# TRAP-Based Techniques for Recognition of Noisy Speech*

František Grézl and Jan Černocký

Speech@FIT, Brno University of Technology, Czech Republic
{grezl,cernocky}@fit.vutbr.cz

**Abstract.** This paper presents a systematic study of performance of TempoRAl Patterns (TRAP) based features and their proposed modifications and combinations for speech recognition in noisy environment. The experimental results are obtained on AURORA 2 database with clean training data. We observed large dependency of performance of different TRAP modifications on noise level. Earlier proposed TRAP system modifications help in clean conditions but degrade the system performance in presence of noise. The combination techniques on the other hand can bring large improvement in case of weak noise and degrade only slightly for strong noise cases. The vector concatenation combination technique is improving the system performance up to strong noise.

## 1 Introduction

In recent years, Temporal Pattern (TRAP) based feature extraction has become popular and especially systems combining TRAP with conventional parameters such as MFCC or PLP exhibit good performances [1].

Unlike mostly used features which are based on full spectrum with short time context, temporal pattern (TRAP) features are based on narrow band spectrum with long time context. These features are derived from temporal trajectory of spectral energy in frequency bands in two steps: First, critical band trajectory is turned into band-conditioned class posteriors estimates using nonlinear transformations — neural net. Second, overall class posteriors estimates are obtained by merging all band-conditioned posteriors. The merging is done by another neural net. Overall class posteriors transformed into form required by a standard GMM-HMM decoder are called TRAP (or TRAP-based) features. The fact that the first step of TRAP processing happens in frequency bands should make TRAP-based features robust in frequency selective noise. In [2], TRAP features were tested with the Qualcomm-ICSI-OGI features [3] on the AURORA 2 database.

Since, numerous modifications of TRAP features were proposed and tested. The concatenation of several critical bands was tested in [4]. In addition, the Principal Component Analysis (PCA) was performed on concatenated vectors and the resulting bases

* This work was partly supported by European projects Caretaker (FP6-027231), by Grant Agency of Czech Republic under project No. 102/05/0278 and by Czech Ministry of Education under project No. MSM0021630528. The hardware used in this work was partially provided by CESNET under projects No. 119/2004, No. 162/2005 and No. 201/2006.

V. Matoušek and P. Mautner (Eds.): TSD 2007, LNAI 4629, pp. 270–277, 2007.
© Springer-Verlag Berlin Heidelberg 2007

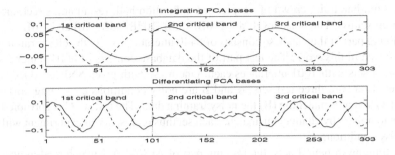

**Fig. 1.** Integrating and differentiating PCA bases

were used for dimensionality reduction. It was observed that the PCA bases have similar shapes as the Discrete Cosine Transform (DCT) bases. Further, the majority of the PCA bases for three concatenated critical band energy trajectories have shapes which perform integration and differentiation of individual critical bands (see Fig. 1). In bases performing integration, all parts corresponding to individual bands have similar shape. In bases performing differentiation, the part corresponding to the middle band is close to zero and the shapes for border bands have opposite phases.

In [5] integrating and differentiating of critical bands is applied directly on the critical band spectrogram prior to the temporal pattern selection, creating so called modified temporal pattern (MTRAP). It was shown that one modification (integration or differentiation) itself does not achieve the performance of the basic TRAP system. The combination of two MTRAP systems or MTRAP and basic TRAP system is necessary. Possible combinations are examined in [6] showing the effectiveness of simple vector concatenation technique where temporal patterns from differently modified critical band spectrograms are concatenated on the input of band-conditioned neural net. However, all results are obtained on a small task (digits) on clean telephone speech and the robustness of the proposed improvements to noise was not verified.

We made efforts to evaluate the proposed techniques on noisy speech from Aurora 2 database while having only clean training data to see whether these techniques are beneficial also in noisy conditions. The description of experimental setup is given in section 2. The following sections then give the overview of used techniques and results. Section 3 introduces TRAP-based feature extraction, section 4 summarizes the multiband system and system with critical band spectrogram modification, and section 5 describes the combinations of TRAP systems. Conclusions are given in section 6.

## 2  Experimental Setup

The AURORA 2 database was designed to evaluate speech recognition algorithms in noisy conditions. The framework was prepared as contribution to the ETSI STQ-AURORA DSR Working Group [7]. The database consists of connected digits task (11 words) spoken by American English speakers. A selection of 8 different real-world noises has been added to the speech with different signal to noise ratio (SNR). The noises are suburban train, crowd of people, car, exhibition hall, restaurant, street, airport and train station. The noise levels are 20dB, 15dB, 10dB, 5dB, 0dB and -5dB.

The training part of the database consists of 8440 utterances. Only the clean training scenario is used in this work. The test part of the database consists of 4004 sentences divided into 4 sets with 1001 utterances each. One noise with given SNR is added to each subset. There are three test sets: **A** and **B** are noisy conditions containing noises matching (A) and non-matching (B) the noisy training data. The test set **C** is corrupted, in addition to noises, by channel mismatch. Each set thus represents an experiment with unique noisy conditions.

For the training of neural nets, the training part of AURORA 2 was forced-aligned using models trained on OGI-Stories database [8]. OGI-Stories were also added to the neural net training set to enrich the phoneme context (in digits, phonemes are occurring in the same context). The target 21 phonemes are those which occur in digits utterances including silence. Other phonemes are not used for training but they create context in TRAP vectors.

The reference recognizer shipped with AURORA was used. The number of results per experiment is given by number of SNR and number of noises. To be able to compare the results from different experiments, we report an average word error rate (WER) for given SNR.

## 3    TRAP-Based Feature Extraction

To obtain the critical band energy trajectory, we have to get the critical-band spectrogram first. This is done by segmentation of the speech into 25 ms frames spaced by 10 ms. Then the power spectrum is computed from each speech frame and integrated by 15 Bark-scaled trapezoidal filters. Finally, logarithm is taken.

In such critical-band spectrogram, the TRAP vector is selected as 101 consecutive frames (center frame +/− 50 frames context) in a given frequency band. The TRAP vector is mean and variance normalized and weighted by Hamming window. In case the PCA or DCT dimensionality reduction is desired, matrix multiplication follows. The resulting vector is then converted into band-conditioned class posteriors by a band-specific **band probability estimator** – a three layer neural net trained to classify the input vector in one of the 21 phonetic classes. All band-conditioned posterior estimates are then concatenated in one vector. Before presenting this vector to the **merger probability estimator** to obtain the overall class posteriors, negative logarithm is taken. Merger probability estimator is also a three layer neural net. Target classes are the same 21 phonemes as for band probability estimator. The block diagram of the TRAP system which converts the critical-band spectrogram to phoneme posteriors estimates is shown in Fig. 2.

The TRAP-based features are obtained from phoneme posteriors obtained by taking the logarithm and PCA decorrelation. This features form an input to the AURORA 2 GMM-HMM recognizer.

The resulting features are denoted *basic TRAP* and obtained results are shown in Tab. 1. There is no dimensionality reduction (matrix multiplication) of TRAP vector in basic TRAP features.

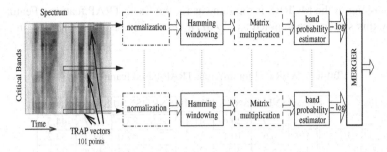

**Fig. 2.** TRAP system for converting the critical band spectrogram to phoneme posteriors. After post-processing, the resulting features are used in standard GMM-HMM recognizer.

## 4 Modifications

### 4.1 Multi-band TRAP System

The multi-band system was proposed in [4]. Three adjacent bands are used as input to the band probability estimator. Frequency shift between two band probability estimators input is one band. Features obtained by this system are denoted as *3b TRAP* and the results are shown on the 3rd line of Tab. 1.

In [4], it was also shown that the dimensionality reduction of concatenated TRAP vectors can further improve the recognition accuracy. We used the neural net training data to compute the PCA bases. The input 303 point long vector was reduced to 150 points. Features obtained by this system are denoted as *3b TRAP + PCA* and the results are shown on the 4th line of Tab. 1.

### 4.2 Critical-Band Spectrogram Modification

According to study presented in [4], it is possible to replace the PCA bases by bases created by concatenating the DCT bases in integrating or differentiating manner (see Fig. 1). In [5], this integration and differentiation was applied directly on the critical band spectrogram using so called *modifying operators*. It was also shown, that replacing the system with integration of critical band spectrogram by the *Basic TRAP system* does not hurt the system performance but rather brings slight improvement. Therefore, we will stick with basic TRAP and differentiation of the critical band spectrogram systems.

The frequency differentiating (FD) operator is a column vector $FD = [1, 0, -1]^T$. The modified critical band spectrogram (MCRBS) is computed as projection of the operator on the original spectrum (CRBS). One point of MCRBS in given time $t$ and in given frequency band $f$ is computed as

$$MCRBS(t, f) = \sum_{i=f-f_c}^{f+f_c} FD(t, i) \times CRBS(i) \qquad (1)$$

where $f_c$ is the frequency context of the FD operator (in our case 1).

The processing of the MCRBS is the same as for the Basic TRAP features. Features obtained by this system are denoted as *FD MTRAP* and the results are shown on last line of Tab. 1.

Table 1. WER [%] for different TRAP-based features

| features | clean | SNR20 | SNR15 | SNR10 | SNR5 | SNR0 | SNR-5 | average |
|---|---|---|---|---|---|---|---|---|
| MFCC | 0.8 | 7.9 | 20.4 | 41.1 | 64.8 | 83.9 | 92.7 | 44.5 |
| basic TRAP | 1.9 | 6.5 | 10.7 | 19.2 | 37.8 | 69.0 | 87.7 | 33.2 |
| 3b TRAP | 1.8 | 5.5 | 9.8 | 20.3 | 42.5 | 74.3 | 89.4 | 34.8 |
| 3b TRAP + PCA | 1.4 | 5.7 | 12.7 | 32.3 | 69.0 | 88.5 | 91.7 | 43.1 |
| FD MTRAP | 2.1 | 7.6 | 15.2 | 33.1 | 63.1 | 84.5 | 90.8 | 42.3 |

Table 2. WER [%] of different Basic TRAP and FD MTRAP system combinations

| combination | clean | SNR20 | SNR15 | SNR10 | SNR5 | SNR0 | SNR-5 | average |
|---|---|---|---|---|---|---|---|---|
| lin ave | 1.5 | 4.4 | 8.9 | 21.7 | 50.7 | 80.5 | 89.8 | 36.8 |
| log ave | 1.3 | 3.8 | 8.2 | 20.6 | 49.4 | 80.4 | 89.6 | 36.2 |
| inv ent th = 1.0 | 1.4 | 3.9 | 7.9 | 18.6 | 44.6 | 78.0 | 89.6 | 34.9 |
| inv ent th = 2.5 | 1.3 | 3.6 | 7.0 | 16.3 | 38.2 | 72.4 | 87.9 | 32.4 |
| vector concat | 1.6 | 4.2 | 7.6 | 15.6 | 36.4 | 70.9 | 88.5 | 32.1 |

## 5    System Combinations

Combination of TRAP system at different levels is examined in [6]. Simple multi-stream combination and vector concatenation techniques were giving the best results. Here, we apply the concatenation techniques on *Basic TRAP* and *FD MTRAP* systems.

### 5.1    Multi-stream Combination

This combination technique combines the final probability estimations from different systems – i.e. outputs of merger probability estimators. The outputs from the TRAP systems are posterior probabilities $P(q_k|\mathbf{x}_t, \theta)$, where the $q_k$ is the $k^{th}$ output class of total $K$ classes, $\mathbf{x}_t$ is the input feature vector at time $t$ and $\theta$ is set of neural net parameters. The systems have the same targets, thus we can use techniques for posterior probability combination. The resulting posterior probability vector for combining $I$ systems will be $\hat{P}(q_k|\mathbf{X}_t, \Theta)$ where $\mathbf{X}_t$ is the set of all input vectors $\mathbf{X}_t = \{\mathbf{x}_t^1, \mathbf{x}_t^2, \dots \mathbf{x}_t^I\}$ and $\Theta = \{\theta^1, \theta^2, \dots \theta^I\}$ is the set of all parameters.

First we performed an **average of output probabilities**, which simply averages the outputs belonging to the same class. This combination of *Basic TRAP* and *FD MTRAP* is denoted as *lin ave* and results are shown on the first line in Tab. 2.

Another possibility is to take the **average of logarithm of output probabilities**, which is equivalent to geometric averaging of the linear posteriors. This multi-stream system combination is denoted as *log ave* and results are shown on the second line in Tab. 2.

Finally, we have explored **entropy based combination** inspired by [9], which is actually a weighted average of output probabilities where weights are estimated for each frame individually.

The entropy of $i^{th}$ system outputs at given time $t$:

$$h_t^i = -\sum_{k=1}^{K} P(q_k|\mathbf{x}_t^i, \theta^i) log_2(P(q_k|\mathbf{x}_t^i, \theta^i)) \qquad (2)$$

can be used as confidence measure of this system. This information is used for weighting the outputs of different systems. The weight for $i^{th}$ system at time $t$ is

$$w_t^i = \frac{1/h_t^i}{\sum_{i=1}^{I} 1/h_t^i} \qquad (3)$$

High entropy means that the posterior probabilities are approaching equal probability for all classes. The stream with high entropy has less discrimination, therefore outputs of such system should be weighted less. The stream with low entropy has higher discrimination and its outputs should be weighted more. This weighting scheme prefers the input stream which has higher disriminability, i.e. is more noise robust.

**Inverse entropy weighting with static threshold** was used in our experiments. If the system entropy at given time is higher than a threshold, the entropy is set to a large value:

$$\tilde{h}_t = \begin{cases} 10000 & : & h_t^i > th \\ h_t^i & : & h_t^i \le th \end{cases} \qquad (4)$$

If both systems have entropy bigger than threshold $th$, both obtain small (but the same) weight and the output will be equal to the average of both systems. If systems have small entropy $< th$, the output will be given by the weighted average. If only one system has entropy $> th$, this output will be suppressed by the small weight and the output will be given by the system with entropy $< th$. We have tuned the threshold value and best results were obtained with $th = 3.5$.

This combination of *Basic TRAP* and *FD MTRAP* is denoted as *inv ent th = val* where *val* is the threshold value. The results are shown in Tab. 2.

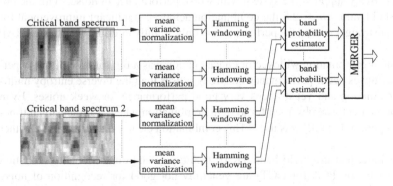

**Fig. 3.** Block diagram of system with vector concatenation

## 5.2   Vector Concatenation

The simple way of combination different feature vectors is to directly concatenate them. The concatenation of the TRAP vectors obtained from different critical band spectrograms is done on the input of band probability estimator. It means that all processing (normalization, windowing, DCT) is done for each vector independently. Fig. 3 shows the processing for system with vector concatenation. The results for vector concatenation system combination are denoted *vector concat* and are shown on last line of Tab. 2.

# 6   Conclusions and Discussions

The results for standard MFCC features are given for comparison on first line in Tab. 1. MFCC features gain better performance in clean conditions but are more vulnerable in noisy conditions compared to basic TRAP features. Basic TRAP are set as a baseline we compare the other TRAP-based techniques to.

The multi-band TRAP system achieves improvement for clean speech and speech with SNR > 10dB. For stronger noises the performance is inferior to basic TRAP features. We explain this behavior by the fact that concatenation of the TRAP vectors form adjacent critical bands spreads the noise from one band to three band probability estimators. Hence, instead of one impaired band-conditioned posterior estimates in case of basic TRAP system, there are three impaired estimates and the overall estimates suffer.

The PCA dimensionality reduction improves the performance on clean speech in agreement with [4], but the deterioration in noisy cases is severe. This is due to the fact that while doing the matrix multiplication, the change of one point in input vector affects all points of output vector. Thus the noise affects the band estimates much more.

The FD MTRAP features has also very poor performance compared to the basic TRAP but we expect them to help in combination.

The *lin ave* and *log ave* multi-stream combination are able to achieve better performance for weak noises with SNR < 10dB, but the results for stronger noises are badly affected by the system which is more vulnerable to the noise. This is – to some extent – solved by the inverse entropy based combination. By increasing the threshold, additional improvement is obtained for smaller SNR (stronger noise), which means that we effectively suppress the system with worse performance in noise. With the optimal threshold value $th = 2.5$ the combination achieves significant improvement for SNR>5. For stronger noises, the performance is only slightly inferior to the basic TRAP system.

The system combination with vector concatenation was a big surprise of our experiments. It achieves better performance on strong noises than the inverse entropy multi-stream combination and yet it keeps very good performance for week noises. Even larger improvement was observed for system combination where 2-dimensional time-frequency operator G2 [10] was used. This combination clearly outperformed all other systems.

We conclude, that the multi-band TRAP techniques and dimensionality reduction of TRAP vector by PCA (or DCT) are generally not good for recognition of noisy speech. It is due the inherent spreading of noise samples to larger area. The multi-stream combination techniques, namely the inverse entropy combination, improve significantly recognition of speech with weak noise (SNR > 5). The vector concatenation technique can bring improvement also in strong noises up to SNR = 0.

# References

1. Chen, B., Zhu, Q., Morgan, N.: Learning long-term temporal features in LVCSR using neural networks. In: Proc. ICSLP 2004, Jeju Island, KR (2004)
2. Jain, P., Hermansky, H., Kingsbury, B.: Distributed speech recognition using noise-robust MFCC and TRAPS-estimated manner features. In: Proc. of ICSLP 2002, Denver, Colorado, USA (2002)
3. Adami, A., Burget, L., Dupont, S., Garudadri, H., Grezl, F., Hermansky, H., Jain, P., Kajarekar, S., Morgan, N., Sivadas, S.: Qualcomm-ICSI-OGI features for ASR. In: Proc. ICSLP 2002, Denver, Colorado, USA (2002)
4. Jain, P., Hermansky, H.: Beyond a single critical-band in TRAP based ASR. In: Proc. Eurospeech 2003, Geneva, Switzerland, pp. 437–440 (2003)
5. Grézl, F., Hermansky, H.: Local averaging and differentiating of spectral plane for TRAP-based ASR. In: Proc. Eurospeech 2003, Geneva, Switzerland (2003)
6. Grézl, F.: Combinations of TRAP-based systems. In: Sojka, P., Kopeček, I., Pala, K. (eds.) TSD 2004. LNCS (LNAI), vol. 3206, pp. 323–330. Springer, Heidelberg (2004)
7. Pearce, D.: Enabling new speech driven servicesfor mobile devices: An overview of the ETSIstandards activities for distributed speech recognition front-ends. In: Applied Voice Input/Output Society Conference (AVIOS2000), San Jose, CA (2000)
8. Cole, R., Noel, M., Lander, T., Durham, T.: New telephone speech corpora at CSLU. In: Proc. of EUROSPEECH 1995, Madrid, Spain, pp. 821–824 (1995)
9. Misra, H., Bourlard, H., Tyagi, V.: New entropy based combination rules in HMM/ANN multi-stream asr. In: Proc. ICASSP 2003, Hong Kong, China (2003)
10. Grézl, F.: Local time-frequency operators in TRAPs for speech recognition. In: Matoušek, V., Mautner, P. (eds.) TSD 2003. LNCS (LNAI), vol. 2807, pp. 269–274. Springer, Heidelberg (2003)

# Intelligibility Is More Than a Single Word: Quantification of Speech Intelligibility by ASR and Prosody*

Andreas Maier[1,3], Tino Haderlein[1], Maria Schuster[1], Emeka Nkenke[2],
and Elmar Nöth[3]

[1] Abteilung für Phoniatrie und Pädaudiologie, Universität Erlangen-Nürnberg
Bohlenplatz 21, 91054 Erlangen, Germany
[2] Mund-, Kiefer- und Gesichtschirurgische Klinik, Universität Erlangen-Nürnberg,
Glückstraße 11, 91054 Erlangen, Germany
[3] Lehrstuhl für Mustererkennung, Universität Erlangen-Nürnberg
Martensstraße 3, 91058 Erlangen, Germany
Andreas.Maier@informatik.uni-erlangen.de

**Abstract.** In this paper we examine the quality of the prediction of intelligibility scores of human experts. Furthermore, we investigate the differences between subjective expert raters who evaluated speech disorders of laryngectomees and children with cleft lip and palate. We use the recognition rate of a word recognizer and prosodic features to predict the intelligibility score of each individual expert. For each expert and the mean opinion of all experts we present the best features to model their scoring behavior according to the mean rank obtained during a 10-fold cross-validation. In this manner all individual speech experts were modeled with a correlation coefficient of at least $r > .75$. The mean opinion of all raters is predicted with a correlation of $r = .90$ for the laryngectomees and $r = .86$ for the children.

## 1 Introduction

Until now speech disorders are evaluated subjectively by an expert listener showing only restricted reliability. For scientific purposes therefore a panel of several expert listeners is needed. For the objective evaluation we developed a new method to quantify speech disorders. In our recent work we evaluated our method with patients whose larynx was removed (laryngectomees) and children with cleft lip and palate (CLP).

By removal of the larynx the patient looses the ability to speak. The patient's breathing is maintained by a detour of the trachea to a hole in the throat—the so called tracheostoma. In order to restore the speech ability of the patient a shunt valve is placed between the trachea and the esophagus. Closure of the tracheostoma forces the air stream from the patient's lungs through the esophagus into the vocal tract. In this way, a tracheoesophageal voice is formed. In comparison to normal voices the quality of such a voice is low [1]. Nevertheless, it is considered as state-of-the-art of substitute voices.

Children with cleft lip and palate suffer from various graduations of speech disorders. The characteristics of these speech disorders are mainly a combination of different articulatory features, e.g. nasal air emissions that lead to nasality, a shift in localization of

---

* This work was supported by the Johannes-und-Frieda-Marohn Stiftung and the Deutsche Forschungsgemeinschaft (German Research Foundation) under grant SCHU2320/1-1.

V. Matoušek and P. Mautner (Eds.): TSD 2007, LNAI 4629, pp. 278–285, 2007.

**Table 1.** Correlations of the individual raters to the mean of the other raters

| laryngectomees | | | children | | |
|---|---|---|---|---|---|
| rater | mean of other raters | | rater | mean of other raters | |
| | $r$ | $\rho$ | | $r$ | $\rho$ |
| rater L | .84 | .82 | rater B | .95 | .92 |
| rater S | .87 | .84 | rater K | .94 | .93 |
| rater F | .80 | .77 | rater L | .94 | .93 |
| rater K | .81 | .83 | rater S | .94 | .92 |
| rater H | .80 | .77 | rater W | .96 | .92 |

articulation (e.g. using a /d/ instead of a /g/ or vice versa), and a modified articulatory tension (e.g. weakening of the plosives /t/, /k/, /p/) [2].

In [1] it was shown that—next to the recognition rate of a speech recognizer—prosodic features also hold information on the intelligibility. In this paper we successfully combine both approaches to enhance the prediction quality of our automatic evaluation system for speech disorders. Furthermore, we investigate the individual differences in intelligibility perception and their relation to the prosodic information.

## 2  Databases

The 41 laryngectomees (mean $62.0 \pm 7.7$ years) with tracheoesophageal substitute voice read the German version of the fable "The North Wind and the Sun". It is phonetically balanced and contains 108 words of which 71 are unique.

The children's speech data was recorded using a German standard speech test (PLAKSS [3]). The test consists of 33 slides which show pictograms of the words to be named. In total the test contains 99 words which include all German phonemes in different positions (beginning, center and end of a word). Additional words, however, were uttered in between the target words, since the children tend to explain the pictograms with multiple words. Informed consent had been obtained by all parents of the children prior to the examination. The database contains speech data of 31 children and adolescents with CLP (mean $10.1 \pm 3.8$ years).

All speech samples were recorded with a close-talking microphone (DNT Call 4U Comfort headset) at a sampling frequency of 16 kHz and quantized with 16 bit. The data were recorded during the regular out-patient examination of the patients. All patients were native German speakers, some of them using a local dialect.

## 3  Subjective Evaluation

Both corpora were evaluated by a panel of five speech experts. The experts rated each turn on a Likert scale between $1 \equiv$ very good and $5 \equiv$ very bad. So a floating point value was computed for each patient to represent his intelligibility, as commonly used for scientific purposes.

In order to compare the scores we computed Pearson's product moment correlation coefficient $r$ and Spearman's correlation coefficient $\rho$. Table 1 shows the agreement of the individual raters to mean of the respective other raters.

## 4   Automatic Speech Recognition System

A word recognition system developed at the (deleted) was used. As features we use mel-frequency cepstrum coefficients 1 to 11 plus the energy of the signal for each 16 ms frame (10 ms frame shift). Additionally 12 delta coefficients are computed over a context of 2 time frames to the left and the right side (56 ms in total). The recognition is performed with semi-continuous Hidden Markov Models (HMMs). The codebook contains 500 full covariance Gaussian densities which are shared by all HMM states. The elementary recognition units are polyphones [4]. The polyphones were constructed for each sequence of phones which appeared more than 50 times in the training set.

For our purpose it is necessary to put more weight on the recognition of acoustic features. So we used only a unigram language model to restrict the amount of linguistic information which is used to prune the search tree.

The training set for the adults' speech recognizer are dialogues from the VERBMO-BIL project [5]. The topic of the recordings is appointment scheduling. The data were recorded with a close–talking microphone with 16 kHz and 16 bit. The speakers were from all over Germany, and thus covered most regions of dialect. However, they were asked to speak standard German. About 80% of the 578 training speakers (304 male, 274 female) were between 20 and 29 years old, less than 10% were over 40. This is important in view of the test data, because the average age of our test speakers is over 60 years; this may influence the recognition results. A subset of the German VERBMOBIL data (11,714 utterances, 257,810 words, 27 hours of speech) was used for the training set and 48 utterances (1042 words) for the validation set (the training and validation corpus was the same as in [6]).

The training set of the children's recognizer contained 53 children with normal speech between 10 and 14 years of age. In order to increase the amount of training data, speech data of adult speakers from VERBMOBIL—whose vocal tract length was adapted to children's speech—were added. Further enhancement of the children's recognizer was done by MLLR adaptation to each speaker as described in [7]. A more detailed description of the recognizer, the training set, and the language model is presented in [8,9].

## 5   Prosodic Features

The prosody module used in these experiments was originally developed within the VERBMOBIL project [5], mainly to speed up the linguistic analysis [11,12]. It assigns a vector of prosodic features to each word in a word hypothesis graph which is then used to classify a word w.r.t., e.g. carrying the phrasal accent and being the last word in a phrase. For this paper, the prosody module takes the text reference and the audio signal as input and returns 37 prosodic features for each word and then calculates the mean, the maximum, the minimum, and the variance of these features for each speaker, i.e. the prosody of the whole speech of a speaker is characterized by a 148-dimensional vector. These features differ in the manner in which the information is combined (cf. Fig. 1):

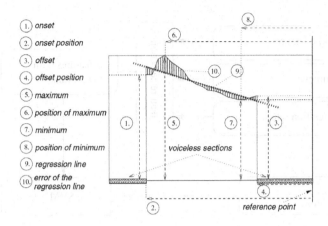

**Fig. 1.** Computation of prosodic features within one word (after [10])

1. onset
2. onset position
3. offset
4. offset position
5. maximum
6. position of maximum
7. minimum
8. position of minimum
9. regression line
10. mean square error of the regression line

These features are computed for the fundamental frequency ($F_0$) and the energy (absolute and normalized). Additional features are obtained from the duration and the length of pauses before and after the respective word. Furthermore jitter, shimmer and the length of voiced (V) and unvoiced (UV) segments are calculated as prosodic features.

## 6   Automatic Evaluation

The automatic evaluation system employs support vector regression (SVR) [13] for prediction of the experts' scores.

As displayed in Fig. 2 we utilize on the one hand the word accuracy (WA) and the word recognition rate (WR) of a speech recognizer.

$$\mathrm{WR} = \frac{C}{R} \times 100\,\%$$

is computed as the percentage of correctly recognized words $C$ and the number of reference words $R$. In addition

$$\mathrm{WA} = \frac{C - I}{R} \times 100\,\%$$

weights the number of wrongly inserted words $I$ in this percentage.

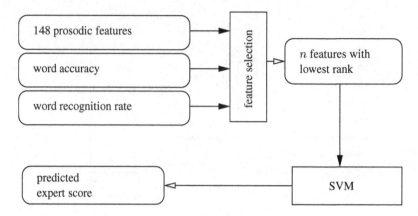

**Fig. 2.** Proposed system for the prediction of the expert scores

On the other hand 148 prosodic features as features for the system. So we obtain 150 features in total. In order to select a subset of the features we applied a simple algorithm based on the multiple regression/correlation analysis [14] (also called "linear regression" in some cases). The algorithm builds—based on the best $n - 1$ subset— all possible sets with $n$ features and picks the set with the best regression to the target value (Here: the mean opinion of the experts). This algorithm returned better features than other feature selection algorithms like correlation-based feature subset selection [15] or consistency subset evaluation [16]. However, the algorithm can select $m - 1$ features at most, where $m$ is the number of subjects in the test set. If a feature was not selected we assigned rank 149.

All evaluations presented here were done in a 10-fold cross validation (CV) manner since the number of patients in each group is rather small. In order to present a feature ranking for the feature selection we computed the mean rank of all CV iterations for each feature. This, however, does not mean, that the particular feature has been selected for all CV iterations.

## 7   Results

The additional use of prosody could enhance the accuracy of the prediction compared to [1] and [9] for the adults' speech data.

Table 2 gives an overview about the quality of the CV prediction on the laryngec-tomees' database. We stopped reporting additional features when the correlation did not increase further. Combination of either the WR or the WA with prosodic features yields improvement in most cases. The prediction of our gold standard—the mean opinion of all experts—is improved by 3.4 % in case of Pearson's $r$ and 4.8 % for Spearman's $\rho$ relatively. Note that the correlations cannot be compared directly to those of Table 1 since these correlations were not computed in cross-validated manner.

Furthermore, prosody is also useful to model the intelligibility perception of each individual expert. As can be seen in Table 2 rater L's intelligibility scores are modeled best by the

**Table 2.** Overview on the prediction performance done by different feature sets on the laryngectomees' database

| feature | mean rank | prediction SVR | | reference raters |
|---|---|---|---|---|
| | | $r$ | $\rho$ | |
| word accuracy | 0 | .87 | .83 | all raters |
| mean $F_0$ of all words | 17 | **.90** | **.87** | all raters |
| word recognition rate | 17.1 | .68 | .67 | rater L |
| word accuracy | 18.6 | .70 | .71 | rater L |
| maximum silence before word | 34.7 | .75 | .76 | rater L |
| mean $F_0$ regression line | 39 | **.77** | **.78** | rater L |
| word recognition rate | 14.9 | **.77** | **.74** | rater S |
| word accuracy | 14.9 | **.76** | **.78** | rater R |
| word accuracy | 16.9 | .71 | .72 | rater K |
| mean silence after word | 23.2 | .69 | .73 | rater K |
| maximum $F_0$ minimum position | 35.5 | .74 | .78 | rater K |
| maximum $F_0$ minimum | 46.6 | **.77** | **.78** | rater K |
| word accuracy | 0 | .76 | .70 | rater H |
| minimum $F_0$ minimum | 30.6 | **.78** | **.72** | rater H |

**Table 3.** Prediction of the experts' scores by different feature sets on the children's database

| feature | mean rank | prediction SVR | | reference raters |
|---|---|---|---|---|
| | | $r$ | $\rho$ | |
| word accuracy | 0 | **.86** | **.84** | all raters |
| word accuracy | 0 | **.86** | **.84** | rater B |
| word accuracy | 14.9 | **.77** | **.76** | rater K |
| word accuracy | 0 | **.78** | **.77** | rater L |
| word recognition rate | 36.4 | **.80** | **.77** | rater S |
| word accuracy | 0 | **.80** | **.80** | rater W |

- word recognition rate,
- the word accuracy,
- the maximum silence before each word, and the
- mean of the of the regression coefficient of the fundamental frequency's slope.

The scores of each individual rater are modeled by these features with a correlation of $r > .75$ and $\rho > .72$. The raters S and R seem to judge the intelligibility only by means of either WR or WA. Their opinion of intelligibility cannot be explained further by means of prosody.

With the children's data no further improvement was obtained by the application of prosodic features for the prediction of experts' scores (cf. Table 3). Although high correlations between single prosodic features and the mean opinion of the experts exist, the best feature for the prediction of the experts is always the word accuracy.

We suppose that the summarization of the prosodic information is too rough for the case of the children. For the laryngectomees the mean, the variance, the minimum, and the maximum of the prosody of all single words seems to be enough. This might be related to the kind of this disorder: most affected are the fundamental frequency and the duration of pauses since the generation of the tracheoesophageal speech is artifical and the speaking with such a voice is exhausting and the speaker has to stop more often and unexpectedly. Both effects seem to reduce the intelligibility.

For the case of children the prosody of the single words of the speech test has to be differentiated more closely: The prosody of the children depends on the difficulty of the target words. We assume that the prosody of the difficult words is more monotonous than the prosody of familiar and simple words, which are also uttered in between the target words. We will examine this aspect more closely in our future work.

## 8  Summary

In this paper we successfully combined prosodic features with the recognition rate of a word recognizer to improve the reliability of the automatic speech intelligibility quantification system. A feature selection using multiple regression analysis yielded a prediction system that computes scores which are very close to the experts' scores ($r =.90$ and $\rho =.87$). For the data of the children no further improvement was obtained by additional prosodic information. However, the quality of the children's prediction system ($r =.87$ and $\rho =.84$) is in the same range as the laryngectomees' prediction system. Therefore, both systems can be used to replace the time and cost intensive subjective evaluations. In addition, the system can be used to investigate the intelligibility perception of the human experts. So a list of features can be computed which models the intelligibility rating of each expert best. The prediction of these single expert models has always a correlation which is above $r > .75$ and $\rho > .74$.

## References

1. Haderlein, T.: Nöth, E., Schuster, M., Eysholdt, U., Rosanowski, F.: Evaluation of Tracheoesophageal Substitute Voices Using Prosodic Features. In: Hoffmann, R., Mixdorff, H. (eds.) Proc. Speech Prosody, 3rd International Conference, Dresden, Germany, TUDpress, pp. 701–704 (2006)
2. Harding, A., Grunwell, P.: Active versus passive cleft-type speech characteristics. Int. J. Lang. Commun. Disord. 33(3), 329–352 (1998)
3. Fox, A.: PLAKSS - Psycholinguistische Analyse kindlicher Sprechstörungen. Swets & Zeitlinger, Frankfurt a.M., now available from Harcourt Test Services GmbH, Germany (2002)
4. Schukat-Talamazzini, E., Niemann, H., Eckert, W., Kuhn, T., Rieck, S.: Automatic Speech Recognition without Phonemes. In: Proceedings European Conference on Speech Communication and Technology (Eurospeech), Berlin, Germany, pp. 129–132 (1993)
5. Wahlster, W. (ed.): Verbmobil: Foundations of Speech-to-Speech Translation. Springer, Berlin (2000)
6. Stemmer, G.: Modeling Variability in Speech Recognition. PhD thesis, Chair for Pattern Recognition, University of Erlangen-Nuremberg, Germany (2005)
7. Gales, M., Pye, D., Woodland, P.: Variance compensation within the MLLR framework for robust speech recognition and speaker adaptation. In: Proc. ICSLP '96. Philadelphia, USA, vol. 3 pp. 1832–1835 (1996)

8. Maier, A., Hacker, C., Nöth, E., Nkenke, E., Haderlein, T., Rosanowski, F., Schuster, M.: Intelligibility of children with cleft lip and palate: Evaluation by speech recognition techniques. In: Proc. International Conf. on Pattern Recognition. Hong Kong, China, vol. 4, pp. 274–277 (2006)

9. Schuster, M., Maier, A., Haderlein, T., Nkenke, E., Wohlleben, U., Rosanowski, F., Eysholdt, U., Nöth, E.: Evaluation of Speech Intelligibility for Children with Cleft Lip and Palate by Automatic Speech Recognition. Int. J. Pediatr. Otorhinolaryngol. 70, 1741–1747 (2006)

10. Kießling, A.: Extraktion und Klassifikation prosodischer Merkmale in der automatischen Sprachverarbeitung. Berichte aus der Informatik. Shaker, Aachen (1997)

11. Nöth, E., Batliner, A., Kießling, A., Kompe, R., Niemann, H.: Verbmobil: The Use of Prosody in the Linguistic Components of a Speech Understanding System. IEEE Trans. on Speech and Audio Processing 8, 519–532 (2000)

12. Batliner, A., Buckow, A., Niemann, H., Nöth, E., Warnke, V.: The Prosody Module. [5], 106–121

13. Smola, A., Schölkopf, B.: A tutorial on support vector regression. In: NeuroCOLT2 Technical Report Series, NC2-TR-1998-030 (1998)

14. Cohen, J., Cohen, P.: Applied Multiple Regression/Correlation Analysis for the Behavioral Sciences. Lawrence Erlbaum Associates, Hillsdale, New Jersey (1983)

15. Hall, M.A.: Correlation-based Feature Subset Selection for Machine Learning. PhD thesis, University of Waikato, Hamilton, New Zealand (1998)

16. Liu, H., Setiono, R.: A probabilistic approach to feature selection - a filter solution. In: 13th International Conference on Machine Learning, pp. 319–327 (1996)

# Appositions Versus Double Subject Sentences – What Information the Speech Analysis Brings to a Grammar Debate

Horia-Nicolai Teodorescu[1,2] and Diana Trandabăț[1,3]

[1] Institute for Computes Science, Romanian Academy
[2] Faculty of Electronics and Telecommunications, Technical University "Gh. Asachi" of Iaşi
[3] Faculty of Computer Science, University "Al. I. Cuza" of Iaşi
hteodor@etc.tuiasi.ro, dtrandabat@info.uaic.ro

**Abstract.** We propose a method based on spoken language analysis to deal with controversial syntactic issues; we apply the method to the problem of the double subject sentences in the Romanian language. The double subject construction is a controversial linguistic phenomenon in Romanian. While some researchers accept it as a language 'curiosity' (specific only to the Asian languages, but not to the European ones), others consider it apposition-type structure, in order to embody its behavior in the already existing theories. This paper brings a fresh gleam of light over the debate, by presenting what we believe to be the first study on the phonetic analysis of double-subject sentences in order to account for its difference vs. the appositional constructions.

## 1 Introduction

Grammatical issues are often controversial and leave space to interpretations. We introduce a new method to help decision in such controversial cases, based on the analysis of speech. The main idea is that two different grammatical structures should have different prosodic interpretations, while instances of a single syntactic structure should have similar correspondences in the speech, all other variables kept constant (speaker, environment etc.). All European languages use appositions to emphasize a specific meaning the speaker wishes to convey. Some languages, like the Japanese, Mandarin, Korean and the Thai languages, use for similar purposes specific constructions, named "double-subject constructions" [5], [6]. (For a detailed analysis of the double subject issue in Asian languages, as well as for an extensive list of references on the topic, see [6]). Such constructions are unknown to most modern European languages, like English or French. In the Romanian linguistic community there has been in recent years a debate on some types of sentences which are considered by several researchers [1], [2] and by us a double-subject construction.

The purpose of this paper is to present a detailed analysis on the contrastive prosodic features of the double-subject sentences and apposition constructions in Romanian. The analysis goes beyond the basic prosody, as represented by pitch values and trajectory; it aims to determine the evolution of higher formants and

V. Matoušek and P. Mautner (Eds.): TSD 2007, LNAI 4629, pp. 286–293, 2007.

temporal patterns. After presenting the different approaches to double-subject sentences in Section 2, we discuss the methodology behind the double-subject corpus creation and its analysis: annotation, acoustic parameters determination, etc. The results of the prosodic analysis are presented in Section 4, before drawing some conclusions and indicating some further directions.

## 2 Double-Subject Sentences in Romanian

The semantic arguments of a predicate (the subject, the direct object and the indirect object) can be doubled, in the Romanian language. While the objects are commonly doubled by clitic pronouns (the doubling is sometimes mandatory, like in *L-am văzut pe Ion,* EN: I saw John), the subjects receive, occasionally, and mainly colloquially, a doubling pronoun (not only in Romanian, as Masahiro [6] shows[1]). The doubling of the subject for the Romanian language is a controversial phenomenon: after having long been considered an apposition, Alexandra Cornilescu [2] has reopened the doubling problem, Verginica Barbu [1] has modeled it using HPSG instruments, but until today, there is no unitary consensus. In this context, supplementary information should be gathered on the specificities of the double-subject constructions contrasted both to the single subject sentences and to sentences which include appositions. Specific phonetic constructions for the three cases would be a significant argument for three independent linguistic constructions. What supplementary information the pronouncing brings, from a descriptive perspective, in double-subject phrases, remains an open question. The present paper partially answers this question.

We provide subsequently a few examples of brief sentences with double subject in the Romanian language. To translate these sentences, we use the symbol Ø to mark the place of the missing doubled subject in the sentences translated in correct English. Examples of sentences with double subject are:

(a)  Vine    ea  mama!
    *Comes  she  mom!  [Mom Ø is coming!]
(b)  „A trecut  el    aşa  un răstimp." (Sadoveanu M.)
    *Passed has  it    thus    a time.  [A time has Ø thus passed.]

The first author proposes that the double-subject sentences convey different meanings, depending on the prosody, for example:

- a neutral pronunciation indicates a non-determination of the time interval.
- a pronunciation accentuating the pronoun "el" (EN: he) indicates that the speaker has an idea about the time interval duration, and that the focus is on the passing of that time, and not on the duration.
- if the sentence is further developed, it can bring a further specification of the interval. For example, in the development „A trecut el aşa un răstimp de lung, încât..."

---

[1] There is no definite explanation why not all languages accept the double-subject structure. For these languages, in most of the cases, the doubling of the subject is realized as an apposition. The Romanian language has both double subject and apposition structures.

(EN: A <u>so long</u> time has thus passed, that…), the duration of the interval is specified in a certain way.

(c)   O şti        el     careva     cum   să rezolve   asta.
      *would know   he     someone    how   to solve     this. [He would know Ø
      how to solve this.]

Different pronunciations may mark either the fact that the speaker does not know who is the person mentioned ( „el"), either that he knows, but has no intention on telling to the audience (when the accent is on „careva", EN: someone), or clearly specifies, by an apposition, who is envisaged, if the sentence is developed as „O şti el careva, <u>Ion</u>, cum să rezolve asta" (EN: He, <u>John</u>, would know how to solve this). Notice that such a sentence, including both apposition and double subject, is a strong argument in favor of the existence of the double subject constructions as a distinct linguistic structure.

For the examples b) and c), the interpretation is that the information must be partially known by the auditorium (knowledge at the generic level, but not at the level of instantiation with a concrete individuality).

(d)   Mama   vine        şi   ea   mai târziu.
      *Mom   is coming   also she   later. [Also mom is coming later.]
(e)   Mama   ştie   ea   ce   face.
      *Mom   knows   she   what   is doing. [Mom knows what she is doing.]

Examples d) and e) are considered by some linguists [1] as constructions with doubled subject, while other authors [2] consider them particular structures of the Romanian language. We intend to compare them to examples a) – c) to see if there are differences in their prosodic realizations.

In this context, we recorded a set of sentences bearing doubled subject for a comparative analysis of the prosody in sentences with doubled/simple subject and appositions, in order to observe the modifications involved by the doubling of the subject. This paper aims to bring clarifications on the change of prosody in double-subject sentences in comparison with simple sentences and appositions.

# 3  Methodology

A principle we propose and use here is that consistent distinctions at the phonetic level between two specific syntactic constructions reflect and represent an argument to distinguish at the syntactical level between the two constructions. In order to realize a correlation between the semantic charge carried by a sentence and the representation of its subject, the five sentences presented in Section 2 have been recorded by 15 speakers. The database is freely accessible on the web site of the Romanian Sounds Archive [8]. The Romanian Sounds Archive contains over 1000 distinct recordings, available in various accuracy and encoding formats (more methodological aspects are given in [10], in this volume).

Apart the archive itself, the site hosts also documentations regarding the description of the technical modalities and conditions (protocols) involved by the realization of the archive. Namely, the database contains two types of protocols:

-   The documentation protocol, which contains the speaker profile (linguistic, ethnic, medical, educational, professional information about the speaker), and a questionnaire regarding the speaker's health, especially concerning the pathologies of the phonating tract.
-   The recording protocol, containing information about the noise acceptable values, the microphone, the soundboard, and the corresponding drivers.

### 3.1  Double-Subject Spoken Database

After subjects have been informed about the objectives of the project, they signed an informed consent according to the Protection of Human Subjects Protocol of the U.S. Food and Drug Administration and to the Ethical Principles of the Acoustical Society of America for Research Involving Human Subjects. The speakers' selection was tributary to the Archive's constraints (the documentation protocol).

The recordings (sound files) corresponding to the simple subject, double-subject, and apposition sentences have been recorded according to the methodology explained in the recording protocol of the Romanian Sound Archive [8]. The recordings were performed using the GoldWave[TM] application [3], with a sampling frequency of 22050Hz [10].

The speakers[2] have recorded several variants of the five sentences mentioned in Section 2; the sentences have been uttered with neutral tone, accentuation of the doubling pronouns, focuses on the words next the pronouns, and respectively the extension of the sentences.

(a)   RO: Vine ea mama!               EN: Mom Ø is coming!
(b)   RO: A trecut el aşa un răstimp    EN: A time has Ø thus passed.
(c)   RO: O şti el careva cum să rezolve asta.
      EN: He would know Ø how to solve this.
(d)   RO: Mama vine şi ea mai târziu.   EN: Also mom is coming later.
(e)   RO: Mama ştie ea ce face.         EN: Mom knows what she is doing.

Corresponding variants of the five mentioned sentences with simple subject and appositions have also been recorded. Every speaker pronounced each sentence three times, following the archive recording protocol (see for details [8]).

### 3.2  Analysis Methodology

We performed the analysis of the double subject in two steps. The first step requires finding and correlating the double sentences parameters with the corresponding simple sentences parameters. The second phase envisages the contrastive analysis between double subject and appositions.

The sentences have been annotated using the Praat[TM] software [7] at phoneme level. Then, the syllable, word, sentence, subject position, and articulation type level

---

[2] Fifteen speakers have been recorded for the double subject analysis. The results discussed in Section 4 consider only seven subjects: subject #1, subject #2, subject #12, subject #13 (female) and subject #5, subject #6, subject #15 (male), selected because they all work in academic/university environment, and should therefore be more familiar with the linguistic structures of the Romanian language.

were easily created. After the annotation, the pitch and the formants (F0–F4) are determined for the sentence vowels and semi-vowels. For a determination as precise as possible, a segment of the vowel fulfilling the following conditions is selected:

-   The selected segment should be a central area, where there are no transitions of the formants to those of the joined phonemes;
-   The formant's frequency should not present big fluctuations. The fluctuations of the formants and their correlation to the double subject will be analyzed in a subsequent stage;
-   The formant's contour should not contain interruptions.

Unfortunately, various analysis tools provide different results. This is due to the fact that there is no single definition for these parameters for non stationary signals (as the speech signal is), various tools using different ad hoc definitions. Therefore, we have used several programs, namely Praat™ [7], Klatt analyzer™ [4], GoldWave™ [3] and WASP™ [11] to determine the acoustic parameters. The obtained results, discussed in the next section, use a mean of the values obtained with the four analysis programs.

## 4  Double-Subject Sentences Analysis

The hypothesis that motivated this analysis is to provide prosodic evidence of the fact that double subject sentences and appositional constructions are two linguistically different phenomena.

We analyzed therefore the values of the formants and duration of the vowels for seven subjects from our database for the constructions:

-   Vine mama (EN: Mom is coming) – simple subject
-   Vine ea mama (EN: Mom Ø is coming) – doubled subject
-   Vine ea, mama (EN: She, mom, is coming) – apposition.

We realize that an analysis over seven subjects can have no claims on generality, but it represents a good start for the pioneering contrastive analysis on the specificities of the Romanian double subject and apposition constructions.

Fig. 1 presents the relative deviation for the sentences "Vine ea mama" vs. "Vine mama" for the seven speakers. The relative deviation $\sigma_r$ was computed as:

$$\sigma_r = \frac{\overline{\Delta F0_k^v}}{\overline{F0^v}} \text{ and } F0_k^v - F0^v = \Delta F0_k^v$$

where $v$ represents each vowel in the sentence, and $k$ each of the seven speakers.

For each subject (1, 2, 5, 6, 12, 13 and 15) we computed the pitch values for double subject sentences (DS) and the corresponding simple subject ones (SS). Thus, the first bar in graph represents DS_1, the double subject sentence for subject 1, the second gar represents the pitch values for simple subject construction (SS) for subject 1, etc.

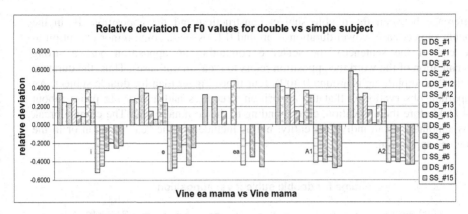

**Fig. 1.** Pitch values for the double subject and simple subject constructions

When comparing simple subject with double subject constructions, an increasing tendency of the F0 values in the simple subject sentences vs. double-subject sentences was observed. The major differences in the pitch values are visible for the vowels in unaccented syllables [see for details 9].

As for the other formants, it looks that they are fluctuating and carries no double subject information. However, they carry information about the speaker [9]. Also, no significant differences were found in the duration comparison.

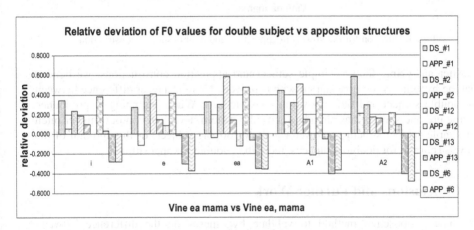

**Fig. 2.** Pitch values for the double subject and apposition constructions

When comparing apposition structures with double subject structures, we must emphasize that the formant values bring no definite difference. Fig. 2 shows the pitch values for the double subject and the apposition sentence for five of the seven considered subjects. The pitch tendency has no obvious pattern. The data recordings we have annotated and analyzed are not sufficient to draw pertinent statistic conclusions. However, our hypothesis on different patterns for different syntactic constructions is confirmed by the duration of the vowel. Fig. 3 shows the significant difference between the double subject construction and the apposition. If, in the double subject case, vowel duration is around 0.100s (with some minor exception to the end of the sentence), the sentence containing an apposition bear a strong accentuation of the word the apposition refers to ("ea" in our case). Thus, the duration of the "ea" diphthong is around 0.400s, four times bigger than for double subject. An important observation is that the apposition structures had a very big pause (about 0.400s) before the apposition, corresponding to the comma. break. The comma break was annotated as an individual entity, not as included in the "ea" pronoun or in the "mama" apposition.

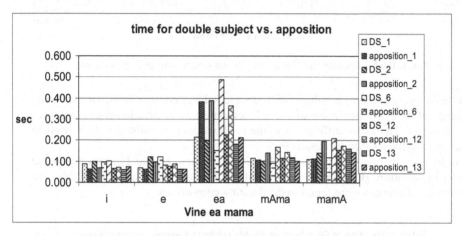

**Fig. 3.** Duration for the vowels in apposition and double subject construction

After analyzing double / simple subject and apposition constructions, we believe that the hypothesis we have started with is proven. There is a clear difference between the double subject and the apposition constructions. When beginning to pronounce a structure, the speaker has already a prosodic pattern: the pitch contour (higher pitch for simple subject structures, lower values for double subject) or the duration of the vowels (normal for double sentences, more than double for appositions).

## 5   Conclusions and Further Work

We have proposed a method to validate hypotheses on the difference between syntactical constructions based on marked differences in the prosody of spoken sentences incorporating such constructions. Specifically, we proposed to use prosodic differences as an argument in deciding when two constructions are different. We have analyzed the influence of the double-subject construction on the prosody in the Romanian language. The analysis involved short sentences which are parallel in the sense that they are identical up to the use of double-subject or apposition constructions.

The main conclusion which can be derived from this preliminary research is that the two syntactic constructions differ in a consistent way from a prosodic point of view. Namely, the word that the apposition explains has duration four times bigger than normal simple subject sentences or double subject constructions. A second conclusion is that the frequency of the pitch and the central frequency of first formant are different in the two constructions, but both the way of changing and the change amplitude depend significantly on the speaker. These differences represent an argument supporting the existence of double subject construction in the Romanian language – the only Latin, moreover the only non-Asian language exhibiting such a construction.

In the future, we will analyze more recordings in order to confirm these findings and to detect an inter-speaker patterning.

## Acknowledgement

This research is part of the Romanian Academy "priority research" topic "Cognitive Systems".

## References

1. Barbu, V.: Double subject constructions in the Romanian language. A HPSG perspective (in Romanian), in Aspects of the Romanian language dynamics (in Romanian), Ed. Universității Bucharest, vol. II, pp. 73–79 (2003)
2. Cornilescu, A.: The Double Subject Construction in Romanian. Notes on the Syntax of the Subject, Revue Roumaine de Linguistique, (3-4), 1–45 (1997)
3. GoldWave – Audio software, http://www.goldwave.com
4. KPE80 – A Klatt Synthesiser and Parameter Editor, http://www.speech.cs. cmu.edu/comp.speech/Section5/Synth/klatt.kpe80.html
5. Kumashiro, Toshiyuki, Langacker, Ronald, W.: Double-subject and complex-predicate constructions. Cognitive Linguistics (14-1), 1–45 (2003)
6. Masahiro, O.: Analyzing Japanese double-subject construction having an adjective predicate. In: Proceedings of the 16th conference on Computational linguistics, Copenhagen, Denmark, vol. 2, pp. 865–870 (1996)
7. Praat: doing phonetics by computer, http://www.praat.org
8. Romanian Sounds Archive, http://www.etc.tuiasi.ro/sibm/omanian_spoken_language/index.htm
9. Teodorescu, H.-N., Feraru, M., Trandabăț, D.: Studies on the Prosody of the Romanian Language: The Emotional Prosody and the Prosody of Double-Subject Sentences. In: Burileanu, Teodorescu (eds.) Advances in Spoken Language Technology, Romanian Academy Press, pp. 171–182 (2007)
10. Teodorescu, H.-N., Feraru, M.: A study on Speech with Manifest Emotions. In: Proc. TSD 2007. LNCS, Springer, Heidelberg (2007) (in this volume)
11. UCL Phonetics & Linguistics, WASP – Waveform Annotations Spectrograms and Pitch, http://www.phon.ucl.ac.uk/resource/sfs/wasp.htm

# Automatic Evaluation of Pathologic Speech
# – from Research to Routine Clinical Use

Elmar Nöth[1], Andreas Maier[1,2], Tino Haderlein[1,2], Korbinian Riedhammer[1],
Frank Rosanowski[2], and Maria Schuster[2]

[1] Universität Erlangen-Nürnberg, Lehrstuhl für Mustererkennung (Informatik 5)
Martensstraße 3, 91058 Erlangen, Germany
noeth@informatik.uni-erlangen.de
http://www5.informatik.uni-erlangen.de
[2] Universität Erlangen-Nürnberg, Abteilung für Phoniatrie und Pädaudiologie
Bohlenplatz 21, 91054 Erlangen, Germany

**Abstract.** Previously we have shown that ASR technology can be used to objectively evaluate pathologic speech. Here we report on progress for routine clinical use: 1) We introduce an easy-to-use recording and evaluation environment. 2) We confirm our previous results for a larger group of patients. 3) We show that telephone speech can be analyzed with the same methods with only a small loss of agreement with human experts. 4) We show that prosodic information leads to more robust results. 5) We show that text reference instead of transliteration can be used for evaluation. Using word accuracy of a speech recognizer and prosodic features as features for SVM regression, we achieve a correlation of .90 between the automatic analysis and human experts.

## 1 Introduction

In speech therapy, objective evaluation of voice and speech quality is necessary for at least 1) patient assessment, 2) therapy control, 3) evaluation of different therapy methods using groups of patients, and 4) preventive screening. Normally, a group of experts rates some aspect of a patient's utterance like intelligibility, nasality, or harshness. This property is typically rated on a five to seven point Likert scale [1], e.g. from 1 = "very high" to 5 = "very low". The average or median of the ratings is then considered as an "objective" rating of this aspect of the patient's voice or speech. However, except for research projects, such a procedure is not done for financial, cost-cutting reasons. Thus, the patient is often evaluated by just one expert, sometimes only in a very crude way, e.g. the expert only distinguishes between "changed" and "unchanged intelligibility". With significant inter- and intra-rater variability, there normally is no objective evaluation of a patient's voice and speech available. Therefore, there is a strong need for an easy to apply, cost-effective, instrumental, and objective evaluation method.

In two research studies [2,3] we showed that for two groups of patients such a method is available, using automatic speech recognition technology: For a group of children with "Cleft Lip and Palate" (CLP) we recorded names of objects shown on pictograms and for a group of patients with tracheoesophageal (TE) substitute voice (after removal of the larynx due to cancer) we recorded a phonetically rich read text.

V. Matoušek and P. Mautner (Eds.): TSD 2007, LNAI 4629, pp. 294–301, 2007.

The recordings were transliterated and rated by a group of speech experts according to different aspects like intelligibility, nasality, and match of breath/sense units on a five-point Likert scale. The average of the intelligibility ratings was compared to the word accuracy (WA) of an automatic speech recognizer (ASR) which was calculated w.r.t. the transliteration. The correlation between these two ratings was 0.84 [3,4] for TE and 0.9 [2] for CLP speech. When we projected the WA to the Likert scale and considered the ASR as an additional rater, the inter-rater agreement to the human raters was in the same range as the inter-rater agreement between the human raters. We can thus conclude that the WA can be used as an objective instrumental evaluation method.

In this paper we want to report on several steps that bring us closer to using ASR in everyday clinical use. In detail we will restrict ourselves to TE patients and will address the following topics which are all important steps towards a routine use of our evaluation methods:

1. Can we create an easy-to-use and easily available interface to our analysis environment?
2. Do our results hold for a larger, more representative group of patients?
3. A very important communication situation for the patient is the communication over the telephone, where other information channels are missing. Can we evaluate speech via this reduced information channel?
4. It is well known that prosody is an important aspect of speech perception. Can prosodic features improve our evaluation results?
5. To show the agreement between human experts and ASR, we carefully translit- erated the utterances as a reference for WA. This would not be done outside of a research study. How well does the ASR rating agree to the human rating, when it is evaluated w.r.t. the reference text rather than the transliteration?

The rest of this paper is organized as follows: In Section 2 we give a short character- istic of TE voice and of our database. In Section 3 we introduce our recognition system and recording environment (topic 1). In Section 4 we try to answer the topics 2-5 named above. We conclude with a discussion and summary.

## 2   TE Voice and Used Database

The TE substitute voice is currently state-of-the-art treatment to restore the ability to speak after laryngectomy [5]: A silicone one-way valve is placed into a shunt between the trachea and the esophagus which on the one hand prevents aspiration and on the other hand deviates the air stream into the upper esophagus during expiration. The up- per esophagus, the pharyngo-esophageal (PE) segment, serves as a sound generator. Tissue vibrations of the PE segment modulate the streaming air and generate the pri- mary substitute voice signal which is then further modulated in the same way as normal speech. In comparison to normal voices the quality of substitute voices is low, e.g. the change of pitch and volume is limited and inter-cycle frequency perturbations result in a hoarse voice [6]. Acoustic studies of TE voices can be found for instance in [7,8].

41 laryngectomees ($\mu = 62.0 \pm 7.7$ years old, 2 female and 39 male) with TE substitute voice read the German version of the text "The North Wind and the Sun",

a fable from Aesop. It is a phonetically rich text with 108 words (71 disjoint) which is often used in speech therapy in German speaking countries. The speech samples were recorded with a close-talking microphone with 16 kHz and 16 bit.

To determine the loss of information due to the telephone channel, we played back the close-talking recordings using a standard PC and loudspeaker in a quiet office environment and placed a telephone headset in front of the loudspeaker, i.e. we created a telephone quality (8 kHz a-law) version of the database. Due to the multiple AD/DA conversions and the different frequency characteristics of the loudspeaker and the microphones we expect the recognition rates to be a lower bound for the recognition rates for real telephone calls.

## 3    The Automatic Speech Analysis System

For the objective measurement of the intelligibility of pathologic speech, we use a hidden Markov model (HMM) based ASR system. It is a state-of-the-art word recognition system developed at the Chair of Pattern Recognition (Lehrstuhl für Mustererkennung) of the University of Erlangen-Nuremberg. In this study, the latest version as described in detail in [9] was used. A commercial version of this recognizer is used in high-end telephone-based conversational dialogue systems by *Sympalog* (www.sympalog.com), a spin-off company of the Chair of Pattern Recognition. As features we use 11 Mel-Frequency Cepstrum Coefficients and the energy of the signal for a 16 ms analysis frame (10 ms shift). Additionally 12 delta coefficients are computed over a context of 2 time frames to the left and the right side (56 ms in total). The recognition is performed with semi-continuous HMMs. The codebook contains 500 full covariance Gaussian densities which are shared by all HMM states. The elementary recognition units are polyphones [10], a generalization of triphones.

The output of the word recognition module is used by our prosody module to calculate word-based prosodic features. Thus, the time-alignment of the recognizer and the information about the underlying phoneme classes (like *long vowel*) can be used by the prosody module. For each word we extract 22 prosodic features over intervals of different sizes, i.e. the current word or the current word and the previous word. These features model F0, energy and duration, e.g. maximum of the F0 in the word pair "current word and previous word". In addition, 15 global prosodic features for the whole utterance are calculated, e.g. standard deviation of jitter and shimmer. In order to evaluate the pathologic speech, we calculate the average, the maximum, the minimum, and the variance of the 37 turn- and word-based features for the whole text to be read. Thus we get 148 features for the whole text. A detailed description of the features is beyond the scope of this paper. We will restrict ourselves to explaining in Section 4 those features which proved to be most relevant for our task. A detailed discussion of our prosodic features can be found in [11,12].

### 3.1    Recognizer Training

The basic training set for our recognizers are dialogues from the VERBMOBIL project [13]. The topic of the recordings is appointment scheduling. The data were recorded with a close-talking microphone with 16 kHz and 16 bit. The speakers were from all over Germany and thus covered most dialect regions. However, they were asked to speak standard German. About 80% of the 578 training speakers (304 male, 274 female) were between 20 and 29 years old, less than 10% were over 40. This is important in view of the test data, because the fact that the average age of our test speakers is more than 60 years may influence the recognition results. A subset of the German VERBMOBIL data (11,714 utterances, 257,810 words, 27 hours of speech) was used for the training set and 48 utterances (1042 words) for the validation set (the training and validation corpus was the same as in [9]).

In order to get a telephone speech recognizer, we downsampled the training set to telephone quality. We reduced the sampling rate to 8 kHz and applied a low-pass filter with a cutoff frequency of 3400 Hz to simulate telephone quality. Thus, we used "the same" training data for the close-talking and telephone recognizer. A loss in evaluation quality will therefore mainly be caused by the different channels, not by different amounts of training data.

In [4], we showed for a corpus of 18 TE speakers that a monophone-based recognizer for close-talking signals produced slightly better agreement with speech experts' intelligibility ratings than a polyphone-based recognizer. We wanted to verify all the results for the larger corpus of 41 TE speakers. Therefore we created four different recognizers: For the 16 kHz and the 8 kHz training data, we created a polyphone-based and a monophone-based recognizer (rows "16/m", "8/m", "16/p", "8/p" in Table 2). After the training, the vocabulary was reduced to the words occurring in the German version of the "The North Wind and the Sun".

### 3.2  Recording Environment

For routine use of our evaluation system, it must be easily and cheaply available from any phoniatric examination room. We created PEAKS (Program for Evaluation and Analysis of all Kinds of Speech disorders), a client/server recording environment. The system can be accessed from any PC with internet access, a browser, a sound card, and Java Runtime Environment (JRE) 1.5.0.6. The texts to be read and pictograms to be named are displayed in the browser. The patient's utterances are recorded by the client and transferred to the server. The ASR system analyzes the data and sends the evaluation results back to the client. The recordings are stored in an SQL database. A secure connection is used for all data transfer. A registered physician can group his patients according to disorder, create new patient entries, create new recordings, analyze patients and groups of patients. The physician has only access to his patients but physicians can share groups of patients. For the telephone data, the patient gets a handout from his physician with a unique ID and the text to be read. The server can be accessed from the public telephone system. PEAKS is used by physicians from 3 clinics of our university, collecting data from patients with CLP, TE voice, epithelium cancer in the oral cavity, and partial laryngectomy. More information can be found at http://www5.informatik.uni-erlangen.de/Research/Projects/Peaks.

## 4   Experimental Results

### 4.1   Subjective Evaluation

A group of 5 voice professionals subjectively estimated the intelligibility of the 41 patients while listening to a play-back of the close-talking recordings. A five-point Likert scale was applied to rate the intelligibility of each recording. In this manner an averaged mark – expressed as a floating point value – for each patient could be calculated. We assigned this mark also to the telephone recordings.

To judge the agreement between the different raters we calculated correlation coefficients and the weighted multi-rater $\kappa$ [14] for the "intelligibility" rating. The average correlation coefficient between a single rater and the average of the 4 other raters was 0.81, the weighted multi-rater $\kappa$ for the 5 raters was 0.45. A $\kappa$ value greater than 0.4 is said to show moderate agreement.

### 4.2   Automatic Evaluation

We applied the two close-talking recognizers and the two telephone speech recognizers to the accordant speech data and calculated the correlation between the WAs and the average of the experts' intelligibility rating. The WA was calculated w.r.t. the reading text and w.r.t. the transliteration. The $\kappa$ values were calculated using the recognizer as a 6th rater. For this we mapped the WAs to the Likert scale, using the thresholds that are given in Table 1.

**Table 1.** Thresholds for mapping the WA of the close-talking (c/t) and the telephone (tel) ASR systems to marks on the Likert scale for rating the intelligibility of the patients

| Mark | 5 | 4 | 3 | 2 | 1 | Mark | 5 | 4 | 3 | 2 | 1 |
|------|-----|------|------|------|------|--------|-----|------|------|------|------|
| WA c/t | < 5 | < 25 | < 40 | < 55 | ≥ 55 | WA tel | < 5 | < 15 | < 25 | < 45 | ≥ 45 |

In a second step we applied a 10-fold cross-validation multi correlation/regression analysis [15] to determine the features with the best average rank among WA and the 148 prosodic features. These features are either global or the average features calculated for words or word pairs (see above and [11,12]).

We used these features and the average expert rating for SVM regression [16,17]. Rounding the SVM regression value to the next integer we again treated the automatic result as a 6th rater and calculated the multi-rater $\kappa$.

The multi correlation/regression analysis chose the following features (in descending order):

– WA always had the the best rank.
– The global F0 mean.
– The variance of the energy maximum.
– The maximum pause duration before a word.
– The mean of the F0 regression coefficient.

In this work, only the first two features were used due to the small size of the test set.

**Table 2.** Evaluation results for the four different recognizers for the 41 patients. The WA is calculated w.r.t. the transliteration (trl) and text reference (ref), $r$ is the correlation between the WA or the SVM regression and the average expert rating. For the description of the recognizers see Section 3.1.

| reco | eval | $\mu_{WA}$ | $r_{WA}$ | $\kappa_{WA}$ | $r_P$ | $\kappa_P$ | reco | eval | $\mu_{WA}$ | $r_{WA}$ | $\kappa_{WA}$ | $r_P$ | $\kappa_P$ |
|------|------|-----------|----------|---------------|-------|------------|------|------|-----------|----------|---------------|-------|------------|
| 16/m | trl | 37.7 | -.85 | .44 | — | — | 16/p | trl | 38.6 | -.89 | .47 | .89 | .48 |
| 8/m | trl | 31.1 | -.78 | .38 | — | — | 8/p | trl | 27.5 | -.84 | .40 | — | — |
| 16/m | ref | 37.7 | -.85 | .44 | — | — | 16/p | ref | 38.6 | -.89 | .46 | .90 | .47 |
| 8/m | ref | 31.0 | -.78 | .38 | — | — | 8/p | ref | 27.5 | -.83 | .41 | — | — |

Table 2 shows the results for the 4 recognizers based on WA and WA in combination with the best prosodic feature (P). Note that $r_{WA}$ is negative, since good speakers have low Likert values and high WAs, while $r_P$ is positive since SVM regression tries to predict the average score of the human raters. Figure 1 shows the SVM regression values vs. the average experts' score as well as the regression line. The result of the 16/p recognizer and the text reference were used for the calculation of the WA.

# 5 Discussion and Summary

In the following we want to discuss the topics addressed in Section 1.

1. The recording environment is highly accepted by our clinical colleagues. One major reason is that there is no installation cost, since practically all examination rooms already have a telephone and a PC with internet access. We are currently expanding the data collection to other German clinics.
2. The results reported on 18 patients in [3,4] were mostly confirmed for the 41 patients. The best correlation (-.89) and $\kappa$ values (.47) were slightly higher than for the 18 patients (-.84 and .43). For the larger corpus, the polyphone-based recognizers produced better and more consistent results than the monophone-based ones. Thus, our assumption that the monophone models are more robust towards the strongly distorted TE speech [4] seems not to hold.
3. The results for the telephone recognizers show that the loss of information due to the telephone channel are acceptable, e.g. from -.89 and .47 for "16/p" to -.84 and .40 for "8/p", respectively. Due to the loss of quality in telephone transmission, the multiple AD/DA conversions, and the different frequency characteristics of the loudspeaker and the microphones, the overall WA for the simulated telephone calls is reduced. Also, the training data of the speech recognizer for the 8 kHz was down-sampled close-talking data and not real telephone data. We chose this way instead of using real telephone training data, since we wanted the telephone recognizer to be trained with the same training data as the recognizer for the close-talking data. Reducing the acoustic mismatch of training and evaluation data might lower the loss of correlation.
4. Adding prosodic features to the evaluation vector increases the correlation to the human experts' scores (from .89 to .90) and makes the analysis more robust. We are currently porting the prosody module to telephone speech.

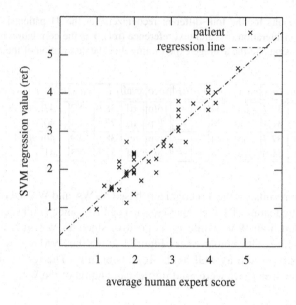

**Fig. 1.** SVM regression value for the 41 recordings in comparison to the average of the experts' intelligibility score

5. The results in Table 2 show that there is practically no difference between the results evaluated against the transliteration and against the reference text. Thus we can do without the cumbersome transliteration.

In conclusion we can say that our evaluation system provides an easy to apply, cost-effective, instrumental, and objective evaluation for TE speech. We are currently enhancing our analysis environment in order to provide a modular platform which can be easily expanded:

- From the medical point of view we can add new intelligibility tests to provide speech evaluation for a larger spectrum of speech disorders. The easy to use graphical user interface allows a fast evaluation of these tests.
- From the technical point of view we are able to plug in different ASR systems in order to provide more flexibility when realizing these new tests.
- Once a new intelligibility test is integrated and validated, it can immediately be used in clinical routine in all clinics participating. Thus PEAKS not only speeds up research studies but also helps to reduce the gap between research and practice.

## Acknowledgments

This work was funded by the Deutsche Krebshilfe (German Cancer Aid) under grant 106266, the Deutsche Forschungsgemeinschaft (German Research Foundation) under grant SCHU2320/1-1, and the Johannes-und-Frieda-Marohn Stiftung. The responsibility for the content of this paper lies with the authors.

# References

1. Likert, R.: A Technique for the Measurement of Attitudes. Archives of Psychology 140 (1932)
2. Schuster, M., Maier, A., Haderlein, T., Nkenke, E., Wohlleben, U., Rosanowski, F., Eysholdt, U., Nöth, E.: Evaluation of Speech Intelligibility for Children with Cleft Lip and Palate by Means of Automatic Speech Recognition. International Journal of Pediatric Otorhinolaryngology 70, 1741–1747 (2006)
3. Schuster, M., Haderlein, T., Nöth, E., Lohscheller, J., Eysholdt, U., Rosanowski, F.: Intelligibility of Laryngectomees' Substitute Speech: Automatic Speech Recognition and Subjective Rating. European Archives of Oto-Rhino-Larngology and Head & Neck 263, 188–193 (2006)
4. Schuster, M., Nöth, E., Haderlein, T., Steidl, S., Batliner, A., Rosanowski, F.: Can you Understand him? Let's Look at his Word Accuracy — Automatic Evaluation of Tracheoesophageal Speech (Vol. 1.) 61–64
5. Brown, D., Hilgers, F., Irish, J., Balm, A.: Postlaryngectomy Voice Rehabilitation: State of the Art at the Millennium. World J. Surg. 27(7), 824–831 (2003)
6. Schutte, H., Nieboer, G.: Aerodynamics of esophageal voice production with and without a Groningen voice prosthesis. Folia Phoniatrica et Logopaedia 54, 8–18 (2002)
7. Robbins, J., Fisher, H., Blom, E., Singer, M.: A Comparative Acoustic Study of Normal, Esophageal, and Tracheoesophageal Speech Production. Journal of Speech and Hearing Disorders 49, 202–210 (1984)
8. Bellandese, M., Lerman, J., Gilbert, H.: An Acoustic Analysis of Excellent Female Esophageal, Tracheoesophageal, and Laryngeal Speakers. and Hearing Research 44, 1315–1320 (2001)
9. Stemmer, G.: Modeling Variability in Speech Recognition. Studien zur Mustererkennung, vol. 19. Logos Verlag, Berlin (2005)
10. Schukat-Talamazzini, E., Niemann, H., Eckert, W., Kuhn, T., Rieck, S.: Automatic Speech Recognition without Phonemes. In: Proc. European Conf. on Speech Communication and Technology. Berlin, vol. 1, pp. 111–114 (1993)
11. Batliner, A., Buckow, A., Niemann, H., Nöth, E., Warnke, V.: The Prosody Module. [13], 106–121
12. Haderlein, T., Nöth, E., Schuster, M., Eysholdt, U., Rosanowski, F.: Evaluation of Tracheoesophageal Substitute Voices Using Prosodic Features. In: Proc. of 3rd International Conference on Speech Prosody, Dresden, pp. 701–704 (2006)
13. Wahlster, W. (ed.): Verbmobil: Foundations of Speech-to-Speech Translation. Springer, Berlin (2000)
14. Davies, M., Fleiss, J.: Measuring agreement for multinomial data. Biometrics 38(4), 1047–1051 (1982)
15. Cohen, J., Cohen, P.: Applied Multiple Regression/Correlation Analysis for the Behavioral Sciences. Lawrence Erlbaum Associates, Hillsdale, New Jersey (1983)
16. Platt, J.: Fast Training of Support Vector Machines using Sequential Minimal Optimization. In: Schölkopf, B., Burges, C., Smola, A. (eds.) Advances in Kernel Methods – Support Vector Learning, pp. 185–208. MIT Press, Cambridge (1999)
17. Smola, A., Schölkopf, B.: A Tutorial on Support Vector Regression. In: NeuroCOLT2 Technical Report Series, NC2-TR-1998-030 (1998)

# The LIA Speech Recognition System: From 10xRT to 1xRT

G. Linarès, P. Nocera, D. Massonié, and D. Matrouf

Laboratoire Informatique d'Avignon, LIA, Avignon, France
{georges.linares, pascal.nocera, dominique.massonie,
driss.matrouf}@lia.univ-avignon.fr

**Abstract.** The LIA developed a speech recognition toolkit providing most of the components required by speech-to-text systems. This toolbox allowed to build a Broadcast News (BN) transcription system was involved in the ESTER evaluation campaign ([1]), on *unconstrained transcription* and *real-time transcription* tasks. In this paper, we describe the techniques we used to reach the real-time, starting from our baseline 10xRT system. We focus on some aspects of the A* search algorithm which are critical for both efficiency and accuracy. Then, we evaluate the impact of the different system components (lexicon, language models and acoustic models) to the trade-off between efficiency and accuracy. Experiments are carried out in framework of the ESTER evaluation campaign. Our results show that the real time system reaches performance on about 5.6% absolute WER whorses than the standard 10xRT system, with an absolute WER (Word Error Rate) of about 26.8%.

## 1 Introduction

The LIA developed a full set of software components for speech-to-text system building, including tools for speech segmentation, speaker tracking and diarization, HMM training and adaptation... The aim of the toolkit is to provide software for transcription system design and implementation. It is composed of two main packages addressing a large part of speech-to-text related tasks. The first one contains software components for segmentation and speaker recognition. It is based on ALIZE toolkit ([2]). The second is dedicated to HMM-based acoustic modeling and decoding. This software environment allowed us to build a Broadcast News (BN) transcription system which was involved in ESTER evaluation campaign. In this paper, we describe the efforts we produced to reach the real time, starting from this baseline BN system.

The core of the transcription toolkit is constituted of a recognition engine (Speeral [3]), which is a stack decoder derived from the A* algorithm. Most of the real-time speech recognition systems used a beam-search approach, since A* performs a depth-first exploration of the search graph. Usually, the main motivation for using A* decoder relies in it's well known capacity in integrating new information sources into the search. In the first part of this paper, we investigate some methods for reaching the real time by using such an asynchronous engine. We propose an architecture for a very fast access to linguistic and acoustic resources, and we show how it can be taken advantage of the specificity of the A* decoder to improve the efficiency of pruning.

V. Matoušek and P. Mautner (Eds.): TSD 2007, LNAI 4629, pp. 302–308, 2007.

In the second part, we present our BN system and we discuss about efficiency issues related to the global transcription strategy. Then, starting from the 10xRT system, we evaluate the system configuration which leads to an optimal trade-off beetwen accuracy and decoding duration. Finally, section 4 provides some conclusion on this work.

## 2   The Speeral Decoder

### 2.1   Search Strategy

A* is an algorithm dedicated to the search of the best path in a graph. It has been used in several speech recognition engines, generally for word-graph decoding. In Speeral, the search algorithm operates on a phoneme lattice, which are estimated by using cross-word and context-dependent HMM.

The exploration of the graph is supervised by an estimate function $F(h_n)$ which evaluates the probability of the hypothesis $h_n$ crossing the node $n$:

$$F(h_n) = g(h_n) + p(h_n) \tag{1}$$

where $g(h_n)$ is the probability of the current hypothesis which results from the partial exploration of the search graph (from the starting point to the current node $n$); $p(h_n)$ is the probe which estimates the probability of the best hypothesis from the current node $n$ to the ending node.

The graph exploration is based on the function of estimate $F()$. Indeed, the stack of hypothesis is ordered on each node according to $F()$. The best paths are then explored firstly. This deep search refines the evaluation of the current hypothesis and low-probability paths are cutted-off, leading to search backtrack. It is clear that precision of the probe function is a key point for search efficiency. We have produced substantial efforts to improve the probe by integrating, as soon as possible, all available information. The Speeral look-ahead strategy is described in the next section.

### 2.2   Acoustic-Linguistic Look-Ahead

As explained previously, the probe function aims to evaluate the probability of each path which have to be developed. The more this approximation is close to the exact one, the soon a decision of leaving or developping a path is taken. Moreover, the CPU-cost of this function is critical while it is used at each node of the search graph. We use a long-term acoustic probe combined with a short-term linguistic look-ahead. The acoustic term is computed from a Viterbi-back algorithm based on context-free acoustic models. This algorithm evaluates the best acoustic probabilities from the end-point to the current one. Of course, the evaluation of all partial paths are performed once, in a first pass. Nevertheless, as explained in the last section, the probe must provide an upper limit of path probabilities. So, the best phoneme sequences are rescored by using *upper-models*. Upper models are context-free models resulting from the aggregation in a large HMM, of all context-dependant states associated to a context-free one, remaining left-Right transition constraints. Hence, the probability of emission given the upper-model is an upper-limit of path-probabilities, given any context-dependent model.

Anticipating the linguistic information (known as LMLA - Language Model Look-Ahead) enables the comparison of competing hypotheses before reaching a word boundary. The probability of a partial word corresponds to the best probability in the list of words sharing the same prefix. The probability of a partial word corresponds to the best probability in the list of words sharing the same prefix:

$$P(W^*|h) = max_i P(W_i|h)$$

where $W^*$ is the best possible continuation word and $h$ the word history (partially present in $g(h_n)$). The lexicon is stored as a PPT (Pronunciation Prefix Tree), each node containing the list of reachable words. To ensure the consistency between linguistically well formed hypotheses and pending ones, linguistic probabilities have to be computed at the same n-gram order. This means doing LMLA also at the 3-gram level. We developed a fast computation and approximation method based on a divide and conquer strategy ([4]). Our approach consists in first comparing the list $W^*$ with the list of available trained 3-grams stored in the LM. The LM is an on-disk tree structure containing lists of word probabilities at each n-gram level. Comparing lists at runtime spares most of the LM back-off computation with low overhead. The LMLA approximation does not affect the results. Moreover we introduced precomputed LMLA probabilities to speed-up the computation of the biggest lists.

### 2.3   Optimizing the Computation of Acoustic Likelihoods

The *a priori* estimation of the contribution of each component in the decoding duration is difficult, as it depends to the search strategy and to the models complexity. Nevertheless, considering the complexity of acoustic models involved in LVCSR systems, the computation of acoustic probabilities may take a large part of the decoding computational cost. [5] estimates this ratio ranges between 30% and 70% in a large vocabulary system. More recent systems use models of millions of parameters; such complexity should not be tractable without any fast calculation method. We produced substantial efforts in optimizing the management of acoustic scoring, by using an efficient caching scheme and an original method for fast-likelihood computation.

**Likelihood Access and Caching.** A* decoding and state tying lead to an asynchronous use of acoustic probabilities, at both the HMM and GMM levels:

- probabilities $P(X_{t,t+d}|H_i)$ of observation sequences $X_{t,t+d}$ given a HMM $H_i$ are firstly computed during first pass of acoustic-phonetic decoding. Rescoring with upper-models require the evaluation of $P(X_t|S_i)$, for each GMM $S_i$ matching to the best phonetic sequence;
- as the search develops a part of the exploration graph, various concurrent hypothesis are evaluated. Each of them corresponds to a phonetic sequence. It is clear that competing word-hypotheses could share some phonetic sub-sequences;
- state tying leads to involve the same state in computation of different HMM probabilities; the architecture of acoustic handler should take advantage of this state sharing.

In order to avoid multiple computation of a likelihoods, we separate clearly the search algorithm and an acoustic handler which is in charge of acoustic probabilities computing and caching. This handler is based on a two-level caching mechanism; as the search algorithm has to score an hypothesis, it requires a probability $P(X_{t,t+d}|H_i)$. The acoustic handler search this score in the level-1 cache (L1); if it is not found, this score is computed by using the Viterbi algorithm and the probabilities of emission $P(X_{t,t+d}|H_i)$. These last ones are searched in the level-2 cache (L2). When the targeted values are not found, they are computed by using a fast likelihood methods which are describe below. L1 and L2 caches are implemented as circular buffers. Moreover, likelihood computation function is written in assembly language, by using SIMD instruction set. Finally, likelihoods are computed on-demand, allowing to limit the computed scores to the ones effectively required by the search and to take benefit from the lexical and linguistic constraints. The figure 1 describes the global architecture of the acoustic handler.

**Fast Likelihood Computation.** Numerous methods for fast likelihood computation have been evaluated the last decades. Most of them rely on Gaussian selection techniques which identify, in the full set of Gaussian, the ones which contribute significantly to the frame likelihood estimate. We developed an original method which guaranties a constant precision $\epsilon$ of the likelihood approximation. This method consists in off-line clustering of Gaussian and in on-line selection of Gaussian clusters.

As proposed in some papers ([6],[5]), Gaussian are clustered by a classical k-means algorithm on the full set of Gaussian, using a minimum-likelihood loss distance. Each center of cluster is a mono-Gaussian model $G_i$ resulting of the merge of all members of the cluster.

The on-line selection process consists in selecting a set of clusters which model the observation neighborhood. It is important to note that, at contrary of classical Gaussian selection methods, the number of selected cluster is variable, according to the considered frame and to the expected precision $\epsilon$.

The clusters are selected by computing the likelihood of the frame knowing each cluster center $G_i$; these likelihoods are used for partitioning clusters into two subsets (tagged *selected* and *unselected* clusters) respecting the rules: (1) each frame likelihood knowing a selected cluster center is greater than each unselected one and (2) the sum of unselected clusters likelihood is lower than an *a priori* fixed precision threshold.

Lastly, *a posteriori* probabilities are computed using only Gaussian belonging to selected clusters. Probabilities of unselected Gaussian are estimated by backing off to the cluster probabilities.

Our experiments have shown that this method allows to decrease the number of computed likelihood by a factor 10 without impacting the WER (Word Error Rate). In adverse acoustic conditions, this method allows to remain a good acoustic precision, since computational cost is increasing.

## 3   Overview of the Broadcast News Transcription System

In addition to intrinsic difficulty of speech recognition, broadcast news transcription adds specific problems related to the signal flow continuity as well as the diversity of

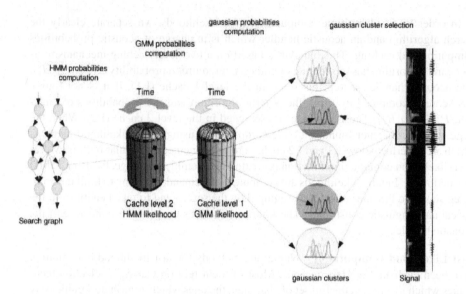

**Fig. 1.** Architecture of Speeral acoustic handler; likelihood are computed on-demand, depending to the path which are effectively developed by the search algorithm; the cache L1 stores acoustic probabilities of an observation sequence given a HMM; the L2 cache stores probabilities of an observation $X_t$ knowing the considered state $S_i$. Finally, the Gaussian probabilities are computed, or approximated by the ones of Gaussian-clusters.

the acoustic conditions. Recognizers require tractable speech segments and high level acoustic information about the nature of segments (speaker identity, recording conditions, etc.). In our system, 2 successive segmentation passes are performed. Speech segments are initially isolated from audio flow; then, a wide/narrow band segmentation identifies telephone segments. We use a method based on a hierarchical classification based of GMM classifiers and morphological rules ([7]). Speaker segmentation is achieved by a fast method ($\simeq$ 0.05xRT) based on mono-Gaussian models and the BIC-based criterion ([8]).

The 10xRT system runs two decoding passes; the first one provides transcript which is used for MLLR-based adaptation. While the same models are involved in the 2 passes, the pruning scheme changes: the first pass takes about 3xRT for about 6xRT for the final one. As we aim to reach the real time, the RT system runs only 1 pass, without any speaker adaptation. In the following, we study how we can reach real time by tuning the acoustic and linguistic models involved in the system. All tests reported in this section were performed on 3 hours extraced from the ESTER development corpus.

### 3.1    Transcription

### 3.2    Acoustic Modelling

We use a classical PLP parametrization; feature vectors are composed of 12 coefficients, plus energy, and first and seconds derivative of these 13 coefficients. At last, we perform a cepstral normalization (mean removal and variance reduction) in a 500ms sliding window. The system uses context-dependent models trained on the 90 hours of Ester transcribed data. State tying is performed by a decision tree algorithm, using acoustic context related questions.

Two sets of speaker-independent acoustic models are used: a large band model and a narrow band model, both gender-dependent. Ester corpus provides a small amount of narrow-bandwidth data; so, narrow-bandwidth models were first trained using filtered large bandwidth data (using a low-pass filter); finally, telephone models were mapped to real narrow-bandwidth data extracted from the Ester train corpus.

The 10xRT system is based on acoustic models set containing 10000 HMM for 3600 emitting states, 64 Gaussian each. Here, we use only the first pass of this system, which is about 3xRT. This acoustic model contains about 230000 Gaussian components. In spite of efficient Gaussian selection, this model is too large for real time decoding. We built two smaller model sets, composed respectively by 3600 states 24 Gaussian each (90k Gaussian, noted 90Kg) and 936 emitting states (60k Gaussian, noted 60Kg). Tests are carried out by using the pruning scheme of the 3xRT system (3xRT); CPU-time is computed on a small server (2.2GHz opteron, only one process dedicated to the test). Results show that the 90k model allows a very significant gain in term of efficiency, since the WER is increased of about 0.5% absolute.The model 330k, composed by 5200 emitting states is too large considering the amount of training data.

**Table 1.** WER of systems according to the number of Gaussian components

| Acoustic model | 60Kg | 90Kg | 230Kg | 330Kg |
|---|---|---|---|---|
| WER | 25.8% | 24.7% | 24.2% | 24.5% |
| RealTime Factor | 1.6 | 1.9 | 2.9 | 3.4 |

### 3.3  Lexical and Language Models

The linguistic resources are extracted from two corpora: newspaper Le Monde from 1987 to 2003 (330 Million words) and ESTER (960K words). The trigram language model was learned on the corpus Le Monde and ESTER training set. It is obtained by a linear combination of three models; the first two were learned on the data of Le Monde 1987-2002 and on Le Monde 2002-2003, and the last on the ESTER corpus. Lastly, these models are mixed into an unique model with interpolation coefficients determined by the ESTER development corpus entropy. The language model used by the 10xRT system is based on a lexicon of 65000 word (named 65Kw)and and language model including 16.7 Million of bigrams and about 20M of trigrams. In order to reduce the computational cost due to the size of lexicon, we built a 20000 word dictionary (named 20Kw) for which the out of vocabulary rate is about 1.2% (0.5% for 65Kw lexicon) results are compared to the ones obtained by using the 65000 word lexicon. The table 2 compares the results obtained by using this two LM, for 60Kg and 90Kg acoustic models. For these tests, the system is configured in one drastic pruning scheme (noted 1xRT), according to the targeted real-time decoding. Results of the system based on 65Kw lexicon and 90Kg acoustic model obtain very good results (24.7%WER), while being under the 2xRT; moreover, the configuration 90Kg and 65Kw represents a very good trade-off which could be reached by using a more recent processor. Finally, by using acoustic model of 90K Gaussian and a lexicon 20000 word, the system runs in about 1.0xRT and reach a WER of 26.8% (cf. table 2).

**Table 2.** WER and real-time factors for Speeral decoding according to the lexicon size, the size of acoustic models, and the pruning scheme

| System | 20Kw 60Kg 1xRT | 65Kw 60Kg 1xRT | 20Kw 90Kg 1xRT | 65Kw 90Kg 1xRT | 65Kw 90Kg 3xRT |
|---|---|---|---|---|---|
| WER | 28.5% | 27.5% | 26.8% | 25.6 | 24.7% |
| Real time factor | 0.7 | 0.9 | 1.0 | 1.3 | 1.9 |

## 4 Conclusion and Perspectives

We presented the main aspects of the LIA real time transcription system. An efficient architecture is proposed and we provide a full methodological framework for fast A* decoding. Our results shows that real-time can be reached while remaining the functional model of our baseline system. This real-time system obtained an absolute WER of 26.8% WER. This system ranked 2 in *real time transcription task* of the ESTER evaluation campaign.

## References

1. Galliano, S., Geoffrois, E., Mostefa, D., Choukri, K., Bonastre, J.F., Gravier, G.: The ESTER Phase II evaluation campaign for the rich transcription of French broadcast news. In: Proc. of the ECSCT (2005)
2. Bonastre, J.F., Wils, F., Meignier, S.: ALIZE, a free toolkit for speaker recognition. In: ICASSP'05, Philadelphia, USA (2005)
3. Nocera, P., Linarés, G., Massonié, D.: Phoneme lattice based a* search algorithm for speech recognition. In: Sojka, P., Kopeček, I., Pala, K. (eds.) TSD 2002. LNCS (LNAI), vol. 2448, Springer, Heidelberg (2002)
4. Massonié, D., Nocéra, P., Linarès, G.: Scalable language model look-ahead for LVCSR. Inter-Speech'05, Lisboa, Portugal (2005)
5. Knill, K., Gales, M., Young, S.: Use of gaussian selection in large vocabulary continuous speech recognition using HMMS. In: Proc. ICSLP'96, Philadelphia, PA, USA, vol. 1, pp. 470–474. Cambridge University, Cambridge (1996)
6. Bocchieri, E.: Vector quantization for the efficient computation of continuous density likli-hood. In: IEEE, Proc ICASSP'93, Speech Research Dept., AT&T Lab., Murray Hill, vol. 2, pp. 692–696. IEEE, Los Alamitos (1993)
7. Nocera, P., Fredouille, C., Linares, G., Matrouf, D., Meignier, S., Bonastre, J.F., Massonié, D., Béchet, F.: The LIA's French broadcast news transcription system. In: SWIM: Lectures by Masters in Speech Processing, Maui, Hawaii (2004)
8. Meignier, S., Moraru, D., Fredouille, C., Bonastre, J.F., Besacier, L.: Step-by-step and inte-grated approaches in broadcast news speaker diarization. In: Odyssey'04. Toledo University, vol. 20, pp. 303–330 (2004)

# Logic-Based Rhetorical Structuring for Natural Language Generation in Human-Computer Dialogue

Vladimir Popescu[1,2], Jean Caelen[1], and Corneliu Burileanu[2]

[1] Laboratoire d'Informatique de Grenoble, Grenoble Institute of Technology, France
{vladimir.popescu, jean.caelen}@imag.fr
[2] Faculty of Electronics, Telecommunications and Information Technology,
University "Politehnica" of Bucharest, Romania

**Abstract.** Rhetorical structuring is field approached mostly by research in natural language (pragmatic) interpretation. However, in natural language generation (NLG) the rhetorical structure plays an important part, in monologues and dialogues as well. Hence, several approaches in this direction exist. In most of these, the rhetorical structure is calculated and built in the framework of Rhetorical Structure Theory (RST), or Centering Theory [7], [5]. In language interpretation, a more recent formal account of rhetorical structuring has emerged, namely Segmented Discourse Representation Theory (SDRT), which alleviates some of the issues and weaknesses inherent in previous theories [1]. Research has been initiated in rhetorical structuring for NLG using SDRT, mostly concerning monologues [3]. Most of the approaches in using and / or approximating SDRT in computer implementations lean on dynamic semantics, derived from Discourse Representation Theory (DRT) in order to compute rhetorical relations [9]. Some efforts exist in approximating SDRT using less expressive (and expensive) logics, such as First Order Logic (FOL) or Dynamic Predicate Logic (DPL), but these efforts concern language interpretation [10]. This paper describes a rhetorical structuring component of a natural language generator for human-computer dialogue, using SDRT, approximated via the usage of FOL, doubled by a domain-independent discourse ontology. Thus, the paper is structured as follows: the first section situates the research in context and motivates the approach; the second section describes the discourse ontology; the third section describes the approximations done on vanilla SDRT, in order for it to be used for language generation purposes; the fourth section describes an algorithm for updating the discourse structure for a current dialogue; the fifth section provides a detailed example of rhetorical relation computation. The sixth section concludes the paper and gives pointers to future research and improvements.

## 1 Introduction

This paper describes a module for rhetorical structure computation, for a natural language generator in human-computer dialogue. More specifically, pragmatic aspects concerning rhetorical coherence of speech turns in dialogue are discussed. It is known that when pragmatic aspects are concerned, total domain or application independence is an elusive goal, since the "pragmatics" of a fragment of speech strongly depends on the context in which that fragment occurs.

V. Matoušek and P. Mautner (Eds.): TSD 2007, LNAI 4629, pp. 309–317, 2007.

The research described in this paper is performed in the framework of a project consisting in the development of a task-oriented dialogue system, designed in a generic manner (i.e., easily customizable to different tasks and applications) and applied to several domains, such as meeting room reservation, in an enterprise, or book reservation, in a library [2]. Our team has already designed and implemented components regarding dialogue management, task planning, semantic parsing of user's requests, and pragmatic interpretation of these. For pragmatic interpretation purposes, SDRT has been adapted and extended for dialogue, in order to integrate the concept of *topic* [2]. However, the component responsible for generating system's responses has been reduced to a template-based generator, thus providing little flexibility and robustness with respect to contextual variations, and lacking pertinence with respect to dialogue dynamics [6].

As for the use of SDRT in NLG, research of L. Danlos or L. Roussarie could be put forward [3]; for the extensions performed on SDRT in order to integrate aspects related to dialogue (however, for interpretation purposes), work of L. Prevot and his team could be mentioned [4]. For approximations or reformulations of SDRT using less "heavy" logics, work of M. Staudacher [10] could be mentioned, for example.

The novelty of the approach described in this paper resides in the usage of a logic formalism less complex (from a computational point of view) than dynamic semantics or logics - first order predicate logic, parameterized by a discourse ontology specifying the scopes of the entities invoked in logic formulas. The advantage of this approach is, at a practical level, that it allows the usage of software tools and environments designed for FOL (for instance, several flavors of PROLOG) in a straightforward manner. Then, from a methodological point of view, the approach proposed is task-independent, since the discourse ontology does not depend on the constraints imposed by a specific application domain; the coupling with the task-specific aspects is made via a task ontology, handled by the task controller in the dialogue system [2], [6]. This allows for an augmented portability of the generation module, thus lowering the costs for the adaptation to a new task.

The following section describes the particular elements chosen for the formal expression of logic formulas in a discourse ontology (that is generic, i.e, task-independent); the third section presents the approximation of a fragment of SDRT using FOL and integrating a set of semantics for the rhetorical relations being used in dialogues; the fourth section describes an algorithm for updating the discourse structure (called Segmented Discourse Structure - SDRS) of a current dialogue; the fifth section provides a detailed example of rhetorical relation computation in the SDRS update process. Finally, the sixth section concludes the paper and provides pointers to further work.

## 2    Task-Independent Discourse Ontology

A logic formula expressed in FOL contains five types of information: (i) connectives: $\wedge, \vee, \neg, \Rightarrow$, (ii) quantifiers: $\forall, \exists$, (iii) objects, and (iv) predicates.

Type (i) entities can link different logic formulas; type (ii) entities can precede type (iii) entities that, in their turn, can be followed by type (iv) entities. Type (iv) entities can be preceded by type (iii) entities.

In order to be able to state in detail the content of the logic formulas expressing utterances, facts and rules in the ontology (described hereafter), one takes into account that the dialogues concerned involve negotiation on time intervals. Thus, a taxonomy of possible *moments of time* is defined, in order to augment the expressive power of FOL. These temporal markers are: (i) $t_\#$ - present conditional, (ii) $t_+$ - future and "new", (iii) $t$ - present, (iv) $t_-$ - past simple and "old", (v) $t_=$ - past perfect, (vi) $t_\pm$ - past conditional, (vii) $t_\mp$ - future in the past, (viii) $t_\exists$ - a unique moment not precisely situated on the time axis, (ix) $t_\forall$ - any moment, eternal, permanent.

As for the discourse ontology for "pragmatic" generation (a.k.a. rhetorical structuring), its elements are represented by entities in a knowledge base. This knowledge base specifies a set of objects, functions and predicates involved in the expression of the *semantics* of the rhetorical relations used, in the framework of SDRT, and is used by the generator to infer the rhetorical relations between utterances[1].

For this knowledge base, a minimal set of predicates is defined: (i) $\in$ - MemberOf, (ii) $\subset$ - SubclassOf, (iii) $\ni$ - ClassOf, (iv) $\supset$ - SuperclassOf, (v) $\cap = \emptyset$ - Disjoint, (vi) $\cup = $ All - ExhaustiveDecomposition, (vii) $(\cap = \emptyset) \wedge (\cup = $ All$)$ - Partition. In order to express **measures**, one defines the predicates: smaller, greater, and equals. For temporal event handling, a specific set of predicates could be defined, as in [8], but in our generator these latter ones are not used, therefore are not mentioned here.

In order to enforce the structure of the ontology so that it remains generic with respect to the task and at the same time flexible and useful, the ontology contains the following particular entities: (i)functions: $SARG()$, $Plan()$; (ii) predicates: answer(), bad_time(), good_time(), emitter(), receiver(), topic(), enounce(), question().

The discourse ontology contains elements (terminals, i.e, without subtypes) whose properties must be specified via a set on axioms, denoting by $\Omega$ the set of values for the object in the task ontology[2], by $K(\alpha)$ the clause logically expressing the semantics of the utterance $\alpha$ and by $t_\alpha$, the moment in time when utterance $\alpha$ is produced. Hence, the specific predicates and functions are: equals($\alpha$, question), equals($\alpha$, enounce), equals($\beta$, confirmation($\alpha$)), equals($\beta$, answer($\alpha$)), topic($\alpha$), emitter($\alpha$), entails $(\alpha, \beta)$, $SARG(\alpha)$, good_time($\Delta t_\beta$), bad_time($\Delta t_\beta$), and $\Delta t_\beta$. For example, the axiom specifying the semantics of topic($\alpha$)is given below:

topic($\alpha$) ::= ExhaustiveDecomposition($i, j; v_i, \omega_j$) $\wedge$ MemberOf($v_i, K(\alpha)$)$\wedge$
MemberOf($\omega_j, \Omega$) $\wedge$ ($\exists k$ : equals($v_k, \omega_j$) $\wedge$ MemberOf($v_k, K(\alpha)$)).

One notices that, unlike the classical SDRT, the notion of topic of an utterance is defined here in terms of sets of objects in the domain ontology, referred to in a determined manner[3] in the utterance. Hence, the topic relations between utterances are computed using the task/domain ontology, handled by the task controller.

---

[1] The utterances are represented as clauses in FOL.

[2] The rhetorical structuring component in the linguistic generator adapts to the task in this manner: the task-dependent aspects are provided by the task manager that handles, in its turn, a task ontology, different from the discourse ontology.

[3] Here, an object is referred to in a determined manner if and only if the logic variable designating the object in the ontology has an allowed value, assigned in the logic clause expressing the semantics of the utterance.

# 3   SDRT Adaptation for Language Generation in Dialogue

For the goals supposed by our project, a subset of SDRT has been chosen, namely 17 rhetorical relations, specified below; these rhetorical relations are due to be used at the "pragmatic" level for rhetorical structuring of the utterances to be generated. These rhetorical relations are grouped in three categories:

– **first-order** relations - relations strongly related to *temporal* aspects in dialogue; these relations will be used in an approximative manner, specific to the characteristics of the type of dialogue concerned (see below); these relations are: Q-Elab, IQAP, P-Corr and P-Elab, with the informal semantics stated in [1];
– **second-order** relations - relations less constrained by the temporal context of the dialogues concerned; by consequence, these relations are used in the generator in the same manner as specified by vanilla SDRT [1]; these relations are: $Background_q$, $Elab_q$, $Narration_q$, QAP, ACK and NEI;
– **third-order** relations - relations specific to monologues and used to relate utterances within a speech turn, generated by one speaker, either the human ($U$), or the machine ($M$); these relations are: Alternation, Background, Consequence, Elaboration, Narration, Contrast and Parallel, with informal semantics defined in [1].

These rhetorical relations are speech act types, reflecting the dependencies between the success of the (performance of the) current speech act and the content of a set of preceding acts in dialogue. Accordingly, a dialogue is considered pragmatically *coherent* if and only if for any utterance there exists at least one connection, via a discourse relation, between it and another speech act belonging to the dialogue history (i.e., to the set of preceding speech turns in dialogue). Such a connection is called SARG ("Speech Act Related Goal").

In language pragmatic interpretation, SARGs are recovered out of the discourse context, whereas in generation, these SARGs are defined by the speech acts *enclosed* in the clauses come from the dialogue controller [2], being thus priorly known. Hence, in language generation the issue is not the identification of the SARGs, but their *placement* in an existing discourse context.

In order to enforce rhetorical coherence by taking into account pragmatic aspects in language generation, we consider only a particular type of dialogues, encountered in meeting room reservation, book reservation in a library, or, more generally, in situations where the agreement on time intervals for the usage of a specified set of resources is essential. Hence, the characteristics of such a dialogue are: (i) the limited scope of the goals, and (ii) the importance of temporal aspects.

As in work reported by Schlangen et al. [9], the SDRT is approximated in that the answers that the system is supposed to generate are restricted to information concerning (i) description of time intervals of availability of a certain book or item, in relation to the interval $\Delta t$, convenient for the user; (ii) the adequacy or non-adequacy of the interval $\Delta t$, i.e., whether it is a "good" time interval or not (in other words, whether in the interval $\Delta t$ the user having made a request to the system may have access to the book he wishes for, or not).

Therefore, the generator ought to produce utterances situating the time interval proposed by the user $U$, $\Delta t$, in terms of appropriate time for the loan (good_time($\Delta t$)), or not appropriate (bad_time($\Delta t$)).

The rules for computing the discourse relations situating the communicative intention determined by the dialogue controller, with respect to discourse context are monotonic in our model, while in vanilla SDRT there are non-monotonic. This choice is motivated by the possibility to avoid thus computationally intensive consistency verifications. These verifications would have been necessary if one had used non-monotonic rules, since the truth value of the semantics inferred via these rules could have changed at each discourse structure update.

The knowledge of the prior general goal of the users in dialogue, in relation to the temporal aspects of the conversation, allows us to state the semantics of the first-order rhetorical relations. For example, the semantics of Q-Elab is specified below:

Q-Elab$(\alpha, \beta)$ ::= equals $(\alpha,$ enounce$) \wedge$ equals$(\beta,$ question$) \wedge \neg$Disjoint$(\Delta t_\beta, SARG(\alpha))$
$\wedge\ (\forall$ answer $(\beta)) \neg \Rightarrow \neg SARG(\alpha)$.

In words, this formula states that the relation Q-Elab can be inferred between utterances $\alpha$ and $\beta$ if and only if $\beta$ is a question so that any answer to it elaborates a plan for satisfying the SARG of $\alpha$.

The semantics of the second-order and third-order rhetorical relations are specified in a similar manner; for space considerations, their semantics are not given here.

## 4   Dialogue SDRS Update

This section presents an algorithm for updating the discourse structure (SDRS) of a dialogue, adding an utterance to it[4]. More specifically, for a clause $K(\alpha)$, come from the dialogue controller, expressing the communicative intention to be put in an utterance $\alpha$, one wishes to find the set $\Re$ of discourse relations relating utterance $\alpha$ to a set $\aleph$ of previous utterances in dialogue.

The algorithm for updating the dialogue SDRS is presented below:

1. for each utterance labeled $\alpha$ to add to the SDRS (and generate):
   (a) read $K(\alpha)$, through a query to the dialogue controller;
   (b) perform initializations: $\aleph(\alpha) \leftarrow \emptyset$, $\Re(\alpha) \leftarrow \emptyset$;
   (c) for each utterance $\beta_i$ in $\bar{\aleph}$ $(i = 1, ...|\bar{\aleph}|)$:
       i. for each rhetorical relation $\rho_j$ known $(j = 1, ...17)$:
          A. retrieve the logic formulas $K(\beta_i)$ and $\Sigma_j$;
          B. compute the truth value of $\Sigma_j\ (\sigma\ (K(\alpha), K(\beta_i)))$ and denote it by $\gamma$;
          C. if $\gamma$ =FALSE, then go to 1.(c).i.;
             else, perform $\Re \leftarrow \Re \cup \{\rho_j\}$ and $\aleph \leftarrow \aleph \cup \{\beta_i\}$ and go to 1.(c).i.;
          D. perform: $\aleph(\alpha) \leftarrow \aleph(\alpha) \cup \aleph$ and $\Re(\alpha) \leftarrow \Re(\alpha) \cup \Re$;
2. compute the truth value of equals($|\aleph(\alpha)|, 0) \wedge$ equals($|\Re(\alpha)|, 0$) and denote it by $\upsilon$:
   (a) if $\upsilon$ =FALSE, then:

---

[4] An utterance is added to a SDRS if and only if at least a rhetorical relation between it and utterances in the dialogue history is found.

     i. add $\alpha$ to the utterances in the current SDRS;
    ii. add $\Re(\alpha)$ to the rhetorical relations in the current SDRS;
(b)  else build a new SDRS having one utterance, $\alpha$, and no rhetorical relations.

In the algorithm specified above, we denoted by $\bar{\aleph}$ the set of *all* the utterances preceding $\alpha$ in the current dialogue and we searched, for each utterance $\beta_i$ in the dialogue history, the rhetorical relations $\rho$ so that $\Sigma_\rho\left(\sigma\left(K(\alpha), K(\beta_i)\right)\right) \neq$ FALSE, where $\sigma$ denotes the permutation operation, $\Sigma_\rho$ denotes the formal semantics of $\rho$, $|X|$ denotes the number of elements in the set $X$, and $\cup$, the set union operator.

This algorithm builds a *complete* discourse structure, in that all the logically possible rhetorical relations between pairs of utterances (out of which at least one is due to be generated by the machine) are found; the possible contradictions involved by this approach (stemming from the fact that a rhetorical relation might be revised by consequent rhetorical relations or future utterances) are alleviated by the monotonicity assumption governing the logic framework used. This leads to the creation of all the potential discourse structures, without finding the most coherent one (in the sense discussed in [1]).

## 5   Rhetorical Relations Computation: Extended Example

In the algorithm presented in the previous section for updating the SDRS for a dialogue, the essential step is 1.(c).i.B; this step will be detailed through an example in this section. More specifically, one will show the manner in which the truth values for the clauses $\Sigma_j\left(\sigma\left(K(\alpha), K(\beta_i)\right)\right)$ are computed, where $i$ spans over the set of utterances, available as logic clauses in a discourse structure, and $j$, over the set of rhetorical relations known.

We consider the following dialogue fragment, in French language (English translations are provided in italics below each utterance):

$U : \alpha$ : Je lirai ce livre lundi.
(*I will read this book on Monday.*)
$M : \beta$ : Est-ce bien pour vous à 14 h ?
(*Is it good for you at 2 P.M.?*)

In fact, the utterance $\alpha$ is come from the user, thus available at the same time as text and as clause expressing the meaning of the utterance[5], while the utterance $\beta$ is available only as a clause, expressing a communicative intention, computed by the dialogue controller [2], [6]; its textual form is due to be determined by the language generator. More specifically, between utterances $\alpha$ and $\beta$ the rhetorical relation Q-Elab can be inferred. However, the generator cannot know it in advance, thus it has to try each of the 17 rhetorical relations used; we suppose, for simplicity, that it is precisely the relation Q-Elab($\alpha$, $\beta$) that is checked. Hence, the following processing stages are performed:

1. Find the clauses expressing the semantics of the utterances $\alpha$ and $\beta$:
   $- \alpha \mapsto K(\alpha) ::= \exists X, Y :$ object$(X) \wedge$ equals$(X, $ book$) \wedge$ agent$(Y) \wedge$ equals$(Y, U) \wedge$
   equals$(t_\alpha, t) \wedge$ read$(Y, X) \wedge$ equals$(\Delta t_\alpha, t_+)$;

---

[5] The clause associated to an utterance come from the user is computed by a semantic parser together with a pragmatic interpreter [6].

$- \beta \mapsto K(\beta) ::= \exists X, Y : \text{object}(X) \wedge \text{equals}(X, \text{book}) \wedge \text{agent}(Y) \wedge \text{equals}(Y, U) \wedge$
$\text{read}(Y, X) \wedge \text{greater}(t_\beta, t_\alpha) \wedge \text{equals}(\Delta t_\beta, '14h') \wedge \text{equals}(\Delta t_\beta, t_+) \wedge$
$\text{equals}(\text{good\_time}(\Delta t_\beta), ?);$

2. Retrieve the semantics of the rhetorical relation currently checked:

$\Sigma_{\text{Q-Elab}} ::= \text{equals}(\alpha, \text{enounce}) \wedge \text{equals}(\beta, \text{question}) \wedge \neg \text{Disjoint}(\Delta t_\beta, SARG(\alpha)) \wedge$
$(\forall \text{ answer}(\beta) \neg \Rightarrow \neg SARG(\alpha));$

3. Expand each entity in the semantics of the rhetorical relation currently checked:

   $- \text{equals}(\alpha, \text{enounce}) ::= \neg \text{equals}(\alpha, \text{question}) \mapsto \forall v : \text{MemberOf}(v, K(\alpha)) \vee \exists \omega :$
   $\text{MemberOf}(\omega, \Omega) \vee \text{equals}(v, \omega);$

   $- \text{equals}(\beta, \text{question}) ::= \exists v' : \text{MemberOf}(v', K(\beta)) \wedge \neg \exists \omega' : \text{MemberOf}(\omega', \Omega) \wedge$
   $\text{equals}(v', \omega');$

   $- SARG(\alpha) ::= \exists X, Y, \theta : \text{object}(X) \wedge \text{equals}(X, \text{book}) \wedge \text{agent}(Y) \wedge \text{equals}(Y, U) \wedge$
   $\text{good\_time}(\theta) \wedge \text{equals}(\theta, \Delta t_\alpha) \wedge \text{greater}(\theta, '\text{lundi}');$

   $- \Delta t_\beta ::= '14h' \wedge '\text{lundi}';$

   $- \forall \text{answer}(\beta) ::= \forall \delta : \text{equals}(\delta, \text{answer}(\beta)) \mapsto \forall \delta : \text{greater}(t_\delta, t_\beta) \wedge \text{equals}(\text{topic}(\delta),$
   $\text{topic}(\beta)) \wedge \forall v'' : \text{MemberOf}(v'', K(\delta)) \Rightarrow \exists \omega'' : \text{MemberOf}(\omega'', \Omega) \wedge \text{equals}(v'', \omega'');$

   $- \neg SARG(\alpha) ::= \forall X, Y, \theta : \text{object}(X) \vee \text{equals}(X \text{book}) \vee \text{agent}(Y) \vee \text{equals}(Y, U) \vee$
   $\text{bad\_time}(\theta) \vee \text{equals}(\theta, \Delta t_\alpha) \vee \text{greater}(\theta, '\text{lundi}');$

   $- \text{topic}(\beta) ::= \text{ExhaustiveDecomposition}(\text{book}, \text{read}, \text{good\_time}('14h'),$
   $\text{good\_time}('\text{lundi}'), t_+);$

   $- \text{good\_time}(\theta) ::= \exists \gamma, \pi : \neg \text{Disjoint}(\text{topic}(\gamma), \text{topic}(\pi)) \wedge \text{smaller}(t_\alpha, t_\pi) \wedge$
   $(\text{SubclassOf}(\theta, \Delta t_\alpha) \vee \text{equals}(\theta, \Delta t_\alpha)) \wedge \pi : \text{equals}(\Delta t_\pi, \theta);$

4. Compute the truth value of each clause between conjunctions, in the semantics of
   the rhetorical relation currently checked:

   $- \text{equals}(\alpha, \text{enounce}) \wedge \text{equals}(\beta, \text{question}) \rightarrow \text{TRUE}$; this expression is obtained
   substituting the semantics of $\alpha$ in the semantics of the clause shown above and
   exploring in the discourse ontology *and* in the task ontology ($\Omega$) all the possible
   values for the variables involved;

   $- \neg \text{Disjoint}(\Delta t_\beta, SARG(\alpha)) \rightarrow \text{TRUE}$; this results directly from the clause:
   $\text{SubclassOf}(\Delta t_\beta, \Delta t_\alpha) \wedge \text{SubclassOf}(\Delta t_\alpha, SARG(\alpha));$

   $- \forall \text{answer}(\beta) \neg \Rightarrow \neg SARG(\alpha) \mapsto \forall \delta : \text{equals}(\delta, \text{answer}(\beta)) ::= \forall \delta : \text{greater}(t_\delta, t_\beta) \wedge$
   $\text{equals}(\text{topic}(\delta), \text{ExhaustiveDecomposition}(\text{book}, \text{read}, \text{good\_time}('14h'),$
   $\text{good\_time}('\text{lundi}'), t_+)) \wedge \forall v'' : \text{MemberOf}(v'', K(\delta)) \Rightarrow \exists \omega'' : \text{MemberOf}(\omega'', \Omega) \wedge$
   $\text{equals}(v'', \omega'') \neg \Rightarrow \forall X, Y : \text{object}(X) \vee \text{equals}(X, \text{book}) \vee \text{agent}(Y) \vee \text{equals}(Y, U) \vee$
   $\text{bad\_time}(\Delta t_\beta) \vee \text{equals}(\Delta t_\beta, \Delta t_\alpha) \vee \text{greater}(\Delta t_\beta, '\text{lundi}');$

   the last formula was obtained substituting the predicates and functions by their explicit definitions, and the variable $\theta$, by $\Delta t_\beta$; then, by *skolemization* of the variables
   in this formula, one has:

   $\forall \text{ answer}(\beta) \neg \Rightarrow \neg SARG(\alpha) \mapsto \forall \delta : \text{equals}(\delta, \text{answer}(\beta)) \mapsto \forall \delta : \text{greater}(t_\delta, t_\beta) \wedge$
   $\text{equals}(\text{topic}(\delta), \text{ExhaustiveDecomposition}(\text{book}, \text{read}, \text{good\_time}('14h'),$
   $\text{good\_time}('\text{lundi}'), t_+)) \wedge \forall v'' : \text{MemberOf}(v'', K(\delta)) \Rightarrow \text{MemberOf}(\omega_0, \Omega) \wedge \text{equals}(v'',$
   $\omega_0) \neg \Rightarrow \forall X, Y : \text{object}(X) \vee \text{equals}(X, \text{book}) \vee \text{agent}(Y) \vee \text{equals}(Y, U) \vee \text{bad\_time}(\Delta t_\beta)$
   $\vee \text{equals}(\Delta t_\beta, \Delta t_\alpha) \vee \text{greater}(\Delta t_\beta, '\text{lundi}');$

   then, universal quantifiers are eliminated and the possible paths in the task ontology
   are explored, obtaining non-contradiction, hence the value TRUE.

This example shows that it is possible, in principle, to compute a rhetorical relation connecting a current utterance (due to be generated in textual form), available only as a logic formula, to a dialogue in progress, using FOL and a task-independent discourse ontology. The computational costs of the algorithm proposed is yet to be evaluated using a dialogue corpus (acquired by the Wizard of Oz method [2], such as the PVE - "Portail Vocal pour l'Entreprise" corpus [6], [11]).

## 6   Conclusions and Further Work

This paper has presented a rhetorical structuring component of a natural language generator for human-computer dialogue. The pragmatic and contextual aspects are taken into account communicating with a task controller providing domain and application-dependent information, structured in a task ontology. In order to achieve the goal of computational feasibility and genericity, SDRT has been strongly adapted to natural language generation in dialogue. Thus, a discourse ontology has been defined and several axioms structuring it have been specified; moreover, specific predicates and functions have been given a formal account. Then, using this ontology, a set of semantics, in first-order predicate logic, has been specified for a fragment of SDRT. The advantage of this approach resides in the possibility to readily use software tools designed for FOL and in the relative simplicity of the formal statement; this latter point allows for straightforward extensions or customizations to different types of dialogue (e.g. tutoring dialogue). The implementation of the rhetorical structuring component (described in this paper) in ISO PROLOG is currently under way.

However, several improvements could be brought to current processing stages, mostly with respect to the computational cost involved and secondly regarding the precision and reliability of the rhetorical relation computation process. The reduction of discourse structure updating time could be achieved by limiting the dialogue history, for a current utterance, to $N$ previous utterances; psycholinguistic studies motivate a choice of $N = 7$ [8]. The goal of augmenting the reliability of the inference could be achieved by limiting the number of candidate rhetorical relations, for a given pair of utterances; this could be done by using speech acts to characterize the utterances from an illocutionary point of view [2], [11]. Thus, for a given pair of utterances, a mapping is built between the corresponding pair of speech acts and the set of possible rhetorical relations connecting them; in this respect, a study of our team has already been done [11].

## References

1. Asher, N., Lascarides, A.: Logics of Conversation. Cambridge University Press, Cambridge (2003)
2. Caelen, J., Xuereb, A.: Interaction et pragmatique, Editions Hermés, Paris (2007)
3. Danlos, L., El-Ghali, D.: A complete integrated NLG system using AI and NLU tools. In: Proceedings of COLING'02, Taiwan (2002)
4. Maudet, N., Muller, Ph., Prevot, L.: Tableaux conversationnels en SDRT, Actes de TALN 2004, Fès, Maroc (2004)

5. McTear, M.F.: Spoken Language Technology: Enabling the Conversational User Interface, ACM Computing Surveys, 34(1) (2002)
6. Nguyen, H.: Dialogue homme–machine : Modélisation de multisession, PhD Thesis, Joseph Fourier University, Grenoble, France (2005)
7. Reiter, E., Dale, R.: Building Natural Language Generation Systems. Cambridge University Press, Cambridge (2000)
8. Russell, S., Norvig, P.: Artificial Intelligence: A Modern Approach. Prentice Hall, Englewood Cliffs (2003)
9. Schlangen, D., et al.: Resolving Underspecification using Discourse Information. In: Proceedings of BI-DIALOG 2001, Springer, Heidelberg (2001)
10. Staudacher, M.: SDRT Reformulated using DPL, Term Paper, Bielefeld University (2005)
11. Xuereb, A., Caelen, J.: Actes de langage et relations rhétoriques en dialogue homme–machine, Séminaire Logique et Dialogue, France (2004)

# Text-Independent Speaker Identification Using Temporal Patterns

Tobias Bocklet, Andreas Maier, and Elmar Nöth

University of Erlangen Nuremberg, Chair for Pattern Recognition,
Martensstr.3, 91058 Erlangen, Germany
Andreas.Maier@informatik.uni-erlangen.de

**Abstract.** In this work we present an approach for text-independent speaker recognition. As features we used Mel Frequency Cepstrum Coefficients (MFCCs) and Temporal Patterns (TRAPs). For each speaker we trained Gaussian Mixture Models (GMMs) with different numbers of densities. The used database was a 36 speakers database with very noisy close-talking recordings. For the training a Universal Background Model (UBM) is built by the EM-Algorithm and all available training data. This UBM is then used to create speaker-dependent models for each speaker. This can be done in two ways: Taking the UBM as an initial model for EM-Training or Maximum-A-Posteriori (MAP) adaptation. For the 36 speaker database the use of TRAPs instead of MFCCs leads to a frame-wise recognition improvement of 12.0 %. The adaptation with MAP enhanced the recognition rate by another 14.2 %.

## 1 Introduction

The extraction of speaker-dependent information out of the voice of the user, so that a person can be identified or additional speaker specific information is obtained, is an important task these days. Speaker-dependent information is the identity of a speaker, the language, the age, the gender, or the channel he or she is calling from.

These pieces of information about the identity of the speaker or specific characteristics of the person are helpful for several applications. Identification of a person can be used to allow or restrict a person the use of certain services or the access to certain places. In these cases the user does not need to have a password, an account, a personal identification number (PIN), or a door-key anymore. The access is granted or denied only by the person's voice. It is also possible to perform the identification process in a secure way over the telephone.

In our approach each speaker is modeled by a *Gaussian Mixture Model* (GMM). To train the system first of all a Universal-Background-Model (UBM) is created comprising the complete amount of training data. This is achieved by the EM-Algorithm. The UBM is then used to create a speaker model in two ways: Either EM-Training is performed and the UBM is needed as an initial speaker model or Maximum-A-Posteriori (MAP) adaptation is applied, where the UBM is combined with the speaker-dependent training data. We used two different features in this work: *Mel Frequency-Cepstrum-Coefficients* (MFCCs), which extract the features over a short temporal context (16 ms)

V. Matoušek and P. Mautner (Eds.): TSD 2007, LNAI 4629, pp. 318–325, 2007.

and *TempoRAl Patterns* (TRAPs). TRAPs calculate the features over a very long temporal context (150 ms).

For training and evaluation we employed a database provided by the company ME-DAV (www.medav.com). The database is called SET-M. The Verbmobil [1] database was used to generate transformation matrices for the dimension reduction of the TRAPs by *Linear Discriminant Analysis* (LDA). These two databases are presented in the following.

## 2   Databases

### 2.1   SET-M

The SET-M-corpus contains speech recordings of 36 persons, each of them reading two newspaper articles. The texts are semantically different. One text is a newspaper article dealing with computer viruses, the other article is about children who have attention deficit disorder (ADD). The data was recorded by a close-talking microphone. In order to simulate telephone quality, the data was $\mu$-law coded. Additionally it was artificially corrupted by convolution with Gaussian noise. In total 84 min of speech was available, recorded with a sample rate of 22kHz and re-sampled to 16kHz. The computer virus text was used for training, the other one for testing. The total amount of the training set was 45 min and the length of the test set was 39 min respectively.

### 2.2   Verbmobil

The Verbmobil (VM) database (see [1]) is a widely used speech collection. We used a German subset of the whole corpus which was already investigated in [2]. The scenario of the corpus is human-human communication with the topic of business appointment scheduling. It contains in total 27.7 hours of continuous speech by 578 speakers of which 304 were male and 274 were female. The size of the vocabulary is 6825 words. On average each of the 12,030 utterances contains 22 words. The data of this corpus was transliterated and a forced alignment was performed. This produced phonetic labels for each speech frame. These labels are then utilized to train the transformation matrix of the Linear Discriminant Analysis which is used to reduce the dimension of our TRAPs from 556 to 24.

## 3   Applied Methods

### 3.1   Features

As features the commonly used *Mel Frequency Cepstrum Coefficients* (MFCCS) and *TempoRAl Patterns* (TRAPs) are employed. MFCCs calculate the features on a short temporal context but they take the complete frequency domain into consideration. TRAPs examine each frequency band of the recordings separately over a very long temporal context.

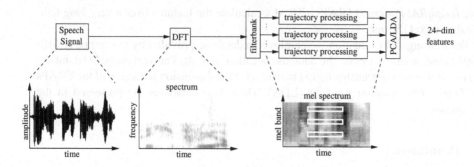

**Fig. 1.** Feature extraction for Temporal Patterns

**Mel Frequency Cepstrum Coefficients.** The 24 dimensional MFCCs consist of 12 static and 12 dynamic components. The 12 static features are composed by the spectral energy and 11 cepstral features. Furthermore the 12 dynamic features are calculated as an approximation of the first derivative of the static features using a regression line over 5 time frames. The time frames are computed for a period of 16 ms with a shift of 10 ms.

**Temporal Patterns.** The TRAPs we used in this work are quite similar to the original approach of Hermansky ([3]). The main difference of our approach are the time trajectories and their processing. Fig. 1 shows the complete extraction method. The time trajectories consider a long temporal context (150 ms) of 18 mel-bands. These mel-bands are generated by a convolution of the spectrum with triangular filter-banks. Each trajectory is smoothed by a Hamming window and transformed by application of the discrete Fast Fourier Transformation afterwards. These magnitudes in the frequency domain are then filtered by canceling all frequencies except the interval from 1 to 16Hz. A detailed explanation can be found in [4]. The fusion of the trajectories combined with a dimension reduction is not performed by neural networks, as in the original paper, but by concatenation of the high-dimensional features and application of either *Linear Discriminant Analysis* (LDA) or *Principal Component Analysis* (PCA) afterwards. The result of this dimension reduction were 24-dimensional features, as in case of MFCCs.

To train the transformation matrices of the LDA transform, labeled data was needed. We decided to use the Verbmobil database, because the data of this corpus was already transliterated and forced aligned. This produced labels in form of 47 German phonetic classes.

### 3.2 Classifier Specifications and Test Phase

In this work the speakers are modeled by *Gaussian Mixture Models* (GMMs) as described in [5]. Each speaker $\lambda$ is modeled by $M$ unimodal weighted Gaussian Distributions:

$$p(\boldsymbol{x} \mid \lambda) = \sum_{i=1}^{M} w_i p_i(\boldsymbol{x}). \qquad (1)$$

with

$$p_i(\boldsymbol{x}) = \frac{1}{(2\pi)^{D/2} \mid \boldsymbol{K_i} \mid^{1/2}} e^{-(1/2)(\boldsymbol{x}-\mu_i)^T \boldsymbol{K_i}^{-1}(\boldsymbol{x}-\mu_i)} \qquad (2)$$

where $\mu_i$ denotes the mean vector and $\boldsymbol{K_i}$ the covariance matrix of the Gaussians. Unlike [5] we used full covariance matrices in our work, because preliminary comparisons showed a slight advantage of fully occupied matrices. The number of densities is varied from 16 to 2048 in $2^x$ steps. For classification a standard Gaussian Classifier is used. The classifier calculates for each feature vector of a specific speaker an allocation probability for each speaker model. This is done for all speech frames of one utterance. Then the probabilities of each model are accumulated. The model which achieved the highest value is expected to be the correct one.

### 3.3   Training

In Fig. 2 the general procedure of the training phase is shown. After feature extraction a *Universal Background Model* (UBM) is generated. Therefore, we comprised all the available training data. Then either a standard EM-Training or MAP adaptation [6,7] was applied to derive speaker-dependent models.

The EM-algorithm consist of the E-step (Eq. 3) where the A Posteriori probabilities of a feature vector $\boldsymbol{x_t}$ for every mixture $i$ is calculated.

$$p(i \mid \boldsymbol{x_t}) = \frac{\omega_i p_i(\boldsymbol{x_t})}{\sum_{j=1}^{M} \omega_j p_j(\boldsymbol{x_t})}. \qquad (3)$$

$p(i \mid \boldsymbol{x_t})$ is then used in the M-Step to reestimate the components of the new speaker model $\lambda'$:

$$\text{Mixture weights: } w_i' = \frac{1}{T} \sum_{t=1}^{T} p(i \mid \boldsymbol{x_t}) \qquad (4)$$

$$\text{Mean values: } \mu_i' = \frac{\sum_{t=1}^{T} p(i \mid \boldsymbol{x_t})\boldsymbol{x_t}}{\sum_{t=1}^{T} p(i \mid \boldsymbol{x_t})} \qquad (5)$$

$$\text{Covariance matrices: } \boldsymbol{K_i'} = \frac{\sum_{t=1}^{T} p(i \mid \boldsymbol{x_t})}{\sum_{t=1}^{T} p(i \mid \boldsymbol{x_t})}(\boldsymbol{x_t} - \mu_i')(\boldsymbol{x_t} - \mu_i')^T \qquad (6)$$

$(\boldsymbol{x_t} - \mu_i')^T$ in Eq. 6 describes the transposed mean subtracted feature vector. After the M-step the model $\lambda$ is replaced by the new estimated model $\lambda'$.

The MAP-adaptation also uses (Eq. 3) to estimate $p(i \mid \boldsymbol{x_t})$ out of the UBM parameters and the speaker-dependent feature vectors $\boldsymbol{x_t}$. The weight $(\tilde{\omega}_i)$, mean $(\tilde{\mu}_i)$ and variance $(\tilde{\boldsymbol{K}}_i)$ parameters of each mixture $i$ are computed by:

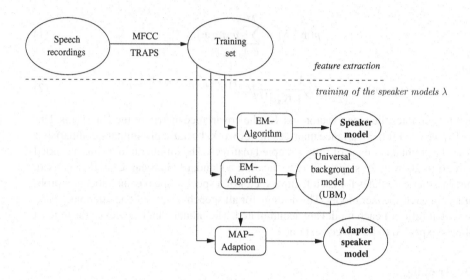

**Fig. 2.** General proceeding of the training phase

$$\tilde{\omega}_i = \sum_{t=1}^{T} p(i \mid x_t) \tag{7}$$

$$\tilde{\mu}_i = \sum_{t=1}^{T} p(i \mid x_t) x_t \tag{8}$$

$$\tilde{K}_i = \sum_{t=1}^{T} p(i \mid x_t) x_t x_t^T \tag{9}$$

Finally these newly calculated statistics are combined with the UBM statistics to create the parameters for the adapted density $i$: $\hat{\omega}_i, \hat{\mu}_i, \hat{K}_i$ (see [6,7]):

$$\hat{\omega}_i = [\alpha_i \tilde{\omega}_i / T + (1 - \alpha_i)\omega_i]\gamma \tag{10}$$

$$\hat{\mu}_i = \alpha_i \tilde{\mu}_i + (1 - \alpha_i)\mu_i \tag{11}$$

$$\hat{K}_i = \alpha_i \tilde{K}_i + (1 - \alpha_i)(K_i + \mu_i \mu_i^T) - \mu_i \mu_i^T \tag{12}$$

The adaptation coefficient $\alpha_i$ is defined as:

$$\alpha_i = \frac{n_i}{n_i + \tau}, \tag{13}$$

where $\tau$ has to be selected by the user. In preliminary experiments we distinguished the best value to be 50 for our database. (Eq. 10) contains the scale factor $\gamma$, which normalizes the sum of all new estimated a priori probabilities $\hat{\omega}_i, i \in 1, ..M$ to 1.

Both algorithms take the UBM as an initial model and for each single speaker one speaker-distinguishing model is created. The difference between EM-Training and

MAP adaptation is, that MAP adaptation calculates the parameters of the speaker-dependent Gaussian mixtures in only one iteration step and combines them with the UBM-parameters.

## 4  Experiments and Results

In preliminary experiments we investigated the best TRAPs parameters. For the preliminary experiments we used speaker models with 64 Gaussian densities and standard EM-Training. The parameters we varied were the context (15 or 30), the use of filtered and normal TRAPs and the application of PCA or LDA alternatively. For the SET-M database we used a context of 15 and filtered TRAPs. The feature reduction was performed by LDA, because it outperformed the PCA approach.

**Table 1.** Frame-wise (fw) and speaker-level (sp) recognition results achieved on the SET-M corpus

| | EM-Training | | | | | | | | MAP | | | | | | | |
| | MFCC | | | | TRAPs | | | | MFCC | | | | TRAPs | | | |
| Density | fw | 100f | 500f | sp | fw | 100f | 500f | sp | fw | 100f | 500f | sp | fw | 100f | 500f | sp |
|---|---|---|---|---|---|---|---|---|---|---|---|---|---|---|---|---|
| 32 | 21.6 | 76.4 | 79.7 | 92 | 24.2 | 70.2 | 86.7 | 92 | 24.8 | 80.7 | 91.7 | 92 | 27.6 | 76.3 | 90.6 | 97 |
| 64 | 19.2 | 67.5 | 79.7 | 83 | 23.2 | 68.8 | 85.1 | 92 | 24.4 | 81.4 | 92.6 | 92 | 26.9 | 77.0 | 90.2 | 97 |
| 128 | 16.7 | 65.6 | 77.0 | 86 | 22.4 | 66.5 | 84.0 | 92 | 23.4 | 81.2 | 92.1 | 97 | 24.9 | 73.9 | 86.7 | 92 |
| 256 | 14.5 | 61.7 | 74.6 | 83 | 21.5 | 60.5 | 83.4 | 92 | 22.3 | 80.4 | 91.7 | 94 | 22.4 | 73.3 | 85.8 | 92 |
| 512 | 12.5 | 36.4 | 39.6 | 33 | 21.2 | 52.6 | 63.2 | 67 | 19.9 | 78.5 | 91.9 | 100 | 20.2 | 72.6 | 86.7 | 92 |
| 1024 | 12.0 | 48.2 | 59.5 | 64 | 16.1 | 34.0 | 36.7 | 36 | 16.8 | 73.7 | 90.4 | 97 | 16.9 | 67.4 | 86.9 | 94 |
| 2048 | 9.5 | 20.7 | 19.5 | 14 | 15.9 | 31.0 | 34.4 | 31 | 12.4 | 64.5 | 82.3 | 97 | 15.1 | 64.4 | 88.4 | 94 |

Table 1 shows the recognition results for the EM-Training and the MAP adaptation. It contains the results for both features: MFCC and TRAPs. *fw* denotes the recognition result, reached when deciding for each frame separately (frame-wise) and *sp* denotes the recognition results of the classification of all vectors of one speaker (speaker-level). The columns named *100f* and *500f* contain the recognition results after classification with 100 and 500 frames each (no overlap).

In the case of EM-Training we observed a maximal frame-wise recognition rate of 24.16 % with TRAPs features and 32 Gaussian mixtures. The maximal recognition rate for the *speaker* decision was 91.67 %. For the SET-M corpus the MAP adaptation outperforms the EM-Training. In the case of MAP adaptation the highest frame-wise recognition result (27.58 %) was achieved by 32-dimensional speaker models and TRAPs. The maximal value in case of *speaker* decision (100 %) was accomplished with 512-dimensional models and MFCCs.

In Fig. 3 we plotted the recognition results dependent on the amount of test feature vectors. Therefore, we classified all data of the test speakers after a given number of frames (no overlap). One can see, that the slope of the recognition results is almost zero when more than 500 frames are used.

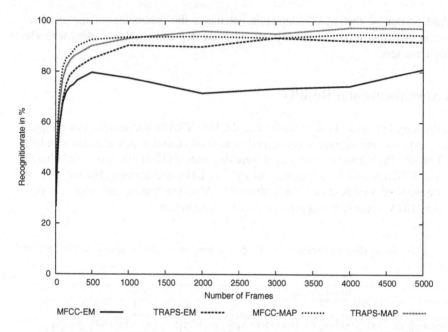

**Fig. 3.** Speaker recognition results dependent on the amount of test feature vectors

## 5   Discussion

Using TRAPs in case of text-independent speaker recognition can improve the recognition results, especially if the recordings are very noisy, like the database of this paper. So the recognition could be improved from 21.57 % to 24.84 % (12.0 %) in case of frame-wise recognition. Due to the fact, that the amount of training data was very low in this database, the use of more Gaussian mixtures even decreased the frame-wise recognition result. The *speaker* recognition reaches its maximum (91.67 %) at speaker models with 32 densities.

The training with MAP-adaptation improves the frame-wise recognition rate by another 14.2 % from 24.84 % to 27.58 %. But an increase of the number of Gaussian densities does not achieve an improvement of the recognition rate of the TRAPs. On speaker-level MFCCs obtain better results than TRAPs, if a larger number of densities is chosen.

Therefore, we conclude that TRAPs have a better recognition rate on frame-level due to the larger context and that the properties of a speaker can be modeled with TRAPs using fewer Gaussian densities than MFCCs. We will examine this aspect further in future experiments.

## 6   Summary

In this paper we evaluated a system for speaker-independent speaker recognition. We used 2 different kinds of features: MFCCs and TRAPs. Both analyze the spectrum of a given recording. MFCCs examine the complete frequency domain on a short temporal context and TRAPs calculate features by analyzing different frequency bands over a longer time period. For the training we created a UBM by standard EM-Training on all the available training data. To build one model for every speaker we took this UBM as an initial model and applied EM-Training or MAP adaptation respectively. For this step we only used the speaker-dependent training data. For the evaluation of our system we performed experiments on the SET-M database.

We improved the frame-wise recognition result by 12.0 % when using TRAPs instead of MFCCS. The application of MAP adaptation improved the frame-wise recognition results by additional 14.2 %. The *speaker* recognition result also was increased and for 512 Gaussian densities 100 % were reached.

# References

1. Wahlster, W.: Verbmobil: Foundations of Speech-to-Speech Translation. Springer, Berlin (2000)
2. Stemmer, G.: Modeling Variability in Speech Recognition. PhD thesis, Chair for Pattern Recognition, University of Erlangen-Nuremberg, Germany (2005)
3. Hermansky, H., Sharma, S.: TRAPS – classifiers of temporal patterns. In: Proc. International Conference on Spoken Language Processing (ICSLP), Sydney, Australia (1998)
4. Maier, A., Hacker, C., Steidl, S., Nöth, E., Niemann, H.: Robust Parallel Speech Recognition in Multiple Energy Bands. In: Kropatsch, W.G., Sablatnig, R., Hanbury, A. (eds.) Pattern Recognition. LNCS, vol. 3663, pp. 133–140. Springer, Heidelberg (2005)
5. Reynolds, D.A., Rose, R.C.: Robust Test-Independent Speaker Identification Using Gaussian Mixture Speaker Models. IEEE Transaction on Speech and Audio Processing 3, 72–83 (1995)
6. Gauvain, J., Lee, C.: Maximum A-Posteriori Estimation for Multivariate Gaussian Mixture Observations of Markov Chains. IEEE Transactions on Speech and Audio Processing 2, 291–298 (1994)
7. Reynolds, D.A., Quatieri, T.F., Dunn, R.B.: Speaker Verification Using Adapted Gaussian Mixture Models. Digital Signal Processing, 19–41 (2000)

# Recording and Annotation of Speech Corpus for Czech Unit Selection Speech Synthesis*

Jindřich Matoušek[1] and Jan Romportl[2]

[1] University of West Bohemia, Faculty of Applied Sciences,
Department of Cybernetics, Univerzitní 8, 306 14 Plzeň, Czech Republic
jmatouse@kky.zcu.cz
[2] SpeechTech, s.r.o., Morseova 5, 301 00 Plzeň, Czech Republic
jan.romportl@speechtech.cz

**Abstract.** The paper gives a brief summarisation of preparation and recording of a phonetically and prosodically rich speech corpus for Czech unit selection text-to-speech synthesis. Special attention is paid to the process of two-phase orthographic annotations of recorded sentences with regard to their coherence.

## 1 Introduction

Quality of synthetic speech produced by a concatenation-based synthesis system crucially depends on the quality of its acoustic unit inventory. Several factors contribute to the quality of the acoustic unit inventory, such as speech corpus from which the units are extracted, the type of the units (i.e. phone, diphone, triphone etc.), labelling accuracy, the number of instances per each unit, prosodic richness of each unit etc.

A process of the speech corpus preparation involves several steps like text collection preprocessing, sentence selection according to specified criteria, recording by a suitable speaker and orthographic annotation with its revision. This paper summarises such steps in the preparation of a new speech corpus for the Czech TTS system ARTIC [1] and specially pays attention to the orthographic annotation and its revision.

The new speech corpus is intended to provide enough data (approx. five thousand sentences) for robust unit selection text-to-speech synthesis as well as for prosodic-syntactic parsing and explicit prosody modelling. Special care is thus given to assuring segmental and supra-segmental balance of recorded sentences together with exact correspondence with their orthographic form.

## 2 Selection of Sentences

The sentences have been selected from a large collection of Czech newspaper texts covering various domains like news, sport, culture, economy, etc. Further discussion on suitability of such kind of texts is provided in [2] together with a detailed description of sentence preprocessing.

---

* Support for this work was provided by the Ministry of Education of the Czech Republic, project No. 2C06020, and the EU 6th Framework Programme IST-034434.

V. Matoušek and P. Mautner (Eds.): TSD 2007, LNAI 4629, pp. 326–333, 2007.

From the total number of 524,472 sentences a selection algorithm has automatically chosen approximately five thousand sentences so as the resulting selection was phonetically and prosodically (or segmentally and supra-segmentally) balanced. The question has arisen, whether the sentences should be balanced *naturally* or *uniformly* – as [2] shows, we have decided for the latter, i.e. frequency of all segmental units (diphones) should be uniform (obviously, this cannot be fulfilled but at least it ensures rare units to appear as frequently as possible). In addition to this, further restrictions on the selection process were imposed: the sentences shorter than 3 and longer than 30 words were excluded; 3,500 sentences had to be declaratory, 900 interrogative ("yes/no") questions, 310 supplementary ("wh-") questions and 311 exclamatory or imperative sentences. Moreover, other sentences have been selected "by hand" (due to requirements for specific contexts) so that the final total number was 5,139.

Both segmental and supra-segmental balancing processes were carried out by a greedy algorithm maximising diphone entropy of selected sentences. Prosodic richness was introduced into this process by creating six variants of each diphone based on its position within a prosodic structure of a given sentence [3]. The selection algorithm then treated these variants as different units and thus maximised entropy also among them, which basically lead to better deployment of various prosodic contexts in the selected sentences [2].

## 3   Recording of the Corpus

Unlike speech recognition tasks, where some kind of noise depending on the environment the speech recogniser will run in is almost always desirable [4], high-quality noise-free recordings are required for concatenative speech synthesis. Hence, our corpus was recorded in a soundproof studio. An AKG C 3000B large-diaphragm cardioid condenser microphone with a pop filter installed to reduce the force of air puffs emerging from bilabial plosives and other strongly released stops was used. A high fidelity capture card capable of up to 96 kHz AD conversion was utilised. For our purposes, 48 kHz AD conversion has been actually performed. Glottal signal measured by an electroglottograph device was recorded along with the speech signal. The glottal signal is suitable for the detection of glottal closure instants (also called pitch-marks) which are used for accurate pitch contours estimation, pitch-synchronous speech synthesis, very precise voiced/unvoiced signal detection, or smooth concatenation of speech segments in unit selection speech synthesis [1,5].

A female voice talent possessing a pleasant voice, good voice quality and professional recording experience was chosen to record the corpus. The recording ran in a sentence-by-sentence manner. The speaker was instructed to read each sentence naturally but with no emotions and no amount of expressiveness. She was also asked to speak clearly and to keep her normal speaking rate and volume. Being aware of the importance to keep the recordings consistent both in phonetic and prosodic (within the framework of symbolic prosody description [5]) terms, an expert in acoustic phonetics and orthoepy supervised the recordings; his job was to check the consistency of recordings and also the constancy of speaker's voice quality and pronunciation. The average duration of a recording session was about 4 hours which resulted in about 13 recording sessions.

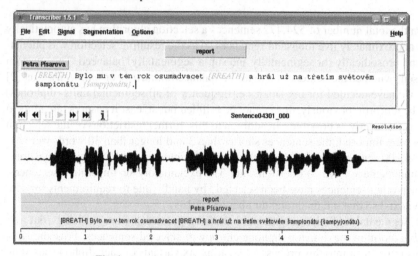

**Fig. 1.** A typical window of the Transcriber

# 4   Annotation of the Corpus

To know the correspondence between the speech signals and their linguistic representations on orthographic (and later on phonetic) level, the orthographic annotation of each recorded sentence is necessary. As the very precise annotation is very important for corpus-based speech synthesis (where the annotation serves as a base for indexing large speech unit inventories, and any misannotation often causes glitches in the synthesised speech), the annotation process was divided into two phases. In the first phase the recordings were transcribed by a skilled annotator and the "initial" annotation (ANN1) was obtained in this way. In the second phase the annotation ANN1 was revised and possibly corrected by another annotator ("revised" annotation ANN2).

The annotations were done using the special annotation software Transcriber, a tool for assisting the creation of speech corpora [6] (see Figure 1). It makes it possible to manually segment, label and transcribe speech signals for later use in automatic speech processing. Transcriber is freely available from Linguistic Data Consortium (LDC) web site http://www.ldc.upenn.edu/.

## 4.1   Annotation Rules

During the annotation process, each sentence is transcribed in the way it was really pronounced. Unlike the "prescribed" sentences selected by the sentence selection algorithm in Section 2 (denoted as "patterns" and marked as ANN0 hereafter), the really uttered sentences can contain mispronunciation, unintelligible pronunciation, missing or extra words, various non-speech events like breathing or clicking, and very rarely also some kinds of noises (in our case mostly caused by a failure of the recording system). The rules used for the annotation of the corpus were adopted from [4,7] (where they were utilised for the purposes of speech recognition) and are listed here (if specified further in the examples, P means pattern sentences (i.e. what should have been read) and R means what was actually read):

**Table 1.** List of non-speech events and their brief description

| Event | Description |
|---|---|
| BREATH | audible breath |
| CLICK | extraordinary mouth click |
| NO-SILENCE | no leading or trailing silence |
| UNINTELLIGIBLE | unintelligible pronunciation |
| NOISE | noise |
| NO-SPEECH | no speech present in the signal |

1. For each sentence, the annotator is given an orthographic pattern of the sentence, i.e. what the speaker was expected to read. These are the sentences selected in Section 2.

   E.g.: V Tatrách začíná univerziáda.

2. Non-speech events and noises are indicated by a descriptor enclosed in square brackets. The descriptors contain only capitalised alphabetic characters and dashes. The list of the descriptors used during the annotation is given in Table 1. The descriptor is placed at the point at which the non-speech event occurred. Two non-speech events caused by the speaker are distinguished: [BREATH], indicating an "audible" breath, and [CLICK], marking a loud extraordinary mouth click.

   E.g.: [BREATH] Využívání služby je jednoduché a pohodlné.

3. Each recording should start and end with a silence. If not, a special "non-speech-event-like" mark [NO-SILENCE] must be put on the appropriate place (either at the beginning or the end of the sentence).

   E.g.: Další důležitou změnou je věk [NO-SILENCE].

4. The conventions in Czech written texts are abided by (including punctuation) – each sentence starts with a capitalised word, all other words except for proper names (e.g. Josef) and acronyms (e.g. NATO) are transcribed with low-case letters.

   E.g.: V Kosovu jsou vojáci NATO, kteří střeží bezpečnost.

5. Everything uttered is transcribed as words, including numbers or dates. Again, rules for correct writing of Czech numbers are followed.

   E.g. (P): Skončil až někde na 163. místě.

   E.g. (R): Skončil až někde na sto šedesátém třetím místě.

6. If any word was pronounced differently from the given pattern (either as another meaningful word or a non-sense word) and the mispronunciation was clear and intelligible, the original word must be replaced with the really uttered word (the non-sense word must be enclosed by * to indicate that it is not a typo).

   E.g. (P): V minulých dvou dnech si zdříml jen příležitostně.

   E.g. (R): V minulých dvou letech si *zdríml* jen příležitostně.

7. If any word was pronounced differently from the given pattern and the mispronunciation was not intelligible (e.g. stammering, hesitation or surprise when uttering a word, a word corrupted by the recording system failure, etc.), non-speech event [UNINTELLIGIBLE] is placed in front of the word.

   E.g.: V ruce žmoulá [UNINTELLIGIBLE] kapesník.

8. The pronunciation of some words (especially the foreign ones) does differ from its written form and does not obey the Czech pronunciation rules [8]. Such words must be followed by a "commentary" notation, containing the "pronunciation form" of the word, and will be referred as exceptions henceforth.

   E.g.: V Tatrách začíná univerziáda {unyverzyjáda}.

9. Abbreviations are transcribed as they were spelled, also using the "commentary" notation, e.g. IBM can be transcribed as "aj bí em", "i b@ m@", or "í bé em", where "@" stands for a reduced vowel (schwa).

   E.g.: V Kosovu jsou jednotky KFOR {káfor} z mise OSN {ó es en}.

10. Although the recordings were made in a soundproof studio with a high-quality recording system, one must always take some possible noises into account. Indeed, sometimes (very rarely) there was a failure of the recording system causing some portions of the recordings to sound like a buzzing. Such portions of the signal (either silence, a single word, or a sequence of words) must be denoted by a special descriptor [NOISE] (in the case of a sequence of words, the beginning of the noise event is denoted by [NOISE>] and the end by [<NOISE], or by descriptor [NO-SPEECH] if the speech signal was completely missing.

   E.g.: To je také důvod, proč [NOISE>] píšu [<NOISE].

Since there were relatively many recordings with [NO-SILENCE],[NO-SPEECH] or [NOISE] events (around 5 %), sentences with these events were recorded once more.

## 4.2   The 1st Phase of Annotation

For the 1st-phase annotation, the "prescribed" sentences (ANN0) selected by the algorithm briefly described in Section 2 were used as patterns. Following the annotation rules described in Section 4.1, the annotation ANN0 was modified by the first annotator. As a result, ANN1 annotation was obtained. Approximately 72% of all transcribed sentences and 96% of all words were identical in ANN0 and ANN1 (as there were no non-speech events available in ANN0, they were not counted in during the comparison). As the exceptional words were not marked as exceptions in ANN0 (see Section 4.1, rule No. 8), most of the differences were the exceptions themselves. The results of the comparison are shown in Table 2 in section ANN0-ANN1. The results after suppressing the influence of exceptions (by supplementing ANN0 with "pronunciation forms" of the exceptions from a dictionary of exceptions – note that not all exceptions were actually present in the dictionary and that some exceptions were mistyped in ANN1 because the relative occurrences of different words increased) are shown in section ANN0-ANN1*.

## 4.3   The 2nd Phase of Annotation – Revision

Being aware of the importance of the precise annotation of the source speech data for corpus-based speech synthesis, all annotations were subject of a revision. The revision ANN2 was made by another annotator – she used ANN1 annotations and corrected them if needed. Approximately 96% of all sentences and more than 99% of all words

**Table 2.** Comparison of the pattern (ANN0), initial (ANN1) and revised (ANN2) annotations (relative occurrences in percents) and percentage of words and sentences which were equal in the comparisons. Sections with * denote that ANN0 was supplemented with "pronunciation forms" of the exceptions.

| Differences | ANN0-ANN1 | ANN0-ANN1* | ANN0-ANN2 | ANN0-ANN2* | ANN1-ANN2 |
|---|---|---|---|---|---|
| Missing exceptions | 2.60 | 0.13 | 2.77 | 0.16 | 0.17 |
| Different words | 0.79 | 0.83 | 0.88 | 0.94 | 0.09 |
| Extra words | 0.07 | 0.07 | 0.08 | 0.08 | 0.08 |
| Missing words | 0.05 | 0.05 | 0.05 | 0.05 | 0.03 |
| Words OK | 96.49 | 98.92 | 96.22 | 98.77 | 99.62 |
| Sentences OK | 72.32 | 87.13 | 70.83 | 87.15 | 96.24 |

**Table 3.** Differences between words and their examples as annotated in ANN1 and revised in ANN2. Percentage is shown within all differences.

| Typo | Perch. [%] | ANN1 | ANN2 |
|---|---|---|---|
| TYPO1 | 47.37 | blondýna | blondýnka |
| LAST | 19.30 | jak | jako |
| LENGTH | 14.04 | benzinů | benzínů |
| TYPO2 | 10.53 | jak | jako |
| MISP | 8.77 | Jankulovski | Jarkulovski |

were found to be the same in both annotations. The comparison of both annotations is given in Table 2 in section ANN1-ANN2 and comprises also non-speech events.

Both the words missed in ANN1 ("missing words") and deleted in ANN2 ("extra words") were mostly non-speech events (about 73%, or 82% respectively). The rest were mostly monosyllabic words. The differences between words in both annotations ("different words") are summarised in Table 3. They typically consist in:

- the last letter of a word was missing or extra (LAST);
- a vowel letter was shortened or lengthened (LENGTH);
- a word was mistyped in ANN1 as another meaningful word (TYPO1);
- a word was mistyped in ANN1 as a non-sense word (TYPO2);
- a word had been pronounced as a non-sense word but the original transcription from ANN0 was left in ANN1 (MISP).

As for the exceptional words not marked as exceptions in the initial annotations (ANN1), six types of exceptions were observed and are analysed in Table 4:

- words containing "s" were pronounced with [z] (S-Z);
- consonant [j] was inserted between [i] and a vowel (INS-J);
- "d", "t", "n" were pronounced as non-palatal consonants [d], [t], [n] when followed by "i" (typical for foreign words in Czech, DTN) [8];

**Table 4.** Missing exceptions and their examples. Percentage is shown within all differences. PRON stands for the "pronunciation form" of a word.

| Exceptions | Perch. [%] | Word | PRON |
|---|---|---|---|
| S-Z | 28.18 | Klausem | Klauzem |
| INS-J | 23.64 | policie | policije |
| DTN | 20.00 | politika | polityka |
| LEN | 12.73 | Rudolfinum | Rudolfínum |
| OTHER | 9.09 | pokeru | pokru |
| DBL | 6.36 | Gross | Gros |

- words containing a short vowel were pronounced with a corresponding long vowel (LEN);
- words containing a double letter were pronounced with the corresponding single consonant (DBL);
- the other words (OTHER).

The final revised annotations comprise 62,332 running words (7.60% of them being non-speech events and 2.62% being exceptions as defined by rule No. 8 in Section 4.1) in 5,139 sentences. The lexicon made from the annotations contains 17,630 different words, 0.02% of which being non-speech events and 6.11% being exceptions.

## 5   Conclusion

We have briefly summarised the whole process of creation of the new Czech speech corpus for unit selection text-to-speech synthesis together with the requirements posed on it, as well as the aims this corpus has been intended with. The emphasis was actually mainly placed on the important step of the orthographic annotations carried out as a two-phase process, where the importance of the second annotation phase (i.e. revision) has been discussed.

The Table 2 clearly shows the improvement of the annotation coherence and its correspondence with the speech data after the annotation revision. Although the difference between the numbers of correct words in both annotation steps (ANN1 and ANN2) may first seem rather insignificant (from the point of view of ANN2 there were 99.62% words correctly annotated in ANN1), but concerning the total number of word tokens from the corpus (62,332), the 2nd-phase annotation has corrected 237 words which would cause – if being left uncorrected – fairly noticeable problems in resulting synthesised speech because wrongly assessed segments from these words would be repetitively used in the concatenation process during unit selection (since generally the whole corpus is used at once).

As can be further seen, careful classification and annotation of non-speech events and speaker's mistakes is of great importance too. The column ANN0-ANN2* from the

Table 2 can thus be regarded as a very rough "measure" of how the speaker was correct and precise in recording. Indeed this value comprises possible errors in the source text (ANN0) corrected by the speaker herself and perhaps also mistakes that have been made (or unnoticed) both in ANN1 and ANN2, but the core is definitely constituted by the differences in what the speaker was supposed to read and what has actually read.

# References

1. Matoušek, J., Tihelka, D., Romportl, J.: Current State of Czech Text-to-Speech System ARTIC. In: Sojka, P., Kopeček, I., Pala, K. (eds.) TSD 2006. LNCS (LNAI), vol. 4188, pp. 439–446. Springer, Heidelberg (2006)
2. Matoušek, J., Romportl, J.: On Building Phonetically and Prosodically Rich Speech Corpus for Text-to-Speech Synthesis. In: Proc. Computational Intelligence. San Francisco, U.S.A, pp. 442–447 (2006)
3. Romportl, J.: Structural Data-driven Prosody Model for TTS Synthesis. In: Proc. Speech Prosody. Dresden, Germany, pp. 549–552 (2006)
4. Radová, V., Psutka, J.: Recording and Annotation of the Czech Speech Corpus. In: Sojka, P., Kopeček, I., Pala, K. (eds.) TSD 2000. LNCS (LNAI), vol. 1902, pp. 319–323. Springer, Heidelberg (2000)
5. Tihelka, D., Matoušek, J.: Unit Selection and its Relation to Symbolic Prosody: a New Approach. In: Proc. Interspeech. Pittsburgh, U.S.A., pp. 2042–2045 (2006)
6. Barras, C., Geoffrois, E., Wu, Z., Liberman, M.: Transcriber: development and use of a tool for assisting speech corpora production. Speech Communication 33, 1–2 (2000)
7. Psutka, J., Radová, V., Müller, L., Matoušek, J., Ircing, P., Graff, D.: Large Broadcast News and Read Speech Corpora of Spoken Czech. In: Proc. Eurospeech. Ålborg, Denmark, pp. 2067–2070 (2001)
8. Psutka, J., Müller, L., Matoušek, J., Radová, V.: Talking with Computer in Czech. Academia, Prague (in Czech) (2006)

# Sk-ToBI Scheme for Phonological Prosody Annotation in Slovak

Milan Rusko[1], Róbert Sabo[1], and Martin Dzúr[2]

[1] Department of Speech analysis and Synthesis of Institute of Informatics of Slovak Academy of Science, Dúbravská cesta 9, 845 07 Bratislava, Slovakia
{milan.rusko,robert.sabo}@savba.sk
[2] Department of Slovak Literature and Literary Science of Philosophical faculty of Comenius University, Gondova 2, 818 01 Bratislava, Slovakia
martindzur@gmail.com

**Abstract.** Research and development in speech synthesis and recognition calls for a phonological intonation annotation scheme for the particular language. Inspired by the successful ToBI (Tones and Break Indices) for American English [1] and GToBI [2] for German, this paper introduces a new intonation annotation scheme for Slovak, Sk-ToBI. In spite of the fact that Slovak prosodic rules differ from those of English or German, we decided to follow the main principals of ToBI and to define a special Slovak version of Tones and Break Indices annotation scheme. The speech material belonging to different styles, which was used for the preliminary study of accents in Slovak is shortly described and the conventions of Sk-ToBI annotation are presented.

## 1 Motivation

Prosody analysis and processing represent an inevitable part of current automatic speech processing systems. A phonological intonation annotation scheme is needed for this purpose. As there was no such scheme available for Slovak, we decided to create a system for intonation labeling based on the ideas of the ToBI annotation. Our definition is based on previous research and findings in the field of Slovak phonetics and phonology but it also draws inspiration from foreign conventions for prosody labeling (e.g. GToBI).

The most common convention for prosody labeling is ToBI (Tones and Break Indices) that was set up by a team of American researchers on the basis of the Pierrehumbert's model of intonation and presented in 1992 [3]. Although Slovak does not have the same prosodic features as English, we have drawn inspiration from this convention as far as the use of its labels and tiers is concerned but we have adapted it to Slovak prosody. We had to determine basic features and rules of the Slovak intonation, which enabled us to create a set of essential types of pitch accents and their combinations that can be found in the spoken form of this language.

Slovak is a Slavic language. It is a stress language with fixed stress on the first syllable. As far as we know this is the first attempt to create a phonological prosody annotation scheme for this language.

V. Matoušek and P. Mautner (Eds.): TSD 2007, LNAI 4629, pp. 334–341, 2007.

There are various tonal realizations of utterances in Slovak. We will outline especially the ones belonging to the neutral style of standard Slovak, and thus to create a basis for setting up intonation labeling conventions that we have called Slovak Tones and Break Indices (Sk-ToBI). This system is used not only for labeling pitch accents (tones) but also for marking the intermissions (breaks), and gives so information on time segmentation of the speech. The Miscellaneous tier allows for annotation of disfluences, nonverbal speech displays of the speaker and some supralinguistic and extralinguistic phenomena possibly influencing the intonation.

Regional accents of Slovak have their accent structure very different from that of standard Slovak and they are not included in this study.

## 2 Speech Material

As a basic source of knowledge for our study we used the theoretical works on the Slovak phonology (e.g. [4] and [5]). This study was then followed by a research on the recorded speech material. Our speech sources consisted of several speech databases:

1. "intonation part" of the speech synthesis database which was designed to cover all the basic types of sentence intonation contours in Slovak [6] (artificial, read material),
2. speeches of politicians from the Slovak parliament (rhetoric style),
3. TV news (reportage style) [7],
4. TV debates "Pod lampou" (live dialogues),
5. database of recorded puppet plays of a traditional puppeteer Bohuslav Anderle (artistic style with very expressive speech).

Approximately one hour from every of the databases was listened through by two researchers with degree in linguistics. Every utterance, that seemed to contain new unseen features, was analysed in detail and checked for suitability of Sk-ToBI for its annotation. New features were added to the definition and the unused features were excluded from the annotation scheme.

## 3 Sk-ToBI: Break Index Tier

Break indices are used to rate the degree of juncture between words and between the final word and the silence. They are marked after all words. All junctures - including those after fragments and filled pauses - must be assigned an explicit break index value [8,9].

Similarly to TOBI we use in the Slovak system indices 0-4 and signs "?" and "-" for questionable parts.

### 3.1. Description of Break Indices

"0" – marks the junction that is imperceptible in the case of neutral pronunciation e.g. for word boundaries in clitic groups (see Fig. 1. "a predsa").

"1" – marks the break that occurs at "normal" phrase-medial word boundaries.

"2" – marks a strong disjuncture marked by a pause or virtual pause, but with a well-formed tune continuing across the juncture [8,9]. It can be described as a slightly unnatural prosodic disfluency and it often occurs in a case of a short interruption (pause) of the utterance and also in some principle clauses (e.g. *"Vojdúc do izby utrel si nohy" after the word "izby" ("Having entered the room he cleaned his feet" after the word "room".)).*

"3" – marks the typical boundary strenght at intermediate phrase.

"4" – marks the break at the end of the utterance.

## Uncertainty Marks

"-" – is used when an annotator is not sure, which break index value to apply and hesitates between two indices. In this case it is recommended to use the higher index from the two and the mark „-" should be added. For example when one cannot decide between the index „1" and „2", the sign "2-" should be used.

"?" – is used when an annotator cannot determine the break index at all.

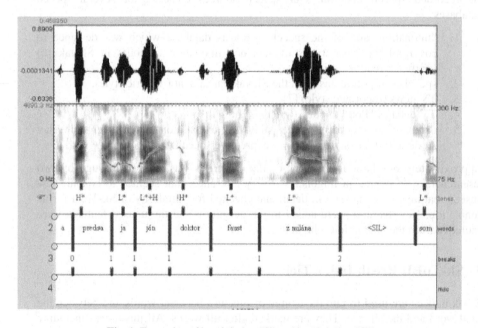

**Fig. 1.** Examples of break index "0" and break index "2"

The picture shows a use of the break index "0" in the clitic group "a predsa" and a longer unnatural pause after the break index "2" (strong disjuncture with no tonal markings). The example is taken from the performance of a puppeteer. The annotated utterance says "a predsa ja, Ján doktor Faust z Milána" ("…nevertheless me, Jan doctor Faust from Milan…").

# 4 Sk-ToBI: Tone Tier

To mark the tones and pitch accents we have adopted the signs used in English ToBI. We have introduced a new type of accent (*!H-*). We have excluded bitonal accents H+!H* and L+H* and also H+L* (used in GToBI) from the Sk-ToBI system:

Single tone accents:
*H\**    high pitch accent,
*L\**    low pitch accent,
*!H\**   an accent pitched approximately in the middle of the range of the melodic contour (this accent can follow after the same accent (*!H\**) or after the high accent (*H\**).

Bitonal accent:
*L\*+H*   low pitch accent with raise to high target after accented syllable.

Boundary tones:
*%H*    onset at the beginning of the speaker's utterance with a very high pitch,
*H-*    ending of the intermediate phrase with a high low pitch (before a break of type „3"or with combination with final boundary (H-H% or H-L%) tone before „4") ,
*L-*    ending of the intermediate phrase with a pitch (before a break of type „3" or with combination with final boundary (L-L% or L-H%) tone before „4"),
*!H-*   ending of the intermediate phrase approximately in the middle of the utterance pitch contour boundaries (before a break index „3")
*H%*    ending of an intonation phrase with high pitch - anticadence   (break index „4"),
*L%*    ending of the intonation phrase with low pitch - conclusive cadence (before a break index „4").

Auxiliary mark:
*HiF0*   highest pitch level of the speaker's utterance.

**Uncertanties.** Similarly to the break index tier, the annotator can use in this tier the mark "?". When he cannot exactly determine the tone "H" or "L", the tone should be marked as "X?*", "X?-", "X?%" or *? as the transcriber is not certain even there is a pitch accent.

## 4.1 Possible Realizations of Pitch Contours in Slovak

Ends of intermediate phrases are traditionally labelled within English, German and other ToBI systems by a set of symbols (L-, H-) and ends of intonation phrases by (L%, H%). The first group is applied to label the tone pitch immediately before the end of the intermediate phrase and the second one indicates the end of utterance with a conclusive cadence (falling intonation) or anticadence (rising intonation). The first group is used with break indices type "3" and in combination with the second group with break indices type "4".

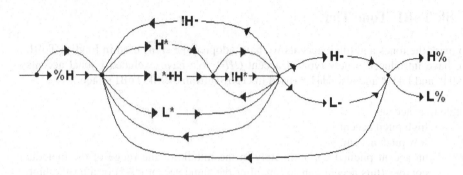

**Fig. 2.** The diagram of possible sequences of tones and pitch accents in the Slovak sentence

The course of intonation contour at the end of the intermediate phrase (index type "3") is called semicadence, and the three typical boundary tones it can reach are shown in the examples in Table 1.

**Table 1.** Three possible types of semicadence at the end of intermediate phrase

| | | |
|---|---|---|
| L- | Low pitched semicadence – typical for paratactic clauses. | *Ležím s knihou v posteli a nezunuje sa mi.*<br>H\* L\* L\* L-L\* L-L%<br>(I lie in a bed with a book and it doesn't bother me.) |
| !H- | Semicadence placed within the middle pitch range – typical for the most of Slovak hypotactic clauses but also within longer simple sentences. | *Chcete sa s ním stretnúť a dať mu ten dar?*<br>H\* H\* L\* !H- L\* H\* H-H%<br>(Do you want to meet him and give him that gift?) |
| H- | High pitched semicadence – typically occurs in complex questions and in some types of hypotactic clauses (e.g. disjunctive clauses). | *Už si ho nevšímala, ani mu neodpovedala.*<br>H\* L\* H- L\* L L\* L-L%<br>(She was no longer noticing him nor she was answering to him.) |

As far as the end of intonation phrase is concerned, the course of intonation contour at its end is cadence or anticadence and the four typical boundary tone combinations it can reach are shown in the examples in Table 2.

It is necessary to be aware of the natural fall of the pitch that results from decreasing air pressure in lungs during the utterance. This decline is, however, not marked by annotation.

Cases such as H-H% and L-H% refer to anticadence and L-L% and H-L% refer to conclusive cadence.

**Table 2.** Four types of realization of the end of the intonation phrase

| L-L% | End of the utterance by conclusive cadence (typical for declarative sentences). | *Ležím s knihou v posteli a nezunuje sa mi.*<br><br>H* L* L* L-L*  L-L%<br><br>(I lie in a bed with a book and it doesn't bother me.) |
|---|---|---|
| H-L% | High pitched end of the utterance with conclusive cadence (typical for declarative sentences, or exclamatory sentences). | *V y n e s     s m e t i!*<br><br>L*+H    H*    H-L%<br><br>(Empty the trash bin!) |
| L-H% | End of the utterance with low placed pitch with anticadence. (e.g. yes – no question). | *Neprišiel?*<br>L-H%<br><br>L*<br><br>(Hasn't he come?) |
| H-H% | End of the utterance with high pitch and anticadence (e.g. yes–no question). | *Chcete sa s ním stretnúť a dať mu ten dar?*<br><br>H* H* L* !H- L* H* H-H%  H*<br><br>(Do you want to meet him and give him that gift?) |

## 4.2 Specific Features of Sk-ToBI

Slovak is a language with constant position of word stress[1] on the first syllable. That is why tone accent occurs on the first syllable and bitonal accents such as L+H*, H+!H* or H+L* need not be included in the annotation system.

# 5 Sk-ToBI: Miscellaneous Tier

The miscellaneous (MSC) tier is designated for transcribing different types of disfluences that can influence prosody. For example pause, breath, crooning (filled pauses), cough, laugh, etc. The set of marks of this tier is not closed and it can be broadened in the case of need.

# 6 Sk-ToBI: Orthographic Tier

The orthographic tier is a straightforward transcription of all of the words in the utterance, in ordinary Slovak orthography. From the common practice we have adopted a rule to sign longer silent segments as "<SIL>" in this tier.

---

[1] Our conception is based on the neutral pronunciation of Slovak. In case of some dialects, especially the eastern ones, it will be necessary to use other models.

# 7  Conclusion

Our paper presents a phonological prosody annotation scheme for Slovak, Sk-ToBI. We have analyzed a speech material of different speaking styles (read sentences, TV news and reportages, TV debates, political speeches from parliament and recorded plays of a puppeteer, representing artistic style with very expressive speech). We found the most frequent pitch contours occurring in Slovak utterances. Following the ideas of ToBI annotation, we have defined a specialized Tones and Break Indices annotation scheme for Slovak. We are aware of the fact that every particular realization of the utterance and its consequent annotation will always depend on the individual speaker and annotator and so it can differ from standard situations considered in this paper. It is however necessary for the annotators to have a basic set of labels and annotation conventions well defined, to agree with these conventions and to follow them. We believe that current definition of Sk-ToBI gives a serious basis for prosody research and opens a possibility to start a development of prosody-driven and expressive speech oriented applications in Slovak, mainly in the area of analysis, recognition [10] and synthesis of human speech. Basic rules of SK-ToBI as they were presented in this paper represent a system compatible to a high degree with the original definition of ToBI. Results of verification of completeness and practical usability of this scheme as well as examination of inter-annotator agreement will be published in forthcoming papers.

## Acknowledgments

This work was funded by the Slovak Agency for Science, VEGA, grant No. 2/2087/22.

## References

1. Silverman, M., et al.: ToBI: A standard for labeling English prosody. In: Proceedings of the 2nd International Conference of Spoken Language Processing, Banff, pp. 867–870 (1992)
2. Baumann, S., Grice, M., Benzmüller, R.: GToBI–a phonological system for the transcription of German intonation. In: Proceedings Prosody 2000, Speech Recognition and Synthesis Workshop, Cracow, pp. 21–28 (2000)
3. Pierrehumbert, J.B.: The Phonology and Phonetics of English Intonation. PhD dissertation, IULC edition. MIT, Cambridge (1980)
4. Král', Á.: Pravidlá slovenskej výslovnosti, SPN, Bratislava (1988)
5. Sabol, J., Král', Á.: Fonetika a fonolgia, SPN, Bratislava (1989)
6. Rusko, M., Darjaa, S., Trnka, M., Cerak, M.: Slovak Speech Database for Experiments and Application Building in Unit-Selection Speech Synthesis. In: Sojka, P., Kopeček, I., Pala, K. (eds.) TSD 2004. LNCS (LNAI), vol. 3206, pp. 457–464. Springer, Heidelberg (2004)
7. Zibert, J., Mihelic, F., et al.: The COST278 Broadcast News Segmentation and Speaker Clustering Evaluation - Overview, Methodology, Systems, Results. In: Proceedingds of the 9th European Conference on Speech Communication and Technology Interspeech 2005, Lisbon, pp. 629–632 (2005)

8. Beckman, M.E., Gayle, M.A.: Guidelines for ToBI labelling. Version 2.0 (February 1994), http://www.ling.ohio-state.edu/research/phonetics/E_ToBI/
9. Hirschberg, J., Beckman, M.: The ToBI annotation conventions (1994), http://www.ling.ohio-state.edu/~tobi/ame_tobi/annotation_conventions.html
10. Cernak, M., Wellekens, Ch.J.: Emotional aspects of intrinsic speech variabilities in automatic speech recognition. In: SPECOM 2006, 11th International Conference Speech and Computer, Saint-Petersburg, Russia, pp. 405–408 (2006)

# Towards Automatic Transcription
## of Large Spoken Archives in Agglutinating Languages
## – Hungarian ASR for the MALACH Project

Péter Mihajlik[1], Tibor Fegyó[1,2], Bottyán Németh[1,2], Zoltán Tüske[1],
and Viktor Trón[3]

[1] Department of Telecommunications and Media Informatics,
Budapest University of Technology and Economics,
1117, Budapest, Magyar tudósok krt. 2, Hungary
[2] AITIA International, Inc., 1039 Budapest, Czetz János u. 48-50, Hungary
[3] University of Edinburgh, 2 Buccleuch Place, EH8 9LW Edinburgh, United Kingdom
{mihajlik,fegyo,bottyan,tuske}@tmit.bme.hu, v.tron@ed.ac.uk

**Abstract.** The paper describes automatic speech recognition experiments and
results on the spontaneous Hungarian MALACH speech corpus. A novel
morph-based lexical modeling approach is compared to the traditional word-
based one and to another, previously best performing morph-based one in terms
of word and letter error rates. The applied language and acoustic modeling
techniques are also detailed. Using unsupervised speaker adaptations along with
morph based lexical models 14.4%-8.1% absolute word error rate reductions
have been achieved on a 2 speakers, 2 hours test set as compared to the speaker
independent baseline results.

## 1 Introduction

The MALACH project (Multilingual Access to Large Spoken Archives) aims at
providing improved access to the archived testimonies of Holocaust survivors. The
testimonies were given in 32 languages and a considerable subpart (more than 2,000
hours) consists of Hungarian interviews. Our current aim was to automatically
transcribe Hungarian MALACH speech data at good accuracy. According to our
knowledge no other research groups have previously published spontaneous
conversational Hungarian LVCSR (Large Vocabulary Continuous Speech
Recognition) results.

As presented in [1], Hungarian is a morphologically very rich language. The main
source of word form variability is agglutination but inflections of verbs and nouns
also commonly occur. A pioneering attempt to LVCSR of Hungarian is described in
[2], where rule-based morphological segmentation and analysis is integrated for read
newspaper speech recognition. Unfortunately, the applied morphological analyzer is
not available publicly and no explicit comparison of morph- and word-based ASR
approaches was made.

Though Hungarian is considered as a phonologically writing language and
grapheme to phoneme rules are applied successfully in a variety of ASR tasks [1],

V. Matoušek and P. Mautner (Eds.): TSD 2007, LNAI 4629, pp. 342–349, 2007.
© Springer-Verlag Berlin Heidelberg 2007

some issues regarding pronunciation modeling of spontaneous speech still remain. Similarly to other languages, significant deletions can happen in spontaneously uttered words. Also, foreign and some traditional words' pronunciations must be modeled via exception pronunciation dictionaries. Moreover, pronunciation modeling is not entirely straightforward, if morph-based language modeling is intended to be used but annotation is based on words.

In the following we present our work and results related to the large vocabulary spontaneous Hungarian MALACH speech recognition task.

# 2 Database

As already introduced, more than two thousand hours of MALACH speech is recorded, however, only 31 hours of this speech were transcribed so far and from now on we will refer to this specific part as the MALACH Hungarian speech corpus.

## 2.1 Speech and Transcription Data

Speech was recorded with a sampling frequency of 44.1 kHz in common environments (usually survivors' homes). Given the nature of the interviews, MALACH speakers are typically aged and their speech is sometimes incoherent, rich in disfluencies and occasionally strongly accented. However, some of the interviewees speak close to the standard way.

During the transcription process, orthographic and phonemic variants were noted in parallel if pronunciation could not be automatically derived correctly from orthography. Phonological co-articulations were not considered during transcription, but foreign words, spellings, partially pronounced "broken" words, etc. were marked explicitly.

## 2.2 Training and Test Sets

For *training*, 15-minute segments from 104 speakers were transcribed, starting from 30th minute of each interview (yielding a total of 26 hours). For *test*, a 5 hour set was defined with variable length of transcribed data from 10 other speakers.

The test set was partitioned into several subsets. Matched subset for *speaker independent* recognition is defined as the collection of test utterances resulted after the 15th minute of the testimonies (about half of test recordings). Weakly-matched subset containing many named entities is defined as the complement of the matched subset, i.e., utterances from the first 15 minutes.

Test subsets are also defined for *speaker adapted* ASR. One female and one male speakers' 1-1 hours data are used for unsupervised speaker adaptations and tests.

# 3 Lexical Modeling Approaches

The first task was to select the basic lexical units. These units are used in language modeling and also, the same units are mapped to basic acoustic elements (phonemes) through pronunciation modeling.

The obvious language modeling problem of agglutinative languages, i.e., the high number of rare word forms, is generally alleviated by using morphemes instead of words as basic units. A fundamental question is how to segment the words into morphemes. A rule-based approach is used for newspaper reading in Hungarian [2]. In [3] morpheme-like lexical units are obtained by rule-based tools, but they perform in Basque ASR worse than word-based ones. [4] successfully applies morphs resulted from statistics and rule-based analysis for Korean LVCSR. Considerable error reductions are reported in [5] using a statistics-based unsupervised morph segmentation tool called "Morfessor Baseline" [6] for Finnish read newspaper speech recognition. Additionally, good results are achieved using the same statistical segmentation tool for Estonian and Turkish read speech recognition [7]. Finally, the improved, Morfessor Categories ML algorithm performed as best in the Morpho Challenge 2005 project [8] for Finnish and Turkish LVCSR.

In this study the following lexical modeling approaches are compared on the MALACH Hungarian task in term of ASR accuracies:

- *Word lexical modeling.* A traditional approach that is used most commonly in other language (English, Czech, Slovak, Russian, etc.) MALACH LVCSR tasks [9], [10], where simple words act as basic lexical units. This approach served as *baseline*.
- *Statistics-based morph lexical modeling.* In [11] best ASR results on the Hungarian MALACH task are obtained when the unsupervised *Morfessor Categories ML* algorithm [12] was used for the derivation of basic (morph) lexical units.
- *Combined, rule- and statistics-based morph lexical modeling.* A novel approach is applied here for morph segmentations. The technique is the combination of the rule-based Hunmorph [13] and the statistics-based Morfessor Baseline [6] methods. Essentially the MDL (Minimum Description Length) principle is applied for the unsupervised disambiguation of rule-based morph segmentations. This approach will be called as *Hunmorph MDL* in the rest of the paper.

### 3.1 The Application of Morph Segmentations

Each morph segmentation tool was applied to the Hungarian MALACH manually transcribed training text as follows.

1. All text tokens from the preprocessed (lower cased, etc.) training text are collected into a list.
2. Spellings, foreign words and any non-word tokens are removed.
3. Morph segmentation is applied on the filtered list resulting in a word to morph dictionary.
4. Word break symbols (#) are inserted into the training text.
5. Training text words found in the word to morph dictionary are replaced by (white space separated) sequences of morphs.

In this way word-based training text is transformed into a sequence of morphs in a broader sense where (possibly agglutinated) foreign words, word break symbols, spellings, hesitations and tagged morphs etc. are treated equally as simple morphs.

The recognition results obtained using various lexical models are displayed on Table 1. In the followings we briefly describe the applied language and acoustic modeling techniques.

# 4 Language Modeling

For language modeling only training data transcripts (200K running words) were used. Both word- and morph-based 3-gram language models were built with modified, interpolated Kneser-Ney smoothing using the SRILM toolkit [14]. Other orders of n-gram models (n=2, 4) were tried but found suboptimal in terms of recognition accuracy and/or speed (results not displayed). No language model pruning was applied.

### 4.1 Word-Based LM

Word-based language models were taught on the simplest "traditional" way. No classes or any supplementary or unnecessary information (e. g. word break symbols) were used.

The number of word forms occurred only once in the training text in the proportion of all word forms (vocabulary size) was 62%. This clearly shows the difficulties of using word-based LM-s in case of agglutinative languages and relatively small training text databases.

### 4.2 Morph-Based LM

Hungarian – similarly to many other agglutinative languages – writes words. However, if the lexical units are morphs then the output of the recognizers will be a sequence of morphs, too, that is not straightforward to convert to word sequence without supplementary information. Since reliable morphological analysis is generally not available for the morphs we have decided to use a simple statistical approach found in [5] to overcome the problem.

The essence of the approach is the introduction and application of explicit word break symbols (#), which are considered as "normal" morphs in language modeling and so the restoration of word forms from the ASR output is trivial.

A disadvantage of this technique is that the effective history of morph n-gram models is shortened. This is especially true if the average number of morphs per word is low (close to 1). Regarding the MALACH Hungarian data the average number of morphs per word was close to 1.6 at each morph segmentation approach.

# 5 Acoustic Modeling

Similarly to other languages, acoustic modeling of Hungarian means pronunciation modeling, i.e., mapping of lexical units to phone like units and acoustic modeling of the basic phone units. In the experiments we used a simplified phoneme set as basic acoustic models, i.e., long and short consonants were not differentiated.

### 5.1 Pronunciation Modeling

**Word Pronunciations.** Word to phoneme mapping was performed using simple rules for the majority of training text words [1]. Foreign, traditional, etc. words' transcriptions were obtained explicitly from the manual annotations if given. Weighted alternative pronunciations were used only for those orthographic words, which occurred in the training transcriptions with more than one different explicitly annotated pronunciation as exceptions.

**Morph Pronunciations.** Theoretically morphs can be mapped to phonemes using the same grapheme to phoneme rules like words. Exceptionally pronounced morphs would need, however, special treatment since the exception list is based on words. Additionally, morph segmentation of words may be resulted in phonemically incorrect grapheme splittings. (Hungarian has many phonemes, which are written with more than one grapheme.)

We used an approximation to obtain morph to phoneme mappings. First, the word exception pronunciation dictionary is simply applied to the *morphs* of the training text. Then remaining morphs which were not found in the word exception dictionary are mapped to phonemes using grapheme to phoneme rules [1].

Though the method allows incorrect pronunciations, as well, practically it enables the usage of morph-based language models.

### 5.2 Phoneme Acoustic Modeling

The same phoneme acoustic models were applied for word and morph lexical units. Decision-tree state clustered cross-word triphone models with approximately 3000 HMM states were trained using ML estimation [15]. Three state left-to-right HMMs were applied with 10 mixture components per state. PLP-based acoustic features were obtained by using a modified HTK front-end [10]. No phone durational models were used.

## 6   Experimental Results

One-pass decoding was performed by an enhanced version of the decoder mentioned in [16] with a Real-Time Factor (RTF) of 2-4 on a 3GHz CPU. Previously, the recognition network was constructed and optimized offline on the triphone-level based on the AT&T FSM toolkit [17].

### 6.1   Speaker Independent ASR Results

In the speaker independent tests all the 5 hours length test data – divided into matched and weakly matched subparts – were used. Acoustic models were trained with all the training data (26 hours, 104 speakers).

Letter Error Rates (LER's) were calculated as well, because in case of morphologically rich languages they sometimes can be more reliably used for evaluations than Word Error Rates (WER's).

**Table 1.** Speaker independent ASR results on the MALACH Hungarian spontaneous speech corpus. Word and letter error rates are in [%], best values are italicized.

| Lexical modeling approach | Vocabulary size | Weakly matched | | Matched | |
|---|---|---|---|---|---|
| | | WER | LER | WER | LER |
| Word – baseline | 20K | 56.15 | 28.66 | 52.96 | 25.57 |
| Morph – Morfessor Cat. ML | 5.5K | *55.94* | *28.17* | 51.07 | 24.53 |
| Morph – Hunmorph MDL | 7K | 56.01 | 28.26 | *50.90* | *24.39* |

It can be seen that morph-based approaches consistently outperform the word-based baseline. As Table 1 shows, improvements are lager in the Matched subset – about 2% WER absolute – although they seem minor considering the magnitudes of absolute error rates. Nevertheless, the results are comparable to the MALACH ASR systems for other languages even though the Hungarian training data are four times smaller than the other databases.

## 6.2 Speaker-Wise ASR Results with Unsupervised Speaker Adaptations

We made speaker-wise tests to investigate the effect of acoustic model adaptations on various lexical modeling approaches. For unsupervised speaker adaptation a three-pass method was applied on the female (F) and male (M) speakers' data mentioned in 2.2.

In the first pass speaker independent (denoted with SI) ASR results were used as transcripts for the first, 32 class – formally supervised – MLLR adaptations [15] resulting in SA1 acoustic models. Then speech recognition was performed with these speaker adapted acoustic models (per speaker). Second, speaker dependent ASR results were used in a second adaptation round, which resulted in refined speaker adapted acoustic models denoted with SA2. These models were applied in the third, final recognition pass.

**Table 2.** Speaker-wise ASR results on the MALACH Hungarian spontaneous speech corpus. Three stages of speaker adaptations are evaluated. SI: no adaptations, SA1: one iteration, SA2: two iterations. Word and letter error rates are in [%], best values are italicized.

| Lexical modeling approach | Acoustic models | F speaker | | M speaker | |
|---|---|---|---|---|---|
| | | WER | LER | WER | LER |
| Word – baseline | SI | 51.86 | 21.70 | 49.09 | 23.73 |
| Word – baseline | SA1 | 46.23 | 17.32 | 43.94 | 19.67 |
| Word – baseline | SA2 | 46.41 | 17.23 | 43.69 | 19.60 |
| Morph – Morfessor Cat. ML | SI | 47.10 | 19.25 | 48.05 | 23.26 |
| Morph – Morfessor Cat. ML | SA1 | 38.06 | 13.79 | 41.86 | 18.61 |
| Morph – Morfessor Cat. ML | SA2 | 37.85 | *13.64* | 41.54 | 18.58 |
| Morph – Hunmorph MDL | SI | 46.07 | 18.59 | 48.00 | 23.41 |
| Morph – Hunmorph MDL | SA1 | 37.56 | 13.84 | 41.45 | 18.42 |
| Morph – Hunmorph MDL | SA2 | *37.46* | 13.86 | *40.98* | *18.31* |

As Table 2. shows, the morph lexical models obtained with the Hunmorph MDL approach perform as best nearly under all circumstances. It can easily be noticed that the improvements from the baselines are much larger than in the previous experiments. However, high speaker variability can also be observed in the improvements. Finally, it is acknowledged that the language independent Morfessor Categories ML based technique performed nearly as well as the language specific Hunmorph MDL.

## 7  Conclusions

The first attempts to automatically transcribe spontaneous Hungarian were introduced. We can conclude that spontaneous LVCSR of such a morphologically rich language can effectively be improved using appropriate morph lexical, language and pronunciation models. As presented, lexical modeling does not necessarily need to be language dependent if the given language is agglutinative. Furthermore, in can be noticed that the improvements are higher after speaker adaptations of acoustic models. Finally, we would expect further error rate reductions from the application of discriminative training techniques, which is our future plan.

## Acknowledgments

The authors gratefully acknowledge the cooperation of Josef Psutka, Pavel Ircing, Jindrich Matousek, Vlasta Radová and Josef J. Psutka from the Department of Cybernetics at the University of West Bohemia in Plzeň. Thanks to Bill Byrne for initiating the Hungarian MALACH project.

The project was supported by the Survivors of the SHOAH Visual History Foundation and the Johns Hopkins University in Baltimore (Agreement JHU 8202–48279), by the National Science Foundation (Grant No. IIS-0122466), and by the National Office for Research and Technology (GVOP–3.1.1–2004/05–0385/3.0).

## References

1. Szarvas, M., Fegyó, T., Mihajlik, P., Tatai, P.: Automatic Recognition of Hungarian: Theory and Practice. International Journal of Speech Technology 3, 277–287 (2000)
2. Szarvas, M., Furui, S.: Finite-State Transducer based Modeling of Morphosyntax with Applications to Hungarian LVCSR. In: Proceedings of ICASSP, Hong Kong, China, pp. 368–371 (2003)
3. López, K., Graña, M., Ezeiza, N., Hernández, M., Zulueta, E., Ezeiza, A., Tovar, C.: Selection of Lexical Units for Continuous Speech Recognition of Basque. In: Sanfeliu, A., Ruiz-Shulcloper, J. (eds.) CIARP 2003. LNCS, vol. 2905, pp. 244–250. Springer, Heidelberg (2003)
4. Kwon, O.–W., Park, J.: Korean large vocabulary continuous speech recognition with morpheme-based recognition units. Speech Communication 39(3-4), 287–300 (2003)

5. Hirsimäki, T., Creutz, M., Siivola, V., Kurimo, M., Pylkkönen, J., Virpioja, S.: Unlimited vocabulary speech recognition with morph language models applied to Finnish. Computer, Speech and Language 20(4), 515–541 (2006)
6. Creutz, M., Lagus, K.: Unsupervised Morpheme Segmentation and Morphology Induction from Text Corpora Using Morfessor 1.0., Publications in Computer and Information Science, Report A81, Helsinki University of Technology (March 2005)
7. Kurimo, M., Puurula, A., Arisoy, E., Siivola, V., Hirsimaki, T., Pylkkonen, J., Alumae, T., Saraçlar, M.: Unlimited vocabulary speech recognition for agglutinative languages, HLT–NAACL, New York, USA (2006)
8. Kurimo, M., Creutz, M., Varjokallio, M., Arisoy, E.: Saraçlar, Murat.: Unsupervised segmentation of words into morphemes - Morpho Challenge 2005, Application to Automatic Speech Recognition. In: Interspeech 2006. Pittsburgh, USA (September 17-21, 2006)
9. Ramabhadran, B., Juang, J., Picheny, M.: Towards Automatic Transcription of Large Spoken Archives - English ASR for the MALACH Project. In: International Conference on Acoustics, Speech, and Signal Processing, Genf. (2003)
10. Psutka, J., Ircing, P., Psutka, J.V., Hajic, J., Byrne, W.J., Mírovský, J.: Automatic transcription of Czech, Russian, and Slovak spontaneous speech in the MALACH project. In: Interspeech–2005, pp. 1349–1352 (2005)
11. Mihajlik, P., Fegyó, T., Tüske, Z., Ircing, P.: A Morpho-graphemic Approach for the Recognition of Spontaneous Speech in Agglutinative Languages – like Hungarian. In: Interspeech 2007. Antwerp, Belgium (August 27-31, 2007)
12. Creutz, M., Lagus, K.: Inducing the Morphological Lexicon of a Natural Language from Unannotated Text. In: Proceedings of AKRR'05, Espoo, Finland (June 15-17, 2005)
13. Trón, V., Németh, L., Halácsy, P., Kornai, A.: Gyepesi, Gy., Varga, D.: Hunmorph: open source word analysis. In: Proc. ACL 2005, Software Workshop, pp. 77–85 (2005)
14. Stolcke, A.: SRILM: – an extensible language modeling toolkit. In: Proc. Intl. Conf. on Spoken Language Processing, Denver, pp. 901–904 (2002)
15. Young, S., Ollason, D., Valtchev, V., Woodland, P.: The HTK book (for HTK version 3.2.) Cambridge (2002)
16. Fegyó, T., Mihajlik, P., Szarvas, M., Tatai, P., Tatai, G.: VOXenter – Intelligent voice enabled call center for Hungarian. In: EUROSPEECH-2003, Genf., pp. 1905–1908 (2003)
17. Mohri, M., Pereira, F., Riley, M.: Weighted Finite-State Transducers in Speech Recognition. Computer Speech and Language 16(1), 69–88 (2002)

# Non-uniform Speech/Audio Coding Exploiting Predictability of Temporal Evolution of Spectral Envelopes*

Petr Motlicek[1,2], Hynek Hermansky[1,2,3], Sriram Ganapathy[1,3], and Harinath Garudadri[4]

[1] IDIAP Research Institute,
Rue du Simplon 4, CH-1920, Martigny, Switzerland
{motlicek,hynek,ganapathy}@idiap.ch
[2] Faculty of Information Technology, Brno University of Technology,
Božetěchova 2, Brno, 612 66, Czech Republic
[3] École Polytechnique Fédérale de Lausanne (EPFL), Switzerland
[4] Qualcomm Inc., San Diego, California, USA
hgarudad@qualcomm.com

**Abstract.** We describe novel speech/audio coding technique designed to operate at medium bit-rates. Unlike classical state-of-the-art coders that are based on short-term spectra, our approach uses relatively long temporal segments of audio signal in critical-band-sized sub-bands. We apply auto-regressive model to approximate Hilbert envelopes in frequency sub-bands. Residual signals (Hilbert carriers) are demodulated and thresholding functions are applied in spectral domain. The Hilbert envelopes and carriers are quantized and transmitted to the decoder. Our experiments focused on designing speech/audio coder to provide broadcast radio-like quality audio around $15 - 25$kbps. Obtained objective quality measures, carried out on standard speech recordings, were compared to the state-of-the-art 3GPP-AMR speech coding system.

## 1 Introduction

State-of-the-art speech coding techniques that generate toll quality speech typically exploit the short-term predictability of speech signal in the $20 - 30$ms range [1]. This short-term analysis is based on the assumption that the speech signal is stationary over these segment durations. Techniques like Linear Prediction (LP), which is able to efficiently approximate short-term power spectra by Auto-Regressive (AR) model [2], are applied.

However, speech signal is quasi-stationary and carries information in its dynamics. Such information is not adequately captured by short-term based approaches. Some considerations that motivated us to explore novel architectures are mentioned below:

---

* This work was partially supported by grants from ICSI Berkeley, USA; the Swiss National Center of Competence in Research (NCCR) on "Inter active Multi-modal Information Management (IM)2"; managed by the IDIAP Research Institute on behalf of the Swiss Federal Authorities, and by the European Commission 6th Framework DIRAC Integrated Project.

V. Matoušek and P. Mautner (Eds.): TSD 2007, LNAI 4629, pp. 350–357, 2007.
© Springer-Verlag Berlin Heidelberg 2007

- When the signal dynamics are described by a sequence of short-term vectors, many issues come up, like windowing, proper sampling of short-term representation, time-frequency resolution compromises, etc.
- There are situations where LP provides a sub-optimal filter estimate. In particular, when modeling voiced speech, LP methods can be adversely affected by spectral fine structure.
- The LP based approaches do not respect many important perceptual properties of hearing (e.g., non-uniform critical-band representation).
- Conventional LP techniques are based on linear model of speech production, thus have difficulties encoding non-speech signals (e.g., music, speech in background, etc.).

Over the past decade, research in speech/audio coding has been focused on high quality/low latency compression of wide-band audio signals. However, new services such as Internet broadcasting, consumer multimedia, or narrow-band digital AM broadcasting are emerging. In such applications, new challenges have been raised, such as resiliency to errors and gaps in delivery. Furthermore, many of these services do not impose strict latency constraints, i.e., the coding delay is less important as compared to bit-rate and quality requirements.

This paper describes a new coding technique that employs AR modeling applied to approximate the instantaneous energy (squared Hilbert envelope (HE)) of relatively long-term critical-band-sized sub-band signals. It has been shown in our earlier work that based on approximating the envelopes in sub-bands we can design very low bit-rate speech coder giving intelligible output of synthetic quality [3]. In this work, we focus on efficient coding of residual information (Hilbert carriers (HCs)) to achieve higher quality of the re-synthesized signal. The objective quality scores based on Itakura-Saito (I-S) distance measure [4] and Perceptual Evaluation of Speech Quality (PESQ) [5] are used to evaluate the performance of the proposed coder on challenging speech files sampled at 8kHz.

The paper is organized as follows: In Section 2, a basic description of the proposed encoder is given. In Section 3, the decoding-side is described. Section 4 describes the experiments we conducted to validate the approach using objective quality measurements.

## 2   Encoding

New techniques utilizing LP to model temporal envelopes of input signal have been proposed [6,7]. More precisely, HE (squared magnitude of an analytic signal), which yields a good estimate of instantaneous energy of the signal, can be parameterized by Frequency Domain Linear Prediction (FDLP) [7]. FDLP represents frequency domain analogue of the traditional Time Domain Linear Prediction (TDLP) technique, in which the power spectrum of each short-term frame is approximated by the AR model.

The FDLP technique can be summarized as follows: To get an all-pole approximation of the squared HE, first the Discrete Cosine Transform (DCT) is applied to a given audio segment. Next, the autocorrelation LP technique is applied to the DCT transformed signal. The Fourier transform of the impulse response of the resulting all-pole model approximates the squared HE of the signal.

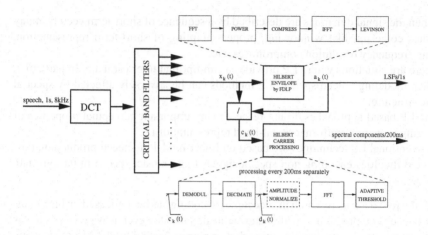

**Fig. 1.** Simplified structure of the proposed encoder

Just as TDLP fits an all-pole model to the power spectrum of the input signal, FDLP fits an all-pole model to the squared HE of the signal. As discussed later, this approach can be exploited to approximate temporal envelope of the signal in individual frequency sub-bands. This presents an alternate representation of signal in the 2-dimensional time-frequency plane that can be used for audio coding.

### 2.1   Parameterizing Temporal Envelopes in Critical Sub-bands

The graphical scheme of the whole encoder is depicted in Fig. 1. First, the signal is divided into 1000ms long temporal segments which are transformed by DCT into the frequency domain, and later processed independently. In order to avoid possible artifacts at segment boundaries, 10ms overlapping is used.

To emulate auditory-like frequency selectivity of human hearing, we apply $N_{BANDs}$ Gaussian functions ($N_{BANDs}$ denotes number of critical sub-bands), equally spaced on the Bark scale with standard deviation $\sigma = 1$ bark and center frequency $F_k$, to derive sub-segments of the DCT transformed signal. FDLP technique is performed on every sub-segment of the DCT transformed signal (its time-domain equivalent obtained by inverse DCT is denoted as $x_k(t)$, where $k$ denotes frequency sub-band). Resulting approximations of HEs in sub-bands are denoted as $a_k(t)$.

### 2.2   Excitation of FDLP in Frequency Sub-bands

To reconstruct the signal in each critical-band-sized sub-band, the additional component – Hilbert carrier (HC) $c_k(t)$ is required (residual of the LP analysis represented in time-domain). Modulating $c_k(t)$ with approximated temporal envelope $a_k(t)$ in each critical sub-band yields the original $x_k(t)$ (refer [8] for mathematical explanation).

Clearly, $c_k(t)$ is analogous to excitation signal in TDLP. Utilizing $c_k(t)$ leads to perfect reconstruction of $x_k(t)$ in sub-band $k$ and, after combining the sub-bands, in perfect reconstruction of the overall input signal.

**Processing Hilbert Carriers (HCs):** For convenience in processing and encoding, we need the sub-band carrier signals to be low-pass. This can be achieved by demodulating $c_k(t)$ (shifting Fourier spectrum of $c_k(t)$ from $F_k$ to 0 Hz). Since modulation frequency $F_k$ of each sub-band is known, we employ standard procedure to demodulate $c_k(t)$ through the concept of *analytic signal* $z_k(t)$. $z_k(t)$ is the complex signal that has zero-valued spectrum for negative frequencies. To demodulate $c_k(t)$, we perform scalar multiplication $z_k(t).c_k(t)$. Demodulated carrier in each sub-band is low-pass filtered and down-sampled. Frequency width of the low-pass filter as well as the down-sampling ratio is determined using the frequency width of the Gaussian window (the cutoff frequencies correspond to 40dB decay in magnitude with respect to $F_k$) for a particular critical sub-band. The resulting time-domain signal (denoted as $d_k(t)$) represents demodulated and down-sampled HC $c_k(t)$. $d_k(t)$ is a complex sequence, because its Fourier spectrum is not conjugate symmetric. Perfect reconstruction of $c_k(t)$ from $d_k(t)$ can be done by reversing all the pre-processing steps.

Since HCs $c_k(t)$ are quite non-stationary, they are split into 200ms long sub-segments (10ms overlap for smooth transitions) and processed independently.

**Encoding of Demodulated HCs:** Temporal envelopes $a_k(t)$ and complex valued demodulated HCs $d_k(t)$ carry the information necessary to reconstruct $x_k(t)$. If the original HE is used to derive $d_k(t)$, then $|d_k(t)| = 1$, and only the phase information from $d_k(t)$ would be required for perfect reconstruction. However, since FDLP yields only approximation of the original HEs, $|d_k(t)|$ in general will not be perfectly flat and both components of complex sequence are required.

The coder implemented is an "adaptive threshold" coder applied on Fourier spectrum of $d_k(t)$, independently in each sub-band, where only the spectral components having magnitudes above the threshold are transmitted. The threshold is dynamically adapted to meet a required number of transmitted spectral components (described later in Section 4.1). The quantized values of magnitude and phase for each selected spectral component are transmitted.

## 3 Decoding

In order to reconstruct the input signal, the carrier $c_k(t)$ in each critical sub-band needs to be re-generated and then modulated by temporal envelope $a_k(t)$ obtained using FDLP.

A graphical scheme of the decoder, given in Fig. 2, is relatively simple. It inverts the steps performed at the encoder. The decoding operation is also applied on each (1000ms long) input segment independently. The decoding steps are: (a) Signal $d_k(t)$ is reconstructed using inverse Fourier transform of transmitted complex spectral components. $d_k(t)$ is then up-sampled to the original rate and modulated on sinusoid at $F_k$ (i.e., its Fourier spectrum is frequency-shifted and post-processed to be conjugate symmetric). This results in the reconstructed HC $c_k(t)$. (b) Temporal envelope $a_k(t)$ is reconstructed from transmitted AR model coefficients. The temporal trajectory $x_k(t)$ is obtained by modulating $c_k(t)$ with $a_k(t)$.

The above steps are performed in all frequency sub-bands. Finally: (a) The temporal trajectories $x_k(t)$ in each critical sub-band are projected to the frequency domain by

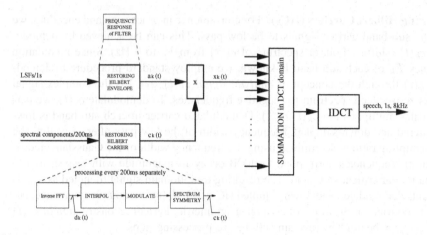

**Fig. 2.** Simplified structure of the proposed decoder

DCT and summed. (b) A "de-weighting" window is applied to compensate for the effect of Gaussian windowing of DCT sequence at the encoder. (c) Inverse DCT is performed to reconstruct 1000ms long output signal (segment). Fig. 3 shows time-frequency characteristics of the proposed coder for a randomly selected speech sample.

## 4   Experiments

All experiments were performed with speech signals sampled at $F_s = 8$kHz. We used decomposition into $N_{BANDs} = 13$ critical sub-bands, which roughly corresponds to partition of one sub-band per bark.

FDLP approximating HE in each frequency sub-band $a_k(t)$ is represented by Line Spectral Frequencies (LSFs). Previous informal subjective listening tests, aimed at finding sufficient approximations $a_k(t)$ of temporal envelopes, showed that for coding the 1000ms long audio segments, the "optimal" AR model is of order $N_{LSFs} = 20$ [3].

### 4.1   Objective Quality Tests on HC

We used Itakura-Saito (I-S) distance measure [4] as a simple method together with ITU-T P.862 PESQ objective quality tool [5] to adjust the threshold values on Fourier spectrum of $d_k(t)$ for reconstructing HCs $c_k(t)$ at the decoder.

These measures were used to evaluate performance as a function of variable number of Fourier spectral components for the reconstruction of $c_k(t)$ (this number is always constant over all sub-bands) while fixing all other parameters.

The performance was tested on a sub-set of TIMIT – speech database [9], containing 380 speech sentences sampled at $F_s = 8$kHz. A total of about 20 minutes of speech was used for the experiments.

I-S measure was performed on short-term frames (30ms frame-length, 7.5ms frame-skip). Encoded sentences were compared to original sentences measuring the I-S distance between them. The lower values of I-S measure indicate smaller distance and

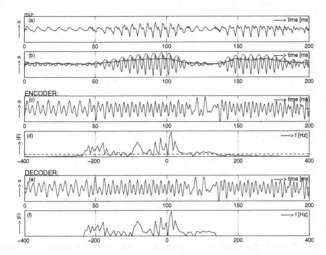

**Fig. 3.** Time-Frequency characteristics generated from randomly selected speech sample: (a) 200ms segment of the input signal. (b) $x_3(t)$ sequence (frequency sub-band $k = 3$, center frequency $F_3 = 351$Hz ), thin upper line represents original HE, solid upper line represents its FDLP approximation. (c) Original HC $c_3(t)$. (d) Magnitude Fourier spectral components of the demodulated HC $d_3(t)$, the solid line represents the selected threshold. (e) Reconstructed HC $c_3(t)$ in the decoder. (f) Magnitude Fourier spectral components of $d_3(t)$ post-processed by adaptive threshold.

better speech quality. As suggested in [10], to exclude unrealistically high spectral distance values, 5% of frames with the highest I-S distances were discarded from the final evaluation. This method ensures a reasonable measure of overall performance.

PESQ scores were also computed for the reconstructed signal. The quality estimated by PESQ corresponds to the average user perception of the speech sample under assessment PESQ – MOS (Mean Opinion Score).

Fig. 4 shows the mean I-S distance value as well as mean PESQ score computed over all TIMIT DB sub-set as a function of the number of Fourier spectral components used to reconstruct spectrum of demodulated HC $d_k(t)$ in each critical sub-band. Both objective quality measures show marked improvement when the number of spectral components is increased from 30 to 80.

We repeated the above objective tests with 3GPP-Adaptive Multi Rate (AMR) speech codec at 12.2kbps [11] on the same database, and show the results in Fig. 4. The results indicate that if $d_k(t)$ is reconstructed from $\sim 65$ Fourier spectral components (in each critical sub-band, per 200ms), the proposed coder achieves similar performance to AMR codec with respect to the chosen objective measures. Informal subjective results showed that the speech quality was comparable to that of AMR 12.2, while the quality for music signals was noticeably better.

In this paper, we do not discuss the quantization block and entropy coder. However, in additional informal experiments, LSFs describing temporal envelopes $a_k(t)$ as well as the selected spectral components of $d_k(t)$ were quantized (split VQ technique). These

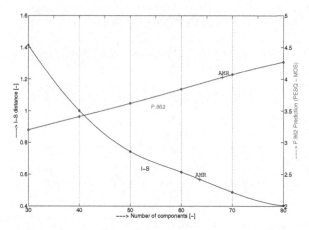

**Fig. 4.** Global mean I-S distance measure of the proposed coder as a function of the number of Fourier spectral components used to reconstruct $d_k(t)$ in each critical sub-band. "+" marks the performance of the 3GPP-AMR speech codec at 12.2kbps.

preliminary experiments show the promise of a coder in encoding speech and music signals at an average bit-rate of $15 - 25$kbps.

## 5   Conclusions

A novel variable bit-rate speech/audio coding technique based on processing relatively long temporal segments of audio signal in critical-band-sized sub-bands has been proposed and evaluated. The coder architecture allows to easily control the quality of reconstructed sound and the final bit-rate, thus making it suitable for variable bandwidth channels. The coding technique representing input signal in frequency sub-bands is inherently more robust to losing chunks of information, i.e., less sensitive to dropouts. This can be of high importance for any Internet protocol service.

We describe experiments focused on efficient representation of excitation signal for the proposed FDLP coder. Such parameter setting does not indeed correspond to "optimal" approach (e.g., we use uniform spectral parameterization of Hilbert carriers in all sub-bands, uniform quantization of LSFs, simple Gaussian decomposition, etc). All these would be the direction of future research in improving the proposed coder. To convert the proposed speech/audio coding technique into a real application, formal subjective tests need to be made both on speech and music recordings.

## References

1. Spanias, A.S.: Speech Coding: A Tutorial Review. In: Proc. of IEEE, vol. 82(10) (October 1994)
2. Makhoul, J.: Linear Prediction: A Tutorial Review. In: Proc. of IEEE, vol. 63(4) (April 1975)

3. Motlicek, P., Hermansky, H., Garudadri, H., Srinivasamurthy, N.: Speech Coding Based on Spectral Dynamics. In: Sojka, P., Kopeček, I., Pala, K. (eds.) TSD 2006. LNCS (LNAI), vol. 4188, Springer, Heidelberg (2006)
4. Quackenbush, S.R., Barnwell, T.P., Clements, M.A.: Objective Measures of Speech Quality. Advanced Reference Series. Prentice-Hall, Englewood Cliffs, NJ (1988)
5. ITU-T Rec. P.862: Perceptual Evaluation of Speech Quality (PESQ), an Objective Method for End-to-end Speech Quality Assessment of Narrowband Telephone Networks and Speech Codecs, ITU, Geneva, Switzerland (2001)
6. Herre, J., Johnston, J.H.: Enhancing the performance of perceptual audio coders by using temporal noise shaping (TNS), in 101st Conv. Aud. Eng. Soc. (1996)
7. Athineos, M., Hermansky, H., Ellis, D.P.W.: LP-TRAP: Linear predictive temporal patterns. In: Proc. of ICSLP, Jeju, S. Korea, pp. 1154–1157 (October 2004)
8. Schimmel, S., Atlas, L.: Coherent Envelope Detector for Modulation Filtering of Speech. In: Proc. of ICASSP, Philadelphia, USA, vol. 1, pp. 221–224 (May 2005)
9. Fisher, W.M., et al.: The DARPA speech recognition research database: specifications and status. In: Proc. DARPA Workshop on Speech Recognition, pp. 93–99 (February 1986)
10. Hansen, J.H.L., Pellom, B.: An Effective Quality Evaluation Protocol for Speech Enhancement Algorithms. In: Proc. of ICSLP, Sydney, Australia, vol. 7, pp. 2819–2822 (December 1998)
11. 3GPP TS 26.071: AMR speech CODEC, General description,
    http://www.3gpp.org/ftp/Specs/html-info/26071.htm

# Filled Pauses in Speech Synthesis: Towards Conversational Speech

Jordi Adell[1], Antonio Bonafonte[1], and David Escudero[2]

[1] Universitat Politècnica de Catalunya, Barcelona 08034, Spain
{jadell,antonio}@gps.tsc.upc.edu
http://www.talp.upc.edu
[2] Universidad de Valladolid, 47011 Valladolid, Spain
descuder@infor.uva.es
http://www.infor.uva.es

**Abstract.** Speech synthesis techniques have already reached a high level of naturalness. However, they are often evaluated on text reading tasks. New applications will request for conversational speech instead and disfluencies are crucial in such a style. The present paper presents a system to predict filled pauses and synthesise them. Objective results show that they can be inserted with 96% precision and 58% recall. Perceptual results even shown that its insertion increases naturalness of synthetic speech.

## 1 Introduction

Speech synthesis has already reached high naturalness, mainly due to the use of effective techniques such us unit selection-based systems [1] or other new rising technologies [2] based on the analysis of huge speech corpora. The main application of speech synthesis has been focused by now on read style speech as it can be assessed that read style is the most generalist style to be extrapolated to any other situation. But nowadays and future applications of text to speech (TTS) systems like film dubbing, robotics, dialogue systems, or multilingual broadcasting demand a variety of styles as the users expect the interface to do more than just reading information.

If synthetic voices want to be integrated in future technology, they must simulate the way people talk instead the way people read. This objective has been already tackled in several manners such as emotional speech synthesis [3], voice quality modelling [4] or even pronunciation variants [5]. In our opinion style is more important; it is desirable synthetic speech to be more conversational-like rather than reading-like speech. Therefore, we claim it is necessary to move from *reading* to *talking* speech synthesisers.

Both styles differ significantly from each other due to the inclusion of a set of a variety of prosodic resources affecting to the rhythm of the utterances. Disfluencies are one of these resources defined as phenomena that interrupt the flow of speech and do not add propositional content to an utterance [6]. Disfluencies are very frequent in normal speech [7] so that it is possible to hypothesise the need to include these prosodic events to approximate to talking speech synthesis. We have already presented experiences on synthesising disfluencies in TTS systems in previous works [8], now we present a work to predict where the disfluencies must be placed in the text.

V. Matoušek and P. Mautner (Eds.): TSD 2007, LNAI 4629, pp. 358–365, 2007.

In the goal of integrating disfluencies in conversational speech synthesis, it is not only important that the system is able to pronounce disfluencies, but also to give the system capabilities to predict them. Previous experiences in pauses prediction [9] lead us to give to this process a relevant role because wrong predictions can decrease dramatically the quality of the synthetic voices. Due to the complexity of the disfluencies phenomenon we focus on the simplest and more frequent type of disfluency: filled pauses. Filled pauses, in contrast with other disfluencies like repairs or repetitions, are easy to synthesise and permit the use of similar algorithms as the ones already used to predict pauses.

First, in Section 2 the framework is set, both the prediction algorithm and the corpus are described. Next section explains the synthesis method used for filled pauses. Finally, Section 4 presents objective and perceptual evaluation results and in Section 5 conclusions are discussed.

## 2 Prediction of Filled Pauses

In the framework of disfluent speech synthesis, the synthesis of filled pauses is essential. In some applications, such as dialogue systems, filled pauses position could already be given to the synthesiser. However, in some other applications such as speech translation, text is given in spoken style but with no disfluencies in it. Is in this kind of applications where filled pauses prediction makes its contribution. Filled pauses can be inserted, for example, in between the two utterances of a repetition: *no, uh no I won't go!* or after an acknowledge phrase: *well, uh I will go!*. The proposed approach to this task can be carried on due to the increasing availability of large corpora. Mainly for recognition purposes, many spoken corpora are currently being build. Machine learning techniques can take advantage from this corpora in order to perform tasks as the one proposed here.

### 2.1 Classifier Description

The algorithm presented here is based on a combination of language modelling plus a decision tree. Both are well known machine learning techniques. The focus of the work done is on choosing proper features with respect to the task.

The decision tree performs a binary classification task. It classifies each word in the text whether it has a filled pause following it or not. The aim of the algorithm is to decide whether a filled pause has to be inserted or not after each word present in the text. The decision tree does the classification based on feature vectors which mainly contain language model probabilities and POS tags.

First of all, a language model is trained [10]. Afterwards, the text is tagged with Part-of-Speech (POS) labels [11]. Using the tagged text and the language model a feature vector is generated for each word. A special tag is used as sentence beginning and is treated as a word in order to allow insertions at beginning of sentences.

### 2.2 Feature Sets

Several sets of features have been tested in order to see which one is better suited for the filled pause insertion task. Three classes of features have been used: text-based, ngram-based and a disfluency-based one. Text-based features are: $w_i$ and $pos_i$, where $w_i$ is the actual word to which is applied the decision tree, $pos_i$ is the Part-of-Speech of word $w_i$. Ngram-based features are: $p(w_i|h_i)$ and $p(FP|h_{i+1})$ where $p(w_i|h_i)$ is the probability of word $w_i$ given the history $h_i$ and the language model, and $p(FP|h_i)$ is the probability of a filled pause given previous history $h_i$. Finally, the disfluency-based feature is binary and indicates whether the filled pause is to be inserted in between the two utterances of a repetition or not and will be referred as *repeat*. Repetitions of one and two words have been considered. The four sets used in the experiments are:

- *Set1*: $w_i$, $pos_i$, $pos_{i-1}$, $pos_{i+1}$, $p(w_i|h_i)$, $p(w_{i+1}|h_{i+1})$ and $p(FP|h_i)$;
- *Set2*: $w_i$, $pos_{i-2}$, $pos_{i-1}$, $pos_i$, $pos_{i+1}$, $pos_{i+2}$, $p(w_i|h_i)$, $p(w_{i+1}|h_{i+1})$, $p(w_{i+2}|h_{i+2})$ and $p(FP|h_i)$;
- *Set3*: $w_i$, *repeat*, $pos_i$, $pos_{i-1}$, $pos_{i+1}$, $p(w_i|h_i)$, $p(w_{i+1}|h_{i+1})$ and $p(FP|h_i)$;
- *Set4*: $pos_i$, $pos_{i-1}$, $pos_{i+1}$, $p(w_i|h_i)$, $p(w_{i+1}|h_{i+1})$ and $p(FP|h_i)$;

Ngram-based features have been chosen based on the work of [12], where they claim that filled pauses can be used by the listener in order to identify whether what is going to be said afterwards will be hard to understand or not. Since a filled pause mainly points out speaker problems on finding the desired words, it suggests to the listener that following words will be hard to understand because they have been hard to produce. This reasoning has lead us to add language model probabilities as features. Lower probabilities of the actual and next word will indicate an uncommon expression and therefore it might have been difficult to generate.

POS values have been used here, in order to give the machine learning technique (i.e. decision tree) the possibility to evaluated syntactic structures. Different context lengths have been tested too. Shorter one in Set number 1 and larger one in Set number 2.

Filled pauses can also be in the editing phase of a disfluency [13]. This is the reason for including the feature *repeat* in feature set number 3.

The inclusion of the word itself makes the algorithm very slow since all questions about each possible word has to be tested. In order to avoid this practical problem, we have added to our system the concept of *candidate*. A candidate is a word that allows a filled pause to be placed after it. Therefore, no filled pauses will be placed after words that are not considered candidates. The set of candidates is chosen by sorting words in the training data set. The number of times they preceded a filled pause is used as sorting criteria. This list is then truncated and top most words are considered candidates. Moreover, few candidates are enough to consider a high number of filled pauses. For example, ten candidates can cover 53% of filled pauses in the corpus used in this paper.

In order to test whether the candidates-based approach leads to an improvement, the last data-set has been added to the experiments. *Set4* is equal to *Set1* but without feature $w_i$, only $pos_i$ is known from words preceding a filled pause.

## 2.3   Corpora

The algorithm presented in Section 2.1 is data-driven. Therefore, corpora is needed to train it. The corpus used in the present work has been collected within the LC-STAR project. It consists on lab recorded "spontaneous" conversations about the tourist information topic. Several volunteers were asked to perform a conversation over a telephone playing several roles such as a costumer of a hotel. No guidance on the specific subject was given. It is highly spontaneous, it contains 317,000 words. In addition to filled pauses, it contains disfluencies such as restarts or repetitions. It has been manually transcribed at word level mainly for recognition purposes.

It is a Spanish corpus of 64 speakers. It contains 317,000 words and 5,700 filled pauses (i.e. 1.8%). The vocaulary size is 11,000 words and 1,173 out of them preceded a filled pause at leas once.

## 3  Rhythm and Synthesis of Filled Pauses

In previous works [8] we implemented a set of rules to predict duration and F0 contours of filled pauses and repetitions. Those rules where expected to be useful for the unit selection TTS system in order to look up the corresponding units in the inventory to be concatenated to compose the final sequence. As far as it concerns to filled pauses, the main result of the study was to show that filled pauses are very stable in duration (we decided to use a constant value) and in frequency (the lowered mean value of the preceding and the following word). But results where not satisfactory in terms of the quality of the synthetic output of the TTS system because the rules implemented did not consider several important aspects concerning to the rhythm imposed by the disfluencies that must be taken into account.

In this section we present a study based on a corpus of 65 sentences recorded from a male speaker specially for disfluent speech synthesis. We differentiate here between the rhythm of the whole sentence (rhythTot), the rhythm preceding the filler pause (rhythPre), the one following the filled pause (rhythPost) and the duration of the filled pause itself. The mean syllable duration is used as a measure of the rhythm. Table 1 shows values for the 48 filled pauses that appear in the corpus.

As can be observed in Table 1 there are no significant differences ($P > 0.05$) between the rhythm of the whole sentence, the preceding and following to the filled pause. However, the filled pause duration is significantly higher than the rest ($P < 0.05$). Therefore, we can conclude that the insertions of filled pauses does not modify the rhythm of the utterance. Also the duration of the syllable just before the filled pause has been studied. As can be observed again in Table 1 there are significant differences between the sentence rhythm and this syllable length. This result lead us to conclude that it exists a lengthening of the syllable immediately before the filled pause with respect to the global rhythm of the sentence.

**Table 1.** Study of filled pauses rhythm, calculated over the 48 realisations that appear in the corpus. Units are *ms*.

|                    | FPdur | rhythmTot | rhythmPre | rhythmPost | lastSyl |
|--------------------|-------|-----------|-----------|------------|---------|
| Average            | **320** | 190       | 180       | 190        | **360** |
| Standard Deviation | 160   | 30        | 50        | 40         | 50      |

As the TTS system does not include these considerations, at the moment we decided to enter disfluencies manually in a pool of sentences to test the benefits of the inclusion of filled paused on generating more expressive synthetic speech. The procedure followed was first to generate free of filled pauses synthetic speech and then to insert the filled pause in the place indicated by the algorithm explained in previous sections. We select the disfluency to be inserted (emm, uhh) from a list of them taking also into account its F0 values and the rules devised in [8]. By using the PSOLA features of the praat program [14] the duration of the syllables and the duration of filled pauses are adapted to the new situation. These modifications are done following conclusions presented in this section.

## 4   Evaluation

### 4.1   Prediction

In order to evaluate the prediction algorithm, the corpus (see Section 2.3) has been split into three sets: training, development and test sets. Which consisted on 80%, 10% and 10% of the whole database respectively. A set of language models and decision trees where trained for several amounts of candidates. Moreover, the best set of models in the development set where chosen to be applied to the test set. F-measure was chosen as the optimisation function.

**Table 2.** Summary of experiments for each set of features proposed. Best systems on development set are shown. F-measure corresponds to test-set.

|  | Set1 | Set2 | Set3 | Set4 |
|---|---|---|---|---|
| Motivation | small context | large context | *repeat* feature | no-candidates |
| F-measure | **86** | **86** | 84 | 85 |
| Language model order | 2 | 2 | 2 | 4 |
| Number of candidates | 40 | 40 | 30 | 0 |

Table 2 shows F-measure values over the test for best systems on the development set and for each set of features proposed. There are not big differences between feature sets. The use of the *repeat* feature slightly decreases the F-measure. In contrast, the use of candidates slightly improves results, moreover, its computational cost is much lower.

Table 3 shows detailed results for the winner system based on Set1. The algorithm classifies each word as preceding ($FP$) or not ($\overline{FP}$) a filled pause. It can be observed how the algorithm presented here inserts filled pauses with 96% precision. However, it only adds 57% of filled pauses existing in the test set.

**Table 3.** Classification Results of best system. Precision and Recall are shown as well as F-measure for each class. Second order language model and 20 candidates.

| Confusion Matrix | $\overline{FP}$ | $FP$ | Recall | Precision | F-measure |
|---|---|---|---|---|---|
| $\overline{FP}$ | **30,991** | 12 | 99.9% | 99.2% | 99 |
| $FP$ | 254 | **347** | **57.7%** | **96.7%** | **82.3** |

Presented results lead us to claim that the algorithm presented can be used to learn from conversational corpus in order to generate them afterwards.

## 4.2 Perceptual Evaluation

Disfluent speech synthesis is a relatively recent research line. The evaluation of such systems is thus an unsolved issue. While speech synthesis evaluation is still a hot research topic; emotional, disfluent, ... speech synthesis even add more difficulties to the evaluation process. However, qualitative as well as quantitative perceptual evaluations are necessary to allow us to extract conclusions that can lead our future research work.

In order to evaluate the quality of the inserted filled pauses as well as of its synthesis, some sentences extracted from the test set described in Section 4.1 were synthesised with and without disfluencies. The unit-selection synthesiser from Universitat Politècnica de Catalunya was used [15], and filled pauses were inserted in the audio using the methodology described in Section 3.

The test consisted on a set of 6 audio pairs. Each pair consisted on a sentence synthesised with and without filled pauses. Three of them where randomly chosen from the set of sentences the algorithm matched the reference (i.e. inserted a filled pause in a place where the reference contained one). The rest where chosen from the set the algorithm inserted a filled pauses in a place where the reference did not contain any (i.e. the algorithm did not match the reference). Each evaluator had to answer to 5 questions for each pair. Three of them related to naturalness and adequacy of the voice for a dialogue system:

- **Q1** Do you think that filled pauses make the voice *(more|equal|less)* natural?
- **Q2** Do you think that filled pauses make the voice *(more|equal|less)* suitable for a dialogue?
- **Q3** Do you think that filled pauses make the voice *(more|equal|less)* human-like?

and two questions related to the position of filled pauses and quality of their synthesis:

- **Q4** Do you think that filled pauses are *(correctly|incorrectly)* pronounced?
- **Q5** Do you think that filled pauses are *(correctly|incorrectly)* placed?

Answers to first three questions where given values $[more = 1, equal = 0, less = -1]$ respectively. The questionnaire was answered by 21 evaluators which had few or no relation with speech synthesis plus 4 evaluators that are speech experts. Results are shown in Table 4.

**Table 4.** Perceptual results. Mean values are given and variance in brackets for Q1, Q2 and Q3. For Q4 and Q5 the percentage of evaluator that answer *"correct"* is given.

| question | Q1 | | Q2 | | Q3 | | Q4 | | Q5 | |
|---|---|---|---|---|---|---|---|---|---|---|
| | *NoExp.* | *Exp.* | *NoExp.* | *Exp.* | *NoExp.* | *Exp.* | *NoExp.* | *Exp.* | *NoExp.* | *Exp.* |
| *match* | 0.4(0.6) | 0.6(0.2) | 0.08(0.7) | 0.3(0.5) | 0.4(0.5) | 0.7(0.4) | 71% | 91% | 71% | 83% |
| *no-match* | 0.5(0.6) | 0.2(0.7) | 0.6(0.4) | 0.6(0.6) | 0.3(0.5) | 0.8(0.1) | 78% | 100% | 82% | 83% |

In Table 4 it can be observed how mean values show that the sentences presented in the test have been considered more natural (0.4 to 0.6) and more human-like (0.4 to 0.8). They have been considered slightly more suitable for a dialogue.

**Table 5.** Qualitative summary of answers to the open question: *Do you think that it is interesting that in human-machine interactions speech synthesisers produce a more human-like voice?*

| Opinion | Evaluators | Reason |
|---|---|---|
| In favour | 22 | *human-like speech is easier to understand* |
| Does not matter | 2 | *Not necessary but can be understood.* |
| Against | 1 | *Communication with machines has to be simple.* |

On the other hand, experts and non-experts agree that filled pauses where mainly correctly placed and correctly pronounced. As can be seen in Table 4, experts evaluated synthesis with higher values than non-experts, it might be due to the fact that they are more used to speech synthesis quality.

In addition to these five questions an open question has been included in the test. It was: *Do you think that it is interesting that in human-machine interactions speech synthesisers produce a more human-like voice?* and answers to this question have been summarised in Table 5. These comments, thus, support our claim that talking speech synthesis is worth further research.

## 5 Conclusion

In the present paper we have described a system that is able to insert filled pauses in a text. It has been evaluated objectively against a reference corpus. Results have shown that filled pauses can be inserted with a precision of up to 96%. These results, lead us to the conclusion that filled pauses can be correctly inserted in a text by means of simple machine learning techniques. Furthermore, perceptual results support the conclusion coming up from objective results. Even sentences that do no correspond with the reference have been evaluate as correct. It has also been shown that the use of filled pauses in speech synthesis can increase the naturalness of the speech and that conversational speech is somehow desired by users. This encourages future work on disfluent speech synthesis including repetitions, restarts, etc.

## Acknowledgements

This work has been partially founded by the Spanish government, under grant TEC2006-13694-C03 and by Consejería de Educación under project JCYL (VA053A05).

## References

1. Mostefa, D., Garcia, M.N., Hamon, O., Moreau, N.: Deliverable 16: Evaluation report. Technical report, ELDA (2006)
2. Bennett, C.L., Black, A.W.: The blizzard challenge 2006. In: Proceedings of Blizzard Challenge 2006 Workshop, Pittsburgh, PA (2006)
3. Shröder, M.: Emotional Speech Synthesis: A Review. In: Proceedings of Eurospeech. Aalborg, Denmark, vol. 1, pp. 561–564 (2001)

4. Gobl, C., Bennet, E., Chasaide, A.N.: Expressive synthesis: How crucial is voice quality. In: Proceedings of IEEE Workshop on Speech Synthesis, Santa Monica, California, pp. 91–94 (2002)
5. Werner, S., Hoffman, R.: Pronunciation variant selection for spontaneous speech synthesis - A summary of experimental results. In: Proc. of International Conference on Speech Prosody, Dresden, Germany (2006)
6. Tree, J.E.F.: The effects on of false starts and repetitions on the processing of subsequent words in spontaneous speech. Journal of Memory and Language 34, 709–738 (1995)
7. Tseng, S.C.: Grammar, Prosody and Speech Disfluencies in Spoken Dialogues. PhD thesis, Department of Linguistics and Literature, University of Bielefeld (1999)
8. Adell, J., Bonafonte, A., Escudero, D.: Disfluent speech analysis and synthesis: a preliminary approach. In: Proc. of 3th International Conference on Speech Prosody, Dresden, Germany (2006)
9. Agüero, P.D., Bonafonte, A.: Phrase break prediction: a comparative study. In: Proc. of XIX Congreso de la Sociedad Española para el procesamiento del Lenguaje Natura, Alcala de Henares, Spain (2003)
10. Stolcke, A.: SRILM - an extensible language modeling toolkit. In: Proc. Intl. Conf. Spoken Language Processing, Denver, Colorado (2002)
11. Carreras, X., Chao, I., Padró, L., Padró, M.: Freeling: An open-source suite of language analyzers. In: Proc. of the 4th International Conference on Language Resources and Evaluation (LREC'04), Lisbon, Portugal (2004)
12. Tree, J.E.F.: Listeners' uses of um and uh in speech comprehension. Memory & Cognition 29(2), 320–326 (2001)
13. Shriberg, E.E.: Preliminaries to a Theory of Speech Disfluencies. PhD thesis, Berkeley's University of California (1994)
14. Boersma, P., Weenink, D.: Praat: doing phonetics by computer (version 4.3.04) (2005), http://www.praat.org/
15. Bonafonte, A., Agüero, P.D., Adell, J., Pérez, J., Moreno, A.: Ogmios: The UPC text-to-speech synthesis system for spoken translation. In: TC-STAR Workshop on Speech-to-Speech Translation, Barcelona, Spain, pp. 199–204 (2006)

# Exploratory Analysis of Word Use and Sentence Length in the Spoken Dutch Corpus

Pascal Wiggers and Leon J.M. Rothkrantz

Man–Machine Interaction Group
Delft University of Technology
Mekelweg 4, 2628 CD Delft, The Netherlands
p.wiggers@tudelft.nl, l.j.m.rothkrantz@tudelft.nl

**Abstract.** We present an analysis of word use and sentence length in different types of Dutch speech, ranging from conversations over discussions and formal speech to read speech. We find that the distributions of sentence length and personal pronouns are characteristic for the type of speech. In addition, we analyzed differences in word use between male and female speakers and between speakers with high and low education levels. We find that male speaker use more fillers, while women use more pronouns and adverbs. Furthermore, gender specific differences turn out to be stronger than differences in language use between groups with different education levels.

## 1 Introduction

It is well-known that language use varies in different situations, for example conversational speech differs from read speech. In addition language use also depends on social factors such as gender and education level [1]. To cope with this variation speech and language processing systems are often developed for a specific domain. A deeper understanding of variations in language use can help us to identify new domains in which existing systems can be used with minor changes and to develop more general models and systems that incorporate knowledge of language variation, such as personalized interfaces or adaptive language models for speech recognition [2] which is the ultimate goal of our research.

In this paper we use an exploratory data-mining approach to investigate how the type of speech influences word use in a large corpus and we identify features of the text that allow us to determine the type of speech. We also look at the influence of the speakers gender and education level on language variation in spontaneous speech.

The corpus used is the Spoken Dutch Corpus (Corpus Gesproken Nederlands or CGN) [3]. A corpus of almost nine million words of standard Dutch as spoken in the Netherlands and Flanders. The corpus is subdivided in 15 components that contain different types of speech, ranging from spontaneous conversations to more formal speech such as sermons and read speech. Table 1 gives an overview of the components. It lists the total number of words in each category. We focus on the recordings made in the Netherlands, leaving the Flemish recordings out of consideration[1]. The last column of Table 1 gives the number of words per category for the recordings from the Netherlands.

---

[1] However, most of the results in this paper also hold for the Flemish part of the corpus.

V. Matoušek and P. Mautner (Eds.): TSD 2007, LNAI 4629, pp. 366–373, 2007.

**Table 1.** Overview of the Spoken Dutch Corpus

| Component | words | NL |
|---|---|---|
| a. Spontaneous conversations (face-to-face) | 2,626,172 | 1,747,789 |
| b. Interviews with teachers of Dutch | 565,433 | 249,879 |
| c. Spontaneous telephone dialogs (switchboard) | 1,208,633 | 743,537 |
| d. Spontaneous telephone dialogs (mini disc) | 853,371 | 510,204 |
| e. Simulated business negotiations | 136,461 | 136,461 |
| f. Interviews/ discussions/debates (broadcast) | 790,269 | 539,561 |
| g. (political) Discussions/debates/meetings | 360,328 | 221,509 |
| h. Lessons recorded in the classroom | 405,409 | 299,973 |
| i. Live (e.g. sports) commentaries (broadcast) | 208,399 | 130,377 |
| j. News reports/reportages (broadcast) | 186,072 | 90,866 |
| k. News (broadcast) | 368,153 | 285,298 |
| l. Commentaries/columns/reviews (broadcast) | 145,553 | 80,167 |
| m. Ceremonial speeches/sermons | 18,075 | 5,565 |
| n. Lectures/seminars | 140,901 | 61,834 |
| o. Read speech | 903,043 | 551,624 |

In this paper we are mainly interested in words and part-of-speech (POS) tags that are most characteristic for a particular subset of the data. To operationalize this notion we use log-likelihood ratios [4] defined by:

$$G^2 = 2 \sum_i O_i \ln(\frac{O_i}{E_i}) \tag{1}$$

Where $O_i$ is the frequency observed in a cell of a contingency table and $E_i$ is the corresponding expected value:

$$E_i = \frac{N_j \sum_i O_i}{\sum_j N_j} \tag{2}$$

$N_j$ is the total frequency in a column of the table and $i$ ranges over all elements in a row of the table. The log-likelihood ratio can be thought of as a measure of surprise and is related to the $\chi^2$-test, but more reliable than the latter in the face of sparse data. The values of $G^2$ are approximately $\chi^2$ distributed. It can be used in the hypothesis testing framework as a measure of statistical significance. However, as pointed out by [5] among others, the null hypothesis that subsets are drawn at random from the same underlying population is almost always defeated when looking at linguistic phenomena because language use clearly is not random. Outside the hypothesis testing framework likelihood ratios are nevertheless useful as a way to rank words in order of distinctiveness as shown in [5,6,7].

## 2  Related Work

There is a large body of work in corpus analysis, e.g. [8] investigates social differentiation defined by factors as gender, age and social group, in the use of English vocabulary using the British national corpus. They also look at the difference between spoken and written text. [5] is concerned with comparing corpora and looks at the difference between male and female word use, as does [9] for Swedish. The CGN is also used in [10] and [11]. In [10] gender differences in speech rate and the use of fillers and nouns is investigated. They find that male speakers use more fillers and female speakers use more pronouns. The work of [11] uses multivariate analysis to investigate lexical richness. They find that more formal speech (components $k$ and $m$ of the corpus) has high lexical richness, while spontaneous speech does not. In addition, they note that female speech has lower lexical richness than male speech. Hypothesizing that women elaborate longer on a particular topic than men would.

## 3 Type of Speech

To investigate differences between types of speech we analyze the differences in sentence length, part-of-speech and word distributions between the components of the corpus. For word distributions we remove all words that occur less than five times in the whole corpus to remove or at least moderate topic dependency.

### 3.1 Sentence Length

We expect sentence length to be a good indicator of the type of speech, therefore we analyze sentence lengths in the subcorpora. Table 2 summarizes the results. As one would expect sentences in spontaneous speech (components $a$, $c$ and $d$) are shorter on average than sentences in formal speech. Lectures ($n$) and political debates ($g$) contain the longest sentences on average. When taking a closer look we find that for all components the sentence length distributions are highly skewed, sentences can occasionally get very long. Therefore, the median and inter quartile range (IQR) are better indicators of sentence distribution than mean and sample deviation.

Fig. 1 shows the distributions (up to sentence length 30) for several components. The distributions of components $b$ and $c$ are characteristic for spontaneous speech, about a quarter of all sentences are single word sentences containing fillers, yes/no answers and backchannels. For the remaining lengths there is a peak around length 6. Debates ($g$) show a more uniform distribution over sentence lengths while news $k$ has a nicely balanced bell-shaped distribution. It makes one wonder whether anchormen and news reporters think about sentence length when writing their texts.

### 3.2 Part of Speech

We ranked POS-categories using likelihood ratios and found that fillers, nouns, determiners, adverbs and pronouns best differentiate between subcorpora. Fig. 2 compares the relative frequencies of the components for several POS-tags. It can be observed that fillers and adverbs as well as incomplete words are much more common in spontaneous speech while nouns and determiners are characteristic of more formal and read speech.

**Table 2.** Sentence length statistics per category of the CGN

| Comp. | avg. | s.dev. | mode | median | IQR | Comp. | avg. | s.dev. | mode | median | IQR |
|---|---|---|---|---|---|---|---|---|---|---|---|
| a | 6.42 | 6.70 | 1 | 5 | 8 | i | 9.90 | 9.72 | 4 | 7 | 8 |
| b | 11.34 | 11.09 | 1 | 8 | 13 | j | 12.20 | 9.98 | 1 | 10 | 12 |
| c | 6.37 | 6.76 | 1 | 4 | 8 | k | 13.46 | 5.39 | 12 | 13 | 7 |
| d | 6.80 | 7.03 | 1 | 5 | 8 | l | 13.04 | 10.65 | 7 | 10 | 11 |
| e | 8.59 | 9.85 | 1 | 5 | 11 | m | 11.76 | 7.22 | 10 | 10 | 9 |
| f | 11.29 | 11.45 | 1 | 8 | 13 | n | 28.15 | 21.16 | 19 | 23 | 24 |
| g | 20.05 | 17.20 | 1 | 16 | 19 | o | 11.51 | 8.61 | 5 | 9 | 10 |
| h | 7.98 | 7.72 | 1 | 6 | 9 | | | | | | |

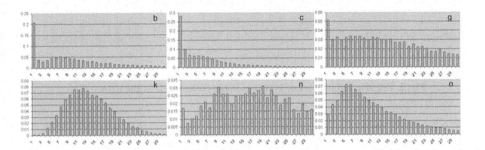

**Fig. 1.** Sentence length distributions for several components of the CGN

Pronouns occur less in broadcast categories; in particular news contains little pronouns. Note that the relative number of verbs is more or less equal for all components.

### 3.3 Words

Words that based on likelihood ratios differentiate most between components are: *ja* (yes), *de* (the), *het* (the), *uh* (filler), *voor* (for), *xxx* (incomprehensible), *ik* (me), *van* (of), *nee* (no), *je* (you), *nou* (well), *in* (in), *haar* (her), *hij* (he), *u* (you, polite). Articles, fillers and (personal) pronouns make up most of this list. We further investigated the words in these classes. Fig. 3 compares relative frequencies of some words in the components of the corpus. The personal pronouns are very good indicators of the type of speech. The first person singular *ik* and its abbreviated form *'k* are most used in spontaneous speech. The colloquial second person singular *je* is used slightly more often in spontaneous speech than in formal and read speech. It is used most often in lessons. An explanation for this is that in contemporary Dutch *je* is also used as an indefinite pronoun. The second person singular *jij* is used most in discussions and debates, whereas the polite form *u* is used most often in interviews and formal speech. Third person singular is used most in read speech (stories). The word *ze* which can mean either she or they occurs relatively often in spontaneous speech. It is interesting to note that *ie* the slang version of 'he' occurs relatively often in sports commentaries. Finally, *wij* (we) is used most prominently in debates and formal speech.

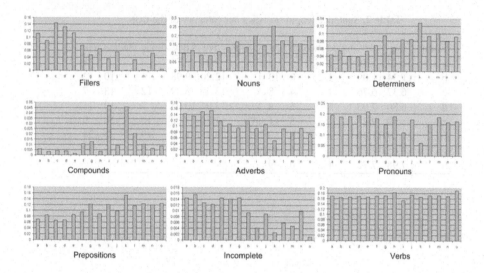

**Fig. 2.** Relative frequencies of POS-tags in the components of the CGN

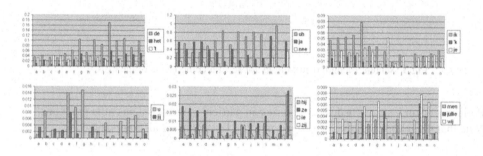

**Fig. 3.** Distributions of common words in components of the CGN

## 4    Gender

To analyze the influence of gender on word use we selected a subset of the corpus from components *a,c* and *d,* the most spontaneous speech. We select at random an equal number of words from an equal number of male and female speakers in order to reduce the influence of text length.

It turns out that there is little difference in sentence length between male and female speakers. Men use slightly longer sentences than women on average as shown in Table 3, but there is also more variation in sentence length for male speakers.

In Table 4 the frequencies of POS-categories for male and female speaker are shown, ranked by likelihood ratio. As was also found by [10] on a slightly different subset of the corpus, male speakers use more fillers. In addition male speaker use more determiners and compound names and do not complete words more often. Female speakers use

**Table 3.** Summary statistics of sentence length for male and female speakers

|        | average | sample dev. | mode | median | IQR |
|--------|---------|-------------|------|--------|-----|
| male   | 8.8     | 9.1         | 1    | 6      | 10  |
| female | 7.5     | 7.8         | 1    | 5      | 8   |

more adverbs and pronouns. Detailed investigation shows that women especially use more personal pronouns (relative frequencies of 0.092 for male speakers vs. 0.1 for female speakers) this is in line with results found by [12,9].

**Table 4.** Relative frequencies of POS-tags for male and female speakers ordered by likelihood ratios

| part of speech   | male    | female  | llr   | part of speech   | male   | female  | llr    |
|------------------|---------|---------|-------|------------------|--------|---------|--------|
| Fillers          | 0.129   | 0.117   | 645.3 | Nouns            | 0.0971 | 0.092   | 156.2  |
| Adverbs          | 0.139   | 0.151   | 579.7 | Verbs            | 0.165  | 0.171   | 119.8  |
| Determiners      | 0.0423  | 0.0367  | 438.4 | Conjunctions     | 0.0643 | 0.068   | 115.5  |
| Incomplete words | 0.01501 | 0.0127  | 211.6 | Adjectives       | 0.0604 | 0.0636  | 92.91  |
| Compounds        | 0.0056  | 0.0043  | 199.6 | Prepositions     | 0.0701 | 0.0668  | 88.01  |
| Pronouns         | 0.1886  | 0.1959  | 183.2 | Numerals         | 0.0132 | 0.0121  | 43.31  |
| Foreign words    | 0.00311 | 0.00216 | 181.4 | Incomprehensible | 0.0071 | 0.0067  | 12.84  |

Before collecting statistics on word use hapax legomena are removed from the vocabulary. The left side of Table 5 shows the ten words with the highest log-likelihood values that are most often used by women. The right side of Table 5 shows the ten words with the highest log-likelihood that are more often used by men. The tendency of women to use more personal pronouns and for men to use more fillers can also be observed from those lists.

**Table 5.** Relative frequencies of high likelihood ratio words used most often by women (left) and men (right)

| word            | male    | female   | llr   | word             | male      | female    | llr    |
|-----------------|---------|----------|-------|------------------|-----------|-----------|--------|
| oh (filler)     | 0.00732 | 0.009991 | 434.8 | uh (filler)      | 0.03669   | 0.02596   | 1990   |
| ze (them, she)  | 0.0075  | 0.00981  | 324.4 | de (the)         | 0.01816   | 0.01517   | 285.2  |
| zei (said)      | 0.0011  | 0.00195  | 275.8 | xxx              | 0.02219   | 0.01927   | 219.5  |
| leuk (nice)     | 0.00163 | 0.00261  | 240.2 | een (a)          | 0.01873   | 0.01642   | 161.6  |
| want (because)  | 0.00444 | 0.00595  | 230.4 | we (we)          | 0.006424  | 0.005572  | 63.55  |
| toen (then)     | 0.00315 | 0.00434  | 198.5 | in (in)          | 0.01015   | 0.009105  | 60.24  |
| heel (very)     | 0.00327 | 0.00442  | 180.7 | hum (filler)     | 0.0003717 | 0.0001934 | 59.7   |
| zo (therefore)  | 0.0077  | 0.00927  | 154   | effe (short time)| 0.001028  | 0.0007182 | 57.61  |
| ik (I)          | 0.0259  | 0.0287   | 145.9 | enzovoort (etc.) | 7.01E-05  | 9.58E-06  | 54.19  |
| echt (really)   | 0.00348 | 0.0045   | 136.7 | is (is)          | 0.01649   | 0.01525   | 51.31  |

We also look at the type-token ratios for male and female speakers. The type-token ratio is defined as the number of different words (types) divided by the total number of words (tokens) and can be seen as a measure of lexical richness [11]. A lower type-token ratio means a relatively smaller number of different words. Unfortunately, the type-token ratio is text length dependent. The longer a text, the lower the type-token ratio will be [11]. This is why we used randomly selected equal size subsets for male and female speakers. The ratios found were 0.0195 for male speakers and 0.0187 for female speakers. This confirms the results of [11] where a different procedure for data selection and different subsets of the CGN corpus were used.

## 5   Education Level

Finally, we look at the influence of education level on word use. We define two education levels, high (college or university) and low. We used the same subcorpus of spontaneous speech as for the analysis of gender influence and applied the same procedure to obtain equally sized subsets. Contrary to what we expected we find that there is no significant difference in sentence length for these two groups. The differences in POS-tag distributions are also smaller than for gender. Table 6 lists the POS-tag with the highest likelihood-ratios. It turns out that the differences for nouns, determiners and fillers is not significant (by any standard).

The difference in type-token-ratio is bigger than for gender. The ratio is 0.031 for the high education group and 0.028 for low education level. The overall differences in word use between the two groups is small. The low education group uses more slang and abbreviations, while the high education group uses more determiners.

**Table 6.** Relative frequency of POS-tags with the highest likelihood-ratios for education level

| part of speech | high | low | llr | part of speech | high | low | llr |
|---|---|---|---|---|---|---|---|
| Adjectives | 0.0641 | 0.0588 | 223.5 | Conjunctions | 0.067 | 0.065 | 63.52 |
| Foreign words | 0.00294 | 0.00207 | 139.7 | Verbs | 0.167 | 0.171 | 53.17 |
| Numerals | 0.0123 | 0.0138 | 84.95 | Incomprehensible | 0.0069 | 0.0077 | 47.12 |

## 6   Conclusions

In this paper we analyze differences in word use between different types of speech as well as gender specific word use. Rather than starting with a set of hypothesis we use an exploratory approach to get more insight in the data, that will allow us to formulated hypothesis, in the form of language models, about word use in the future.

We find clear differences in sentence length between different types of speech and show that the distributions of determiners, fillers and especially personal pronouns are quite different for different types of speech.

It is also shown that word use is gender specific. Differences between higher and lower educated speakers are less clear. In the future we want extend our analysis to other speaker characteristics such as age and region of living and look at the distributions of $n$-grams and grammar.

# References

1. Boves, T., Gerritsen, M.: Inleiding in de sociolinguistiek. Uitgeverij Het Spectrum, Utrecht (1995)
2. Wiggers, P., Rothkrantz, L.J.: Topic-based language modeling with dynamic bayesian networks. In: the Ninth International Conference on Spoken Language Processing (Interspeech 2006 ICSLP) Pittsburgh, Pennsylvania, pp. 1866–1869 (September 2006)
3. Schuurman, I., Schouppe, M., Hoekstra, H., van der Wouden, T.: Cgn, an annotated corpus of spoken dutch. In: Proceedings of the 4th International Workshop on Linguistically Interpreted Corpora (LINC-03), Budapest, Hungary (April 14, 2003)
4. Dunning, T.: Accurate methods for the statistics of surprise and coincidence. Computational Linguistics 19(1), 61–74 (1993)
5. Kilgarri, A.: Comparing corpora. International Journal of Corpus Linguistics 37, 1–37 (2001)
6. Rayson, P., Garside, R.: Comparing corpora using frequency profiling (2000)
7. Daille, B.: Combined approach for terminology extraction: lexical statistics and linguistic.ltering. Technical Report 5, Lancaster University (1995)
8. Rayson, P., Leech, G., Hodges, M.: Social di.erentiation in the use of english vocabulary: some analyses of the conversational component of the british national corpus. s. International Journal of Corpus Linguistics 2(1), 133–152 (1997)
9. Harnqvist, K., Christianson, U., Ridings, D., Tingsell, J.G.: Vocabulary in interviews as related to respondent characteristics. Computers and the Humanities 37, 179–204 (2003)
10. Binnenpoorte, D., Bael, C.V., Os, E.D., Boves, L.: Gender in everyday speech and language: a corpus-based study. In: Interspeech 2005, pp. 2213–2216 (2005)
11. van Gijsel, S., Speelman, D., Geeraerts, D.: Locating lexical richness: a corpus linguistic, sociovariational analysis. In: J.M.V., et al. (eds.) Proceedings of the 8th International Conferene on the statistical analysis of textual data (JADT), Besanon, France, pp. 961–971 (2006)
12. Argamon, S., Koppel, M., Fine, J., Shimoni, A.R.: Gender, genre, and writing style in formal written texts. Text 23(3) (2003)

# Design of Tandem Architecture
# Using Segmental Trend Features*

Young-Sun Yun[1] and Yunkeun Lee[2]

[1] Dept. of Information and Communication Engineering,
Hannam University, Daejeon, Republic Of Korea
ysyun@hannam.ac.kr
[2] Spoken Language Processing Team,
ETRI, Daejeon, Republic of Korea
yklee@etri.re.kr

**Abstract.** This paper investigates the tandem architecture (TA) based on segmental features. The segmental feature based recognition system has been reported to show better results than the conventional feature based system in previous studies. In this paper we tried to merge the segmental feature with the tandem architecture which uses both hidden Markov models and neural networks. In general, segmental features can be separated into the trend and location. Since the trend means variation of segmental features and since it occupies a large portion of segmental features, the trend information was used as an independent or additional feature for the speech recognition system. We applied the trend information of segmental features to TA and used posterior probabilities, which are the output of the neural network, as inputs of the recognition system. Experiments were performed on Aurora2 database to examine the potentiality of the trend feature based TA. The results of our experiments verified that the proposed system outperforms the conventional system on very low SNR environments. These findings led us to conclude that the trend information on TA can be additionally used for the traditional MFCC features.

## 1 Introduction

HMM has been widely used in various areas for a long time owing to its easy implementation, flexible modeling capability, and high performance. However, it has been reported that the HMM does not effectively represent the temporal dependency of speech signals because of the weakness of its assumptions. Various studies have been done to mitigate the weakness by adopting segmental models [3,5] or trajectory approaches [4,6]. These models use the segmental features instead of the frame features or the regression function of frame features.

Various studies have been conducted on combining the Neural Network (NN) and HMM or Gaussian Mixture Model (GMM) approach within a single system, expecting to potentially combine the advantages of both systems. Under the influence of these approaches, a simple variant, tandem approach has been presented [10,11,12], in which

---

* He was a visiting researcher at ETRI from May 2006 to Feb. 2007.

V. Matoušek and P. Mautner (Eds.): TSD 2007, LNAI 4629, pp. 374–381, 2007.
© Springer-Verlag Berlin Heidelberg 2007

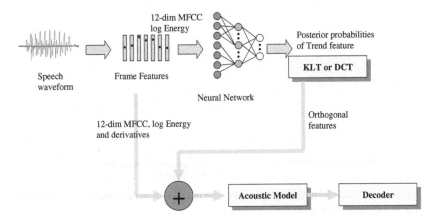

**Fig. 1.** System overview of proposed tandem architecture : Trend features are used as stand alone features or augmented feature of generally used MFCCs and their derivatives

a NN classifier is first trained to estimate the posterior probabilities of phone sets. These probabilities are then used as inputs of the conventional HMM systems.

In this paper, we present a new method of TA (Tandem Architecture) that uses the segmental information instead of the phone label as has been done in the previous studies. The proposed approach estimates the posterior probabilities of each predetermined trend information extracted from segmental features rather than subword units such as phoneme-like units. The trajectories, in general, can be separated into two parts: trend and location. The trend corresponds to the type of variation and the location means the smoothed value of a trajectory at current time. If the trajectory is a linear system, the trend represents the slope, and if it is a quadratic system, the trend shows the parabolic tendency. Since the applied system is a parametric trajectory system[9], the trend and the location can be separated without difficulty and the trend can easily be used for the inputs of the speech recognition system.

## 2   Design of Tandem Architecture

The general system flow of TA has been well illustrated by Hermansky [10]. In our system, the structure is modified to include the segmental information. The segmental features are used to replace general MFCCs or they are additively used with the general features. Either combined with general features or independently used, the orthogonal features obtained by KLT (Karhunen Loéve Transform) were fed into a HMM like speech recognition system. The overall system is shown in Fig. 1.

In the proposed architecture, 12 dimensional MFCCs and log Energy are employed as inputs of NN (Neural Network) whose outputs emit the posterior probabilities of predetermined trend features. The types of trend features are categorized by the trend quantization codebook. The posterior output probabilities are orthogonalized by KLT or DCT (simplified KLT) and merged into streams corresponding to HMM inputs. The rest of them are similar to the well-known HMM based speech recognition system.

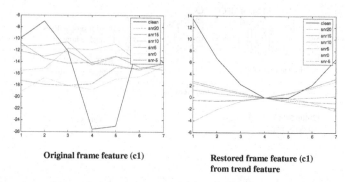

**Original frame feature (c1)**          **Restored frame feature (c1)**
                                        **from trend feature**

**Fig. 2.** Comparison of frame features and restored frame features from the trend on various SNR, c1 means the first cepstral dimension

To obtain the orthogonal features using NN and KLT, we consider two flows. One flow is the set of frames that are directly used for inputs of NN, and the other is the frame features restored from the trend feature. Comparison of both types of features is shown in Fig. 2 at various SNR environments.

In the first method, the segmental features are used only for finding the codeword indexes from the trend quantization codebook. On the other hand, in the second approach, the segmental features are obtained from the set of frames, and the restored frame features become the inputs of NN. This means removal of trajectory estimation error or smoothing of frame features. The segmental features are easily obtained from the frame features, in both approaches, by simple pseudo inverse multiplication as explained in Section 3. The difference between the two methods is whether the inputs of NN are noisy features or not. In the former, the flow is simple but the NN is not well converged. The NN acts like a simple classifier for the clean trajectory features in the latter approach. The KLT matrix is obtained from the eigenvectors of NN posterior outputs for given training patterns in common. The training process is illustrated in Fig. 3.

## 3   Segmental-Feature

In the segmental feature system, the input speech signals, which are transformed to trajectories based on well-known speech features, are transferred to the classification module. Deng has proposed a parametric approach for a non-stationary state HMM where polynomial trend functions are used as a time varying means of the output Gaussian distributions in the HMM states[1]. In another trajectory method[2], Gish modeled each feature dimension of a speech segment as a polynomial regression function. These features are called *segmental features* because the features are extracted from segments that correspond to the set of frames. In contrast to Gish's approach, Yun et al. has proposed features with the fixed length segment[7,8,9].

In this section, the parametric approach is explained. To express the fixed segment with time indexes, the segmental features can be expressed as

$$\mathbf{C}_t = \mathbf{Z}\mathbf{B}_t + \mathbf{E},\tag{1}$$

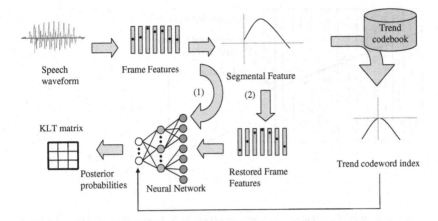

**Fig. 3.** Training process: NN is to be learned from (1) the original set of frame features or from (2) the restored frames features of trend information. KLT matrix is obtained from the outputs of NN.

where $C_t$ and $B_t$ are the speech segment and trajectory coefficients at time $t$. In this equation, the segmental feature is extracted from the successive frame features using the design matrix $Z$. Each frame is represented by a $D$ dimensional feature vector, and $Z$ and $B_t$ are $N \times R$ design matrix and $R \times D$ trajectory coefficient matrix, respectively. $E$ finally denotes the residual error which is assumed to be independent and identically distributed.

Since errors are supposed to be independent and identically distributed, trajectory coefficient matrix $\hat{B}_t$ can be easily obtained by *a linear regression* or the following matrix equation:

$$\hat{B}_t = [Z'Z]^{-1} Z'C_t. \tag{2}$$

## 4 Trend Features

A segment is modeled as a polynomial trajectory of fixed duration. The trajectory is obtained by the sequence of feature vectors of speech signals, and it can be divided into trend and location. The trend indicates the variation of consequent frame feature vectors, while the location points to the positional difference of trajectories [9](Fig. 4).

### 4.1 Separation of Trend and Location

The trajectory can be rewritten by the linear regression function. For each feature dimension, the following polynomial is considered:

$$y_{\tau,i} = b_{1,i}z_{\tau,1} + b_{2,i}z_{\tau,2} + b_{3,i}z_{\tau,3} + \cdots + b_{R,i}z_{\tau,R}, \ 1 \leq i \leq D, \tag{3}$$

where $y_{\tau,i}$ means the cepstral features of $i$th dimension of $\tau$th frame in a segment, $b_{r,i}$ is $r$th trajectory coefficient, and $z_{\tau,r}$ is the element of the design matrix $Z$.

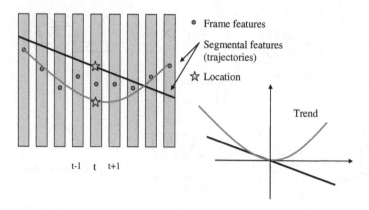

**Fig. 4.** Conceptual drawing of frame features, segmental features (trajectories), and trend features

From the above equation, the first column element of the design matrix is one, i.e. $z_{\tau,1} = 1[9]$. Therefore, $b_{1,i}$ means a *location* (intercept) on cepstral feature domain, while the remains of the equation are related to the segmental variation, e.g. *trend*.

If the above polynomial function is considered by matrix equation, the first row of the trajectory coefficient matrix $b_1$ means the $D$-dimensional location and the remains of rows are considered as the $(R - 1) \times D$-dimensional trend. The coefficient matrix $T_t$ for the trend can be defined as follows:

$$\mathbf{T}_t = \begin{bmatrix} \mathbf{b}_2^t \cdots \mathbf{b}_R^t \end{bmatrix}^T. \tag{4}$$

## 4.2   Trend Clustering

Trend clustering algorithm is similar to the algorithm of well-known vector quantization. However, the distance measure has to be modified to compare two trends. The Euclidean distance is modified to reflect the trend characteristics as follows:

$$D(\mathbf{T}_i, \mathbf{T}_j) = \frac{1}{N} \sum_{\tau=1}^{N} \{\tilde{\mathbf{z}}_\tau (\mathbf{T}_i - \mathbf{T}_j)\} \{\tilde{\mathbf{z}}_\tau (\mathbf{T}_i - \mathbf{T}_j)\}', \tag{5}$$

where $\tilde{\mathbf{z}}_\tau$ is the row vector of design matrix which excludes the first column value, $\mathbf{T}_i, \mathbf{T}_j$ are trend coefficient matrices.

## 5   Experiments and Results

To explore the characteristics of the proposed architecture, we examined the frame correct rates of NN at first. The frame correct rates represent the number of segments that are correctly classified by NN. It means whether or not the maximum posterior probabilities of NN match the given trend indexes. The segmental feature is obtained from 7 frames whose dimensions are 13, i.e. 12 MFCC and log energy. The inputs of NN

**Table 1.** Frame correct rates of the neural network on Aurora2 clean training database to examine its performance. Training segments are randomly selected about 5% from the recommended clean training database. The remaining segments of the database are used for evaluation.

|        | Training set (about 5%) | Test set (about 95%) |
|--------|:-----------------------:|:--------------------:|
| Best   | 71.24                   | 60.60                |
| Top 5  | 94.82                   | 94.44                |
| Top 10 | 98.79                   | 98.81                |

consist of 78 feature dimensions (6 frames times 13 dimensions except the center frame because the center frame means the location information and only the trend information is used in our approach) and the NN has 160 hidden nodes and 64 output nodes. The NN is trained using the backpropagation algorithm. To involve the trend information of the segmental features in the TA, the second approach in the training process (Fig. 3) is selected for fast convergence of NN and for simplicity.

The results are shown in Table 1. About 5%of the clean training set, which is recommended from ETSI, is randomly selected for NN training because the total number of segments is so huge that the network is not well converged. Correct tests were conducted on the data used for training the NN as well as the data that were not used in the training. From the experiments, we considered that about 5% training data is sufficient to train NN because Top 10 candidates almost contain the correct segmental information.

The second experiments are performed for evaluation of the proposed architecture and trend features. The TA using trend features are tested for *set c* of Aurora2 test database in which the noises of "subway" and "street" are artificially added at several SNRs after filtered with a "MIRS" characteristics to simulate the influence of terminals with different characteristics such as telephones. The various combinations of feature dimensions are tested with trend features or not. From the results, when the trend information is added to the widely used features (MFCC and log energy, their derivatives) the proposed methods shows better results than the traditional feature systems at SNR 5, 0, and -5 environments. Since the current classification performance of the NN is far from our satisfactory, if more studies on NN or KLT are done, the suggested method maybe shows the better performances.

# 6 Summary

We have proposed a TA that uses segmental features represented by the polynomial regression function. The trajectory representing the segmental features can be separated into the trend and the location; the trend shows the variation of segmental features, and the location indicates the reference point that is corresponding to segment mid-point value. The proposed method uses the quantized trend of trajectories for inputs of neural network, which omits the posterior probabilities of the given trend information. To evaluate the performance of TA based on trend features, the experiments are done on Aurora2 database. From the experimental results, even though the NN's performance is

**Table 2.** Comparison of performances using 12 MFCC and log energy, their 1st derivatives, combined features, concatenated features by trend information

| Noisy Env. | | B13[1] | $\Delta^2$ | TI13[3] | B26 | B13+TI13 | B39 | B26+TI13 |
|---|---|---|---|---|---|---|---|---|
| | clean1 | 95.70 | 97.94 | 82.10 | 99.08 | 97.39 | 99.02 | 98.43 |
| | SNR 20 | 59.13 | 93.83 | 80.81 | 93.31 | 84.28 | 94.29 | 92.05 |
| | SNR 15 | 43.23 | 84.80 | 72.46 | 86.49 | 69.88 | 87.60 | 84.31 |
| Subway | SNR 10 | 27.82 | 61.34 | 58.67 | 71.08 | 51.43 | 73.23 | 68.59 |
| | SNR 5 | 14.25 | 34.51 | 37.43 | 42.59 | 33.01 | 49.62 | 43.44 |
| | SNR 0 | 7.86 | 19.16 | 18.33 | 17.90 | 18.05 | 23.89 | 17.78 |
| | SNR -5 | 7.55 | 12.16 | 11.18 | 9.64 | 10.75 | 10.68 | 9.18 |
| | clean2 | 95.28 | 98.00 | 82.98 | 98.73 | 97.37 | 99.00 | 98.61 |
| | SNR 20 | 71.16 | 94.50 | 79.99 | 92.90 | 90.11 | 95.10 | 94.59 |
| | SNR 15 | 58.98 | 87.94 | 69.77 | 85.97 | 80.68 | 88.72 | 87.97 |
| Street | SNR 10 | 43.95 | 67.29 | 54.32 | 68.50 | 62.70 | 72.55 | 74.21 |
| | SNR 5 | 29.63 | 39.33 | 31.98 | 44.65 | 42.02 | 46.86 | 50.48 |
| | SNR 0 | 14.81 | 21.07 | 16.60 | 20.37 | 21.80 | 22.01 | 26.18 |
| | SNR -5 | 9.67 | 11.52 | 10.16 | 10.43 | 12.36 | 10.97 | 13.66 |

not good, we can see the possibility of TA based on trend feature. If the more studies are done for the NN's output normalization or KL transform, the better results may be achieved.

## Acknowledgment

We would like to thank the reviewers for their valuable remarks and suggestions.

## References

1. Deng, L.: A generalized hidden Markov model with state-conditioned trend functions of time for the speech signal. Signal Processing 27, 65–78 (1992)
2. Gish, H., Ng, K.: A segmental speech model with application to word spotting. In: Proc. of Int. Conf. on Acoustics, Speech and Signal Proc. vol. II, pp. 447–450 (1993)
3. Gales, M.J.F., Young, S.J.: Segmental Hidden Markov Models. In: Proc. of European Conf. on Speech Comm. and Tech., pp. 1579–1582 (1993)
4. Gish, H., Ng, K.: Parametric trajectory models for speech recognition. In: Proc. of Int. Conf. on Spoken Lang. Proc. vol. I, pp. 466–469 (1996)
5. Ostendorf, M., Digalakis, V., Kimball, O.A.: From HMMs to Segment Models: A Unified View of Stochastic Modeling for Speech Recognition. IEEE Tr. on Speech and Audio Processing 4(5), 360–378 (1996)

---

[1] B# means the baseline system along to the number of feature dimensions (13 : 12MFCC+logE, 26 : 12MFCC+logE+$\Delta$, 39 : 12MFCC+logE+$\Delta$ + $\Delta^2$).

[2] $\Delta$ means the first derivatives of B13(12MFCC+logE).

[3] TI13 means the posterior probabilities of trend information on tandem architecture and 13-dimensional features are selected by KL transforms.

6. Holmes, W.J., Russell, M.J.: Probabilistic trajectory segmental HMMs. Computer Speech and Language 13, 3–37 (1999)
7. Yun, Y.S., Oh, Y.H.: A Segmental-Feature HMM for Speech Pattern Modeling. IEEE Signal Processing Letters 7(6), 135–137 (2000)
8. Yun, Y.S., Oh, Y.H.: A Segmental-Feature HMM for Continuous Speech Recognition Based On a Parametric Trajectory Model. Speech Communication 38(1), 115–130 (2002)
9. Yun, Y.S.: Sharing Trend Information of Trajectory in Segmental Feature HMM. In: Proc. of Int. Conf. On Spoken Language Proc., Denver, Colorado, USA, pp. 2641–2644 (2002)
10. Hermansky, H., Ellis, D., Sharma, S.: Tandem connectionist feature extraction for conventional HMM systems. In: Proc. of Int. Conf. on Acoustics, Speech and Signal Proc., Istanbul, Turkey, pp. 1635–1638 (2000)
11. Ellis, D.W.P., Singh, R., Sivadas, S.: Tandem acoustic modeling in large-vocabulary recognition. In: Proc. of Int. Conf. on Acoustics, Speech and Signal Proc., Salt Lake City, USA, pp. 517–520 (2001)
12. Sivadas, S., Hermansky, H.: Generalized Tandem Feature Extraction. In: Proc. of Int. Conf. on Acoustics, Speech and Signal Proc. vol. II, pp. 56–59 (2003)

# An Automatic Retraining Method for Speaker Independent Hidden Markov Models

András Bánhalmi, Róbert Busa-Fekete, and András Kocsor

Research Group on Artificial Intelligence of the Hungarian Academy of Sciences
and of the University of Szeged,
H-6720 Szeged, Aradi vértanúk tere 1., Hungary
{banhalmi,busarobi,kocsor}@inf.u-szeged.hu

**Abstract.** When training speaker-independent HMM-based acoustic models, a lot of manually transcribed acoustic training data must be available from a good many different speakers. These training databases have a great variation in the pitch of the speakers, articulation and the speed of talking. In practice, the speaker-independent models are used for bootstrapping the speaker-dependent models built by speaker adaptation methods. Thus the performance of the adaptation methods is strongly influenced by the performance of the speaker-independent model and by the accuracy of the automatic segmentation which also depends on the base model. In practice, the performance of the speaker-independent models can vary a great deal on the test speakers. Here our goal is to reduce this performance variability by increasing the performance value for the speakers with low values, at the price of allowing a small drop in the highest performance values. For this purpose we propose a new method for the automatic retraining of speaker-independent HMMs.

## 1 Introduction

The probabilistic models for speech recognition are normally trained on a large amount of manually segmented data samples that contain utterances recorded from many speakers. While these speaker-independent models usually can attain a higher average accuracy on most speakers, speaker-dependent models which are trained only on the utterances of a given speaker are much more efficient in the recognition task for this specific speaker. The problem with developing speaker-dependent systems is that large amounts of speech training data for each speaker is usually unavailable. Thus in order to achieve the efficiency of the speaker dependent model, various techniques have been proposed.

Two main approaches are the transformation of the incoming feature vectors (eg. by VTLN or CMN) and the modification of the parameters of the speaker-independent acoustic models (speaker adaptation techniques). The VTLN (vocal tract length normalization) method normalizes the spectrum of the input spoken data by converting it as if all the samples had been pronounced with the same vocal tract length [1], [2], [3], while the basic CMN (cepstral mean normalization) method converts the cepstral coefficients of the input data in such a way that the samples for each speaker have the same mean value [4] [5]. The other common approach for adjusting the speaker

V. Matoušek and P. Mautner (Eds.): TSD 2007, LNAI 4629, pp. 382–389, 2007.

**Table 1.** The weighted k-means method

| Input: | A set of M points (X) and nonnegative weights (W) |
|--------|----------------------------------------------------|
| 1 | Choose a number of k centroids from X randomly |
| 2 | Repeat the next section |
| 3 | Using the given cluster centres, assign each point to the cluster with the nearest center |
| 4 | Replace each cluster center by the weighted mean of the points in the cluster |
| 5 | For each point, move it to another cluster, if that would lower the total energy function. If a point is moved, update the cluster centres of the two affected clusters. |

independent models to better approximate the performance of speaker dependent models is speaker adaptation. In classical HMM-based systems, various speaker-adaptation techniques have been used with great success [6], [7], [8], [9], [10]. These techniques are based on the modification of the parameters of the speaker-independent system to maximize the likelihood (ML) of the adaptation data of the new speaker.

In this paper we propose a new method for improving the average accuracy after adaptation. A fundamental problem of automatic adaptation techniques is the faulty automatic segmentation, as it induces false adaptation-training data. To create a more accurate automatic segmentation, we applied an iterative method which works only on the train database, and tries to minimize the error of the segmentation. Surprisingly, not only the average accuracy of the adapted model seems to improve, but the non-adapted speaker independent model as well. The reason for this could be that by using our method the underrepresentation of certain kinds of data is reduced. And if the data representation is better, then the accuracy of the recognition and the adaptation should be better.

## 2   The Retraining Procedure

The iterative retraining method we introduce here begins with a normal Viterbi training. For each iteration the last retrained HMM is used to automatically re-segment the train database. After automatic segmentation the differences between the automatically segmented boundary positions and the manually segmented boundary positions are computed for each phoneme in the train database. The next HMM is then trained by weighting the training data according to these differences. When weighting the data, the weight of a segment is multiplied by the weight of the previous iterations. After all the weights for all the phonemes in the train database have been computed, the weighted retraining procedure begins with weighted K-Means clustering. This method is described in more detail in Table 1. The weighted mean for one cluster is computed with the following formula:

$$mean(X, W) = \frac{\sum\limits_{i=1}^{N} W_i X_i}{\sum\limits_{i=1}^{N} W_i}, \tag{1}$$

**Table 2.** The retraining method

| Input: | A manually segmented multi-speaker train database |
|---|---|
| 1 | Train the initial model (M) with the normal train method |
| 2 | W=[1,...,1] |
| 3 | Repeat the next section |
| 4 | Do automatic segmentation on the train database, using the M model |
| 5 | Compute the differences of the automatically and manually segmented boundaries for each phoneme in the train database |
| 6 | The difference value ($D_i$) for a frame ($x_i$) is computed by the sum of the two boundary differences of the phoneme segment containing the given frame. |
| 6 | Compute the weights ($W'$) for each frame with a monotone growing concave function to ensure a lower sensitivity of the weighting. $$W' = 1 + log(D + 1)$$ |
| 7 | $W = W \otimes W'$ |
| 8 | Train new HMM model with weighting (using W): |
| 9 | Initialize the model using weighted k-means, weighted mean, and weighted covariance |
| 10 | Train the new model (M) using the weighted Viterbi training method |

where X contains the data points of one cluster, N is the number of the points in this cluster, and W contains their nonnegative weights. The weighted energy function of one cluster is defined as:

$$E(X, W) = \sum_{i=1}^{N} W_i (X_i - mean(X, W))^2, \tag{2}$$

where the total weighted energy of the database is the sum of the weighted energies for each cluster.

After applying this bootstrapping procedure, the usual Viterbi training is used with the modification of computing the means and the covariances with weighting. The exact formula for the weighted mean was given previously, for the variance vector of data X with weights W (using Hadamard product):

$$var(X, W) = \frac{\sum_{i=1}^{N} W_i (X_i - mean(X, W)) \otimes (X_i - mean(X, W))}{\sum_{i=1}^{N} W_i} \tag{3}$$

In order to give a more precise description, the pseudo code of the iterative weighting method is given in the Table 2 below.

# 3    The Continuous Speech Recognizer and Automatic Segmentation

The continuous speech recognizer was developed within the framework of a medical dictation project. This project was initiated by two university departments with financial support from state funds [11] [12]. Here the recognizer is based on a multi-stack Viterbi decoding and n-gram language models. All the hypotheses in a stack with the same phonetic transcriptions are merged and handled as one. Available cutting parameters are the size of the stacks, the maximal probability difference from the first hypothesis (with the highest probability), the maximal number of the new hypotheses after extension with phonemes, the maximal number of the hypotheses to be extended with phonemes, the step size of the extension with a new phoneme, the step size of the extension with a new word, along with a few less important parameters. Only the maximal probability values are computed for the states at Viterbi decoding, and some Gaussian computing speed-up techniques are applied. This continuous speech recognizer is quite effective for real-time recognition tasks with medium-sized or small-sized dictionaries.

The automatic segmentation part is incorporated in the continuous speech recognizer. Thus in this methodology the automatic segmentation is not a distinct procedure, but actually uses the same technique as the speech recognizer during testing. Thus the automatic segmentation is done using a multi-stack Viterbi decoding, and the same cutting parameters are available as mentioned above.

# 4    Experiments and Results

For training purposes speech signals from 100 speakers (26 women and 74 men) were recorded and manually segmented. The ages of the speakers were between 13 and 72. Each speaker uttered 6 long sentences (average 16 words per sentence) and 12 distinct words. The database contains about 10800 words (85300 phonemes), the total length of the speech signals is about 100 minutes. For adaptations and tests, the speech samples of 5 speakers (3 women and 2 men) were recorded. Each speaker uttered the same 86 sentences (613 words) for the adaptation, and 20-20 different paragraphs of medical reports [13] for tests. The total length of the recorded speech samples for adaptation was about 6 minutes and for the tests it was about 15 minutes per speaker.

For each test, the language model was the same word 3-gram model. The Hidden Markov models for 33 phonetic classes had 3 states, and each state contained 3 mixture components. All the parameters of the continuous speech recognizer were set to same value for all the test cases, and were set so that the recognition was about 1.5 times faster than real-time. Similarly, the parameters for the automatic segmentation method were set so that it was approximately 3 times faster, than real-time.

For adaptation the MAP method was used modifying only the mean values of each mixture component [14]:

$$\mu_{new} = \frac{N}{N+\alpha}\left(\frac{1}{N}\sum_{i=1}^{N} x_i\right) + \frac{\alpha}{N+\alpha}\mu_0, \tag{4}$$

The parameter $N$ represents the number of examples ($x_i$) for the given mixture component, while the parameter $\alpha$ controls the speed of changing the mean of the mixture. The $\mu_0$ represents the initial mean value.

The performances obtained from the tests on 5 different speakers (M1, M2, W1, W2, W3) are listed in Table 3 below, while the corresponding word error rate reductions are given in Table 4 below. These results were obtained using HMMs trained after 40 iterations. The results show that the lower performance values improved by using the weighting method, while the two maximal value remained the same or fell by a few percent. The average word error rate reductions show a more significant improvement in performance after the adaptation method was applied.

**Table 3.** Accuracies for 5 speakers, obtained after 40 iterations. Adaptation was done with automatic segmentation using both the 'unweighted' and the 'weighted' HMMs.

| | Test database | | | | | Average |
|---|---|---|---|---|---|---|
| | Men | | Women | | | |
| | M1 | M2 | W1 | W2 | W3 | |
| normal | 97.57% | 98.53% | 83.42% | 95.33% | 82.99% | 91.57% |
| weighted | 97.57% | 98.02% | 86.12% | 97.74% | 86.35% | 93.16% |
| ad. normal | 94.43% | 95.82% | 92.42% | 98.07% | 97.60% | 95.67% |
| ad. weighted | 98.43% | 96.70% | 94.52% | 98.39% | 97.67% | 97.14% |

**Table 4.** Relative word error rate reductions. Here 'n.' means the normal, 'w.' means the weighted training method.

| | M1 | M2 | W1 | W2 | W3 | Average |
|---|---|---|---|---|---|---|
| n. → w. | 0.00% | -34.69% | 16.28% | 51.61% | 19.75% | 18.98% |
| ad. n. → ad. w. | 71.81% | 21.05% | 27.70% | 16.58% | 2.92% | 34.03% |
| n. → ad. n. | -129.22% | -184.35% | 54.28% | 58.67% | 89.89% | 48.62% |
| n. → ad. w. | 35.39% | -124.49% | 66.95% | 65.52% | 66.11% | 86.30% |

Figure 1 shows, how the accuracy and the total sum of boundary differences vary in the first 40 iterations. It can be seen that there is some negative correlation between the difference measure, and the average accuracy. The exact correlation value is actually -0.6. In Figure 2 the accuracies on the 5 speakers vary overall together, but this tendency is less typical for speakers with higher initial performance values. The correlation values between the accuracy of the speakers are between 0.4 and 0.7, except for the two speakers with the worst initial accuracy, whose value is 0.84.

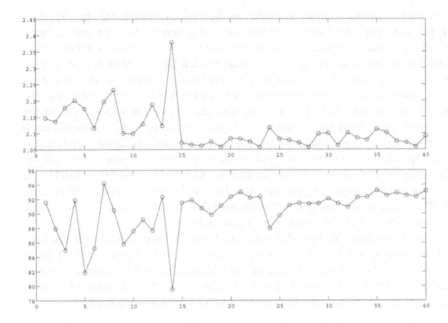

**Fig. 1.** The upper picture shows the total sum of boundary differences (in units of $10^6$), while the lower picture shows the average accuracy on 5 test speakers. The x-axis shows the number of training iterations performed.

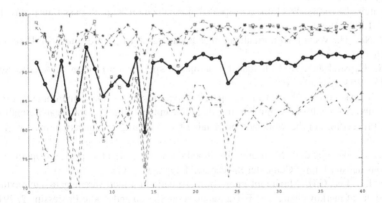

**Fig. 2.** The accuracies on 5 test speakers. The solid line shows the average accuracy. The x-axis shows the number of training iterations performed.

# 5   Conclusions and Future Work

The main challenge of continuous speech recognition is to improve the efficiency of the system still further. To achieve this numerous suggestions and experiments have been made. But here we tried to tackle the problem at its root that is we attempted to improve the recognition accuracy by modifying the training method of the base speaker-independent system. The results show that the total difference between the automatically and the manually segmented boundaries correlates with the total word error rate. By minimizing this difference we could create Hidden Markov Models with an even higher average accuracy. We constructed a simple iterative method for this minimization task, and then we got very promising results for the word error rate reduction. The increase in the accuracy is even more noticeable after speaker adaptation. The reason for this might be because the the efficiency of the adaptation methods depends much more on the performance of the base Hidden Markov Model. On the one hand there is a direct dependency: the better the initial model is, the better the adapted one will be. On the second hand there is an indirect dependency too: the better the base model is, the more accurate the automatic segmentation will be, so the adapted model will perform better. Thus one can see, that by reducing the total difference by only 4.7%, the average accuracy can increase by 19%, and after adaptation by 34%.

We plan to continue this line of research, that is to develop methods that can raise speech recognition performance. Our next piece of work will be a new method which uses the boundary differences to initialize the Hidden Markov Models in a new way. Instead of weighting the training data, we would like to use the difference information to iteratively cluster the training data into clusters with an increasing number.

# References

1. Lee, L., Rose, R.C.: Speaker normalisation using efficient frequency warping procedures. In: Proc. ICASSP96 (1996)
2. McDonough, J., Byrne, W., Luo, X.: Speaker normalization with all-pass transforms. In: Proc. ICSLP98 (1998)
3. Pitz, M., Ney, H.: Vocal tract normalization as linear transformation of mfcc. In: Proc. EuroSpeech2003 (2003)
4. Furui, S.: Cepstral analysis technique for automatic speaker verification. J. Acoust. Soc. Amer. 55, 1204–1312 (1974)
5. Kitaoka, N., Akahori, I., Nakagawa, S.: Speech recognition under noisy environments using spectral subtraction with smoothing of time direction and real-time cepstral mean normalization. In: Proceedings of the Workshop on Hands-Free Speech Communication, pp. 159–162 (2001)
6. Leggetter, C., Woodland, P.: Maximum likelihood linear regression for speaker adaptation of continuous density hmms. Computer Speech and Language 9, 171–185 (1995)
7. Gauvain, J.L., Lee, C.H.: Maximum a posteriori estimation for multivariate gaussian mixture observations of markov chains. IEEE Transactions on Speech and Audio Processing 2, 291–298 (1994)
8. Sankar, A., Lee, C.: A maximum-likelihood approach to stochastic matching for robust speech recognition. IEEE Trans. on Speech and Audio Processing 3, 190–202 (1996)
9. Digalakis, V., Rtischev, D., Neumeyer, L.: Speaker adaptation using constrained reestimation of gaussian mixtures. IEEE Trans. on Speech Audio Processing, 357–366 (1995)
10. Diakoloukas, V., Digalakis, V.: Maximum-likelihood stochastic-transformation adaptation of hidden markov models. IEEE Trans. on Speech Audio Processing 2, 177–187 (1999)
11. Vicsi, K., Kocsor, A., Teleki, C., Tóth, L.: Hungarian speech database for computer-using environments in offices (in Hungarian). In: Proc. of the 2nd Hungarian Conf. on Computational Linguistics, pp. 315–318 (2004)

12. Bánhalmi, A., Kocsor, A., Paczolay, D.: Supporting a Hungarian dictation system with novel language models (in Hungarian). In: Proc. of the 3rd Hungarian Conf. on Computational Linguistics, pp. 337–347 (2005)
13. Banhalmi, A., Paczolay, D., Toth, L., Kocsor, A.: First results of a hungarian medical dictation project. In: Proc. of IS-LTC, pp. 23–26 (2006)
14. Thelen, E.: Long term on-line speaker adaptation for large vocabulary dictation. In: Proc. of IEEE ICSPL, pp. 2139–2142. IEEE Computer Society Press, Los Alamitos (1996)

# User Modeling to Support the Development of an Auditory Help System

Flaithrí Neff, Aidan Kehoe, and Ian Pitt

University College Cork, Cork, Ireland
{fn2, ak2, i.pitt}@cs.ucc.ie

**Abstract.** The implementations of online help in most commercial computing applications deployed today have a number of well documented limitations. Speech technology can be used to complement traditional online help systems and mitigate some of these problems. This paper describes a model used to guide the design and implementation of an experimental auditory help system, and presents results from a pilot test of that system.

## 1 Introduction

Guided by research, online help systems have evolved over the past forty years and now include features such as full text search supported by natural language query, a focus on task-oriented material and tutorials incorporating multimedia elements. Empirical studies show that well-designed online help can be effective [1]. However, these approaches have some well documented limitations which are outlined below.

**Hands/eyes busy:** In some applications the user's hands and eyes are busy e.g. playing computer games, object manipulation in drawing applications, etc. Pausing the application, or the introduction of an overlay assistance window, can result in considerable disruption to the user's work.

**Context switching:** Applications often open an additional window to display assistance information to users. Studies on traditional computing applications show that switching between the application and the user assistance window can be problematic, especially for novice users [2].

**Limited display real estate:** Display of help material on portable devices with small display sizes and limited resolution is difficult.

**Paradox of the active user:** Carrol and Rosson [3] observed that when users encounter a problem they are reluctant to break away from their current task to consult the available help material.

The incorporation of an embedded help pane to display context-sensitive help within an application has been shown to be effective in addressing some of these problems [4], and has been implemented in successful commercial applications including Intuit Quicken and Microsoft Money. Use of an embedded assistance pane is obviously not an option in scenarios of limited display real estate.

V. Matoušek and P. Mautner (Eds.): TSD 2007, LNAI 4629, pp. 390–397, 2007.
© Springer-Verlag Berlin Heidelberg 2007

We have developed an experimental auditory online help system that uses speech technology to provide help within the context of an application, aiming to mitigate some of the problems associated with traditional online help systems. Our system has been designed to complement existing online help methods, and work within the constraints of commercially available online help and speech technologies.

## 2 Auditory Help System Model

The success of our auditory help system within what is a predominantly visual environment depends greatly on effective interaction between the machine and the human user's cognitive processes. We hope to show that by using the Auditory Help User Interface Model (AHUIM), proposed by Neff and Pitt (figure 1), we will gain a deeper understanding of the cognitive factors involved when applying our auditory help system. This, in turn, will help us to avoid design flaws by allowing us to anticipate the data flow of auditory (speech and non-speech) information during a specific task.

The model focuses on four primary stages of human auditory cognition.

- The sensory input stage: rules of auditory scene analysis [5] are applied.
- The attention stage: rules based on the notion of focused and peripheral attention mechanisms. [6]
- The memory stage: rules based on the changing-state hypothesis [7], [8].
- The schema stage: rules based on the idea of top-down influences and user experience [9].

We utilize the model to identify potential areas of interference between speech and non-speech streams within our online help system. Interference can occur at early, automated stages (streaming and attention) as well as at a more sophisticated level such as at stages of top-down processing and memory encoding.

Using our auditory help system, which was developed in accordance with the model, we performed a pilot study. We presented context-specific help topics to subjects for a primarily graphic application in a visually dominated environment i.e. Microsoft Paint running on Windows XP.

### 2.1 Multiple Streams

Our use of speech in the auditory help system is primarily to communicate specific information to the user while the use of non-speech sound is primarily to relay non-specific, structural information.

Although not exhibited in the model itself, we acknowledge that a certain level of interaction between auditory streams and concurrent visual content may also occur [10], [11]. One specific visual feature is of particular concern, and that is the reading of menu items in an application. This task begins as a visual one but is converted by the perceptual system into an auditory form as sub-vocal rehearsal. Sub-vocal rehearsal has the potential to disrupt the memory encoding of auditory help information due to its occupation of the top-down rehearsal process. However, menu items are short and concise and the drain on the rehearsal process is negligible.

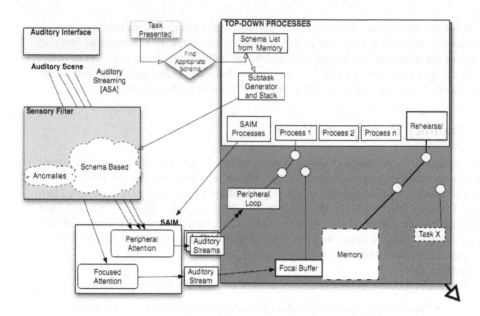

**Fig. 1.** Auditory Help User Interface Model (AHUIM)

The AHUIM shows how concurrent auditory information is automatically segregated into streams based on the rules of Auditory Scene Analysis [5]. The initial sensory mechanism is usually very effective at individualizing the many sonic attributes and assigning them to their proper physical sources. The schema filter further reduces the number of auditory streams to only the most relevant (namely the speech stream and the supporting sonification). However, beyond this stage the possibility of interaction between speech and non-speech sound increases.

## 2.2  Attention

The first stage where interference may occur between speech and non-speech streams is during the attention process, referred to as the Stream Attention and Inhibition Manager (SAIM) in the model. Only one stream at a time has the privilege to receive Focused Attention. The other concurrent streams are demoted to Peripheral Attention. This has important implications for our auditory help system as it is the focused attention attribute that relays information to the Focal Buffer, which in turn has exclusive access to the memory component. The most important stream in our auditory help system is the speech stream since this contains specific information. Therefore, non-speech sound must be either designed in such a way as not to remove focused attention from speech, or else it should not occur concurrently with speech. Even when employing the second method, the attention mechanism seems to need time to readjust to a new stream [6], and if no allowance is made for this the beginning of the new stream may be lost. This was something we found in our initial design and therefore we utilize non-speech sound only when vitally necessary and only at key structural points such as the very beginning of the document and for bullet points etc.

Although the Peripheral Attention facility does not have any access to the Focal Buffer, and is therefore not encoded, it can still disrupt top-down processes that may be required by the stream in Focused Attention. Rules at this point of the model are based primarily on the Changing-State Hypothesis [8] and the notion that interference is not content oriented but rather process oriented in the auditory domain. Jones et al focus on the top-down process of seriation to demonstrate this hypothesis. Consequently, in our implementation of the auditory help system, we try to avoid concurrent tasks that share top-down processes. For example, any acoustic signal that changes over time (including speech) requires the top-down seriation process, whether it is in Focused Attention or Peripheral Attention. In our implementation of the auditory help topics, sonified (non-speech) features are short and are as acoustically unvaried as possible. This rule is applied especially if speech is concurrent or adjacent. A similar approach to top-down disruption is applied to the memory component of the model.

## 2.3 Memory

Most visual displays provide the user with a form of external memory by providing accurate information on screen so there is no need for the user to store it in human memory. It is obvious from the AHUIM that information communicated via auditory means lacks external memory. In the auditory domain we rely heavily on the dynamic involvement of human working memory. To achieve a higher degree of efficiency from human memory, we need to avoid perceptual interaction and interference between speech and non-speech sounds at this critical stage. As depicted in the model, a coinciding task (such as sub-vocal rehearsal) may also require the top-down rehearsal process. This may temporarily steal memory encoding time from our speech stream.

In our auditory help system, the memory encoding of specific information via the speech stream is heavily reliant on the Rehearsal Process. The temporary re-allocation of the Rehearsal Process to another task will have a negative impact on the accuracy of the speech information encoded into memory. In our study, this becomes a concern in two scenarios and so user control over the auditory output becomes important. The first scenario relates to sub-vocal rehearsal of text-based menu items. A need to consult menu items is a factor in some of the help topics. Therefore, the user should have control over the auditory output so that only one information type is pulling on rehearsal resources at a time (either the sub-vocal rehearsal or the speech stream). The second scenario involves speech overwriting speech. In other words, a line of speech relaying one bullet point may need time to be engraved (rehearsed) into memory before a second line of speech is heard. This especially relates to trailing notes in the original visual version of the help files. In the visual domain, trailing notes often complement preceding information. However, when directly converted to speech via simple TTS, these notes often cause the more important preceding information to be lost (overwritten in memory). This was an important issue in our auditory help system design and a re-working of the textual content was needed before an aural rendition of the data commenced.

# 3  Auditory Help System

The model outlined in section 2 of the paper guided the decisions on audio channel configuration, and the use of speech and non-speech sounds. As well as considering technical constraints it was essential to consider cognitive load and usability issues associated with providing auditory help within an application context. Figure 2 shows an overview of the system. The standard application audio remains directed to the system speakers. The user hears the auditory help system output on the headset earphone. The user can use the headset microphone to issue voice commands to control access (play, pause, continue, stop) to the help system i.e. the user hears the standard application audio through the speakers, and interacts with the help system via the headset. Microsoft Speech Application Interface on Windows was used for the initial implementation.

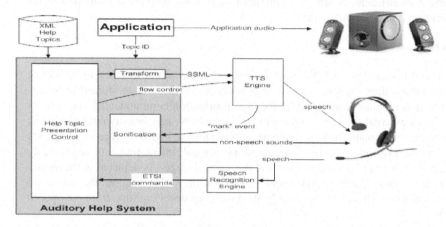

**Fig. 2.** Auditory Help System Overview

We have chosen to use speech synthesis, rather than human speech recordings. For mainstream online help systems the option of using human speech recordings is not realistic for cost reasons [12]. As illustrated in Figure 2 the speech synthesis markup language (SSML) "mark" element is used to trigger playback of non-speech sounds in synchronization with the synthesized speech output. The TTS (Text-to-Speech) engine notifies the help system when a structural mark element is encountered and the appropriate non-speech sounds are generated in a synchronized manner with the synthesized speech.

## 3.1  Categories of Help Topics

It is important to consider the cognitive load and usability issues associated with providing auditory help within an application context. Procedural information, simple instructional help, and basic conceptual information topics can be presented effectively using speech. Application of the minimalist instruction approach outlined by Carroll [13] typically results in shorter task-oriented topics suitable for presentation aurally. These types of topics can address the "what is this?" and "what is next?" style questions that can form a significant portion of user help requests [14]. As a result, for the pilot study we deliberately used this type of shorter task-oriented help topic.

Help topics targeted at advanced users often contain complex sentence structures and diagrams. Reference information typically includes elements such as lists, tables, etc. These types of information are difficult to present effectively using speech alone.

## 3.2 Pilot Study Results

Using a group of 10 first-year arts students, we performed a pilot study using the auditory help system. The aim of the study was to evaluate the effectiveness of help topics when the information was presented aurally to the subjects. Each of the subjects was required to perform four tasks in the Microsoft Paint application. While the tasks themselves were fairly straightforward, each task used somewhat obscure and non-intuitive functionality of the application, and thus it required the subject to review the supporting help material. The help material for each task was made directly available to the subjects (i.e. they did not have to search the help material to find the specific topic). Two of the help topics were presented to the subjects visually i.e. using the standard HTML Help view. The other two help topics were presented aurally i.e. using speech synthesis engine "ATT DTNV1.4 Mike" with complementary non-speech sounds.

There was a surprisingly low successful task completion rate for the first test (70%). We believe the reason for this is that some of the subjects did not read the help material carefully i.e. they quickly scanned the topic text but did not read it thoroughly. For subsequent tests we observed that the subjects reviewed the help topics (both visual and aural) much more carefully, probably because they better appreciated that the help topic material did actually contain the information required to help them complete the task.

| Test Description | Help | Task Completion |
|---|---|---|
| 1. Change foreground/background color | Visual | 70% |
| 2. Change size of image to 10x10 cm | Aural | 100% |
| 3. Turn grid lines on | Visual | 90% |
| 4. Draw horizontal and 45 degree lines | Aural | 80% |

In general the subjects were able to comprehend and successfully execute the short task-oriented help topics that were presented aurally. The use of non-speech sounds, played in-synch with the synthesized speech to represent help topic structural information, did not impair the subjects understanding of the topics. While all of the subjects considered the quality of the speech synthesis to be good, the tests highlighted a number of problems unique to the help topics presented aurally including:

**Pronunciation:** The speech synthesis engine used an American-style pronunciation of some words (e.g. "attribute") that subjects found difficult to understand.

**User Interface References:** Help topics often contain material that references specific user interface elements e.g. menu items, toolbars, etc. The HTML Help visual presentation of these user interface element references is done in a variety of ways including use of an icon, use of capitalization, quotes, etc. In the auditory version of the help topics we did use the SSML EMPH tag to highlight these user interface references. In spite of this use of word emphasis, several of the subjects still had problems relating the emphasized words to the appropriate user interface elements.

**Clarity:** Some help topics contained material that could be easily misinterpreted when presented using speech, and more explicit language would be required to clarify the aural presentation e.g. the fragment "holding down SHIFT while dragging" should be changed to "holding down the SHIFT key while dragging". Such lack of clarity in the aural help topic resulted in a lower successful task completion rate for test 4.

These types of issues resulted in the majority of the subjects having to listen to the auditory help topics multiple times to comprehend them (e.g. subjects required an average of 1.6 listens for the topic in test 2, and an average of 2.2 listens for the topic in test 4). These issues could be addressed by applying textual layout guidelines for speech rendering [15]. Such guidelines should result in faster task completion times and improved successful task completion rates.

## 4  Summary

An auditory help system can complement traditional online help systems, and it offers possibilities to mitigate some of the challenges associated with those systems. There are significant issues to be considered in design and implementation of such a system including: technological constraints of speech components; perceptual interaction between visual and auditory information; and perceptual interaction between speech and non-speech sounds within the auditory help system.

Fischer [16] highlights the importance of saying "the right thing at the right time in the right way", and this is especially important for speech-based interfaces. The model outlined in section 2 of the paper has been important in guiding the design and implementation of this experimental auditory help system. The pilot study has demonstrated that speech technology can be effective in presenting short task-oriented help topics to users aurally within the context of an application. We are planning a number of larger scale studies to expand on this work.

## References

1. Hackos, J.T., Stevens, D.: Standards for online communication. Wiley, New York (1997)
2. Kearsley: Online help systems, Ablex Publishing, Norwood, NJ (1988)
3. Carroll, J.M., Rosson, M.B.: Paradox of the active user. interfacing thought: cognitive aspects of human-computer interaction. MIT Press, Cambridge, MA (1987)
4. Grayling, T.: Usability Test of Two Browser-based Embedded Help Systems. Journal of the Society of Technical Communication 49(2), 193–209 (2002)
5. Bregman, A.S.: Auditory Scene Analysis: The Perceptual Organization of Sound. MIT Press, Cambridge (1990)

6.  Wrigley, S.N., Brown, G.J.: A model of auditory attention. Technical Report CS-00-07, Speech and Hearing Research Group, University of Sheffield (2000)
7.  Jones, D.M., Macken, W.J.: Irrelevant Tones Produce an Irrelevant Speech Effect: Implications for Phonological Coding in Working Memory. Journal of Experimental Psychology: Learning, Memory, and Cognition 19(2), 369–381 (1993)
8.  Jones, D.M., Madden, C., Miles, C.: Privileged access by irrelevant speech to short-term memory: The role of changing state. Quarterly Journal of Experimental Psychology 44A, 645–669 (1992)
9.  Baddeley, A.D.: Your Memory – A User's Guide. Prion, London (1996)
10. Bonnel, A.M., Hafter, E.R.: Divided attention between simultaneous auditory and visual signals. Perception & Psychophysics 60(2), 179–190 (1998)
11. Johnson, J.A., Zatorre, R.J.: Attention to Simultaneous Unrelated Auditory and Visual Events: Behavioral and Neural Correlates. Advance Access (2005)
12. Davison, G., Murphy, S., Wong, R.: The use of eBooks and interactive multimedia as forms of technical documentation. In: ACM Conf. on Design of Communication, pp. 108–115 (2005)
13. Carroll, J.M.: The Nurnberg Funnel: designing minimalist instruction for practical computer skill. The MIT Press, Cambridge, Massachusetts (1990)
14. Roesler, A.W., McLellan, S.G.: What help do users need?: taxonomies for on-line information needs & access methods. In: Proc. SIGCHI Conf. on Human factors in computing systems, pp. 437–441 (1995)
15. Kehoe, A., Pitt, I.: Designing Help Topics for use with Text-To-Speech. In: ACM Conf. on Design of Communication, USA (2006)
16. Fischer, G.: User Modeling in Human–Computer Interaction. User Modeling and User-Adapted Interaction 11(1-2), 65–86 (2001)

# Fast Discriminant Training of Semi-continuous HMM

G. Linarès and C. Lévy

Laboratoire Informatique d'Avignon, LIA, Avignon, France
{georges.linares,christophe.levy}@lia.univ-avignon.fr

**Abstract.** In this paper, we introduce a fast estimate algorithm for discriminant training of semi-continuous HMM (Hidden Markov Models).

We first present the *Frame Discrimination* (FD) method proposed in [1] for weight re-estimate. Then, the weight update equation is formulated in the specific framework of semi-continuous models. Finally, we propose an approximated update function which requires a very low level of computational resources.

The first experiments validate this method by comparing our fast discriminant weighting (FDW) to the original one. We observe that, on a digit recognition task, FDW and FD estimate obtain similar results, when our method decreases significantly the computational time.

A second experiment evaluates FDW in Large Vocabulary Continuous Speech Recognition (LVCSR) task. We incorporate semi-continuous FDW models in a Broadcast News (BN) transcription system. Experiments are carried out in the framework of ESTER evaluation campaign ([2]). Results show that in particular context of very compact acoustic models, discriminant weights improve the system performance compared to both a baseline continuous system and a SCHMM trained by MLE algorithm.

## 1 Introduction

Last decade, some works deal with reduction of resources required by acoustic models for speech recognition, both in terms of memory occupation and CPU time. Some authors propose approaches based on semi-continuous hidden Markov Models (SCHMM) ([3], [4]). These models consist in a common codebook of Gaussian and state-dependent weight vectors. This full Gaussian tying allows a significant reduction of model complexity. Moreover, it could limit the estimate problems due to the lack of training data.

Training of SCHMM requires both codebook building and weight estimate. This is generally achieved by using Maximum Likelihood (ML) criterion. Nevertheless, discriminative training is now frequently used for acoustic model training. Few works have shown that these methods could bring good results compared to classical MLE, especially when large training corpora are available or on small vocabulary tasks ([5],[1]). Unfortunately, these methods require much more CPU resources than classical MLE training.

This paper is focuses on the use of frame-discrimination based training of SCHMM. We present a technique for fast training of discriminative semi-continuous HMM. We formulate the weight update equation presented by [1] in the specific framework of semi-continuous models. Then, we investigate SCHMM particularities which could improve FD-based training.

V. Matoušek and P. Mautner (Eds.): TSD 2007, LNAI 4629, pp. 398–405, 2007.

In the first section, we introduce the standard update rule and we re-formulate it in the case of semi-continuous HMM. Then, we propose an approximation which dramatically reduces the training costs. In the second section, we perform an experimental evaluation of our training method. SCHMM are estimated using the standard update equation and the fast one. Both systems are compared to a SCHMM estimated by MLE algorithm, on a digit recognition task. In the third section, fast update rule is evaluated on a real-time LVCSR task, in the experimental framework of the ESTER evaluation campaign ([2]). The FDW training performance is compared with a SCHMM trained by MLE and with a classical continuous HMM.

Lastly, we conclude and propose further developments of the proposed method.

## 2  Fast Discriminant Weighting for SCHMM Training

Acoustic model estimate based of the maximization of mutual information has been largely studied for last few years. The general principle of this approach is to reduce the error rate by maximizing the likelihood gap between the good transcripts and the bad ones. The search of $\lambda$ parameters is performed by maximizing the MMIE objective function $F_{mmie}$:

$$F_{mmie}(\lambda) = \sum_{r=1}^{R} log \frac{P_\lambda(O_r|M_{w_r})P(w_r)}{\sum_{\tilde{w}} P_\lambda(O_r|M\tilde{w})P(\tilde{w})} \tag{1}$$

where $w_r$ is the correct transcript, $M_w$ the model sequence associated with the word sequence $w$, $P(w)$ the linguistic probabilities and $O_r$ an observation sequence. The denominator of the objective function sums the acoustico-linguistic probabilities of all possible hypothesis.

One of the main difficulties in the parameter estimate is the complexity of the objective function (and the derived update rules) which requires a scoring of all bad paths for the denominator evaluation. In order to reach reasonable computational cost, several methods have been presented in the literature. Authors proposed to use word or phone lattices ([6]), specific acoustic model topology ([1]), etc. In spite of the complexity decrease reached by these techniques, discriminative training still requires very large computing resources, significantly more than MLE training.

In the particular case of semi continuous HMM, only the weight update is required. Moreover, the sharing of Gaussian components allows a significant decrease of the training duration. This is mainly due to the size of the codebook, which contains generally less parameters than classical continuous models. Moreover, sharing the gaussian components over the states could allow a direct selection of discriminant components. We highlight this point by developing a *frame discrimination* method proposed in [1]. In this paper, the authors propose an frame-based method for discriminative training. A weight update rule is proposed wich aims to improve the low stability of gradiant-descent based weight estimate rules. The proposed technique consists in finding the weights $\tilde{c}_{jm}$ to maximize the auxilary function:

$$F_c = \sum_{j,m} \left[ \gamma_{jm}^{num} log(\tilde{c}_{jm}) - \frac{\gamma_{jm}^{den}}{c_{jm}} \tilde{c}_{jm} \right] \tag{2}$$

where $\gamma_{jm}^{num}$ and $\gamma_{jm}^{den}$ are the occupancy rates estimated respectively on positive examples (corresponding to a correct decoding situation, noted $num$) and on negative examples ($den$) ; $c_{jm}$ is the weight of the component $m$ of state $j$ at the previous step and $\tilde{c}_{jm}$ is the updated weight $(j, m)$.

By optimizing each term of that sum holding all other weights, the convergence can be reached in few iterations. Each term of the previous expression is convex. Therefore, the update rule can be directly calculated using the equation:

$$\tilde{c}_{jm} = \frac{\gamma_{jm}^{num}}{\gamma_{jm}^{den}} c_{jm} \tag{3}$$

where $\gamma_{jm}^k$ is the probability of being in the component $m$ of state $j$; this probability is estimated on the corpus $\Omega_k$ which consists in all frames associated with the state $k$. Therefore, the occupation rate can be expressed using the likelihood functions $L()$:

$$\gamma_{jm}^k = \sum_{X \in \Omega^k} \frac{L(X|S_j)}{\sum_i L(X|S_i)} \frac{c_{jm} L(X|G_{jm})}{L(X|S_j)} \tag{4}$$

$$\gamma_{jm}^k = \sum_{X \in \Omega^k} c_{jm} \frac{L(X|G_{jm})}{\sum_i L(X|S_i)} \tag{5}$$

By isolating the likelihood of the frame $X$ knowing the state $S_k$ in the denominator, we obtain:

$$\gamma_{jm}^k = \sum_{X \in \Omega^k} c_{jm} \frac{L(X|G_{jm})}{L(X|S_k) + \sum_{i \neq k} L(X|S_i)} \tag{6}$$

In semi-continuous models, components $G_{jm}$ are state-independent. Denoting $\epsilon_k = \sum_{i \neq k} L(X|S_i)$, the occupation ratio can be formulated as:

$$\frac{\gamma_{jm}^{num}}{\gamma_{jm}^{den}} = \frac{\sum_{X \in \Omega^j} \frac{L(X|G_{jm})}{L(X|S_j) + \epsilon_j}}{\sum_l \sum_{X \in \Omega^l} \frac{L(X|G_{lm})}{L(X|S_l) + \epsilon_l}} \tag{7}$$

By assuming $\epsilon \simeq 0$, numerator and denominator of previous ratio are reduced to the update function of classical EM weight estimate. Then, previous equation can be approximated by:

$$\frac{\gamma_{jm}^{num}}{\gamma_{jm}^{den}} = \frac{c_{jm}}{\sum_l c_{lm}} \tag{8}$$

By combining this heuristic to equation 3, we obtain the weight update formula:

$$\tilde{c}_{jm} = \frac{c_{jm}^2}{\sum_l c_{lm}} \tag{9}$$

Weight vectors are normalized after each iteration.

Therefore, this training technique uses the Gaussian sharing properties of SCHMM to estimate discriminative weights directly from MLE weights, without any additional likelihood calculation. Comparing to classical MMIE training functions, neither search algorithm nor lattice computation are required for denominator evaluation. So, this method permits a model estimate while involving a computational cost equivalent to the one required by MLE training. Nevertheless, this technique is based on the assumption that $\epsilon_i$ are state-independent (cf. Equation 7). The *a priori* validation of such an assumption seems to be difficult, especially due to the particular form of equation 7, where the $\epsilon_i$ quantities contribute both to the numerator and to the denominator of the cost function.

# 3    Contrastive Experiments on Digit Recognition Task

The first experiments are conducted on French digit recognition task. We use the plate-form described in [7] for embedded speech recognition. Here, the small amount of available task-specific data for model training allows a full evaluation of weight update rule, without any approximation based on partial decoding.

## 3.1    Training Strategy

The recognition engine is designed for embeded recognition systems using only few hardware resources and small amount of task-specific training data. The training strategy proposed in [7] consists in using two corpora. The first one is a large database used to estimate of a generic model. The second is a GMM trained with the classical EM algorithm. In a second step, this GMM is adapted (using Maximum A Posteriori technique) to the targeted context using the task-specific database. Lastly, state-dependent parameters (i.e. weight vectors in the case of semi-continuous models) are estimated on the task-specific corpus.

## 3.2    Acoustic Model Training

In our first experiment, the generic GMM is trained on BREF120 corpus ([8]). This database contains about 100 hours of speech pronouced by 120 speakers. Test and adaptation sets are extracted from BDSON database. These corpora consist respectivly in 2300 (test) and 700 (train) digit utterances.

Several codebooks are trained by varying the Gaussian number from 216 to 1728. For each of them, a SCHMM is estimated by the MLE algorithm. Then, two sets of discriminant models are trained, using either the initial weight estimator (denoted FD) or our fast estimator (FDW).

## 3.3    Results

Table 1 shows the results obtained by MLE and the two discriminative approaches using various codebook sizes. The two last methods obtain very close results, slightly better for fast estimator. Nevertheless, differences are not really significant since they remain inside the confidence interval (0.6% large in these tests). We can notice that both the weighting methods outperform significantly the standard MLE training, except for the model based on the smallest codebook whit only 216 Gaussian components. Approaches based on mixture weighting assume that some Gaussian components carry state-specific or phone-specific information; this assumption is probably not satisfied here, since the acoustic space is covered by a very small set of Gaussian components.

Globally, our experiments show that the proposed approximation has no negative impact on system performance while reducing the computational cost required by MMIE training. Moreover, full MMIE estimate (including mean and variance update) has been evaluated on similar tasks in literature. Experiments have shown good performance compared to MLE approaches ([5]). Our results suggest that frame-discrimination based weighting could be also efficient in such conditions, while preserving both model compacity and computational costs.

**Table 1.** Digit Error Rate (DER) according to the Gaussian number of the codebook (# GAUSS) for semi-continuous models estimated using MLE, initial FD and fast FD (FDW) algorithms. These experiments were performed on a test set of 2300 digit utterances extracted from the French BDSON database.

| # GAUSS | MLE | FD | FDW |
|---------|-------|-------|-------|
| 216 | 3.91% | 4,26% | 4,09% |
| 432 | 3.09% | 2,70% | 2,39% |
| 864 | 2.74% | 2,48% | 2,57% |
| 1728 | 2.74% | 2,30% | 2,00% |

## 4 Fast Discriminant Weighting on LVCSR Task

### 4.1 Experimental Framework

Evaluation on LVCSR task has been conducted using the real-time broadcast news recognition system developed at the LIA ([9]), in the framework of ESTER evaluation campaign. The aim of ESTER campaign was the evaluation of broadcast news transcription systems. Data were recorded from radiophonic shows in various conditions (studio and telephone speech, live and recorded shows, ...). The train corpus is made up of 200 hours of transcripted speech, but no restriction about the amount and the source of training data was imposed. Here, we evaluate discriminant semi-continuous HMM using the recognition engine we involved in the real-time transcription task. This system is based on an A* search algorithm working on a phone lattice. For the real-time task, this system uses trigram language models, a vocabulary of 25000 word and context-dependent acoustic models. These models contain 3000 HMM sharing 936 emiting states for a global number of free parameters of 4.8 millions. Real-time is reached by using fast linguistic look-ahead and on-demand Gaussian selection. In spite of the relatively low CPU resources required by this recognizer, the memory space used does not fit the limited capacities of light systems like mobile phones.

Here we evaluate compact acoustic models based on discriminant SCHMM, by comparing a baseline continuous HMM, SCHMM trained by MLE algorithm and SCHMM trained by our fast FDW estimator.

## 4.2   SCHMM Estimate

**Baseline System and Codebook Building.**  As discussed in introduction, one of the major interest of semi-continuous HMM lies in its compacity, which could be critical for embedded systems. In order to evaluate our method on LVCSR task while respecting compacity constraint, we use first a relatively small continuous model. This baseline model (named C-CI) is context-independent and each state is composed at most of 64 Gaussian components. Feature vectors contain 12 PLP (Perceptual Linear Prediction) coefficients, the energy and first and second derivatives of these 13 parameters. Lastly, cepstral normalization (mean subtraction and variance reduction) is performed in a sliding window of 500 ms.

C-CI model is trained using a classical EM algorithm, on the 200 hours of speech from ESTER train corpus.

Lastly, this baseline model is also used for codebook building. All its 7000 Gaussian components are collected and grouped into a common codebook which will be used for the semi-continuous HMM training, with both MLE and MMIE algorithm.

**MLE Weight Update.**  Given a Gaussian codebook, the memory occupation required by each state-dependent GMM remains very low. We use a set of 10000 models sharing 3500 emiting states. The tying is performed using the decision tree method. All state-dependent weight vectors are then estimated using MLE algorithm. At this point, we obtain a first context-dependent SCHMM (SC0) estimated by MLE and based on the codebook of 7000 Gaussian components and on 3500 state-dependent weight vectors of 7000 coefficients.

Then, vector sizes are reduced by removing the Gaussian components for which weights are lower than a fixed threshold $\gamma$. Setting $\gamma$ at $10^{-5}$, the model contains 580000 weights. This number corresponds to a mean of 121 Gaussian components per state. This first model (SC-MLE-L) contains finally 1100000 parameters (weights and codebook). This is much more than the complexity of the original continuous model C-CI which has been used for codebook building.

In order to compare semi-continuous and continuous models for a fixed complexity, we also build a more compact SCHMM. By setting $\gamma$ threshold to 0.3, we obtain a model (SC-MLE-C) containing 610000 free parameters.

**Updating Discriminant Weights.**  Using large amount of training data and large vocabulary systems, the full estimate of objective function denominator is not reachable in a reasonable time. So, we use our fast estimator.

FDW weights are calculated directly from SC0 model using equation 9. Then, weight vectors are reduced by applying the method used for the SCHMM models which were estimated by MLE. By setting $\gamma$ at the value used for SC-MLE-L model ($10^{-5}$), we obtain a model (SC-FDW) containing a total of 80000 weights (and 615000 parameters), corresponding to an average of 22 Gaussian per state. Therefore, the Gaussian selection based on discriminant weighting is dramatically more efficient than using MLE models. This first result enlights the coding redundancy of MLE models, where the main part of Gaussian components is not state-specific. This point is an expected consequence of MLE approaches. On the other hand, it can be also due to the low precision of the initial GMM, for which complexity remains low regarding both the targeted task and the corpus variability.

### 4.3  Results

Table 2 presents the results obtained on 4 hours of French BN shows. FDW models (SC-FDW) outperform the continuous model C-CI of 2,2% WER (absolute) while demonstrating a similar level of complexity. On the other hand, the compact MLE model (SC-MLE-C) obtains low results. By saving all significative weights (SC-MLE-L model), the complexity reaches to 1 million parameters while the WER is reduced to 34.9%. It seems to be clear that the forced reducing of SC-MLE-C leads to a dramatic loss of relevant information.

FDW semi-continuous model improves performance for highly compact models; nevertheless, context-dependent continuous models allow a best decoding: the model involved in the ESTER evaluation campaign the on real-time transcription task obtains a 26.8% WER on this corpus, using 7.1 million free parameters (about 8 times more than our discriminant SCHMM).

**Table 2.** Word error rate (WER) and complexity (expressed as a number of free parameters) of context-independent continuous HMM (C-CI), semi-continuous HMM trained by the fast frame discrimination algorithm, compact SCHMM trained by MLE (SC-MLE-C), large SCHMM trained by MLE (SC-MLE-L). Experiments carried out on 4 hours of French broadcast news from ESTER corpus.

|       | C-CI  | SC-FDW | SC-MLE-C | SC-MLE-L |
|-------|-------|--------|----------|----------|
| WER   | 41.8% | 39.6%  | 48.8%    | 34.9%    |
| # PAR | 544k  | 615k   | 610k     | 1100k    |

## 5  Conclusion

We have proposed a fast training algorithm for discriminant training of SCHMM. We have formulated the *frame discrimination* update rules in the particular case of semi-continuous models and we have proposed an heuristic which allows to computed directly the discriminant weights from MLE ones, without any requirement of train database decoding.

The experiments conducted on a digit recognition task validate this fast approximation: performance is close to that obtained using the initial updating rule. Thus, discriminant models outperform significantly MLE models except for extremely small codebook.

Our first experiments on LVCSR task show that SCHMM models could also take benefit of the proposed technique of fast discriminant weighting.

Moreover, the weighting method is usually not considered as a critical point in the global scheme of discriminant training. Our experiments show that as gaussian are estimated by using MLE criterion, the discriminant capacity of HMM could be significantly improved by using discriminant weights.

# References

1. Woodland, P., Povey, D.: Large scale discriminative training for speech recognition. In: ISCA ITRW Automatic Speech Recognition: Challenges for the Millenium, Paris, pp. 7–16 (2000)
2. Galliano, S., Geoffrois, E., Mostefa, D., Choukri, K., Bonastre, J.F., Gravier, G.: The ester phase ii evaluation campaign for the rich transcription of French broadcast news. In: Proc. of the European Conf. on Speech Communication and Technology (2005)
3. Huang, X., Alleva, F., Hon, H.W., Hwang, M.Y., Rosenfeld, R.: The SPHINX-II speech recognition system: an overview. Computer Speech and Language 7(2), 137–148 (1993)
4. Vaich, T., Cohen, A.: Comparison of continuous-density and semi-continuous hmm in isolated words recognition systems. In: EUROSPEECH'99, pp. 1515–1518 (1999)
5. Normandin, Y., Cardin, R., Mori, R.D.: High-performance digit recognition using maximum mutual information. 2, 299–311 (1994)
6. Aubert, X., Ney, H.: Large vocabulary continuous speech recognition using word graphs. In: Proc. ICASSP '95, pp. 49–52 (1995)
7. Lévy, C., Linarès, G., Nocera, P., Bonastre, J.: 7 in Digital Signal Processing for In-Vehicle and Mobile Systems 2. In: Abut, H., Hansen, J.H.L., Takeda, K. (eds.) Embedded mobile phone digit-recognition, Springer Science, Heidelberg (2006)
8. Lamel, L., Gauvain, J., Eskénazi, M.: BREF, a large vocabulary spoken corpus for French. In: Proceedings of European Conference on Speech Communication and Technology (Eurospeech'1991), Gênes, Italie, pp. 505–508 (1991)
9. Nocera, P., Fredouille, C., Linarés, G., Matrouf, D., Meignier, S., Bonastre, J.F., Massonié, D., B'ehet, F.: The LIA's French broadcast news transcription system. In: SWIM: Lectures by Masters in Speech Processing, Maui, Hawaii (2004)

# Speech/Music Discrimination Using Mel-Cepstrum Modulation Energy*

Bong-Wan Kim[1], Dae-Lim Choi[1], and Yong-Ju Lee[2]

[1] Speech Information Technology and Industry Promotion Center,
Wonkwang University, Korea
{bwkim, dlchoi}@sitec.or.kr
[2] Division of Electrical Electronic and Information Engineering,
Wonkwang University, Korea
yjlee@wonkwang.ac.kr

**Abstract.** In this paper, we propose Mel-cepstrum modulation energy (MCME) as an extension of modulation energy (ME) for a feature to discriminate speech and music data. MCME is extracted from the time trajectory of Mel-frequency cepstral coefficients (MFCC), while ME is based on the spectrum. As cepstral coefficients are mutually uncorrelated, we expect MCME to perform better than ME. To find out the best modulation frequency for MCME, we make experiments with 4 Hz to 20 Hz modulation frequency, and we compare the results with those obtained from the ME and the MFCC based cepstral flux. In the experiments, 8 Hz MCME shows the best discrimination performance, and it yields a discrimination error reduction rate of 71% compared with 4 Hz ME. Compared with the cepstral flux (CF), it shows an error reduction rate of 53%.

## 1 Introduction

There often arises a need to discriminate speech signals from music signals as automatic speech recognition (ASR) systems come to be applied to multimedia domains. Speech/music discriminator (SMD) can be used as a preprocessor to ASR systems to exclude music portions from the multimedia data before transcription.

To improve the performance of SMD, the selection of appropriate feature parameter(s) is very important. Thus, many feature parameters have been proposed to discriminate speech/music signals. Those parameters can be roughly classified into time domain, spectral domain and cepstral domain parameters.

In the time domain, zero crossing rate [1,2] and short term energy [3] have been proposed. In the spectral domain, the balancing point of the spectral power distribution (spectral centroid) and the 2-norm of the frame-to-frame spectral amplitude difference vector (spectral flux, SF) have been used [1]. And 4 Hz modulation energy (ME) has also used to discriminate speech/music signals [1] because speech has a characteristic energy modulation peak around the 4 Hz syllabic rate [4]. In the cepstral domain, the

---

* This work was supported by the Korea Research Foundation Grant funded by the Korean Government (MOEHRD) (The Regional Research Universities Program/Center for Healthcare Technology Development).

V. Matoušek and P. Mautner (Eds.): TSD 2007, LNAI 4629, pp. 406–414, 2007.

method which uses the frame-to-frame cepstral difference or the cepstral flux (CF) has been proposed [5]. CF can be viewed as a cepstral domain extension of SF. Among other domains pulse metric [1] which reflects rhythms of audio signals and entropy and dynamism [6] which use phoneme recognition results have also been proposed. Also features which use consonant, vowel and silence segment information obtained from a phoneme recognizer have been proposed [7].

In general, it is known that SF is less effective than CF because SF is more sensitive even to minute spectral variation as a result of its difference computation of all filterbank results, whereas, in the case of CF, only the lower part of cepstral components which represents spectral envelops is used in its difference computation. Like SF, ME uses all of the filterbank outputs in the calculation. As the spectral energies in adjacent bands are highly correlated, the use of ME would not be expected to work well.

Therefore in this paper, we propose to use Mel-cepstrum modulation energy (MCME), which is a Mel-cepstrum domain extension of ME, as a feature parameter to discriminate speech/music signals. As cepstral coefficients are mutually uncorrelated, we expect MCME to perform better than ME.

The paper is organized as follows. In Sect. 2, we describe ME and MCME. In Sect. 3, we describe databases used in the experiments, and in Sect. 4, we deal with the results of the experiments. Our conclusion follows in Sect. 5.

## 2  ME and MCME

### 2.1  ME

Let $X[n, k]$ be the DFT of $n^{th}$ frame of audio signal. Then, we can obtain magnitude modulation spectrum (MMS) by taking a second DFT of the time sequence of the $k^{th}$ DFT coefficients as:

$$MMS[n, k, q] = \sum_{p=0}^{P-1} |X[n + p, k]| e^{-j2\pi qp/P} \qquad (1)$$

where $n$ is a frame index of audio signal, $k$ is an index of frequency axis in the first DFT, $q$ is an index of frequency axis in the second DFT and $P$ is the size of second DFT. Lower $q'$ indicates slower spectral changes while higher $q'$ indicates faster spectral changes. Speech has faster spectral changes than music signal due to its sequential alternation of voiced and unvoiced sounds, so modulation spectrum can be used as a feature to discriminate speech/music signals.

To calculate ME, in general, the result of the first DFT is not directly used but the result of Mel-filterbank analysis is used as the input to the second DFT. As speech has a characteristic energy modulation peak around the 4 Hz syllabic rate [4], only 4 Hz (or 4 Hz to 8 Hz) ME has been used as a feature to discriminate between speech and musical signals [1,2,8,9,10].

In this paper the ME is defined as:

$$ME[n, q] = \frac{\frac{1}{M} \sum_{m=0}^{M-1} |FMS[n, m, q]|^2}{\frac{1}{P} \sum_{p=0}^{P-1} log(E[n + p])} \qquad (2)$$

where $FMS[n, m, q]$ is modulation spectrum calculated on the time trajectory of filter-bank output, $M$ is the order of filterbank and $E[n]$ is the short term energy of the $n^{th}$ frame of source signal.

The denominator for normalization is slightly different from the one in the literature [1,2] where normalized ME is extracted by calculating the ratio between the target mod-ulation (4 Hz) component and the sum of all modulation components and adding the ratios over all the filterbank bands. We simply use the short term energy of source signal to reduce dynamic range of target modulation energy. With the preliminary experiment, we found that the proposed method shows better performances than the conventional method. We think that the conventional method's deteriorating performances are due to its sensitivity to variation of modulation spectrum as it uses the sum of all modu-lation components as a denominator for normalization. Furthermore, the conventional method has the disadvantage that it needs recalculation of all modulation components per frame, but the proposed method uses simple short term energy of source signal which is calculated only once when MFCC or filterbank results are calculated.

## 2.2 MCME

Let $C[n, l]$ be the real cepstrum of the DFT $X[n, k]$. We can obtain a cepstrally smoothed estimate of spectrum, $S[n, k]$, by using a rectangular low quefrency lifter which retains only the lower part of cepstral coefficients as $C[n, l]$ is a real symmetric sequence.

$$logS[n, k] = C[n, 0] + \sum_{l=1}^{L-1} 2C[n, l]cos(2\pi lk/K) \tag{3}$$

Using (3), we can obtain a cepstally smoothed estimate of MMS by taking the second DFT of $logS[n, k]$ over $P$ points and considering the $q^{th}$ coefficients as follows:

$$MMS'[n, k, q] = \sum_{p=0}^{P-1} log(S[n+p, k])e^{\frac{-j2\pi pq}{P}}$$

$$= \sum_{p=0}^{P-1} C[n+p, 0]e^{\frac{-j2\pi pq}{P}} + \sum_{l=1}^{L-1} \frac{cos(2\pi kl)}{K} \sum_{p=0}^{P-1} 2C[n+p, l]e^{\frac{-j2\pi pq}{P}}. \tag{4}$$

Using the last part of (4), Mel-cepstrum modulation spectrum (MCMS) is defined as:

$$MCMS[n, l, q] = \sum_{p=0}^{P-1} C[n+p, l]e^{-j2\pi pq/P}. \tag{5}$$

From (4) and (5), we find that $MMS'[n, k, q]$ is a linear transformation of $MCMS[n, l, q]$. As MFCC coefficients are mutually uncorrelated, we expect MCMS to perform better than MMS.

Tyagi and his collegues introduced the MCMS to be used as the dynamic feature of MFCC for ASR and proved its effectiveness compared with delta parameter and

RASTA PLP in the noise environment [12,13] as even in the presence of acoustic interference the temporal characteristics of speech appear to be less variable than the static characteristics [11].

Thus, in this paper we propose to use modulation energy based on the MCMS as a feature parameter to discriminate speech/music signals. MCME, proposed in this paper, is defined as:

$$MCME[n, q] = \frac{\frac{1}{L}\sum_{l=0}^{L-1} |MCMS[n, l, q]|^2}{\frac{1}{P}\sum_{p=0}^{P-1} log(E[n + p])}. \tag{6}$$

4 Hz ME and 4 Hz MCME extracted from music signal and speech signal are depicted in Fig. 1, where the thick lines are for ME. Comparing (a) and (b), we can find that ME and MCME have higher values for speech signal than for music signal. And we can see that MCME has a higher value than ME for both signals, and the former shows more significant difference between speech and music signals than the latter.

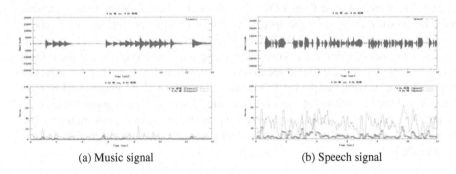

(a) Music signal                           (b) Speech signal

**Fig. 1.** Comparison of 4 Hz ME and 4 Hz MCME

# 3    Databases

## 3.1    Databases

**Read speech and clean music data set (RDCD).** This set is composed of read speech data instead of spontaneous speech and music data extracted from music CDs. As ME and MCME are related to the syllabic rate, we include Korean, English and Chinese sentences in the speech data to avoid language dependency. A total of 2,120 Korean sentences, uttered by 20 speakers, from CleanSent01 DB [14] (about 189 minutes), a total of 1,680 English sentences from the test set of the TIMIT Acoustic-Phonetic Continuous Speech Corpus [15] (about 86 minutes) and a total of 1,000 Chinese sentences, uttered by 10 speakers, from Chinese03 DB [14] (about 62 minutes) are included in this set. For music data, a total of 100 musical pieces (about 420 minutes) from "RWC - Music Database: Music Genre" (RWCMG) [16] and a total of 100 pieces (about 406 minutes) from "RWC - Music Database: Popular Music" [16] are included in this set. RWCMG has 10 main genre categories and 33 subcategories.

**Broadcast data set (BRDC).** When the training set is constructed from RDCD, the deterioration of discrimination performances is expected with the test data in the different environment, such as spontaneous speech and music data with channel noises, even though the performances are good on the data in the same environment. Therefore we have included internet broadcast data to verify the performances on the data in the different environment. We recorded 3 music programs from an FM radio channel which are simultaneously broadcasted via internet by Korean Broadcasting System (KBS) [17] in Korea. We captured the programs through the realtime internet receiver software provided by KBS. We have excluded advertisement parts from the data, but included the parts with some background music while the program announcer is speaking. Due to the nature of the programs, host's conversation with guests, sound of laughing and prosodic speech were naturally included in the data. We have manually segmented data into speech and music parts before discrimination test. The speech data is about 70 minutes long and the music data in the BRDC set is about 87 minutes long.

## 3.2 Organization of Training and Test Set

Training data were selected from RDCD set. We have used 1,060 Korean sentences uttered by 10 speakers to train the speech model, and 33 pieces, 1 piece per 1 subcategory, selected randomly from RWCMG to train the music model. 16 pieces of music have lyrics of artist(s). Full data excepting training data were used as test data. Data sizes of training and test data are respectively about 236 minutes and 1,084 minutes in length. All the data were sampled at 16 kHz sampling rate in 16 bit linear PCM format. All music and broadcast data for test were split into 15 second long segments in average to avoid the effects which data length may have on discrimination performances. The numbers of speech sentences and 15 second long segments for discrimination test are shown in Table 1.

**Table 1.** Number of data for discrimination test

| Test set | | # data |
|---|---|---|
| | Speech | 3,740 |
| RDCD | Music | 2,649 |
| | Total | 6,389 |
| | Speech | 265 |
| BRDC | Music | 339 |
| | Total | 604 |
| | Speech | 4,005 |
| Total | Music | 2,988 |
| | Total | 6,993 |

# 4   Experimental Results and Discussion

## 4.1   Experimental Setup

We have used single mixture Gaussian mixture model (GMM) as a discriminator, and the HTK [18] for the experimental platform. We have used 25 ms Hamming window and 10 ms frame shift to extract 24 filterbank analysis results and 12 MFCC results. The order of filterbank 24 is like that of the filterbank for extracting 12 MFCC results. As frame shift is in the 10 ms unit, the second level sampling rate to extract ME and MCME is 100 Hz. We have used 25 point (= 250 ms long) bandpass filter based on DFT and 10 frame shift (= 100 ms frame shift) to extract ME and MCME. In this way, we have finally got 10 features per second.

## 4.2 Results and Discussion

To verify the effectiveness of MCME compared with ME, we have carried out discrimination experiments in the range of 4 Hz to 20 Hz modulation energy. The results are shown in Table 2. ME shows its best performance at 4 Hz, but its performances deteriorate as modulation frequency changes, especially over 8 Hz. The overall performance of MCME is better than ME's and stabler in spite of changes in modulation frequency. MCME yields the best accuracy at modulation frequency of 8 Hz. MCME yields discrimination error reduction rate of 72% in average compared with ME. Especially, 8 Hz MCME yields reduction rate of 71% compared with 4 Hz ME, and reduction rate of 74% compared with 8 Hz ME. To illustrate the fact that 8 Hz MCME yields better performance than 4 Hz MCME, we depict boxplots of 4 Hz ME, 4Hz MCME and 8 Hz MCME generated from training data in Fig. 2. We can see that speech and music data are separated more clearly in MCME than in ME. Although the means of speech data

**Table 2.** Speech/Music discrimination error rate (%) of ME and MCME

| Feature | Test set | | Modulation Frequency (Hz) | | | | | Average |
|---|---|---|---|---|---|---|---|---|
| | | | 4 | 8 | 12 | 16 | 20 | |
| ME | RDCD | Speech | 0.0 | 0.1 | 0.1 | 0.0 | 0.3 | 0.10 |
| | | Music | 1.8 | 2.5 | 5.2 | 6.2 | 6.3 | 4.40 |
| | | Average | 0.90 | 1.30 | 2.65 | 3.10 | 3.30 | 2.25 |
| | BRDC | Speech | 17.4 | 10.9 | 13.6 | 27.9 | 42.3 | 22.42 |
| | | Music | 3.2 | 3.8 | 4.1 | 2.9 | 2.7 | 3.34 |
| | | Average | 10.30 | 7.35 | 8.85 | 15.40 | 22.50 | 12.88 |
| | Total | Speech | 1.1 | 0.8 | 0.9 | 1.9 | 3.1 | 1.56 |
| | | Music | 2.0 | 2.7 | 5.1 | 5.8 | 5.9 | 4.30 |
| | | Average | 1.55 | 1.75 | 3.00 | 3.85 | 4.50 | 2.93 |
| MCME | RDCD | Speech | 0.0 | 0.0 | 0.0 | 0.0 | 0.1 | 0.02 |
| | | Music | 1.3 | 0.5 | 1.2 | 1.4 | 1.4 | 1.16 |
| | | Average | 0.65 | 0.25 | 0.60 | 0.70 | 0.75 | 0.59 |
| | BRDC | Speech | 0.4 | 0.8 | 0.4 | 1.9 | 5.7 | 1.84 |
| | | Music | 4.7 | 3.8 | 4.1 | 4.1 | 5.0 | 4.34 |
| | | Average | 2.55 | 2.30 | 2.25 | 3.00 | 5.35 | 3.09 |
| | Total | Speech | 0.0 | 0.0 | 0.0 | 0.1 | 0.5 | 0.12 |
| | | Music | 1.7 | 0.9 | 1.6 | 1.7 | 1.8 | 1.54 |
| | | Average | 0.85 | 0.45 | 0.80 | 0.90 | 1.15 | 0.83 |

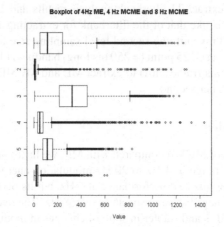

1: 4 Hz ME (speech), 2: 4 Hz ME (music), 3: 4 Hz MCME (speech),
4: 4 Hz MCME (music), 5: 8 Hz MCME (speech), 6: 8 Hz MCME (music)

**Fig. 2.** Boxplots of 4 Hz ME, 4 Hz MCME and 8 Hz MCME

**Table 3.** Speech/Music discrimination error rate(%) of MFCC based CF, compared with 4 Hz ME and 8 Hz MCME

| Test set | | CF | 4 Hz ME | 8 Hz MCME |
|---|---|---|---|---|
| RDCD | Speech | 0.1 | 0.0 | 0.0 |
| | Music | 1.5 | 1.8 | 0.5 |
| | Average | 0.80 | 0.90 | 0.25 |
| BRDC | Speech | 0.4 | 17.4 | 0.8 |
| | Music | 4.1 | 3.2 | 3.8 |
| | Average | 2.25 | 10.30 | 2.30 |
| Total | Speech | 0.1 | 1.1 | 0.0 |
| | Music | 1.8 | 2.0 | 0.9 |
| | Average | 0.95 | 1.55 | 0.45 |

and music data in 8 Hz MCME have lower values than in 4 Hz MCME, 8 Hz MCME has narrower range of fluctuation. In the case of music, the range of outliers in 8 Hz MCME is more significantly narrowed than in 4 Hz MCME. Taking these into consideration, we think that 8 Hz MCME is a good feature parameter to discriminate speech and music signals.

We have carried out another experiment to compare MCME with CF which is frequently used in SMD that does not use ME based feature. CF, used in this paper, is defined as:

$$CF[n] = \frac{1}{P-1} \sum_{p=1}^{P-1} \sqrt{\sum_{l=0}^{L-1} (C[n+p,l] - C[n,k])^2} \qquad (7)$$

where $P$ is the frame size to calculate cepstral differences and $L$ is the order of cepstrum. Although LPC cepstrum is used to extract CF in [5], we have used MFCC and we set 25 for $P$ and 12 for $L$ for fair comparison with MCME. The results are shown in Table 3. Although CF shows good performance compared with ME, it shows worse performance than the proposed MCME. As the discrimination error rate reduction of 53% of 8 Hz MCME when compared with CF shows, the effectiveness of the proposed MCME seems to be clear.

# 5  Conclusion

In this paper, we have proposed to use MCME to improve the performance of ME. MCME is extracted from the time trajectory of MFCC, while ME is based on the spectrum. As cepstral coefficients are mutually uncorrelated, MCME performs better than ME. In the experiments, 8 Hz MCME shows the best discrimination performance, and it yields discrimination error reduction rate of 71% compared with 4 Hz ME. Compared with the MFCC based CF, it shows error reduction rate of 53%. Although MCME shows good performance for speech data, it yields slightly worse performance for music data. In near future, we expect to be able to improve discrimination accuracy by adding feature(s) which takes characteristics of music data into consideration.

# References

1. Scheirer, E., Slaney, M.: Construction and evaluation of a robust multifeature speech/music discriminator. In: Proc. ICASSP-97, vol. 2, pp. 1331–1334 (1997)
2. Lu, L., Jiang, H., et al.: A robust audio classification and segmentation method. In: Proc. 9th ACM Multimedia, pp. 203–211 (2001)
3. Saunders, J.: Real-time discrimination of broadcast speech/music. In: Proc. ICASSP-96, vol. 2, pp. 993–996 (1996)
4. Houtgast, T., Steeneken, H.J.M.: The modulation transfer function in room acoustics as a predictor of speech intelligibility. Acoustica 28, 66–73 (1973)
5. Asano, T., Sugiyama, M.: Segmentation and classification of auditory scenes in time domain. In: Proc. IWHIT98, pp. 13–18 (1998)
6. Ajmera, J., McCowan, I., et al.: Speech/music discrimination using entropy and dynamism features in a HMM classification framework. Speech communication 40(3), 259–430 (2003)
7. Žibert, J., Pavešić, N., Mihelič, F.: Speech/non-speech segmentation based on phoneme recognition features. EURASIP Journal on Applied Signal Processing, Article ID 90945 2006(6), 1–13 (2006)
8. Pinquier, J., Rouas, J.L.: A fusion study in speech/music classification. In: Proc. ICME '03, vol. 1, pp. I-409–412 (2003)
9. Karnebäck, S.: Discrimination between speech and music based on a low frequency modulation feature. In: Proc. Eurospeech-2001, pp. 1891–1894 (2001)
10. Eronen, A., Klapuri, A.: Musical instrument recognition using cepstral coefficients and temporal features. In: Proc. ICASSP '00, vol. 2, pp. II753–II756 (2000)
11. Kingsbury, B.E.D., Morgan, N., et al.: Robust speech recognition using the modulation spectrogram, Speech Communication, 25(1-3) (1998)
12. Tyagi, V., McCowan, I., et al.: On Factorizing Spectral Dynamics for Robust Speech Recognition. In: Proc. Eurospeech-2003, pp. 981–984 (2003)

13. Tyagi, V., McCowan, I., et al.: Mel-cepstrum modulation spectrum (MCMS) features for robust ASR. In: Proc. ASRU '03, pp. 399–404 (2003)
14. SiTEC (Speech Information Technology and Industry Promotion Center), http://www.sitec.or.kr
15. Garofolo, John, S., et al.: TIMIT Acoustic-Phonetic Continuous Speech Corpus, Linguistic Data Consortium (LDC), Philadelphia, USA
16. Goto, M.: Development of the RWC Music Database. In: Puntonet, C.G., Prieto, A.G. (eds.) ICA 2004. LNCS, vol. 3195, Springer, Heidelberg (2004)
17. KBS (Korea Broadcasting System), http://www.kbs.co.kr
18. HTK (Hidden Markov Model Toolkit), http://htk.eng.cam.ac.uk

# Parameterization of the Input in Training the HVS Semantic Parser

Jan Švec[1], Filip Jurčíček[1], and Luděk Müller[2]

[1] Center of Applied Cybernetics, University of West Bohemia,
Pilsen, 306 14, Czech Republic
{honzas,filip}@kky.zcu.cz
[2] Department of Cybernetics, Faculty of Applied Sciences,
University of West Bohemia,
Pilsen, 306 14, Czech Republic
muller@kky.zcu.cz

**Abstract.** The aim of this paper is to present an extension of the hidden vector state semantic parser. First, we describe the statistical semantic parsing and its decomposition into the semantic and the lexical model. Subsequently, we present the original hidden vector state parser. Then, we modify its lexical model so that it supports the use of the input sequence of feature vectors instead of the sequence of words. We compose the feature vector from the automatically generated linguistic features (lemma form and morphological tag of the original word). We also examine the effect of including the original word into the feature vector. Finally, we evaluate the modified semantic parser on the Czech Human-Human train timetable corpus. We found that the performance of the semantic parser improved significantly compared with the baseline hidden vector state parser.

## 1 Introduction

This article concerns statistical semantic parsing which we interpret as a search process of the sequence of semantic tags $C = (c_1, c_2, \ldots, c_t)$ that maximizes the a posteriori probability $P(C|W)$ given the input sequence of words $W = (w_1, w_2, \ldots, w_t)$. This process can be described as:

$$C^* = \arg\max_C P(C|W) = \arg\max_C P(W|C)P(C) \tag{1}$$

where the probability $P(C)$ is called the semantic model and the probability $P(W|C)$ is called the lexical model. The simplest implementation of this process is Pieraccini's semantic finite state tagger (FST) which was used for the ATIS task [1]. The main disadvantage of this tagger is its inability to capture either long distance dependencies or a hierarchical structure of the processed utterance.

Several attempts has been proposed to overcome these disadvantages. In [2] He and Young described a HMM based parser with a hidden state variable implemented as a stack and where the state transitions are modeled using pushdown operations on this stack. The stack (called also vector state) stores the information about the hierarchical

V. Matoušek and P. Mautner (Eds.): TSD 2007, LNAI 4629, pp. 415–422, 2007.

structure of a processed sequence of words. They called their parser the hidden vector state (HVS) parser [2].

The HVS parser is able to decode the hierarchical structure of the input sequence because it approximates the pushdown automaton. To estimate the parser parameters one needs a training data set provided with a structured semantic annotation. This structured semantic annotation forms a semantic tree. The nodes of the semantic tree are labeled with semantic concepts, which are considered to be basic units of particular meaning. The annotation must define an ordering relation between the nodes of the semantic tree. The order of nodes must correspond to the order of words in the underlying sentence. As a result, any semantic tree can be expressed as a sequence of vectors containing semantic concepts.

To alleviate the annotation stage the training data could be only weakly annotated by so-called abstract semantic trees. The abstract semantic tree does not provide the explicit word alignment between the nodes of the tree and the words of the sentence. Due to this merit the abstract annotation represents a robust annotation scheme to the intent to obtain a high inter-annotator agreement. We write the abstract semantic tree in the parenthesized form, which express both the parent/child relationships and the ordering of nodes in the semantic tree. For example, the sentence *Jede nějaký spěšný vlak do Prahy kolem čtvrté odpoledne?*[1] could be annotated with the semantic tree showed in Fig. 1. The corresponding abstract semantic tree is DEPARTURE( TO( STATION ), TIME ). Both the semantic and the abstract semantic trees can be converted into a sequence of vectors as it is shown at the bottom of the figure.

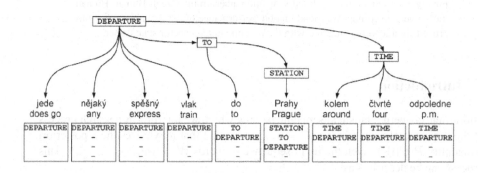

**Fig. 1.** An example of a full semantic parse tree with the corresponding stack sequence

Later in this paper, we present an extension of the original HVS parser which we call the *input parameterization of the HVS parser*. The input of the original HVS parser is a pure sequence of words $W$. However, if the parser has some additional information, it can take advantage of it and use a sequence of feature vectors $S = (s_1, s_2, \ldots, s_T)$ where for every input word $w_t$ one feature vector $s_t = (s_t[1], \ldots, s_t[N])$ is computed. For example, the feature vector can be composed of some prosodic and linguistic features computed for the word $w_t$. In our experiments, we used the combination of the

---

[1] Lit. translation: *Does any express train go to Prague around four p.m.?*.

input word, its lemma, and morphological tag generated by an automatic morphology analyzer.

This article is organized in the following manner: in Section 2 we shortly describe the original HVS parser. In Section 3 we propose a novel method for the input parameterization of the HVS parser. Section 4 provides experimental results and finally, Section 5 concludes this paper.

## 2   The HVS Parser

The HVS parser is an approximation of a pushdown automaton. This is mainly due to the limited stack depth and a reduced set of allowed stack operations. In other words, the HVS parser is the generalization of the finite state tagger. The HVS parser has a larger state space in comparison with Pieraccini's parser and state transitions are modeled using pushdown operations. The HVS parser is able to better capture the hierarchical structure typical of natural language.

The original HVS parser proposed by He and Young uses two stack operations: popping zero to four concepts off the stack and pushing exactly one new concept onto the stack. These operations are implemented in the semantic model which is given by:

$$P(C) = \prod_{t=1}^{T} P(pop_t \mid c_{t-1}) P(c_t[1] \mid c_t[2], c_t[3], c_t[4]) \tag{2}$$

where the hidden variable $pop_t$ is the stack shift operation and takes values in the range $0, \ldots, 4$ and the hidden variable $c_t = (c_t[1], \ldots, c_t[4])$ is the vector state (the stack) of the HVS model. The depth of the stack is chosen to be the maximal depth of semantic trees found in the training data. We observed that the stack of at most four concepts is quite sufficient in all experiments described in Section 4. The concept $c_t[1]$ is the preterminal concept of the word $w_t$ and the concept $c_t[4]$ is the root of semantic tree. The value of $pop_t$ represents the count of concepts to be popped off the stack at time $t$. The value $pop_t = 0$ means that no concept is popped off the stack so the stack $c_t$ grows by one new concept $c_t[1]$. Values greater than zero lead to popping $pop_t$ concepts off the stack and to pushing one new concept $c_t[1]$ onto the stack.

The lexical model imposes an additional constraint on the stack sequence by allowing only such stack sequences that correspond to the input word sequence. The original HVS parser can process one word $w_t$ at time $t$. The lexical model is given by:

$$P(W|C) = \prod_{t=1}^{T} P(w_t|c_t[1, \ldots 4]) \tag{3}$$

Starting with the definition of the HVS parser, we extend the lexical model so that it accepts a sequence of feature vectors (parameterized input) on its input.

## 3   Input Parameterization

The input parameterization extends the HVS parser into a more general HVS parser with the input feature vector (HVS-IFV parser). This parser works on a sequence of feature vectors $S = (s_1, \ldots, s_T)$ instead of a sequence of words $W$. The feature vector is defined as $s_t = (s_t[1], s_t[2], \ldots, s_t[N])$. To every word $w_t$, we assign the fixed number $N$ of features. If we use the feature vector $s_t$ instead of the word $w_t$ in Eq. 3, the lexical model changes as follows:

$$
\begin{aligned}
P(S|C) &= \prod_{t=1}^{T} P(s_t \mid c_t) \\
&= \prod_{t=1}^{T} P(s_t[1], s_t[2], \ldots s_t[N] \mid c_t)
\end{aligned}
\tag{4}
$$

Using the chain rule, we can rewrite this equation into the form:

$$
P(S|C) = \prod_{t=1}^{T} \prod_{i=1}^{N} P(s_t[i] \mid s_t[1, \ldots i-1], c_t)
\tag{5}
$$

To minimize the data sparsity problem, we used the conditional independence assumption between the features $s_t[i]$ and $s_t[j]$, $i \neq j$ given the concept $c_t$. This kind of assumption is also used for example in the naive Bayes classifier to avoid the curse of dimensionality problem. The lexical model of the HVS-IFV parser is then given by:

$$
P(S|C) = \prod_{t=1}^{T} \prod_{i=1}^{N} P(s_t[i] \mid c_t)
\tag{6}
$$

Because the conditional independence assumption is hardly expected to be always true, we need to modify the search process defined in Section 1. Let's assume that for example we have the sequence of the feature vectors $S = (s_t[1], s_t[2])_{t=1}^{T}$ where $s_t[1]$ is equal to $s_t[2]$ for every $t$. The search process is given by:

$$
\begin{aligned}
C^* &= \arg\max_{C} \prod_{t=1}^{T} \left[ \prod_{i=1}^{2} P(s_t[i] \mid c_t) \right] P(pop_t \mid c_{t-1}) P(c_t[1] \mid c_t[2], c_t[3], c_t[4]) \\
&= \arg\max_{C} \prod_{t=1}^{T} \left[ P(s_t[1] \mid c_t) \right]^2 P(pop_t \mid c_{t-1}) P(c_t[1] \mid c_t[2], c_t[3], c_t[4])
\end{aligned}
$$

As we can see, the lexical model probability $P(S|C)$ is exponentially scaled with the factor 2 and it causes imbalance between the lexical and the semantic model depending on the dimension $N$ of the feature vector. Therefore, we use the scaling factor $\lambda$ to compensate the imbalance caused by the use of feature vectors:

$$
C^* = \arg\max_{C} P(S|C) P^\lambda(C)
\tag{7}
$$

The HVS-IFV parser is defined by equations 2, 6, and 7. To find the optimal value of $\lambda$, we use a grid search over the finite set of values to maximize the concept accuracy measure defined in Section 4.2. The grid search is performed on the development data.

# 4   Experiments

The semantic parsers evaluated in this article were trained and tested on the Czech human-human train timetable (HHTT) dialogue corpus [3]. The HHTT corpus consists of 1,109 dialogues completely annotated with the abstract semantic annotations. Both operators and users have been annotated. The corpus comprises 17,900 utterances in total. The vocabulary size of the whole corpus is 2,872 words. There are 35 semantic concepts in the HHTT corpus. The dialogues were divided into training data (798 dialogues - 12,972 dialogue acts, 72%), development data (88 dialogues - 1,418 dialogue acts, 8%), and test data (223 dialogues - 3,510 dialogue acts, 20%). The development data were used for finding the optimal value of the semantic model scaling factor.

The training of the HVS parser is divided into three parts: 1) initialization of the semantic and lexical models, 2) estimation of parameters of the semantic and lexical models, 3) smoothing the probabilities of the semantic and lexical models. All probabilities are initialized uniformly. To estimate the parameters of the semantic and the lexical model, it is necessary to use the expectation-maximization (EM) algorithm because abstract semantic annotations does not provide fully annotated parse trees. There are no explicit word level annotations. After training the parameters we smooth all three probabilities using the back-off model.

We evaluated our experiments using two measures: semantic accuracy ($SAcc$) and concept accuracy ($CAcc$). We could not use the PARSEVAL measures [4] because they rely on availability of full parse trees of the test data. As we already mentioned above, the HHTT corpus has no explicit word level annotation.

## 4.1   Semantics Accuracy

Two semantic annotations are considered equal only if they exactly match each other. Exact match is very tough standard because under the exact match the difference between semantics ARRIVAL( TIME, FROM( STATION ) ) and ARRIVAL( TIME, TO( STATION ) ) is equal to the difference between semantics ARRIVAL( TIME, FROM( STATION ) ) and DEPARTURE( TRAIN_TYPE ). The semantic accuracy of a hypothesis is defined as

$$SAcc = \frac{E}{N} \cdot 100\% \tag{8}$$

where $N$ is the number of evaluated semantic annotations and $E$ is the number of hypothesis semantic annotations which exactly match the reference.

## 4.2   Concept Accuracy

Similarity scores between the reference and the hypothesis semantics can be computed also by the tree edit distance algorithm [5]. The tree edit distance measures the similarity between two trees by comparing subtrees of both the reference and the hypothesis annotations.

The tree edit distance algorithm computes the minimum number of substitutions, deletions, and insertions required to transform one tree into another one and uses the dynamic programing. The operations act on the tree nodes and modify the tree by changing the parent/child relationships of the tree. The tree edit distance is convenient for measuring similarity between two abstract semantic annotations because it does not rely on the alignment of the annotation and the underlying word sequence. We define the concept accuracy as

$$CAcc = \frac{N - S - D - I}{N} \cdot 100\% \tag{9}$$

where $N$ is the number of concepts in the reference semantic annotation and $S$, $D$, and $I$ are the numbers of substitutions, deletions, and insertions, respectively, in the minimum-cost alignment of the reference and the hypothesis semantic annotation.

### 4.3 Morphological Analysis

Every utterance of the HHTT training corpus was automatically processed using the morphological analyser and the tagger from Prague Dependency Treebank (PDT) [6]. The morphological analysis for every input word generated a lemma and a morphological tag which were then used as features in the parameterized input of the HVS parser. The morphological tag is represented as a string of several symbols, each for one morphological category: part-of-speech, detailed part-of-speech, gender, number, case, possessor's gender, possessor's number, person, tense, degree of comparison, negation, and voice. We have also done some experiments with reduced morphological tags (we removed some morphological categories from the morphological tag) but these experiments bring no improvement in concept accuracy. Because Czech has very rich inflection, the lemmatization of input sentence reduced the vocabulary. The vocabulary size for every input feature is shown in Table 1.

### 4.4 Results

Table 1 shows the performance of the original HVS parser described in Section 2 with words, lemmas, and morphological tags on its input. The original HVS parser with words on its input was chosen as the baseline. To measure the statistical significance, we used the paired $t$-test (NIST MAPSSWE test) and the difference was taken as significant if $p$-value of this test was $< 0.01$.

We combined these features and passed their combination to the input of HVS-IFV parser described in Section 3. The performance on the development data set is shown in Table 2. We can see, that the HVS-IFV parser whose feature vector contains besides words also their lemmas yields better performance. Table 3 reports the performance comparison of the original HVS parser and the best HVS-IFV parser evaluated both on the development and the test data sets.

**Table 1.** The performance and the vocabulary size of the HVS parser with different inputs evaluated on the development data

| Input feature | Vocab. size | $CAcc$ | $SAcc$ | $p$-value |
|---|---|---|---|---|
| words (baseline) | 696 | 67.0 | 52.8 | |
| lemmas | 551 | 68.3 | 53.4 | $< 0.01$ |
| morph. tags | 225 | 42.2 | 30.5 | $< 0.01$ |

**Table 2.** The performance of the HVS-IFV parser with different inputs evaluated on the development data

| Input features | SAcc | CAcc | p-value |
|---|---|---|---|
| words, m. tags | 69.3 | 55.3 | < 0.01 |
| lemmas, m. tags | 70.3 | 56.2 | < 0.01 |
| words, lemmas | 73.1 | 58.2 | < 0.01 |
| words, lemmas, m. tags | 72.2 | 58.1 | < 0.01 |

**Table 3.** The performance of the baseline HVS and developed HVS-IFV parsers. The parsers were evaluated on the test and the development data.

| Parser Type | Test data | | | Development data | | |
|---|---|---|---|---|---|---|
| | CAcc | SAcc | p-value | CAcc | SAcc | p-value |
| HVS (baseline) | 64.9 | 50.4 | | 67.0 | 52.8 | |
| HVS-IFV (words, lemmas) | 69.4 | 57.0 | < 0.01 | 73.1 | 58.2 | < 0.01 |

# 5 Conclusion

In this work, we presented a modification of the HVS parser. The proposed HVS-IFV parser is able to parse the sequence of feature vectors. We started with the original lexical model of the HVS parser and we used the assumption of conditional independence of the input features to simplify the parser's lexical model. We showed that the conditional independence assumption affects the search process. We used the exponential scaling of the semantic model probability to correct the imbalance between the lexical and the semantic model of the HVS-IFV parser.

We processed the training corpus with the automatic morphology analyzer and tagger. We used original words, their lemmas, and morphological tags as the input features of the HVS-IFV parser. The best performance was achieved with the feature vector composed of words and lemmas. This approach significantly increases the performance of the semantic parser. We believe that the improvements come mainly from the ability of lemmas to cluster the original word vocabulary into classes with the same meaning and in this way make the model more robust. However, we found that in some cases it is still useful to include the words into the feature vector because the words help to distinguish the ambiguous cases where two words with the different meaning has the same lemma. In total, $SAcc$ was significantly increased from 50.4% to 57.0% and $CAcc$ from 64.9% to 69.4% measured on the test data.

## Acknowledgments

This work was supported by the Ministry of Education of the Czech Republic under project No. 1M0567 (CAK).

# References

1. Hemphill, C.T., Godfrey, J.J., Doddington, G.R.: The ATIS spoken language systems pilot corpus. In: Proceedings of DARPA Speech and Natural Language Workshop, Hidden Valley, PA, USA, pp. 96–101 (1990)
2. He, Y., Young, S.: Hidden vector state model for hierarchical semantic parsing. In: Proceedings of ICASSP, Hong Kong (2003)
3. Jurčíček, F., Zahradil, J., Jelinek, L.: A Human-Human Train Timetable Dialogue Corpus. In: Proceedings of EUROSPEECH, Lisboa, Portugal (2005)
4. Black, E., Abney, S., Flickinger, D., Gdaniec, C., Grishman, R., Harrison, P., Hindle, D., Ingria, R., Jelinek, F., Klavans, J., Liberman, M., Marcus, M., Roukos, S., Strzalkowski, S.T.: A procedure for quantitatively comparing the syntactic coverage of english grammars. In: Proceedings of the 1990 DARPA Speech and Natural Language Workshop, Pacific Grove, CA, pp. 306–311 (1990)
5. Klein, P.: Computing the edit-distance between unrooted ordered trees. In: Bilardi, G., Pietracaprina, A., Italiano, G.F., Pucci, G. (eds.) ESA 1998. LNCS, vol. 1461, pp. 91–102. Springer, Heidelberg (1998)
6. Hajič, J.: Disambiguation of Rich Inflection (Computational Morphology of Czech). Karolinum Press, Charles University, Prague (2001)

# A Comparison Using Different Speech Parameters in the Automatic Emotion Recognition Using Feature Subset Selection Based on Evolutionary Algorithms

Aitor Álvarez, Idoia Cearreta, Juan Miguel López, Andoni Arruti,
Elena Lazkano, Basilio Sierra, and Nestor Garay

Computer Science Faculty (University of the Basque Country)
Manuel Lardizabal 1, E-20018 Donostia (Gipuzkoa), Spain
aalvarez031@ikasle.ehu.es

**Abstract.** Study of emotions in human-computer interaction is a growing research area. Focusing on automatic emotion recognition, work is being performed in order to achieve good results particularly in speech and facial gesture recognition. This paper presents a study where, using a wide range of speech parameters, improvement in emotion recognition rates is analyzed. Using an emotional multimodal bilingual database for Spanish and Basque, emotion recognition rates in speech have significantly improved for both languages comparing with previous studies. In this particular case, as in previous studies, machine learning techniques based on evolutive algorithms (EDA) have proven to be the best emotion recognition rate optimizers.

## 1 Introduction

Affective computing, a discipline that develops devices for detecting and responding to users' emotions [20], is a growing research area [22]. The main objective of affective computing is to capture and process affective information with the aim of enhancing and naturalizing the communication between the human and the computer.

Development of affective systems is a challenge that involves analysing different multimodal data sources. A large amount of data is needed in order to include a wide range of emotionally significant material. Affective databases are a good chance for developing such applications, either for affective recognizers or either for affective synthesis.

This papers presents a study aimed at giving a new step towards researching relevant speech parameters in automatic speech recognition area. Based on the same classification techniques used in [1], where basic speech parameters were used, a wide group of new parameters have been used in order to analyse if emotion recognition rates have improved. This work has also served to check the efficiency of these new parameters in the study of emotions.

After a brief review on related work, corpus, speech parameters and machine learning techniques used for this study are detailed. Achieved experimental results are shown next. Finally, some conclusions and future work are highlighted.

V. Matoušek and P. Mautner (Eds.): TSD 2007, LNAI 4629, pp. 423–430, 2007.

## 2 Related Work

Affective resources, such as affective databases, provide a good opportunity for training affective applications, either for affective synthesis or for affective recognizers based on classification via artificial neural networks, Hidden Markov Models, genetic algorithms, or similar techniques (e.g., [2, 8]). There are some references in the literature that present affective databases and their characteristics. [4] carried out a wide review of affective databases. Other interesting reviews are the ones provided in [11] and [16].

Most of these references are related to English, while other languages have less resources developed, especially the ones with relatively low number of speakers. This is the case of Standard Basque. To our knowledge, the first affective database in Standard Basque is the one presented by [18]. Concerning to Spanish, the work of [12] stands out.

This type of databases usually record information such as images, sounds, psychophysiological values, etc. RekEmozio database is a multimodal bilingual database for Spanish and Basque [17], which also restores information came from processes of some global speech parameters extraction for each audio recording. Some of these parameters are prosodic features while others are quality features.

Machine Learning (ML) paradigms take a principal role in some works related to emotion recognition that can be found in literature. [5] presented a good reference paper. The Neural Networks Journal devoted special issue to emotion treatment from a Neural Networks perspective [23]. The work by [4] is related with this paper in the sense of using a Feature Selection method in order to apply a Neural Network to emotion recognition in spoken English, although both the methods to perform the FSS and the paradigms are different. In this line it has to be pointed out the work by [10] which uses a reduced number of emotions and a greedy approach to select the features.

## 3 Study of Automatic Emotion Recognition Relevant Parameters Using Machine Learning Paradigms

### 3.1 Corpus

The emotions used were chosen based on Ekman's six basic emotions [6], and neutral emotion was added. RekEmozio database has been used for this study. Its characteristics are described in [17].

### 3.2 Emotional Feature Extraction

One of the most important questions for automatic recognition of emotions in speech is which features should be extracted from the voice signal. Previous studies show that it is difficult to find specific voice features that could be used as reliable indicators of the emotion present in the speech [14]. In order to improve the results of previous works, new parameters have been calculated using the same recordings.

These parameters have been collected from the work carried out of [7] and consist of a total of 91. Parameters are divided in this way:

### 3.2.1  Prosodic Features
Many parameters have been computed that model the F0, energy, voiced and unvoiced regions, pitch derivative curve and the relations between the features as proposed in [3, 21].

- F0 based features: Values of the F0 curve in the voiced parts. Maximum, position of the maximum, minimum, position of the minimum, mean, variance, regression coefficient and its mean square error are computed.
- Energy: maximum and its position curve in the whole utterance, minimum and its relative time position, mean, variance, regression coefficient and its mean square error values are computed.
- Voiced/unvoiced regions based features: F0 value of the first and last voiced frames, number of regions with more than three successive voiced and unvoiced frames, amount of voiced and unvoiced frames in the utterance, length of the longest voiced and unvoiced regions, ratio of number of voiced and voiced regions and the ratio of number of voiced and unvoiced frames in the whole utterance are computed.
- Pitch contour derivative based features: Based on the derivative of the F0, maximum, minimum, mean, variance, regression coefficient and its mean square error values are computed.
- Relations among features: Mean, variance, mean of the maximum, variance of the maximum, mean of the pitch ranges and mean of the flatness of the pitch based on every voiced region pitch values are computed. The pitch increasing and decreasing in voiced parts and in the whole utterance is also measured as well as the mean of the voiced regions duration. Many features related with the energy among the voiced regions have been taken account, like global energy mean, vehemence, mean of the flatness and tremor in addition to others.

### 3.2.2  Quality Features
Features have been computed related with the voice quality parameters, such as formants, energy band distribution, harmonicity to noise ratio and active level in speech.

- Formant frequency based features: Mean of the first, second and third formant frequencies and their bandwidths among all voiced region are computed as well as the maximum and the range of the second formant ratio.
- Energy band distribution: The four frequency bands used are the following: 0 Hz to F0 Hz, 0Hz to 1000 Hz, 2500 Hz to 3500 Hz and 4000 Hz to 5000 Hz. The energy contained in the corresponding band is calculated for all voiced parts and divided by the energy over all frequencies of the voiced parts of utterance. The longest region is calculated and the energy values are also computed in that region as well as rate and relative energy contained in voiced regions and energy over all utterance.

- <u>Harmonicity to noise ratio</u>: The ratio of the energy of the harmonic frames to the energy of the remaining part of the signal is computed. In this sense, maximum harmonicity, minimum, mean, range and standard deviation have been analysed.
- <u>Active level features</u>: Maximum, minimum, mean and variance of the speech active level among the voiced regions are computed.

### 3.3 Machine Learning Standard Paradigms Used

The models that have been constructed to solve this problem of classification made up of 91 speech related values for each sample are constructed for the previous study, while the label value is one of the seven emotions identified. Therefore, *Decision Trees, Instance-Based Learning* and *Naive Bayes* compose the sort of classifiers.

Most of the supervised learning algorithms perform rather poorly when faced with many irrelevant or redundant features (depending on the specific characteristics of the classifier). Bearing it in mind, Feature Subset Selection (FSS) [15] is applicated, as it can be reformulated as follows: *given a set of candidate features, select the 'best' subset in a classification problem.* In this case, the 'best' subset will be the one with the best predictive accuracy. The FSS proposes additional methods to reduce the number of features so as to improve the performance of the supervised classification algorithm.

For applicating FSS, an Estimation of Distribution Algorithm (EDA) [19] has been used having model accuracy as fitness function. It is necessary to clarify that given the number of dimensions of the classification problem (search space for EDA is $2^{91}=2,4758E27$), it is not possible to consider all the possibilities.

## 4   Experimental Results

The above mentioned methods have been applied over the crossvalidated data sets using the MLC++ library [13]. Each dataset corresponds to a single actor. Experiments were carried out with and without FSS in order to extract the accuracy improvement introduced by the feature selection process and then comparated with the results obtained in the previous study. The first two tables show classification results obtained using the whole set of variables, for Basque and Spanish languages respectively. First column presents used Machine Learning paradigms (as Decision Trees classifiers, ID3 and C4.5 paradigms; as Instance-Based Learning classifiers, IB paradigm; and finally, Naive Bayes classifier (NB) have been used). Last column presents the total average for each classifier. Each remaining column represents a female (Fi) or male (Mi) actor, and mean values corresponding to each classifier/gender are also included.

Results do not seem very impressive. In fact, as it can be see in Figure 1, results from previous studies are not improved, except for one case. For both Basque and Spanish languages, ID3 best classifies emotions for female actresses, as well as for male actors for Basque. IB appears as the best classifier for Spanish for male actors.

**Table 1.** 10-fold crossvalidation accuracy for Basque using the whole variable set

|      | Female |       |       |       | Male  |       |       |       |       | Total |
|------|--------|-------|-------|-------|-------|-------|-------|-------|-------|-------|
|      | F1     | F2    | F3    | mean  | M1    | M2    | M3    | M4    | mean  |       |
| IB   | 34.00  | 42.91 | 33.91 | 36.94 | 56.18 | 41.00 | 36.91 | 36.82 | 42.73 | 40.25 |
| ID3  | 49.45  | 45.91 | 46.78 | **47.38** | 54.27 | 44.00 | 51.45 | 49.45 | **49.79** | **48.75** |
| C4.5 | 42.73  | 40.09 | 42.73 | 41.85 | 60.36 | 39.55 | 48.45 | 37.82 | 46.55 | 44.54 |
| NB   | 39.82  | 31.00 | 46.45 | 39.09 | 60.36 | 29.91 | 36.91 | 41.44 | 42.16 | 40.84 |

**Table 2.** 10-fold crossvalidation accuracy for Spanish using the whole variable set

|      | Female |       |       |       |       |       | Male  |       |       |       |       |       | Total |
|------|--------|-------|-------|-------|-------|-------|-------|-------|-------|-------|-------|-------|-------|
|      | F1     | F2    | F3    | F4    | F5    | mean  | M1    | M2    | M3    | M4    | M5    | mean  |       |
| IB   | 36.46  | 41.92 | 41.92 | 43.64 | 33.64 | 39.52 | 30.00 | 36.46 | 44.55 | 36.46 | 30.00 | **35.49** | 37.51 |
| ID3  | 38.18  | 47.27 | 55.45 | 43.64 | 44.55 | **45.82** | 24.55 | 40.00 | 50.00 | 46.36 | 34.55 | 39.09 | **42.46** |
| C4.5 | 42.73  | 48.18 | 50.91 | 50.91 | 45.45 | 47.64 | 21.82 | 39.09 | 46.36 | 48.18 | 27.27 | 36.54 | 42.00 |
| NB   | 34.55  | 34.45 | 40.91 | 32.73 | 31.82 | 34.89 | 20.91 | 39.09 | 40.00 | 35.45 | 21.82 | 31.45 | 33.17 |

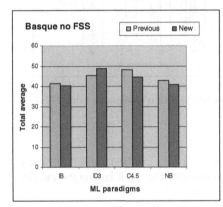
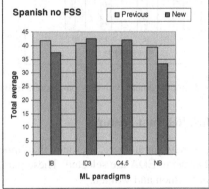

**Fig. 1.** Comparison between previous and new results in all classifiers using the whole variable set

Results obtained after applying FSS are more appealing, as it can be seen in Tables 3 and 4. In fact, there is a substantial improvement in all cases with regard to previous results. IB classifier appears once again as the best paradigm for all categories, both female and male, and Basque and Spanish languages. Moreover, as it can be seen in Figure 2, accuracies outperform previous ones between 6 and 10%. It must also be highlighted once more that FSS improves the well classified rate for all ML paradigms.

**Table 3.** 10-fold crossvalidation accuracy for Basque using FSS

|      | Female |       |       |       | Male  |       |       |       |       | Total |
|------|--------|-------|-------|-------|-------|-------|-------|-------|-------|-------|
|      | F1     | F2    | F3    | mean  | M1    | M2    | M3    | M4    | mean  |       |
| IB   | 72.55  | 79.73 | 62.27 | **71.52** | 91.36 | 73.00 | 77.82 | 71.82 | **78.50** | **75.50** |
| ID3  | 71.00  | 71.73 | 66.64 | 69.79 | 78.73 | 65.82 | 72.64 | 66.91 | 71.03 | 70.50 |
| C4.5 | 67.73  | 75.91 | 68.09 | 70.58 | 76.73 | 65.82 | 69.91 | 68.91 | 70.34 | 70.44 |
| NB   | 73.00  | 77.73 | 63.36 | 71.36 | 89.45 | 67.27 | 66.18 | 65.36 | 72.07 | 71.76 |

Table 4. 10-fold crossvalidation accuracy for Spanish using FSS

| | Female | | | | | | Male | | | | | | Total |
|---|---|---|---|---|---|---|---|---|---|---|---|---|---|
| | F1 | F2 | F3 | F4 | F5 | **mean** | M1 | M2 | M3 | M4 | M5 | **mean** | |
| IB | 72.73 | 72.73 | 80.91 | 76.36 | 64.55 | **73.46** | 58.18 | 72.73 | 76.36 | 70.00 | 62.73 | **68.00** | **70.73** |
| ID3 | 67.27 | 75.45 | 73.64 | 72.73 | 68.18 | 71.45 | 51.82 | 63.64 | 76.36 | 69.09 | 59.09 | 64.00 | 67.72 |
| C4.5 | 70.91 | 75.45 | 74.55 | 64.55 | 66.36 | 70.36 | 54.55 | 63.64 | 80.91 | 66.36 | 56.36 | 64.36 | 67.35 |
| NB | 75.45 | 73.64 | 68.18 | 67.27 | 64.55 | 69.82 | 50.00 | 60.00 | 76.36 | 68.18 | 58.18 | 62.54 | 66.18 |

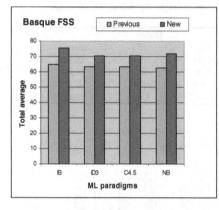 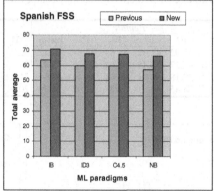

**Fig. 2.** Improvement with new parameters in Basque and Spanish languages using FSS

## 4.1 Most Relevant Features

In case of new parameters, the ones that appear most times in best subsets for each actor after using EDA are selected. These parameters consider both languages and are different for men and women.

- Most used parameters for men (Basque and Spanish languages): *regression coefficient for F0 and its mean square error, variance of the pitch values over the voiced regions, mean of the pitch means in every voiced regions, variance of the pitch means in every voiced regions, mean of the flatness of the pitch for every voiced regions, relation between the maximum of the energy in all voiced regions and the maximum of the utterance, maximum of the energy curve, slope coefficient of the regression line for the energy curve and its mean square error, variance of the energy curve, F0 value for the first voiced frame, number of voiced frames in the utterance, maximum, mean and variance of the active level in the speech signal over the voiced regions.*
- Most used parameters for women (Basque and Spanish languages): *mean F0 value calculated over the voiced regions of the utterance, relation between the maximum of the energy in all voiced regions and the maximum of the utterance, tremor or number of zero-crossings over a window of the energy curve derivative, maximum and minimum of the energy curve in the whole utterance, regression coefficient for the energy curve values and its mean square error, mean and variance of the energy values over the whole utterance, ratio of number of voiced frames vs.*

*number of all frames, mean of the harmonicity to noise ratio, mean and variance of the active level in speech signal in every voiced regions.*

## 5   Conclusions and Future Work

Affective databases have been very useful for developing affective computing systems, being primarily used for training affective recognition systems. RekEmozio database is being used to training some automatic recognition systems applied to the localization where authors make their research.

Emotion recognition rates have improved using the parameters defined in this paper, but it must also be taken into account that such improvement has been achieved after applying EDA for FSS. In the future, new voice features related to emotions are expected be taken into account, with the aim of improving current results. Another option in order to improve emotion recognition rate would be to merge all parameters in one single classification.

This paper also describes how results obtained by Machine Learning techniques applied to emotion classification can be improved automatically by selecting an appropriate subset of classifying variables by FSS. Classification accuracies, although not very impressive yet, are clearly improved over the results obtained using the full set of variables. Still, an analysis of the features selected by FSS is required as an effort to extract meaningful information from selected set. Merging or combining information from multiple sources by means of a multiclassifier model [9] can help obtaining better classification accuracies.

## Acknowledgements

The involved work has received financial support from the Department of Economy of the local government "Gipuzkoako Foru Aldundia" and from the University of the Basque Country.

## References

1. Álvarez, A., Cearreta, I., López, J.M., Arruti, A., Lazkano, E., Sierra, B., Garay, N.: Feature Subset Selection based on Evolutionary Algorithms for automatic emotion recognition in spoken Spanish and Standard Basque languages. In: Sojka, P., Kopeček, I., Pala, K. (eds.) TSD 2006. LNCS (LNAI), vol. 4188, pp. 565–572. Springer, Heidelberg (2006)
2. Athanaselis, T., Bakamidis, S., Dologlou, I., Cowie, R., Douglas-Cowie, E., Cox, C.: Asr for emotional speech: clarifying the issues and enhancing performance. Neural Networks 18, 437–444 (2005)
3. Batliner, A., Fisher, K., Huber, R., Spilker, J., Nöth, E.: Desperately Seeking Emotions: Actors, Wizards, and Human Beings. In: Cowie, R., Douglas-Cowie, E., Schröder, Manuela (Hrsg.) Proc. ISCA Workshop on Speech and Emotion: A Conceptual Framework for Research Newcastle, Northern Ireland, pp. 195–200 (September 2000)

4. Cowie, R., Douglas-Cowie, E., Cox, C.: Beyond emotion archetypes: Databases for emotion modelling using neural networks. Neural Networks 18, 371–388 (2005)
5. Dellaert, F., Polzin, T., Waibel, A.: Recognizing Emotion in Speech. In: Proc. of ICSLP'96 (1996)
6. Ekman, P., Friesen, W.: Pictures of facial affect. Consulting Psychologist Press, Palo Alto, CA (1976)
7. Emotion Recognition in Speech Signal: Retrieved (March 30, 2007), from http://lorien.die.upm.es/partners/sony/main.html
8. Fragopanagos, N.F., Taylor, J.G.: Emotion recognition in human-computer interaction. Neural Networks 18, 389–405 (2005)
9. Gunes, V., Menard, M., Loonis, P., Petit-Renaud, S.: Combination, cooperation and selection of classiers: A state of the art. International Journal of Pattern Recognition 17, 1303–1324 (2003)
10. Huber, R., Batliner, A., Buckow, J., Noth, E., Warnke, V., Niemann, H.: Recognition of emotion in a realistic dialogue scenario. In: Proc. ICSLP'00, pp. 665–668 (2000)
11. Humaine: Retrieved (March 26, 2007) (n.d.), from http://emotion-research.net/
12. Iriondo, I., Guaus, R., Rodríguez, A., Lázaro, P., Montoya, N., Blanco, J.M., Bernadas, D., Oliver, J.M., Tena, D., Longhi, L.: Validation of an acoustical modelling of emotional expression in Spanish using speech synthesis techniques. In: SpeechEmotion'00, pp. 161–166 (2000)
13. Kohavi, R., Sommerfield, D., Dougherty, J.: Data mining using MLC++, a Machine Learning Library in C++. International Journal of Artificial Intelligence Tools 6(4), 537–566 (1997), http://www.sgi.com/Technology/mlc/
14. Laukka, P.: Vocal Expression of Emotion. Discrete-emotions and Dimensional Accounts. Acta Universitatis Upsaliensis. Comprehensive Summaries of Uppsala Dissertations from the Faculty of Social Sciences, Uppsala 141, 80 (2004) ISBN 91-554-6091-7
15. Liu, H., Motoda, H.: Feature Selection for Knowledge Discovery and Data Mining. Kluwer Academic Publishers, Dordrecht (1998)
16. López, J.M., Cearreta, I., Fajardo, I., Garay, N.: Validating a multimodal and multilingual affective database. In: To be published in Proceedings of HCI International, Springer, Heidelberg (2007)
17. López, J.M., Cearreta, I., Garay, N., de Ipiña, K.L., Beristain, A.: Creación de una base de datos emocional bilingüe y multimodal. In: Redondo, M.A., Bravo, C., Ortega, M. (eds) Proceedings of the 7th Spanish Human Computer Interaction Conference, Interaccion'06, Puertollano, pp. 55–66 (2006)
18. Navas, E., Hernáez, I., Castelruiz, A., Luengo, I.: Obtaining and Evaluating an Emotional Database for Prosody Modelling in Standard Basque. In: Sojka, P., Kopeček, I., Pala, K. (eds.) TSD 2004. LNCS (LNAI), vol. 3206, pp. 393–400. Springer, Heidelberg (2004)
19. Pelikan, M., Goldberg, D.E., Lobo, F.: A Survey of Optimization by Building and Using Probabilistic Models. Technical Report 99018, IlliGAL (1999)
20. Picard, R.W.: Affective Computing. MIT Press, Cambridge, MA (1997)
21. Tato, R., Santos, R., Kompe, R., Pardo, J.M.: Emotional space improves emotion recognition. In: Hansen, J.H.L., Pellom, B. (eds.) Proceedings of 7th International Conference on Spoken Language Processing (ICSLP'02 – INTERSPEECH'02). Denver, Colorado, USA, pp. 2029–2032 (2002)
22. Tao, J., Tan, T.: Affective computing: A review. In: Tao, J., Tan, T., Picard, R.W. (eds.) ACII 2005. LNCS, vol. 3784, pp. 981–995. Springer, Heidelberg (2005)
23. Taylor, J.G., Scherer, K., Cowie, R.: Neural Networks. special issue on Emotion and Brain 18(4), 313–455 (2005)

# Benefit of Maximum Likelihood Linear Transform (MLLT) Used at Different Levels of Covariance Matrices Clustering in ASR Systems*

Josef V. Psutka

University of West Bohemia, Faculty of Applied Sciences,
Department of Cybernetics,
Univerzitní 8, 306 14 Plzeň, Czech Republic
psutka_j@kky.zcu.cz

**Abstract.** The paper discusses the benefit of a Maximum Likelihood Linear Transform (MLLT) applied on selected groups of covariance matrices. The matrices were chosen and clustered using phonetic knowledge. Results of experiments are compared with outcomes obtained for diagonal and full covariance matrices of a baseline system and also for widely used transforms based on Linear Discriminant Analysis (LDA), Heteroscedastic LDA (HLDA) and Smoothed HLDA (SHLDA).

## 1 Introduction

The absolute majority of current LVCSR systems work with acoustic models based on hidden Markov Models (HMMs). Output distributions tied to the states of the model and expressed by multidimensional Gaussian distributions (simply by "Gaussians") or, more exactly, by the mixtures of Gaussians are considered to be a fundamental attribute of this concept. The application of a mixture of Gaussians for modeling an output distribution results from an effort to both catch the possible non-Gaussian nature of density functions which are associated with a particular state and model mutual correlations of elements in feature vectors.

Since ideally all output distributions should be computed for each incoming feature vector, it is useful notably for real-time applications to reduce the huge amount of computations which increase with a size of the dimension of a feature space and also with the number of Gaussians. To reduce the computation burden associated with evaluating output distributions, we can apply some of the following techniques:

- To execute decorrelation of feature vectors and to use diagonal rather a full covariance matrices (CMs) for the modeling of output distributions. For these purposes, usually some orthogonal transform based on the DCT (Discrete Cosine Transform) or the NPS (Normalization of Pattern Space) is applied [1].

* This paper was supported by the AVCR, project no. 1QS101470516., GACR, project no. 102/05/0278 and the project of the EU $6^{th}$ FP no. IST-034434.

V. Matoušek and P. Mautner (Eds.): TSD 2007, LNAI 4629, pp. 431–438, 2007.

- To reduce the dimension of pattern space using the projection of feature vectors from the original space to the space with lower dimension. A typical approach is based on PCA (Principal Component Analysis), LDA (Linear Discriminant Analysis) or HLDA (Heteroscedastic LDA).
- To change over from the triphone- to the monophone-based structure of HMMs where an influence of suppressed dependencies among features is alleviated mainly by enhancing the number of Gaussians in the individual states of monophone models. LVCSR systems with a triphone-based structure work typically with 30 up to 120 thousand Gaussians, whereas systems working with monophone-based structure use from 5 to 15 thousand Gaussians [2].

Naturally, there are many other clever approaches which speed up computations or choose only relevant states with associated Gaussians for evaluations. Generally, it is possible to say that both the pass from the triphone- to the monophone-based concept on the one hand and the various transformation techniques on the other hand decrease the number of computations, but they simultaneously bring about increasing the word error rate ($WER$) in comparison with using full CMs. Moreover, in case of transforms applied in a level of feature vectors, it is usually unfeasible to find the only one transformation which could decorrelate all elements of feature vectors of all states.

Recently, new approaches have been designed which alleviate the above-mentioned increasing the $WER$, while simultaneously preserving a relatively high computation efficiency. These techniques, known as Maximum Likelihood Linear Transform (MLLT) [3], [4] and Semi-Tied Covariance (STC) [7], suppose one transformation matrix to be tied with a group of covariance matrices belonging to (an) individual state(s) of a HMM (i.e. a set of transform matrices is used for the accomplishment of the transformation).

This paper describes experiments which were performed using the MLLT applied on selected groups of CMs (i.e. multiple application of a MLLT principle), which should be decorrelated. The covariance matrices were chosen and clustered using phonetic knowledge. The results of experiments are compared with outcomes obtained for diagonal and full covariance matrices and also for transforms based on LDA, HLDA and SHLDA (Smoothed HLDA) [5].

## 2   Maximum Likelihood Linear Transform (MLLT)

Current ASR systems use HMMs with continuous parameters which are represented for each state by a Gaussian Mixture Model (GMM). A standard GMM with parameters given by $\Theta = \{c_j, \boldsymbol{\mu}_j, \boldsymbol{\Sigma}_j\}_{j=1}^M$ it is of the form

$$p(\boldsymbol{x}|\Theta) = \sum_{j=1}^M c_j \, N(\boldsymbol{x}; \boldsymbol{\mu}_j, \boldsymbol{\Sigma}_j), \qquad (1)$$

where $M$ is the number of components in a mixture, $c_j$ is the $j$-th component weight satisfying requirements: $c_j \geq 0$, $\sum_{j=1}^M c_j = 1$, $\boldsymbol{\mu}_j$ is the mean value of the $j$-th component and $\boldsymbol{\Sigma}_j$ is a square covariance matrix of the $j$-th component of the rank $n$. As was mentioned in Section 1, almost all currently used systems work with the CM of a diagonal form. In comparison with a full covariance concept, this approach has an evident advantage especially owing to lower computational and storage burden and also due to a robust parameter estimation. However, the diagonal concept of CMs can be fully beneficial only on condition that elements of the feature vector are mutually independent. The MLLT introduces a new form of a CM which allows sharing a few full covariance matrices over many distributions. Instead of having a distinct CM for every component in the recognizer, each CM consists of two elements: a non-singular linear transformation matrix $W^T$ shared over a set of components, and the diagonal elements in the matrix $\Lambda_j$. The inverse covariance (precision) matrix $\Sigma_j^{-1}$ is then of the form

$$\Sigma_j^{-1} \approx W \Lambda_j W^T = \sum_{k=1}^{n} \lambda_j^k w_k w_k^T, \qquad (2)$$

where $\Lambda_j$ is a diagonal matrix with entries $\Lambda_j = \text{diag}(\lambda_j) = \text{diag}(\lambda_j^1, \lambda_j^1, \ldots, \lambda_j^n)$ and $w_k^T$ is the $k^{th}$ row of the transformation matrix $W^T$. Estimations of model parameters could be performed by the maximum likelihood approach. We look for such a set of parameters $\hat{\Theta}$, which satisfies the equation

$$\hat{\Theta} = \underset{\Theta}{\text{argmax}} \sum_{i=1}^{N} \log\ p(x_i|\Theta). \qquad (3)$$

The solution of this equation cannot be determined in an analytical form. However, we can use an iterative procedure based on the EM algorithm.

## 2.1   Levels of CMs Clustering

The MLLT (or perhaps better the multiple-MLLT) proposed in this article supposes the transform matrix $W^T$ to be generally searched for each of $R$ selected groups of CMs. For a better insight into this problem Fig. 1 illustrates examples of using transformation matrices $W_r^T$ in a HMM structure of an ASR system. A diagram shows five levels of a tree structure based on a successive division of a whole set of GMMs. In fact, this division is based on phonetic knowledge and supposes the 3-state HMM of a triphone to be used as a basic unit for acoustic modeling.

It is evident that the whole acoustic model "$\bullet - \bullet + \bullet$" consists of a set of particular triphone models "$\bullet - ? + \bullet$". In a simplified way all triphones with the same centre phoneme can be considered to be a generalized model of a monophone, e.g. the phoneme "A" depicted in the form "$\bullet - A + \bullet$". Considering different frequency of occurrence of individual triphones in training data it is pertinent owing to a robust estimation of statistics of less frequent units to cluster phonetically similar states of models (usually using a phonetic decision tree). It means that each state of one generalized monophone can include several clustered representatives each of them is described by an individual mixture (GMM) equipped with $M$ components. Individual components of a GMM can be described by probability density functions of the form $N(x; \mu_{\Phi(\alpha)\_\beta\_\gamma}, \Sigma_{\Phi(\alpha)\_\beta\_\gamma})$, where $x$ is the feature vector, $\mu_{\Phi(\alpha)\_\beta\_\gamma}$, is the mean value and $\Sigma_{\Phi(\alpha)\_\beta\_\gamma}$ is the covariance matrix of the $\gamma$-th component of the $\beta$-th representative (mixture) of the state $\alpha$ of the phoneme $\Phi$.

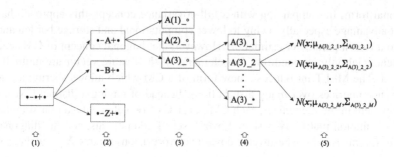

**Fig. 1.** Levels of CMs clustering

As was described above there are basically five different levels of CMs clustering, which are indicated in Fig. 1:

- $MLLT(1)$: There is only one transformation matrix for all components of all GMMs (mixtures) of all states of all generalized monophones.
- $MLLT(2)$: There is a separate transformation matrix for all components of all mixtures of all states of an individual generalized monophone (it means that each generalized monophopne has its own transformation matrix).
- $MLLT(3)$: In this case, a separate transformation matrix is connected to all components of all mixtures belonging to an individual state of a given generalized monophone.
- $MLLT(4)$: There is a separate transformation matrix for all components of a given mixture which belongs to an individual state of a given generalized monophone.
- $MLLT(5)$: Each component in all mixtures of HMMs has its own transformation matrix. This approach is equivalent to the case of GMMs equipped with the full CMs, because each symmetric positive definite matrix has clear decomposition to the diagonal and transformation matrices (eigen values and eigen vectors).

### 2.2 Practical Application of the MLLT for Different Levels of CMs Clustering

Let us suppose $R$ groups of CMs to be selected in a given level of clustering. It means that each group of CMs will share its own transformation matrix $W_r^T$, $r = 1, \ldots, R$. As was mentioned above, to estimate parameters of HMMs including parameters of transformation matrices $W_r^T$, $r = 1, \ldots, R$, we will have to use the EM algorithm. This algorithm supposes the construction of an optimization function $Q(\bar{\Theta}, \Theta)$, which can be expressed in the form [6]

$$
Q(\bar{\Theta}, \Theta) = \sum_{r=1}^{R} \sum_{(s,j) \in \Xi^{(r)}} \sum_{t} \gamma_{sj}(t) \log \left[ c_{sj} \, N(o(t); \mu_{sj}, \Lambda_{sj}, W_r^T) \right] =
$$

$$
= \sum_{r=1}^{R} \sum_{(s,j) \in \Xi^{(r)}} \sum_{t} \gamma_{sj}(t) \{ \log c_{sj} - \frac{n}{2} \log(2\pi) + \frac{1}{2} \log \det \left( W_r \, \Lambda_{sj} \, W_r^T \right) - \quad (4)
$$

$$
- \frac{1}{2} \left( o(t) - \mu_{sj} \right)^T W_r \, \Lambda_{sj} \, W_r^T \left( o(t) - \mu_{sj} \right) \},
$$

where $\bar{\boldsymbol{\Theta}} = \{\bar{\Theta}_r\}_{r=1}^R$ is a set of parameters of the old and $\boldsymbol{\Theta} = \{\Theta_r\}_{r=1}^R$ parameters of the new system. Let us remark that $\Theta_r = \{c_{sj}, \boldsymbol{\mu}_{sj}, \boldsymbol{\Lambda}_{sj}, \boldsymbol{W}_r^{\mathrm{T}}\}_{(s,j)\in\Xi^{(r)}}$ have their own transformation matrix $\boldsymbol{W}_r^{\mathrm{T}}$. $\Xi^{(r)}$ is a set of couples $(s, j)$, where $s$ corresponds to the state and $j$ is an index of a component. In fact $s$ matches the state of an original triphone model which is, in a sense of above described clustering process, consecutively clustered in various ways (Note: Because each state of an original triphone model was always represented by one mixture (GMM), we can alternatively see the index $s$ as a label of a concrete mixture in the system of all mixtures representing the whole HMM of a task); $\gamma_{sj}(t)$ is the a posteriori probability of the $j$-th component and the $s$-th state given an observation of the feature vector $o(t)$ and the set of parameters $\boldsymbol{\Theta}$; $\boldsymbol{\Lambda}_{sj}$ is the diagonal matrix, $\boldsymbol{\mu}_{sj}$ is the mean value and $c_{sj}$ is the a priori probability of the $s$-th state and the $j$-th component.

Let us have a look at the set $\Xi^{(r)}$ from the point of view of the variant $MLLT(2)$. In this case the $R$ would be equal to the number of phonemes (in the phoneme alphabet) and one set $\Xi^{(r)}$ would consist of all components of all states which are connected with all triphones containing the same centre phoneme.

Searching for maximum of $Q(\bar{\boldsymbol{\Theta}}, \boldsymbol{\Theta})$ respecting $c_{sj}$, $\boldsymbol{\mu}_{sj}$ and $\boldsymbol{\Sigma}_{sj}$ results in expected equations

$$\hat{c}_{sj} = [\sum_t \gamma_{sj}(t)] \, [\sum_j \sum_t \gamma_{sj}(t)]^{-1} \tag{5}$$

$$\hat{\boldsymbol{\mu}}_{sj} = [\sum_t \gamma_{sj}(t)o(t)] \, [\sum_t \gamma_{sj}(t)]^{-1} \tag{6}$$

$$\hat{\boldsymbol{\Sigma}}_{sj} = [\sum_t \gamma_{sj}(t)(o(t) - \hat{\boldsymbol{\mu}}_{sj}) \, (o(t) - \hat{\boldsymbol{\mu}}_{sj})^{\mathrm{T}}] \, [\sum_t \gamma_{sj}(t)]^{-1} \tag{7}$$

Maximization of (4) simultaneously with respect to $\boldsymbol{\Lambda}_{sj}$ and $\boldsymbol{W}_r^{\mathrm{T}}$ is possible only in quite trivial cases. For that reason, the maximum has to be estimated by a special iterative procedure [6], [7]. This technique consists firstly in optimization of $\boldsymbol{\Lambda}_{sj}$ at fixed $\boldsymbol{W}_r^{\mathrm{T}}$ and in the next step, on the contrary, the $\boldsymbol{W}_r^{\mathrm{T}}$ is optimized at the fixed $\boldsymbol{\Lambda}_{sj}$, which results in the estimate

$$\hat{\boldsymbol{\Lambda}}_{sj} = \left[\mathrm{diag}\,(\boldsymbol{W}_r^{\mathrm{T}} \hat{\boldsymbol{\Sigma}}_{sj} \boldsymbol{W}_r)\right]^{-1}, \tag{8}$$

where $(s, j) \in \Xi^{(r)}$, $\hat{\boldsymbol{\Sigma}}_{sj}$ is the estimate of a covariance matrix of the $j$-th component and the $s$-th state. An iterative procedure for an estimate of an individual row of the transformation matrix $\boldsymbol{W}_r^{\mathrm{T}}$ can be written in the form of

$$(\tilde{\boldsymbol{w}}_k^{(r)})^{\mathrm{T}} = (c_k^{(r)})^{\mathrm{T}} \, (\boldsymbol{G}_k^{(r)})^{-1} \sqrt{\frac{T}{(c_k^{(r)})^{\mathrm{T}}(\boldsymbol{G}_k^{(r)})^{-1}(c_k^{(r)})}}, \tag{9}$$

where

$$\boldsymbol{G}_k^{(r)} = \sum_{(s,j)\in\Xi^{(r)}} \sum_t \hat{\lambda}_{sj}^k \gamma_{sj}(t) \left[o(t) - \hat{\boldsymbol{\mu}}_{sj}\right] \left[o(t) - \hat{\boldsymbol{\mu}}_{sj}\right]^{\mathrm{T}}, \tag{10}$$

$k = 1, \ldots, n$; $n$ is a dimension of the feature space, $T$ is a number of speech feature vectors, $\hat{\lambda}_{sj}^k$ is the $k$-th diagonal element of the matrix $\hat{\boldsymbol{\Lambda}}_{sj}$, $(\tilde{\boldsymbol{w}}_k^{(r)})^{\mathrm{T}}$ is the estimate of the $k$-th row of the transformation matrix $\boldsymbol{W}_r^{\mathrm{T}}$, $(c_k^{(r)})^{\mathrm{T}}$ is the $k$-th row of the cofactor matrix to the transformation matrix $\boldsymbol{W}_r^{\mathrm{T}}$, $\gamma_{sj}(t)$ is the a posteriori probability of the $j$-th component and the $s$-th state given an observation of a feature vector $o(t)$ at the time $t$ and a set of parameters $\boldsymbol{\Theta}$. Let us describe one pass through the modified EM algorithm:

1. Using up-to-date set of parameters we compute a posteriori probabilities $\gamma_{sj}(t)$ for all components $j$, all states $s$ and all times $t$ (this point corresponds to a standard step of the Baum-Welch procedure).
2. On the basis of $\gamma_{sj}(t)$ we estimate for all $j$ and $s$ a priori probabilities $\hat{c}_{sj}$ (5), mean values $\hat{\boldsymbol{\mu}}_{sj}$ (6) and covariance matrices $\hat{\boldsymbol{\Sigma}}_{sj}$, see (7).
3. Using current transformation matrices $\boldsymbol{W}_r^{\mathrm{T}}$ we compute new estimates of all diagonal matrices $\hat{\boldsymbol{\Lambda}}_{sj}$ according to (8).
4. We estimate all transformation matrices $\boldsymbol{W}_r^{\mathrm{T}}, r = 1, \ldots, R$. To accomplish this step, a special iterative procedure must be run. For individual transformation matrices we compute:
   - the matrices $\boldsymbol{G}_k^{(r)}$ according to (10), where $k = 1, \ldots, n$ and $n$ is the dimension of a feature space,
   - the cofactor matrix $\boldsymbol{W}_r^{\mathrm{COF}}$ to the current estimate of the transformation matrix $\boldsymbol{W}_r^{\mathrm{T}}$,
   - the individual rows of the transformation matrix $\tilde{\boldsymbol{W}}_r^{\mathrm{T}}$ according to (9).
   
   Then we update $\boldsymbol{W}_r^{\mathrm{T}}$ and continue the iterative procedure from the point of 4a) until the convergence is reached.
5. We replace the old estimate of the parameters $\bar{\boldsymbol{\Theta}}$ by the new $\boldsymbol{\Theta}$ ones and repeat by the step 1 until the convergence of the EM algorithm is reached.

As soon as the transform matrices $\boldsymbol{W}_r^{\mathrm{T}}$, $r = 1, \ldots, R$, are estimated, we can use them for transformation of feature vectors $o(t)$ to the new feature space. Generally, the feature vector $o(t)$ should be put as many transformations

$$o(t)^{(r)}(t) = \boldsymbol{W}_r^{\mathrm{T}}\, o(t), \quad \text{for } r = 1, \ldots, R, \tag{11}$$

as many groups $R$ of CMs are in a given task clustered. The resulting likelihood function $\log N(o(t); \boldsymbol{W}_r^{\mathrm{T}}\boldsymbol{\mu}_{sj}, \boldsymbol{\Lambda}_{sj}^{-1})$ must then be normalized by a likelihood compensation term, which is proportional to $\log \det \boldsymbol{W}_r^{\mathrm{T}}$.

## 3   Results of Experiments

All the experiments were performed using a speech data set of telephone quality. The corpus consists of Czech read speech transmitted over a telephone channel. One thousand speakers were asked to read various sets of 40 sentences. The digitization of an input analog telephone signal was provided by a telephone interface board DIALOGIC D/21D at 8 kHz sample rate and converted to the mu-law 8 bit resolution. The telephone test set consisted of 100 sentences randomly selected from utterances of 100 different

**Table 1.** Results of comparative experiments

|  | # of transf. matrices | WER[%] | #of estimated parameters |
|---|---|---|---|
| DIAG | - | 14.03 | ≈2.33M |
| LDA dim 26 | 1 | 13.74 | ≈1.68M |
| HLDA dim 25 | 1 | 13.81 | ≈1.62M |
| SHLDA dim 25 | 1 | 13.96 | ≈1.62M |
| MLLT(1) | 1 | 11.68 | ≈2.33M |
| MLLT(2) | 43 | 11.64 | ≈2.38M |
| MLLT(3) | 129 | 11.46 | ≈2.50M |
| MLLT(4) | 4044 | 11.29 | ≈7.57M |
| FULL | - | 9.18 | ≈22.70M |

speakers who were not included in the telephone training database. The lexicon in all test tasks contained 475 different words. Since several words had multiple different phonetic transcriptions, the final vocabulary consisted of 528 items. There were no OOV words.

The front-end of the ASR system is based on the MFCC parameterization. Feature vectors consisted of 12 static + 12 delta + 12 delta-delta = 36 coefficients (including the zeroth cepstral coefficient representing the signal energy) were computed with a frame spacing of 10ms. The number of coefficients in a feature vector and the number of band-pass filters ($f$ =15) applied in the frequency axis (0÷4kHz) is a result of a thorough analysis and extensive experimental works in which robust setting of the MFCC-based parameterization was searched for [6]. Cepstral mean normalization (CMN) was used to reduce the effect of constant channel characteristics. No variance or vocal tract length normalization (VTLN) was applied in these experiments.

The basic speech unit in all the experiments was a triphone. Each individual triphone is represented by a 3-state HMM; each state is provided by a mixture of 8 components of a multivariate Gaussians. In all recognition experiments a language model based on zero-grams was applied so that the influence of individual transforms could be better judged. For that reason, the perplexity of the task was 528.

The goal of following experiments was to explore how an application of different levels of CMs clustering influences recognition results (WER). In order to judge a benefit of the Maximum Likelihood Linear Transform used at different levels of CMs clustering, we also performed comparative experiments on a standard baseline ASR system with CMs of the diagonal form (DIAG) and with the system working with the full CMs (FULL). In addition, several comparative tests were also made with widely used linear transforms based on LDA, HLDA and SHLDA. In these experiments we searched for the lowest dimension of the feature space which yielded lower or equal WER than the solution based on DIAG. In the case of SHLDA we set the smoothing factor $\alpha$ on a recommended value, which is $\alpha = 0.8$ [5]. The Results of all the experiments are itemized in Table 1 together with information about the number of parameters that have to be estimated.

## 4 Conclusions

The Maximum Likelihood Linear Transform (MLLT) applied at different levels of co-variance matrices clustering (i.e. multiple applications) is a very effective technique of parameter decorrelation, which overcomes the widely used Linear Discriminant Analysis (LDA), Heteroscedastic LDA (HLDA) and also SHLDA (Smoothed HLDA) and approaches the application with the full CMs. However, results shown in Table 1 indicate that the growth of a number of transform matrices also causes the growth of a number of parameters that should be estimated, as well as the increase in the number of computations, especially during enumeration of output distributions. Therefore, the tradeoff between computational burdens (e.g. a case of real time applications) and the WER is necessary. We can also consider a different concept of CMs clustering which would not be based on phonetic knowledge but rather directly on real data, i.e. the "geometrical" form of covariance matrices. This concept could bring the same or better results simultaneously with decreasing the number of transformation matrices.

## References

1. Psutka, J.V., Müller, L.: Comparison of various decorrelation techniques in automatic speech recognition. Jour. of Syst., Cyb. and Inf. 5(1), 27–30 (2007)
2. Psutka, J.V., Müller, L.: Building Robust PLP-based Acoustic Module for ASR Application. In: Proc. of the 10th SPECOM'2005, Greece, pp.761–764 (2005)
3. Olsen, P.A., Gopinath, R.A.: Modeling inverse covariance matrices by basis expansion. IEEE Trans. on Speech and Audio Proc. 12(1), 272–281 (2004)
4. Visweswariah, K., Axelrod, S., Gopinath, R.A.: Acoustic modeling with mixtures of subspace constrained exponential models. In: Proc. of the 7th Eurospeech'2003, Geneva, Switzerland, pp. 2613–2616 (2003)
5. Burget, L.: Combination of Speech Features Using Smoothed Heteroscedastic Linear Discriminant Analysis. In: Proc. of the 8th ICSLP'2004, Jeju, Korea, pp. 2549–2552 (2004)
6. Psutka, J.V.: Techniques of parameterization, decorrelation and dimension reduction in ASR systems. Thesis. Depart. of Cybernetics, Univ. of West Bohemia, Pilsen (in Czech) (2007)
7. Gales, M.J.F.: Semi-Tied Covariance Matrices for Hidden Markov Models. IEEE Transactions on Speech and Audio Processing 7(3), 272–281 (1999)

# Information Retrieval Test Collection for Searching Spontaneous Czech Speech⋆

Pavel Ircing[1], Pavel Pecina[2], Douglas W. Oard[3], Jianqiang Wang[4],
Ryen W. White[5], and Jan Hoidekr[1]

[1] University of West Bohemia, Faculty of Applied Sciences,
Department of Cybernetics
Univerzitní 8, 306 14 Plzeň, Czech Republic
{ircing, hoidekr}@kky.zcu.cz
[2] Charles University, Institute of Formal and Applied Linguistic
Malostranské náměstí 25, 118 00 Praha, Czech Republic
pecina@ufal.mff.cuni.cz
[3] University of Maryland, College of Information Studies/UMIACS
College Park, MD 20742, USA
oard@umd.edu
[4] State University of New York at Buffalo,
Department of Library and Information Studies
Buffalo, NY 14260, USA
jw254@buffalo.edu
[5] Microsoft Research
One Microsoft Way, Redmond, WA 98052, USA
ryenw@microsoft.com

**Abstract.** This paper describes the design of the first large-scale IR test collection built for the Czech language. The creation of this collection also happens to be very challenging, as it is based on a continuous text stream from automatic transcription of spontaneous speech and thus lacks clearly defined document boundaries. All aspects of the collection building are presented, together with some general findings of initial experiments.

## 1 Introduction

The very essence of an information retrieval (IR) system is to satisfy user's information needs, expressed by a query submitted to the system. The degree of user satisfaction is of course inherently subjective and therefore there is a need for some form of (automatic) quantitative evaluation of the system effectiveness. Such evaluation is usually performed on a defined test collection which includes a representative set of documents, a representative set of topics (formalized information needs) and, most importantly, judgments of the relevance of each document to each topic. The process of relevance assessment is extremely labour-intensive, and perhaps explains why no large-scale Czech IR collection has so far been available (at least according to the authors' knowledge).

⋆ This work was supported by projects MSMT LC536, GACR 1ET101470416, MSM0021620838 and NSF IIS-0122466.

V. Matoušek and P. Mautner (Eds.): TSD 2007, LNAI 4629, pp. 439–446, 2007.
© Springer-Verlag Berlin Heidelberg 2007

During the course of the MALACH project, which aims to improve access to the large multilingual spoken archives using advanced ASR and IR techniques [1], the need for an IR test collection arose quite naturally. The archives in question consist of the digitized videotaped testimonies given by the survivors and witnesses of the Holocaust and the ultimate goal of the project is to allow potential users to watch the passages relevant to their queries. In order to facilitate the IR itself, the soundtrack from the testimonies must be transformed into text using an Automatic Speech Recognition (ASR) decoder.

The test collection for English was created first [2]. However, a significant subset of the English interviews (approx. 4,000 of them) was manually subdivided into topically coherent segments, equipped with a three-sentence summary and indexed with keywords selected from a pre-defined thesaurus. The test collection was then built using these manually annotated data. No such manual segmentation was available in the case of the Czech interviews and this fact not only made the Czech IR more difficult but shifted the very nature of the task - the goal for Czech IR experiments is to identify appropriate replay start points rather than to select among pre-defined segments. Nevertheless, some form of manual indexing was performed even on the Czech data, as described in the following section. Both English and Czech collection were used as the reference corpora in the Cross-Language Speech Retrieval (CL-SR) track at the CLEF-2006 evaluation campaign (http://www.clef-campaign.org/).

## 2   Collection

### 2.1   "Documents"

The collection consists not of documents but rather of (354) interviews; a continuous text stream coming from an ASR decoder. To be precise, there are actually four text streams, generated by two different ASR systems (for details about the first one see [3], the second one is described in [4]), each of them providing the transcription of both the left and right stereo channels. Each of those channels was recorded from a separate microphone, one placed on the interviewer and one on the interviewee, thus yielding acoustic signals with non-negligible differences.

These ASR transcripts are further accompanied with:

– English thesaurus terms that were manually assigned with one-minute granularity by subject-matter experts. Two broad types of thesaurus terms are present, with some describing concepts and others describing locations. The location terms are most often pre-combined with a time clause, which reflects the fact that political boundaries and place names sometimes changed over the time frame described by interviewees in this collection. Unlike the English collection, where the keywords are associated with entire indexer-defined segments, the thesaurus terms in the Czech collection are used as onset marks—they appear only once at the point where the indexer recognized that a discussion of a topic or a location-time pair had started; continuation and completion of discussion are not marked.
– Automatically produced Czech translations of the English thesaurus terms. These were created using the following resources: (1) professional translation of about 3,000 thesaurus terms that were selected to provide broad coverage of the constituent words, (2) volunteer translation of about 700 additional thesaurus terms,

and (3) a custom-built machine translation system that reused words and phrases from manually translated thesaurus terms to produce additional translations [5]. Some words (e.g., foreign place names) remained untranslated when none of the three sources yielded a usable translation.

One simple way of automatically identifying points within an interview at which replay should be started is to divide the interview into passages, and then to return the start time of the passages that best match a query. Very short passages would not contain enough words to reliably match the query, so we arbitrarily chose roughly 400 words as a passage length.[1] Non-overlapped passages of that length would yield a larger temporal granularity than would be desirable, so a 67% overlap was used (i.e., passage start times were spaced about 133 words apart). In practice, this makes the minimum temporal granularity for start times about 75 seconds, roughly five times larger than the temporal granularity of the relevance assessments for this test collection (see Section 2.3).

The 11,377 resulting passages can then be treated as "documents" for which the following fields are provided:

- DOCNO. The specification for the start time of a segment, in the "VHF[interview]-[seconds]" format required for scoring.
- INTERVIEWDATA. The first name and last initial of the interviewee; additional names (e.g., maiden names and aliases) may also be present. This field is the same for every passage that is drawn from the same interview.
- ENGLISHMANUKEYWORD. This field was intended to contain thesaurus terms that has been manually assigned to a time that fell within the segment, but a script bug resulted in inclusion of thesaurus terms from earlier in the interview. Terms found in this field are therefore not useful in this version of the collection.
- CZECHMANUKEYWORD. The automatically produced English translations of the thesaurus terms in the ENGLISHMANUALKEYWORD field. Because of the errors in that field, the CZECHMANUALKEYWORD field is similarly unusable.
- ASRSYSTEM. Usually 2006, which was the more recent (and hence the more accurate) of the two ASR systems. In the rare instances when no words were produced by the 2006 system, this value is 2004. The 2004 system had been designed to transcribe colloquial Czech. In the 2006 system, lexical substitution was used to generate formal Czech.
- ASRTEXT. The words produced by the ASR channel that produced the largest number of words for that passage (usually this is the channel assigned to the interviewee).
- CHANNEL. The stereo ASR channel that was automatically chosen (left or right).
- ENGLISHAUTOKEYWORD. English terms from the same thesaurus that were automatically assigned based on words found in the ASR stream. A $k$-NN classifier was trained for this purpose using English data (manually assigned thesaurus terms and manually written segment summaries) and run on the automatically generated English translations of the Czech ASRTEXT (produced using a probabilistic dictionary - see [6] for details). Because the classifier was trained on data in which thesaurus

---

[1] Specifically, the sum of the word durations in each passage, excluding all silences, is exactly 3 minutes.

terms were associated with segments rather than start points, the natural interpretation of an automatically assigned thesaurus term is that the classifier believes that the indicated topic is associated with the words spoken in the given passage.

- CZECHAUTOKEYWORD. Automatically produced Czech translations of the thesaurus terms in the ENGLISHAUTOKEYWORD field.

For example:

```
<DOC>
<DOCNO>VHF10325-1080.34</DOCNO>
<INTERVIEWDATA>Alexej H...</INTERVIEWDATA>
<ENGLISHMANUKEYWORD>social relations in prisons</ENGLISHMANUKEYWORD>
<CZECHMANUKEYWORD>společenská vztahy v vězení</CZECHMANUKEYWORD>
<ASRSYSTEM>2006</ASRSYSTEM>
<CHANNEL>left</CHANNEL>
<ASRTEXT> PĚKNĚ TAKŽE NĚKDY I TY I TY HLÍDAČI NE </s> <s> TO MYSLÍ
POSLOUCHALI POSLOUCHALI TO CO ZPÍVÁM HO NECHAL CELKEM ASI TO ZPÍVAL ONI NÁS JAKO NE
</s> <s> KDE MÁTE NA MYSLI </s> <s> NO TAM JSME
BYLI ASI </s> <s> DO JARA ROKU ČTYŘICET </s> <s> NO TO SEM NÁS
VOZILI NA NA NA SILNICI A NÁM SE PODAŘILO NĚKOLIKA ...</ASRTEXT>
<ENGLISHAUTOKEYWORD>fate of loved ones | living conditions in the
camps | Poland 1941 (June 21) - 1944 (July 21) | Germany 1945
(January 1 - May 7) ...</ENGLISHAUTOKEYWORD>
<CZECHAUTOKEYWORD>osudy blízkých | životní podmínky v táborech |
Polsko 1941 (21. červen) - 1944 (21. červen) | Německo 1945
(1. leden - 7. květen) ...</CZECHAUTOKEYWORD>
</DOC>
```

## 2.2  Topics

Currently there are 115 topics specified for the collection. All of them were originally constructed in English and then translated into Czech and in some cases adapted in order to increase the number of relevant passages in the collection (this was usually done by removing geographic restrictions from the topic). As for the original English topics, they were mostly compiled from the real requests made by scholars, educators and documentary film makers to the administrators of the archive.

The topics are represented in the well-known TREC-style format as shown in the following example:

```
<top>
<num>1225</num>
<title>Osvobození Buchenwaldu a Dachau </title>
<desc>Výpovědi svědků osvobození koncentračních táborů Buchenwald
a Dachau.</desc>
<narr>Relevantní materiál by měl zahrnovat příběhy přeživších nebo
osvoboditelů popisující tyto události. Osvobození jiných táborů
není relevantní.</narr>
</top>
```

where the `<title>`, `<desc>` (description) and `<narr>` (narrative) fields gradually provide more detailed specification about the user's request. Both Czech and English versions of the topics are available for searching the Czech collection to the CLEF participants.

## 2.3  Relevance Judgments

Relevance judgments are prepared within the CLEF evaluation campaign at Charles University in Prague. In 2006 the judgments were completed for a total of 29 topics by five domain experts. The assessors were Czech native speakers with a good knowledge of English. They were working 20 hours a week in average for a period of five months. The two-sided relevance assessment process was performed in two phases supported by an advanced search system designed especially for this task at the University of Maryland in College Park. Full Czech ASR transcript of the best audio channel and the manually assigned keywords from English thesaurus were indexed as overlapping passages as described in Section 2.1.

**Search-guided Assessment.** Each assessor processed one topic at a time. The assessor started with an individual topic research using external resources (such as books, encyclopedias, and web pages) followed by a presentation of the topic to the other assessors

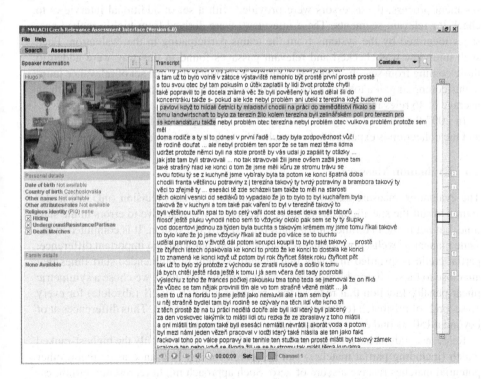

**Fig. 1.** Screenshot of the assessment interface

and discussion aimed at detailed specification of the topic relevance that all assessors agreed on. Then the assessor iterated between formulating the topic-related queries (using either ASR or thesaurus terms, or both) and searching the collection for interviews containing potentially relevant passages.

Each promising interview was displayed in a detailed view providing an interactive search capability, showing the Czech ASR transcript, English thesaurus terms, and the possibly relevant passages identified by graphical depiction of the retrieval status value. The assessor could scroll through the interview, search using either type of thesaurus terms, and also replay the audio from any point in order to identify the start time and end time of the relevant periods by indicating points on the transcript. The mGAP measure (see Section 2.4) employs only the start times (converted to 15-second granularity), however both start and end times are available for future research. A screenshot from the interface used by the assessors is shown in Fig. 1.

At least five relevant passages were required to minimize the effect of quantization noise on the computation of mGAP. Twenty nine such topics were distributed to the CLEF 2006 participants as evaluation topics. Each participating team employed its existing information retrieval systems on these topics and submitted maximum of five official runs.

**Highly-ranked Assessment.** Following completion of the search-guided relevance assessment process, the assessors were provided with a set of additional interviews to check for relevant segments. These interviews were derived from highly ranked passages identified by the systems from the teams participating in the evaluation. Each such interview was checked and relevant passages found in this way were added to those coming from search-guided assessment to produce the final set of relevance judgments comprising of a total of 1,322 start times for relevant passages identified with an average of 46 relevant passages per topic.

In 2007 the currently ongoing relevance assesment process follows the same rules and the collection is expected to be enriched by judgments for new 25-30 topics.

## 2.4   Evaluation Measure

The evaluation measure called mean Generalized Average Precision (mGAP) is designed to suit the specific needs of this collection — it is sensitive to errors in the start time, but not in the end time, of passages retrieved by the system. It is computed in the same manner as well-know mean average precision, but with one important difference: partial credit is awarded in a way that rewards system-recommended start times that are close to those chosen by assessors. After a simulation study, we chose a symmetric linear penalty function that reduces the credit for a match by 0.1 (absolute) for every 15 seconds of mismatch (either early or late) (see [7] for details). Thus differences at or beyond a 150 second error are treated as a no-match condition.

Relevance judgments are drawn without replacement so that only the highest-ranked match (including partial matches) can be scored for any relevance assessment; other potential matches receive a score of zero. Such approach might represent a pitfall, especially in the case of using the overlapping passages. Specifically, the start time of the highest-ranking passage that matches (however poorly) a passage start time in the relevance judgments will "use up" that judgment. Subsequent passages in which the same matching terms were present would then receive no credit at all, even if they were closer matches than the highest-ranking one.

## 3   Initial Experiments

The first experiments were performed on the set of 11,377 passages (see Section 2.1), using quite a simple document-oriented IR system based on the *tf.idf* model. Detailed description of the experiments can be found in [8], here we will only summarize the most important findings:

- The artificially created "documents", however not topically coherent, are usable for initial experiments with the described collection as the system is indeed able to identify a significant number of relevant starting points.
- The best result was achieved when only the ASRTEXT field was indexed. We knew that the manually assigned keywords were misaligned, but the poor performance of the indexes involving automatic keywords was surprising. Manual examination of a few CZECHAUTOKEYWORD fields indeed indicates a low density of terms that appear as if they match the content of the passage, but additional analysis will be needed before we can ascribe blame between the transcription, classification and translation stages in the cascade that produced those keyword assignments.
- Proper linguistic preprocessing seems to be indispensable for good performance of the Czech IR system - both lemmatization and stemming boosted the performance almost by a factor of two in comparison with the runs using the original word forms.

## 4   Conclusion and Future Work

The presented test collection constitutes a valuable resource for facilitating research into IR for the Czech language. As was already mentioned, the collection is going to be further enriched for this year's CLEF campaign. Moreover, the assessors at Charles University are preparing also the document-oriented, text-based Czech collection for the CLEF Ad-Hoc track. Once these collections are completed, we will have a rather rich set of resources for experiments with Czech IR systems. The development of such systems is our current top priority — so far we have employed only the standard (document-oriented) IR approaches, not reflecting the specific nature of the collection described in this paper and taking into account the properties of the Czech language only to a limited extent.

## References

1. Byrne, W., Doermann, D., Franz, M., Gustman, S., Hajič, J., Oard, D., Picheny, M., Psutka, J., Ramabhadran, B., Soergel, D., Ward, T., Zhu, W.J.: Automatic Recognition of Spontaneous Speech for Access to Multilingual Oral History Archives. IEEE Transactions on Speech and Audio Processing 12(4), 420–435 (2004)

2. Oard, D., Soergel, D., Doermann, D., Huang, X., Murray, G.C., Wang, J., Ramabhadran, B., Franz, M., Gustman, S.: Building an Information Retrieval Test Collection for Spontaneous Conversational Speech. In: Proceedings of SIGIR 2004, Sheffield, UK, pp. 41–48 (2004)
3. Shafran, I., Byrne, W.: Task-Specific Minimum Bayes-risk Decoding Using Learned Edit Distance. In: Proceedings of ICSLP 2004, Jeju Island, South Korea, pp. 1945–1948 (2004)
4. Shafran, I., Hall, K.: Corrective Models for Speech Recognition of Inflected Languages. In: Proceedings of EMNLP 2006, Sydney, Australia, pp. 390–398 (2006)
5. Murray, C., Dorr, B.J., Lin, J., Hajič, J., Pecina, P.: Leveraging Reusability: Cost-effective Lexical Acquisition for Large-scale Ontology Translation. In: Proceedings of ACL 2006, Sydney, Australia, pp. 945–952 (2006)
6. Olsson, S., Oard, D., Hajič, J.: Cross-Language Text Classification. In: Proceedings of SIGIR 2005, Salvador, Brazil, pp. 645–646 (2005)
7. Liu, B., Oard, D.: One-Sided Measures for Evaluating Ranked Retrieval Effectiveness with Spontaneous Conversational Speech. In: Proceedings of SIGIR 2006, Seattle, Washington, USA, pp. 673–674 (2006)
8. Ircing, P., Oard, D., Hoidekr, J.: First Experiments Searching Spontaneous Czech Speech. In: Proceedings of SIGIR 2007, Amsterdam, The Netherlands (2007)

# Inter-speaker Synchronization in Audiovisual Database for Lip-Readable Speech to Animation Conversion

Gergely Feldhoffer, Balázs Oroszi, György Takács, Attila Tihanyi, and Tamás Bárdi

Faculty of Information Technology,
Péter Pázmány Catholic University,
Budapest, Hungary
{flugi, oroba, takacsgy, tihanyia, bardi}@digitus.itk.ppke.hu

**Abstract.** The present study proposes an inter-speaker audiovisual synchronization method to decrease the speaker dependency of our direct speech to animation conversion system. Our aim is to convert an everyday speaker's voice to lip-readable facial animation for hearing impaired users. This conversion needs mixed training data: acoustic features from normal speakers coupled with visual features from professional lip-speakers. Audio and video data of normal and professional speakers were synchronized with Dynamic Time Warping method. Quality and usefulness of the synchronization were investigated in subjective test with measuring noticeable conflicts between the audio and visual part of speech stimuli. An objective test was done also, training neural network on the synchronized audiovisual data with increasing number of speakers.

## 1 Introduction

Our language independent and direct speech to facial animation conversion system was introduced in 1, 2, and 3. The main system components are the acoustic feature extraction, the feature point coordinate vector calculation and running a standard MPEG4 face animation model as it is shown in Figure 1. The input speech sound is sampled at 16 bit/48 kHz and then acoustic feature vectors based on Mel-Frequency Cepstrum Coefficients (MFCC) are extracted from the signal. The feature vectors are sent to the neural network (NN), which computes a special weighting vector that is a compressed representation of the target frame of the animation.

A conceptual element of our system is to process only continuous acoustical and facial parameters. The traditional solutions transform the continuous process of speech into a discrete set of language elements like phonemes and visemes. The second part of the traditional systems converts discrete text or phoneme strings into animated faces. The modular structure and the separated training of the elements mean the main benefit of such discrete element based systems. [4],[5] Their problems are the accumulated error and the lost original temporal and dynamic structure.

One of the benefits of our direct solution is the reservation of the original temporal and energy structure of the speech process. Thus the naturalness of rhythm is guarantied. Further benefit of our solution is a relatively easy implementation in the

V. Matoušek and P. Mautner (Eds.): TSD 2007, LNAI 4629, pp. 447–454, 2007.

environment of limited computational and memory resources. A rather promising feature of our system is the language independent operational capacity.

The single speaker version works quite well. Deaf persons can recognize about 50% of the words correctly. The neural network has been trained by speech parameters at the input and principal components of the facial parameters at the output and parameter pairs have originated from the same speaker 1.

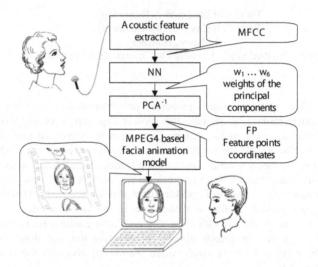

**Fig. 1.** Structure of the direct speech to facial animation conversion system

In this paper we report on the results of speaker variation tests in our system. The main technical problem has been caused by timing differences in the utterances of different speakers. In traditional discrete systems the labeling of small units by phoneme codes can eliminate the timing problems. The training and testing procedures are based on phoneme size segments 1, 2.

Dynamic Time Warping algorithm (DTW) is a well refined procedure in time matching of two utterances and widely used in the isolated word recognition systems. This algorithm does not need an exact segmentation within complete utterances. So a special version of DTW has been applied in our program.

Several theoretical questions have been raised in the investigation of multiple speaker situations. Hard of hearing persons stated that the lip-readability of speakers is very different due to the lip level articulation, while the intelligibility of their voice is similar to that of the hearing people. How can we optimally train the neural network? Whether train by a high number of speakers representing the average population or train by a carefully selected group of lip-readable persons? We do not need to transform a speech signal to its original face movement. Rather we need a transformation to one easily lip-readable face. How can we test the efficiency of time matching? How can we measure objectively the performance of the system changing the speakers?

## 2 Database Construction

The traditional audio-visual databases are elaborated for testing by hearing people. Our database was constructed according to our special specifications.

Professional interpreters/lip-speakers were invited to the record sessions. Their speech tempo and articulation style is accommodated to the communication with hard of hearing persons. Their articulation emphasizes visible distinctive features of speech.

Sentences from a popular Hungarian novel were selected. The selection differs from the phonetically balanced criteria. The rare phonetic elements and visually confusable phoneme pairs were represented on higher level, than their average probability. The records consist of 100 sentences.

The head of speakers have been softly fixed to reduce the motion of the head. We used commercially available video cameras with 720x576 resolutions, 25 fps PAL format video – which means 40ms for audio and video frames. The video recordings have concentrated only on the area of the mouth and vicinity to let maximum resolution. The text was visible on a big screen behind the camera. The camera produced portrait position picture to maximize the resolution on the desired area of the face.

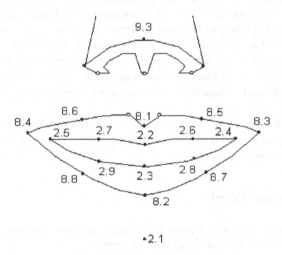

**Fig. 2.** The applied MPEG-4 feature point (FP) subset

Yellow markers were used on the nose and chin of speakers as reference points. Red lipstick emphasized the lip color for easier detection of lip contours. The records were stored on digital tape and then copied into a PC.

The records were prepared in an acoustically treaded room by a dynamic microphone. The input speech sound is sampled at 16 bit/48 kHz. The speakers could repeat sentences in case of wrong pronunciations. The audio and video files were synchronized by "papapapa" phrases recorded at the beginning and end of files. The closures end opening of explosives are definite both in audio and video files.

The video signal was processed frame by frame. The first step was the identification of yellow dots on the nose and on the chin based on their color and brightness. The color values of the red lips were manually tuned to get the optimal YUV parameters for identification of internal and external lip contours. On the shape of lips the left most and right most points identified the MPEG-4 standard FP-s 8.4 and 8.3. The further FP coordinates were determined at the cross points of halving vertical lines and lip contours. FPs around the internal contour were located similarly. An extra module calculated the internal FPs in the cases of closed lips. The FP XY coordinates can be described by a 36 element vector frame by frame. The first 6 principal component values (PCA) were used to compress the number of video features.

The input speech was pre-emphasis filtered then in 40 ms frames 1024 element FFT with Hamming windows were applied to gain spectrum vectors. Next step calculated 16 Mel-Frequency Cepstrum Coefficients (MFCC) in each frame.

### 2.1 The DTW Procedure

The temporal matching has the following task: the frame "i" in the audio and video records of Speaker A has a corresponding frame "j" in the records of speaker B. The correspondence means the most similar acoustic features and lip position parameters. Each speaker read the same text so several corresponding frames are evident for example at sentence beginnings. The time warping algorithm in the isolated word recognition can provide a suboptimal matching between frame series. [6]

The voice records were used for warping. Voice frames are characterized by MFCC feature vectors. The distance metric in the DTW algorithm was the sum of MFCC coefficient absolute differences. The total frame number in records A and B might be different.

In the database 40 sentences of 5 speakers were warped to the files of all other speakers. This warp is represented by indexing the video frames for 40 ms audio windows as possible jumps or repeats.

## 3   Experiments and Results

Subjective tests were used to measure the quality of time warping in cases of changing of speakers in audio and visual parts of records.

An objective measure, the average error of calculated FP coordinates related to natural FP parameters of speaker A was investigated as a function of number of speakers involved in the training. The neural net has 16 input nodes, 20 hidden nodes and 6 output nodes. The input parameters are the MFCC parameter values of the voice. The output parameters are the principal component values of the FP coordinates. So the system calculates face animation parameters from the voice features. The training errors express the average value of distances of the calculated FP coordinates from the target FP coordinates. The voice might be from speaker A, B, C and D and the target FP coordinates are the elements of PF coordinates are from the video records of speaker A.

### 3.1 Subjective Testing of the Time Warping

The quality test of matching of the records needs a subjective assessment. For this reason we prepared a test sequence of voice and video records of feature points. The audio part and video parts were taken randomly from the same records and other cases the voice parts were from different speakers and the video frames were warped to the audio frames spoken by the different speakers. In the case of ideal warping the test persons could not differentiate the audio-video pairs whether they were from the same records or warped different records. In the test they expressed the opinion on a scale: 5- surely identical, 4-probably identical, 3-uncertain, 2-probably different, 1-surely different the origin of the voice and video.

### 3.2 Results of the DTW Tests

21 test persons watched the 15 audio-video pairs in random order. The results are summarized in Figure 3.

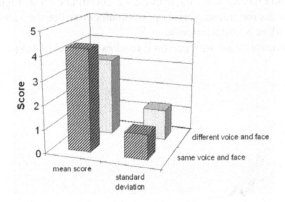

**Fig. 3.** Results of the subjective DTW tests. The opinion score values: 5- surely identical, 4-probably identical, 3-uncertain, 2-probably different, 1-surely different the origin of the voice and video.

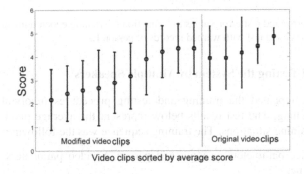

**Fig. 4.** Ratings (average score and deviation) of video clips grouped by modified and original video data, sorted by average score

The cases when the audio and video parts of the records were from the same persons the average score was 4.2. The average opinion expressed, that the test persons can differentiate original and warped pairs at some level.

The same time the warped pairs have a score of 3.2 so it is in between probably identical and uncertain value. This score value proves that the warping is good because the result is not on the "different" side.

The video clips were rearranged in a way that we put into one group the modified and in other group the original ones during the evaluation process. The results are in Figure 4. The standard deviation values are overlapped within the two groups.

### 3.3 Training and Testing the System by a Single Speaker

In this experiment the records of 5 speakers were studied. Matrix Back-propagation algorithm was used to train the neural network. Speaker A, B, C and D were used for the training of the neural net and speaker E was used for testing. In this section the records of speaker A was used for training and Speaker E for testing. The average training error is a good indicator for the average FP coordinate error. Figure 5 shows, that after 10.000 epochs the training error approximately reaches the value below one pixel during independent training processes.

Meantime the test error value with person E reaches the level of 1.5 pixels.

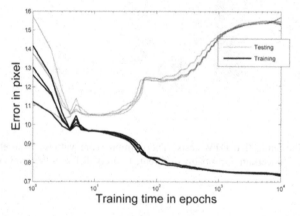

**Fig. 5.** The training error and test error values as a function of training epoch number. Training with records of person A and test with warped records of person E.

### 3.4 Training and Testing the System by Multiple Speakers

In this section we repeated the training and testing procedures involving several speakers in the training. The test results below represent the average pixel errors in case of different training situations. The training sequence was the following:

- person A voice parameters as input and person A video parameters as target values of the net
- the previous + person B voice parameters to A as input and warped A video parameters as target values of the net

- the previous + person C voice parameters to A as input and warped A video parameters as target values of the net
- the previous + person D voice parameters to A as input and warped A video parameters as target values of the net

The training and testing error values are shown in Figure 6. More and more persons involved in the training caused higher training error and lower testing error values. The ratio of testing error and training error is diminished from more than 100% to 7%.

### 3.5  Discussion

Voice features of four speakers were involved in the training of the neural net. The average pixel error of the calculated FP coordinates related to the original FP coordinates of speaker A. The values of FP test error (testing with speaker E whose data were not involved in the training) decrease from 1.5 to 1.0 by increasing the number of speakers from 1 to 4. In the case we involved more speakers in the training there was no considerable decrease in FP error. The training error increases because of the higher variations in the training set.

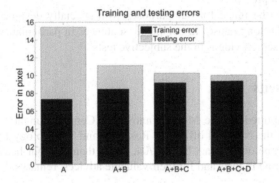

**Fig. 6.** Variation of training and testing errors involving 1-4 speakers in the training procedure

The decreased value of the test errors proves that the time warping works well. The test error value in case of 4 speakers in the training is only 7% higher then the training error. This means that four speakers represent well the speaker variations in the system.

It is possible to express the error in ratio of the domains. On a PAL screen which contains only a face from the eyes to the neck, the average domain size of a feature point coordinate is about 40 pixels. 1 pixel error means about 2.5% error this way. The reason of using pixel as a unit of error is the database recording method. The error of FP calculations has 2 pixels in the training material. This calculation error is the ±1 pixel uncertainty of the automatic detection algorithm. The error range of the calculated FP coordinates from the voice parameters and the detection uncertainty error are comparable.

# 4  Conclusions

The speaker dependent variations in the direct calculation of face animation parameters from the voice parameters can be treated by the applied methodology. DTW is a convenient solution to compensate the lack of phoneme level in multi-personal issues of speech-to-animation conversion.

The subjective tests proved that the time warping based on voice parameters can map well the speech process of speaker A into the speech process of speaker B who has the same text. The level of testing error is lower than the critical error which disturbs the lip reading for hard of hearing persons 1. So the training of the conversion system performs the speaker independent criteria on the required level.

The increasing of the number of feature points around the inner contour of lips improved the readability of the face animation.

It is easier to implement and train systems by simple DTW matching instead of phoneme level labeling, which is a time consuming manual work.

Holding to continuous speech processing is good to support potential language independency also.

The direct subjective test with hard of hearing persons will be done in the next phase of our research project.

DTW method can be tuned for our purposes with specially designed cumulative distance formulas, for long pauses in composite sentences in particular, which caused the errors of lowest scored videos in the subjective tests.

# Acknowledgements

This work was supported by the Mobile Innovation Center, Hungary. The authors would like to thank the National office for Research and Technology for supporting the project in the frame of Contract No 472/04. Many thanks to our hearing impaired friends for participating in tests and for their valuable advices, remarks.

# References

1. Takács, G., Tihanyi, A., Bárdi, T., Feldhoffer, G., Srancsik, B.: Speech to Facial Animation Conversion for Deaf Customers. In: Proceedings of EUSIPCO Florence Italy (2006)
2. Takács, G., Tihanyi, A., Bárdi, T., Feldhoffer, G., Srancsik, B.: Signal Conversion from Natural Audio Speech to Synthetic Visible Speech. In: Proceedings of International Conference on Signals and Electronic Systems, Lodz, Poland, vol. 2, p. 261 (2006)
3. Takács, G., Tihanyi, A., Bárdi, T., Feldhoffer, G., Srancsik, B.: Database Construction for Speech to Lip-readable Animation Conversion. In: Proceedings of ELMAR Zadar, Croatia p. 151 (2006)
4. Granström, B., Karlsson, I., Spens, K.-E.: SYNFACE – a project presentation. In: Proc of Fonetik 2002, TMH-QPSR, vol. 44, pp. 93–96 (2002)
5. Johansson, M., Blomberg, M., Elenius, K., Hoffsten, L.E., Torberger, A.: Phoneme recognition for the hearing im-paired. TMH-QPSR, Fonetik 44, 109–112 (2002)
6. Rabiner, L.R., Juang, B-H.: Fundamentals of speech recognition (1993)

# Constructing Empirical Models
# for Automatic Dialog Parameterization

Mikhail Alexandrov[1], Xavier Blanco[1], Natalia Ponomareva[2], and Paolo Rosso[2]

[1] Universidad Autonoma de Barcelona, Spain
[2] Universidad Politecnica de Valencia, Spain

**Abstract.** Automatic classification of dialogues between clients and a service center needs a preliminary dialogue parameterization. Such a parameterization is usually faced with essential difficulties when we deal with politeness, competence, satisfaction, and other similar characteristics of clients. In the paper, we show how to avoid these difficulties using empirical formulae based on lexical-grammatical properties of a text. Such formulae are trained on given set of examples, which are evaluated manually by an expert(s) and the best formula is selected by the Ivakhnenko Method of Model Self-Organization. We test the suggested methodology on the real set of dialogues from Barcelona railway directory inquiries for estimation of passenger's politeness.

## 1 Introduction

### 1.1 Problem Setting

Nowadays, dialogue processing is widely used for constructing automatic dialogue systems and for improving service quality. By the word "dialogue" we mean a conversation between a client and a service center, and by the word "processing" we mean a classification of clients. Politeness, competence, satisfaction, etc. are very important characteristics for client classification but its formal estimation is quite difficult due to the high level of subjectivity. Thus, these characteristics usually are not taken into account or they are estimated manually [1].

In this work, we aim to construct an empirical formula to evaluate the mentioned characteristics, which is based on:

(i) objective lexical-grammatical indicators related to a given characteristic;
(ii) subjective expert opinion about dialogues.

The selection of lexical-grammatical indicators depends on expert experience. However, some simple indicators are often obvious, e.g. polite words for estimation of politeness, "if-then" expressions for the estimation of competence, or objections for estimation of a level of satisfaction. The technical problem is to find an appropriate tool for revealing such indicators and include this linguistic tool into the automatic process of dialogue parameterization.

V. Matoušek and P. Mautner (Eds.): TSD 2007, LNAI 4629, pp. 455–463, 2007.

Subjective expert opinion(s) may be obtained by means of manual evaluation of a set of dialogues. For this, a fixed scale is taken and each dialogue is evaluated in the framework of this scale. Usually symmetric normalized scale [-1,1] or positive normalized scale [0,1] is considered.

In order to construct an empirical formula we use the Inductive Method of Model Self-Organization (IMMSO) proposed by Ivakhnenko [7]. This method allows to select the best formula from a given class using the training and the control sets of examples.

For definiteness, in this paper we consider only client's politeness. And it should be emphasized that we have no aim to find the best way for numerical estimation of politeness. Our goal is only to demonstrate how one may transform the lexical-grammatical properties of a text and the subjective expert opinion to these numerical estimations.

The paper is organized as follows. Section 2 describes the linguistic factors that should be taken into account in the formula to be constructed. Section 3 shortly describes the Ivakhnenko method. Section 4 contains the results of experiments. Conclusions and future work are drawn in Section 5.

In order to construct an empirical formula we use the Inductive Method of Model Self-Organization (IMMSO) proposed by Ivakhnenko [7]. This method allows to select the best formula from a given class using the training and the control sets of examples.

For definiteness, in this paper we consider only client's politeness. And it should be emphasized that we have no aim to find the best way for numerical estimation of politeness. Our goal is only to demonstrate how one may transform the lexical-grammatical properties of a text and the subjective expert opinion to these numerical estimations.

The paper is organized as follows. Section 2 describes the linguistic factors that should be taken into account in the formula to be constructed. Section 3 shortly describes the Ivakhnenko method. Section 4 contains the results of experiments. Conclusions and future work are drawn in Section 5.

## 1.2  Related Works

The existing automatic tools related with the estimation of politeness only detect polite (impolite) expressions in dialogues but do not give any numerical estimation of the level of politeness [2,3]. And it can be easy explained: such estimations are too subjective. In the work [11], some formal factors of politeness are proposed and the empirical formula based on these factors is constructed. Nevertheless this formula was not properly justified: it was given in advance and fitted to data.

The Ivakhnenko method (it is better to say 'approach'), mentioned above has many applications in Natural Sciences and Techniques [7]. It has been applied in Computational Linguistics for constricting empirical formulae for testing word similarity [4,9].

## 2  Models for Parameter Estimation

### 2.1  Numerical Indicators

The model to be constructed represents a numerical expression, which depends on various indicators of politeness of a given text and determines a certain level of politeness. This level is measured by a value between 0 and 1, where 0 corresponds to a regular politeness, and 1 corresponds to the highest level of politeness. We do not consider any indicators of impoliteness, although in some cases it should be done.

In this paper we take into account the following 3 factors of politeness: the first greeting ($G$), polite words ($W$) and polite grammar forms ($V$). As examples of polite words such well-known expressions as "please", "thank you", "excuse me", etc. can be mentioned. We considered verbs in a subjunctive mood as the only polite grammar forms, e.g. "could you", "I would", etc.

We take into account the following two circumstances:

(i) The level of politeness does not depend on the length of dialogue. It leads to the necessity to normalize a number of polite expressions and polite grammar forms on the length of dialogue. The dialogue's length here is the number of client's phrases.

(ii) The level of politeness depends on the number of polite words and polite grammar forms non-linearly: the greater number of polite words and grammar forms occur in a text the less contribution new polite words and grammar forms give. It leads to the necessity to use any suppressed functions as the logarithm or the square root, etc.

Therefore, we consider the following numerical *indicators* of politeness:

$$G = \{0,1\}, \quad W = Ln(1 + N_w/L), \quad V = Ln(1 + N_v/L), \tag{1}$$

where $N_w$, $N_v$ are a number of polite words and polite grammar forms respectively and $L$ is a number of phrases.

It is evident that: a) $W = V = 0$, if polite words and polite grammar forms do not occur; b) $W = V = Ln(2)$, if polite words and polite grammar forms occur in every phrase. All these relations are natural and easy to understand.

### 2.2  Example

Here we demonstrate how the mentioned indicators are manifested and evaluated. Table 1 shows the example of dialogue (the records are translated from Spanish into English). Here *US* stands for a user and *DI* for a directory inquire service. This example concerns the train departure from Barcelona to other destinations both near the Barcelona and in other provinces of Spain.

Table 2 shows the results of parameterization of this dialogue and its manual estimation by a user.

**Table 1.** Example of a real dialogue between passengers and directory inquires

| | |
|---|---|
| *US*: **Good evening**. **Could you** tell me the schedule of trains to Zaragoza for tomorrow?<br>*DI*: For tomorrow morning?<br>*US*: Yes<br>*DI*: There is one train at 7-30 and another at 8-30<br>  *US*: And later?<br>  *DI*: At 10-30<br>  *US*: And till the noon?<br>  *DI*: At 12<br>*US*: **Could you** tell me the schedule till 4 p.m. more or less?<br>  *DI*: At 1-00 and at 3-30<br>  *US*: 1-00 and 3-30<br>  *DI*: hmm, hmm <SIMULTANEOUSLY><br>  *US*: And the next one?<br>  *DI*: I will see, one moment. The next train leaves at 5-30 | *US*: 5-30<br>*DI*: hmm, hmm < SIMULTANEOUSLY ><br>*US*: Well, and how much time does it take to arrive?<br><br>*DI*: 3 hours and a half<br>*US*: For all of them?<br>*DI*: Yes<br>*US*: Well, **could you** tell me the price?<br>*DI*: 3800 pesetas for a seat in the second class<br>*US*: Well, and what about a return ticket?<br>*DI*: The return ticket has a 20% of discount<br>*US*: Well, so, it is a little bit more than 6 thousands, no?<br><br>*DI*: Yes<br>*US*: Well, **thank you very much**<br>*DI*: Don't mention it, good bye |

**Table 2.** Parameterized dialogue

| First greeting $g$ | Number of polite words $Nw$ | Number of polite grammar forms $Nv$ | Indicator $G$ | Indicator $W$ | Indicator $V$ | Estimation |
|---|---|---|---|---|---|---|
| *Yes* | 1 | 3 | 1 | 0.07 | 0.21 | 0.75 |

In our work all factors $G$, $w$, $v$ are detected by means of the *NooJ* resource [10]. Early for the same goal we used morphological analyzers described in [6]. The *NooJ* is a linguistic tool to locate morphological, lexical and syntactic patterns used for raw texts processing. The results of *NooJ* work were fixed in a file for further processing by Ivakhnenko method.

## 2.3 Numerical Models

Having in view the three factors described above the following series of polynomial models can be suggested for automatic evaluation of the level of politeness:

Model 1:  $F(G,W,V) = A_0$

Model 2:  $F(G,W,V) = C_0 G$

Model 3:  $F(G,W,V) = A_0 + C_0 G$

Model 4:  $F(G,W,V) = A_0 + C_0 G + B_{10} W + B_{01} V$

Model 5:  $F(G,W,V) = A_0 + C_0 G + B_{10} W + B_{01} V + B_{11} VW$ $\qquad(2)$

Model 6:  $F(G,W,V) = A_0 + C_0 G + B_{10} W^2 + B_{01} V^2$

Model 7:  $F(G,W,V) = A_0 + C_0 G + B_{11} VW + B_{20} W^2 + B_{02} V^2$

Model 8:  $F(G,W,V) = A_0 + C_0 G + B_{10} W + B_{01} V + B_{11} VW + B_{20} W^2 + B_{02} V^2$

etc.

Here: $A_0, C_0, B_{ij}$ are undefined coefficients. It is easy to see that all these models are the polynomials with respect to the factors $W$ and $V$. Such a representation is enough general for various function $\psi(W,V)$ and this a reason of its application. Of course, one can suggest the other type of models.

## 3    The Ivakhnenko Method

### 3.1    The Contents of the Method

Inductive method of model self-organization (IMMSO) was suggested and developed by Ivakhnenko and his colleagues at 80s. This method allows to determine the model of optimal complexity, which well describe a given experimental data. Speaking 'model' we mean a formula, equation, algorithm, etc.

This method does not require any a priori information concerning distribution of parameters of objects under consideration. Just for this reason the Ivakhnenko method proves to be very effective in the problems of Data and Text Mining. Nevertheless it should be said that if such a priori information exists then the methods of Pattern Recognition will give better results.

This method has one restriction: it cannot find the optimal model in any continuous class of models because its work is based on the competition of the models. So this method is titled as an *inductive* one. The main principle of model selection is the principle of stability: the models describing different subsets of a given data set must be similar.

Here are the steps of the Ivakhnenko method

(1) An expert defines a sequence of models, from the simplest to more complex ones.
(2) Experimental data are divided into two data sets: training data and control data, either manually or using an automatic procedure.
(3) For a given kind of model, the best parameters are determined using, first, the training data and, then, the control one. For that any internal criteria of concordance between the model and the data may be used (e.g., the least squares criterion).
(4) Both models are compared on the basis of any external criteria, such as the criterion of regularity, or criterion of unbiasdness, etc. If this external criterion achieves a stable optimum, the process is finished; otherwise, more complex model is considered and the process is repeated from the step (3).

Why we expect the external criterion to reach any optimum? The fact is the experimental data are supposed to contain: (a) a regular component defined by the model structure and (b) a random component-noise. A simplified model does not react to noise, but simultaneously it does not reflect the nature of objects. Otherwise, a sophisticated model can reflect very well real object behavior but simultaneously such a model will reflect a noise. In both cases the values of the penalty function (external criterion) are large. The principle of model *self-organization* consists in that an external criterion passes its minimum when the complexity of the model is gradually increased.

## 3.2 Application of Method

There are two variants of the Ivakhnenko method:

    I Combinatorial Method
    II Method of Groupped Arguments

In the first case all variants of model are considered step-by-step. And in the second one the models are filtered [8]. In this work we use only the first method and consequently consider all 8 models presented in the section 2.3.

Parameters of the concrete model are determined by means of the least square method. For this we fix one of the models (2) and construct the system of lineal equations for a given set of dialogs:

$$F(g_i, w_i, v_i) = P_i, \quad i=1,..,N \tag{3}$$

Here: $g,w,v$ are the factors, $P_i$ are the manual estimations of dialog, $N$ is the number of dialogs. For example, the dialog described above forms the following equation for the 4$^{\text{th}}$ model: $A_0 + C_0 + 0.07B_{10} + 0.021\,B_{01} = 0.75$.

The system (3) is a system of lineal equations with respect to undefined coefficients. It can be solved by the least square method. It should take into account that the number of equations must be several times more then the number of parameters to be determined. It allows to filter a noise in the data. Speaking 'noise' we mean first of all fuzzy estimations of politeness.

According to IMMSO methodology for the series of models starting with the first model from (2) any external criterion is calculated and checked whether it achieved an optimal point. Depending on the problem different forms of this criterion can be proposed [8]. In our case we use the criterion of regularity. It consists in the following:

- model parameters (coefficients $A_0$, $C_0$, etc.) are determined on the training data set
- this model is applied to control data set and 'model' politeness is calculated
- the relative difference between the model politeness and the manual politeness of an expert is estimated

All these actions can be reflected by the following formula

$$K_r = \frac{\sqrt{\sum_N \left(P_i\,(T) - P_i\right)^2}}{\sqrt{\sum_N \left(P_i\right)^2}} \tag{4}$$

where $P_i(T)$ are the 'model' estimations of politeness on the control data set, that is the left part of the equations (3), $P_i$ are the manual estimations of dialogs from the control data set, $N$ is the number of dialogs in control data set. It should emphasize that the model parameters are determined on the training data set.

# 4 Experiments

The data we used in our experiments represent a corpus of 100 person-to-person dialogues of Spanish railway information service. The short characteristics of the corpus (length of talking, volume of lexis) are described in [5]. From the mentioned corpus of dialogues we took randomly $N = 15$ dialogues for training data set and $N=15$ dialogues for control data set. The level of politeness was estimated manually in the framework of scale [0, 1] with the step 0.25. Table 3 represents the part of data used for the experiments.

**Table 3.** Example of data used in the experiments

| G | W | V | $W^2$ | WV | $V^2$ | Manual estimation |
|---|---|---|---|---|---|---|
| 1 | 0.134 | 0.194 | 0.0178 | 0.0259 | 0.0377 | 1 |
| 0 | 0.111 | 0.057 | 0.0124 | 0.0064 | 0.0033 | 0.75 |
| 1 | 0.000 | 0.074 | 0.0000 | 0.0000 | 0.0055 | 0.25 |
| 1 | 0.000 | 0.031 | 0.0000 | 0.0000 | 0.0009 | 0 |
| 1 | 0.000 | 0.118 | 0.0000 | 0.0000 | 0.0139 | 0.75 |
| 1 | 0.043 | 0.043 | 0.0018 | 0.0018 | 0.0018 | 0.5 |
| 1 | 0.000 | 0.000 | 0.0000 | 0.0000 | 0.0000 | 0.25 |
| 1 | 0.043 | 0.083 | 0.0018 | 0.0035 | 0.0070 | 0.5 |
| 0 | 0.000 | 0.074 | 0.0000 | 0.0000 | 0.0055 | 0 |
| 1 | 0.134 | 0.069 | 0.0178 | 0.0092 | 0.0048 | 1 |

We tested all 8 models (2) and calculated the criterion of regularity (4). The results are presented in the Table 4.

**Table 4.** Criterion of regularity

| Model- | Model- | Model- | Model- | Model- | Model- | Model- | Model- |
|---|---|---|---|---|---|---|---|
| 0.505 | 0.567 | 0.507 | *0.253* | 0.272 | 0.881 | 1.875 | 0.881 |

It is easy to see, that the lineal model is a winner. The fact that the model reflects only trend could be explained by imperfectness of a given class of models and/or a high level of noise. Joining together all 30 examples we determined the final formula as:

$$F(g, w, v) = -0.14 + 0.28G + 3.59W + 3.67V \qquad (5)$$

This formula provides 24% of relative mean square root error.

In order to evaluate the sensibility of results to the volume of data the same calculations were completed on the basis 10 dialogs taken for training and 10 data taken for control. We considered only first 4 models: more complex models needs more data. The results presented in the Table 5 show that the dependence on the volume is insignificant with respect to the behavior of external criterion.

**Table 5.** Criterion of regularity for shorten data set

| Model- | Model- | Model- | Model- |
|--------|--------|--------|--------|
| 0.497  | 0.503  | 0.502  | 0.319  |

## 5  Conclusions and Future Work

In this paper, we suggested the simple methodology for automatic estimation of various 'fuzzy' dialogue characteristics, which have a large level of subjectivity. We applied this methodology for the estimation of politeness. The constructed formula correctly reflects the contribution of selected factors of politeness: all factors have positive coefficients. The obtained error is comparative with the step of the manual dialogue estimation.

In the future, we intend to consider more complex empirical models for estimation of politeness, culture and competence, satisfaction.

## Acknowledgements

The authors thank Dr. Alexander Gelbukh from the National Polytechnic Institute of Mexico for his valuable notes and help in this investigation.

This work has been partially supported by MCyT TIN2006-15265-C06-04 research project.

## References

1. Alexandrov, M., Sanchis, E., Rosso, P.: Cluster analysis of railway directory inquire dialogs. In: Matoušek, V., Mautner, P., Pavelka, T. (eds.) TSD 2005. LNCS (LNAI), vol. 3658, pp. 385–392. Springer, Heidelberg (2005)
2. Alexandris, C., Fotinea, S.E.: Discourse particles: Indicators of positive and non-positive politeness in the discourse structure of dialog systems for modern greek. Intern. J. for Language Data Processing Sprache Datenverarbeitung, 1-2, 19–29 (2004)
3. Ardissono, L., Boella, C., Lesmo, L.: Indirect speech acts and politeness: A computational approach. In: Proceedings of the 17th Cognitive Science Conference, pp. 113–117 (1995)
4. Blanco, X., Alexandrov, M., Gelbukh, A.: Modified Makagonov's Method for Testing Word Similarity and its Application to Constructing Word Frequency List. In: Advances in Natural Language Processing, Mexico, pp. 27–36 (2006)
5. Bonafonte, A.: Desarrollo de un sistema de dialogo oral en dominios restringidos (in Spanish). In: I Jornadas en Tecnologia de Habla (2000)
6. Gelbukh, A., Sidorov, G.: Approach to construction of automatic morphological analysis systems for inflective languages with little effort. In: Gelbukh, A. (ed.) CICLing 2003. LNCS, vol. 2588, pp. 215–220. Springer, Heidelberg (2003)

7. Ivakhnenko, A.: Inductive method of model self-organization of complex systems (in Russian), Tehnika Publ. (1982)
8. Ivakhnenko, A.: Manual on typical algorithms of modeling (in Russian). Tehnika Publ. (1980)
9. Makagonov, P., Alexandrov, M.: Constructing empirical formulas for testing word similarity by the inductive method of model self-organization. In: Advances in Natural Language Processing. Mexico, pp. 239–247 (2004)
10. NooJ description: http://www.nooj4nlp.net
11. Ponomareva, N., Blanco, X.: Example-based empirical formula for politeness estimation in dialog processing. In: Proceedings of 10th Intern. Workshop on NooJ applications (to be published, 2007)

# The Effect of Lexicon Composition
# in Pronunciation by Analogy

Tasanawan Soonklang[1], R.I. Damper[1], and Yannick Marchand[2]

[1] School of Electronics and Computer Science
University of Southampton
Southampton SO17 1BJ, UK
ts03r,rid@ecs.soton.ac.uk
[2] Institute for Biodiagnostics (Atlantic)
1796 Summer Street, Suite 3900, Halifax
Nova Scotia, Canada B3H 3A7
Yannick.Marchand@nrc-cnrc.gc.ca

**Abstract.** Pronunciation by analogy (PbA) is a data-driven approach to phonetic transcription that generates pronunciations for unknown words by exploiting the phonological knowledge implicit in the dictionary that provides the primary source of pronunciations. Unknown words typically include low-frequency 'common' words, proper names or neologisms that have not yet been listed in the lexicon. It is received wisdom in the field that knowledge of the class of a word (common versus proper name) is necessary for correct transcription, but in a practical text-to-speech system, we do not know the class of the unknown word *a priori*. So if we have a dictionary of common words and another of proper names, we do not know which one to use for analogy unless we attempt to infer the class of unknown words. Such inference is likely to be error prone. Hence it is of interest to know the cost of such errors (if we are using separate dictionaries) and/or the cost of simply using a single, undivided dictionary, effectively ignoring the problem. Here, we investigate the effect of lexicon composition: common words only, proper names only or a mixture. Results suggest that high-transcription accuracy may be achievable without prior classification.

## 1 Introduction

Text-to-phoneme conversion is an integral part of several important speech technologies. The main strategy to determine pronunciation from spelling is to look up the word in a dictionary (or 'lexicon', or 'lexical database') to retrieve its pronunciation, since this is straightforward to implement and yields ~100% accuracy. However, the set of all words of a language is unbounded, so is not possible to list them all. Missing words typically include low-frequency 'common' words, neologisms and proper names, i.e., of people, streets, companies, etc. Thus, there must be a backup strategy for pronouncing unknown words not in the dictionary.

One of the most successful backup strategies (vastly superior to expert rules [1]) is pronunciation by analogy (PbA), which exploits the phonological knowledge implicit in the dictionary of known words to generate a pronunciation for an unknown word. So

V. Matoušek and P. Mautner (Eds.): TSD 2007, LNAI 4629, pp. 464–471, 2007.

far, many variants of PbA have been proposed and evaluated with different lexicons. With very few exceptions, previous works using PbA assumed that any missing words tend to be neologisms and so have used a lexicon of common words only. Yet there is a general consensus in the field that knowledge of word class (common word versus proper name) is essential to high-accuracy pronunciation.

In practice, when encountering an unknown word in the input to a text-to-speech (TTS) system, we would not know if it is a proper name or a common word. It should be possible to develop techniques for automatic classification, but these will never be entirely error-free. Therefore, one of several aspects to investigating the performance of PbA is whether or not it makes a difference when the system infers a pronunciation by analogy with a lexicon containing: (1) known common words only, (2) known proper names only, or (3) a mix of common words and proper names.

If high accuracy can be obtained in case (3), then automatic classification of unknown words (with attendant potential for errors) might be avoided. Since PbA infers pronunciations using lexical words most similar (in an analogical sense) to the unknown word, there is a reasonable chance of this. In the best case, the pronunciation of a proper name will be inferred predominantly by analogy with proper names in the dictionary, whereas the pronunciation of a common word will be inferred predominantly by analogy with common words in the dictionary, without having to separate the lexical entries into the two classes in advance. In this paper, we test this possibility, focusing on the effect that lexicon composition has on pronunciation accuracy for PbA.

## 2   Pronunciation by Analogy

An early, influential PbA system was PRONOUNCE, described by Dedina and Nusbaum [2]. Since then, there have been many variants, e.g., [3,4,5,6,7,8], more or less based on PRONOUNCE. The variant of PbA used in this work features several enhancements to PRONOUNCE as detailed in [8]. The pronunciation of an unknown word is assigned by comparing a substring of the input to a substring of words in the lexicon, gaining a phoneme set for each substring that matches, and then assembling the phoneme sets together to construct the pronunciation. As depicted in Figure 1, this process is comprised of four components briefly described as follows.

### 2.1   Aligned Lexical Database

PbA requires a dictionary in which the letters of each word's spelling are aligned in one-to-one fashion with the phonemes (possibly including "nulls") of the corresponding pronunciation. We use the algorithm of Damper et al. [9] for this.

### 2.2   Substring Matching

Substring matching is performed between the input letter string and dictionary entries, starting with the initial letter of the input string aligned with the end letter of the dictionary entry. If common letters in matching positions in the two strings are found,

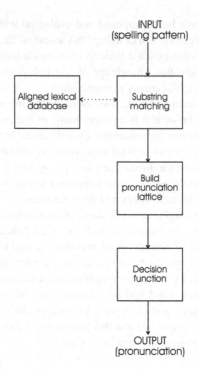

**Fig. 1.** Dedina and Nusbaum's PRONOUNCE

their corresponding phonemes (according to the prior alignment) and information about their positions in the input string are used to build a pronunciation lattice, as detailed next. One of the two strings is then shifted relative to the other by one letter and the matching process continued, until the end letter of the input string aligns with the initial letter of the dictionary entry. This process is repeated for all entries in the dictionary.

### 2.3   Building the Pronunciation Lattice

The pronunciation lattice is a directed graph in which information on matching substrings is used to construct nodes and arcs in the lattice for the particular input string. A lattice node represents a matched letter, $L_i$, at some position, $i$, in the input. The node is labelled with its position $i$ in the string and the phoneme corresponding to $L_i$ in the matched substring, $P_{im}$ say, for the $m$th matched substring. An arc is labelled with the phonemes intermediate between $P_{im}$ and $P_{jm}$ in the phoneme part of the matched substring and the frequency count, increasing by one each time the substring with these phonemes is matched during the search through the lexicon. If the arcs correspond to bigrams, the arcs are labelled only with the frequency. The phonemes of the bigram label the nodes at each end. Additionally, there is a *Start* node at position 0, and an *End* node at position equal to the length of the input string plus one.

## 2.4   Decision Function

Finally, the decision function finds the complete shortest path(s) through the lattice from *Start* to *End*. The possible pronunciations for the input correspond to the output strings assembled by concatenating the phoneme labels on the nodes/arcs in the order that they are traversed. In the case of only one candidate pronunciation corresponding to a unique shortest path, this is selected as the output. If there are tied shortest paths, then the five strategies of heuristic scoring of candidate pronunciations proposed in [8] and [10] are used, and combined by rank fusion to give a final result.

# 3   Lexical Databases

Two publicly-available dictionaries have been used in this work: the British English Example Pronunciation (BEEP) of common words and the Carnegie-Mellon University Dictionary (CMU) of common words and proper names. The former is intended to document British English pronunciations, whereas the latter contains American English pronunciations. We have also studied proper-name and common-word subsets of CMU and mixtures of BEEP and CMU.

## 3.1   BEEP

BEEP is available as file `beep.tar.gz` from `ftp://svr-ftp.eng.cam.ac.uk/comp.speech/dictionaries/`. It contains approximately 250,000 word spellings and their transcriptions. After removing some words that contain non-letter symbols and/or words with multiple pronunciations, the number of words used in this work is 198,632. The phoneme set for BEEP consists of 44 symbols.

## 3.2   CMU Dictionary

CMU contains both common words and proper names, and their phonemic transcriptions. The phoneme set for CMU contains 39 symbols. The latest version (CMU version 0.6) can be downloaded from `http://www.speech.cs.cmu.edu/cgi-bin/cmudict`. There are some duplicate words, some containing non-letter symbols and some where the pronunciations obviously does not match the spelling. These were removed to leave 112,102 words. CMU can be partitioned into two subsets as follows.

**Proper Name Subset.** There is no single, easily-available list of proper names and their pronunciations. However, a proper-name dictionary can be developed by using a list of proper names (without pronunciations) together with the standard CMU version 0.6. The list of names can be downloaded as file `cmunames.lex.gz` from `http://www.festvox.org`. It includes the most frequent names and surnames in the USA and their pronunciations [11], from a wide variety of origins. The procedure was simply to extract from CMU pronunciations for the names on the first list. (Note, however, that some names on this list were not found in CMU.) We refer to this extracted subset as Names. The number of proper names in Names is 52,911.

**Common Word Subset.** After extracting the proper names from CMU as above, the remaining words form the common word subset of 59,191 words. We call this dictionary Com.

Finally, we have used a 'Mixture' dictionary; a combination of the BEEP and Names dictionaries. Because different phoneme sets are used by the two dictionaries, we need to collapse the larger of the two (BEEP) onto the smaller (CMU), so that there is a uniform inventory of phonemes. This is the process of harmonisation [1]. Precise details of the harmonisation scheme are omitted for the sake of space.

## 4   Results

Performance was evaluated using a leave-one-out strategy. That is, each word was removed in turn from the dictionary and a pronunciation derived by analogy with the remaining words. Results are reported in terms of words correct, i.e., the number of words for which all phonemes of the transcription exactly match all the phonemes of the corresponding word in the lexicon. Stress assignment has been ignored for simplicity.

Table 1 shows the results of PbA with BEEP, Names and the Mixture dictionary for all combinations of the three dictionaries as test set and lexical database. It should be noted that all entries are significantly different from one another (binomial tests, one-tailed, $p \sim 0$). As can be seen, best results for a given test-set dictionary are achieved when the same dictionary is used as the lexical database. Much higher accuracy is achieved when BEEP is used as the test set and lexical database (87.50% words correct) than when Names is used as the test set and lexical database (68.35% words correct). This is to be expected in view of the diversity of origin of the proper names and different degrees of asssimilation into English [12,13], making their pronunciations harder to infer. Cross-lexicon testing leads to a very large deterioration in performance. Although it is tempting to think that this indicates that proper name transcription is a harder problem than common name transcription, the difference could be due primarily or solely to the different sizes of lexicon, since PbA transcription accuracy is a strong function of dictionary size, increasing as the size of dictionary increases [14].

Using the Mixtures dictionary as both test set and lexical database reflects the practical situation in which no attempt is made to classify the word class, merely treating all words as from the same class. Here the relevant result is 78.08% words correct, a long way below the performance when words from BEEP are pronounced by analogy with the entire BEEP dictionary. Note that a simple weighted linear sum of the BEEP/BEEP and Names/Names results (where the weights are the proportions of the two classes of word) would predict a result of 83.5% words correct, some way above the 78.08% result actually obtained. In effect, this weighted linear sum forms an upper bound on the performance that could be obtained if we had a perfect means of identifying the class of any input word.

In the results of the previous paragraph, the Mixture dictionary is of course heterogeneous, consisting of a British English lexicon of common words (whose phoneme set has had to be harmonised to CMU) and an American English dictionary of proper names. This was done to have the largest possible dictionaries. We have also studied

**Table 1.** Percentage words correctly transcribed by PbA for BEEP, Names and Mixture dictionaries

|  | Lexicon | | |
|---|---|---|---|
| Test set | BEEP | Names | Mixture |
| BEEP | 87.50 | 15.93 | 83.62 |
| Names | 23.57 | 68.35 | 55.08 |
| Mixture | 73.34 | 26.62 | 78.08 |

**Table 2.** Percentage words correctly transcribed by PbA with Com, Names and CMU dictionaries

|  | Lexicon | | |
|---|---|---|---|
| Test set | Com | Names | CMU |
| Com | 75.67 | 28.20 | 75.94 |
| Names | 38.63 | 68.35 | 51.10 |
| CMU | 64.36 | 39.18 | 72.13 |

the performance of PbA when the three dictionaries (common words, proper names and mixture) are homogeneous, all being derived from CMU. That is, we have used Com, Names and CMU as the three dictionaries. In this case, CMU acts as the dictionary of 'mixtures' (containing both common words and proper names.) Table 2 shows the corresponding results.

Here, the pattern of results is quite similar except for the case of common words tested against the full CMU dictionary. The Com vs. Com result of 75.67% words correct is not significantly different from the Com vs. CMU result of 75.94% words correct (binomial test, two-tailed, $p = 0.876$). That is, extending the lexical database from Com to CMU when testing Com did *not* lead to any deterioration in performance, unlike the corresponding BEEP/Mixture case where there was a large deterioration. We are inclined to believe that the difference is due to the inhomogeneity of the latter (Mixture) dictionary, and the avoidance of harmonisation for Com/CMU. Thus, we give more credence to the results of Table 2 than to those of Table 1.

The very positive Com/CMU result is intriguing. Why does Com vs. CMU, where there is a partial mismatch of the test set and lexicon, perform as well as Com vs. Com, where there is not? It cannot be because proper names are similar in some way to common words with respect to pronunciation by analogy, since this interpretation is denied by all the other results. For instance, there is a huge drop in performance (binomial test, one-tailed, $p \sim 0$) when testing Names against the full CMU dictionary, indicating that proper names have some special characteristics different from common words, as expected from their diversity. The most likely explanation is that PbA is somehow successful in forming strong analogies between common test words and common words in the CMU lexicon, while analogies between these words and the proper names in the lexicon (i.e., the 'wrong' class) are much weaker. This interpretation is currently under investigation.

Let us turn finally to the result of most practical interest; that is, the comparison of Com vs. Com with CMU vs. CMU. This reflects the situation where we have a single, undivided lexicon in the TTS system. Here, the relevant figures are 75.67% and 72.13% words correct, respectively. This latter figure is almost exactly what we would predict from a weighted linear sum of the Com vs. Com and Name vs. Names results. This is an important finding, since it constitutes compelling evidence for independent errors for the two different classes of word. It supports the working hypothesis of strong analogies between test words of a particular class and lexical entries of the same class and weak analogies between test words of a particular class and lexical entries of the other class. If correct, this means there would be no advantage to attempting automatic inference of input-word class, since the analogy process itself.

## 5   Conclusions

Pronunciation by analogy has been described and tested with different lexicon compositions: common words only, proper names only, and a mixture of the two. Although we attempted to exploit the existence of the large BEEP dictionary, the attempt was complicated by the absence of a list of proper names and their pronunciations for British English. Thus, we believe that our most credible results are those for American English using the CMU dictionary, and common-word and proper-name subsets thereof. In this case, excellent performance has been obtained on the mixture, comparable to that on common names alone. This intriguing result suggests that there may be no need for automatic word class categorisation (common word versus proper name) to be attempted, with its attendant dangers of mis-classification. This interpretation is greatly strengthened by the observation that the result when testing all available words is almost exactly that predicted by a linear sum of the individual word accuracies, weighted by the relative proportions of common words and proper names, respectively, in CMU. As this prediction is based on assuming independence of errors for the two classes of word, it can be viewed as an upper bound on performance for a mixed lexicon.

## References

1. Damper, R.I., Marchand, Y., Adamson, M.J., Gustafson, K.: Evaluating the pronunciation component of text-to-speech systems for English: A performance comparison of different approaches. Computer Speech and Language 13(2), 155–176 (1999)
2. Dedina, M.J., Nusbaum, H.C.: Pronounce: A program for pronunciation by analogy. Computer Speech and Language 5(1), 55–64 (1991)
3. Sullivan, K.P.H., Damper, R.I.: Novel-word pronunciation: A cross-language study. Speech Communication 13(3-4), 441–452 (1993)
4. Federici, S., Pirrelli, V., Yvon, F.: Advances in analogy-based learning: False friends and exceptional items in pronunciation by paradigm-driven analogy. In: Proceedings of International Joint Conference on Artificial Intelligence (IJCAI'95) Workshop on New Approaches to Learning for Natural Language Processing, Montreal, Canada, pp. 158–163 (1995)
5. Yvon, F.: Grapheme-to-phoneme conversion using multiple unbounded overlapping chunks. In: Proceedings of Conference on New Methods in Natural Language Processing (NeMLaP-2'96), Ankara, Turkey, pp. 218–228 (1996)

6. Damper, R.I., Eastmond, J.F.G.: Pronunciation by analogy: Impact of implementational choices on performance. Language and Speech 40(1), 1–23 (1997)

7. Pirrelli, V., Yvon, F.: The hidden dimension: A paradigmatic view of data-driven NLP. Journal of Experimental and Theoretical Artificial Intelligence 11(3), 391–408 (1999)

8. Marchand, Y., Damper, R.I.: A multistrategy approach to improving pronunciation by analogy. Computational Linguistics 26(2), 195–219 (2000)

9. Damper, R.I., Marchand, Y., Marsters, J.-D.S., Bazin, A.I.: Aligning text and phonemes for speech technology applications using an EM-like algorithm. International Journal of Speech Technology 8(2), 149–162 (2005)

10. Damper, R.I., Marchand, Y.: Information fusion approaches to the automatic pronunciation of print by analogy. Information Fusion 71(2), 207–220 (2006)

11. Font Llitjós, A.: Improving pronunciation accuracy of proper names with language origin classes, Master's thesis, Carnegie Mellon University, Pittsburgh, PA (2001)

12. Vitale, T.: An algorithm for high accuracy name pronunciation by parametric speech synthesizer. Computational Linguistics 17(3), 257–276 (1991)

13. Spiegel, M.F.: Proper name pronunciations for speech technology applications. International Journal of Speech Technology 6(4), 419–427 (2003)

14. Soonklang, T., Damper, R.I., Marchand, Y.: Effect of lexicon size on pronunciation by analogy of English, submitted to Interspeech 2007, Antwerp, Belgium (August 2007)

# Festival-si: A Sinhala Text-to-Speech System

Ruvan Weerasinghe, Asanka Wasala, Viraj Welgama, and Kumudu Gamage

Language Technology Research Laboratory, University of Colombo School of Computing,
35, Reid Avenue, Colombo 00700, Sri Lanka
arw@ucsc.cmb.ac.lk,
{awasala,vwelgama,kgamage}@webmail.cmb.ac.lk

**Abstract.** This paper brings together the development of the first Text-to-Speech (TTS) system for Sinhala using the Festival framework and practical applications of it. Construction of a diphone database and implementation of the natural language processing modules are described. The paper also presents the development methodology of direct Sinhala Unicode text input by rewriting letter-to-sound rules in Festival's context sensitive rule format and the implementation of Sinhala syllabification algorithm. A Modified Rhyme Test (MRT) was conducted to evaluate the intelligibility of the synthesized speech and yielded a score of 71.5% for the TTS system described.

## 1 Introduction

In this paper, we describe the implementation and evaluation of a Sinhala text-to-speech system based on the diphone concatenation approach. The Festival framework [1] was chosen for implementing the Sinhala TTS system. The Festival Speech Synthesis System is an open-source, stable and portable multilingual speech synthesis framework developed at the Center for Speech Technology Research (CSTR), of the University of Edinburgh.

TTS systems have been developed using the Festival framework for different languages including English, Japanese [1], Welsh [12], [2], Turkish [9], and Hindi [5], [8], Telugu [3], [5], among others. However, no serious Sinhala speech synthesizer has been developed this far. This is the first known documented work on a Sinhala text-to-speech synthesizer. The system is named "Festival-si" in accordance with common practice.

The rest of this paper is organized as follows: Section 2 gives an overview of the Sinhala phonemic inventory; Section 3 explains the diphone database construction process; the implementation of natural language processing modules is explained in section 4. Section 5 discusses the potential applications while Section 6 presents an evaluation of the current system. The work is summarized and future research directions and improvements are discussed in the last section.

## 2 Sinhala Phonemic Inventory

Sinhala is one of the official languages of Sri Lanka and the mother tongue of the majority - 74% of its population. Spoken Sinhala contains 40 segmental phonemes;

V. Matoušek and P. Mautner (Eds.): TSD 2007, LNAI 4629, pp. 472–479, 2007.

14 vowels (/i/, /i:/, /e/, /e:/ ,/æ/, /æ:/, /ə/, /ə:/, /u/, /u:/, /o/, /o:/, /a/, /a:/) and
26 consonants as classified below in Table 1 [4].

**Table 1.** Spoken Sinhala Consonant Classification

|  | Labial | Dental | Alveolar | Retroflex | Palatal | Velar | Glottal |
|---|---|---|---|---|---|---|---|
| Stops | p b | t d |  | ʈ ɖ |  | k g |  |
| Affricates |  |  |  |  | tʃ dʒ |  |  |
| Pre-nasalized voiced stops | ᵐb | ⁿd |  | ⁿɖ |  | ᵑg |  |
| Nasals | m |  | n |  | ɲ | ŋ |  |
| Trill |  |  | r |  |  |  |  |
| Lateral |  |  | l |  |  |  |  |
| Fricatives | f v | s |  |  | ʃ |  | h |
| Approximants |  |  |  |  | j |  |  |

## 3 Diphone Database Construction

This section describes the methodology adopted in the construction of Sinhala
diphone database.

Prior to constructing the diphone database, the answers to the following two
questions were investigated [9]: What diphone-pairs exist in the language? What
carrier words should be used?. Generally, the number of diphones in a language is
roughly the square of the number of phones. Therefore, 40 phonemes for Sinhala
identified in section 2 suggest roughly 1600 diphones should exist. The first phase
involved the preparation of matrices mapping all possible combinations of consonants
and vowels; i.e. CV, VC, VV, CC, _V, _C, C_ and, V_. Here '_' denotes a short
period of silence. In the second phase, redundant diphones were marked to be omitted
from the recording. Due to various phonotactic constraints, not all phone-phone pairs
occur physically. (for instance, diphone "ᵐb-ᵑg" never occurs in Sinhala). All such
non-existent diphones were identified after consulting a linguist. Finally,
1413 diphones were determined.

The third phase involved in finding the answer to the second question; What carrier
words should be used?; In other words, to compile set of words each containing an
encoded diphone. Following the guidelines given in the Festvox manual [1] it was
decided to record nonsense words containing the targeted diphone. These nonsense
words were embedded in carrier sentences including four other nonsensical context
words. A care was taken when coining these nonsensical words, so that these words
act in accordance with phonotactics of the Sinhala language. The diphone is extracted
from the middle syllable of the middle word, minimizing the articulatory effects at the
start and end of the word. Also, the use of nonsensical words helped the speaker to
maintain a neutral prosodic context. The output of the third phase was 1413 sentences.

The fourth phase involved recording the sentences. A native professional male speaker chosen for recording practiced the intelligibility and pronunciation of the sentences. He was advised to maintain a constant pitch, volume, and fairly constant speech rate during the recording. In order to maintain all of the above stated aspects, recordings were limited to two 30-minute sessions per day. At each session, 100 sentences were recorded on average.

Recording was done in a professional studio with an optimum noise free environment. Initially the sentences were directly recorded to Digital Audio Tapes, and later transferred into wave files, redigitising at 44.1 kHz/16 bit quantization.

The most tedious and painstaking tasks were carried out in the fifth phase where the recordings were split into individual files, and diphone boundaries hand-labeled using the speech analysis software tool 'Praat'[1] . Afterwards, a script was written to transform Praat text-grid collection file into diphone index file (EST) as required by Festival [1].

The method for synthesis used in this project is Residual Excited Linear Predictive Coding (RELP Coding). As required by this method, pitch marks, Linear Predictive Coding (LPC) parameters and LPC residual values had to be extracted for each diphone in the diphone database. The script `make_pm_wave` provided by speech tools [1] was used to extract pitch marks from the wave files. Then, the `make_lpc` command was invoked in order to compute LPC coefficients and residuals from the wave files [1]. Having tested synthesizing different diphones, several diphones were identified problematic. An analysis of the errors revealed that most were due to incorrect pitch marking caused by the use of default parameters when extracting the pitch marks. The accurate parameters obtained by analyzing samples of speech with Praat were set in the scripts as per the guidelines given in [8]. Moreover, it was realized that lowering the pitch of the original wave files resulted in a more lucid speech. A proprietary software tool was used to lower the recorded pitch, and normalize it in terms of power so that all diphones had an approximately equivalent power. Subsequently, modified `make_pm_wave` and `make_lpc` scripts were used to extract the necessary parameters from the wave files. These overall post-processing steps significantly improved the voice quality.

A full listing of the scripts used for recording and creating the diphone database is available for download from http://www.ucsc.cmb.ac.lk/ltrl/projects/si.

## 4  Natural Language Processing Modules

When building a new voice using Festvox [1], templates of the natural language processing modules required by Festival are automatically generated as Scheme files. The NLP modules should be customized according to the language requirements. Hence, the language specific scripts (phone, lexicon, tokenization) and speaker specific scripts (duration and intonation) can be externally configured and implemented without recompiling the system [1], [9]. The NLP related tasks involved when building a new voice are [1], [5]: defining the phone-set of the language,

---

[1] Available from: http://www.praat.org

tokenization and text normalization, incorporation of letter-to-sound rules, incorporation of syllabification rules, assignment of stress patterns to the syllables in the word, phrase breaking, assignment of duration to phones and generation of f0 contour.

## 4.1 The Phone Set Definition

The identified phone-set for Sinhala in section 2 is implemented in the file festvox/ ucsc_sin_sdn_phoneset.scm. The proposed set of symbol scheme is found to be a versatile representation scheme for Sinhala phone-set. Along with the phone symbols, features such as vowel height, place of articulation and voicing are defined. Apart from the default set of features, new features that are useful in describing Sinhala phones are also defined. e.g. whether a consonant is pre-nasalized or not. These features will prove extremely useful when implementing prosody.

## 4.2 Tokenization and Text Normalization

The default text tokenization methodology implemented in Festival (which is based on whitespace, and punctuation characters) is used to tokenize Sinhala text. Once the text has been tokenized, text normalization is carried out. This step converts digits, numerals, abbreviations, and non-alphabetic characters into word sequence depending on the context. Text normalization is a non trivial task. Therefore, prior to implementation, it was decided to analyze running text obtained from a corpus. Text obtained from the category *"News Paper > Feature Articles > Other"* of the *UCSC Sinhala corpus BETA* was chosen due to the heterogeneous nature of these texts and hence better representation of the language in this section of the corpus[2]. A script was written to extract sentences containing digits from the text corpus. The issues were identified by thoroughly analyzing the sentence. Strategies to address these issues were devised. A function is implemented to convert any number (decimal or integer up to 1 billion) into spoken words.

In Sinhala, the conversion of common numbers is probably more complicated when compared to English. In certain numerical expressions, the number may be concluded from a word suffix. e.g. 5 pahen *(out of five)*, 1  paləvæni *(1st)*. Such expressions are needed to be identified by taking into consideration the added suffix in a post-processing module. A function is implemented to expand abbreviations into full words. Common abbreviations found by the corpus analysis are listed, but our architecture allows easy incorporation of new abbreviations and corresponding words. In some situations, the word order had to be changed. For example, 50% must be expanded as " " - sijəjət ə panəha *(percent hundred)*, 50m should be expanded as mi:t ər panəha *(meters fifty)*. All above mentioned functions are called effectively by analyzing the context, and then accurate expansions are obtained

The tokenization and text normalization modules are implemented in festvox/ ucsc_sin_sdn_tokenizer.scm and capable of normalizing elements such as numbers,

---

[2] This accounts for almost two-thirds of the size of this version of the corpus.

currency symbols, ratios, percentages, abbreviations, Roman numerals, time expressions, number ranges, telephone numbers, email addresses, English letters and various other symbols.

## 4.3 Letter-to-Sound Conversion

The letter-to-sound module is used to convert an orthographic text into its corresponding phonetic representation. Sinhala being a phonetic language has an almost one-to-one mapping between letters and phonemes.

We implemented the grapheme to phoneme (G2P) conversion architecture proposed by Wasala et al. in [10]. In this architecture, the UTF-8 textual input is converted to ASCII based phonetic representation defined in the Festival. This process takes place at the user-interface level. Owing to the considerable delay experienced when synthesizing the text, it was decided to re-write the above G2P rules in the Festival's context sensitive format [1]. The rules were re-written in UTF-8 multi-byte format following the work done for Telugu language [3]. The method was proven to work well causing no delay at all. The 8 rules proposed in [10] expanded up to 817 rules when re-written in context sensitive format. However, some frequently encountered important words were found incorrectly phonetized by these rules. Hence, such words along with their correct pronunciation forms were included in Festival's addenda, a part lexicon. The letter-to-sound rules and lexicon are implemented in festvox/ucsc_sin_lexi.scm.

Festival's UTF-8 support is still incomplete; however, we believe the above architecture as the best to deal with Unicode text input in Festival over other proposed methods [12], [10].

## 4.4 Syllabification and Stress Assignment

Instead of Festival's default syllabification function `lex.syllabify.phstress` based on sonority sequencing profile [1], a new function (`syllabify 'phones`) is implemented to syllabify Sinhala words. In this work, syllabification algorithm proposed by Weerasinghe et al. [11] is implemented. This algorithm is reported to have 99.95% accuracy [11]. The syllabification module is implemented in festvox/ucsc_sin_sdn_syl.scm.

## 4.5 Phrase Breaking Algorithm

The assignment of intonational phrase breaks to the utterances to be spoken is an important task in a text-to-speech system. The presence of phrase breaks in the proper positions of an utterance affects the meaning, naturalness and intelligibility of speech. There are two methods for predicting phrase breaks in Festival. The first is to define a Classification and Regression Tree (CART). The second and more elaborate method of phrase break prediction is to implement a probabilistic model using probabilities of a break after a word based on the part of speech of the neighboring words and the previous word [1]. However, due to the unavailability of a Part-of-Speech (POS), and a POS tagger for Sinhala, probabilistic model cannot be constructed yet. Thus, we opted for the simple CART based phrase breaking algorithm described in [1]. The algorithm is based on the assumption that phrase boundaries are more likely between content words and function words. A rule is defined to predict a break if the current word is a content word and the next is seemingly a function word and the current word is more than 5 words from a punctuation symbol.

This algorithm, initially developed for English, has proved to produce reasonable results for Sinhala as well. The phrasing algorithm is defined in festvox/ucsc_sin_sdn_phrase.scm.

### 4.6 Prosodic Analysis

Prosodic analysis is minimal in the current system and will be implemented in the future. The major challenge for building prosody for Sinhala is the lack of a POS tagset, POS tagger and tagged text corpus. An experiment was carried out to adapt CART trees generated for an English voice prosody (f0 & duration) modules into Sinhala. The CART trees were carefully modified to represent the Sinhala Phone-set. The phone duration values were also hand modified to incorporate natural phone durations. The above steps resulted in more natural speech when compared to the monotonous speech produced before incorporating them. These adapted modules (cmu_us_kal_dur.scm, cmu_us_kal_int.scm) are incorporated to the Festival-si system.

## 5  Integration with Different Platforms

Festival offers a powerful platform for the development and deployment of speech synthesis systems. Since most Linux distributions now come with Festival pre-installed, the integration of Sinhala voice in such platforms is very convenient. Furthermore, following the work done for Festival-te, the Festival Telugu voice [3], the Sinhala voice developed here was made accessible to GNOME-Orca[3] and Gnopernicus[4] - powerful assistive screen reader software for people with visual impairments.

Motivated by the work carried out in the Welsh & Irish Speech Processing Resources (WISPR) project [12], steps were taken to integrate Festival along with the Sinhala voice into the Microsoft Speech Application Programming Interface (MS-SAPI) which provides the standard speech synthesis and speech recognition interface within Windows applications [13]. As a result of this work, the MS-SAPI compliant Sinhala voice is accessible via any speech enabled Windows application. We believe that the visually impaired community would be benefited at large by this exercise owing to the prevalent use of Windows in the community. The Sinhala voice also proved to work well with Thunder[5] a freely available screen reader for Windows. This will cater to the vast demand for a screen reader capable of speaking Sinhala text. It is noteworthy to mention that for the first time the print disabled community in

---

[3] Available from: http://live.gnome.org/Orca
[4] Available from: http://www.baum.ro/gnopernicus.html

Sri Lanka will be able to work on computers in their local language by using the current Sinhala text-to-speech system.

## 6  Evaluation

Text-to-speech systems can be compared and evaluated with respect to intelligibility, naturalness, and suitability for used application [6]. As the Sinhala TTS system is a general-purpose synthesizer, a decision was made to evaluate it under the intelligibility criterion.

A Modified Rhyme Test (MRT) [6], [9] was designed to test the Sinhala TTS system. The test consists of 50 sets of 6 one or two syllable words which makes a total set of 300 words. The words are chosen to evaluate phonetic characteristics such as voicing, nasality, sibilation, and consonant germination. Out of 50 sets, 20 sets were selected for each listener. The set of 6 words is played one at the time and the listener marks the synthesized word. The overall intelligibility of the system from 20 listeners is found to be 71.5%. According to the authors' knowledge, this is the only reported work in the literature describing the development of a Sinhala text-to-speech system, and more importantly the first Sinhala TTS system to be evaluated using the stringent Modified Rhyme Test.

## 7  Conclusions and Future Work

In this paper we described the development and evaluation of the first TTS system for Sinhala language based on the Festival architecture. The design of a diphone database and the natural language processing modules developed have been described.

Future work will mainly focus on improving the naturalness of the synthesizer. Work is in progress to improve the prosody modules. A speech corpus containing 2 hours of speech has been already recorded. The material is currently being segmented, and labeled. We are also planning to improve the duration model using the data obtained from the annotated speech corpus. A number of other ongoing projects are aimed at developing a POS tag set, POS tagger and a tagged corpus for Sinhala.

Further work will focus on expanding the pronunciation lexicon. At present, the G2P rules are incapable of providing accurate pronunciation for most compound words. Thus, we are planning to construct a lexicon consisting of compound words along with common high frequency words found in our Sinhala text corpus, which are currently incorrectly phonetized.

## Acknowledgement

This work was made possible through the PAN Localization Project, (http://www.PANL10n.net) a grant from the International Development Research Center (IDRC), Ottawa, Canada, administered through the Center for Research in Urdu Language Processing, National University of Computer and Emerging Sciences, Pakistan.

---

[5] Available from: http://www.screenreader.net/

# References

1. Black, A.W., Lenzo, K.A.: Building Synthetic Voices, Language Technologies Institute, Carnegie Mellon University and Cepstral LLC. Retrieved from (2003), http://festvox.org/bsv/
2. Jones, R.J., Choy, A., Williams, B.: Integrating Festival and Windows. InterSpeech 2006. In: 9th International Conference on Spoken Language Processing, Pittsburgh, USA (2006)
3. [Error during LaTeX to Unicode conversion]
4. Karunatillake, W.S.: An Introduction to Spoken Sinhala, 3rd edn. M.D. Gunasena & Co. ltd. 217, Olcott Mawatha, Colombo, 11 (2004)
5. Kishore, S.P., Sangal, R., Srinivas, M.: Building Hindi and Telugu Voices using Festvox. In: Proceedings of the International Conference On Natutal Language Processing 2002 (ICON-2002), Mumbai (2002)
6. Lemmetty, S.: Review of Speech Synthesis Technology, MSc. thesis, Helsinki University of Technology (1999)
7. Louw, A.: A Short Guide to Pitch-Marking in the Festival Speech Synthesis System and Recommendations for Improvement. Local Language Speech Technology Initiative (LLSTI) Reports. Retrieved from (n.d.), http://www.llsti.org/documents.htm
8. Ramakishnan, A.G., Bali, K., Talukdar, P.P., Krishna, N.S.: Tools for the Development of a Hindi Speech Synthesis System. In: 5th ISCA Speech Synthesis Workshop, Pittsburgh, pp. 109–114 (2004)
9. Salor, Ö., Pellom, B., Demirekler, M.: Implementation and Evaluation of a Textto- Speech Synthesis System for Turkish. In: Proceedings of Eurospeech-Interspeech 2003, Geneva, Switzerland, pp. 1573–1576 (2003)
10. Wasala, A., Weerasinghe, R., Gamage, K.: Sinhala Grapheme-to-Phoneme Conversion and Rules for Schwa Epenthesis. In: Proceedings of the COLING/ACL 2006 Main Conference Poster Sessions, Sydney, Australia, pp. 890–897 (2006)
11. Weerasinghe, R., Wasala, A., Gamage, K.: A Rule Based Syllabification Algorithm for Sinhala. In: Dale, R., Wong, K.-F., Su, J., Kwong, O.Y. (eds.) IJCNLP 2005. LNCS (LNAI), vol. 3651, pp. 438–449. Springer, Heidelberg (2005)
12. Williams, B., Jones, R.J., Uemlianin, I.: Tools and Resources for Speech Synthesis Arising from a Welsh TTS Project. In: Fifth Language Resources and Evaluation Conference (LREC), Genoa, Italy (2006)
13. Microsoft Corporation.: Microsoft Speech SDK Version 5.1. Retrieved from (n.d.), http://msdn2.microsoft.com/en-s/library/ms990097.aspx

# Voice Conversion Based on Probabilistic Parameter Transformation and Extended Inter-speaker Residual Prediction*

Zdeněk Hanzlíček and Jindřich Matoušek

University of West Bohemia, Faculty of Applied Sciences,
Department of Cybernetics,
Univerzitní 8, 306 14 Plzeň, Czech Republic
zhanzlic@kky.zcu.cz, jmatouse@kky.zcu.cz

**Abstract.** Voice conversion is a process which modifies speech produced by one speaker so that it sounds as if it is uttered by another speaker. In this paper a new voice conversion system is presented. The system requires parallel training data. By using linear prediction analysis, speech is described with line spectral frequencies and the corresponding residua. LSFs are converted together with instantaneous $F_0$ by joint probabilistic function. The residua are transformed by employing residual prediction. In this paper, a new modification of residual prediction is introduced which uses information on the desired target $F_0$ to determine a proper residuum and it also allows an efficient control of $F_0$ in resulting speech.

## 1 Introduction

The aim of voice conversion is to transform an utterance pronounced by a source speaker so that it sounds as if it is spoken by a target speaker.

In [1] and [2], some initial experiments on voice conversion were presented. In this paper, a new voice conversion system is described. This system requires parallel training data which is analysed by using pitch-synchronous linear prediction. Thus, speech frames are represented by line spectral frequencies (LSFs), corresponding residua and instantaneous fundamental frequency. LSFs are converted together with instantaneous $F_0$ by using a joint probabilistic function. The residua are transformed by employing residual prediction – a method which estimates a suitable residuum for a given parameter vector. In this paper, a new extension of this method is introduced which uses information on the desired target $F_0$ to determine a proper residuum and it also allows an efficient control of $F_0$ in resulting speech.

This paper is organized as follows. Section 2 describes speech data used in our experiments and methods used for its analysis, synthesis and time-alignment. Section 3 deals with parameter transformation for voiced and unvoiced speech. Section 4 describes the extended inter-speaker residual prediction. In Section 5, the performance of our conversion system is evaluated. Finally, Section 6 concludes the paper and outlines our future work.

---

* Support for this work was provided by the Ministry of Education of the Czech Republic, project No. 2C06020, and the EU 6th Framework Programme IST-034434.

V. Matoušek and P. Mautner (Eds.): TSD 2007, LNAI 4629, pp. 480–487, 2007.

## 2   Speech Data

Speech data for our experiments were recorded under special conditions in an anechoic chamber. Along with the speech signal, the glottal signal (EGG) was recorded to ensure more robust pitch-mark detection and $F_0$ contour estimation.

Firstly, one female speaker recorded the reference utterances – a set of 55 short sentences. All sentences were in the Czech language. Subsequently, four other speakers (two males and two females) listened to these reference utterances and tried to repeat them in the reference speaker's style. This should guarantee better pronunciation and prosodic consistency among all speakers.

### 2.1   Speech Analysis and Synthesis

Our voice conversion system employs pitch-synchronous linear prediction (LP) analysis. Each voiced frame is two pitch long with one pitch overlap. Pitch-marks are extracted from the EGG signal. Unvoiced frames are 10 msec long with 5 msec overlap. LP parameters are represented by their line spectral frequencies (LSFs), which are converted by employing a probabilistic function (e.g. [3] or [5]). Residual signal is represented by its amplitude and phase FFT-spectra, which are transformed by using residual prediction (e.g. [5] or [6]).

The reconstruction of speech is performed by a simple OLA method. For analysis and synthesis we employ special weight windows; in both cases a square root of Hann window is used. This is a trade-off between efficacious speech description and smooth frame composition on condition of correct speech reconstruction.

### 2.2   Speech Data Alignment and Selection

To find the conversion function properly, the training data has to be correctly time-aligned. This is performed by the dynamic time warping (DTW) algorithm. For each frame, the feature vector consists of delta-LSFs and V/U flag whose value is 1 for voiced and 0 for unvoiced frame.

After time-alignment, some suspicious data have to be excluded from the training set, because they probably correspond to incorrect time-alignment caused e.g. by prosodic or pronunciation mismatch:

- pairs composed of one voiced and one unvoiced frame
- long constant sections (horizontal or vertical) of warping function
- frame pairs with a very low energy or with too different energy values
- frame pairs with too different values of normalized $F_0$

## 3   Parameter Transformation

Parameters (LSFs) are transformed using a probabilistic conversion function based on the description of training data with a Gaussian mixture model (GMM). The conversion function is determined for voiced and unvoiced speech separately. Although unvoiced speech is supposed to be unimportant for speaker identity perception, the conversion of unvoiced speech proved good on transitions between voiced and unvoiced speech. Without unvoiced speech transformation, some unusual source speaker's glimmer was noticed in the converted utterances.

### 3.1 Simple LSF Transformation

This approach to parameter transformation was proposed by Stylianou et al. [3] and later improved by Kain et al. [5]. However, we used it only for the conversion of unvoiced speech. The interrelation between source and target speaker's LSFs ($x$ and $y$, respectively) is described by a joint GMM with $Q$ mixtures

$$p(x,y) = \sum_{q=1}^{Q} \alpha_q \mathcal{N}\left\{ \begin{bmatrix} x \\ y \end{bmatrix}; \mu_q, \Sigma_q \right\}. \tag{1}$$

All unknown parameters (mixture weights $\alpha_q$, mean vectors $\mu_q$ and covariance matrices $\Sigma_q$) are estimated by employing the expectation-maximization (EM) algorithm. The mean vectors $\mu_q$ and covariance matrices $\Sigma_q$ consist of blocks which correspond to source and target speaker's components

$$\mu_q = \begin{bmatrix} \mu_q^x \\ \mu_q^y \end{bmatrix} \qquad \Sigma_q = \begin{bmatrix} \Sigma_q^{xx} & \Sigma_q^{xy} \\ \Sigma_q^{yx} & \Sigma_q^{yy} \end{bmatrix}. \tag{2}$$

The transformation function is defined as a conditional expectation of target $y$ given source $x$

$$\tilde{y} = E\{y|x\} = \sum_{q=1}^{Q} p(q|x)\left[ \mu_q^y + \Sigma_q^{yx}\left(\Sigma_q^{xx}\right)^{-1}(x - \mu_q^x) \right], \tag{3}$$

where $p(q|x)$ is the conditional probability of mixture $q$ given source $x$

$$p(q|x) = \frac{\alpha_q \mathcal{N}\{x; \mu_q^x, \Sigma_q^{xx}\}}{\sum_{i=1}^{Q} \alpha_i \mathcal{N}\{x; \mu_i^x, \Sigma_i^{xx}\}}. \tag{4}$$

### 3.2 Combined LSF and F$_0$ Transformation

This extension of the aforementioned simple LSF transformation was introduced by En-Najjary at al. [7]; however, the implemented system employed the Harmonic plus Noise Model of speech production.

This method exploits the interdependency between LSFs and instantaneous $F_0$; they are converted together by using one transformation function. Formally, new variables are introduced

$$\chi = \begin{bmatrix} 10^2 \cdot x \\ f_x \end{bmatrix} \qquad \psi = \begin{bmatrix} 10^2 \cdot y \\ f_y \end{bmatrix}. \tag{5}$$

A simple composition of LSFs and instantaneous $F_0$ would be unsuitable because the importance of particular components would not be well-balanced. This is the reason for introducing the weighting factor $10^2$; this value was experimentally selected and performs well for all speaker combinations. In [7], the balancing of components is solved by the normalization of fundamental frequency.

Again, the joint distribution of $\chi$ and $\psi$ is estimated using EM algorithm

$$p(\chi, \psi) = \sum_{q=1}^{Q} \alpha_q \mathcal{N} \left\{ \begin{bmatrix} \chi \\ \psi \end{bmatrix} ; \mu_q = \begin{bmatrix} \mu_q^\chi \\ \mu_q^\psi \end{bmatrix}, \Sigma_q = \begin{bmatrix} \Sigma_q^{\chi\chi} & \Sigma_q^{\chi\psi} \\ \Sigma_q^{\psi\chi} & \Sigma_q^{\psi\psi} \end{bmatrix} \right\} \qquad (6)$$

and the conversion function is defined as the conditional expectation of target $\psi$ given source $\chi$.

$$\tilde{\psi} = E\{\psi|\chi\} = \sum_{q=1}^{Q} p(q|\chi) \left[ \mu_q^\psi + \Sigma_q^{\psi\chi} (\Sigma_q^{\chi\chi})^{-1} (\chi - \mu_q^\chi) \right]. \qquad (7)$$

The resulting vector $\tilde{\psi}$ is decomposed into LSFs $\tilde{y}$ and instantaneous $F_0$ $\tilde{f}_y$ which is further used in the extended residual prediction method (see Section 4).

## 4  Residual Prediction

Residual prediction is a method which allows the estimation of a suitable residuum for a given parameter vector. It would be unsatisfactory to use the original source speaker's residua because the residual signal still contains significant information on speaker identity, mainly in voiced speech.

In voice conversion framework (see e.g. [6] or [5]), the residual prediction is traditionally based on probabilistic description of source speaker's cepstral parameter space – with a GMM. For each mixture of this model, a typical residual signal is determined; it is represented by its amplitude and phase residual spectrum. Naturally, this method is only used for voiced frames. In unvoiced speech, residua are adopted from source speech without any modification.

In [1] and [2], a new approach to residual prediction – so-called inter-speaker residual prediction – was introduced. In comparison with the traditional residual prediction, the cardinal difference is that the target speaker's residua are estimated directly by using the source speaker's parameter vectors. Moreover, the source speaker's parameter space is described in a non-probabilistic manner.

In this paper, a new extension of this method is proposed which uses information on the desired instantaneous $F_0$ during the selection of a suitable residuum and facilitates a simple and efficient control of $F_0$ in the transformed speech.

### 4.1  Training Stage

A non-probabilistic description of source LSF space is used. Source LSFs are clustered into $Q$ classes by employing the binary split k-means algorithm; a reasonable value of $Q$ is about 20. Each class $q$ is represented by its LSF centroid $\bar{x}_q$. The pertinence of parameter vector $x_n$ ($n = 1, 2, \ldots N$) to class $q$ ($q = 1, 2, \ldots Q$) can be expressed by the following weight

$$w(q|x_n) = \frac{\left[ d(\bar{x}_q, x_n) \right]^{-1}}{\sum_{i=1}^{Q} \left[ d(\bar{x}_i, x_n) \right]^{-1}} \qquad (8)$$

All training data are uniquely classified into these classes. For each class $q$, a set $R_q$ of pertaining data indices is established

$$R_q = \{k; 1 \leq k \leq N \wedge w(q|x_k) = \max_{i=1...Q} w(i|x_k)\}. \tag{9}$$

Thus all data $x_r$ for $r \in R_q$ belongs into class $q$. Within each parameter class $q$, the data is divided into $L_q$ subclasses according to their instantaneous $F_0$. The number of subclasses $L_q$ differs for particular parameter classes $q$. Each $F_0$ subclass is described by its central frequency $\bar{f}_q^\ell$ ($q$-th LSF class, $\ell$-th $F_0$ subclass) and the set of data belonging into this subclass is defined as a set $R_q^\ell$ of corresponding indices

$$R_q^\ell = \{k; k \in R_q \wedge d(\bar{f}_q^\ell, f_k) = \min_{i=1...L_q} d(\bar{f}_q^i, f_k)\}. \tag{10}$$

For each $F_0$ subclass, a typical residual amplitude spectrum $\hat{r}_q^\ell$ is determined as the weighted average of amplitude spectra belonging into this subclass

$$\hat{r}_q^\ell = \frac{\sum_{n \in R_q^\ell} r_n w(q|x_n)}{\sum_{n \in R_q^\ell} w(q|x_n)}. \tag{11}$$

Although all FFT-spectra cover the same frequency range given by the sampling frequency $f_s$, their lengths in samples are different because they correspond to the lengths of pitch-synchronously segmented frames. Thus all spectra have to be interpolated to the same length; cubic spline interpolation is used and the target length equals to the average length of all spectra within particular subclasses.

Similarly, the typical residual phase spectrum $\hat{\varphi}_q^\ell$ is determined. However because of phase warping problem, it is not calculated but it is only simply selected

$$\hat{\varphi}_q^\ell = \varphi_{n^*} \qquad n^* = \arg\max_{n \in R_q^\ell} w(q|x_n). \tag{12}$$

The selected residual phase spectrum should be interpolated to the same length as amplitude spectrum. To avoid the phase warping problem, nearest neighbour interpolation is used.

## 4.2   Transformation Stage

In the transformation stage, the desired target instantaneous fundamental frequency $\tilde{f}_n$ has to be known for each voiced frame; it is obtained by combined LSF & $F_0$ transformation (see Section 3.2).

The target residual amplitude spectrum $\tilde{r}_q$ is calculated as the weighted average over all classes. However, from each class $q$ only one subclass $\ell_q$ is selected whose centroid $\bar{f}_q^{\ell_q}$ is the nearest to the desired fundamental frequency $\tilde{f}_n$

$$\tilde{r}_n = \sum_{q=1}^{Q} \hat{r}_q^{\ell_q} w(q|x_n) \qquad \ell_q = \arg\min_{\ell=1...L_q} d(\bar{f}_q^\ell, \tilde{f}_n) \tag{13}$$

The target residual phase spectrum is selected from the parameter class $q^*$ with the highest weight $w(q|x_n)$ from the F0 subclass $\ell^*$ with the nearest central frequency $\bar{f}_{q^*}^{\ell^*}$.

$$\tilde{\varphi}_n = \hat{\varphi}_{q^*}^{\ell^*} \qquad q^* = \underset{q=1...Q}{\arg\max}\, w(q|x_n)$$

$$\ell^* = \underset{\ell=1...L_{q^*}}{\arg\min}\, d(\bar{f}_{q^*}^{\ell}, \tilde{f}_n) \tag{14}$$

The resulting amplitude and phase FFT-spectra have to be interpolated to the length given by the desired F0 $\tilde{f}_n$. The speech quality deterioration caused by this interpolation should not be significant, because the length of predicted residuum is very close to the target length.

## 5   Experiments and Results

In this section, the assessment of the described conversion system is presented. In the first subsection, mathematical evaluation of LSF and F0 transformation is presented. The second subsection deals with subjective evaluation by listening tests.

In all experiments, the conversion from the reference speaker to all other speakers was performed. 40 utterances were used for training and 15 different utterances for the assessment.

### 5.1   Objective Evaluation – LSF and F0 Transformation

The performance of LSF transformation can be expressed by using the performance index $I_{LSF}$

$$I_{LSF} = 1 - \frac{E(\tilde{y}, y)}{E(x, y)}, \tag{15}$$

where $E(x, y)$ is the average Euclidean distance between LSFs of 2 time-aligned utterances $x = \{x_1, x_2, \ldots x_N\}$ and $y = \{y_1, y_2, \ldots y_N\}$

$$E(x, y) = \frac{1}{N} \sum_{n=1}^{N} (x_n - y_n)^{\top}(x_n - y_n). \tag{16}$$

The higher value of performance index signifies the better conversion performance (maximum value is 1).

Similarly, the F0 transformation can be evaluated by performance index $I_{F_0}$

$$I_{F_0} = 1 - \frac{E(\tilde{f}_y, f_y)}{E(f_x, f_y)}, \tag{17}$$

or it can be also simply assessed by using average Euclidean distance $E(\tilde{f}_y, f_y)$ between transformed $\tilde{f}_y$ and target $f_y$ (the result is in Hz).

Results are stated in Table 1. They are presented separately for each speaker to expose that the outcomes are speaker dependent.

**Table 1.** Mathematical evaluation of LSF and $F_0$ transformation performance

| Target speaker | Male 1 | Male 2 | Female 1 | Female 2 |
|---|---|---|---|---|
| LSF performance index (voiced speech) | 0.412 | 0.335 | 0.317 | 0.344 |
| LSF performance index (unvoiced speech) | 0.316 | 0.254 | 0.237 | 0.217 |
| $F_0$ performance index | 0.764 | 0.836 | 0.510 | 0.336 |
| Default $F_0$ distance [Hz] (source – target) | 50.64 | 68.12 | 30.38 | 21.76 |
| Final $F_0$ distance [Hz] (transformed – target) | 11.97 | 11.15 | 14.89 | 14.49 |

Though some performance indices were proposed which should facilitate more complex transformation assessment (e.g. spectral performance index), they do not often correspond to the real speech quality and resulting speaker identity as it is perceived by people. Thus, the best way of evaluating a voice conversion system in a complex way is listening tests.

### 5.2 Subjective Evaluation – Speaker Discrimination Test

An extension of standard ABX test was used. 10 participants listened to triplets of utterances: original source and target (A and B in a random order) and transformed (X). They made decisions whether X sounds like A or B and rate their decision according to the following scale

1. X sounds like A
2. X sounds rather like A
3. X is halfway between A and B
4. X sounds rather like B
5. X sounds like B

For unified result interpretation, cases when A was from target and B from source speaker were reversed. Thus all results correspond to the case when A is source and B target utterance. Then the higher rating signifies the more effective conversion. Average rating for female-to-male conversion was 4.36 (i.e. listeners were sure of the target speaker identity) and for female-to-female conversion was 3.54 (i.e. the identity was closer to the target speaker, but it was not so persuasive).

## 6   Conclusion and Future Work

In this paper a new voice conversion system was introduced which is based on probabilistic transformation of LSF and fundamental frequency and which utilizes the extended inter-speaker residual prediction for determination of proper residual signals.

Speaker discrimination tests revealed that the identity of converted speech is closer to the target speaker. However in cases of similar source and target voices (female-to-female conversion), the decision was not definite. This was probably caused by insufficient speaking style consistency and a small amount of training data. Generally, all speakers had some difficulty reproducing the reference utterances; they focused on mimicking but their speech lost its fluency and naturalness. Thus, in our future work we will concentrate on approaches which do not require parallel training data.

# References

1. Hanzlíček, Z., Matoušek, J.: First Steps towards New Czech Voice Conversion System. In: Sojka, P., Kopeček, I., Pala, K. (eds.) TSD 2006. LNCS (LNAI), vol. 4188, pp. 383–390. Springer, Heidelberg (2006)
2. Hanzlíček, Z.: On Residual Prediction in Voice Conversion Task. In: Proceedings of the 16th Czech-German Workshop on Speech Processing, ÚRE AVČR, Prague, Czech Republic, pp. 90–97 (2006)
3. Stylianou, Y., Cappé, O., Moulines, E.: Continuous Probabilistic Transform for Voice Conversion. IEEE Transactions on Speech and Audio Processing 6(2), 131–142 (1998)
4. Kain, A., Macon, M.W.: Design and Evaluation of Voice Conversion Algorithm Based on Spectral Envelope Mapping and Residual Prediction. In: Proceedings of ICASSP'01 (2001)
5. Kain, A.: High Resolution Voice Transformation. Ph.D. thesis, Oregon Health & Science University, Portland, USA (2001)
6. Sündermann, D., Bonafonte, A., Ney, H., Höge, H.: A Study on Residual Prediction Techniques for Voice Conversion. In: Proceedings of ICASSP'05, pp. 13–16 (2005)
7. En-Najjary, T., Rosec, O., Chonavel, T.: A Voice Conversion Method Based on Joint Pitch and Spectral Envelope Transformation. In: Proceedings of Interspeech 2004 - ICSLP, pp. 1225–1228 (2004)

# Automatic Czech – Sign Speech Translation*

Jakub Kanis and Luděk Müller

University of West Bohemia, Faculty of Applied Sciences,
Department of Cybernetics
Univerzitní 8, 306 14 Pilsen, Czech Republic
{jkanis,muller}@kky.zcu.cz

**Abstract.** This paper is devoted to the problem of automatic translation between
Czech and SC in both directions. We introduced our simple monotone phrase-
based decoder - **SiMPaD** suitable for fast translation and compared its results
with the results of the state-of-the-art phrase-based decoder - **MOSES**. We compare the translation accuracy of handcrafted and automatically derived phrases
and introduce a "class-based" language model and post-processing step in order
to increase the translation accuracy according to several criteria. Finally, we use
the described methods and decoding techniques in the task of SC to Czech automatic translation and report the first results for this direction.

## 1 Introduction

In the scope of this paper, we are using the term Sign Speech (SS) for both the Czech
Sign Language (CSE) and Signed Czech (SC). The CSE is a natural and adequate communication form and a primary communication tool of the hearing-impaired people in
the Czech Republic. It is composed of the specific visual-spatial resources, i.e. hand
shapes (manual signals), movements, facial expressions, head and upper part of the
body positions (non-manual signals). It is not derived from or based on any spoken
language. CSE has basic language attributes, i.e. system of signs, double articulation,
peculiarity and historical dimension, and has its own lexical and grammatical structure.
On the other hand the SC was introduced as an artificial language system derived from
the spoken Czech language to facilitate communication between deaf and hearing people. SC uses grammatical and lexical resources of the Czech language. During the SC
production, the Czech sentence is audibly or inaudibly articulated and simultaneously
with the articulation the CSE signs of all individual words of the sentence are signed.

The using of written language instead of spoken one is a wrong idea in the case of
the Deaf. Hence, the Deaf have problems with the majority language understanding
when they are reading a written text. The majority language is the second language
of the Deaf and its use by the deaf community is only particular. Thus the majority
language translation to the sign speech is highly important for better Deaf orientation
in the majority language speaking world. Currently human interpreters provide this
translation, but their service is expensive and not always available. A full dialog system

---

* Support for this work was provided by the GA of the ASCR and the MEYS of the Czech
Republic under projects No. 1ET101470416 and MŠMT LC536.

(with ASR and Text-to-Sign-Speech (TTSS) [1] systems on one side (from spoken to sign language) and Automatic-Sign-Speech-Recognition (ASSR) and TTS systems on second side (from sign to spoken language)) represents a solution which does not intent to fully replace the interpreters, but its aim is to help in everyday communication in selected constraint domains such as post office, health care, traveling, etc. An important part of TTSS (conversion of written text to SS utterance (animation of avatar)) and ASSR systems is an automatic translation system which is able to make an automatic translation between the majority and the sign language.

In rest of this paper we describe our phrase-based translation system (for both directions: Czech to SC and SC to Czech). We compare the translation accuracy of a translation system based on phrases manually defined in the process of training corpus creation (phrases defined by annotators) with accuracy of the system based on phrases automatically derived from the corpus in the training process of Moses decoder [2]. In addition, we introduce a "class-based" language model based on the semantic annotation of the corpus and the post-processing method for Czech to SC translation.

## 2   Phrase-Based Machine Translation

The machine translation model is based on the noisy channel model scheme. When we apply the Bayes rule on the translation probability $p(\mathbf{t}|\mathbf{s})$ for translating a sentence $\mathbf{s}$ in a source language into a sentence $\mathbf{t}$ in a target language we obtain:

$$\mathrm{argmax}_t p(\mathbf{t}|\mathbf{s}) = \mathrm{argmax}_t p(\mathbf{s}|\mathbf{t}) p(\mathbf{t})$$

Thus the translation probability $p(\mathbf{t}|\mathbf{s})$ is decomposed into two separate models: a translation model $p(\mathbf{s}|\mathbf{t})$ and a language model $p(\mathbf{t})$ that can be modeled independently. In the case of phrase-based translation the source sentence $\mathbf{s}$ is segmented into a sequence of $I$ phrases $\bar{\mathbf{s}}_1^I$ (all possible segmentations has the same probability). Each source phrase $\bar{s}_i, i = 1, 2, ..., I$ is translated into a target phrase $\bar{t}_i$ in the decoding process. This particular $ith$ translation is modeled by a probability distribution $\phi(\bar{s}_i|\bar{t}_i)$. The target phrases can be reordered to get more precise translation. The reordering of the target phrases can be modeled by a relative distortion probability distribution $d(a_i - b_{i-1})$ as in [3], where $a_i$ denotes the start position of the source phrase which was translated into the $ith$ target phrase, and $b_{i-1}$ denotes the end position of the source phrase translated into the $(i - 1)th$ target phrase. Also a simpler distortion model $d(a_i - b_{i-1}) = \alpha^{|a_i - b_{i-1} - 1|}$ [3], where $\alpha$ is a predefined constant, can be taken. The best target output sentence $\mathbf{t_{best}}$ for a given source sentence $\mathbf{s}$ then can be acquired as:

$$\mathbf{t_{best}} = \mathrm{argmax}_t p(\mathbf{t}|\mathbf{s}) = \prod_{i=1}^{I} [\phi(\bar{s}_i|\bar{t}_i) d(a_i - b_{i-1})] p_{LM}(\mathbf{t})$$

Where $p_{LM}(\mathbf{t})$ is a language model of the target language (usually a trigram model with some smoothing usually built from a huge portion of target language texts). Note, that more sophisticated model in [3] uses more probabilities as will be given in Section 4.1.

## 3    Tools and Evaluation Methodology

### 3.1    Data

The main resource for the statistical machine translation is a parallel corpus which contains parallel texts of both the source and the target language. Acquisition of such a corpus in case of SS is complicated by the absence of the official written form of both the CSE and the SC. Therefore for all our experiments we use the Czech to Signed Czech (CSC) parallel corpus ([4]).

The CSC corpus contains 1130 dialogs from telephone communication between customer and operator in a train timetable information center. The parallel corpus was created by semantic annotation of several hundreds of dialog and by adding the SC translation of all dialogs. A SC sentence is written as a sequence of CSE signs. The whole CSC corpus contains 16 066 parallel sentences, 110 033 running words and 109 572 running signs, 4082 unique words and 720 unique signs. Every sentence of the CSC corpus has assigned the written form of the SC translation, a type of the dialog act, and its semantic meaning in a form of semantic annotation. For example (we use English literacy translation) for Czech sentence: *good day I want to know how me it is going in Saturday morning to brno* we have the SC translation: *good_day I want know how _ _ go in Saturday morning to brno* and for the part: *good day* the dialog act: *conversational_domain="frame" + speech_act="opening"* and the semantic annotation: *semantics="GREETING"*. The dialog act: *conversational_domain="task" + speech_act="request_info"* and semantic annotation: *semantics="DEPARTURE(TIME, TO(STATION))"* is assigned to the rest of the sentence. The corpus contains also handcrafted word alignment (added by annotators during the corpus creation) of every Czech – SC sentence pair. For more details about the CSC corpus see [4].

### 3.2    Evaluation Criteria

We use the following criteria for evaluation of our experiments. The first criterion is **Sentence Error Rate (SER):** It is a ratio of the number of incorrect sentence translations to the number of all translated sentences. The second criterion is **Word Error Rate (WER):** This criterion is adopted from ASR area and is defined as the Levensthein edit distance between the produced translation and the reference translation in percentage (a ratio of the number of all deleted, substituted and inserted produced words to the total number of reference words). The third criterion is **Position-independent Word Error Rate (PER):** it is simply a ratio of the number of incorrect translated words to the total number of reference words (independent on the word order). The last criterion is **BLEU** score ([5]): it counts modified n-gram precision for output translation with respect to the reference translation. A lower value of the first three criteria and a higher value of the last one indicate better i.e. more precise translation.

### 3.3    Decoders

We are using two different phrase-based decoders in our experiments. The first decoder is freely available state-of-the-art factored phrase-based beam-search decoder - **MOSES** ([2]). Moses can work with factored representation of words (i.e. surface form, lemma, part-of-speech, etc.) and uses a beam-search algorithm, which solves a problem of the exponential number of possible translations (due to the exponential number of possible alignments between source and target translation), for efficient decoding. The training tools for extracting of phrases from the parallel corpus are also available, i.e. the whole translation system can be constructed given a parallel corpus only. For language modeling we use the SRILM[1] toolkit.

The second decoder is our simple monotone phrase-based decoder - **SiMPaD**. The monotonicity means using the monotone reordering model, i.e. no phrase reordering is permitted. In the decoding process we choose only one alignment which is the one with the longest phrase coverage (for example if there are three phrases: $p_1$, $p_2$, $p_3$ coverage three words: $w_1$, $w_2$, $w_3$, where $p_1 = w_1 + w_2$, $p_2 = w_3$, $p_3 = w_1 + w_2 + w_3$, we choose the alignment which contains phrase $p_3$ only). Standard Viterbi algorithm is used for the decoding. SiMPaD uses SRILM[1] language models.

## 4 Experiments

### 4.1 Phrases Comparison

We compared the translation accuracy of handcrafted phrases with the accuracy of phrases automatically derived from the CSC corpus. The handcrafted phrases were simply obtained from the corpus. The phrase translation probability was estimated by the relative frequency ([3]):

$$\phi(\bar{s}_i|\bar{t}_i) = \frac{count(\bar{s}_i, \bar{t}_i)}{\sum_{\bar{s}_i} count(\bar{s}_i, \bar{t}_i)}$$

We used training tools of Moses decoder for acquiring the automatically derived phrases. The phrases were acquired from Giza++ word alignment of parallel corpus (word alignment established by Giza++[2] toolkit) by some heuristics (we used the default heuristic). There are many parameters which can be specified in the training and decoding process. Unless otherwise stated we used default values of parameters (for more details see Moses' documentation in [2]). The result of training is a table of phrases with five probabilities of the translation model: phrase translation probabilities $\phi(\bar{s}_i|\bar{t}_i)$ and $\phi(\bar{t}_i|\bar{s}_i)$, lexical weights $p_w(\bar{s}_i|\bar{t}_i)$ and $p_w(\bar{t}_i|\bar{s}_i)$ (for details see [3]) and phrase penalty (always equal $e^1 = 2.718$).

For comparison of results we carried out 20 experiments with various partitioning of data to the training and test set. The average results are reported in Table 1. The first column shows results of Moses decoder run on the handcrafted phrase table (phrase translation probability $\phi(\bar{s}_i|\bar{t}_i)$ only - HPH). The second column comprises results of Moses with automatically derived phrases (again phrase translation probability $\phi(\bar{s}_i|\bar{t}_i)$ only APH_PTS) and the third column contains results of Moses with automatically derived phrases and with all five translation probabilities (APH_ALL). The same language model was used in all three cases. The best results are in boldface. We used the standard sign test for the statistical significance determination. All results are given on the level of significance $= 0.05$.

---

[1] Available at http://www.speech.sri.com/projects/srilm/download.html
[2] Available at http://www.isi.edu/~och/GIZA++.html

**Table 1.** The translation results for comparison of handcrafted and automatically derived phrases

|        | HPH | APH_PTS | APH_ALL |
|--------|-----|---------|---------|
| SER[%] | 45.30 ± 2.40 | **44.21 ± 2.92** | 45.30 ± 2.40 |
| BLEU   | 65.17 ± 1.83 | 67.47 ± 1.92 | **68.77 ± 1.72** |
| WER[%] | 20.74 ± 1.31 | 17.42 ± 1.23 | **16.37 ± 1.02** |
| PER[%] | 11.78 ± 0.77 | 12.01 ± 0.89 | **11.22 ± 0.82** |

The results show that the automatically acquired phrases have the same or better translation accuracy than the handcrafted ones. The best result we got for automatically acquired phrases with full translation model (all five translation probabilities used in decoding process APH_FULL). However, there is a difference in size of phrase tables. The table of automatically acquired phrases contains 273 226 items (phrases of maximal length 7, the whole corpus) while the table of handcrafted phrases contains 5415 items only. The size of phrase table affects a speed of translation, the smaller table the faster decoding.

### 4.2 "Class-Based" Language Model

As well as in the area of ASR, there are problems with out-of-vocabulary words (OOV) in automatic translation area. We can translate only words which are in the translation vocabulary (we know their translation to the target language). By the analysis of the translation results we found that many OOV words are caused by missing a station or a personal name. Because the translation is limited to the domain of dialogs in train timetable information center, we decided to solve the problem of OOV words similarly as in work [6], where the class-based language model was used for the real-time closed-captioning system of TV ice-hockey commentaries. The classes of player's names, nationalities and states were added into the standard language model in this work. Similarly, we added two classes into our language model - the class for all known station names: STATION and the class for all known personal names: PERSON. Because the semantic annotation of corpus contains station and personal names, we can simply replace these names by relevant class in training and test data and collect a vocabulary of all station names for their translation (the personal names are always spelled). Table 2 describes the results of comparison of both decoders with and without "class-based" language model. We carried out 20 experiments with a various partitioning of data to the training and test set. The standard sign test was used for statistical significance determination. The significantly better results are in boldface. In the first column there are the results of SiMPaD with a trigram language model of phrases (SiMPaD_LMP) and in the second one the results of SiMPaD with a trigram "class-based" language model of phrases (SiMPaD_CLMP). Because SiMPaD uses the table of handcrafted phrases (no more than 5.5k phrases), the used language model is based on phrases too. In the third and fourth column there are results of the Moses decoder (the phrase table of automatically acquired phrases was used) with the trigram language model (Moses_LM) and with the trigram "class-based" language model (Moses_CLM).

**Table 2.** The results for comparison of decoding with and without "class-based" language model

|  | SiMPaD_LMP | SiMPaD_CLMP | Moses_LM | Moses_CLM |
|---|---|---|---|---|
| SER[%] | $44.84 \pm 1.96$ | $\mathbf{42.11 \pm 2.16}$ | $45.30 \pm 2.40$ | $\mathbf{42.94 \pm 2.20}$ |
| BLEU | $67.92 \pm 1.93$ | $\mathbf{70.68 \pm 1.73}$ | $68.77 \pm 1.72$ | $\mathbf{71.17 \pm 1.69}$ |
| WER[%] | $16.02 \pm 1.08$ | $\mathbf{14.61 \pm 0.96}$ | $16.37 \pm 1.02$ | $\mathbf{15.07 \pm 1.00}$ |
| PER[%] | $13.30 \pm 0.91$ | $\mathbf{11.97 \pm 0.80}$ | $11.22 \pm 0.82$ | $\mathbf{9.94 \pm 0.77}$ |

The "class-based" language model is better than the standard word-based one in all cases, for both decoders and in all criteria. The perplexity of language model was reduced to about 29 % on average in the case of phrase-based models (SiMPaD) and about 28 % in the case of word-based models (Moses), from $44.45 \pm 4.66$ to $31.60 \pm 3.38$ and from $38.69 \pm 3.79$ to $27.99 \pm 2.76$, respectively. The number of OOV words was reduced to about 53 % on average for phrase-based and about 63 % for word-based models (from 1.80 % to 0.85 % and from 1.43 % to 0.54 %, respectively).

### 4.3 Post-processing Enhancement

We found out that for translation from the Czech to the SC we can obtain even better result when we use an additional post-processing method. Firstly, we can remove the words which are omitted in translation process (they are translated into 'no translation' sign respectively) from the resulting translation. Anyway, to keep these words in training data gives better results (more detailed translation and language models). Secondly, we can substitute OOV words by a finger-spelling sign. Because the unknown words are finger spelled in the SC usually. The results for SiMPaD and Moses (suffix _PP for post-processing method) are in Table 3.

**Table 3.** The results of post-processing method in Czech $\Longrightarrow$ Signed Czech translation

|  | SiMPaD_CLMP | SiMPaD_CLMP_PP | Moses_CLM | Moses_CLM_PP |
|---|---|---|---|---|
| SER[%] | $42.11 \pm 2.16$ | $\mathbf{40.59 \pm 2.06}$ | $42.94 \pm 2.20$ | $\mathbf{41.97 \pm 2.20}$ |
| BLEU | $70.68 \pm 1.73$ | $\mathbf{73.43 \pm 1.78}$ | $71.17 \pm 1.69$ | $\mathbf{73.64 \pm 1.84}$ |
| WER[%] | $14.61 \pm 0.96$ | $\mathbf{14.23 \pm 1.06}$ | $15.07 \pm 1.00$ | $\mathbf{14.73 \pm 1.16}$ |
| PER[%] | $11.97 \pm 0.80$ | $\mathbf{9.65 \pm 0.78}$ | $9.94 \pm 0.77$ | $\mathbf{8.67 \pm 0.73}$ |

### 4.4 Czech to SC Translation

The same corpus, methods and decoders as for Czech to SC translation can be used for the inverse translation direction, i.e. from SC to Czech. The results for the SC to Czech translation are reported in Table 4. The second and the fourth columns contain results for test data where we kept also the words with 'no translation' sign and that were omitted in Czech to SC translation (suffix _WL). Finally, in the first and the third columns there are results for real test data (i.e. without the words with 'no translation' sign in the Czech to SC translation direction) (suffix _R).

**Table 4.** The results for Signed Czech $\Longrightarrow$ Czech translation

|  | SiMPaD_CLMP_R | SiMPaD_CLMP_WL | Moses_CLM_R | Moses_CLM_WL |
|---|---|---|---|---|
| SER[%] | 67.84 ± 1.56 | 57.08 ± 1.74 | 64.07 ± 2.71 | 51.74 ± 2.28 |
| BLEU | 39.55 ± 1.45 | 53.04 ± 1.24 | 50.23 ± 1.80 | 61.97 ± 1.80 |
| WER[%] | 36.15 ± 1.06 | 25.36 ± 0.78 | 29.65 ± 1.16 | 20.41 ± 1.04 |
| PER[%] | 33.04 ± 0.99 | 22.21 ± 0.71 | 26.00 ± 1.04 | 16.28 ± 0.85 |

Of course, the results for the test data containing also the words with 'no translation' sign are better, because there are more suitable words which should be in the resulting translation. Hence, for a better translation it is suitable to include the information on omitted words into the translation model. The Moses's results are better than SiMPaD's, because the word-based language model is more suitable for SC – Czech translation than phrase-based one (both trained on the corpus only).

## 5    Conclusion

We compared the translation accuracy of handcrafted and automatically derived phrases. The automatically derived phrases have the same or better accuracy than the handcrafted ones. However, there is a significant difference in the size of phrase tables. The table of automatically acquired phrases is more than 50 times larger than the table of handcrafted phrases. The size of phrase table affects a speed of the translation. We developed our decoder SiMPaD which uses handcrafted phrase table and some heuristics (monotone reordering and alignment with the longest phrase coverage) to speed up the translation process. We compared the SiMPaD's results with the state-of-the-art phrase-based decoder Moses. We found that the SiMPaD's results are fully comparable with the Moses's results while SiMPaD is almost 5 times faster than the Moses decoder.

We introduced "class-based" language model and post-processing method which improved the translation results from about 8.1 % (BLEU) to about 27.4% (PER) of relative improvement in case of SiMPaD decoder and from about 7.1% (BLEU) to about 22.7 % (PER) of relative improvement in case of Moses decoder (the relative improvement is measured between the word-based model - _LM(P) and the class-based model with post-processing - _CLM(P)_PP).

The same corpus, methods and decoders as for Czech to SC translation we used for SC to Czech translation and obtained first results for this translation direction. The experiment showed that it would be important to keep in some way the information on words that have 'no translation' sign in the Czech to SC translation direction to get better translation results.

## References

1. Krňoul, Z., Železý, M.: Translation and Conversion for Czech Sign Speech Synthesis. In: Proceedings of 10th International Conference on TEXT, SPEECH and DIALOGUE TSD 2007, Springer, Heidelberg (2007)

2. Koehn, P., et al.: Moses: Open Source Toolkit for Statistical Machine Translation. In: Annual Meeting of the Association for Computational Linguistics (ACL), demonstration session, Prague, Czech Republic, (June 2007)
3. Koehn, P., et al.: Statistical Phrase-Based Translation, HLT/NAACL (2003)
4. Kanis, J., et al.: Czech-Sign Speech Corpus for Semantic Based Machine Translation. In: Sojka, P., Kopeček, I., Pala, K. (eds.) TSD 2006. LNCS (LNAI), vol. 4188, pp. 613–620. Springer, Heidelberg (2006)
5. Papineni, K.A., et al.: Bleu: a method for automatic evaluation of machine translation, Technical Report RC22176 (W0109-022), IBM Research Division, Thomas J. Watson Research Center (2001)
6. Hoidekr, J., et al.: Benefit of a class-based language model for real-time closed-captioning of TV ice-hockey commentaries. In: Proceedings of LREC 2006, Paris: ELRA, pp. 2064–2067 (2006) ISBN 2-9517408-2-4

# Maximum Likelihood and Maximum Mutual Information Training in Gender and Age Recognition System*

Valiantsina Hubeika, Igor Szöke, Lukáš Burget, and Jan Černocký

Speech@FIT, Brno University of Technology, Czech Republic
xhubei00@stud.fit.vutbr.cz,
{szoke, burget, cernocky}@fit.vutbr.cz

**Abstract.** Gender and age estimation based on Gaussian Mixture Models (GMM) is introduced. Telephone recordings from the Czech SpeechDat-East database are used as training and test data set. Mel-Frequency Cepstral Coefficients (MFCC) are extracted from the speech recordings. To estimate the GMMs' parameters Maximum Likelihood (ML) training is applied. Consequently these estimations are used as the baseline for Maximum Mutual Information (MMI) training. Results achieved when employing both ML and MMI training are presented and discussed.

## 1 Introduction

Estimation of gender and age is an open topic in the speech processing field. When gender estimation is a simple task with two classes, age estimation is a lot more complicated due to non-linearity in changing of voice during aging. It is difficult to define precisely a border between two adjoined age groups. In this work, age groups were created experimentally according to limited amount of the available data. Nevertheless, the achieved results are optimistic.

Previously carried out studies [1], [2], [3] proved that it is possible to estimate gender and age of an unknown speaker only by listening to a low quality recording of his/her voice, such as from an analogue telephone line. Subjective gender estimation by human listeners shows very high performance. The accuracy of estimation is almost 100 %. However, estimation is not as accurate in case of children and elderly people.

During aging, changes of voice are continuous, therefore precise age estimation is unfeasible even by human listeners. Studies [1], [3] show that the accuracy in case of subjective age estimation by human listeners depends on several factors. Estimation is more precise using long sentences instead of isolated words. An important fact is that

---

* This work was partly supported by European projects AMIDA (IST-033812) and Caretaker (FP6-027231), by Grant Agency of Czech Republic under project No. 102/05/0278 and by Czech Ministry of Education under project No. MSM0021630528. The hardware used in this work was partially provided by CESNET under projects No. 119/2004, No. 162/2005 and No. 201/2006. Lukáš Burget was supported by Grant Agency of Czech Republic under project No. GP102/06/383.

V. Matoušek and P. Mautner (Eds.): TSD 2007, LNAI 4629, pp. 496–501, 2007.

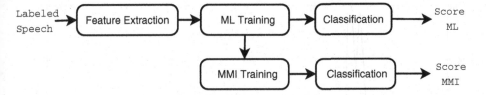

**Fig. 1.** Structure of the recognizer

voice of an atypical speaker can seem to be far younger or far older than he or she actually is. When using whole sentences in case of typical speakers, the error is mostly not greater than 10 years. In case of short isolated words from atypical speakers, the error can rise up to 40 years [3].

This work presents automatic gender and age estimation from telephone speech recordings based on Gaussian Mixture Models (GMM) which are proven to be a powerful tool often employed in text-independent classification tasks. The GMM parameters are estimated using ML training [4] and following MMI training [5]. The paper is organized as follows: Section 2 introduces the approach. Experiments are described in sections 3 and 4. Finally, the results are summed up in section 5.

## 2  Architecture of the Recognizer

The basic structure of the recognizer is shown in figure 1. The HTK toolkit [4] and STK toolkit from Speech@FIT[1] for HMM-based speech processing are used. Perl, C Shell and Awk scripts are used to process the data and evaluate the results.

The Czech SpeechDat-East database[2], used in the experiments, contains telephone speech recordings (8 kHz / 8 bit) from 1052 Czech speakers. 12 phonetically rich phrases from each speaker are used. The data are divided into training and test sets, that are mutually disjoint. The training set amounts to 81% of all data and consists of recordings from speakers aged 9 to 79 years. The remaining 19 % is the test set which consists of recordings from speakers aged 12 to 75 years. Distribution of single ages in the database is presented in figure 2. In both, training and test set, gender is covered equally. Altogether, 10207 recordings are used as training set and 2397 as test set. According to the transcription files, a lot of data contain speaker, background and channel noises. Only about 12 % of all the available data are considered as clear.

### 2.1  Feature Extraction – Mel-Frequency Cepstral Coefficients

Speech is divided to frames with a sampling window of 25 ms with a shift of 10 ms. From every frame, 12 MFCC and either the energy (age estimation) or the log-energy (gender estimation) are extracted [4]. The first order and the second order time derivatives are concatenated with the base static coefficients. The final feature vector has 39 coefficients.

---

[1] http://www.fit.vutbr.cz/research/groups/speech/stk.html
[2] http://www.fee.vutbr.cz/SPEECHDAT-E

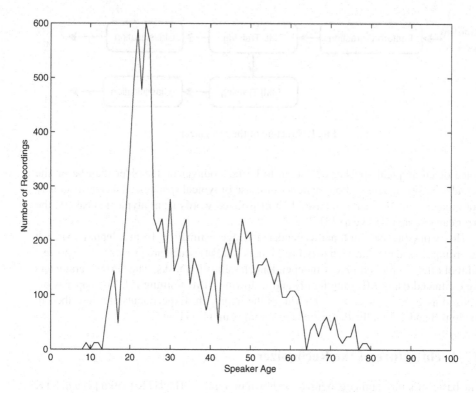

**Fig. 2.** Distribution of Single Ages in the Czech SpeechDat-East Database

## 2.2 Models' Training

Gaussian mixture models are used to represent distributions of cepstral features of gender and age classes. When using GMM, the recognition process is divided to two subproblems: estimation of the parameters of GMM using a set of training samples and following classification using trained models [4]. First, models' parameters (means and covariance matrices) are estimated using Maximum Likelihood (ML) training technique [4]. ML training determines GMM's parameters that maximize the likelihood of the given data samples by estimating means and covariance matrices from all the data for given class. When models are ML trained, they are used as the starting point for discriminative training [5]. Discriminative training is an approach used to maximize the probability of correct decision and to minimize the classification error. MMI objective function is:

$$F_{MMI}(\lambda) = \sum_{r=1}^{R} \log \frac{p_\lambda(O_r|s_r)^{K_r} P(s_r)}{\sum_{\forall s} p_\lambda(O_r|s)^{K_r} P(s)} \qquad (1)$$

where $p_\lambda(O_r|s_r)$ is likelihood of $r$-th training segment, $O_r$, given the correct transcription (gender or age) of the segment, $s_r$, and model parameters, $\lambda$. $R$ is the number of training segments and the denominator is the overall probability density, $p_\lambda(O_r)$. The prior probabilities, $P(s_r)$ and $P(s)$, are considered to be equal for all classes and are

**Table 1.** Age groups with spans of 25 years and the amount of used training and test recordings

|  | Young | Middle Aged | Elderly |
|---|---|---|---|
| Range | 9..30 | 31..55 | 56..79 |
| Training Set | 4259 | 3333 | 969 |
| Test Set | 1125 | 984 | 276 |

**Table 2.** Age groups with spans of 5 years and the amount of used training and test recordings

|  | Group | | | | | | | | | | | | |
|---|---|---|---|---|---|---|---|---|---|---|---|---|---|
|  | 1 | 2 | 3 | 4 | 5 | 6 | 7 | 8 | 9 | 10 | 11 | 12 | 13 |
| Age from | 9 | 16 | 21 | 26 | 31 | 36 | 41 | 46 | 51 | 56 | 61 | 66 | 71 |
| Age to | 15 | 20 | 25 | 30 | 35 | 40 | 45 | 50 | 55 | 60 | 65 | 70 | 79 |
| Training Set | 84 | 999 | 2507 | 1113 | 838 | 599 | 720 | 1020 | 755 | 514 | 287 | 202 | 192 |
| Test Set | 48 | 237 | 624 | 240 | 252 | 96 | 336 | 204 | 72 | 144 | 84 | 24 | 24 |

dropped. Usually, segment likelihood, $p_\lambda(O_r|s)$, is computed as multiplication of frame likelihoods incorrectly assuming statistical independence of feature vectors. The factor $0 < K_r < 1$ can be considered as a compensation for underestimating segment likelihoods caused by this assumption. This compensation factor is experimentally set to 0.01. MMI objective function (1) can be increased by re-estimating model parameters using extended Baum-Welch algorithm [6].

## 3   Gender Estimation

This work shows that automatic estimation proves to be almost as accurate as in case of subjective estimation by human listeners (see the introduction). When using all the available data and 30 Gaussian components (further adding of Gaussians shows no improvement of the result) in each gender GMM trained by ML, the accuracy is 94.64 %. With MMI re-estimation of the models' parameters, the accuracy went up to 97.41 %. Further improvement was achieved by filtering the training data. When utterances containing noise, mis-pronunciations or other defects (according to the transcription files) are discarded, the accuracy increased up to 98.25 %.

## 4   Age Estimation

Age groups are formed according to the limited size of the database. The first experiment is performed to estimate which age category the speaker belongs to. Three groups are defined with spans of 25 years (table 1). The aim of the second experiment is to estimate the age more precisely. Classification was done using 13 age groups with 5 year spans (table 2).

When models are ML trained, the accuracy using groups of 25 years is 49.60 %. The most accurate estimation is obtained for speakers belonging to the group of young

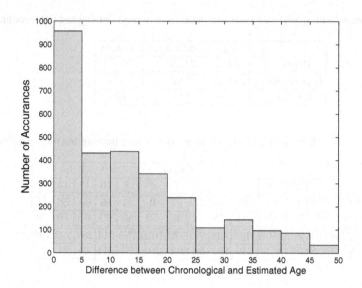

**Fig. 3.** Estimation Error Rate when Using Groups with Spans of 5 Years

people (56.62 %). The age of elderly people is estimated with the greatest error (only 28.26 % accuracy). The estimation accuracy in case of the middle aged speakers is 47.56 %. This difference in accuracy is caused by non-uniform distribution of single ages in the database, where 50 % of all recordings belong to young people, 39 % belongs to middle aged people and only 11 % belongs to elderly people.

When the models' parameters are MMI re-estimated, the accuracy of classification is 60.13 %. Correct classification in case of young speakers is done for 78.49 % of all utterances. In case of old people, the accuracy decreased to 17.39 % (due to low amount of training data). For middle aged people, the estimation was correct in 52.13 % of all cases.

After, the data were divided to groups with the ranges of 5 years (Tab. [3]) and 13 GMMs were ML trained. The average difference between chronological age and estimated age is 13.71 years. After MMI training, this difference went down to 11.38 years. Maximum difference between chronological and estimated age is 50 years (1 % of all cases). For 48 % of all cases, the estimation error is not greater then 10 years. A histogram of the estimation error is presented in figure 3.

## 5  Conclusion

An acoustic recognition system for gender and age estimation was presented. For gender estimation, the accuracy is high and satisfies the expectations.

The age is estimated with errors comparable to subjective human age estimation (errors of 10 years is commonly supposed as standard) although the models of some groups are trained on relatively small amount of data. The training is negatively influenced by large amount of disturbed data contained in the training set. Also, data from atypical speakers affect correct parameter estimation of the models which impairs correct estimation of models' paramteres. A possible solution would be an iterative reduction of outliers in the training data, we are however limited by the its relatively small size.

We have shown that the MMI training increased the accuracy. While the ML training tends to cover the whole regions uniformly by Gaussians, MMI probably concentrates less on the border regions (for example 10 and 11 years) which can not be reliably distinguished anyway, and models better the central parts of age groups.

# References

1. Cerrato, L., Falcone, M., Paoloni, A.: Subjective age estimation of telephonic voices. Speech Communication 31(2), 107–112 (2000)
2. Minematsu, N., Sikeguchi, K., Hirose, K.: Performance improvement in estimating subjective ageness with prosodic features. Speech Prosody (2002)
3. Schotz, S.: A perceptual study of speaker age. Technical report, Lund University (2001)
4. Young, S., Evermann, G., Gales, M., Hain, T., Kershaw, D.: The HTK book (2005)
5. Povey, D.: Discriminative Training for Large Vocabulary Speech Recognition. PhD thesis, Cambridge University (2004)
6. Matejka, P., Burget, L., Schwarz, P., Černocký, J.: Brno University of Technology System for NIST 2005, Language Recognition Evaluation. In: Proceedings of Odyssey 2006: The Speaker and Language Recognition Workshop, pp. 57–64 (2005)

# Pitch Marks at Peaks or Valleys?*

Milan Legát, Daniel Tihelka, and Jindřich Matoušek

University of West Bohemia, Faculty of Applied Sciences,
Department of Cybernetics,
Univerzitní 8, 306 14 Plzeň, Czech Republic
{legatm, dtihelka, jmatouse}@kky.zcu.cz

**Abstract.** This paper deals with the problem of speech waveform polarity. As
the polarity of speech waveform can influence the performance of pitch marking
algorithms (see Sec. 4), a simple method for the speech signal polarity determination is presented in the paper. We call this problem peak/valley decision making,
i.e. making of decision whether pitch marks should be placed at peaks (local maxima) or at valleys (local minima) of a speech waveform. Besides, the proposed
method can be utilized to check the polarity consistence of a speech corpus, which
is important for the concatenation of speech units in speech synthesis.

## 1 Introduction

The modern pitch-synchronous methods of speech processing rely on a knowledge of
the moments of glottal closure in speech signals. These moments are called glottal closure instants (GCIs) or pitch marks, if we speak about their location in speech. They
are usually used in pitch-synchronous speech synthesis methods (e.g. PSOLA or some
kinds of sinusoidal synthesis), where they ensure that speech is synthesized in a consistent manner. Knowing the position of pitch marks, a very accurate estimation of $f_0$
contour could be obtained and utilized in a number of speech analysis and processing
methods.

The problem of pitch marking has been tackled by several approaches including
wavelet-based analysis [1], application of nonlinear system theory [2] and many methods based on or similar to autocorrelation analysis and/or thresholding[3]. Before any
pitch marking algorithm is employed, it needs to be decided whether the pitch marks
should be placed at peaks or at valleys of a speech waveform. As we have found out
during our experiments, this decision is very important for the performance of the
pitch marking algorithm in terms of its accuracy and robustness. In [4] the problem
of peak/valley decision making is solved by comparing of the $f_0$ contour calculated
using AMDF (Average Magnitude Difference Function) and $f_0$ contours derived from
valley and peak based pitch mark sequences. The decision depends on the deviation
between these contours.

Though this method gives quite reliable results, there are some disadvantages. First,
the estimation of $f_0$ contour is time-consuming. Second, $f_0$ estimation is an error prone

* This work was supported by the Ministry of Education of the Czech Republic, project
No. 2C06020, and the EU 6th Framework Programme IST-034434.

task and the errors in the contour, even if it is filtered, can affect the peak/valley decision. In this paper, we propose a simple method based on the confrontation of peaks and valleys of a speech waveform.

This paper is organized as follows. Section 2 serves to describe the proposed method. In Section 3, we briefly discuss the effects of the speech signal polarity on the synthetic speech. In Section 4, we describe our experiments to demonstrate the performance of the proposed method. Section 5 gives the conclusions of this paper.

# 2 The Proposed Method

## 2.1 Motivation

During the development of our pitch marking algorithm [5] we observed large variation in its performance. We have found out that this was due to the polarity mismatch present in our speech corpus [6]. This mismatch is illustrated in Fig. 1, where two segments of two different sentences are shown.

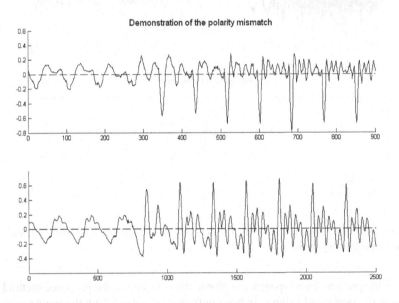

**Fig. 1.** Polarity mismatch. In the upper part there is a segment of the Sentence1 (negative polarity), while in the lower part there is a segment cut from the Sentence2 (positive polarity).

This observation led to the idea of development of peak/valley decision making method.

## 2.2 Description

Before we employ an automatic algorithm, the typical $f_0$ of the given speaker needs to be estimated. As the single speaker is recorded during corpus generation, this task can be very simply accomplished manually. Once we obtain the typical $f_0$ of the speaker, we can use this information as an input of the automatic peak/valley decision making method.

The proposed method can be summarized as follows. First, the speech waveform should be pre-processed. The aim of pre-processing is to reduce higher frequencies present in unvoiced segments and an extraneous noise. We accomplish this task by low-pass filtering by 23-rd order FIR filter with the cutoff frequency 900 Hz. The parameters of the filter were set ad hoc. This filter removes high frequencies and saves the valleys and peaks in voiced segments (See Fig. 2). The next step of pre-processing is the signal scaling. The aim of the scaling is to obtain a signal with zero mean value. This is necessary for later stages of the algorithm.

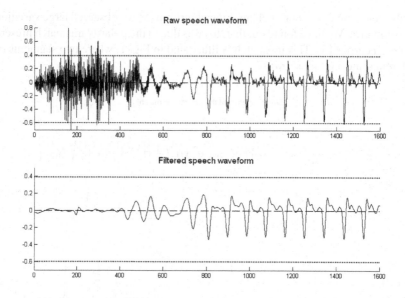

**Fig. 2.** Raw and filtered speech waveform. The dotted lines serve to illustrate how the noise can influence the peak/valley decision.

Having the pre-processed speech waveform, the next step of the proposed method is to confront the peaks and valleys. In this confrontation we use both the pre-processed speech waveform (*speech*) and its absolute value (*abs_speech*):

$$abs\_speech = |speech| . \tag{1}$$

The method can be summarized as follows:

1. Reset the counters *peak_count* and *valley_count*.
2. Find global maximum of the *abs_speech*. Denote its time coordinate as $t_m$.
3. If the position of this maximum corresponds with the position of the peak in *speech*, increment the counter *peak_count*, otherwise *valley_count* is incremented.

4. Set the value of *abs_speech* to zero in the range:

$$[t_m - 2/3 * f_0, t_m + 2/3 * f_0], \tag{2}$$

where $f_0$ is the estimate of speaker's typical value of the fundamental frequency. The length of this range was set experimentally.
5. Repeat steps 2, 3 and 4 until the *rms* value of the *abs_speech* is lower than $thresh *$ $rms\_speech$, where *rms_speech* is the *rms* value of the signal *speech*. The range of the constant $thresh$ is $[0.2, 0.7]$, the higher this value is the faster the peak/valley decision is made.

In fact, the above mentioned algorithm confronts the peaks and valleys in terms of their amplitudes. For the final peak/valley decision, we also calculate the overall energy above *e_above* and below zero *e_below* of the signal *speech*. The values of these energies are used as auxiliary predictors. If the value of the counter *peak_count* is higher than *valley_count* and *e_above* is higher than *e_below*, peaks are decided to be convenient for pitch marks placement and vice versa. If the values of counters are not in accordance with the values of energies, the decision is made solely on the basis of the values of the counters, but in this case it is marked as uncertain.

## 3   Discussion

In this section, we would like to address the issue of signal polarity unification during corpus recording. We have recently recorded a new speech corpus [6]. Although we placed emphasis on keeping recording conditions equal through all the recording sessions, we found several recording sessions to have speech signal with inverted polarity (See Fig. 1) – we are still examining the causes of this, but it may be related to the assembling/dissembling of the recording devices due to sharing the recording room with other projects.

The need to unify the polarity of the speech signal of all recorded phrases is obvious. As the pitch marks are placed only either at minima or maxima, phase mismatches will occur when speech units taken from signals with inverted polarity are concatenated, no matter how precisely pitch marks are detected; this is illustrated in Fig. 3. Synthetic speech will then contain audible "glitches" at such concatenation points [7], which we also confirmed in [8].

## 4   Experiments and Results

Rather than the experiments, in the first part of this section the results of the practical utilization of the proposed method are shown. The method was employed to check and unify the polarity of the newly recorded speech corpus [6]. The results were more than satisfactory - 98.14% of correct decisions, 1.36% of correct but uncertain decisions and only 0.5% of errors. It means that in 99.5% of cases the automatic method makes the same polarity decision as human would make. For all erroneous decisions the values of counters *peak_count* and *valley_count* were almost equal, so that these errors can be very simply detected or corrected by the setting of a threshold.

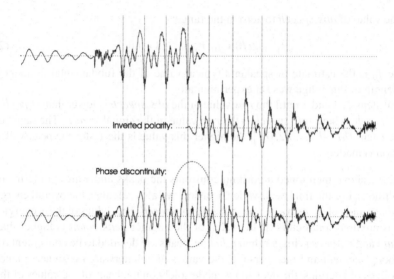

**Fig. 3.** Phase mismatch when units with different speech polarity are concatenated. Pitch-marks are placed at the negative amplitudes (valleys) of speech signal.

Besides, we have designed an experiment to find out how the peak/valley decision influences the performance of a pitch marking algorithm. For this purpose we have used the pitch marking method described in [5]. We have tested the performance of this algorithm depending on the type of pitch mark positions – either peaks(local maxima) or valleys (local minima) of the speech waveform. The experiment was conducted in three languages – Czech (CZ-M male and CZ-F female), Slovak (SK-F) and German (GE-M). In 8 sentences (i.e. about 7.000 pitch marks) the pitch marks were placed manually to test the performance of the automatic pitch marking method. Two reference pitch mark sets were used for testing – peak-based pitch marks and valley-based pitch marks. The summary of the results can be seen in Tab. 1. The average loss of accuracy if the pitch marks were placed into incorrect positions (i.e. placing to peaks if the polarity is negative and vice versa) was 8.6%.

**Table 1.** Summary of experiment results. The values in the table are accuracies of automatic pitch marking in percents. "Peak" means peak-based pitch marks, "Valley" means valley-based pitch marks. The polarity of tested sentences was negative.

|      | Peak  | Valley |
|------|-------|--------|
| CZ-M | 88.18 | 98.10  |
| CZ-F | 87.20 | 97.74  |
| SK-F | 88.21 | 97.19  |
| DE-M | 86.04 | 91.01  |

# 5    Conclusion

In this paper, we have addressed the problem of speech waveform polarity. We have proposed a simple method for speech signal polarity checking. This method can be used for peak/valley decision making (i.e. the decision whether pitch marks should be placed in local maxima (peaks) or local minima (valleys) of the speech waveform), which is the first step before any pitch marking algorithm is employed. We have shown in our experiments how the peak/valley decision can influence the performance of the pitch marking algorithm. The decrease in the accuracy of the automatic pitch marking algorithm was 8.6% in our experiments.

Moreover, the proposed method can be used for checking of the recorded speech corpus in terms of its polarity consistence, as we have experienced the speech signal polarity mismatch during corpus recording.

# References

1. Sakamoto, M., Saito, T.: An Automatic Pitch-Marking Method using Wavelet Trasform. In: Proc. INTERSPEECH, Beijing, China, vol. 3, pp. 650–653 (2000)
2. Hagmüller, M., Kubin, G.: Poincaré pitch marks. Speech Communication 48, 1650–1665 (2006)
3. Matoušek, J., Romportl, J., Tihelka, D., Tychtl, Z.: Recent Improvements on ARTIC: Czech Text-to-Speech System. In: Proc. INTERSPEECH. Jeju, Korea, pp. 1933–1936 (2004)
4. Lin, C.-Y., Roger Jang, J.-S.: A Two-Phase Pitch Marking Method for TD-PSOLA Synthesis. In: Proc. INTERSPEECH. Jeju, Korea, pp. 1189–1192 (2004)
5. Legát, M., Matoušek, J., Tihelka, D.: A Robust Multi-Phase Pitch-Mark Detection Algorithm. In: Proc. INTERSPEECH. Antwerp, Belgium (accepted, 2007)
6. Matoušek, J., Romportl, J.: On Building Phonetically and Prosodically Rich Speech Corpus for Text-to-Speech Synthesis. In: Proc. Computational Intelligence. San Francisco, U.S.A., pp. 442–447 (2006)
7. Huang, X., Acero, A., Hon, H-W.: Spoken Language Processing: a Guide to Theory. Algorithm and System Development Microsoft research. In: PTR 2001, Ch. 16, p. 829. Prentice-Hall, Englewood Cliffs (2001)
8. Tihelka, D., Matoušek, J.: The Analysis of Synthetic Speech Distortions. In: proceedings of 14th Czech–German Workshop. Prague, pp. 124–129 (2004)

# Quality Deterioration Factors
# in Unit Selection Speech Synthesis

Daniel Tihelka, Jindřich Matoušek, and Jiří Kala

University of West Bohemia, Faculty of Applied Sciences,
Department of Cybernetics,
Univerzitní 8, 306 14 Plzeň, Czech Republic
{dtihelka, jmatouse, jkala}@kky.zcu.cz

**Abstract.** The purpose of the present paper is to examine the relationships between target and concatenation costs and the quality (with focus on naturalness) of generated speech. Several synthetic phrases were examined by listeners with the aim to find unnatural artefacts in them, and the mutual relation between the artefacts and the behaviour of features used in given unit selection algorithm was examined.

## 1 Introduction

The quality of synthetic speech, especially its naturalness, represents an issue which still remains open. Although the *unit selection* approach, or the modern *HMM–based synthesis*, are able to reach fairly high level of naturalness, there are still unnatural artefacts of a different nature (depending also on the synthesis method) occurring in synthesized speech when input text comes from the unlimited domain.

The paper evaluates unnatural artefacts (also called *glitches*) perceived in speech generated by unit selection incorporated in our TTS ARTIC [1,2,3], and attempts to reveal the relation between those artefacts and the behaviour of features used for target and concatenation costs computing. There have been a number of papers written, mostly focusing on the relation of human perception to spectral subcost of concatenation cost only – [4,5,6,7,8,9,10] belong among those most important. However, we aim to look at the whole unit selection scheme globally, to examine all the expected or potential factors which can cause the perceived artefacts (for the terminology defining the terms *artefact* and *factor* see [11] or short reminder in Section 2.1). The problem is the greater number of possible causes of artefacts in the unit selection, as well as usual co-incidence of several factors leading to perceived artefact.

The paper is organised as follows. Section 2 describes special evaluation methodology which has to be used for this task, covering also the related terminology. Section 3 is focused on the procedure of artefact collection, defines the set of expected causes (factors) of artefacts, and describes, how the artefacts and factors were related together. The results and the conclusions are presented in Section 4 and Section 5 respectively.

V. Matoušek and P. Mautner (Eds.): TSD 2007, LNAI 4629, pp. 508–515, 2007.

## 2   Evaluation Method

It is obvious that the outlined aims of the paper require the human evaluation of synthetic speech, i.e. listening tests. However, as the standard listening tests (e.g. MOS, CCR or different kinds of decision tests) usually focus either on overall quality (e.g. how close to natural the synthetic phrase is), or on one kind of artefact (e.g. does a particular concatenation point sound continuous or not), none of them is suitable for the proposed purposes. We, therefore, used special methodology designed in [11]. Although the methodology was first used for the evaluation of our single–instance version of our TTS, it is applicable for the larger set of evaluation types and/or system types. The disadvantage of the approach is its labour input, but to our knowledge it is the most straightforward way of analyzing the collected unnatural artefacts.

### 2.1   Terminology

Let us review terminology for this kind of evaluation, as established in [11]. An *artefact* is a term denoting every event or place in synthetic speech which annoys human listeners and causes distortions in synthetic speech being perceived as unnatural; it can also be distributed to several consecutive phones. Artefacts are represented by various kinds of glitches, clicks, cracks, murmuring, speech or prosody discontinuities, machine-like or artificial sounds and so on.

In [11] we also divided artefacts into *fragments*, as the artefacts determined by listeners overlapped in many ways due to the natural fuzziness of human perception. However, we do not explicitly build fragments in the present evaluation, as described in Section 3.1.

Finally, it is clear that artefacts are caused by a coincidence of what is carried out in the TTS algorithm before speech is created. Therefore, the set of actions which can influence the synthetic speech is usually explicitly defined, being called the set of *factors*.

## 3   Evaluation Procedure

The whole evaluation can be divided into several steps. Synthetic speech must be evaluated by the listeners to collect the set of artefacts. The set of factors which are apriori expected must also be determined from the knowledge of synthesis algorithm (although the methodology enables adding further artefacts revealed during the evaluation). Finally, the occurrences of the factors are searched in all phones assigned to the artefacts.

### 3.1   Artefact Collection

Five listeners did the listening to 50 synthetic phrases generated by our unit selection version (described in [2,3]), with the aim to determine artefacts which they found sounding unnatural – the length of each phrase was about 10 seconds. The set of most expected artefacts was predefined ad-hoc, however, each listener could create "new" artefact tag, when none of predefined made sense regarding his/her impression of the perceived distortion. Contrary to [11], listeners were not required to mark artefacts as regions, but only to label the subjectively most prominent phone around which an arte-fact was perceived. It was decided due to the nature of unit selection approach to con-catenate continuous (thus natural) sequences of units, which are whole expected to be affected by the artefact.

As the secondary aim of the evaluation is to acquire enough practical experience to tune the artefact collection process so that it is as easy as possible, the listeners included the authors and Ph.D. students focusing their research activity on speech synthesis. The future plan is to involve ordinary people (as potential users of the evaluated TTS) in the procedure.

The source corpus used for the synthesis of evaluated phrases phrases contains 5,000 phonetically balanced utterances, giving approximately 12.5 hours of speech. It was read by a semi-professional female speaker (with radio news broadcasting experience) in a relatively consistent news-like style, and recorded together with glottal signal in an office at our department. The corpus was segmented automatically, using HMM-based approach [12,13]. Although the corpus was not originally designed for the unit selection approach, it was found to be suitable enough for the experiments with this approach, until a more appropriate corpus is recorded [14].

### 3.2 Factors in Unit Selection

In [11], where the same evaluation method was used for the single-candidate version of our TTS, the definition of factors was much easier, based on the well-known prob-lematic parts – need of signal modification, sensitivity to segmentation inaccuracy and spectral discontinuity between candidates. On the contrary, artefacts considered in the present paper are caused almost purely by the choice of inappropriate unit in the se-lection algorithm (when no further signal processing is carried out, which is our case), no matter how the "suitability" is measured. In the present paper we, therefore, limit ourselves to the analysis of factors intuitively expected to decrease speech quality, and factors directly joined to target and concatenation costs features:

**Segmentation inaccuracy (SI).** In the case of unit selection with diphone units, the ab-solute boundary misplacements is not such a problem as it is in the single-candidate system (due to boundary shift [15], triphones are not taken into account yet [16]). Instead, we look for units which are missegmented as a whole – especially se-quences as [la], [ij], [mn] and/or similar tend to have one of phones very short. To determine if the factor was presented in artefact, the segmentation of units had to be checked manually – there is no automatic measure available yet (we found out HMM score not to be useful very much). Once there is such, the missegmented boundaries could automatically be corrected or the selection can avoid those "bad" units, and this factor would not have to be considered.

**Spectral discontinuity (SD).** The measure of spectral smoothness is the standard part of concatenation cost. Although there were many experiments with various mea-sures compared to human perception carried out [6,7,8,9,10], Euclidean distance between MFCC vectors is still often utilized (also in our case) thanks to its simplic-ity. However, as our experience is increasingly showing us that more appropriate measure of perceived smoothness must be established, we manually checked the discontinuities of spectra around concatenation points for units related to artefact. Manual inspection is also necessary here, just due to the fact that MFCC seems not to be good predictor of perceived discontinuities.

**Target features mismatch (*TFM*).** This meta-factor covers the whole range of factors related to target cost, as described in [2,3]. Its aim is to reveal the relation of artefact occurrences to the mismatch of features used to measure the suitability of units to express the required prosody (or communication function). Obviously, it expects that the features used really describe measured requirements (and mismatch then causes artefact), which does not necessarily have to be true. Therefore, we work on a new approach of determining target features for unit selection. The given *TFM* factors are collected automatically during the synthesis of evaluated phrases.

**Concatenation features mismatch (*CFM*).** It is also a meta-factor covering all factors used to the measure of concatenation smoothness, see [2,3]. The difference from *SD* factor is that while *SD* handles the real occurrence of spectral discontinuity in synthetic speech (is/is not), those factors are supposed to provide a relation of artefact occurrences to the behaviour of features which are expected to ensure both spectral and prosodic smoothness (differences around join point). The *CFM* factors are also collected automatically.

Although listed factors are related to the evaluated TTS, we are convinced that the results will tend to display similar results for each TTS system using similar features, features with similar behaviour, or with similar range of values.

### 3.3   Factors Assignment

As described in Section 3.1, the listeners were asked to mark only the most prominent phone of an artefact perceived. As it can be supposed that artefacts naturally occur around concatenation points or on very short unit sequences of one or two phones (except suprasegmental artefacts like inappropriate communication function, which are not considered by the paper), and as listeners are usually unable to determine the exact phone/phones of artefact either (due to fuzzy nature of perception the place cannot usually be even defined), the sequence of phones affected by each artefact was defined aposteriori as follows. The units preceding the phone holding artefact label were examined until the continuous sequence of at least 3 units was found; similarly for the units following the phone (in the context of [11], such regions can be considered as fragments). In those regions, only the units around *sequence break* were further analysed, as illustrated by the following example:

```
...
mJ    Sentence0457
Ji    Sentence0457 >> region beginning (including the diphone)
ix    Sentence0457
xo    Sentence0457      s1
ov    Sentence4569      s1,s2
vj    Sentence1775        s2
je    Sentence1775
```

```
eC   Sentence1775           s3
Ct   Sentence2666           s3
ti   Sentence2666
iR   Sentence2666   << region end
Ri   Sentence2666
...
```

As described in Section 3.2, some of factors needed manual inspection of phones around sequence breaks – it cannot be done automatically, simply due to the lack of appropriate automatic measure, which, if it exists, could directly be used by selection algorithm to avoid the examined artefacts. Due to the laboriousness of the inspection, only 9 phrases were fully evaluated, which may seem to be a small number for significant results. However, the analysed phrases tend to display very similar tendencies, as shown in Section 4. On the other hand, the manual inspection revealed that some units around sequence breaks were chosen from phrases differing in the voice quality – it defined additional factor *VQ*.

## 4   Results

There were 90 artefacts collected in the 9 analysed phrases (some determined by more than one listener), and 127 unique sequence breaks were found in the artefact regions. Let us note that there were 316 sequence breaks in total in the analysed phrases, and 9 artefacts did not contain any phrase break at all.

The first part of results covers factors which acquired binary values present/absent in the assignment procedure – *SI*, *SD*, *VQ*, and factors from *TFM* set. The special treatment was, however, required for "position in prosodic word" feature, which has been designed not to provide any sharp delimitation of the feature, as described in [2]. The feature was "sharpened" by splitting the position into 5 equally spaced regions (beginning, beginning-middle, middle, etc.), where match/mismatch could be determined. As the target cost is measured independently for units $i$ and $i + 1$ around each sequence break, the results are collected separately for both units; however, as expected, the results are very similar. Let us also note that context mismatch is considered if either left of right context of unit does not fit. Results show that the segmentation inaccuracy cannot still be neglected, as it was found in almost 24% of sequence breaks. Even worse

**Table 1.** The occurrences of factors with binary present/absent decision values in all sequence breaks (127). Features from *TFM* set are listed separately for units $i$ and $i + 1$.

| factors | occurrences | percentage |
|---|---|---|
| SI | 30 | 23.62 |
| SD | 42 | 33.07 |
| VQ | 36 | 28.35 |
| context $i/i + 1$ | 107/106 | 84.25/83.46 |
| prosodeme $i/i + 1$ | 7/ 8 | 5.51/ 6.30 |
| word pos. $i/i + 1$ | 43/ 39 | 33.86/30.71 |

situation is for spectral discontinuity, despite subcost aiming to deal with it. A similar situation is for word position mismatch, but we need to further analyse the failure cases to be able to draw a conclusion. The most frequently occurring factor is context mismatch; however, as it is the least important feature (the least weighted in selection) the precise match is yielded in favour of target features defined to be more important.

The second part of the results is focused on *CFM* factors. We compared the behaviour of individual concatenation subcosts through all sequence breaks (note that concatenation cost is 0 aside sequence breaks), shown in Table 2. Let us note that due to the fact that the concatenation features acquire continuous values, they cannot be evaluated as the previous factors. To do so, we would have to define a threshold which would split the values to correct/failure sets. However, the threshold can, in fact, be chosen randomly, each choice giving different results.

**Table 2.** The mean, standard deviation and median values of *CFM* factors for all sequence breaks. The $F_0$ values were measured only at the sequence breaks of voiced units $i$ and $i + 1$.

| features | mean value | std. dev. | median |
|---|---|---|---|
| $F_0$ difference [Hz] | 6.10 | 5.23 | 4.71 |
| $F_0$ cost | 0.10 | 0.08 | 0.07 |
| intensity cost | 0.06 | 0.07 | 0.04 |
| spectral (MFCC) cost | 0.69 | 0.22 | 0.66 |

**Table 3.** The behaviour of MFCC cost in relation to the occurrences of *SD* factor and the voice of unit transitions

| voice $i, i + 1$ | SD occurrence | number | mean value | std. dev. | median |
|---|---|---|---|---|---|
| voiced | y | 38 | 0.70 | 0.20 | 0.67 |
| voiced | n | 46 | 0.58 | 0.20 | 0.54 |
| unvoiced | y | 4 | 0.77 | 0.09 | 0.77 |
| unvoiced | n | 39 | 0.82 | 0.22 | 0.84 |

In addition, we compared the behaviour of spectral measure using Euclidean distance with MFCC in relation to *SD* factors, which is shown in Table 3. It can be seen that the MFCC-related subcost has the largest contribution to the concatenation cost, and it therefore significantly influences the choice of units to concatenate. However, it does not seems to be able to measure spectral discontinuity very well – the average subcosts are relatively close each other (considering also standard deviation), whether *SD* occurred or not. Moreover, there are more than 26% of distances in voiced sequence breaks without *SD* higher than the average of distances in voiced breaks without *SD*. For the unvoiced sequence breaks the numbers are even closer (which, however, was expected due to the noisy character of unvoiced sounds).

The last part of the results covers the brief comparison of the behaviour of both target and concatenation costs. While target cost acquired the mean value 0.11 (with std.dev. 0.09) for units $i$ and 0.11 (std.dev. 0.08) for units $i + 1$ around all sequence breaks, the concatenation cost acquired the mean value 0.82 (with std.dev. 0.24), which is almost 8-times higher. In unit selection modules using features similar to ours, it is, therefore, the concatenation cost (and MFCC cost as its sub-part) which decides which units to concatenate.

## 5    Conclusion

Although the speech generated by our TTS has been evaluated as "close to natural" [3], the present paper showed that there are still a number of issues to focus on. First of all is the use of different measure for concatenation smoothness which would follow human perception much more closely (to avoid *SD* factors). There is a also need to reduce the mismatch of target features, e.g. the scaling of context features which is required to express the perceived defect of substitution instead of the difference of phone labels. Moreover, we work on a new method of target features design based on the analysis-by-synthesis approach; we expect that it will be able to provide us with the set of features observed in both natural speech and its synthetic variant. The new costs will also need to be better balanced, and finally, the work on segmentation precision will continue. Further analysis of features aside artefacts and/or sequence breaks will yet also be carried out, to get behaviour of the features in speech regions with and without perceived distortions.

There is also the need to tune the evaluation procedure, as we plan to carry out such tests involving ordinary people. Moreover, the assignment procedure needs be simplified not to be so laborious. The evaluation will also be very important for the new unit selection corpus which we are preparing [14].

## Acknowledgements

This research was supported by the GAČR 102/06/P205, and by the EU 6th Framework Program no. IST-034434. Our thanks are also due to Zdeněk Hanzlíček and Milan Legát, who participated on the artefact assessments.

## References

1. Matoušek, J., Romportl, J., Tihelka, D., Tychtl, Z.: Recent Improvements on ARTIC: Czech Text-to-Speech System. In: Proceedings of Interspeech 2004 - ICSLP, Jeju Island, Korea, vol. III, pp. 1933–1936 (2004)
2. Tihelka, D., Matoušek, J.: Unit selection and its relation to symbolic prosody: a new approach. In: Proceedings of Interspeech 2006 – ICSLP. Pittsburgh, USA, pp. 2042–2045 (2006)
3. Tihelka, D.: Symbolic Prosody Driven Unit Selection for Highly Natural Synthetic Speech. In: Proceedings of Interspeech 2005 – Eurospeech. Lisbon, Portugal, pp. 2525–2528 (2005)
4. Bellegarda, J.R.: A Novel Discontinuity Metric for Unit Selection Text-to-Speech Synthesis. In: Proceeding of 5th ISCA Speech Synthesis Workshop. Pittsburgh, pp. 133–138 (2004)
5. Syrdal, A.K., Conkie, A.D.: Data-Driven Perceptually Based Join Costs. In: Proceeding of 5th ISCA Speech Synthesis Workshop. Pittsburgh, pp. 49–54 (2004)

6. Vepa, J., King, S.: Join Cost for Unit Selection Speech Synthesis. In: Text to Speech Synthesis: new Paradigms and Advances, pp. 35–62. Prentice Hall PTR, New Jersey (2004)
7. Vepa, J., King, S.: Kalman Filter-Based Join Cost For Unit-Selection Speech Synthesis. In: Proceedings of the 8th European Conference on Speech Communication and Technology Interspeech 2003 – Eurospeech. Geneva, Switzerland, pp. 293–296 (2003)
8. Kavai, H., Tsuzaki, M.: Acoustic Measures vs. Phonetic Features as Predictors of Audible Discontinuity in Concatenative Speech Synthesis. In: Proceedings of IEEE International Conference on Acoustics, Speech, and Signal Processing – ICASSP '04, Quebec, Canada, vol. 1, pp. 657–660 (2004)
9. Donovan, R.E.: A new Distance Measure for Costing Spectral Discontinuitie In Concatenative Speech Synthesis. In: Proceedings of 4th ISCA Tutorial and Research Workshop on Speech Synthesis, Scotland (2001)
10. Klabbers, E., Veldhuis, R.: On the Reduction of Concatenation Artefacts in Diphone Synthesis. In: Proceedings of International Conference on Spoken Language Processing ICSLP 98, Sydney, Australia vol. 6, pp. 2759–2762 (1998)
11. Tihelka, D., Matoušek, J.: Revealing the most Significant Deterioration Factors in Single Candidate Synthetic Speech. In: Proceedings of SPECOM 2005. Greece, pp. 171–174 (2005)
12. Matoušek, J., Tihelka, D., Psutka, J.: Automatic Segmentation for Czech Concatenative Speech Synthesis Using Statistical Approach with Boundary-Specific Correction. In: Proceedings of the 8th European Conference on Speech Communication and Technology Interspeech 2003 – Eurospeech. Geneva, pp. 301–304 (2003)
13. Matoušek, J., Tihelka, D., Psutka, J.: Experiments with Automatic Segmentation for Czech Speech Synthesis. In: Matoušek, V., Mautner, P. (eds.) TSD 2003. LNCS (LNAI), vol. 2807, pp. 287–294. Springer, Heidelberg (2003)
14. Matoušek, J., Romportl, J.: Recording and Annotation of Speech Corpus for Czech Text-to-Speech Synthesis. LNCS (LNAI). Springer, Heidelberg (2007)
15. Clark, R.A.J., Richmond, K., King, S.: Festival 2 – Build Your Own General Purpose Unit Selection Speech Synthesizer. In: Proceedings of ISCA Speech Synthesis Workshop. Pittsburgh, pp. 173–178 (2004)
16. Tihelka, D., Matoušek, J.: Diphones vs. Triphones in Czech Unit Selection TTS. In: Sojka, P., Kopeček, I., Pala, K. (eds.) TSD 2006. LNCS (LNAI), vol. 4188, pp. 531–538. Springer, Heidelberg (2006)

# Topic-Focus Articulation Algorithm on the Syntax-Prosody Interface of Romanian

Neculai Curteanu[1], Diana Trandabăț[1,2], and Mihai Alex Moruz[1,2]

[1] Institute for Computer Science, Romanian Academy, Iaşi Branch
[2] Faculty of Computer Science, University "Al. I. Cuza" of Iaşi
{curteanu, dtrandabat, mmoruz}@iit.tuiasi.ro

**Abstract.** We propose in this paper an implementation of the Prague School's TFA (Topic-Focus Articulation) algorithm to support the Romanian *prosody design*, relying on the experience with FDG (Functional Dependency Grammar) and SCD (Segmentation-Cohesion-Dependency) parsing strategies for the classical, *i.e.* predication-driven, but Information Structure (IS) non-dependent, syntax. As contributions worth to be mentioned are: **(a)** Outlining the *functional* and *hierarchical* organization of linguistic *markers and structures* within SCD and FDG local-global parsing, on both sides of the *syntax-prosody interface* of Romanian. **(b)** Pointing out the relationship between classical (IS-free) syntactic structures, IS (topic-focus, communicative dynamism) depending textual spans, and the corresponding *prosodic intonational* units. **(c)** *Adapting* and *implementing* the TFA *algorithm* for *the first time* to Romanian prosodic structures, to be continued with TFA sentence-level refinements, its rhetorical-level extension, and embedding into local-global *linking algorithms*.

## 1 A Tribute to Eva Hajičová's Speech

Our basic ideas on developing a sound prediction for (Romanian) prosody received (indirectly) a strong, both theoretical and computational support from Eva Hajičová's recent speech at her ACL Lifetime Achievement Award [8]. This short introduction is intended to support the golden issues of functional and hierarchical marking and markers for linguistic categories and structures, and the (often hidden) way which their boundaries (especially the intonational ones) are settled by. The mappings that project the categories together with their boundaries (markers), structures, and their orderings (either for markers or structures) on the language interfaces, particularly on the syntax-phonology/prosody interface, represent major goals for our present and forthcoming approaches.

For the local (sentence-level) prosody prediction, we grant the phonological approach derived from the intensive use of *information structure* (IS) and its intimate relationship to sentence-level discourse and prosody. Prague School's TFA [8], [9] and Steedman's CCG (Combinatory Categorial Grammar) [14] are paying the due importance to the close relationship between IS, sentence structure, and prosody. Valuable approaches for Romanian on this line come from [7], [15], [12], [6]. Göbbel [7] gives evidence on the relationship between *argument structure* of the Romanian

V. Matoušek and P. Mautner (Eds.): TSD 2007, LNAI 4629, pp. 516–523, 2007.
© Springer-Verlag Berlin Heidelberg 2007

sentence, word order, and the *focus* (F) *projection* into the prosody of the Romanian sentence, using the phonological theories of E. Selkirk [13] *et al.* for English and German. The approach in [7] is, to our knowledge, a *first effective application* of a phonologic discourse theory (this subsuming a syntactic phonology) at sentence-level for Romanian. An interesting discussion on the syntax-phonology interface for Italian (holding also for Romanian) *restructuring* (*i.e.* modal, aspectual, and motion) *verbs* is proposed by Monachesi [12: Chap.5].

## 2 The *IS*-Driven Mapping on the Syntax-Prosody Interface

The present paper wants to propose, along with results on adapting and implementing TFA algorithm to Romanian and a procedure to ascribe adequate intonational units to Romanian FDG-SCD parses, the image of a general parsing strategy, in parallel for both text and speech on the syntax-prosody interface. The structure of this interface consists of three main components: a lattice-type organization of textual *marker* (boundary) *classes*, with the role of syntactic structure delimiting and dependency-establishing (for the IS-driven *syntactic side* of the interface); a similar set of marker classes for the *prosody side*; and an homomorphic general mapping between the two main structures with the property of preserving *subsumption* between the involved (sub)structures: the IS-driven *syntax subsumes* the *prosody*. The SCD marker classes and FX-bar projection structures [4], [5] are concrete steps on this road, taken for both Romanian and English (for a discussion on the *non-isomorphism* between syntax and phonology, see [12 :Chap.5]).

Here are, apart from [8], [9], some few hints supporting the way that markers and structures on text and speech are put into correspondence on the syntax-prosody interface. Proposals in [13] concerning the computation of a phrase stress in the context of *informational focus* (F), *contrastive focus* (FOC), and Chomsky's *phase* [3] entail that FOC, F, F-projection rules, and the organization of sentences into *phases,* all play a crucial role in defining sentence-level patterns of phrase stress and prosody. The same is true for the relationship between spoken, intonational cues, and syntactic-rhetoric textual boundaries used in spoken and written discourse segmentation.

Chiu-yu Tseng *et al.* [16] specifies that spoken discourse prosody is constituted from *prosodic phrase groups,* whose phonologic boundaries are *prosodic fillers* represented by the prosodic word markers and prosodic phrase-discourse markers. Tseng's prosody modelling relies on the RST discourse parsing of English, intensively using the discourse markers (*e.g.* Daniel Marcu's work [11]), references closely related to SCD markers and our own approach to Romanian discourse parsing [4]. The use of text discourse theories to predict the prosody at the global level could better suggest the interplay between prosodic intonational units, proven to be difficult to determine for Romanian [1].

Steedman's CCG observes similar ideas to our image on syntax-prosody interface: (*a*) *Intonation* coincides with surface structure in the sense that all *intonational boundaries* coincide with *syntactic boundaries*. (*b*) The reverse *does not* hold: not all surface syntactic boundaries are *explicitly* marked by intonational boundaries. (*c*) The majority of IS boundaries go unmarked by explicit intonational boundaries; IS boundaries may be marked by other, *more subtle,* articulator markers. (*d*) IS constituents and their semantic interpretation provide the logical forms that discourse semantic functions apply to, the boundaries of these IS constituents *lining up* with the intonational-structure boundaries, when these are present. (*e*) Phrasal intonation and intonational contours are defined in terms of pitch accent(s) tones and boundary tones. *Boundary tones* mark the *right-hand* side *boundary* of the *prosodic phrase*, differing therefore from the *syntactic* and *discourse markers*, which bound the *left-hand* side of a *textual phrase*.

Our present belief is that, for devising an adequate prosody prediction for Romanian, it is necessary to assimilate and adapt the Prague School's TFA [9] and Steedman's CCG experience [14], [2] for the *local prosody* design, compatible with the local-level spoken discourse theories on IS. This should be coupled with the *global-level prosody* approach based on prosodic intonational units (such as Tseng's *prosodic phrase groups* [16]), compatible with the global spoken discourse through, for instance, RST textual discourse [11] and *global linking* algorithm [17, Ch. 7].

Apart from being used as input for the TFA algorithm in Section 3, the FDG parsing [10] is also used for determining the prosodic entities of any given sentence, using intonationally-driven principles (Ghini's *mapping algorithm* [12]) and the following empirically acquired rules (for Romanian):

**Clause acquisition:** every subtree that has as its root a verb forms a clause, together with all of its descendents, with the exception of the verbal descendents which are themselves heads of VG (Verbal Groups) and their subtrees;

**Main clause acquisition:** all verbs on the highest level in the FDG tree of a sentence are heads of the main clauses;

**IP** (*Intonational Phrase*) **acquisition:** each main clause, together with all its subordinate clauses, forms an IP;

**ip** (*intermediate phrase*) **acquisition:** each clause is an *ip*. There are cases when *ip* structures are interrupted by speech break markers (punctuation marks, coordinative conjunctions, etc.)

**RU** (*Rhythmical Unit*) **acquisition:** (*i*) A parent and its first adjacent successor form together an RU (it does not matter whether the successor stands at the left or the right of the parent). The RU also contains the first adjacent descendent of the successor, recursively, to the leaf; (*ii*) Non-adjacent successors are each in separate RUs, in a similar way as in (*i*); (*iii*) If a parent has two or more adjacent successors, the first successor forms an RU with its parent and the other successors form an RU by itself, except the case when the successors have no other descendent (are all leaves for the parent). In this case, all the successors enter in the parent's RU.

An important remark for the RU assignment is that clitics and reflexive pronouns, included in the VG, are considered to be on the same level with the verb.

As an example, the intonational units computed for the FDG trees in Fig. 1 are:

```
[[[O'Brien apucă]_RU [sticla]_RU [de gât]_RU]_ip]_IP [[[şi  turnă      în
*O'Brien   grips  the bottle  by the neck     and  poured  into
pahare]_RU [un    lichid   de un    roşu     aprins]_RU]_ip]_IP.
glasses   a     liquid   of a    red      bright.
```

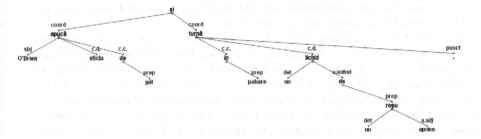

**Fig. 1.** FDG tree example for the sentence "*O'Brien apucă sticla de gât şi turnă în pahare un lichid de un roşu aprins.*" (EN: *O'Brien gripped the bottle neck and poured a bright red liquid into glasses.*).

## 3  Topic-Focus Computing for Romanian

This paper takes a first step to design an adequate syntax-prosody interface for Romanian, by adapting and applying the Prague School's TFA (Topic-Focus Articulation) algorithm [9]. The purpose of this algorithm is to compute the topic-focus (theme-rheme) for a given sentence to obtain, on the basis of sentence IS syntax, a better acquisition of the sentence intonational units, and subsequently, an improved naturalness for the assignment of tone and tune phrases to the established intonational units of the sentence.

The starting status of our prosody prediction system (Fig. 2) is the morphologically-syntactically tagged text, with SCD markers and FDG dependencies. From the FDG parsing tree, two directions are derived: (1) grouping the FDG

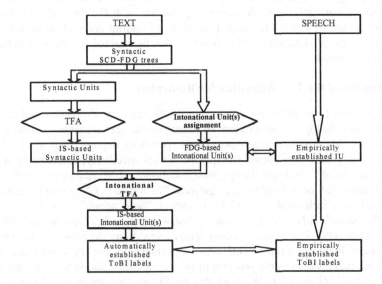

**Fig. 2.** The architecture of the prosody prediction system

branches into syntactic constituents, and (2) detecting intonational units from the FDG tree using the acquisition rules presented above.

The *Topic-Focus Articulation* (TFA) *Algorithm* receives as input the syntactic units of a sentence, SCD markers (*e.g.* quantifiers, case markers, or *predicational* feature), and certain semantic features. The outcome consists of IS-based units, which are actually syntactic constituents with topic-focus values. The obtained IS-syntactic structures are used to re-arrange the intonational units according to their relevance for the speech dynamics, leading to IS-based intonational units and syntax. Applying transformation rules as those proposed by Steedman's CCG for the local level or those drafted by RST-based prosody [16], [11] for the global level, the corresponding ToBI labels could be more adequately established.

The right-hand side of Fig. 2 presents the empirical annotations of the spoken text. The manually determined intonational units and the ToBI labels will be compared to the automatically acquired ones, in order to evaluate and adjust the TFA procedure against the (currently empirical) annotated gold corpus.

We selected for speech recording approx. 150 sentences from George Orwell's novel "1984". This text was selected because we already have an important number of morphological-syntactic annotations for that corpus (morphology, POS, NP-VP groups, SCD markers, etc.). At syntactic level, we added for each sentence an FDG-tree. At phonological level, each spoken sentence was listened to by at least two annotators who added ToBI tone labels and phonological entity boundaries to the spoken texts. This is how the annotators identified the IUs in the sentence discussed above:

$$[[[O'Brien]_{RU} \quad [apucă \quad sticla \quad de \quad gât]_{RU}]_{ip}]_{IP} \quad [[și \quad turnă \quad în$$
$$pahare]_{RU}]_{ip} \quad [[un \quad lichid]_{RU} \quad [de \quad un \quad roșu \quad aprins]_{RU}]_{ip}]_{IP}.$$

One can easily see that there are several discrepancies between the manually and automatically IU annotation, such as the splitting of the (grammatical) subject and the VG (verbal group) in separate rhythmic units. This contrastive analysis helps us to improve the intonational units' detection, compatible with the intonational-driven *mapping algorithm* of Ghini [12] (*e.g.* possibility of adding preposition or length–based exception cases), keeping in mind however that the manual annotation has been empirically performed.

### 3.1 Application of the TFA Algorithm for Romanian

The input for the TFA algorithm consists of FDG trees for Romanian sentences. Besides the morphological annotation of each word, the semantic features of constituents are also classified according to their specificity degrees: (i) general – low specificity, contextually non-bound; (ii) specific – high specificity, contextually non-bound; (iii) indexical – mid-specificity, contextually-bound. Examples of specificity degrees for temporal complements are: **general** "niciodată" (EN: *never*), "mereu" (EN: *always*) etc.; **indexical** "astăzi" (EN: *today*), "anul acesta" (EN: *this year*); **specific** "22 iunie" (EN: *22nd of June*), "într-o frumoasă zi de aprilie" (EN: *a beautiful April day*). The TFA algorithm utilizes specificity degrees only for verbs and temporal / locative complements. An important element in computing the sentence topic-focus values is the position of the verbal direct and indirect arguments within the *Systemic Order (SO)*. We posit that the SO for Romanian, which is used in our implementation of TFA, is the same as the English SO [9]:

```
Time - Actor - Addressee - Objective - Origin - Effect -
Manner - Directional.from - Means - Directional.to - Locative
```

Figure 3 presents an example of the algorithm application, where the parsing tree is described in a LISP-style syntax and the output of the procedure consists of the topic-focus assessment for the verb and all its complementations.

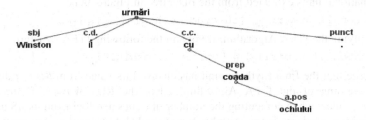

**Fig. 3a.** FDG tree for the sentence *"Winston îl urmări cu coada ochiului."* (EN: *Winston followed him from the tail of the eye.*).

```
Input:
verb (topic(f), sem(interm), label(il urmari),
ltree(act(topic(f), det(1), so(1), surf(np), label
(Winston))), rtree(mod(topic(f), det(1), so(8), surf(np),
label (coada), ltree(prep(cu)), rtree(apos(ochiului)))))
Output:
(Winston)_T (îl urmări)_{T/F} (cu coada ochiului)_{T(F)}
```

**Fig. 3b.** TFA algorithm input (classical syntax) and output (IS-syntax)

## 3.2 Intonational TFA Algorithm

After computing topic/focus for the syntactic constituents, the natural step to follow is to assign topic-focus values to the intonational units computed from the FDG tree. We have called this process the Intonational TFA Algorithm (ITFA), and conceived it as a consistent extension of the Prague TFA. The output of the TFA algorithm is post-processed in order to define the *t(f)* value: if the clause contains an IS unit *a* labeled as *t(f)*, and no other IS unit in that clause bears a focus, then the label of the IS unit *a* changes into *f* focus; in all other instances, the IS unit labeled as *t(f)* will be changed into *t/f* value. This processing helps to define the ITFA algorithm for merging intonational units with IS-based syntactic units.

```
if (IU unit == IS unit)
#the IU and the IS have exactly the same elements
    topic/focus(IU unit) := topic/focus(IS unit)
else if (IU unit contains more than one IS unit)
  if (all are t/f)
      topic/focus(IU unit) := t/f
  else
  count occurrences of topic and focus respectively
# an IS having t/f yields a topic and a focus values
      if (dominant topic/focus can be determined)
          topic/focus(IU unit) := dominant topic/focus
      else
```

#dominant *topic/focus* cannot be determined, for instance if we have one focus and one topic
```
    topic/focus(IU unit) := t(f)
else if (IS unit contains more than one IU unit)
  headIU := the IU containing the head of the IS unit
  topic/focus(headIU) := topic/focus(IS unit)
  topic/focus(rest of IUs) := t
```

The intonational units extracted from the FDG tree in Figure 3a is:

[[[Winston îl urmări]$_{RU}$ [cu coada ochiului]$_{RU}$]$_{ip}$]$_{IP}$.

When applying the ITFA Algorithm, we obtain the following ITFA structure:

[[[(Winston îl urmări)$_T$]$_{RU}$ [(cu coada ochiului)$_{T/F}$]$_{RU}$]$_{ip}$]$_{IP}$.

One can see that the final rhythmic unit has changed its value from *t(f)* to *f* due to the post-processing of the ITFA Algorithm. Also, the RU *"Winston îl urmări"* becomes *topic* since, when computing the number of topics and foci from the IS units, we have two topics and one focus. Another run of the ITFA Algorithm is exemplified using the sentence *"Pe o parte a feţei, carnea i se înnegrea"*. From the FDG tree presented in Fig. 4, the following FDG-based intonational units have been extracted:

[Pe o parte]$_{RU}$ [a feţei]$_{RU}$, [carnea i se înnegrea]$_{RU}$.

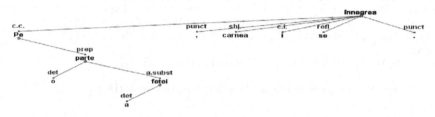

**Fig. 4.** FDG tree for the sentence *Pe o parte a feţei, carnea i se înnegrea* (EN: *On one side of the face, the flesh was turning dark*)

By applying the TFA algorithm, the following IS-based syntactic units have been computed:

(Pe o parte a feţei)$_{T/F}$, (carnea)$_T$ (i se înnegrea)$_F$.

Lastly, when applying the ITFA Algorithm over those two structures, we obtain the following (re)arrangement of the IS information over the intonational units:

[[[(Pe  o  parte)$_T$]$_{RU}$  [(a  feţei)$_{T/F}$]$_{RU}$,  [(carnea  i  se înnegrea)$_{T(F)}$]$_{RU}$]$_{ip}$]$_{IP}$.

## 4  Conclusion

TFA theory and algorithm is a procedure intervening exactly on the core of syntax-prosody interface, with semantic, discursive, and pragmatic surgery tools, bringing clearer rules for establishing the accented-deaccented IS components on the prosodic intonational units of a sentence, with the aim of assigning proper phonological stress (ToBI labels). Finding the set(s) of rules by which one could evaluate *the focused* elements of a spoken utterance is *one* of the keys for an adequate *prosody prediction*. This is the main challenge of this paper. Meanwhile, the contrastive analysis realized on the 150 Romanian sentences (from "1984") corpus for IS-parsing and intonational units' detection helped us to adapt and design (I)TFA algorithms for Romanian prosody prediction and a better understanding of the syntax-prosody interface.

# References

1. Apopei, V., Jitca, D.: A Set of Intonational Category for Romanian Speech and Text Annotation. In: Proceedings ECIT2006 – 4th European Conference on Intelligent Systems and Technologies, Iasi, Romania, pp. 117–124 (2006)
2. Calhoun, S., Nissim, M., Steedman, M., Brenier, J.: A Framework for Annotating Information Structure in Discourse. In: Proceedings of the Workshop on Frontiers in Corpus Annotation II: Pie in the Sky, Ann Arbor, Michigan, pp. 45–52 (2005)
3. Chomsky, N.: Derivation by phase. In: Kenstowicz, M. (ed.) Ken Hale: A life in language, pp. 1–52. MIT Press, Cambridge, MA (2001)
4. Curteanu, N.: Local and Global Parsing with Functional (F)X bar Theory and SCD Linguistic Strategy (I.+II.), Computer Science Journal of Moldova, Academy of Science of Moldova, 14(1) (40), 74–102, 14(2) (41), 155–182 (2006)
5. Curteanu, N., Trandabăţ, D.: Functional (F)X bar Projections for Local and Global Text Structures. The Anatomy of Predication. Revue Roumaine de Linguistique, Romanian Academy Editorial House, Bucharest, (4) (to be published, 2006)
6. Curteanu, N., Trandabăţ, D., Moruz, M.A.: Syntax-Prosody Interface for Romanian within Information Structure Theories. In: Burileanu, C., Teodorescu, H.-N. (eds.) Proceedings of 4th Conference on Speech Technology and Human - Computer Dialogue (SpeD-2007), Romanian Academy Editorial House, Bucharest, pp. 217–228 (2007)
7. Göbbel, E.: On the Relation between Focus, Prosody, and Word Order in Romanian. In: Quer, J., Schroten, J., Scorretti, M., Sleeman, P., Verheugd, E. (eds.) Romance Languages and Linguistic Theory 2001, pp. 75–92 (2003)
8. Hajičová, E.: Old Linguists Never Die, They Only Get Obligatorily Deleted. Computational Linguistics 43(4), 457–469 (2006)
9. Hajičová, E., Skoumalova, H., Sgall, P.: An Automatic Procedure for Topic-Focus Identification. Computational Linguistics 21(1), 81–94 (1995)
10. Järvinen, T., Tapanainen, P.: A dependency parser for English. Technical Report TR-1. Helsinki: University of Helsinki, Department of General Linguistics (1997)
11. Marcu, Daniel: The Theory and Practice of Discourse Parsing and Summarization. The MIT Press, Cambridge, MA (2000)
12. Monachesi, Paola: The Verbal Complex in Romance. A Case Study in Grammatical Interfaces. In: Oxford Studies in Theoretical Linguistics, Oxford University Press, Oxford (2005)
13. Selkirk, E., Kratzer, A.: Focuses, Phases and Phrase Stress. Ms., University of Massachusetts at Amherst (2005)
14. Steedman, M.: Information structure and the Syntax-Phonology Interface. In: Linguistic Inquiry, vol. 34, pp. 649–689. MIT Press, Cambridge (2000)
15. Teodorescu, H.-N.: A Proposed Theory in Prosody Generation and Perception: The Multi-dimensional Contextual Integration Principle of Prosody. In: Trends in Speech Technology, Romanian Academy Publishing House, pp. 109–118 (2005)
16. Tseng, C.-Y., San, Z.-Y., Chang, C.-H., Tai, C.-H.: Prosodic Fillers and Discourse Markers-Discourse Prosody and Text Prediction. In: The Second International Symposium on Tonal Aspects of Languages (TAL 2006), La Rochelle, France, pp. 109–114 (2006)
17. Valin, V., Robert, Jr., D.: Exploring the syntax-semantics interface. Cambridge University Press, Cambridge (2005)

# Translation and Conversion
# for Czech Sign Speech Synthesis*

Zdeněk Krňoul and Miloš Železný

University of West Bohemia, Faculty of Applied Sciences,
Department of Cybernetics
Univerzitní 8, 306 14 Pilsen, Czech Republic
{zdkrnoul,zelezny}@kky.zcu.cz

**Abstract.** Recent research progress in developing of Czech Sign Speech synthesizer is presented. The current goal is to improve a system for automatic synthesis
to produce accurate synthesis of the Sign Speech. The synthesis system converts
written text to an animation of an artificial human model. This includes translation of text to sign phrases and its conversion to the animation of an avatar. The
animation is composed of movements and deformations of segments of hands,
a head and also a face. The system has been evaluated by two initial perceptual
tests. The perceptual tests indicate that the designed synthesis system is capable
to produce intelligible Sign Speech.

## 1 Introduction

For human-computer interaction, speech-based communication systems become more
and more widely used. However, people with hearing or speech disorders cannot use
such systems. Hence, computer communication systems for aurally or speech disabled
people must be based on alternative means, such as Sign Speech. As one part of such
communication system, a system for automatic synthesis of the Sign Speech provides
speech utterances in these dialogs. A sign of the Sign Speech has two components:
the manual one and the non-manual one. The non-manual component is expressed by
a gesture of a face, a movement and a position of a head and other parts of the upper
half-body. The manual component is expressed by shapes, movements and positions of
hands. In principle, there are two ways how to synthesize any sign language utterance
by a computer system: a data driven approach or generation of movements from the
symbolic formulation (a symbolic system).

The data-driven approach is based on capture of body movements. This synthesis
process uses specific movements derived from signing person directly. Second variant uses artificial composition of base movements. The final Sign Speech utterance is
achieved by concatenation of relevant isolated signs. There is an analogy with the spoken form of a given language.

Second approach appears to be more general and appropriate. It uses a trajectory
generator of the manual component, which employs some symbolic system. However,

---

* Support for this work was provided by the Grant Agency of Academy of Sciences of the Czech
Republic, project No. 1ET101470416.

V. Matoušek and P. Mautner (Eds.): TSD 2007, LNAI 4629, pp. 524–531, 2007.

conversion of symbols to really naturally looking animation is a complicated task. Robotic-like movements are often perceived. As a part of the non-manual component, synchronous synthesis of lip articulation (talking head) is necessary for good overall intelligibility. For this purpose, the goal is not to simulate the musculature of the face or inner mouth organs but to produce sufficient support for lip-reading. Lip articulation should be then controlled by several visual parameters which directly determine required deformations.

For synthesis of the manual component of the Sign Speech we designed solution which is based on the symbolic representation of signs. For this purpose, we have chosen the HamNoSys notation system (Hamburg Notation System for Sign Languages[1]) from several notation systems. This notation system has precise description of signs and is usable for the interpretation in the computer system.

## 2 Sign Speech Synthesis

Our synthesis system has two parts: the translation and the conversion subsystem (Figure 1). The translation system transfers Czech written text to its textual representation in the Sign Speech (textual sign representation). The conversion system then converts this textual sign representation to animation of the artificial human model (avatar). Resulting animation then represents the corresponding utterance in the Sign Speech.

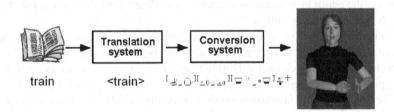

train          &lt;train&gt;

**Fig. 1.** Schema of the Sign Speech synthesis system

### 2.1 Translation System

The translation system is an automatic phrase-based translation system. A Czech sentence is divided into phrases and these are then translated into corresponding Sign Speech phrases. The translated words are reordered and rescored using language model at the end of the translation process. In our synthesizer we use our own implementation of the simple monotone phrase-based decoder - **SiMPaD** [1]. Monotonicity means that there is no reordering model. In the decoding process we choose only one alignment with the longest phrase coverage (i.e. if there are three phrases: $p_1$, $p_2$, $p_3$ coverage three words: $w_1$, $w_2$, $w_3$, where $p_1 = w_1 + w_2$, $p_2 = w_3$, $p_3 = w_1 + w_2 + w_3$, we choose the alignment which contains phrase $p_3$ only). Standard Viterbi algorithm is used for decoding. SiMPaD uses a trigram language model with linear interpolation.

---

[1] Available at www.sign-lang.uni-hamburg.de/projects/HamNoSys.html

## 2.2  Conversion System

The conversion system produces two main components: the manual and non-manual component of a sign. The manual component represents movements, orientations and shapes of hands. The non-manual component is composed of remaining movements of the upper half-body, face gestures and also articulation of lips and inner mouth organs. Symbolic representation was designed to solve the animation problem of the manual component and the upper half-body movements. We use HamNoSys 3.0 for this purpose. This notation is deterministic and suitable for processing of the Sign Speech in a computer system. Synthesis of lip articulation is provided by our talking head subsystem.

## 3  Sign Speech Editor

Methodology of symbolic notation allows precise and extensible description of a sign usable for avatar animation. However, composition of many symbols to a correct string is very difficult. For better coverage of HamNoSys notation features, we improved the SLAPE editor [2]. New functions have been inserted to reach easy transcription. The editor allows notation of two hand movements observed in symmetric signs. Notation of a movement modality is added, too. Modality means a style of a movement (speed, tensity etc.) or a repetition of a movement. The editor allows also notation of a precise location as a contact of the dominant hand with the upper half-body or the second hand. Furthermore, the editor uses avatar animation as a feedback for validation of the string of symbols and easier transcription.

Work with the editor is intuitive. Even those, who are not familiar with a sign language, can use it. Building or correction of signs can be also made in accordance with video records of a signing character. During transcription process, a sign is created by choosing pictures, which represent particular parts of a human body and its basic movements. The final string of HamNoSys symbols is finally saved. For a further synthesis purpose, these strings are stored in the symbol vocabulary.

## 4  Process of Continuous Sign Speech Synthesis

The synthesis process is based on feature frames and animation trajectories. Firstly, our synthesis system automatically generates sequences of feature frames. These feature frames are then transformed into trajectories. Each trajectory controls relevant rotation of bone joints or deformation of triangular meshes directly. The synthesis process involves analysis of symbols for isolated signs, frame processing and final concatenation of isolated signs. The analysis includes parsing of a symbol string into a tree structure and processing of symbols [2]. Next, the sequences of frames are generated for each isolated sign. The final sequence of frames which describes continuous speech is built using the concatenation technique in accordance with the textual sign representation of input text. The illustration of continuous animation is in Figure 2.

The analysis of symbols allows the computation of trajectories only for hands and upper half-body. Trajectories for lip articulation and face gestures are produced by the "talking head" subsystem separately.

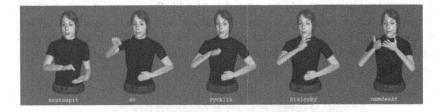

**Fig. 2.** Example of continuous animation

## 4.1  Analysis of Symbols and Trajectories for Isolated Sign

The goal of the analysis of symbols is to transform a particular isolated sign into feature frames. The analysis of the HamNoSys symbols leads to the description based on a context-free grammar. We have constructed over 300 parsing rules. If a sign is accepted by a parser, the algorithm creates a parse tree. A path to each leaf node determines types of symbol processing of the particular symbol. Processing is carried out by several tree walks that reduce the number of nodes. In each node of the parse tree, property items of a symbol are joined and blended. For both hands, these items can be divided to three logic parts: a location, motion and a shape. Processing of these items produces several feature frames for each node. As a result of this analysis, we obtain one sequence of frames in the root of the parse tree. It is then transformed into trajectories by the inverse kinematics technique. The frequency of generated frames is implicitly set to 25 frames per second. The non-manual component of body movements can be supplied by computation of relevant trajectories for the upper half-body.

## 4.2  Talking Head Subsystem

Trajectories for face gestures and also for articulation of lips, a tongue and jaws are created by visual synthesis carried out by the talking head subsystem. This visual synthesis is based on concatenation of phonetic units. Any written text or phrase is represented as a string of successive phones. The lip articulatory trajectories are concatenated by the Visual unit selection method [3]. This synthesis method uses an inventory of phonetic units and the regression tree technique. It allows precise coverage of coarticulation effects. In the inventory of units, several realizations of phoneme are stored. Our synthesis method assumes that the lip and tongue shape is described by a linear model. Realization of phoneme is described by 3 linear components for lip shape and 6 components for tongue shape. The lip components represent linear directions for lip opening, protrusion and upper lip raise. The tongue components consist of jaw height, dorsum raise, tongue body raise, tip raise, tip advance and tongue width. The synthesis algorithm performs a selection of appropriate phoneme candidate according to the context information. This information is built from the triphone context, the occurrence of coarticulation resistant component (of lip or tongue) in adjacent phonemes and also from time duration of neighbored speech segments. Final trajectories are computed by cubic spline interpolation between selected phoneme realizations.

The created trajectories should be time-aligned with the timing of acoustic Sign Speech form. This form is produced by an appropriate TTS system. Synthesis of face gesture trajectories is based on the concatenation and the linear interpolation of the neutral face expression and one of the 6 basic face gestures: happiness, anger, surprise, fear, sadness and disgust. The resulting head animation is in Figure 3.

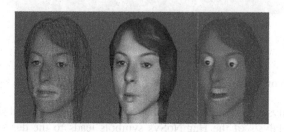

**Fig. 3.** Animation of talking head

### 4.3   Synchrony of Manual and Non-manual Components

Synchrony of the manual and non-manual component is crucial in the synthesis of continuous Sign Speech. The asynchronous components cause overall unintelligibility. The asynchrony should be caused by the different speech rate of spoken and the Sign Speech. We designed an effective solution - a synchrony method at the level of words. This method combines basic concatenation technique with time delay processing. Firstly, for each isolated sign, trajectories from the symbol analysis and trajectories from the talking head subsystem are generated. Time delay processing determines duration of all trajectories and selects the longest synthesized variant. The following step of processing evaluates interpolation time which is necessary for concatenation of particular adjacent isolated signs. This interpolation time ensures the fluent shift of body pose. We select the linear interpolation between the frames on the boundaries of concatenated signs. The interpolation of a hand shape and its 3D position is determined by weight average, the finger direction and palm orientation is interpolated by the extension to the quaternion.

## 5   Animation Model

The animation model tries to produce the Sign Speech by the efficient manner rather than deep understanding to physiological mechanisms. Our animation algorithm employs a 3D geometric animation model of avatar in compliance with H-Anim standard[2]. Our model is composed of 38 joints and body segments. These segments are represented by textured triangular surfaces. The problem of setting the correct shoulder and elbow rotations is solved by the inverse kinematics[3]. There are 7 degrees of freedom for each limb. The rotation of remaining joints and local deformation of the triangular surfaces

---

[2] www.h-anim.org
[3] Available at cg.cis.upenn.edu/hms/software/ikan/ikan.html

allows to set full avatar poses. The deformation of triangular surfaces is primarly used for animation of a face and a tongue model. The surfaces are deformed according to animation schema which is based on the definition of several control 3D points and splines functions [4]. The rendering of the animation model is determined in C++ code and OpenGL.

# 6  Perceptual Evaluation

Two tests on the intelligibility of synthesized Sign Speech have been performed. The goal has been to evaluate the quality of our Sign Speech synthesizer. Two participants who are experts in the Sign Speech served as judges. We used vocabulary of about 130 signs for this evaluation purpose. We completed several video records of our animation and also signing person. The video records of signing person are taken from the electronic vocabulary[4]. The capturing of video records of our animation was under two conditions.

## 6.1  Test A: Isolated Signs

The equivalence test was aimed at the comparison of animation movements of isolated signs with movements of signing person. Video records of 20 pairs of random selected isolated signs were completed. The view on the model of avatar and signing person was from the front. The participants evaluated this equivalence by marks from 1 to 5. The meaning of marks was:

- 1 totally perfect; the animation movements are equivalent to signing person
- 2 the movements are good, the location of hand, shapes or speed of sign are a little different but the sign is intelligible
- 3 the sign is difficultly recognized; the animation includes mistakes
- 4 incorrectly animated movements
- 5 totally incorrect; it is different sign

The results are in top panel of Figure 4. The average mark of participant 1 is 2.25 and of participant 2 is 1.9. The average intelligibility is 70% (marks 1 and 2 indicate the intelligible sign). There was 65% mark agreement between participants. The analysis of signs with lower marks shows that the majority of mistakes is caused by of the symbolic notation rather than inaccuracy in the conversion system. Thus, it is highly important to obtain as accurate symbolic notation of isolated signs as possible.

## 6.2  Test B: Continuous Speech

We created 20 video animation records of short utterances. The view on the avatar animation was here partially from the side. The participants judged the whole Sign Speech utterance. The subtitles (text representation of each sign) were added to the video records. Thus, the participants knew the meaning of the utterance and determined the overall intelligibility. The participants evaluate the intelligibility by marks from 1 to 5. The meaning of marks was:

---

[4] Langer, J. et al.: Znaková zásoba českého znakového jazyka. Palacký Univ. Olomouc.

- 1 the animation shows the signs from subtitles
- 2 good intelligible utterance
- 3 hardly intelligible utterance
- 4 almost unintelligible utterance
- 5 total unintelligible utterance

The results are in bottom panel of Figure 4. All utterances were evaluated by mark 1 or 2. In average, the animation of 70% utterances shows the signs from subtitles. The results indicate that the synthesis of continuous speech is intelligible. The concatenation and synchrony method of isolated signs is sufficient.

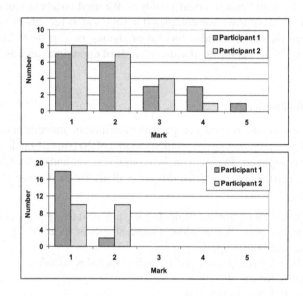

**Fig. 4.** Perceptual evaluation, on top: isolated signs, below: continuous Sign Speech

## 7    Summary and Conclusions

The synthesis system automatically converts written text to the animation of avatar. The synthesis system consists of the translation and conversion system. The translation system translates the Czech written text to its text sign representation. The conversion system then transforms this text sign representation into an animation. For purpose of symbolic notation of signs, the SLAPE editor was improved. The SLAPE editor provides online access to creating and editing of signs. Transcription is intuitive and resulting signs are stored in the vocabulary. For a general purpose, each sign is represented by a string of symbols.

The conversion system uses designed synthesis process of continuous Sign Speech. The synthesis process analyses the string of HamNoSys symbols and creates the sequence of feature frames for isolated signs. Designed concatenation and time delay process allows using separately synthesized trajectories from talking head system together with the trajectories generated from symbolic notations. This trajectory representation of signs is depended on the proportion of animation model (our model is composed of 38 joints and segments of body in this time).

The perceptual tests reveal that the synchrony on the level of word preserves the intelligibility for continuous Sign Speech. But the intelligibility of isolated signs highly depends on symbolic notation of particular signs in the vocabulary. Thus, it is necessary to concentrate on acquisition of precise symbolic notation of isolated signs in future work.

# References

1. Kanis, J., Müller, L.: Automatic Czech – Sign Speech Translation. In: Proceedings of 10th International Conference on TEXT, SPEECH and DIALOGUE TSD 2007, Springer, Heidelberg (2007)
2. Krňoul, Z., Kanis, J., Železný, M., Müller, L., Císař, P.: 3D Symbol Base Translation and Synthesis of Czech Sign Speech. In: Proceedings of SPECOM. St. Petersburg: Anatolya publisher (2006)
3. Krňoul, Z., Železný, M., Müller, L., Kanis, J.: Training of Coarticulation Models using Dominance Functions and Visual Unit Selection Methods for Audio-Visual Speech Synthesis. In: Proceedings of INTERSPEECH 2006 - ICSLP, Bonn (2006)
4. Krňoul, Z., Železný, M.: Realistic Face Animation for a Czech Talking Head. In: Sojka, P., Kopeček, I., Pala, K. (eds.) TSD 2004. LNCS (LNAI), vol. 3206, Springer, Heidelberg (2004)

# A Wizard-of-Oz System Evaluation Study

Melita Hajdinjak and France Mihelič

University of Ljubljana, Faculty of Electrical Engineering,
Laboratory of Artificial Perception, Systems, and Cybernetics,
Tržaška 25, SI-1000 Ljubljana, Slovenia,
{melita.hajdinjak, france.mihelic}@fe.uni-lj.si
http://luks.fe.uni-lj.si/

**Abstract.** In order to evaluate the performance of the dialogue-manager component of a developing, Slovenian and Croatian spoken dialogue system, two Wizard-of-Oz experiments were performed. The only difference between the two experiment settings was in the dialogue-management manner, i.e., while in the first experiment dialogue management was performed by a human, the wizard, in the second experiment it was performed by the newly-implemented dialogue-manager component. The data from both Wizard-of-Oz experiments was evaluated with the PARADISE evaluation framework, a potential general methodology for evaluating and comparing different versions of spoken-language dialogue systems. The study ascertains a remarkable difference in the performance functions when taking different satisfaction-measure sums or even individual scores as the target to be predicted, it proves the indispensableness of the recently introduced *database parameters* when evaluating information-providing dialogue systems, and it confirms the dialogue manager's cooperativity subject to the incorporated knowledge representation.

## 1 Introduction

In order to evaluate the performance of spoken-language dialogue systems, Walker, Litman, Kamm, and Abella [1] proposed PARADISE (PARAdigm for DIalogue System Evaluation), a framework that models user satisfaction as a linear combination of measures reflecting *task success* and *dialogue costs*. Some important PARADISE details and issues were, however, highlighted by Hajdinjak and Mihelič [2].

With the intention of involving the user in all the stages of the design of the natural-language-spoken, weather-information-providing dialogue system [3], Wizard-of-Oz (WOZ) data was collected and evaluated even before the completion of all the system's components. The aim of the first two WOZ experiments [4] was to evaluate the performance of the dialogue-manager component [5,6]. Therefore, while the task of the wizard in the first WOZ experiment was to perform speech understanding and dialogue management, her task in the second WOZ experiment was to perform only speech understanding, and the dialogue-management task was assigned to the newly-implemented dialogue manager.

In the developing spoken dialogue system the slot-filling approach to dialogue management was used, and a special knowledge representation [7], which is consistently

V. Matoušek and P. Mautner (Eds.): TSD 2007, LNAI 4629, pp. 532–539, 2007.

flexible in directing the user to select relevant, available data, was incorporated into the dialogue-management process. Moreover, the dialogue strategy was modeled using *conversational game theory* [5].

## 2 Wizard-of-Oz Experiments

There were 76 and 68 users involved in the first and the second WOZ experiment, respectively. The users were given two tasks – the first task was to obtain a particular piece of weather-forecast information and the second task was a given situation, aim of which was to stimulate them to ask context-specific questions. In addition, the users were given the freedom to ask extra questions. User satisfaction was then evaluated with the user-satisfaction survey introduced by Walker, Litman, Kamm, and Abella [8] and given in table 1, and a comprehensive **User Satisfaction** ($US$) was computed by summing each question's score and thus ranged in value from a low of 8 to a high of 40. The corresponding mean values are given in table 2.

**Table 1.** The user-satisfaction survey used within the PARADISE framework

1. *Was the system easy to understand?* (**TTS Performance**)
2. *Did the system understand what you said?* (**ASR Performance**)
3. *Was it easy to find the message you wanted?* (**Task Ease**)
4. *Was the pace of interaction with the system appropriate?* (**Interaction Pace**)
5. *Did you know what you could say at each point of the dialogue?* (**User Expertise**)
6. *How often was the system sluggish and slow to reply to you?* (**System Response**)
7. *Did the system work the way you expected it?* (**Expected Behaviour**)
8. *From your current experience with using the system, do you think you'd use the system when you are away from your desk?* (**Future Use**)

The task of the wizard in the first WOZ experiment was to simulate Slovenian speech understanding (speech recognition and natural-language understanding) and dialogue management. Croatian speech understanding was not performed since only Slovene users were being involved into the experiment. During the experiment, the wizard was sitting behind the graphical interface, listened to users' queries and tried to mediate an appropriate response, which was being successively followed by the natural-language-generation process and the text-to-speech process.

While the aim of the first WOZ experiment was, first of all, to collect human-computer data, the aim of the second WOZ experiment was to evaluate the newly-implemented dialogue-manager component. Thus, the only difference between the two experiment settings was in the dialogue-management manner, i.e., while in the first experiment dialogue management was performed by a human, the wizard, in the second experiment it was performed by the dialogue manager.

We have claimed [2] that the influence of automatic speech recognition hinders the other predictors from showing significance when evaluating the performance of the dialogue-manager component, excluding the efficiency of its clarification strategies. Only if the user is disencumbered from speech-recognition errors is he/she able to reliably assess the mutual influence of the observable less significant contributors to his/her satisfaction with the dialogue manager's performance. Therefore, in our WOZ experiments speech understanding (i.e., speech recognition and natural-language understanding) was performed by the wizard. As expected, $\kappa$ with mean values near 1 (0.94 and 0.98, respectively) listed in table 2, which was the only predictor reflecting speech-recognition performance, did not show the usual degree of significance in predicting the performance of our WOZ systems (section 3).

## 3　Wizard-of-Oz System Evaluation

The selection of regression parameters is crucial for the quality of the performance equation, and it is usually the result of thorough considerations made during several successive regression analyses. However, following the recommended Cohen's method [9], we first computed the task-success measure

**Kappa coefficient** ($\kappa$), reflecting the wizard's typing errors and unauthorized corrections,

and the dialogue-efficiency costs

**Mean Elapsed Time** ($MET$), i.e., the mean elapsed time for the completion of the tasks that occurred within the interaction, and
**Mean User Moves** ($MUM$), i.e., the mean number of conversational moves that the user needed to either fulfil or abandon the initiated information-providing games.

Second, the following dialogue-quality costs were selected:

**Task Completion** ($Comp$), i.e., the user's perception of completing the first task;
**Number of User Initiatives** ($NUI$), i.e., the number of user's moves initiating information-providing games;
**Mean Words per Turn** ($MWT$), i.e., the mean number of words per user's turns;
**Mean Response Time** ($MRT$), i.e., the mean system-response time;
**Number of Missing Responses** ($NMR$), i.e., the difference between the number of system's turns and the number of user's turns;
**Number of Unsuitable Requests** ($NUR$) and **Unsuitable-Request Ratio** ($URR$), i.e., the number and the ratio of user's initiating moves that were out of context;
**Number of Inappropriate Responses** ($NIR$) and **Inappropriate-Response Ratio** ($IRR$), i.e., the number and the ratio of unexpected system's responses, including pardon moves;
**Number of Errors** ($Error$), i.e., the number of system errors, e.g., interruptions of the telephone connection and unsuitable natural-language sentences;
**Number of Help Messages** ($NHM$) and **Help-Message Ratio** ($HMR$), i.e., the number and the ratio of system's help messages;
**Number of Check Moves** ($NCM$) and **Check-Move Ratio** ($CMR$), i.e., the number and the ratio of system's moves checking some information regarding past dialogue events;

**Number of Given Data** ($NGD$) and **Given-Data Ratio** ($GDR$), i.e., the number and the ratio of system's information-providing moves;

**Number of Relevant Data** ($NRD$) and **Relevant-Data Ratio** ($RDR$), i.e., the number and the ratio of system's moves directing the user to select relevant, available data;

**Number of No Data** ($NND$) and **No-Data Ratio** ($NDR$), i.e., the number and the ratio of system's moves stating that the requested information is not available;

**Number of Abandoned Requests** ($NAR$) and **Abandoned-Request Ratio** ($ARR$), i.e., the number and the ratio of the information-providing games abandoned by the user.

All the mean values of the listed parameters in both WOZ experiments are given in table 2. Those that showed a significant change in value across both WOZ experiments are shaded grey and the corresponding $p$ value is given.

An interesting finding is the relatively high negative correlation (i.e., $-0.53$ in the first experiment and $-0.51$ in the second experiment) between $NUI$ and $MUM$, which reflects the users' ability to adapt to the system's behaviour and to learn how to more efficiently complete the tasks.

Special attention was given to the *database parameters*, i.e., $NGD$, $GDR$, $NRD$, $RDR$, $NND$, and $NDR$, which have been introduced as costs for user satisfaction by Hajdinjak and Mihelič [2]. In our experiments, relying on the extremely sparse and dynamical weather-information source with a time-dependent data structure, it turned out that these parameters can play an important part in predicting the performance of a dialogue system, the performance of its specific components (subsection 3.1), and even in predicting individual user-satisfaction metrics (subsection 3.2).

In addition, table 2 says that in the second WOZ experiment $NRD$ and $RDR$ were almost three times greater than in the first experiment, which confirms the dialogue manager's ability to direct the user to select relevant, available data when his/her explicit request yields no information. Consequently, in the second WOZ experiment, $NND$ was significantly lower ($p < 0.0005$).

## 3.1    Predicting User Satisfaction

Hone and Graham [10] ascertained that the items chosen for the user-satisfaction survey (table 1) introduced within the PARADISE framework were based neither on theory nor on well-conducted empirical research and that the approach of summing all the scores could not be justified. However, in the case of evaluating the performance of a specific component of a dialogue system (e.g., ASR or dialogue-manager performance), we believe [7] that it is more appropriate to take the corresponding, subjective satisfaction-measure set, i.e., the sum of the user-satisfaction scores that are likely to measure the selected aspect of the system's performance.

As the target to be predicted we, therefore, first took **User Satisfaction** ($US$) and afterwards the sum of those user-satisfaction values that (in our opinion) measured the dialogue manager's performance ($DM$) and could, in addition, be well modeled, i.e., the sum of the user-satisfaction-survey scores assigned to **ASR Performance**, **Task Ease**, **System Response**, and **Expected Behaviour**.

**Table 2.** Mean values of the selected parameters in the first (*WOZ1*) and the second (*WOZ2*) WOZ experiment

|  |  | WOZ1 | WOZ2 | p |
|---|---|---|---|---|
| task success | **Kappa koeficient** ($\kappa$) | 0.94 | 0.98 | |
| efficiency | **Mean Elapsed Time** ($MET$)* | 13.76 s | 17.39 s | 0.000 |
| costs | **Mean User Moves** ($MUM$) | 1.48 s | 1.68 s | 0.047 |
| | **Task Completion** ($Comp$) | 0.97 | 0.96 | |
| | **Number of User Initiatives** ($NUI$) | 6.49 | 7.51 | 0.005 |
| | **Mean Words per Turn** ($MWT$) | 9.32 s | 7.56 s | 0.000 |
| | **Mean Response Time** ($MRT$) | 5.13 s | 6.38 s | 0.000 |
| | **Number of Missing Responses** ($NMR$) | 0.60 | 0.75 | |
| | **Number of Unsuitable Requests** ($NUR$) | 0.48 | 0.13 | 0.011 |
| | **Unsuitable-Request Ratio** ($URR$) | 0.08 | 0.02 | |
| | **Number of Inappropriate Responses** ($NIR$) | 0.41 | 0.90 | 0.009 |
| | **Inappropriate-Response Ratio** ($IRR$) | 0.04 | 0.06 | |
| quality | **Number of Errors** ($Error$) | 0.12 | 0.06 | |
| | **Number of Help Messages** ($NHM$) | 0.32 | 0.40 | |
| costs | **Help-Message Ratio** ($HMR$) | 0.03 | 0.03 | |
| | **Number of Check Moves** ($NCM$)* | 0 | 2.19 | 0.000 |
| | **Check-Move Ratio** ($CMR$)* | 0 | 0.16 | 0.000 |
| | **Number of Given Data** ($NGD$) | 4.07 | 4.35 | |
| | **Given-Data Ratio** ($GDR$) | 0.67 | 0.58 | |
| | **Number of Relevant Data** ($NRD$) | 0.70 | 2.06 | 0.000 |
| | **Relevant-Data Ratio** ($RDR$) | 0.10 | 0.28 | 0.005 |
| | **Number of No Data** ($NND$) | 1.67 | 0.94 | 0.000 |
| | **No-Data Ratio** ($NDR$) | 0.22 | 0.12 | |
| | **Number of Abandoned Requests** ($NAR$) | 0.05 | 0.16 | |
| | **Abandoned-Request Ratio** ($ARR$) | 0.01 | 0.02 | |
| | **User Satisfaction** ($US$) | 34.08 | 31.96 | 0.015 |

* Duration of the system's replies is not included.
* In the first WOZ experiment, the wizard did not perform check moves.

After removing about 10% of the outliers in the data from the first WOZ experiment, backward elimination for $F_{out} = 2$ gave the following performance equations:

$$\widehat{\mathcal{N}(US)} = -0.69\mathcal{N}(NND) - 0.16\mathcal{N}(NRD)$$
$$\widehat{\mathcal{N}(DM)} = -0.61\mathcal{N}(NND) + 0.21\mathcal{N}(Comp) - 0.16\mathcal{N}(NRD)$$

with 58% (i.e., $R^2 = 0.58$) and 59% of the variance explained, respectively. To be able to observe the close similarity between these two equations, note that $Comp$ was significant for $US$ ($p < 0.02$), but removed by backward elimination.

In contrast, the data from the second WOZ experiment gave the following performance equations:

$$\widehat{\mathcal{N}(US)} = -0.30\mathcal{N}(CMR) - 0.23\mathcal{N}(MET) + 0.18\mathcal{N}(\kappa)$$
$$\widehat{\mathcal{N}(DM)} = -0.35\mathcal{N}(CMR) + 0.35\mathcal{N}(GDR) + 0.35\mathcal{N}(\kappa) - 0.17\mathcal{N}(ARR)$$

with 26% and 46% of the variance explained, respectively. Again, note that $MET$ was significant for $DM$ ($p < 0.02$) and that $GDR$ and $ARR$ were significant for $US$ ($p < 0.04$), but all removed by backward elimination.

From the coefficients of determination it is clear that $DM$ could be modeled better than $US$, especially in the second experiment. For the second experiment we can even assert that **User Satisfaction** ($US$), as defined within the PARADISE evaluation framework [8], could not be modeled well enough to be appropriate for the prediction of users' satisfaction.

However, let us compare both performance equations predicting $DM$, i.e., ascertain the effect of the dialogue-manager component. The first observation that we make is that none of the predictors is common to both performance equations. All the predictors from the first performance equation (i.e., $NND, Comp, NRD$) were insignificant ($p > 0.1$) for $DM$ in the second experiment. On the other hand, the only predictor from the second performance equation that was significant for $DM$ ($p < 0.004$) in the first experiment, but removed by backward ellimination, was $GDR$.

Unlike the first performance equation with the database parameters $NND$ and $NRD$ as crucial (negative) predictors, the second performance equation clearly shows their insignificance to users' satisfaction with the dialogue manager's performance. Hence it follows that the developed dialogue manager [5,6] with its rather consistent flexibility in directing the user to select relevant, available data does not (negatively) influence users' satisfaction.

Another interesting finding is that, although in the first experiment the users were more comprehensive for quantitative database parameters (i.e., $NGD, NRD, NND$) than for proportional database parameters (i.e., $GDR, RDR, NDR$), in the second experiment it was just the other way round. This finding makes demands upon further empirical research.

### 3.2  Predicting Individual User-Satisfaction Measures

In addition, we thought that it would be very interesting to see which parameters are significant for individual user-satisfaction metrics (table 1). The situation in which the **Task Ease** question was the only measure of user satisfaction, the aim of which was to

**Table 3.** Most significant predictors of the individual user-satisfaction measures in the first (*WOZ1*) and the second (*WOZ2*) WOZ experiment

|  | WOZ1 | WOZ2 |
|---|---|---|
| **TTS Performance** | NND ($p < 0.00005$) | UMN ($p < 0.004$) |
| **ASR Performance** | NND ($p < 0.00005$) | CMR ($p < 0.012$) |
| **Task Ease** | NND ($p < 0.002$) | GDR ($p < 0.02$) |
| **System Response** | NND ($p < 0.0003$) | CMR ($p < 0.0002$) |
| **Expected Behaviour** | NND ($p < 0.00005$) | Comp, RDR, CMR ($p < 0.04$) |

maximize the relationship between elapsed time and user satisfaction, was considered before [11].

First, we discovered that **Future Use** could not be well modeled in the first WOZ experiment and that **User Expertise** and **Interaction Pace** could not be well modeled in the second WOZ experiment, i.e., the corresponding MLR models explained less than 10% of the variance.

Second, the parameters that most significantly predicted the remaining, individual user-satisfaction measures are given in table 3. Surprisingly, in the first experiment, the database parameter $NND$ was most significant for all the individual user-satisfaction measures, but, in the second experiment, it was insignificant for all of them. Moreover, almost all the parameters that were most significant to an individual user-satisfaction measure in the second experiment were insignificant to the same measure in the first experiment. On the one hand, this could indicate that the selected individual user-satisfaction measures really measure the performance of the dialogue manager and consequently illustrate the obvious difference between both dialogue-management manners. On the other hand, one could argue that this simply means that the individual user-satisfaction measures are not appropriate measures of attitude because people are likely to vary in the way they interpret the item wording [10]. Though, due to the huge difference in significance this seems an unlikely explanation.

## 4    Conclusion

In this study we have presented the dialogue-manager evaluation results within the development of a bilingual-spoken dialogue system for weather information retrieval. The data gathered in two WOZ experiments was evaluated with the PARADISE framework. This evaluation resulted in several performance equations predicting different dependent variables, all trying to express the effect of the dialogue-management manner. The observed differences between the derived performance equations make demands upon further empirical research. Not only does a reliable user-satisfaction measure that would capture the performance-measures of different dialogue-system components need to be established, but the reasons for the possible differences between several performance equations also need to be understood and properly assessed.

## References

1. Walker, M.A., Litman, D., Kamm, C.A., Abella, A.: PARADISE: A General Framework for Evaluating Spoken Dialogue Agents. In: Proc. 35th Annual Meeting of the Association of Computational Linguistics, Madrid, Spain, pp. 271–280 (1997)
2. Hajdinjak, M., Mihelič, F.: The PARADISE Evaluation Framework: Issues and Findings. Computational Linguistics 32(2), 263–272 (2006)
3. Žibert, J., Martinčić-Ipšić, S., Hajdinjak, M., Ipšić, I., Mihelič, F.: Development of a Bilingual Spoken Dialog System for Weather Information Retrieval. In: Proc. 8th European Conference on Speech Communication and Technology, Geneva, Switzerland, pp. 1917–1920 (2003)
4. Hajdinjak, M., Mihelič, F.: Conducting the Wizard-of-Oz Experiment. Informatica 28(4), 425–430 (2004)

5. Hajdinjak, M., Mihelič, F.: Information-providing dialogue management. In: Sojka, P., Kopeček, I., Pala, K. (eds.) TSD 2004. LNCS (LNAI), vol. 3206, pp. 595–602. Springer, Heidelberg (2004)
6. Hajdinjak, M., Mihelič, F.: A Dialogue-Management Evaluation Study. Journal of Computing and Information Technology (to appear, 2007)
7. Hajdinjak, M.: Knowledge Representation and Performance Evaluation of Cooperative Automatic Dialogue Systems, Ph. D. Thesis, Faculty of Electrical Engineering, University of Ljubljana, Slovenia (2006)
8. Walker, M.A., Litman, D.J., Kamm, C.A., Abella, A.: Evaluating spoken dialogue agents with paradise: Two case studies. Computer Speech and Language 12(3), 317–347 (1998)
9. Di Eugenio, B., Glass, M.: The Kappa statistic: a second look. Computational Linguistics 30(1), 95–101 (2004)
10. Hone, K.S., Graham, R.: Towards a tool for the Subjective Assesment of Speech System Interfaces (SASSI). Natural Language Engineering: Special Issue on Best Practice in Spoken Dialogue Systems 6(3-4), 287–303 (2000)
11. Walker, M.A., Borland, J., Kamm, C.A.: The utility of elapsed time as a usability metric for spoken dialogue systems. In: Proc. ASRU, Keystone, USA, pp. 317–320 (1999)

# New Measures for Open-Domain Question Answering Evaluation Within a Time Constraint

Elisa Noguera[1], Fernando Llopis[1], Antonio Ferrández[1], and Alberto Escapa[2]

[1] GPLSI. Departamento de Lenguajes y Sistemas Informáticos
Escuela Politécnica Superior. University of Alicante
{elisa,llopis,antonio}@dlsi.ua.es
[2] Departamento de Matemática Aplicada
Escuela Politécnica Superior. University of Alicante
alberto.escapa@ua.es

**Abstract.** Previous works on evaluating the performance of Question Answering (QA) systems are focused on the evaluation of the precision. In this paper, we developed a mathematic procedure in order to explore new evaluation measures in QA systems considering the answer time. Also, we carried out an exercise for the evaluation of QA systems within a time constraint in the CLEF-2006 campaign, using the proposed measures. The main conclusion is that the evaluation of QA systems in realtime can be a new scenario for the evaluation of QA systems.

## 1 Introduction

The goal of Question Answering (QA) systems is to locate concrete answers to questions in collections of text. These systems are very useful for the users because they do not need to read all the document or fragment to obtain a specific information. Questions as: How old is Nelson Mandela? Who is the president of the United States? When was the Second World War? can be answered by these systems. They contrast with the more conventional Information Retrieval (IR) systems, because they treat to retrieve relevant documents to a query, where the query may be a simply collection of keywords (e.g. old Nelson Mandela, president United States, Second World War, ..).

The annual Text REtrieval Conference (TREC[1]), organized by the National Institute of Standards and Technology (NIST), is a serie of workshops designed to advance in the state-of-the-art in text retrieval by providing the infrastructure necessary for large-scale evaluation of text retrieval methodologies. This model has been used by Cross-Language Evaluation Forum (CLEF[2]) in Europe and by the National Institute of Informatics Test Collection for IR Systems (NTCIR[3]) in Asia, which have also studied the cross-language issue. Since 1999, TREC have a specific QA track ([3]). CLEF ([2]) and NTCIR ([1]) have also introduced the QA evaluation. This evaluation consists of given a large number of newspaper and newswire articles, participating systems try to answer

---

[1] http://trec.nist.gov

[2] http://www.clef-campaign.org

[3] http://research.nii.ac.jp/ntcir

V. Matoušek and P. Mautner (Eds.): TSD 2007, LNAI 4629, pp. 540–547, 2007.

a set of questions by analyzing the documents in the collection in a fully automated way.

The main evaluation measures used in these forums are *accuracy* evaluation measure, but different metrics were also considered: *Mean Reciprocal Rank (MRR)*, *K1 measure* and *Confident Weighted Score (CWS)* (for further information about these metrics see [2]).

The motivation of this work is to study the evaluation of QA systems within a time constraint. In order to evaluate the answer time of the systems and compare them, we carried out an experiment in the CLEF-2006 providing a new scenario in order to compare the QA systems. Specifically, we have proposed new measures to evaluate not only the effectiveness of the systems, but also the answer time. As the results achieved by the systems, we can argue that this is a promising step to change the direction of the evaluation in QA systems.

The remainder of this paper is organized as follows: next section presents presents a new proposal of evaluation measures for QA systems. Section 3 describes the experiment carried out in the QA context at CLEF-2006, the evaluation used and the results achieved. Finally, section 4 gives some conclusions and future work.

## 2  New Approaches Evaluating QA Systems

The above mentioned problem can be reformulated in a mathematical way. Let us consider that the answer of each system $S_i$ can be characterized for our purposes by an ordered pair of real numbers $(x_i, t_i)$. The first element of the pair reflects the precision of the answer and the second one the efficiency. In this way, a QA task can be represented geometrically as a set of points located in a subset $D \subseteq \mathbb{R}^2$. Our problem can be solved by giving a method that allows to rank the systems $S_i$ accordingly to some prefixed criterion that take into account both the precision and the efficiency of each answer. This problem is of the same nature as others tackled in Decision theory.

A solution to this problem can be achieved by introducing a total preorder, sometimes referred as total quasiorder, in $D$. Let us remind you that a binary relation $\preceq$ on a set $D$ is a total preorder if it is reflexive, transitive and any two elements of $D$ are comparable. In particular, we can define a total quasiorder on $D$ with the aid of an auxiliary two variables real function $f : D \subseteq \mathbb{R}^2 \rightarrow I \subseteq \mathbb{R}$, in such a way that:

$$(a, b) \preceq (c, d) \iff f(a, b) \leq f(c, b), \forall (a, b), (c, d) \in D. \tag{1}$$

We will refer to this function as a ranking function. One of the advantages of this procedure is that the ranking function contains all the information relative to the chosen criterion to classify the different systems $S_i$. Anyway, let us underline that, since the binary relation $\preceq$ defined in this way is not necessarily an order in $D$, two different elements of $D$ can be equal with respect to the preorder $\preceq$, that is, they are in the same position of the ranking. Mathematically all the elements that are tied in the classification belong to a level curve of the ranking function, that we will call iso-ranking curve. Namely, the iso-ranking curves are characterized by all the elements of $D$ that fulfill the equation $f(x, t) = L$, being $L$ a real number in the image of $f$, $I$. Let us point out that the proposed ranking procedure to evaluate the QA task is of an ordinal type. This means that we should not draw a direct conclusion about the absolute difference of the numerical values of the ranking function for two systems. The only relevant information concerns to their relative position in the relative ranking of the QA task. As a matter of fact, if we consider a new ranking function constructed by composing the initial ranking function with a strictly increasing function, the numerical value assigned to each system is changed but the final ranking obtained is the same as the first one.

In the approach developed in this paper, the precision of the system $S_i$ is given the mean reciprocal rank ($MRR$), so $x_i \in [0, 1]$. The efficiency is measured by considering the answer time of each system, in such a way that a smaller time to answer means a better efficiency of a system. Anyway, to obtain a more suitable scale of representation, we have considered the effective time resulting from dividing the answer time by the maximum answer time obtained in the QA task under consideration, hence we will have that this effective time, denoted as $t_i$, belong to the interval $(0, 1]$. In this way, the accessible region of $\mathbb{R}^2$ is given by the set $D \equiv [0, 1] \times (0, 1]$.

To define a realistic ranking function, it is necessary to require to this function some additional features. These properties are based on the intuitive behavior that our ranking criterion should have to fulfill. For example, as a preliminary approach, we are going to demand the ranking function that:

1. The function $f$ must be continuous in $D$.
2. The supremum of $I$ is given by $\lim_{t \to 0} f(1, t)$. In the case that $I$ is not upper bound, we must have $\lim_{t \to 0} f(1, t) = +\infty$.
3. The infimum of $I$ is given by $f(0, 1)$.

The first condition is imposed for mathematical convenience, although it can be interpreted in terms of some simplified arguments. Namely, this requirement excludes the possibility that if two systems are in different positions in the ranking, any arbitrarily small variation in the precision or the efficiency of one of them changes their relative positions. The second condition is related with the fact that the fictitious system defined by the pair $(1, 0)$ always must be in the first position of the ranking. Finally, the last condition implies that the pair $(0, 1)$ must be in the last position.

## 2.1   Ranking Function Independent of Time ($MRR_2$)

As a first simple example of ranking function, let us consider $MRR_2(x, t) = x$. The preorder induced by this function is closed to the lexicographical order, some times called alphabetic order. For this ranking function we have that:

1. The image of $MRR_2$ is the interval $[0, 1]$.
2. The function $MRR_2$ is continuous in $D$.
3. $\lim_{t \to 0} MRR_2(1, t) = 1$.
4. $MRR_2(0, 1) = 0$.

So, this function fulfills all the previous requirements. On the other hand, the iso-ranking curves of the function are of the form $x = L$, $L \in [0, 1]$ whose representation is a family of vertical segments of length unity (see figure 1). The preorder constructed from this ranking function only takes into account the precision, being unaware of the efficiency of the systems.

## 2.2   Ranking Function with Inverse Temporal Dependence ($MRRT$)

As a second example of ranking function that does take into account the efficiency of the systems, we are going to consider $MRRT(x, t) = x/t$. Let us note that in this case the ranking function is inversely proportional to the time and directly proportional to the precision. In particular, this function verifies the properties:

1. The image of $MRRT$ is the interval $[0, +\infty)$.
2. The function $MRRT$ is continuous in $D$.
3. $\lim_{t \to 0} MRRT(1, t) = +\infty$.
4. $MRRT(0, 1) = 0$.

The associated iso-ranking curves to the function are of the form $x/t = L$, $L \in [0, +\infty)$. Geometrically these curves are a family of segments passing through the point $(0, 0)$ and with slope $1/L$ (see figure 1). In this way, the systems with better efficiency, that is, smaller effective time, obtain for a given value of $x$ a large value of the ranking function. As a matter of fact, both precision and efficiency have the same influence on the ranking function, since a system of values $(x, t)$ is tied with the system $(\alpha x, \alpha t)$ with $0 < \alpha < 1$. On the other hand, although the information of the ranking function is of an ordinal nature, it is desirable that the image of the function is between 0 and 1, since this facilitates an intuitive representation of the values of the ranking function, a condition that this function does not verify either.

## 2.3   Ranking Function with Inverse Exponential-Like with Time Dependence $MRRT_e$

Due to the disadvantages of the previous functions, we propose a new ranking function that depends both on the precision and the efficiency of the system but in which the efficiency has less weight than the precision when evaluating QA systems. Namely, we are going to introduce the ranking function

$$MRRT_e(x, t) = \frac{2x}{1 + e^t}, \tag{2}$$

being $e^t$ the exponential of the effective time. This function fulfills the following requirements:

1. The image of $MRRT_e$ is the interval $[0, 1)$.
2. The function $MRRT_e$ is continuous in $D$.
3. $\lim_{t \to 0} MRRT_e(1, t) = 1$.
4. $MRRT_e(0, 1) = 0$.

The iso-ranking curves of this function are of the form $2x/(1 + e^t) = L$, $L \in [0, 1)$, whose representation is sketched in figure 2. Let us underline that for fictitious efficient systems, that is, those systems that answer instantaneously ($t = 0$), this ranking function coincides with the usual precision classification. Nevertheless, the functional dependence on time modulates the value of $x$, in such a way that when the time grows up the value of the ranking function, for a fixed precision, decreases. Anyway, this modulation is smoother than in the case of the ranking function inversely proportional to time. Moreover, if we consider a given system $S$, we can only tie with it by considering systems whose precision, and efficiency, vary on a particular range, not for any arbitrarily small value of the precision.

## 3    Experiment at QA CLEF-2006

As above is mentioned, we considered the time as a fundamental part in the evaluation of QA systems. In accordance with CLEF organization, we carried out a pilot task at CLEF-2006 whose aim was to evaluate the ability of QA systems to answer within a time constraint. This is an innovative experiment and the initiative is aimed towards providing a new scenario for the evaluation of QA systems. This experiment follows the same procedure that the QA@CLEF-2006, but the main difference is the consideration of the answer time.

In total, five groups took part in this pilot task. The participating groups were: *daedalus* (Spain), *tokyo* (Japan), *priberam* (Portugal), *alicante* (Spain) and *inaoe* (Mexico). All of them participated in the main QA task at CLEF-2006, and have experience researching in QA systems.

### 3.1    Performance Evaluation

In this section we present the evaluation results of the five systems which participated in the realtime experiment. On the one hand, we present precision and the efficiency obtained by these systems. On the other hand, we present the score achieved by them with the different metrics which combine the precision with the efficiency. Tables 1 shows the summary of results for the used metrics (MRR, t, MRRT, $MRRT_e$).

**Table 1.** Evaluation results with the different metrics

| Participant | MRR | rank | t | rank | MRRT | rank | $MRRT_e$ | rank |
|---|---|---|---|---|---|---|---|---|
| daedalus1 | **0.41** | **1°** | 0.10 | 4° | 3.83 | 4° | **0.3881** | **1°** |
| tokyo | 0.38 | 2° | 1.00 | 6° | 0.38 | 6° | 0.2044 | 6° |
| priberam | 0.35 | 3° | **0.01** | **1°** | 32.13 | **1°** | 0.3481 | **2°** |
| daedalus2 | 0.33 | 4° | 0.03 | 3° | 8.56 | 3° | 0.3236 | 3° |
| inaoe | 0.3 | 5° | 0.38 | 5° | 0.78 | 5° | 0.2433 | **4°** |
| alicante | 0.24 | 6° | 0.02 | 2° | 16.23 | 2° | 0.2382 | **5°** |

The precision of the QA systems was evaluated in the experiment with the MRR metric. It is presented in table 1. As above is mentioned, the efficiency of the systems is measured with the answer time (in seconds). In order to normalize the obtained answer times between 0 and 1, we use $t$ as $tsec/tmax$ ($tsec$ is the answer time in seconds and $tmax$ is the highest answer time value of the list).

The overall evaluation of the QA systems, combining precision with the answer time with the $MRR_2$ metric (see section 2) is the same than the evaluation of the MRR (see

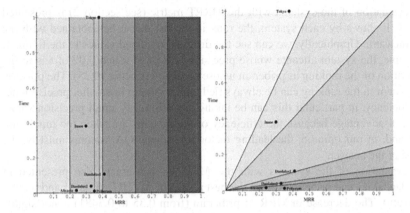

**Fig. 1.** Comparatives of the results obtained by each system with the Lexicographical preorder and $MRRT$ evaluation measures respectively in its iso-ranking function

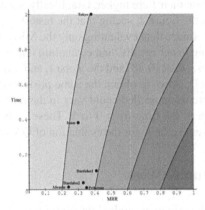

**Fig. 2.** Comparative of the results obtained by each system with the $MRRT_e$ evaluation measure in its iso-ranking function

section 1), because this measure takes into account the precision firstly, and it takes only into account the time if the precision among two or more systems is the same. Graphically, an iso-ranking curve is made up of all the systems with the same value of $x$ and arbitrarily different values of the effective time. That is, the ranking criterion is the same as the usual performance followed up today to evaluate QA systems. The limitations of this procedure, which have motivated this work, are clear if we consider the systems tokio and priberam in figure 1. With the above considered ranking function the system tokyo is in the second position of the ranking and the system priberam in the third one. However, the difference in the precision ($MRR$) of the systems is very small, 0.38 vs. 0.35 , whereas the efficiency of the system priberam is much better than the efficiency of the system tokyo. Therefore, it would be reasonable that the system priberam was preferred to the system tokyo. This is impossible with this kind of ranking functions, since they are independent of time.

The evaluation of the systems with the MRRT metric (see section 2) is presented in table 1. It shows by each system, the rank in the list that it has obtained with the MRRT measure. Graphically, we can see the different obtained values in the figure 1. For example, the system alicante whose precision is 0.24 of $x$ and 0.02 of $t$ is in the same position of the ranking as priberam whose precision is better (0.35). The position of any system in the ranking can be always tied with a system of smaller precision but larger efficiency, in particular this can be taking any arbitrarily small precision value. This is a disadvantage because the efficiency of the systems are taken too much into account and, in our opinion, the leading factor to evaluate QA systems must be the precision of the answer.

The evaluation of the systems with the $MRRT_e$ measure is also presented in table 1. daedalus1 and priberam obtain the best results of $MRRT_e$ (0.3881 and 0.3481 respectively). The decrease in MRR of priberam (from 0.35 to 0.3481) is not significant because it has a short answer time (76 seconds), just like alicante (from 0.24 to 0.2382). Nevertheless, the $MRRT_e$ of daedalus1 reduces the MRR (from 0.41 to 0.3881), because it has a upper answer time (549 seconds). Finally, inaoe and tokyo are significantly penalized because their $t$ are higher. Graphically, we can compare the different values of $MRRT_e$ in the figure 2, seeing that the fastest systems (priberam and alicante) have a similar performace than evaluating only the MRR. However, inaoe and tokyo are penalized obtaining lower results than evaluating only the MRR. For example, tokyo had the second best MRR (0.38) and the worst $t$, and it is penalized being the last in the ranking of $MRRT_e$. Also, to obtain the same position in the ranking as the system $S \equiv (0.4, 0.2)$ the precision needed could vary in the range from 0.36 to 0.67, corresponding to a variation of the time from 0 to 1. These particularities makes the ranking function $MRRT_e$ very suitable for the evaluation of QA systems in realtime.

## 4   Conclusions and Future Work

Mainly, the evalution of QA systems is studied deeply in three known evaluation forums: TREC, CLEF and NTCIR. But, these forums are only focused on evaluating the precision of the systems, and they do not evaluate their efficiency (we consider the answer time of the system as measure of efficiency). Mostly, this evaluation entails accurate systems but slowly at the same time. For this reason, we studied the evaluation of QA systems taking into account the answer time.

For the evaluation of the QA systems, we proposed three measures ($MRR_2$, MRRT, $MRRT_e$) to evaluate them within a time constraint. These measures are based on the Mean Reciprocal Rank (MRR) and the answer time. As preliminary results, we show that $MRRT_2$ only takes into account the precision and the measure MRRT takes into account too much the time. We have solved this inconvenience proposing a new measure called $MRRT_e$. It also combines the MRR with the answer time, but it is based on an exponential function. It penalizes the systems that has a higher answer time. In conclusion, the new measure $MRR_e$ allows classify the systems considering the precision and the answer time.

Futhermore, we carried out a task in the CLEF-2006 in order to evaluate QA systems within a time constraint. This is the first evaluation of QA systems in realtime. It has allowed to stablish a methodology and criterion for the evaluation of QA systems in a new scenario. Fortunately, this exercise did receive a great attention by both organizers and participants, because of seventeen groups were interested into participating and the exercise and the presentation of the results in the workshop were very successful.

Finally, the future directions that we plan to undertake are to take into account more variables as the hardware used by the systems, as well as to insert new control parameters in order to give more significance to efficiency or precision.

## Acknowledgments

This research has been partially supported by the framework of the project QALL-ME (FP6-IST-033860), which is a 6th Framenwork Research Programme of the European Union (EU), by the Spanish Government, project TEXT-MESS (TIN-2006-15265-C06-01) and by the Valencia Government under project number GV06-161.

## References

1. Fukumoto, J., et al.: Question Answering Challenge (QAC-1). In: Proceedings of the Third NTCIR Workshop (2002)
2. Magnini, B., et al.: Overview of the CLEF 2006 Multilingual Question Answering Track. In: WORKING NOTES CLEF 2006 Workshop (2006)
3. Voorhees, E.M., Dang, H.T.: Overview of the TREC 2005 Question Answering Track. In: TREC (2005)

# A Methodology for Domain Dialogue Engineering with the Midiki Dialogue Manager

Lúcio M.M. Quintal[1] and Paulo N.M. Sampaio[2]

[1] Madeira Tecnopolo,
902120-105 Funchal, Portugal
lquintal@madeiratecnopolo.pt
[2] University of Madeira,
9000-105 Funchal, Portugal
psampaio@uma.pt

**Abstract.** Implementing robust Dialogue Systems (DS) supporting natural interaction with humans still presents challenging problems and difficulties. In the centre of any DS is its Dialogue Manager (DM), providing the functionalities which permit a dialogue to move forward towards some common goal in a cooperative interaction. Unfortunately, there are few authoring tools to provide an easy and intuitive implementation of such dialogues. In this paper we present a methodology for dialogue engineering for the MIDIKI DM. This methodology bridges this gap since it is supported by an authoring tool which generates XML and compilable Java representations of a dialogue.

## 1 Introduction

Human interaction with a machine in a natural way, ie, using the same modalities and senses we use in ordinary human-human communication (which include speech, gestures, emotions, etc) is called multimodal (MM) communication. Computer systems which implement such type of interaction are called MM Dialogue Systems (MMDS). One of the main components of a MMDS is the DM. A DM is the core component of a DS and it controls the flow of the interaction with a user, based on a dialogue model. The input to the DM is a formal representation of the input of the user, which in this context is called an input *dialogue move* (a *dialogue move* is a logic predicate that represents the information content and a communicative intent of an utterance or portion of an utterance). The DM interprets this input based on a set of internal rules. The output of the DM will be also a formal representation of the same type, which then is converted to a human readable form. The DM accesses the dialogue history if necessary, to resolve incomplete entries or to refer to previous inputs or outputs. In summary, its main function is to maintain a representation of the current state of the dialogue, managing its progress.

A DM in itself is also a system and is not easily configured. Proper setup of a DM for a dialogue requires knowledge from fields such as linguistics, distributed computing, specific domain knowledge, database access, etc. Congregating all those contributions by means of only programming and scripting languages is not at all an

V. Matoušek and P. Mautner (Eds.): TSD 2007, LNAI 4629, pp. 548–555, 2007.
© Springer-Verlag Berlin Heidelberg 2007

easy job and, apart from expensive commercial solutions, there are no clear methodologies supporting dialogue authoring in DMs. One of the efforts in providing such tools is related to the work of Staffan Larsson and his team, at Gothenburg University, where the Trindikit DM was created [1,2]. On the commercial side, the authoring tool Loquendo SDS Studio for the Loquendo Spoken DS is a representative example [3]. Unfortunately, there is a lack of authoring tools for domain dialogues and the existing ones are not intuitive or accessible to the users. This paper presents a methodology to simplify the development of generic dialogues. The methodology is based on the requirements of the Midiki DM [4] for dialogue authoring. Midiki (MITRE DIalogue Kit) was implemented at MITRE Corporation in 2004 by Carl Burke and others and is available as Open Source. Midiki is based on the Information State Update (ISU) theory of dialogue management [1,5], which identifies the relevant aspects of information in dialogue, how they are updated, and how updating processes are controlled. Midiki was created to provide relatively sophisticated DMs while minimizing investment in development and porting. Using Midiki, developers can alleviate the effort of complex implementations using a simpler dialogue toolkit, while still achieving similar results for appropriate tasks, when compared to the more complex systems [6]. Although Midiki provides some support to debugging, there are no design tools available for it. The authoring methodology proposed in this work provides an easy to use and intuitive technique for domain dialogue authoring, facilitating the prototyping of dialogue systems based on Midiki.

The next section presents the main characteristics of the proposed methodology for the authoring of domain dialogues for MIDIKI DM. Section 3 presents some conclusions of this work.

## 2   The Methodology Proposed for Domain Dialogue Engineering

Despite Midiki being a system with an excellent balance between complexity, flexibility and adaptability to different domains, it does not provide much aid in terms of authoring development. Indeed, the manual available offers many technical references, which the newcomer to DSs will probably find too difficult to understand and put to practice. The methodology we propose brings dialogue implementation closer to the developer, by providing him a tool for rapid application prototyping. At the same time, by proposing an XML [7] representation of the most relevant components of the dialogue, we separate as much as possible conceptual issues from concrete implementation language details, facilitating re-utilization and portability of dialogues. The number of components involved in a dialogue implementation depends on each concrete implementation, but they are in general of two main types: *declarative* and *procedural*. The *declarative* classes are related to the definition of *plans* (a *plan* provides the basic structure of a dialogue and is composed of actions), to the sentences to use and terms to recognize (in the interaction of the system with the user) and also to the data structures. The *procedural* classes are related to the implementation and/or parameterization of code which must be executed in order for the dialogue to run under a Java environment. Midiki, in its standard version, uses exclusively Java classes. Within our methodology we propose an XML representation of the most relevant components of a dialogue system in terms of authoring. The corresponding

DTDs (Document Type Definition) have been specified and only compliant XML files will be generated by the supporting tool. The use of XML presents several advantages, such as (i) Easiness in creating, editing and reading the components of a domain, with the support of an authoring tool; (ii) Facility in converting dialogues to other XML representations, compatible with other dialogue managers; (iii) Independence to the programming language used. This feature facilitates and promotes re-utilization.

The following section describes the *declarative* components of the methodology.

## 2.1   Declarative Components in the Specification of a Dialogue

The *declarative* components are composed by the *domain*, the *lexicon* and the *cells*:

- The *domain* implements the core strategic constituents of a dialogue: *plans*, *strategies* and *attributes* (a *strategy* is equivalent to an action in the *plan*).
- A component with lexical content (the *lexicon*), which defines mappings between user inputs (text or utterances converted into text) and input moves to the system. The *lexicon* also defines the textual content to present to the user, related to output *moves* of the system.
- *Cell* type components (as defined by Midiki), which are responsible for the implementation of the data structures supporting *plans* (such data structures are called *contracts*). Typically there is one *cell* for each main *plan*, ie, each *plan* is based in a supporting *contract*.

All the three components are specified in XML and converted to Java with our tool. The diagram depicted in figure 1 presents the structure of the *declarative* components of a Midiki dialogue (*domain*, *lexicon* and *cell*). These components are presented in the following sections.

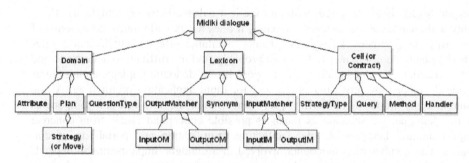

**Fig. 1.** The different elements composing the *declarative* part of a dialogue implementation (*InputOM* stands for input to *OutputMatcher* and *InputIM* stands for input to *InputMatcher*)

The first phase of the work for the creation of a domain dialogue in a DM consists in a classical analysis of requirements and design of a solution, considering the dialogue system as the technological platform of support to the interaction of the user with the service to implement. Our methodology describes the steps that compose the second phase of the work, in a total of 13 steps, related to the specification of the *declarative* and *procedural* components. Once the author has created a new project in the tool, he may build each component (*domain, lexicon* and *cells*) by adding and parameterizing the elements which compose it. The system guides the author whenever is possible and if he wants it so. Table 1 describes all the 9 steps that are followed in our methodology for the *declarative* components.

**Table 1.** List of steps in the methodology and their relation with the *declarative* components

| Components | Steps |
|---|---|
| Domain | Step 1 - Construction of plans with strategies |
| Domain | Step 2 - Declaration of attributes and synonyms |
| Domain | Step 3 - Association of strategies with types of attributes |
| Lexicon | Step 4 - Specification of output matchers |
| Lexicon | Step 5 - Specification of input matchers |
| Cells | Step 6 - Specification of strategy types |
| Cells | Step 7 - Specification of queries |
| Cells | Step 8 - Specification of methods |
| Cells | Step 9 - Inclusion of references to Java classes of the handlers |

Even though the author of a dialogue may follow the most convenient order to accomplish the steps proposed, the methodology can guide him from the beginning to the end of the process. The organization proposed consists in the most intuitive and logic sequence of steps, resulting from our experience in analyzing and using Midiki.

### 2.1.1 Specification of the Domain

The component which implements the core functions of the *domain* is described by the first three steps, which are common to any dialogue implementation independently of the *domain* considered:

**Step 1** - *Initialization of tasks/plans* (XML element *initializeTasks*): consists in the instantiation of the *contracts* and an instantiation of the *plan* structures. It is under *initializeTasks* that plans are constructed through the sequencing of *strategies* (which are equivalent to actions), like for ex. *findout, inform,* etc. The following example presents a simplified partial view of *initializeTasks*, with a focus on *plan* elements (the *strategies*):

```
<plan id="orderTripPlan">
  <strategy name="destinationCity" type="Findout"></strategy>
  <strategy name="tripCost" type="Inform"></strategy>
</plan>
```

The previous XML segment shows a plan named *orderTripPlan* and two *strategies* in that plan; initiated by their order in that plan (representing an usual sequence of occurrence). The *findout strategy* will produce one or more *ask dialogue moves*. Each *ask move* makes the system present a question to user (one or more times, until a valid answer is received and associated with the *findout strategy*). The lexical content for the *ask* move will be retrieved from the *lexicon* component. The second *strategy* presented, an *inform*, produces some information to be presented to the user, possibly including the result of some previous *query* call to a database (the *queryCall strategy* is not showed). Due to space limitation we can not present examples with complete XML specifications for all the steps of the methodology.

**Step 2** - *Initialization of attributes* (XML element *initializeAttributes*): consists in declaring the list of possible values permitted in an answer, corresponding to permitted values for inputs of the user. Each of those categories or types of values will contain the set of values allowed as concrete answers to the questions posed to the user. For instance, an attribute named "paymentMethods" could contain the values {credit, transfer, check}. We then consider the definition of lists of synonyms for the domain attributes or other key words that might occur in an answer from the user. Within the methodology and tool support, this section (element *initializeSynonyms*) is constructed in the domain interface, but the XML generated is in fact included in the lexicon component. The definition of synonyms simplifies the definition of attributes and permits to decrease the number of input matches necessary to catch different lexical forms of the same answer (all with the same semantic). For example, for *synonyms* of type "yesno", a *synonym* for "yes" could contain the values {correct, agreed, affirmative} and a *synonym* for "no" could contain the values {never, impossible}.

**Step 3** - The last step in specifying the *domain* consists in *compiling a list of the actions in the plans*. This step consists in the *initialization of questions* (XML element *initializeQuestions*). Here, the *strategies* are associated with the categories of attributes declared in *initializeAttributes*. All *strategies* which imply questions from the system to the user are included in the list as well as the type of answer for those questions. For instance, a *strategy* named "howToTravel" may be associated with an attribute of type "meansOfTransport".

### 2.1.2 Specification of the Lexicon

The *lexicon* is the second component to be implemented in the methodology. The component which includes the lexical content is described by steps 4 and 5, common to any dialogue implementation, independently of the domain dialogue being considered. In this section we present those two steps:

**Step 4** – This step is related to the *mapping of specific system output moves and the corresponding utterances for the user* (ex. text strings). This is specified by XML elements *outputMatcher*. Those strings are generated by the agent *generateAgent*, based on the textual data contained in the lexicon component. The definition of *outputMatchers* is related to actions of type *findout* (which imply an *ask* dialogue move) and to actions of type *inform*. For instance, an output *move ask*, related to a *findout* *strategy* named "destinationCity" could provide the following output to the user: "To which city would you like to travel?". *Inform* outputs may also include values resulting from database queries.

**Step 5** - Is related to the *mapping of identified tokens or* key words from the utterances of the user input *with concrete dialogue moves for the system*. For instance, an user input "one way" may imply a *move* of type *answer* which, if related to a *strategy* named "wantReturn", can associate it with a value "no" (*strategy* "wantReturn" could previously have produced a question to user, such as: "Would you like a return ticket?" and the answer to that question would be set to "no". Even if the question had not been produced yet, the answer to it would be given in advance. *InputMatcher* XML elements are used for this mapping. The primary usage of the *inputMatchers* will be to intercept common user inputs, which are expected in concrete situations and contexts (mainly when he is answering questions), belonging to the domain for which the system was developed. Thus, the system may react in a predefined manner to usual expressions of the users.

### 2.1.3  Specification of the Contracts/Cells

The supporting data structures of the Midiki DM are the *contracts*. The Java classes which implement the *contracts* are the *cells*, with classes of *contactCell* type. The *contracts* define the signature of the data structures supporting the execution of plans in Midiki. A *plan* is associated to a *contract*, which is implemented in a *cell*. In our proposal, this representation is also described in XML and converted to the native Java format of Midiki. This is the third and last *declarative* component of a dialogue.

**Step 6 -** Is related to the *specification of strategy types* (which may be seen as *plan attributes*). It consists in the inclusion of the list of *strategies* and the types of values to which they shall be binded. For instance, a *strategy* named "howToTravel" might be associated with type "meansOfTransport". In general, the types to choose from will be those defined in *initializeAttributes* (in step 2). This step corresponds XML element *addAttributes*.

**Step 7 -** Is related to the *specification of queries* (corresponds to XML element *addQueries*). The author must include the *queries* which are part of the *strategies* of the *plan*. The result of a *query* may be binded to an answer to a *strategy*. These results are then usually sent to the output by output *moves* of type *inform* which follows the call to a *query* within the *plan*. For example, a query named "flightDetails" may produce a value in an internal variable *nrOfPlace* which will be passed to an *inform move* for being presented to the user.

**Step 8 -** Is related to the *specification of methods* (corresponds to XML element *addMethods*). It consists in the inclusion of the *methods* which are part of the actions of the *plan*. The binding of results from *methods* to *strategies* is identical to the way it is done with *query* results. The difference between a *method* and a *query* in Midiki is that a *query* returns a value and does not change the *information state* (which is composed by a set of *contracts*) whereas a *method* returns a value and changes some item in the *information state*.

**Step 9 -** Is related to the *inclusion in a list of handlers, of the names of queries and methods* with the corresponding filenames of Java classes (specification of XML element *initializeHandlers*). It is a matter of indicating, for each *query* or *method*, which executable Java classes implements the corresponding access to data (if it is a *query*) or which access data but also change the information state (if it is a *method*).

For example, a query named "flightDetails" may have a Java class "flightDetailsHandler.java" containing its handler code.

## 2.2 Procedural Components in the Specification of a Dialogue

Upon completion of all the steps related to the *declarative* components, we must specify and implement the *procedural* components, which are related to the *handlers* (step 10) and the appropriate configuration of the classes *tasks* (step 11), *domain-Agent* (step 12) and *dm* (which stands for dialogue manager in this context and corresponds to step 13). The implementation of the *handlers* in itself (step 10) usually corresponds to a work of programming (in Java, in the case of Midiki). Steps 11, 12 and 13 have been automated by our tool, which automatically generates the corresponding Java classes. The technology used to automatically generate the Java source code is *Java Emitter Templates (JET)*, which is part of *Eclipse Modelling Framework Project (EMF)* [9]. The four steps related to the procedural components are:

- **Step 10** - Implementation of classes of type *handler*, in variable quantity, responsible for answers to *queries* or *methods*. For simple prototypes, where access to data sources may be simulated, it is possible to automate the generation of Java code for *handlers* however, in a general manner, that code is specific to each concrete implementation and should be written by a programmer.
- **Step 11** - Parameterization of the *Tasks*, which constitutes a source of information supporting the *domain*. It creates a list of contracts.
- **Step 12** - Parameterization of *DomainAgent*, responsible for the initialization of various *contracts* and responsible for the connection of the *lexicon* and the *domain* to the *information state*. In the class *domainAgent* must be included information of initialization of the *contracts* and the *handlers*. It is the agent *domainAgent* which establishes the connection of the different components to the *information state*, including the *lexicon* and the *domain*.
- **Step 13** - Parameterization of the *"dialogue manager executive"* Java class. This class is responsible for the initialization of the different agents and system interfaces (like GUI interfaces for example). It references the agents which make up the dialogue manager for the dialogue implemented.

## 3    Conclusions

Currently, there are two DMs available to the general community, Midiki and Trindikit. Both are available as Open Source, but Trindikit requires a commercial Prolog interpreter to run. None of the two provide software application support for the design and implementation of dialogues. Our proposal is a first step towards filling that gap and providing such support to novice developers working with DSs and with DMs in particular. As for future work, we expect to improve these tools (ex. including native support to consistency checking of answers); to integrate Midiki with a complete Open Agent Architecture (OAA) [8] environment, including robust voice recognition and synthesis modules and facilitated support to multimodal interaction.

# References

1. Larsson, S., Traum, D.: Information state and dialogue management in the TRINDI Dialogue Move Engine Toolkit. In: Natural Language Engineering Special Issue on Best Practice in Spoken Language Dialogue Systems Engineering, pp. 323–340, pages 18. Cambridge University Press, U.K (2000)
2. Larsson, S., Berman, A., Hallenborg, J., Hjelm, D.: Trindikit 3.1 Manual. Department of Linguistics, Gothenburg University (2004)
3. Loquendo Spoken Dialog System 6.0 (White Paper), Version 1.0. Loquendo (July 2005), URL http://www.loquendo.com/en/index.htm
4. Burke, C.: Midiki User's Manual, Version 1.0 Beta. The MITRE Corporation (2004), URL http://midiki.sourceforge.net/
5. Traum, D., Larsson, S.: The Information State Approach to Dialogue Management. In: Smith, Kuppevelt (eds.) Current and New Directions in Discourse & Dialogue, pp. 325–353, pages 28. Kluwer Academic Publishers, Dordrecht (2003) (to appear)
6. Burke, C., Doran, C., Gertner, A., Gregorowicz, A., Harper, L., Korb, J., Loehr, D.: Dialogue complexity with portability? Research directions for the Information State approach. In: Proceedings of the HLT-NAACL 2003 Workshop on Research Directions in Dialogue Processing (2003)
7. Sperberg-McQueen, C.M., Thompson, H.: XML Schema, W3C Architecture Domain (April 2000), URL http://www.w3.org/XML/Schema.html
8. Martin, D., Cheyer, A., Moran, D.: The Open Agent Architecture: A framework for building distributed software systems. Applied Artificial Intelligence: An International Journal. 13(1-2). (January-March 1999)
9. Eclipse Modeling Framework Project (EMF): URL http://www.eclipse.org/modeling/emf/

# The Intonational Realization of Requests in Polish Task-Oriented Dialogues

Maciej Karpinski

Institute of Linguistics, Adam Mickiewicz University
ul. Miedzychodzka 5, 60-371 Poznan, Poland
maciej.karpinski@amu.edu.pl

**Abstract.** In the present paper, the intonational realization of *Request Action* dialogue acts in Polish map task dialogues is analyzed. The study is focused on the *Request External Action* acts realized as single, well-formed intonational phrases. Basic pitch-related parameters are measured and discussed. Nuclear melodies are described and categorized, and some generalizations are formulated about their common realizations. Certain aspects of the grammatical form and phrase placement in the dialogue flow are also taken into account. The results will be employed in comparative studies and in the preparation of glottodidactic materials, but they may also prove useful in the field of speech synthesis or recognition.

## 1 Introduction

The intonational realization of dialogue acts (conversational acts, dialogue moves) has been widely studied for both linguistic and technological purposes in a number of languages (e.g., [5], [12], [15], [17]). However, requests (orders, instructions) still remain relatively less explored in this respect, which may result from the fact that their forms are strongly heterogeneous and deeply influenced by numerous cultural factors, including the rules of politenes, the relation between dialogue participants, the extralinguistic situation, the aim and character of the dialogue under study. Nevertheless, the ability to appropriately identify acts of this category is crucial for almost any spoken dialogue system. As in other cases, intonational cues may prove useful here [15], [17], [21]. The ability to use intonationally adequate forms of requests is also a crucial component of linguistic and communicative competence that should be included in foreign language teaching programs.

Few works have been confessed to the issues of the intonational realization of dialogue acts in Polish. In the existing monographical studies of the Polish intonation (e.g., [18]), only some indirect hints can be found. A number of findings on the role of intonation in Polish task-oriented dialogues have been presented in the monography [8] and in some papers by the team of *Pol'n'Asia* project (e.g., [9], [10]). Nevertheless, this study is probably the first to focus on the intonational realization of the Request-type dialogue acts in Polish.

V. Matoušek and P. Mautner (Eds.): TSD 2007, LNAI 4629, pp. 556–563, 2007.

## 2 Selection of Recordings

Ten Polish map task dialogue sessions were selected for the present study from *PoInt* corpus [9]. The speakers were young educated people of similar social status. The number of male and female instruction givers was equal and the total length of the analysed recordings reached 2 hours and 21 minutes. The recordings were made in a sound-treated room with high-quality digital equipment (condenser microphones, professional CD recorders). The corpus was labelled using a dedicated four-dimensional categorization of dialogue acts [7], the key dimension for the present study being "External Action Control". The final shape of that system was influenced by a number of works, including [1], [3], [12], [13], [14], [19], [20]. All the realizations of dialogue acts primarily aimed at triggering extralinguistic actions (*Request External Action, REA*) of the conversational partner were initially taken into consideration. The acts realized as statements describing the route on the map were excluded from further analysis because it proved impossible to differentiate between purely descriptive statements and utterances aimed at triggering extra-linguistic actions. Out of 433 realization of the *REA* that met basic quality conditions (no overlaps, no unfinished phrases), 274 signals in which *RAE* was realized as a single intonational phrase were selected for a closer perception-based scrutiny and instrumental analyzes. Intraphrasal pauses and minor disfluencies were accepted as long as they did not disrupted the melody of the phrase (as it may happen in the case of longer pauses which may sometimes involve "pitch resets").

The actual intonational contour is normally biased by the dialogue situation (e.g., turn-taking-related phenomena) and by the intentions of the speaker (e.g., the intention of continuing or finishing a contribution). While it is intuitionally clear and we can easily notice the tendency of closing a sequence of rising phrases with a falling phrase, it proves difficult to unequivocally distinguish the phrases produced with the intention of immedate continuation by the same speaker from those intended as turn-closing ones.

In order to (at least partially) control the abovementioned phenomena, the extracted utterances were divided into three categories:

- *IC*: immediate continuation (the phrase under study is immediately followed by another utterance by the same speaker);
- *NDC*: no interruption, delayed continuation (the phrase under study is followed by an interval of silence and then by another phrase by the same speaker);
- *IDC*: interrupted; delayed continuation (the phrase under study is followed by an utterance by the other speaker and then, usually, by another utterance by the same speaker).

For future use with larger material, more detailed labels regarding the character of the addressee's response were also applied.

While unfinished, heavily distorted and hardly intelligible phrases were excluded from instrumental analyses, they still formed a substantial part of the corpus and they certainly need further study, demanding a different methodological approach.

# 3 Grammatical Forms of REA Dialogue Acts

Polish is an inflexional language, characterized by a very flexible word order. The "default" grammatical way of expressing *REA* acts is based on the use of verbs in their imperative forms. However, in many social situations this may be perceived as offensive or at least inappropriate. Therefore, other forms are present in everyday communication. Among the extracted utterances realizing the *REA* dialogue act, five main categories were distinguished:

1. *DC2S*: Declarative clause, 2$^{nd}$ pers. sing. (e.g., You are going straight ahead to that tree.);
2. *DC1P*: Declarative clause, 1$^{st}$ pers. plur. (e.g., We are going straight ahead towards that tree.);
3. *IVC*: Imperative clause, using an imperative verb form (e.g., Go straight ahead.);
4. *GC*: Expressions with no verb (e.g., Straight ahead.);
5. *MVC*: Modal verb clauses (mostly based on some equivalents of "must", "to have to" or "should"; e.g., You must go straight ahead.).

# 4 Intonational Realizations of *REA* Dialogue Acts

## 4.1 Instrumental and Perception-Based Analyses

For the three groups utterances (as defined in 2), their basic pitch parameters were obtained automatically, using dedicated Praat [2] scripts. To increase the objectivity of the nuclear melody labelling, the traditional procedure of "impressionistic analysis" was supported by Prosogram, a software tool for tonal perception modelling [16]. Prosogram was used to decide whether a given portion of pitch contour should be tagged as flat, rising or falling when it was not obvious for the listener.

## 4.2 Nuclear Melody

The following forms of the nuclear melody realization were found in the examined realizations of *REA* acts:

- *rise* (the pitch rises gradually or jumps within a single syllable or a sequence of syllables);
- *high* (high pitch is produced without perceivable gradual rise; mostly in the phrases in which the nucleus fell on the ultimate syllable);
- *fall* (including low tone for one-syllable words in the ultimate position);
- *low* (low pitch level reached without perceivable gradual fall; typically occurs when the nucleus is on the ultimate syllable);
- *fall-rise* (a complex contour; can be realized within a single syllable or a number of syllables);
- *rise-fall* (as above);
- *flat* (no perceivable pitch fluctuation).

While the above categories, defined on the sub-phonological (perception-based, [6]) level, bear resemblance to the pitch fluctuation types mentioned in [18], some differences also occur. No definite claims are made here about the phonological relevance of our findings.

Tables 1 and 2 show the shares of the above nuclear melody classes in the three types of contextual settings (as defined in 2) and in the three types of grammatical realizations (as defined in 3).

**Table 1.** The proportions of the nuclear melody categories in the entire analysed data set and in the three categories of contextual settings

| Heading level | Entire set [%] | IC [%] | NDC [%] | IDC [%] |
|---|---|---|---|---|
| fall | 42 | 42 | 47 | 39 |
| rise | 35 | 36 | 31 | 37 |
| fall-rise | 18 | 14 | 15 | 21 |
| high | 2 | 6 | 2 | 1 |
| rise-fall | 2 | 2 | 2 | 0 |
| flat | 1 | 0 | 4 | 1 |
| low | 0 | 0 | 0 | 1 |

Although the proportion of *falls* was higher than of *rises,* there was a substantial number of *fall-rise* nuclear melodies, and probably at least some of them may be counted as implementations of rises. However, since objective differentiation based on the properties of the signal was impossible, rises and falling-rising contours were regarded as separate categories. A typical realization of a "plain" *rise* is shown in Figure 1. The *fall-rise* category was most frequently realized in the way shown in Figure 2. The relative pitch range of the *rise* was roughly twice as high as the relative pitch range of the preceeding *fall. Rise-fall* category was represented very sparsely.

*High* and *low* categories reffer to the cases where the nuclear syllable was the final one in the phrase and there was no perceivable pitch movement within that syllable, but the tone was distinctly higher or lower than the average for a given speaker.

It was noticed that when the nucleus was on the penultimate syllable (unmarked, "default" situation for Polish), the pitch change in *falls* tended to be realized mostly on the nuclear syllable, while in *rises* – mostly on the postnuclear, final syllable of the phrase.

Some utterances produced with *rise* and especially with *fall-rise* nuclear melody were additionally marked with a subjectively prolonged realization of the syllable that contained the main portion of the pitch movement. Since the same speakers in other utterances realized much briefer rises with similar pitch ranges, the explanation based on the idea of creating "more space" for the pitch movement was not satisfactory. An additional perception test was carried out in order to determine whether such contours convey any additional information or meaning.

Ten signals containing the above described features were randomly mixed with ten other signals produced by the same speakers and of possibly similar content but realized without those features. The task of the subjects (fourteen young educated native speakers of Polish) resolved itself in marking which of three pragmatic/semantic features ("insisting", "explanatory", and "expressing doubt") can be attributed to the presented signal. All of the signals initially selected as "unusually prolonged" were marked as "explanatory" by at least 12 out of 14 subjects. Moreover, the same signals were frequently marked as "insisting", but rarely as "expressing doubt", while the others were rarely marked as "explanatory" and generally categorized in a less consistent way.

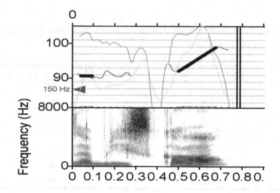

**Fig. 1.** Prosogram and spectrogram of the "plain rise" nuclear melody region in the utterance /za'blisko/ ((don't come) *too close*). The pitch axis is scaled in semitones.

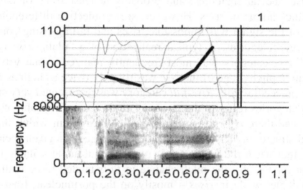

**Fig. 2.** Prosogram and spectrogram of the region of a *fall-rise* shaped nuclear melody in the utterance /'fpravo/ ((go) *to the right*)

### 4.3  Pitch Parameters

For the analyzed signals, the pitch frequency range within the nuclear melody (*PR_NM*) was calculated as the difference between the global maximum and minimum values, and the relative pitch frequency range within the nuclear melody (*RPR_NM*) was calculated as $(f_{0\,max} - f_{0\,min})/f_{0\,max}$. It was found that the mean values of *RPR_NM* differed significantly for *DCIP* and the other two groups (Table 3). The same applied to the average values of the total relative $f_0$ range (within the entire signal, *RPR_T*, calculated similarly to *RPR_NM*) for *DC1P* phrases was higher than for the remaning two types of phrases under study: 50% for *DC1P* vs. 42% for *DC2S* and 38% for *IVC*.

**Table 2.** The mean values of the relative pitch frequency range (as percentages; standard deviation for the sample given in brackets) for the three types of grammatical realizations

| gram. realization category | within the entire utterance [%] *RPR_T* | within the nuclear melody [%] *RPR_NM* |
|---|---|---|
| *DC1P* | 50 (15) | 36 (14) |
| *DC2S* | 42 (12) | 30 (13) |
| *IVC* | 38 (12) | 31 (15) |

It may be suggested that in the case of *DC2S* and especially for *DC2P,* the imperative meaning must be stressed with a wider pitch movement, while for *IVC* it is conveyed mostly by the imperative form of the verb (cf. [4] for a similar trade-off observed for polar and detailed questions in Polish). This phenomenon was not so obvious in the case of *RPR_NM* probably because other strong factors influenced its shape and limited available range of pitch movement.

**Table 3.** The proportions of the nuclear melody varieties in the three categories of grammatical realizations of the *REA* dialogue act

| nuclear melody category | *DC2S* [%] | *DC1P* [%] | *IVC* [%] |
|---|---|---|---|
| *fall* | 47 | 30 | 37 |
| *rise* | 39 | 24 | 35 |
| *fall-rise* | 13 | 35 | 17 |
| *Rise-fall* | 1 | 2 | 0 |
| *high* | 0 | 4 | 7 |
| *low* | 0 | 2 | 0 |
| *flat* | 0 | 2 | 4 |

## 5   Discussion and Conclusions

For the *REA* dialogue acts, the mean value of the total relative pitch frequency range was the highest in the acts grammatically realized as declarative clauses in the 1st *pers. plur.* (cf. Table 2). This tendency was also noticeable for the mean value of the relative $f_0$ range within the nuclear melody. The values of the relative pitch range within phrases were slightly lower than for those obtained for three types of questions coming from the same corpus and utterred the same speakers, on the average, slightly lower than for questions (43% vs. 47%). The value obtained for *Polar Questions* (50%) was higher than for *Wh-questions* (44%) which can be explained by the fact that in the former group, intonation was the only (besides the context) cue to interrogativity, while in the latter, interrogativity was also cued by the *wh-* word [4], [11]. In general, however, the lowest mean value of the $f_0$ range, both in the entire phrase and within the nuclear melody range, was found in the phrases of the *IMC* type, i.e. those based on the verb in the imperative mode. As it was found for questions, the lack of lexical cues may be compensated by more prominent intonational cues. One may offer a similar explanation for *REA* dialogue acts in the Polish language: The lack of lexical and syntactic cues to imperativity is normally compensated by intonational cues.

It was also found that the *IDC* (interrupted realization, delayed continuation) class of utterances was marked by the highest value of the $f_0$ range both within the entire phrase and within the nuclear melody. In this case, a hypothesis may be put forward that the wide pitch fluctuation range serves as a means to upkeep the listener's attention and to suggest that that the utterances is still being continued.

As mentioned at the beginning of the present text, *Requests* form a rich, heterogeneous group which is hard to grasp and define without simplifications. A relatively variable realizations and the number of potentially influential external and internal factors should make us cautious about further inferences, generalizations and interpretations even when statistics shows significant differences between the means for most of the abovementioned contrasts. However, some of the tendencies sketched in the present text seem clear enough to be introduced into the Polish language teaching in order to enhance the communicative competence of the learners. Simultaneously, these contrasts may prove useful in the field of speech technology, serving as a complementary cues in the detection and synthesis of request-type utterances.

As the next step, the results obtained for the Polish language will be compared to the respective results for other languages, including Korean, Thai and Vietnamese. It may become necessary to extend the analyzed material for each language, to rebuild the grammatical categorizations and to modify the approach to the description of intonation in order to grasp and systematically describe universal and language-specific cues to directivity.

## Acknowledgments

This study is a part of the Pol'n'Asia project funded by the Polish Ministry of Science and Higher Education (project code H01D 006 27).

## References

1. Alexandersson, J., Buschbeck-Wolf, B., Fujinami, T., Kipp, M., Koch, S., Maier, E., Reithinger, N., Schmitz, B.: Dialogue Acts in VERBMOBIL-2, 2nd edn. (Deliverable) (1998)
2. Boersma, P., Wenink, D.: Praat. Doing Phonetics by Computer (A computer program; version 4.5) (2005)
3. Carletta, J., Isard, A., Isard, S., Kowtko, J., Doherty-Sneddon, J., Anderson, A.: HCRC dialogue structure coding manual, Human Communications Research Centre, University of Edinburgh, Edinburgh, HCRC TR–82 (1996)
4. Grabe, E., Karpiński, M.: Universal and Language-specific Aspects of Intonation in English and Polish. In: Grabe, E., Wright, D. (eds.) Oxford University Working Papers in Linguistics, Philology & Phonetics, pp. 31–44 (2003)

5. Grice, M., Savino, M.: Map Task in Italian. Asking Questions about New, Given and Accessible Information. Catalan Journal of Linguistics 2, 153–180 (2003)
6. Jassem, W.: Classification and organization of data in intonation research. In: Braun, A., Masthoff, H.R. (eds.) Phonetics and its Applications. Festschrift for Jens-Peter Köster. Wiesbaden: Franz Steiner Verlag, pp. 289–297 (2002)
7. Juszczyk, K., Karpinski, M., Klesta, J., Szalkowska, E., Wlodarczak, M.: Categorization and Labelling of Dialogue Acts in Pol'n'Asia Project (accepted for presentation at Poznan Linguistic Meeting 2007) (2007)
8. Karpinski, M.: Struktura i intonacja polskiego dialogu zadaniowego. Wydawnictwo Naukowe UAM, Poznan (2006)
9. Karpinski, M.: The form and function of selected non-lexical and quasi-lexical units in Polish map task dialogues. In: Proceedings of SASR Conference in Krakow (2005)
10. Karpinski, M., Klesta, J.: The Project of Intonational Database for the Polish Language. In: Puppel, St., Demenko, G. (eds.) Prosody 2000. Poznan: Faculty of Modern Languages and Literature, UAM, pp. 113–119 (2001)
11. Karpinski, M., Szalkowska, E.: On intonation of question-type dialogue moves in Korean and Polish task-oriented dialogues: Spontaneous speech analysis using perception modelling. In: Speech Signal Annotation, Processing and Synthesis Conference, Poznań (2006)
12. Kowtko, J.: The function of intonation in task-oriented dialogue. PhD dissertation presented at the University of Edinburgh (1996)
13. Kowtko, J.C., Isard, S.D., Doherty, G.M.: Conversational Games within Dialogue. HCRC: Human Communication Research Centre Working Papers, Edinburgh (1993)
14. Larsson, S.: Coding Schemas for Dialogue Moves. Göteborg University (1998)
15. Mast, M., Kompe, R., Harbeck, S., Kieling, A., Niemann, H., Noeth, E.: Dialog act classification with the help of prosody. In: Proceedings of ICSLP, pp. 1728–1731 (1996)
16. Mertens, P.: The Prosogram: Semi-Automatic Transcription of Prosody based on a Tonal Perception Model. In: Bel, B., Marlien, I. (eds.) Proceedings of Speech Prosody 2004, Nara (Japan) (2004)
17. Shriberg, E., Bates, R., Stolcke, A., Taylor, P., Jurafsky, D., Ries, K., Coccaro, N., Martin, R., Meteer, M., Van Ess-Dykema, C.: Can Prosody Aid the Automatic Classification of Dialogue Acts in Conversational Speech? Language and Speech 41(3-4), 229–520 (1998)
18. Steffen-Batogowa, M.: Struktura przebiegu melodii jezyka polskiego ogólnego. SORUS, Poznan (1996)
19. Traum, D.R.: 20 Questions for Dialogue Act Taxonomies. Journal of Semantics 17(1), 7–30 (1999)
20. Traum, D.R., Hinkelman, E.A.: Conversation acts in task-oriented spoken dialogue. Computational Intelligence (Special Issue on Non-literal Language) 8(3), 575–599 (1992)
21. Wright, H., Poesio, M., Isard, S.: Using High Level Dialogue Information for Dialogue Act Recognition Using Prosodic Features. In: DIAPRO-1999, pp. 139–143 (1999)

# Analysis of Changes in Dialogue Rhythm Due to Dialogue Acts in Task-Oriented Dialogues

Noriki Fujiwara, Toshihiko Itoh, and Kenji Araki

Graduate School of Information Science and Technology,
Hokkaido University, Sapporo, Japan
{fujiwara, t-itoh, araki}@media.eng.hokudai.ac.jp

**Abstract.** We consider that factors such as prosody of systems' utterances and dialogue rhythm are important to attain a natural human-machine dialogue. However, the relations between dialogue rhythm and speaker's various states in task-oriented dialogue have been not revealed. In this study, we collected task-oriented dialogues and analyzed the relations between "dialogue structures, kinds of dialogue acts (contents of utterances), *Aizuchi* (*backchannel/acknowledgment*), *Repeat* and interjection" and "dialogue rhythm (response timing, F0, and speech rate)".

## 1   Introduction

It is widely accepted that the accuracy of speech recognition and the understanding of language as well as the quality of synthesized speech are important to accomplish natural human-machine communication. The abilities of spoken-dialogue systems have recently increased exponentially. Numerous systems have been developed [1] and some of these are being used in practical applications [2]. However, the communication between humans and machines is not as natural as that between humans. Our previous study [3][4] revealed that factors such as prosody of systems' utterances and dialogue rhythm are important to attain a natural human-machine dialogue.

Kitaoka et al.[5] were interested in dialogue rhythm and developed a free-conversation spoken-dialogue system. They achieved this goal by using machine learning only on keywords and acoustic features from human-human dialogues. We were also interested in dialogue rhythm and developed a spoken-dialogue system for task-oriented dialogue. Our system has the same ability as Kitaoka's but it can also tune the acoustic features of system response to those of a user's utterances [6]. This is because the acoustic features (response timing, F0, and speech rate) of speakers' utterances in free-conversation are claimed to become synchronized with those of their partners' utterances along with increased tension in the dialogue [7]. Although the dialogue rhythm in our system did improve, it was not as smooth and natural as that of a human's. The speakers' state (rise in dialogue tension, dialogue act, and emotions) is usually claimed to influence dialogue rhythm. We believe that dialogue acts (contents of utterances) are particularly important factors in dialogue rhythm. However, our system did not use speaker's dialogue acts to attain dialogue rhythm. It is also not natural for spoken-dialogue systems to always tune the acoustic features of responses to speakers'

V. Matoušek and P. Mautner (Eds.): TSD 2007, LNAI 4629, pp. 564–573, 2007.

utterances regardless of dialogue acts. However, there have been no studies that have investigated the relations between dialogue acts and dialogue rhythm, and there have been no studies that have investigated phenomena such as acoustic synchronicity when task-oriented dialogue is concerned. It is therefore necessary to more thoroughly investigate the rhythm of human-human dialogue to achieve a spoken-dialogue system that enables communication like that between humans.

We collected task-oriented human-human dialogue for the present study, and analyzed the relations between dialogue acts and dialogue rhythm, i.e., the response timing, F0, and speech rate.

## 2   Dialogue Corpus

We recorded task-oriented dialogue to analyze human-human dialogue and annotated it with dialogue-act tags and acoustic labels. The details on the process are described below.

### 2.1   Recording Speech Data

There was a total of 17 subjects, who were undergraduate and graduate university students. The dialogue task was a hotel reservation where one section of the dialogue was spoken by two of these subjects. The first subject played the role of a customer who made a reservation. The second subject played the role of an agent, who searched for hotels and confirmed the reservation. There are cases where the same subject played the role of the customer in one dialogue, and played the agent in another dialogue. Customers interacted according to a "situation" prepared beforehand, and adhered to its context as much as possible. We prepared seven situations. Two subjects are separated by a partition, which makes them invisible to each other. They can only communicate by speaking, without gestures or eye contact.

### 2.2   Acoustic Labelling

We detected the beginning and ending of utterances with speech and waveforms, and labelled each utterance "agent_start", "agent_end", "customer_start", and "customer_end" using "Wavesurfer" speech-analysis software [10]. If there was a pause that lasted longer than 300 ms, we regarded it as a border between utterances. Therefore, one dialogue act, as described in Section 2.3, often consisted of more than one utterance.

### 2.3   Dialogue Act Tagging

A dialogue usually consists of more than one exchange, and the structure of an exchange is usually "Initiate-Response-(Follow-up)". Initiate is a component that functions as an appeal to start a new exchange. Response is a component that functions as a reaction to initiate. Follow-up is component that functions to signal that the current exchange has finished, and is often omitted. Each component includes some dialogue acts. We consulted the literature [8][9] when we defined kinds of the dialogue acts and tagged them. The tagging procedure is described below.

1. Making transcriptions from collected dialogue data.
2. Splitting utterance (transcription) of each speaker by one dialogue act.
3. Judging and tagging the kind of dialogue act to the each splitted utterance according to a decision tree [9] proposed by Discourse Tagging Working Group in Japan.

The dialogue acts that we analyzed and discuss in this paper are described below. The numbers in parentheses denote how many times an agent or a customer used a given dialogue act in the dialogue corpus.

Initiate:

- *Wh-question* (For agents:248, For customers:184): A demand for some values or expressions as a response to a question where the speaker has not forecast his or her partner's response.
- *Request* (For agents: 138, For customers: 131): A demand is made for the listener to act, and some response indicating acceptance or rejection is needed.
- *Inform* (For agents: 50, For customers: 7): Expressing an opinion, knowledge, or facts that the speaker believes to be true.
- *Yes-No question* (For agents: 47, For customers: 35): Answers "Yes" or "No" to a question when the speaker cannot predict his or her partner's response.
- *Confirm* (For agents: 547, For customers: 55): A question is asked by a speaker who can make a prediction or has knowledge about his or her partner's response.

Response:

- *Answers* (For agents: 270, For customers: 242): Utterance that provides content to the demand in *Wh-questions*.
- *Positive* (For agents: 127, For customers: 563): An affirmative response to a *Yes-No question* and acceptance of a demand, request, or preposition.
- *Negative* (For agents: 5, For customers: 28): A negative response to a *Yes-No question* and rejection of a demand, request, or preposition.

Follow-up:

- *Understand* (For agents: 165, For customers: 168): Expressing that the goal of an exchange has been achieved after a response.

Aizuchi & Repeat:

- *Aizuchi* (*backchannel/acknowledgment*) (For agents: 295, For customers: 870): Aizuchi signifies the partner's speech has been heard or the next utterance is prompted (its function is not a definite answer but rather a lubricant to enable smoother conversation).
- *Repeat* (For agents: 187, For customers: 38): An utterance that repeats important words (keywords) included in the preceding speaker's utterances.

Here, *Aizuchi* and *Repeat* are basically included in the Follow-up. However, the contributions of these dialogue acts to dialogue rhythm are considered to differ from Follow-up and we have dealt with *Aizuchi* and *Repeat* as other dialogue acts. Fig. 1 shows an example of such a dialogue tagged with dialogue acts, and Table 1 lists the information from our dialogue corpus.

C: I would like to reserve a hotel in Tokyo.                          [*Request*]
A: A hotel in Tokyo?                                                  [*Confirm*]
C: Yes.                                                               [*Positive*]
A: Can I have your name, please?                                     [*Wh-question*]
C: Suzuki.                                                            [*Answer*]
A: Can I have your telephone number, please?                        [*Wh-question*]
C: 012, (Yes. [*Aizuchi*]) 345, (345. [*Repeat*]) 6789. (6789. [*Repeat*]) [*Answer*]
A: Thank you.                                                        [*Understand*]

("C" denotes "Customer" and "A" denotes "Agent".)

**Fig. 1.** Example of tagged dialogue acts

**Table 1.** Dialogue data

| # of dialogues | | 50 |
|---|---|---|
| # of subjects | | 17 |
| # of utterances | agent | 3844 |
| | customer | 3510 |
| # of dialogue acts | agent | 2215 |
| | customer | 2520 |
| Ave. dialogue duration | | 4 min. 57 sec. |

# 3  Analysis

We investigated the response timing, F0, and speech rate as factors contributing to dialogue acts. The response timing was calculated by subtracting the ending of a previous dialogue act from the beginning of a current one. We calculated the average log (F0), which was calculated by dividing the sum of log (F0) by the number of frames for analysis except for voiceless frames. We used "ESPS/waves+" speech analysis software [11] to estimate F0. The speech rate was calculated by dividing the number of morae by the duration of the utterance.

## 3.1  Analysis of Relations Between Initiate, Response, and Follow-Up

We analyzed the dialogue rhythm of utterances based on Initiate, Response, and Follow-up. Table 2 lists the averages and standard deviations for response timing, the average log (F0), and the speech rate of utterances based on Initiate, Response and Follow-up. The response timing for Response is earlier than Initiate's, and Response's speech rate is slower than Initiate's. Initiate is basically an utterance to start a new exchange and a new topic, and to dominate and manage the flow of dialogue to achieve a task. We therefore considered that Initiate needed a longer time to think about what to say than Response, and the admissible pause to commence speaking could be extended. Response is a reaction to Initiate, except in situations when information is being retrieved,

**Table 2.** Response timing, average log (F0), and speech rate for Initiate, Response, and Follow-up

|  |  | Initiate | Response | Follow-up |
|---|---|---|---|---|
| Response timing | Ave. | 1.06 | 0.44 | 0.50 |
| [*sec*] | SD | 1.00 | 0.74 | 0.81 |
| Ave. of | Ave. | 4.71 | 4.73 | 4.61 |
| log(F0) | SD | 0.24 | 0.29 | 0.20 |
| Speech rate | Ave. | 9.09 | 8.81 | 9.60 |
| [*mora/sec*] | SD | 2.06 | 3.02 | 3.61 |

Response's thinking time was shorter than Initiate's. In addition, as utterances based on Response were affected by temporal restrictions based on the exchange structure and real time, the admissible pause to begin speaking must be short. However, as Response was more frequently included in important content than Initiate, its speech rate slowed to attract his or her partner's attention and convey content accurately. Moreover, the response timing for Follow-up is almost the same as for Response. The reason is the same as for Response which was described above. However, Follow-up's average log (F0) was lower and its speech rate was much faster than the others. The reason for this is that utterances based on Follow-up were considered to be optional utterances and the speaker thought that they were less important than the others.

## 3.2   Analysis Comparing Dialogue Acts

We analyzed the dialogue rhythm of utterances based on all dialogue acts (see Figs. 2, 3, and 4). First, we describe the results for turn-taking which is frequent in a dialogue and highly contributive to dialogue rhythm. In turn-taking, there is a tendency for response timing to occur early in the order of "dialogue acts belonging to Initiate", "dialogue acts belonging to Response (except *Positive*)" and "dialogue acts belonging to Follow-up". The response timing only for *Positive* occurs especially early. This is because most utterances based on *Positive* are responses to *Confirm*, which is a dialogue act to confirm information on the current task and it can easily be predicted during the process of achieving a task. Therefore, we regard utterances that are based on *Positive* to be responses that require hardly any thinking time (yes/no responses). There are therefore many overlaps in this dialogue act. There are tendencies for the response timing of the other dialogue acts, which often treat new or important information, such as *Wh-question* and *Request*, to be delayed and to only occasionally overlap. However, there are tendencies for the response timing of the dialogue act in easily predicted utterances, such as *Confirm* and *Yes-No question*, to speed up and frequently overlap.

There are significant differences in almost all combinations of dialogue acts ($p < 0.01$) according to the results of the t-test for average log (F0). The results also show that F0 is easily affected by dialogue acts. There are basic tendencies in dialogue acts where utterances are predictable or expected by a partner, such as *Positive* and

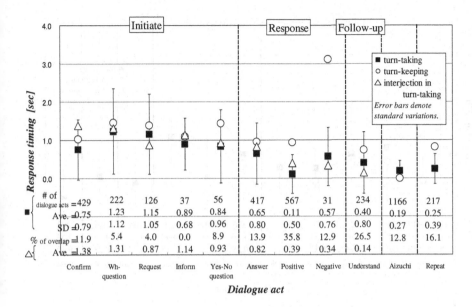

| | Confirm | Wh-question | Request | Inform | Yes-No question | Answer | Positive | Negative | Understand | Aizuchi | Repeat |
|---|---|---|---|---|---|---|---|---|---|---|---|
| # of dialogue acts | 429 | 222 | 126 | 37 | 56 | 417 | 567 | 31 | 234 | 1166 | 217 |
| ■ Ave. | 0.75 | 1.23 | 1.15 | 0.89 | 0.84 | 0.65 | 0.11 | 0.57 | 0.40 | 0.19 | 0.25 |
| SD | 0.79 | 1.12 | 1.05 | 0.68 | 0.96 | 0.80 | 0.50 | 0.76 | 0.80 | 0.27 | 0.39 |
| % of overlap | 1.9 | 5.4 | 4.0 | 0.0 | 8.9 | 13.9 | 35.8 | 12.9 | 26.5 | 12.8 | 16.1 |
| △ Ave. | 1.38 | 1.31 | 0.87 | 1.14 | 0.93 | 0.82 | 0.39 | 0.34 | 0.14 | | |

Dialogue act

**Fig. 2.** Response timing for all dialogue acts

*Understand*, for average log (F0) to become low, and in dialogue acts where utterances are important or unexpected by a partner, for average log (F0) to become high (tendency for emphasis).

The speech rate of dialogue acts that often include new, unexpected, or important information is slow. Utterances based on the dialogue acts which modulate the dialogue rhythm are very fast. Although we thought that the thinking time for *Negative* was almost the same as or a little later than that for *Positive*, *Negative*'s response timing was delayed, its average log (F0) was very high, and its speech rate was slow. We considered that this was because utterances based on *Negative* were important in the sense that they differed from the partner's expectation; therefore, the speaker emphasized them on purpose. In the case of dialogue acts belonging to Response, if "utterances which level of importance is high" were assumed to be "utterances with late response timing", there were strong correlations between response timing and F0 ($r=0.84$), between response timing and speech rate ($r=-1.00$), and between speech rate and F0 ($r=-0.86$). Briefly, there was a tendency by utterances which level of information was high for their response timing to be delayed, their F0 to become high and speech rate to slow; they were often emphasized. In the case of dialogue acts belonging to Initiate, there were strong correlations between response timing and F0 ($r=0.97$) but there were no correlations between other combinations.

Finally, in the case of relations for response timing, F0, and speech rate in each dialogue act, turn-keeping (when a partner does not take turn) appears to have the same tendencies as turn-taking. However, a close analysis reveals that the response timing is later, the average log (F0) is lower and the speech rate is faster for turn-keeping when compared with turn-taking.

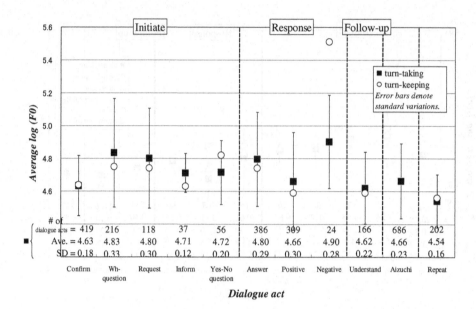

**Fig. 3.** Average log(F0) for all dialogue acts

## 3.3   Analysis Comparing Aizuchi and Repeat

The averages and standard deviations of response timing for *Aizuchi* and *Repeat* were almost the same, and their standard deviations were much smaller than those for the other dialogue acts. The reason for this is that the function to modulate dialogue rhythm is strong and there are stringent constraints about response timing, as is the case for *Aizuchi* and *Repeat*; therefore, a human may respond almost reflexively using various features as acoustic ones. The speech rate for *Repeat* is almost the same as that for *Confirm's*, which is semantically the same but *Repeat's* F0 is much lower than that for the other dialogue acts. The reason for this is that *Repeat* involves implicit confirmation by repeating the keyword(s) included in the partner's utterance, but a speaker intentionally lowers *Repeat's* F0 not to disturb his or her utterances. As above, in the case of utterances that function to modulate dialogue rhythm, if they include information to convey to a partner, such as *Repeat*, their speech rate is normal and their F0 is significantly lowered in order not to disturb partner's utterances. If one partner's utterances have no information to be conveyed (as in *Aizuchi*), their F0 is lowered and speech rate becomes very fast in order not to disturb another partner. This is how dialogue rhythm is being preserved.

## 3.4   Response Timing for Interjections (Filled Pause)

The average of response timing for interjections is 0.94 and its standard deviation is 0.97. We analyzed the relations between the average of response timing for all dialogue acts and the average of response timing for interjections of all dialogue acts (see

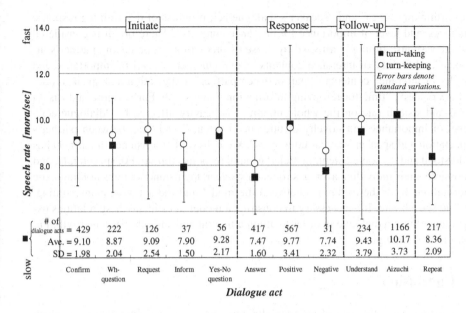

**Fig. 4.** Speech rate for all dialogue acts

triangles in Fig. 2), and we obtained some very interesting results. There is a strong correlation between the average of response timing for all dialogue acts and the average of response timing for interjections ($r=0.73$). The equation of regression line is $y = 0.92x + 0.14$, where $x$ denotes the average of response timing for all dialogue acts and $y$ denotes the average of an interjection's response timing for all dialogue acts. The interjection's response timing for the same thinking state, such as *Positive* and *Negative*, are almost the same, and the interjection's average of response timing for almost all dialogue acts are later than the average of response timing for all dialogue acts. We consider that there is an admissible pause for thinking time (response timing) in all dialogue acts, and the speaker utters interjection when he or she has not decided what to say yet in the admissible pause or he or she has predicted that his or her utterance will be later than the admissible pause.

### 3.5 Toward Accomplishment of Natural and Smooth Human-Machine Communication

From the results described in this section, we took the following into consideration in a task-oriented dialogue between two persons. We considered that there was an admissible pause (thinking time) in all dialogue acts. The pause is determined by kinds of dialogue act (including their exchange structure types) and a state of dialogue structure (turn-taking / turn-keeping). And then, the speaker utters interjection when he or she has not decided yet what to say in the admissible pause or he or she has predicted that his or her utterance will be later than the admissible pause. In the case of actual each response timing, we consider that it is determined by relations between "an admissible pause (thinking time)" and "speaker's dialogue act, progress of determining response sentences, and level of importance for utterance contents". Therefore, it is considered that the response timing is affected by "ease of predicting the preceding partner's utterance", "difficulty of utterance contents", "dialogue act", "level of importance and novelty of utterance contents", "gap between partner's expectation and speaker's actual utterances", "time for retrieving informations" and so on. Furthermore, a decision of F0 and speech rate on the whole utterance is heavily affected by "dialogue act", "level of importance and novelty of utterance contents" and "gap between partner's expectation and speaker's actual utterances". We believe that a listener (a partner) has understood and, to some extent, modeled general response timing, F0, speech rate, and if possible, their individual averages for each speaker, the listener obtains nonlinguistic informations from the difference between the model and the actual response timing, F0, and speech rate. It is necessary to construct a model to estimate these relations using human-human dialogue in order to attain a natural and smooth dialogue rhythm in task-oriented dialogues.

## 4    Conclusion

In this study, we collected task-oriented dialogues and analyzed the relations between "dialogue structures, kinds of dialogue acts, *Aizuchi*, *Repeat*, and interjection" and "dialogue rhythm (response timing, F0, and speech rate)". We gained important knowledge for achieving a smooth and natural spoken-dialogue in task-oriented dialogue system.

Future work is to investigate the shift and synchronization of dialogue rhythm in a task-oriented dialogue, and the relations between dialogue rhythm and the current speaker's dialogue act taking the partner's previous dialogue act into consideration. Moreover, using the results, we plan to improve our spoken-dialogue system in order to enable smoother and more natural communication on the level comparable to human's.

## References

1. Raymond, C., Esteve, Y., Bechet, F., De Mori, R., Damnati, G.: Belief confirmation in spoken dialog systems using confidence measures. In: Proc. of ASRU 2003, pp. 150–155 (2003)
2. Pouteau, X., Krahmer, E., Landsbergen, J.: Robust spoken dialogue management for driver information systems. In: Proc. of Eurospeech'97, pp. 2207–2210 (1997)
3. Yamada, S., Itoh, T., Araki, K.: Linguistic and Acoustic Features Depending on Different Situations - The Experiments Considering Speech Recognition Rate. In: Proc. of INTERSPEECH 2005, pp. 3393–3396 (2005)
4. Yamada, S., Itoh, T., Araki, K.: Is Voice Quality Enough? - Study on How the Situation and User's Awareness Influence the Utterance Features. In: Proc. of INTERSPEECH 2006, pp. 481–484 (2006)
5. Kitaoka, N., Takeuchi, M., Nishimura, R., Nakagawa, S.: Response Timing Detection Using Prosodic and Linguistic Information for Human-friendly Spoken Dialog Systems. Journal of JSAI, SP-E 20(3), 220–228 (2005)
6. Shoji, K., Takahashi, M., Ibara, S., Itoh, T., Araki, K.: Spoken Dialog System considered Rhythm and Synchronized Tendency of Conversation. (in Japanese) IPSJ SIG Tecnical Reports, SLP-61, pp. 43–48 (May 2006)

7. Nagaoka, C., Komori, M., Nakamura, T.: The interspeaker influence of the switching pauses in dialogue (in Japanese) The Japanese Journal of Ergonomics 38(6), 316–323 (2002)
8. Ichikawa, A., Araki, M., Kashioka, H., et al.: Evaluation of Annotation Schemes for Japanese Discourse. In: Proc. of ACL '99 Workshop on Towards Standards and Tools for Discourse Tagging, pp. 26–34 (1999)
9. Araki, M., Itoh, T., Kumagai, T., Ishizaki, M.: Proposal of a Standard Utterance-Unit Tagging Scheme (in Japanese) Journal of JSAI 14(2), 251–260 (1999)
10. http://www.speech.kth.se/wavesurfer/index.html
11. Software manuals of ESPS/waves+ with EnSigTM, Entropic Research Laboratory, Inc. (1997)

# Recognition and Understanding Simulation for a Spoken Dialog Corpus Acquisition*

F. Garcia, L.F. Hurtado, D. Griol, M. Castro, E. Segarra, and E. Sanchis

Departament de Sistemes Informàtics i Computació (DSIC)
Universitat Politècnica de València (UPV)
Camí de Vera s/n, 46022 València, Spain
{fgarcia, lhurtado, dgriol}@dsic.upv.es,
{mcastro, esegarra, esanchis}@dsic.upv.es

**Abstract.** Since the design and acquisition of a new dialog corpus is a complex task, new methods to facilitate this task are necessary. In this paper, we present a methodology to make use of our previous work within the framework of dialog systems in order to acquire a dialog corpus for a new domain. The main idea is the simulation of recognition and understanding errors in the acquisition of the new dialog corpus. This simulation is based on the analysis of such errors in a previously acquired corpus and the definition of a correspondence table among the concepts and attributes of both tasks. This correspondence table is based on the similarity of semantic meaning and frequencies. Finally, the application of this methodology is illustrated in some examples.

## 1 Introduction

The study and development of spoken dialog systems is an emerging field within the framework of language and speech technologies. The scheme used for the development of these systems usually includes several generic modules that deal with multiple knowledge sources. These modules must cooperate to satisfy user requirements: they must recognize the pronounced words, understand their meaning, manage the dialog, perform error handling, access the databases, and generate an oral answer. Each module has its own characteristics and the selection of the most convenient model varies depending on certain factors (the goal of the dialog, the possibility of manually defining the behavior of the module, or the capability of automatically obtaining models from training samples). The process of designing, implementing, and evaluating a dialog system is increasingly complex. One of the most successful approaches is the statistical approach, which probabilistically models processes that are automatically learned from corpora of real human-computer dialogs [1,2,3,4].

The main reason for using the statistical approach is that we want to estimate a dialog manager that is able to deal with variability in user behavior. These models also reduce development and maintenance costs. Systems with improved portability, more robust

* This work has been partially supported by the Spanish Government and FEDER under contract TIN2005-08660-C04-02, and by the Vicerrectorado de Innovación y Desarrollo of the Universidad Politécnica de Valencia under contract 4681.

V. Matoušek and P. Mautner (Eds.): TSD 2007, LNAI 4629, pp. 574–581, 2007.

performance, and an easier adaptation to other domains can be developed. The success of statistical approaches, however, depends on the quality of the models and the quality of the data. Therefore, the acquisition of the corpora and the definition of the semantic representation for the labeling are processes that are key to obtaining the quality data that is needed to train satisfactory models.

In this paper, we present a methodology to make use of our previous work within the framework of dialog systems in order to acquire a dialog corpus for a new domain. The main idea is the simulation of recognition and understanding errors in the acquisition of the new dialog corpus. This simulation is based on the analysis of the recognition and understanding errors generated when our automatic speech recognition and understanding modules [5] are applied to a previously acquired corpus. To translate these errors to the new corpus, a correspondence table among the concepts and attributes of both tasks is defined. This correspondence is based on the similarity of the semantic meanings and frequencies. This methodology has been applied within the framework of two Spanish projects: DIHANA [6] and EDECAN [7].

The main objective of the DIHANA project was the design and development of a dialog system for access to an information system using spontaneous speech. The domain of the project was the query to an information system about railway timetables and prices by telephone in Spanish. Within the framework of this project, we developed a mixed-initiative dialog system to access information systems using spontaneous speech [8]. The behavior of the main modules that compose the dialog system was based on statistical models that were learned from a dialog corpus that was acquired and labeled within the framework of the DIHANA project.

The main objective of the EDECAN project currently underway is to increase the robustness of a spontaneous speech dialog system through the development of technologies for the adaptation and personalization of the system to the different acoustic and application contexts in which it can be used. Within the framework of this project, we will build and evaluate a fully working prototype of a dialog system for access to an information system using spontaneous speech, as in the DIHANA project. In this case, the domain is the multilingual query to an information system about information and booking of sport activities. For the development of the dialog system, we will use statistical approaches as in the DIHANA project. Therefore, the acquisition of a corpus for the new domain will be necessary, and the proposed methodology will be applied.

A new architecture has been designed for the acquisition of this new corpus. Using this architecture, two Wizards of Oz (WOZ) will take part in the acquisition: one to simulate the behavior of the recognition and understanding modules, and the other to simulate the behavior of the dialog manager.

Section 2 presents the architecture that has been defined for the acquisition of the corpus in the EDECAN project. Section 3 presents our previous work within the framework of the DIHANA project and the semantic representation defined for the EDECAN task. Section 4 presents our methodology for simulating errors and confidence values, and the application of this methodology is illustrated in some examples. Finally, Section 5 presents some conclusions and future work.

## 2   An Architecture for the Acquisition of Corpora

As stated in the introduction, we are working on the construction of corpus-based spoken dialog systems for access to information systems. In our approach, the parameters of the main modules that constitute the dialog system are automatically estimated from data. Therefore, when we want to design a dialog system for a new task, we need a corpus of dialogs for this task.

Following the main contributions of the literature, we made acquisitions using the WOZ technique, that is, acquisitions were made with real users and a simulated dialog system. In the WOZ technique, a person substitutes the machine in almost all the functions. In other words, s/he listens to the user turn and builds the query to the information system and the system frame (a codified system answer). This is done by using a software platform (which, for example, stores the historic information supplied by the user in previous turns, etc.) in order to apply the dialog strategy. The system frame is then converted by the Answer Generator and by the Text-To-Speech modules in the answer to the user.

In the WOZ technique, there is usually only one person who performs all the functions described above. In our experience, this is too much for a single person to do. In this work, we propose working with two WOZ: the understanding simulator and the dialog management simulator. The first one listens to the user and simulates the automatic speech recognition and understanding modules, supplying the simulated user frame. From this frame, the second WOZ performs the dialog manager simulation as described above. The architecture proposed for the acquisition of the new dialog corpus is shown in Figure 1.

The separation of the recognition and understanding function and the dialog management function offers several advantages. The main one is that each WOZ must carry

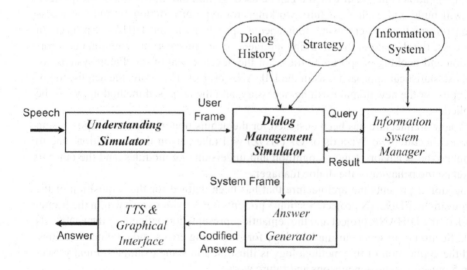

**Fig. 1.** The proposed acquisition schema for the EDECAN corpus

out fewer tasks than before, each WOZ becomes more specialized and the performance of each task improves (if there is a single person doing all the tasks, s/he must manage multiple knowledge sources simultaneously). A separate understanding simulator can better simulate the future automatic understanding module because it only knows (listens to) the user inputs (the system outputs are not listened to). A separate dialog manager simulator can also better simulate the future automatic dialog manager, because their experimental conditions are also the same.

# 3   The DIHANA and the EDECAN Corpora

As in many other dialog systems, the representation of the user and system turns is done in terms of *dialog acts* [9]. The semantic representation chosen for the DIHANA and EDECAN tasks is based on the concept of frame. Therefore, the understanding module generates one or more *concepts* with the corresponding *attributes*.

## 3.1   The Previous Task: DIHANA and the Semantic Representation

One of the objectives of the DIHANA project was the acquisition of a dialog corpus. The DIHANA task consists of a telephone-based information service for trains in Spanish. A set of 900 dialogs was acquired using the standard WOZ technique, and 225 naive speakers collaborated in the acquisition of four dialogs corresponding to different scenarios. The number of user turns was 6,280, and the vocabulary size was 823 different words.

---

M1 Welcome to the railway information system. How can I help you?
U1 Good evening, I want to know timetables from Barcelona to Valencia on April the 24th in the morning.
  HOUR:   *Origin-City* =Barcelona   *Destination-City* =Valencia   *Date* =[2007-24-04]
  *Hour* =morning
M2 There are two trains. The first train leaves at eight twenty-five and the last one leaves at ten thirty. Do you want anything else?
U2 Yes, I would like to know how much the first train costs.
  ACCEPTANCE  PRICE:   *Order-Number* =first
M3 The price of that train is 8.90 €. Do you want anything else?
U3 No, thank you.
  REJECTION
M4 Thanks for using the information system. Have a good journey!

---

**Fig. 2.** A labeled DIHANA dialog (English translation from the original in Spanish). M stands for "Machine turn" and U for "User turn".

**Table 1.** Recognition and understanding of errors in the DIHANA system

|  | Substitutions | Insertions | Deletions |
|---|---|---|---|
| Concepts | 3.63% | 5.33% | 2.33% |
| Attributes | 4.55% | 5.99% | 2.02% |

In this task, we identified six task-dependent concepts: HOUR, DEPARTURE-HOUR, ARRIVAL-HOUR, PRICE, TRAIN-TYPE, TRIP-DURATION, SERVICES and three task-independent concepts: ACCEPTANCE, REJECTION, NOT-UNDERSTOOD. The task-dependent concepts represent the concepts the user can ask for. Each concept has a set of attributes associated to it: *City, Origin-City , Destination-City, Class, Train-Type, Order-Number, Price, Services, Date, Arrival-Date, Departure-Date, Hour, Departure-Hour, Arrival-Hour*. This set represents the restrictions that the user can place on each concept in an utterance. A labeled DIHANA dialog is shown in Figure 2.

A set of experiments was carried out to evaluate the accuracy of our recognition and understanding modules in the DIHANA project [5]. The results obtained for the frame slot accuracy (the number of correctly understood units divided by the number of units in the reference) were: 93.90% for the correct transcriptions without using the recognition module, and 84.20% using both the recognition and understanding modules. Table 1 shows errors for the experiments using recognition and understanding modules in terms of substitutions, insertions and deletions.

### 3.2   The New Task: EDECAN and the Semantic Representation

The new task defined in the framework of the EDECAN project, is a service for the information and booking of sport activities. The service is intended to be used via telephone and via a multimodal information kiosk. We plan to perform the acquisition as explained in Section 2. A kiosk will be set up in a public hall of an education center in our university. A total of 240 dialogs will be recorded by 24 speakers following 15 types of scenarios.

In order to perform this new task, we have obtained a set of 50 person-to-person dialogs that were recorded at the telephone sport service of the University. These dialogs were analyzed and we identified the following: task-independent concepts ACCEPTANCE, REJECTION, NOT-UNDERSTOOD and task-dependent concepts AVAILABILITY, BOOKED, BOOKING, CANCELLATION. The attributes associated to the concepts are: *CourtId, CourtType, Date, Hour, Sport*. An example of a labeled person-to-person dialog is shown in Figure 3.

---

S1   Welcome to the sport service. How can I help you?
U1   I want to know the availability of tennis courts on May the 26th in the evening.
    AVAILABILITY: *Sport* =tennis  *Date* =[2007-26-05] *Hour* =evening
S2   There are three available hours on May the 26th in the evening: from four to five, from six to seven, and from seven to eight. Which do you want to book?
U2   I would like to book from six to seven.
    BOOKING: *Hour* =[18:00-19:00]
S3   I have just booked tennis court number 4 on May the 26th from six to seven in the evening for you. Do you want anything else?
U3   No, thank you.
    REJECTION
S4   Thank you for using the sport service. Goodbye.

---

**Fig. 3.** A labeled EDECAN person-to-person dialog (English translation from the original in Spanish). S stands for "System turn" (the person who attends the service) and U for "User turn".

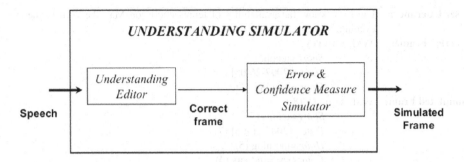

**Fig. 4.** Understanding Simulator for the EDECAN task

## 4  Understanding Simulator

The aim of the understanding simulator in our proposed acquisition architecture is to simulate the behavior of our previous DIHANA recognition and understanding modules. The first WOZ translates the user utterance into a correct user frame using the understanding editor. Then, an error simulator adds errors to the correct frame generating the simulated frame. This process is shown in Figure 4.

The error simulator reproduces the behavior of the recognition and understanding modules developed in the DIHANA project following the error distributions shown in Table 1. To translate these errors to the new corpus, a correspondence table among the concepts and attributes has been manually defined. This correspondence is based on the similarity of the semantic meanings and frequencies. Table 2 shows the correspondences among the concepts and the attributes of the two tasks.

**Table 2.** Correspondence among concepts and attributes of the user turns of the DIHANA and the EDECAN corpora

<table>
<tr><td colspan="2" align="center">CONCEPT correspondence</td><td colspan="2" align="center">ATTRIBUTE correspondence</td></tr>
<tr><td>EDECAN<br>Train Information</td><td>DIHANA<br>Sport Info. & Booking</td><td>EDECAN<br>Train Information</td><td>DIHANA<br>Sport Info. & Booking</td></tr>
<tr><td>AVAILABILITY</td><td>HOUR<br>ARRIVAL-HOUR<br>DEPARTURE-HOUR</td><td>*Sport*</td><td>*City*<br>*Destination-City*<br>*Origin-City*</td></tr>
<tr><td>BOOKING</td><td>PRICE</td><td>*CourtType*</td><td>*Train-Type*<br>*Class*</td></tr>
<tr><td>BOOKED</td><td>TRAIN-TYPE</td><td></td><td></td></tr>
<tr><td>CANCELLATION</td><td>TRIP-DURATION<br>SERVICES</td><td>*CourtId*</td><td>*Order-Number*<br>*Price*<br>*Services*</td></tr>
<tr><td>ACCEPTANCE</td><td>ACCEPTANCE</td><td></td><td></td></tr>
<tr><td>REJECTION</td><td>REJECTION</td><td>*Date*</td><td>*Date*<br>*Arrival-Date*<br>*Departure-Date*</td></tr>
<tr><td>NOT-UNDERSTOOD</td><td>NOT-UNDERSTOOD</td><td></td><td></td></tr>
<tr><td></td><td></td><td>*Hour*</td><td>*Hour*<br>*Arrival-Hour*<br>*Departure-Hour*</td></tr>
</table>

| | |
|---|---|
| **User Utterance:** | I want to know the availability of tennis courts on May the 26th in the evening. |
| **Correct Frame:** | AVAILABILITY:<br>      *Sport* =tennis<br>      *Date* =[2007-26-05]<br>      *Hour* =evening |
| **Simulated Frame:** | AVAILABILITY (8):<br>      *Sport* =tennis (7)<br>      *Date* =[2007-26-05] (7)<br>      *Hour* =morning (5)<br>      *CourtType* =indoors (4) |

**Fig. 5.** An example of applying the error and confidence measure simulator for a user turn (confidence values are shown in brackets)

In a typical DIHANA dialog, the user first asks for the timetable of a trip and then the price of such a trip (as in the example in Figure 2). Analogously, after analyzing the person-to-person dialogs of the sport service at the University, it is usual to ask for the availability of a sport facility and then to book it (as in the example in Figure 3). Therefore, we established a correspondence between HOUR and AVAILABILITY and between BOOKING and PRICE.

The error and confidence measure simulator not only introduces errors into the user frames but also generates a confidence value for each concept and attribute in the simulated frame. This confidence value is calculated using a weighted coefficient that considers whether an error has been introduced in the simulation. Figure 5 shows the effects of applying the error and confidence measure simulator for a user turn.

## 5    Conclusions and Future Work

We have presented a methodology to adapt our previous work within the framework of dialog systems in order to acquire a new corpus for a different domain. This proposal is based on the use of two different WOZ and an error simulator between them.

The first WOZ simulates the combined behavior of the recognition and the understanding modules. S/he listens to the user utterance and transcribes it in terms of semantic user frames. The second WOZ carries out the functions of the dialog manager. S/he receives the user frames, interacts with the information system, and generates the answer of the system in terms of system frames.

The error simulator receives the user frame generated by the first WOZ and returns a modified frame by adding simulated errors. This frame is the input for the second WOZ. The errors introduced by this module simulate the errors introduced by our recognition and understanding modules in our previous project. Moreover, the error simulator calculates a confidence value for each frame slot.

Using an error simulator allows us to achieve two objectives. On the one hand, it allows us to simulate the approximate behavior of our recognition and understanding modules before implementing then. This simulation is based on the analysis of the errors introduced by these modules in our previous project. On the other hand, it is possible to simulate a wide range of application environments by varying the parameters used for the generation of errors. For example, a noisier environment can be simulated by increasing the quantity of errors introduced by the simulator.

Since the two WOZ carry out the correct transcription of their respective modules, each dialog acquired using the proposed methodology is already labeled in terms of understanding frames and dialog management frames. This way, a later labeling phase is not necessary, as is required when a classic approach for the WOZ paradigm is used.

As future work, a more detailed study of the confidence measures used by the error simulator is needed. Once the acquisition is done, we will have real recognition and understanding modules for the new task, and we will be able to compare the behavior of the understanding modules of DIHANA and EDECAN projects to verify whether the correspondence among the semantic representation of the two tasks is appropriate.

# References

1. Potamianos, A., Narayanan, S., Riccardi, G.: Adaptive Categorical Understanding for Spoken Dialogue Systems. IEEE Transactions on Speech and Audio Processing 13(3), 321–329 (2005)
2. Torres, F., Hurtado, L., García, F., Sanchis, E., Segarra, E.: Error handling in a stochastic dialog system through confidence measures. Speech Communication 45, 211–229 (2005)
3. Hurtado, L.F., Griol, D., Segarra, E., Sanchis, E.: A stochastic approach for dialog management based on neural networks. In: Proc. of Interspeech'06-ICSLP, Pittsburgh, pp. 49–52 (2006)
4. Williams, J., Young, S.: Partially Observable Markov Decision Processes for Spoken Dialog Systems. Computer Speech and Language 21(2), 393–422 (2007)
5. Grau, S., Segarra, E., Sanchís, E., García, F., Hurtado, L.F.: Incorporating semantic knowledge to the language model in a speech understanding system. In: IV Jornadas en Tecnologia del Habla, Zaragoza, Spain, pp. 145–148 (2006)
6. Benedí, J., Lleida, E., Varona, A., Castro, M., Galiano, I., Justo, R., López, I., Miguel, A.: Design and acquisition of a telephone spontaneous speech dialogue corpus in Spanish: DIHANA. In: Proc. of LREC'06, Genove, Italy, pp. 1636–1639 (2006)
7. Lleida, E., Segarra, E., Torres, M., Macías-Guarasa, J.I.: EDECAN: sistEma de Diálogo multidominio con adaptación al contExto aCústico y de AplicacióN. In: IV Jornadas en Tecnologia del Habla, Zaragoza, Spain, pp. 291–296 (2006)
8. Griol, D., Torres, F., Hurtado, L., Grau, S., García, F., Sanchis, E., Segarra, E.: A dialog system for the DIHANA Project. In: Proc. of SPECOM'06, S. Petersburgh, pp. 131–136 (2006)
9. Fukada, T., Koll, D., Waibel, A., Tanigaki, K.: Probabilistic dialogue extraction for concept based multilingual translation systems. In: Proc. Int. Conf. on Spoken Language Processing, pp. 2771–2774 (1998)

# First Approach in the Development of Multimedia Information Retrieval Resources for the Basque Context

N. Barroso[1], A. Ezeiza[2], N. Gilisagasti[3], K. López de Ipiña[3],
A. López[3], and J.M. López[3]

University of Basque Country
[1] Department of Systems Engineering and Automation (SEA) Donostia
Aiatek S. Coop. Enteprise.
nora@d-teknologia.com
[2] Department of SEA, Donostia. Ixa taldea. aitzol
ezeiza@ehu.es
[3] Department of SEA, Gasteiz. Computational Intelligence Group
karmele.ipina@ehu.es

**Abstract.** Information Retrieval (IR) applications require appropriate Multimodal Resources to develop all of their components. The work described in this paper is one of the main steps of a broader project that consists in developing a Multimodal Index System for Information Retrieval. The final goal of this part of the project is to create a robust Automatic Speech Recognition System for Basque that also covers the other languages spoken in the Basque Country: Spanish and French. It is widely accepted that the robustness of these systems is directly related to the quality of the resources used during training. Hence, the digital resources for Multilingual Continuous Speech Recognition systems for the three official languages in the Basque Country have to be described.

## 1 Introduction

Information Retrieval (IR) systems provide effective access to on-line multimodal information. Classically IR research has been mainly based on text [1] where the search for any document is carry out by a query that consists of a number of keywords. Nowadays these systems manage multimodal information such as voice, gestures or written information, and those systems require of appropriate Digital Resources [2]. Automatic Indexing of Broadcast News is a topic of growing interest for the mass media in order to take maximum output of their recorded resources. Actually, this is a challenging problem from researchers' point of view due to many unresolved issues such as speaker turn changes and overlapping, different background conditions, large vocabulary, etc. In order to get significant results in this area, high-quality language resources are required. Since the main goal of our project[1] is the

---

[1] This work has been partly developed in the framework of the "Ehiztari" project, funded by the Department of Industry of the Government of the Basque Autonomous Community through the Saiotek program.

V. Matoušek and P. Mautner (Eds.): TSD 2007, LNAI 4629, pp. 582–590, 2007.
© Springer-Verlag Berlin Heidelberg 2007

development of an automatic audio index for Basque and its surrounding languages, which extends our previous work on Broadcast News [3], it is essential to create resources for all the languages spoken by the potential users. Basque, Spanish, and French are used in most of the mass media in the Basque Country, and the three languages have to be taken into account in order to develop an efficient index system. Indeed, the speakers tend to mix words and sentences in these languages in their discourse. Therefore, all the tools (ASR system, NLP system, index system) and resources (digital library, Lexicon, etc.) will be oriented to create a robust multilingual system.

Basque is a Pre-Indo-European language of unknown origin and it has about 1.000.000 speakers. It presents a wide dialectal distribution, including six main dialects: Biscayan, Gipuzkoan, Upper Navarrese, Lower Navarrese, Lapurdian, and Zuberoan. This dialectal variety entails phonetic, phonologic, and morphologic differences. In fact, sometimes the gaps between dialects are so wide that many efforts have been made in the last decades to develop a standard version of Basque called *Batua* that is nowadays in use in almost all the mass media and in the official administration. The Basque Country is situated in the North of Spain and the Southwest of France. There are about 2.75 million Basque citizens in Spain, over 250,000 in France There are also many communities in America as well as in other parts of the world. In Spain, the Basque language (*Euskara*) is spoken in Navarre and in the Basque Autonomous Community, and the latter is the community that most boosts the use of the Basque language and where the University of the Basque Country is located. The Basque language is spoken also in France in the *Département of Pyrénées-Atlantiques*, where French is the only official language. Most works on Natural Language Processing tend to ignore the Basque language spoken in France and its relation with French, since it is difficult to get resources for research or any other goal in this part of the Basque Country.

Most of the mass media in Basque Country use Spanish, French, and/or Basque, and many of them have shown their interest in the development of Index Systems for their media. Thus, the three languages have to be taken into account to develop an efficient Speech Recognition system for Information Retrieval. Many works have been developed with several European languages [4], and specifically French and Spanish have been thoroughly studied [5] [6] but the use of Basque language introduces a new concern: it requires specific Natural Language Processing tools and the resources available are few. Hence, Digital Resources from mass media of the Basque Country have been selected and processed: Videos from the Basque Public Radio and Television (EITB), which emits in Basque and Spanish; Audio files and text scripts from Infozazpi digital radio station that emits in Basque, Spanish and French; and finally, texts extracted from various newspapers also in the three languages.

Section 2 describes the main features of the languages and several evaluations oriented to the development of the indexation tools are carried out. Section 3 details the Language Resources developed. Section 4 deals with the processing and evaluation of the data. Finally, conclusions and future work are summarized in Section 5.

## 2  Analysis of Features Across Language

The analysis of the features of the languages chosen is a crucial issue because they have a clear influence on both the performance of the acoustic decoder and on the vocabulary size of the Index System. On the one hand, the development of ASR multilingual systems based on a single global acoustic model set [7] (combining the sound inventory of all languages) is of increasing importance because of the following advantages: reduced complexity of the system, easier language identification, and efficient bootstrapping of recognition systems in new languages even if only a small amount of training data is available.

**Table 1.** Sound Inventories for Basque, French and Spanish in the SAMPA notation

| Sound Type | Basque | French | Spanish |
|---|---|---|---|
| **Consonant** | | | |
| **Plosives** | p b t d k g c | p b t d k g | p b t d k g |
| **Affricates** | ts ts´ tS | | ts |
| **Fricatives** | gj jj f B T D s s´ S x G Z v h | f v s z S Z | gj jj F B T D s x G |
| **Nasals** | m n J | m n J N | m n J |
| **Liquids** | l L r rr | l R | l L r rr |
| **Vowel glides** | w j | w H j | w j |
| **Vowels** | | | |
| **Vowels** | i e a o u @ | i e E a A O o u  y 2 9 @  e~ a~ o~ 9~, | i e a o u |
| **Indetermination** | - | a-A e~-9~ e-E  o-O 2-9 | - |

**Sounds Inventories**

Table 1 shows the set of sounds for the three languages in the SAMPA (Speech Assessment Methods Phonetic Alphabet) notation, clustered by the type of sound: consonants (Plosives, Affricates, Fricatives, Nasals, Liquids) and vowels (glides, common, indetermination, etc.). All of them have been selected to be used in the indexation system.

- The standard Basque consonant system has 23 obstruents (7 plosives, 3 affricates, and 13 fricatives) and 10 sonorants (1 affricative, 3 nasals, 4 liquids, and 2 semivowel glides). The vowel system comprises 6 oral vowels.
- The standard French consonant system is consists of 12 obstruents (6 plosives and 6 fricatives) and 8 sonorants (3 nasals, 2 liquids, and 3 semivowel glides). The vowel system comprises 12 oral vowels, 4 nasal vowels and several indeterminations such as a-A.
- The standard Spanish consonant system consists of 15 obstruents (6 plosives, 1 africate, and 8 fricatives) and 11 sonorants (1 affricative, 3 nasals, 4 liquids, and 3 semivowel glides). The vowel system comprises 5 oral vowels.

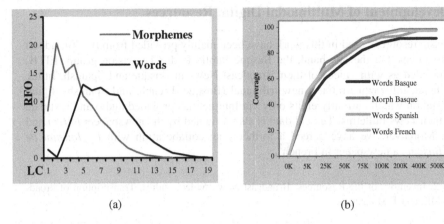

(a)                                                    (b)

**Fig. 1.** (a) Relative Frequency of Occurrence (RFO) of the words and pseudo-morphemes in relation to their Length in Characters (LC) of the STDBASQUE sample. (b) Coverage for the BCNEWS textual sample.

## Vocabulary Size

Basque is an agglutinative language with a special morpho-syntactic structure in the words [8] that may lead to intractable lexis for a LVCSR. A first approach to the problem is to use morphemes instead of words in the system in order to define the vocabulary [9]. This approach has been evaluated over three textual samples analysing the coverage and the Out-of-Vocabulary rate for the words and the morphemic units obtained by the automatic morphological segmentation tool. The tree textual samples consist of about 2M characters, 20.000 sentences, 50.000 words and 20.000 morpheme units.

The first important outcome of this analysis is that the vocabulary size of the pseudo-morphemes is reduced to 60% in all cases in comparison to the vocabulary size of the words. Regarding the unit size, Fig. 1 (a) shows the plot of Relative Frequency of Occurrence (RFO) of the morphemes and words versus their length in characters over the *STDBASQUE* textual sample. Although only 10% of the morphemes have fewer than four characters, such small morphemes have an Accumulated Frequency of about 40% in the databases (the Accumulated Frequency is obtained as the sum of the individual pseudo-morphemes RFO) [9]. To check the validity of the morphemic unit inventory, units having less than 4 characters and having plosives at their boundaries were selected from the texts. They represent some 25% of the total. Such small units are acoustically difficult to recognize and they increase the acoustic confusion and the number of insertions. Finally, Fig. 1 (b) shows the analysis of coverage by Out of Vocabulary rate over the *BCNEWS* textual sample. When pseudo-morphemes are used, the coverage in texts is better and a complete coverage is easily achieved. OOV rate is higher in this sample.

## 3   Development of Multimodal Digital Resources

The basic resources used in this work have been mainly provided from two Broadcast News sources. On the one hand, the Basque Public Radio-Television group (EITB) has provided us with videos of their Broadcast News in Basque and Spanish. In the other hand, Infozazpi irratia, a new trilingual (Basque, French, and Spanish) digital radio station which currently emits only via Internet has provided audio and text data from their news bulletins. Textual data is also provided by the newspapers *Berria* and *Gara*. Moreover, we have agreed a forthcoming collaboration with *Le Journal de Pays Basque*, a newspaper in French[2].

**Table 2.** Inventory of the Resources: Broadcast News (video, audio), Transcription of broadcast-XML, and Text databases

| Language | Broadcast Video | Broadcast Audio | TR. XML-FILE | Text database |
|---|---|---|---|---|
| EU | 6:37 | 18 | 12 | 8M |
| FR | - | 2:58 | 2:58 | 2M |
| ES | 9:35 | 12:34 | 12:34 | 4M |
| Total | 16:12 | 33:12 | 33:12 | 14M |

### Inventory of the Resources

The Inventory of the Resources collected and developed is described in Table 2, and the explanation of the contents is the following:

- *Broadcast news resources:* About 6 hours of video in MPEG4 (WMV 9) format of "Gaur Egun" (Basque) and "Teleberri" (Spanish) programs, the daily broadcast news. Infozazpi irratia provides about 17 hours of broadcast news in the three languages, initially in MP3/96Kps format (this is the format used in their Internet Broadcasting system).
- *Transcription information:* Audio transcription in XML format, containing information about speaker changes, noises and music fragments, and each word's orthographic transcription including word's lemma and Part-Of-Speech disambiguated tags.
- *Textual data:* 2 years of local newspapers in Basque (*Euskaldunon Egunkaria* and *Berria*), in text format. 1 year of newspapers in Spanish provided by *Gara*, in text format. The textual data is completed with the scripts of all the audio recordings from *Infozazpi irratia*. It is worth mentioning that the recordings provided for this work are hourly news bulletins; the bulletins in Basque are about five minutes each, whereas the other bulletins in French and Spanish are about two minutes each. This way, the textual data in Basque is proportionally larger.
- *Lexical resources:* The Lexicons in Basque, French and Spanish have been extracted from the XML transcription files, including phonologic, orthographic, and morphologic information. There are also the correspondent statistical grammars for Basque, French, and Spanish extracted from the textual data.

---

[2] We also would like to thank all the people and entities that have collaborated in the development of this work: UZEI, EITB, Infozazpi irratia, Le journal de Pays Basque, Gara, and Berria.

## Processing of Data

The audio data has been extracted from the MPEG4 video files and it has been stored in WAV format (16 KHz, linear, 16 bits). The MP3 files provided by Infozazpi have been also converted to this standard format for uniform processing reasons. Then the XML label files were created manually, using the *Transcriber* tool [10]. The XML files include information of different speakers, noises, and paragraphs of the broadcast news. Basque XML files include morphologic information such as each word's lemma and Part-Of-Speech tag Table 3.

**Table 3.** Simplified sample of the output of the Transcriber free tool, enriched with morpho-syntactic information of Basque

```
<Sync time="333.439"/>
+horretarako /hortarako/<Word lemma="hori" POS="ADB"/>
+denok /danok/<Word lemma="dena" POS="IZL"/>
lagundu<Word lemma="lagundu" POS="ADI"/>
behar<Word lemma="behar" POS="ADI"/>
dugu<Word lemma="*ukan" POS="ADL"/>
.</Turn>
<Turn mode="spontaneous" fidelity="high" startTime="335.182" endTime="336.065">
<Sync time="335.182"/>
^Batasunak<Word lemma="9batasuna" POS="IZB"/>
```

Using this transcribed information, a Lexicon for each language has been extracted. Additionally, the segments that had a very neatly defined non-neutral emotion have been marked with special tags in order to use this feature in future works on Emotion Recognition Systems. In addition to the data related to the video and audio gathered, there are two independent types of textual resources: The texts extracted from the newspapers and the scripts of the radio. These last resources are very interesting because they are directly related (date, program) with the texts read in the broadcast news in Spanish, French, and Basque. All of them were processed to include morphologic information such as each word's lemma and Part-Of-Speech tag. Using all this information, a Lexicon for each language has been extracted taking into account the context of the words in order to eliminate ambiguity. The Lexicon stores information of each different word appearing in the transcription. Table 4 shows some examples of the Lexicon information.

**Table 4.** Sample of the Lexicon for Basque

| Input | Transcription | Morphological Analysis | LEMA | Morphological segmentation |
|---|---|---|---|---|
| euskaldunena | ewS.'kal.du.ne.'2na | ADJ IZO DEK GEN MG DEK ABS NUMS MUGM | euskaldun | euskaldun=en= |
| | | ADJ IZO DEK GEN NUMP MUGM DEK ABS NUMS | | euskaldun=en= |
| | | ADJ IZO GRA SUP DEK ABS NUMS MUGM | | euskaldun=en=a |
| margolarien | mar.'Go.la.r6i.'2en | IZE ARR DEK GEN NUMP MUGM | margolari | margolari=en |
| | | IZE ARR DEK GEN NUMP MUGM DEK ABS MG | | margolari=en |
| margolaritzan | mar.'Go.la.r6i.'2t&sa | IZE ARR DEK NUMS MUGM DEK INE | margo-laritza | margolaritz=an |
| | mar.'Go.la.r6i.'2t&c~ | | | |
| margolaritza | mar.'go.la.r6i.'2t&sa | IZE ARR; IZE ARR DEK ABS MG | margo-laritza | margolaritza |
| | mar.'go.la.r6i.'2t&c~a | IZE ARR DEK ABS NUMS MUGM | | margolaritza |
| | | | | margolaritz=a |

# 4 Evaluation of the Audio Resources

A preliminary evaluation of digital resources oriented to robust Automatic Multilingual Information Retrieval was carried out, taking into account the future development of the system's key elements. A brief extraction of the result of the evaluation of the database is summarized in Tables 5 and 6.

- *Languages distribution and parallelism.* Tables 5 and 6 show the distribution of the languages in the data provided by the two mass media: EITB and Infozazpi irratia. EITB provides parallel material for Basque and Spanish (about 6 hours each). Most of the material provided by Infozazpi irratia is in Basque but two very interesting parallel samples are also provided in French and Spanish. These parallel sections are not absolutely aligned because they are sometimes written by different reporters, but they have special interest for parallel processing tasks.
- *Speaker segmentation.* The Speaker segmentation in Table 5 shows a substantial difference between the two media. Infozazpi's bulletins are read by only one reporter, while EITB has a wider variety of speakers, being 6,6% of them Non Native Speaker (NNS).
- *Speech-Non Speech automatic segmentation.* Supervised Speech/ Non Speech segmentation (Table 6) shows real differences between digital resources provided by EITB and Infozazpi. The average rate of the former is about 96,5% for both languages but in the latter this average falls to 72% for Basque and to about 45% for the other languages due to longer presence of music. Furthermore, for the EITB database, the speech signal containing background noise is in average of 45,49%: music (10,69%), background speakers (6,04%), white noise (3,59%) and undefined (25,17%). For Infozazpi the background noise is almost always music and it is present in nearly all of the bulletins in French and Spanish.
- *Statistics across languages.* The significant differences across data sets are believed to reflect the multi-style character of the database (Table 6). Number of Words (NW) and Number of Distinct Words (NDW). The number of morphemes has been also obtained for Basque, because of its agglutinative nature.

**Table 5.** Evaluation of the audio resources: Total of Speakers (TS), Non Native Speakers (NNS), Native Speakers (NS), Size (SZ) in Mb, Number Files in WAV Format (NF-WAV), Timeframe in hours (TF-H)

| Media | TS | NNS | NS | SZ-Mb | NF | TF-H |
|-------|-----|-----|-----|-------|-----|-------|
| **Infozazpi** | | | | | | |
| EU | 40 | 0 | 40 | 2440 | 134 | 11:23 |
| ES | 20 | 0 | 20 | 648 | 86 | 2:59 |
| FR | 15 | 0 | 15 | 657 | 86 | 2:58 |
| **EITB** | | | | | | |
| EU | 187 | 14 | 173 | 877 | 7 | 6:37 |
| ES | 175 | 10 | 165 | 870 | 6 | 9:35 |
| **Total** | 437 | 24 | 413 | 5492 | 319 | |

**Table 6.** Evaluation of the audio resources

| Media | %Speech | %Non-Sp | NW | NDW |
|---|---|---|---|---|
| **Infozazpi** | | | | |
| EU | 78 | 22 | 58855 | 5741-W, 3445-M |
| ES | 47 | 53 | 13625 | 2066 |
| FR | 42 | 58 | 20196 | 3494 |
| **EITB** | | | | |
| EU | 97 | 3 | 49185 | 7993-W, 4796-M |
| ES | 96 | 4 | 102301 | 6290 |

## 5 Concluding Remarks

The development and evaluation of appropriate Resources is an important issue for any work involving techniques based on multimodal contents such as Information Retrieval Systems. In the context of this project, Basque, French, and Spanish have to be taken into account because they are official in different parts of the Basque Country, and because of their use in *Infozazpi irratia,* the Basque Public Radio and Television *EITB,* and many other mass media. Therefore, it is very important to lay down the foundations of Multilingual Speech Recognition Systems. In fact, this work tackles this issue from the perspective of making use of a less-resourced language such as Basque against two much more resourced languages such as Spanish and French. Furthermore, this project is an extension of our previous works in this field and it could be considered the first approach to the development of better digital resources for the three official languages spoken in the Basque Country.

## References

1. Frakes, W., Baeza-Yates, R.: Information Retrieval: Data Structures and Algorithms. Prentice-Hall, Englewood Cliffs, N.J (1992)
2. Foote, J.T.: An Overview of Audio Information Retrieval. In: Multimedia Systems, vol. 7(1), pp. 2–11. ACM Press/Springer-Verlag (January 1999)
3. Bordel, G., Ezeiza, A., de Ipiña, K.L., Mendez, M., Peñagarikano, M., Rico, T., Tovar, C., Zuleta, E.: Development of Resources for a Bilingual Automatic Index System of Broadcast News in Basque and Spanish. LREC 2004, vol. III, pp. 881–884 (2004)
4. Vandecatseye, A., et al.: The COST278 pan-European Broadcast News Database. In: Proceedings of LREC 2004, Lisbon (Portugal) (2004)
5. García-Mateo, C., Dieguez-Tirado, J., Docío-Fernández, L., Cardenal-López, A.: Transcrigal: A bilingual system for automatic indexing of broadcast news. LREC (2004)
6. Adda-Decker, M., Adda, G., Gauvain, J., Lamel, L.: Large vocabulary speech recognition in French. Acoustics, Speech, and Signal Processing. In: ICASSP Proceedings, vol. 1, pp. 15–19 (1999)
7. Schultz, T., Waibel, A.: Multilingual and Crosslingual Speech Recognition. In: Proceedings of the DARPA Broadcast News Workshop (1998)

8. Alegria, I., Artola, X., Sarasola, K., Urkia, M.: Automatic morphological analysis of Basque, Literary & Linguistic Computing, vol. 11, pp. 193–203. Oxford Univ. Press, Oxford (1996)
9. de Ipiña, K.L., Graña, M., Ezeiza, N., Hernández, M., Zulueta, E., Ezeiza, A., Tovar, C.: Selection of Lexical Units for Continuous Speech Recognition of Basque. In: Progress in Pattern Recognition, Speech and Image Analysis, pp. 244–250. Springer, Berlin (2003)
10. Barras, C., Geoffrois, E., Wu, Z., Liberman, M.: Transcriber: a Free Tool for Segmenting, Labeling and Transcribing Speech. In: First International Conference on Language Resources and Evaluation (LREC-1998)

# The Weakest Link

Harry Bunt and Roser Morante

Tilburg University, The Netherlands
{bunt,r.morantel}@uvt.nl

**Abstract.** In this paper we discuss the phenomenon of grounding in dialogue using a context-change approach to the interpretation of dialogue utterances. We formulate an empirically motivated principle for the strengthening of weak mutual beliefs, and show that with this principle, the building of common ground in dialogue can be explained through ordinary mechanisms of understanding and cooperation.

## 1 Introduction

Communicating agents are constantly involved in processes of creating and updating a common ground, a set of beliefs that they believe to be shared. The notion of common ground (CG) has been studied from different angles by linguists, logicians, psychologists and computer scientists. Logicians, such as Stalnaker and Lewis, have suggested to define CG in terms of *mutual beliefs*, explained as follows:

(1)  $p$ is a mutual belief of $A$ and $B$ iff:
 - $A$ and $B$ believe that $p$;
 - $A$ and $B$ believe that $A$ and $B$ believe that $p$;
 - $A$ and $B$ believe that $A$ and $B$ believe that $A$ and $B$ believe that $p$;
 and so on *ad infinitum*.

With this notion of CG, the *grounding* of a belief, i.e. the process of adding it to the common ground, is difficult to understand since CG-elements according to (1) contain an infinite amount of information, while only a finite amount of information seems to be obtainable through communication.

In this paper we approach the phenomenon of grounding from a semantic perspective, using the framework of Dynamic Interpretation Theory (DIT)[1]. DIT views the meaning of dialogue utterances in terms of bundles of actions that update both the addressee's and the speaker's information state or 'context model'; these actions are called 'dialogue acts'. Using DIT, we have developed a model of dialogue context update that makes explicit the beliefs that agents hold at every turn of a dialogue. In this paper we argue that grounding in dialogue can be explained by this model, by using a general pragmatic principle for the strengthening of weak beliefs concerning the understanding and acceptance of what is said in a dialogue.

## 2 Views on Grounding

The notion of CG as mutual beliefs has been considered as problematic from a psycholinguistic point of view because of its representational demands. Clark [2](p. 95)

V. Matoušek and P. Mautner (Eds.): TSD 2007, LNAI 4629, pp. 591–598, 2007.

claims that a notion of CG defined as mutual beliefs *"...obviously cannot represent people's mental states because it requires an infinitely large mental capacity"*. While it is true that a mutual belief defined as in (1) seems to have an infinite character, this can be said of any belief. For instance, if $A$ holds a belief that $q$, then he also believes that ($q$ or $r$), for any $r$ that he might consider, due to his capability to perform inferences. Similarly, from the single stored belief that $p$ *is a mutual belief of A and B*, $A$ can infer that $B$ believes that $p$, that $B$ believes that $A$ believes that $p$, and so on *ad infinitum*.[1]

In Clark and Schaefer's model [4] of grounding, participants in a dialogue perform collective actions called *contributions*, divided into an acceptance and a presentation phase, so that every contribution, except for those that express negative evidence, has the role of accepting the previous contribution. A difficulty with this model is that its grounding criterion says that *"the contributor and the partners mutually believe that the partners have understood what the contributor meant"*. So the grounding *process* is conceived in terms of mutual beliefs. However, the central problem of grounding is precisely how mutual beliefs are established. Work based on this model includes its extension to human–computer interaction by Brennan and collaborators [5,6] and the formal theory of grounding by [7].

In his influential computational model of grounding, Traum [8] has introduced separate *grounding acts* which are used to provide communicative feedback and thereby create mutual beliefs. In order for this approach to work, Traum assumes that such communicative acts are always correctly perceived and understood, therefore a dialogue participant does not need feedback about his feedback acts. This is an unwarranted assumption, however. Like any dialogue utterance, an utterance which expresses feedback can suffer from the addressee temporarily being disturbed by the phone, or by an aircraft flying over, or by noise on a communication channel; hence a speaker who performs a grounding act can never be sure that his act was performed successfully until he has received some form of feedback (see also below).

Matheson et al. [9] use elements of Traum's model in their treatment of grounding. They represent grounded and ungrounded discourse units in the information state, and change their status from ungrounded to grounded through grounding acts. The dialogue act `Acknowledgement` is the only grounding act implemented; its effect is to merge the information in the acknowledged discourse unit into the grounded information. They do not deal with cases of misunderstandings or cases where the user asks for acknowledgement. The model keeps only the last two utterances in the information state, so it is not clear what would happen if the utterance to be grounded is more than two utterances back.

# 3   Grounding in DIT

## 3.1   Information Exchange Through Understanding and Adoption

The addition of a belief to a common ground relies on evidence that the belief in question is mutually believed. The nature of such evidence depends on the communicative situation, for instance on whether the participants can see each other, and on whether

---

[1] For the formal logical underpinning of such an inference capability see e.g. [3].

they are talking about something they (both know that they) can both see. We restrict ourselves here to situations where grounding is achieved through verbal communication only, as in the case of telephone conversations, email chats, or spoken human-computer dialogue.

In the DIT framework, information passed from one dialogue participant to another through understanding and believing each other. Understanding is modeled as coming to believe that the preconditions hold which are characteristic for the dialogue acts expressed by that behaviour. For example, if $A$ asks $B$ whether $p$, then as a result of understanding this, $B$ will know that $A$ wants to know whether $p$, and that $A$ thinks that $B$ knows whether $p$. Believing each other leads to what has been called 'belief transfer' [10]. For example, when $A$ has asked $B$ whether $p$, and $B$ answers "Yes", then $A$ may be expected to believe $B$, so from now on $A$ believes that $p$. This is called the *adoption* of information.

To be sure that information is indeed transferred through the mechanisms of understanding and adoption, a speaker needs evidence of correct understanding of his communicative behaviour and of being believed. In order to see how this may happen concretely, consider the following dialogue fragment. $A$ initially contributes utterance 1 expressing an Inform act; let $c_1$ be the precondition that $A$ *believes that p*, with $p$ the propositional content of the act (the information that bus 6 leaves from platform G3). Continued successful communication should lead to both $c_1$ and $p$ at some point being in both $A$'s and $B$'s common ground.

(2) 1. A: Bus 6 leaves from platform G3.
    2. B: Platform G3.
    3. A: That's right.
    4. B: Thank you very much.

In order to come to believe that $p$ is mutually believed, $A$ should have evidence that $B$ understands his utterance 1 and believes its content $p$. $B$'s utterance 2 provides evidence of correct understanding, but not of adoption, since in ]2 he also offers that belief for confirmation. So after 2, $A$ believes that $B$ believes that $A$ believes that $p$, but $A$ does not yet know wether $B$ believes that $p$. $A$'s response 3 tells $B$ that $A$ has understood this, hence it leads to $B$ believing that $A$ believes that $B$ believes that $A$ believes that $p$. $B$'s contribution 4 provides evidence that the previous dialogue acts were performed successfully; therefore, upon understanding utterance 4, $A$ has accumulated the following beliefs:

(3) $A$ believes that $p$
    $A$ believes that $B$ believes that $p$
    $A$ believes that $B$ believes that $A$ believes that $p$
    $A$ believes that $B$ believes that $A$ believes that $B$ believes that $p$
    $A$ believes that $B$ believes that $A$ believes that $B$ believes that $A$ believes that $p$

We see nested beliefs of some depth emerging, but $A$ is still a long way from believing that $p$ is mutually believed – an infinitely long way, in fact.[2] Continuing along this line

---

[2] The construction of the nested beliefs shown in (3) relies on exploiting the cumulative effects of feedback. For a detailed analysis of this phenomenon, called 'feedback chaining', see [11].

obviously does not lead to mutual beliefs in a finite amount of time. One explanation of grounding could perhaps be that human dialogue participants perform a form of induction in order to extend the finite nested beliefs in (3) to infinity, however, we prefer a different explanation.

## 3.2 Strengthening Weak Mutual Beliefs

In natural face-to-face dialogue, the participants give explicit and implicit feedback about their understanding of what is being said by means of facial expressions, head movements, direction of gaze, and verbal elements; speakers thus receive feedback while they are speaking. In situations without visual contact, a speaker often receives no feedback while he is speaking (or typing). This has the effect that, when a speaker has finished a turn, he does not know whether his contribution has been perceived, understood, and accepted. In a situation where "normal input-output" conditions obtain [12], i.e. where participants speak the same language, have no hearing or speaking impairments, use communication channels without severe distortions, and so on, a speaker normally expects that the addressee perceives, understands and believes what is being said.

In the DIT approach to utterance interpretation, such expectations are modeled by the speaker having a doxastic attitude called *weak belief* that the addressee of a dialogue acts believe its preconditions and content to be true. (The most important difference between a weak and a firm belief in that it is not inconsistent to weakly believe that $p$ while at the same time having the goal to know whether $p$. In fact, such a combination forms the preconditions of a Check act.) So after contributing an utterance that expresses a dialogue act with precondition $c_1$, the speaker $A$ has the weak belief that $B$ believes that $c_1$. And similarly, in information-seeking dialogues, assistance dialogues, and other types of cooperative dialogue where the participants are expected to provide correct information about the task at hand, if the utterance offers the information $p$, then the speaker $A$ also has the weak belief that $B$ believes that $p$.

The assumption of being understood and believed is of course not idiosyncratic for a particular speaker, but is commonly made by participants in cooperative dialogue in normal input-output conditions, in particular also by $B$. So $B$ will believe that $A$ makes this assumption, therefore:

(4)  $B$ believes that $A$ weakly believes that $B$ believes that $c_1$.
     $B$ believes that $A$ weakly believes that $B$ believes that $p$.

By the same token, $A$ believes this to happen, hence:

(5)  $A$ believes that $B$ believes that $A$ weakly believes that $B$ believes that $c_1$
     $A$ believes that $B$ believes that $A$ weakly believes that $B$ believes that $p$.

This line of reasoning can be continued *ad infinitum*, leading to the conclusion that, as a result of the assumptions concerning understanding and adoption:

(6)  Both $A$ and $B$ believe that it is mutually believed that $A$ weakly believes that $B$ believes that $c_1$ and that $p$.

This means that, after contributing utterance 1, $A$ will among other things believe the following 'weak mutual beliefs' to be established, 'weak' in the sense that the mutual belief contains a weak belief link:

(7)  a. $A$ believes that it is mutually believed that $A$ weakly believes that $B$ believes that $c_1$.
     b. $A$ believes that it is mutually believed that $A$ weakly believes that $B$ believes that $p$.

More generally, with respect to grounding we may observe that for an agent to ground a belief, what he has to do is not so much extend a set of finitely nested beliefs like (3) to nested beliefs of infinite depth, but to replace the weak belief link in believed mutual beliefs of the form

(8)  $A$ believes that it is mutually believed that $A$ **weakly** believes that $B$ believes $q$

by an ordinary belief link, turning it into

(9)  $A$ believes that it is mutually believed that $A$ believes that $B$ believes $q$

which is equivalent to:[3]

(10)  $A$ believes that it is mutually believed that $q$

So the question is what evidence is necessary and sufficient to strengthen the weakest link in certain 'weak mutual beliefs'.

### 3.3  Empirical Support of the Strengthening Principle

It was suggested above that the evidence behind nested beliefs of the complexity of (3) is not sufficient to establish a mutual belief. That it is a *necessary* condition can be seen from the following example.

(11)  1. A: Bus 6 leaves from platform G3.
      2. B: Platform G3.
      3. A: That's, uh,...., yeah that's right.
      4. B: Excuse me?

With utterance 4, $B$ indicates that he has difficulty understanding utterance 3, which $A$ intended to provide positive feedback on utterance 2. Hence $A$ does not have the evidence required by clause 2 of the SP, and $A$ cannot ground anything at the end of

---

[3] This equivalence depends on the assumption known in epistemic logic as the Introspection axiom. According to this assumption, an agent believes his own beliefs, and in this case an agent also believes that he has a certain goal when he in fact has that goal. A precondition $c_i$ of a dialogue act performed by $A$ is always a property of $A$'s state of beliefs and goals, hence $A$ *believes that* $c_i$ is equivalent to $c_i$. Moreover, all dialogue participants may be assumed to operate according to this assumption, hence $B$ *believes that* $A$ *believes that* $c_i$ is equivalent to $B$ *believes that* $c_i$.

this dialogue fragment. Note that this example illustrates what we said above in relation to Traum's assumption that grounding acts do not require feedback in order to contribute to grounding. Utterance 3 is a counterexample to that assumption.

What evidence is necessary and sufficient for strengthening the weakest link in a weak mutual belief, is an empirical question. The case of (11) represents empirical support for the necessity of evidence of the complexity of (3); we claim that empirical data in fact suggest that the evidence of correct understanding and adoption that supports the beliefs represented in (3) is also *sufficient* for strengthening the weak mutual belief in (6). We express this claim as a pragmatic principle which we call (12) the *Strengthening Principle (SP)*:

(12)  a. A dialogue participant strengthens the weak belief link in a 'weak mutual mutual belief' concerning a precondition of a dialogue act that he has performed, when (1) he believes that the corresponding utterance was correctly understood; (2) he has evidence that: the other participant (a) also believes that, and (b) has evidence that they both have evidence that (1) and (2a) are the case.

   b. Like clause a., replacing "precondition of" by "task-related information, offered by", and replacing "correctly understood" by "believed".

Empirical support of the SP is formed by data showing how the evidence, mentioned in the SP, is created by the various types of dialogue acts and how that influences the grounding of certain beliefs. The dialogue fragments (13) – (16) provide such support.

Example (13) is the fragment (2) continued with utterance 6. According to the SP, $A$ has grounded the information that bus 6 leaves from platform G3 after utterance 4, and assumes $B$ to do the same after utterance 5. Indeed, utterance 6, which provides the already grounded information once more and as such contradicts the SP, would not be a felicitous continuation of the dialogue.

(13)  1. A: The next bus is at 11:02.
      2. B: At 11:02.
      3. A: That's correct.
      4. B: Okay thanks.
      5. A: You're welcome.
      6. A: *So that's at 11:02.

The next example illustrates that it would be infelicitous to continue after utterance 5 by asking whether the information from utterance 1 has come across, which according to the SP has already been grounded. (And this is not due to the change in topic that happens in 2, for the same would occur if utterances 2-3 are replaced by the corresponding ones of example (13).)

(14)  1. A: The next train is at 11:02.
      2. B: And do you know the arrival time?
      3. A: It arrives in Amsterdam at 12.24.
      4. B: Thanks.

5. A: You're welcome.
6. A: *You got that?

In example (15), $B$ continues after the assumed grounding by utterance 5 by expressing doubt about the grounded belief. $B$ could very well express such doubts in his previous turn, as (16) illustrates, but it is too late for that now.

(15)  1. A: The next train is at 11:02.
      2. B: At 11:02.
      3. A: That's correct.
      4. B: Okay thanks.
      5. A: You're welcome.
      6. B: *I thought it would be at 11:08.

(16)  1. A: The next train is at 11:02.
      2. B: At 11:02.
      3. A: That's correct.
      4. B: I thought it would be at 11:08.

Since the only difference between (15) and (16) is the feedback that has been given by utterances 4 and 5, it must be the case that the evidence of underst anding and adoption provided by these utterances makes the difference for grounding.

The support for the SP that we are considering here is empirical in the sense that examples of patterns like (13)-(15) are not found in actual dialogue, whereas we do find examples like (16). The latter is obviously the case; the former is empirically verifiable by searching through dialogue corpora. We have used for this purpose a corpus of air-port information (human-human telephone) dialogues; a corpus of interactive assistance dialogues [13]; and a corpus of train information (human-computer) dialogues [14]. In the human-human dialogues we did not find any counterexamples to the predictions of the SP. In spoken human-computer dialogue, it may be argued that normal input-output conditions do not obtain in view of current limitations of computer speech recognition and understanding. Indeed, in such dialogues we find more abundant feedback than the SP would predict, with unusually frequent explicit feedback from the dialogue system. While this may be reasonable given the system's limited capabilities, this feedback be-haviour is experienced by users as often unnatural and inappropriate. For a more com-plete discussion of how the various types of dialogue acts facilitate, speed up, or delay grounding in dialogue, due to the evidence they provide of the speaker being understood or believed, see [15]).

## 4   Concluding Remarks

In this paper we have presented a simple, empirically based and computationally attrac-tive model of grounding in dialogue. Central to our account is the Strengthening Princi-ple (SP), a pragmatic principle for strengthening weak mutual beliefs, created through the assumed understanding and acceptance of what is said in cooperative dialogue when normal input-ouput conditions obtain. We showed that violations of the SP lead to infe-licitous dialogue behaviour, and checked the predictions of the SP against three corpora of human-human and human-computer dialogues. In human-human dialogues we find no counterexamples, only support for the SP. In spoken human-computer dialogues we find more abundant feedback from the computer than the SP would predict, which can be attributed to violation of the normal input-output conditions assumed by the SP.

The model of grounding, of which we have outlined the theoretical and empirical basis in this paper, has been implemented as part of the Dialogue Manager module in a speech-based information-extraction system (see [16]), proving the consistency and computational feasability of the model.

# References

1. Bunt, H.: Dialogue pragmatics and context speci.cation. In: Bunt, H., Black, W., (eds.) Abduction, Belief and Context in Dialogue. Benjamins, Amsterdam, pp. 81–150 (2000)
2. Clark, H.: Using Language. Cambridge University Press, Cambridge, UK (1996)
3. Colombetti, M.: Formal semantics for mutual belief. Arfti.cial Intelligence 62, 341–353 (1993)
4. Clark, H., Schaefer, E.: Contributing to discourse. Cognitive Science 13, 259–294 (1989)
5. Brennan, S.: The grounding problem in conversations with and through computers. In: Fussell, S., Kreuz, R. (eds.) Social and cognitive psychological approaches to interpersonal communication, pp. 201–225. Lawrence Erlbaum, Hillsdale, NJ (1998)
6. Cahn, J., Brennan, S.E.: A psychological model of grounding and repair in dialog. In: Cahn, J. (ed.) Proc. AAAI FAll Symposium on Psychological Models of Communication in Collaborative Systems, pp. 25–33. AAAI, North Falmouth, MA (1999)
7. Paek, T., Horvitz, E.: Toward a formal theory of grounding. Technical report MSR–TR–2000–40. Microsoft Research, Redmond, WA (2000)
8. Traum, D.: A Computational Theory of Grounding in Natural Language Conversation. PhD Thesis. Dep. of Computer Science, University of Rochester (1994)
9. Matheson, C., Poesio, M., Traum, D.: Modelling grounding and discourse obligations using update rules. In: Proceedings NAACL (2000)
10. Allen, J., Perrault, C.: Analyzing intention in dialogues. Artificial Intelligence 15(3), 143–178 (1980)
11. Bunt, H., Morante, R., Keizer, S.: An emprically based computational model of grounding in dialogue (2007) (Submitted paper)
12. Searle, J.: Speech acts. Cambridge University Press, Cambridge, UK (1969)
13. Geertzen, J., Girard, Y., Morante, R.: The diamond project. In: Proc. 8th Workshop on the Semantics and Pragmatics of Dialogue (CATALOG), Barcelona (2004)
14. OVIS: Dialogue corpus, http://www.let.rug.nl/vannoord/Ovis/
15. Morante, R.: Computing meaning in interaction. PhD Thesis (2007) (forthcoming)
16. Keizer, S., Morante, R.: Dialogue simulation and context dynamics for dialogue management. In: Proc. NODALIDA Conference, Tartu, Estonia (2007)

# A Spoken Dialog System for Chat-Like Conversations Considering Response Timing

Ryota Nishimura[1], Norihide Kitaoka[2], and Seiichi Nakagawa[1]

[1] Department of Information and Computer Sciences, Toyohashi University of Technology,
Japan
{nishimura, nakagawa}@slp.ics.tut.ac.jp
[2] Graduate School of Information Science, Nagoya University, Japan
kitaoka@sp.m.is.nagoya-u.ac.jp

**Abstract.** If a dialog system can respond to a user as naturally as a human, the interaction will be smoother. In this research, we aim to develop a dialog system by emulating the human behavior in a chat-like dialog. In this paper, we developed a dialog system which could generate chat-like responses and their timing using a decision tree. The system could perform "collaborative completion," "*aizuchi*" (back-channel) and so on. The decision tree utilized the pitch and the power contours of user's utterance, recognition hypotheses, and response preparation status of the response generator, at every time segment as features to generate response timing.

## 1 Introduction

Recently, interfaces using automatic speech recognition (ASR) have been developed. In traditional systems, however, there was no reaction to the user during a user utterance, so, the user could not know whether or not the system was hearing the utterance. Therefore, a spoken dialog system gave a *stiff* impression.

In Japanese human-human dialog, well-timed responses such as '*aizuchi*' (sometimes called 'back-channel') and turn-taking make for smooth dialog.

The purpose of this study is to generate a natural response including *aizuchi*, collaborative completions, and turn taking considering response timing. To generate the response timing, we use a decision tree with features related to prosodic information and surface linguistic information. Using this timing generation method, we have been developing a human-friendly spoken dialog system [1]. One of our system's goals is to become very familiar to users so that humans will chat with it.

## 2 Previous Literature on Chat-Like Conversation

The properties of *aizuchi* and turn-taking has been studied [2,3,4,5]. The results indicate that pitch (F0) and power are mainly related to generating *aizuchi* and turn-taking. Some real-time *aizuchi* generation systems developed so far [6,7,8] use pitch (i.e., inverse of fundamental frequency (F0)) and pause duration as features. Some natural turn-taking

V. Matoušek and P. Mautner (Eds.): TSD 2007, LNAI 4629, pp. 599–606, 2007.

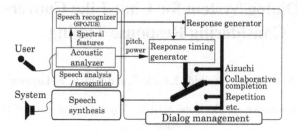

**Fig. 1.** Schematic diagram of dialog system

timing detection systems have also been developed [9,10,11]. Fujie et al. [12], for example, used prosody information, especially F0 and power of the utterance, in order to determine the appropriate timing of the feedback. They use a finite state transducer-based speech recognizer to determine the sentence of the feedback earlier than the end of the utterance. These previous studies dealt with an individual kind of response.

In this paper, we propose a unified approach for generation of various kinds of responses including *aizuchi*, and collaborative completions, considering their timing, mainly based on prosodic information. We previously proposed a method to generate *aizuchi* and turn-taking timing [1], but the approach needed pause detection and thus could not deal with overlapping responses. The new system proposed here is not pause detection-driven, but analyzes user utterances continuously even while the user is speaking. This will enable the system to deal with not only the overlapping *aizuchi* and turn-taking, but also the other responses such as collaborative completion.

## 3   Dialog System

To make spoken dialog systems comply with the above phenomena, we designed the novel system architecture shown in Figure 1. In this section, we introduce an overview of a developed system in the weather information domain.

### 3.1   Speech Analysis and Recognition

The speech recognizer SPOJUS [13] recognizes a user's input. SPOJUS outputs intermediate hypotheses in real-time. We used a vocabulary of 300 words including city names, dates, types of weather, fillers etc., with word class information. Simultaneously, the system analyzes the input to extract prosodic information, such as pitch (F0) and power, using a prosodic analyzer [12,14].

### 3.2   Response Generator

The response generator prepares response sentences using an ELIZA-like method [15] with slot-based history management in addition to recognition hypotheses. Thus, the response generator also serves as a simple dialog manager. A template set of responses is prepared for each dialog act; *aizuchi*, collaborative completions, repetition and other ordinary responses. These templates are used in parallel, so multiple patterns of responses are generated simultaneously.

Our current system deals with '*aizuchi*', 'collaborative completions', 'repetition' and other ordinary responses. Thus, four patterns of response sentences are prepared in parallel. Even while the user is speaking, the response generator continuously updates the responses using the intermediate hypotheses generated by the speech recognizer. Default sentences are also prepared and randomly selected to respond even if no appropriate sentences are prepared by the templates. During a dialog, not only the keywords included in the user utterances but also the current status of the weather extracted from a web site (http://www.imocwx.com/) are kept in the slots and used for response generation.

### 3.3  Response Timing Generator

The response generator only constructs response sentences. To output the sentence, the response timing generator selects an appropriate sentence with the appropriate timing. This timing generator makes a decision to respond or not and which response the system should make using a decision tree. Details are presented in Section 4.

### 3.4  Speech Synthesizer

To output responses by speech, we use the recorded human voice or speech synthesizer voice. GalateaTalk [16] is used for the speech synthesizer, which can controll speaker type, voice tone, speech rate, etc.

## 4  Response Timing Generation

### 4.1  Features for Timing Generation

According to [2] and [5], the contour patterns of pitch and power are related to the timing generation. For example, when pitch and/or power contours of the mora at the end of an utterance follow some proper patterns, the conversational partner's *aizuchi* or turn-taking is triggered. Thus, the first-order regression coefficients of pitch and power sequences in the last three regions of utterances obtained using 55-ms length sliding window with 30-ms overlap (total length is 105 ms) as shown in Figure 2 are used. The longer region also includes the information that triggers responses, so pitch/power contours in the last 500 ms are also used. To describe such patterns, we adopted first-order regression coefficients for 100 ms-length segments with no overlaps. The coefficients of the five continuous segments express the pattern. Such coefficients can be calculated with very small computational cost, and thus the calculation can be done in real time.

'Repetition' and 'collaborative completion' occur when a keyword of the conversation topic is input by the user [17]. When the speaker is afraid the hearer cannot catch up with him/her (imagine that the speaker tells the hearer a telephone number), the speaker divides an utterance into some 'fragments'. In such cases, the hearer often uses 'repetitions' of the fragments to indicate his/her understanding. To imitate this behavior, the timing generator should detect keywords in user utterances. In the recognition results

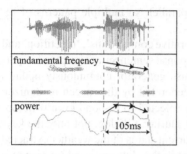

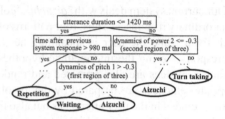

**Fig. 2.** Regression coefficients of fundamental frequency and power at the end of an utterance

**Fig. 3.** Part of the decision tree

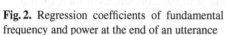

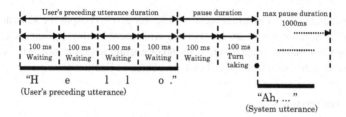

**Fig. 4.** Response timing generated by the decision tree

(or intermediate hypotheses), the attribute of the word is attached to each word, and this information is useful to detect keywords. The task of our system is weather information, so attributes of the keywords include a place-name, date, weather in a topical place, etc. We used the attribute of the last word of hypotheses (or intermediate hypotheses) as a feature.

The following features are used in consideration of the above.

- Duration from the start time of the user's preceding utterance
- Elapsed time from the end of the previous user utterance
- Elapsed time from the end of the previous system utterance
- Pitch/power contour of the last 100 ms (consisting of three values)
- Pitch/power contour of the last 500 ms (consisting of five values)
- Attribute of the last word of the last recognition results (or current intermediate hypotheses)

### 4.2    Response Timing Generation Using Decision Tree

Previously, we proposed a decision tree-based timing generator [1], but it can only treat the response at after the end of user utterances. We modified this method to enable it to generate overlapping responses by scanning every segment whenever the user speaks. A part of the decision tree is shown in Figure 3.

The response timing generator decides response timing as well as the selection of a response sentence from responses prepared by the response generator, using a decision

**Fig. 5.** Example of a dialog between the system and a user

tree based on the features introduced in Section 4.1. The information on whether or not the response contents have been prepared by the response generator is also used as a feature. Features are input to the decision tree every 100 ms. The decision tree selects a dialog act, which the system should do at that moment, from *aizuchi*, collaborative completion, repetition, ordinary response, and *wait*, as illustrated in Figure 4. "*Wait*" means not to output any response. The frequency of the responses except *aizuchi* and repetition is limited to one for one user utterance.

The RWC corpus [18] is used to train the decision tree for *aizuchi*, turn-taking and *wait*. RWC has 48 conversations of about 10-minute durations each for a total of 6.5 hours. It consists of 16,399 utterances. The conversation tasks are 'car sales' and 'overseas trip planning'. The speaker on one side is a professional salesperson, and the questioner / customer on the other side is one of 12 men and women. C4.5 is used for machine learning.

As for the other phenomena; repetitions and collaborative completions, there were not enough training data in the corpus. So we added some rules manually referring to [2,5,8]. For example, 'repetition' occurs when two seconds or more have elapsed from the latest response of the system, and when the last word in the recognition hypothesis is a city name.

The system has some exceptional rules to continue the dialog; for example, the system prompts the user to say something after a long pause (6 seconds in our system). And, when a pause of over 1000 ms occurs after the last user utterance, the system responds to the user without depending on the tree.

## 5   Example of Dialog with the System

An example of dialog with the system is shown in Figure 5. The top shows the user utterances, and the bottom the system responses.

In Figure 5, first the system prompted a start-up utterance. Then, the user said "Hello" and the system also said "Aha, Hello." Next, the system said today's weather to lead the user to the topic of weather. The system obtained the place where the user was (default value) and the weather around there, and kept the information in slots. With the next user utterance "Recently, it often rains, doesn't it?", the system's **collaborative completion** "rains, doesn't it." was **overlapped**. The system detected the keywords/key phrases "saikin (recently)" and "ame (rain)", and knew that it had been raining. So, the

**Table 1.** Evaluation of response timing by subjective evaluation

|          | Timing        | too early | early | good | late | too late | outlier | naturalness(%) |
|----------|---------------|-----------|-------|------|------|----------|---------|----------------|
| aizuchi  | decision tree | 0         | 6     | 61   | 20   | 11       | 2       | 61.0           |
|          | corpus        | 14        | 26    | 58   | 2    | 0        | 0       | 58.0           |
| turn-taking | decision tree | 9      | 26    | 53   | 9    | 2        | 1       | 53.0           |
|          | corpus        | 7         | 31    | 51   | 10   | 0        | 1       | 51.0           |

system predicted that the user would say phrases that meant "it *often* rains" and tried to synchronize to the user with "ooi (many)". The system has some response templates for collaborative completion and activates one of them if the user utterance and the current slot information meet a certain condition written as a decision rule. With the next user utterance "How about the weather in Hamamatsu?", the system detected a keyword "Hamamatsu (city name)" and responded immediately by the way of **repetition**. Then the system replied regarding the weather in Hamamatsu; "It always rains." This dialog contained some chat-like dialog-specific phenomena such as *aizuchi*, repetition, and collaborative completion. Such phenomena often occur in human-human dialogs when the dialogs warm up.

As shown above, our proposed system could work with many kinds of phenomena appearing in natural human-human spoken dialog including overlapping utterances when given appropriate rules, templates and parameters.

## 6   Experiments and Results

### 6.1   Evaluation of Timing Generation

We subjectively evaluated the naturalness of the timing generated by the generator. Here, only *aizuchi* and turn-taking are evaluated, because phenomena with few occurrences such as repetition and collaborative completion have not been sufficiently investigated so far.

To evaluate the timing of the generator, we prepared samples of *aizuchi* and turn-taking whose timing is generated by the decision tree.

We inserted an *aizuchi* extracted from a side of a dialog at the *aizuchi* timing point generated by our timing generator. We also made a samples of turn-taking. Thus, we picked some filled pauses such as "Ettodesune" ("Well ... let's see" in English), which is often employed at the beginning of an utterance, to insert at the time of system formation. Subjects listened to the inserted *aizuchi* with one preceding sentence and evaluated only the timing.

We compared the timing by the generator to that in the corpus. In real dialogs of the corpus, responses may have some meaning consistent with the dialog context and the meaning may make subjects feel natural, especially in the case of turn-taking. To make subjects evaluate only the timing, we also replace the *aizuchi* or filler pauses of the real response with *aizuchi* or a filled pause extracted from other parts of the dialog, as in the case of the generator. The number of samples is 20 for each phenomena. The five subjects heard these sample voices and answered questionnaires (1: too early; 2: early; 3: good; 4: late; 5: too late; and 0: outlier).

The results are shown in Table 1. The "naturalness" in the table indicates the rate of "good." In the table, the naturalness of the decision tree timing is comparable to the naturalness of the corpus (that is, human-human dialog) timing.

### 6.2  Evaluation of Dialog System

The subjects used and evaluated the dialog system with the timing generator. There are four kinds of systems: combinations of using / not using overlap response and using recorded / synthesized voice. After using them, the subjects answered a questionnaire. By comparing recorded voices with synthesized ones, we reveal whether or not there is a difference in the evaluation of timing according to a difference in voice quality. We required to subjects so that subjects focused on the evaluation of "timing" and "overlapping".

According to the results of the questionnaire, two of five subjects prefer the system using *overlap* (including "barge-in"). One said that he could confirm that the system listened to his utterance. As for familiarity with the system, four of five subjects felt the system using *overlap* was 'very good' or 'good' on the basis of a five-grade evaluation from 'very good' to 'very bad'. However, an irrelevant response caused by immature speech recognition and dialog management made subjects feel uncomfortable. When listening to the real responses in the corpus, the subjects felt good for 74% and 80% of *aizuchi* and turn-taking, respectively. Compared with Table 1, this reveals that the contents affect the naturalness of timing. As for speech quality, four of five subjects prefer the recorded human voice. The evaluation of the recording voice is better, even though the timing generator has the same performance. This means that the difference in the naturalness of voice quality influences the evaluation of timing. In fact, the evaluation of timing only is difficult, and the voice quality is also unconsciously related to the evaluation.

## 7  Conclusions

In this paper, we developed a dialog system utilizing real-time response generation and response timing generation, to perform a chat-like friendly conversation. The naturalness of the decision tree-based timing generator was comparable to humans, and the behavior of the dialog system gives a user-friendly impression.

In the future, we will train the decision tree using the dialogs between human and the system. We will also adopt prosodic synchrony to make the system response more natural.

## Acknowledgments

The prosodic analyzer used in our system was designed by Dr. Masataka Goto at the National Institute of Advanced Industrial Science and Technology (AIST), and implemented by Dr. Shinya Fujie at Waseda University.

# References

1. Takeuchi, M., Kitaoka, N., Nakagawa, S.: Timing detection for realtime dialog systems using prosodic and linguistic information. In: Speech Prosody 2004, pp. 529–532 (2004)
2. Koiso, H., Horiuchi, Y., Tutiya, S., Ichikawa, A., Den, Y.: An analysis of turn-taking and backchannels based on prosodic and syntactic features in Japanese map task dialogs. Language and Speech 41(3-4), 291–317 (1998)
3. Geluykens, R., Swerts, M.: Prosodic cues to discourse boundaries in experimental dialogues. Speech Communication 15, 69–77 (1994)
4. Hirschberg, J.: Communication and prosody: functional aspects of prosody. Speech Communication 36, 31–43 (2002)
5. Ohsuga, T., Nishida, M., Horiuchi, Y., Ichikawa, A.: Investigation of the relationship between turn-taking and prosodic features in spontaneous dialogue. In: Proceedings of Eurospeech2005, pp. 33–36 (2005)
6. Ward, N., Tsukahara, W.: Prosodic features which cue back-channel responses in English and Japanese. Journal of Pragmatics 32, 1177–1207 (2000)
7. Okato, Y., Kato, K., Yamamoto, M., Itahashi, S.: Insertion of interjectory response based on prosodic information. In: IEEE Workshop Interactive Voice Technology for Telecommunication Applications (IVTTA-96), pp. 85–88 (1996)
8. Noguchi, H., Den, Y.: Prosody-based detection of the context of backchannel responses. In: Proceedings of ICSLP-98, pp. 487–490 (1998)
9. Sato, R., Higashinaka, R., Tamoto, M., Nakano, M., Aikawa, K.: Learning decision tree to determine turn-taking by spoken dialogue systems. In: ICSLP-02, pp. 861–864 (2002)
10. Hirasawa, J., Nakano, M., Kawabata, T., Aikawa, K.: Effects of system barge-in responses on user impressions. EUROSPEECH-99 3, 1391–1394 (1999)
11. Kamm, C., Narayanan, S., Dutton, D., Ritenour, R.: Evaluating spoken dialogue systems for telecommunication services. In: Eurospeech-97, pp. 2203–2206 (1997)
12. Fujie, S., Fukushima, K., Kobayashi, T.: Back-channel feedback generation using linguistic and nonlinguistic information and its application to spoken dialogue system. In: Interspeech-05, pp. 889–892 (2005)
13. Kai, A., Nakagawa, S.: A frame-synchronous continuous speech recognition algorithm using a top-down parsing of context-free grammar. 257–260 (1992)
14. Goto, M., Itou, K., Hayamizu, S.: A real-time filled pause detection system for spontaneous speech recognition. In: Eurospeech-99, pp. 227–230 (1999)
15. Weizenbaum, J.: ELIZA — a computer program for the study of natural language communication between man and machine. Communications of the ACM 9(1), 36–45 (1965)
16. Kawamoto, S., Shimodaira, H., Nitta, T., Nishimoto, T., Nakamura, S., Itou, K., Morishima, S., Yotsukura, T., Kai, A., Lee, A., Yamashita, Y., Kobayashi, T., Tokuda, K., Hirose, K., Minematsu, N., Yamada, A., Den, Y., Utsuro, T., Sagayama, S.: Open-source software for developing anthropomorphic spoken dialog agent. In: Ishizuka, M., Sattar, A. (eds.) PRICAI 2002. LNCS (LNAI), vol. 2417, pp. 64–69. Springer, Heidelberg (2002)
17. Ishizaki, M., Den, Y.: Danwa to taiwa. Tokyo Daigaku Shuppankai (in Japanese) (2001)
18. Tanaka, K., Hayamizu, S., Yamasita, Y., Shikano, K., Itahashi, S., Oka, R.: Design and data collection for a spoken dialogue database in the real world computing program. In: Proc. ASA-ASJ Third Joint Meeting, pp. 1027–1030 (1996)

# Digitisation and Automatic Alignment
# of the DIALOG Corpus:
# A Prosodically Annotated Corpus of Czech Television
# Debates

Nino Peterek[1], Petr Kaderka[2], Zdeňka Svobodová[2], Eva Havlová[2],
Martin Havlík[2], Jana Klímová[2], and Patricie Kubáčková[2]

[1] Charles University, MFF, Prague
Institute of Formal and Applied Linguistics (ÚFAL)
http://ufal.ms.mff.cuni.cz
[2] The Academy of Sciences of the Czech Republic,
Czech Language Institute (ÚJČ)
http://www.ujc.cas.cz

**Abstract.** This article describes the development and automatic processing of
the audio-visual DIALOG corpus. The DIALOG corpus is a prosodically annotated corpus of Czech television debates that has been recorded and annotated at
the Czech Language Institute of the Academy of Sciences of the Czech Republic.
It has recently grown to more than 400 VHS 4-hour tapes and 375 transcribed
TV debates. The described digitisation process and automatic alignment enable
an easily accessible and user-friendly research environment, supporting the exploration of Czech prosody and its analysis and modelling. This project has been
carried out in cooperation with the Institute of Formal and Applied Linguistics
of Faculty of Mathematics and Physics, Charles University, Prague. Currently
the first version of the DIALOG corpus is available to the public (version 0.1,
http://ujc.dialogy.cz). It includes 10 selected and revised hour-long
talk shows.

## 1 Introduction

The main goal of the digitisation of the DIALOG corpus [1,2] was to develop a computational environment suitable for the structured and fast exploration and research of
prosodic aspects of spoken language. There are many phonetic programs with a deep
prosodic analysis like Praat [3] or ESPS/waves+ [4] that support many levels of annotations and generation of F0, energy and formant plots. These instruments are ideal
for individual research based on a smaller amount of speech material, but in connection with the need for a cooperative prosody database development, the exploration
of a large amount of prosody data and generation of statistics or automatic learning
methods, it was necessary to develop a specific multimedia database system with easy
access to structured data, the Dialogy.Org [5,6]. The basic component of this system is
a corpus of audio-visual data recorded and annotated at the Czech Language Institute
of the Academy of Sciences of the Czech Republic beginning in 1997, the DIALOG
corpus [1,2].

V. Matoušek and P. Mautner (Eds.): TSD 2007, LNAI 4629, pp. 607–612, 2007.

## 1.1  Original Data Storage

Television discussion programmes, such as political debates or entertainment talk shows, are transcribed as plain texts, where speaker turns (incl. speaker overlaps) and basic prosodic events like pauses, final cadences or stress are marked (Table 1).

All processed multimedia data were available only on VHS tapes, which are difficult to access. We have decided to digitise all the tapes into audio-visual files directly accessible from the DIALOG database.

## 1.2  Original Structure of the Transcription

The first paragraph includes the name of the TV debate, the date of recording and the names and initials of the speakers. The rest of the transcription is a sequence of paragraphs representing speakers' turns, each beginning with the speaker's initials.

**Table 1.** Transcription Marks [7] Used in the DIALOG Corpus

| Mark | Description |
|---:|---|
| VM: | speaker identification |
| N1, N2, N3: | unknown speakers |
| Z1, Z2, Z3: | reporters |
| XX: | unrecognised speaker |
| R-: | reporters' speech |
| TH: to já [nevím] | |
| VB: [v té] knize | speech overlaps |
| = | "latching" between turns |
| | (one follows immediately after another) |
| zase | emphatically stressed syllable/word |
| a:le: | phone lengthening/stretching |
| e e: hm ehm mhm em | hesitation and response sounds |
| (.), (..), (...) | short, medium and long pause |
| ? | high rise of final cadence |
| , | slight rise or slight rise and fall of final cadence |
| . | fall of final cadence |
| mon- | unfinished word |
| <uhodil> | speech with laugh |
| (ale) | probable, not precisely recognised word |
| ( ) | not recognised word |
| ((směje se)) | comments of the transcriber |

Turns are speech segments that end with a symbol describing the segment's final tone contour (rising tone, slightly rising or slightly rising-and-falling tone and falling tone). Another labelled prosodic event is stress. Every syllable perceived by the transcriber as emphatically stressed is underscored. The last prosodic event is pause. Three different subjective lengths are labelled.

## 2    Digitisation and Programming of the DIALOG Corpus

In the process of digitisation we have used two parallel methods. VHS tapes, with audio-visual quality usually described as 2MBit/s stream, were replayed by Thomson video player and distributed through the composite signal into a personal computer with an "ATI Wonder ALL" TV grabbing card. This computer recorded all the tapes (usually of the 4-hour length) to a hard drive with the help of "ATI Multimedia Centre" in the MPEG2 video format and 2MBit/s stream (576x486, MP3 sound, 192kBit/s, 44kHz, 16bit). At the same time, the audio signal was recorded onto another computer that was used only for audio recording without loss of data (MS WAV format and 22.05kHz, 16bit, mono). The storage need was about 5GBytes per one 4-hour VHS tape (4.4GB - MP2 audio-video + 600MB WAV audio). We have recorded 150 VHS tapes which currently occupy 750GB storage space. Table 2 lists the main media formats of the DIALOG corpus.

**Table 2.** Media Streams in the DIALOG Corpus

| Stream Description | Formats and Conversion Tools |
|---|---|
| Audio | MP3 (195kBit/s), MP3 (32kBit/s), WAV [8] |
| Audio-visual | MPEG2 (480x576), FLV (320x240) [8] |
| F0 stream | fundamental frequency in voiced regions [9,3] |
| Energy stream | audio dynamic measurements [9,3] |

### 2.1    Text Processing

At present there are 375 shows transcribed in plain text format (with the exception of the underscored stress). To convert this format to the full XML description or a database format, where each word or mark has an identifier and is included in an unambiguous hierarchical structure that is composed of shows and turns, it was necessary to process all the Word DOC files into the XML format using Antiword [10] and then to interpret their annotation, to be easily converted into CSTS ("Czech sentence tree structure") or MySQL table format [11]. CSTS format is necessary for a morphological analyser [12,13] that enriches the transcriptions with POS tags and detailed morphological properties.

### 2.2    Transcription Alignment

All the transcribed data are in the form of plain texts without any time alignment. Only speakers' turns, overlaps and some prosodic events are labelled. The prosodic corpus, to be usable, should have time alignment at the phone level. It enables a fast searching of audio-visual files and computing prosodic statistics over phone strings. Such alignment can be achieved through a technique called forced alignment. It uses HMMs and the Viterbi algorithm to find word or phoneme boundaries, but it is usually restricted only to shorter segments without overlaps, background noise and music because most HMM speech recognition systems consist only of models of context-dependent phonemes, the model of short pauses and the model of longer pauses (even those containing background noise).

The DIALOG corpus has recently grown to more than 400 VHS 4-hour tapes and 375 transcribed TV debates. Therefore it has to be aligned automatically. The solution consists in new models for other non-speech events of arbitrary length (with loop) that can be included into our forced alignment algorithm. These models support the alignment of long speech segments (1 hour in one step) and can handle many non-speech segments. Segments with overlapping speech cannot usually be used for current research because these voices and their F0's are inseparable. We want to have these segments marked (bounded) to exclude probable errors.

### 2.3   Alignment Evaluation

The baseline set of HMM models comes from our Czech speech recognition system [14]. These models have been built using HTK, the hidden Markov model toolkit [15]. The models are continuous density HMMs, trained on approximately 10 hours of read speech from 22 speakers taken from a subset of the CUCFN corpus [14]. The speech feature parametrisation was mel-frequency cepstra, including both delta and delta-delta sub-features; cepstral mean subtraction was applied to all features on a per utterance basis. We work with simpler context-independent models (monophones), which are powerful enough for our alignment task. The first alignment of the whole show occurs with the above-mentioned models. After segmentation into words, the show is divided into turns that serve as the material for the retraining of our acoustic models. After three HTK retraining cycles we arrive at models most suitable for alignment of the show. Such models are very good in the detection of boundaries of turns, words and phones. The evaluation was done on one selected show (Table 3).

**Table 3.** Forced Alignment Evaluation for Show SEDM41 (1 hour)

| Boundary Type | Controlled Segment | Accuracy (%) | Count | # of Misplaced |
|---|---|---|---|---|
| Turns | 1 hour | 99.6 | 460 | 2 |
| Words | 5 min | 98.4 | 906 | 14 |
| Phones | 5 min | 95.7 | 4216 | 181 |

The boundary detection in segments with overlapped speech or with background noise is surprisingly good (but these segments are excluded from the training process). The results of the alignment can be also displayed in the Praat phonetic program [3]. Our alignment process generates labelled files with time marks for turns, words and phones in ESPS/waves+ format [4]. Praat displays all three transcription tiers at once, along with the sound curve, its pitch, intensity contour and many other prosodic details.

## 3   User Interface to the Corpus

One of the aims of the user is to find special prosody contexts in a large amount of material. The search web page of the system Dialogy.Org [6] is designed to enable the easy construction of queries by specifying word forms, lemmas, POS tags, morphological tags, sex of the speaker, name of the show, date of the show, etc. The query form uses the Dialogy.Org system based on the MySQL relational database [11]. The chosen structure of the tables enables the user to track history and calculate statistics of each show and speaker. The relational table structure is described in Figure 1.

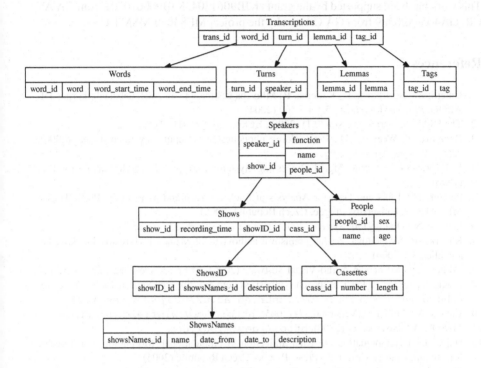

**Fig. 1.** DIALOG Corpus Database Structure

The output of the search query is a list of turns with marked positions matching the search parameters. Every context can be heard and explored in detail by clicking on the head of the selected turn. The summary statistics are optionally printed at the top of the search results page.

At present the first version of the DIALOG corpus is available to the public (version 0.1, http://ujc.dialogy.cz). It includes 10 selected and revised hour-long talk shows.

## 4    Conclusion

The digitised and automatically aligned version of the DIALOG corpus provides easy access to prosodic data with the options of searching for special linguistic contexts, listening to them and computing their statistics. We hope that the larger version of the corpus will provide enough information for further prosodic research and tools for checking linguistic hypotheses and for the preparation of training data of prosodic models usable in text-to-speech and speech recognition systems.

## Acknowledgments

This work has been supported by the grants KJB9061304, KJB900610701 from GA AV ČR, GA405/06/0589 from GA ČR and by the project ME838 of MŠMT ČR.

## References

1. Čmejrková, S., Jílková, L., Kaderka, P.: Mluvená čeština v televizních debatách: korpus DI-ALOG. Slovo a slovesnost 65, 243–269 (2004)
2. The DIALOG corpus (version 0.1) (2006), http://ujc.dialogy.cz
3. Boersma, P., Weenink, D.: Praat: doing phonetics by computer (Version 4.4.X) (2006), http://www.praat.org
4. ESPS/waves+, Entropic Signal Processing System. Entropic Research Laboratory Ltd. (1996)
5. Peterek, N.: Tools and Data for Analysis of Spoken Czech and its Prosody. Ph.D. Thesis, MFF Charles University, Prague, Czech Republic (2006)
6. Peterek, N.: Dialogy.Org System (2006), http://www.dialogy.org
7. Kaderka, P., Svobodová, Z.: Jak přepisovat audiovizuální záznam rozhovoru? Jazykovědné aktuality, 43 (2006)
8. MPlayer and MEncoder Audio-Visual Software (2006), http://www.mplayerhq.hu
9. Black, A., Taylor, P.: Festival speech synthesis system & Edinburgh Speech Tools. University of Edinburgh (1999), http://www.cstr.ed.ac.uk/projects/festival
10. van Os, A.: Antiword (Version 0.37) (2005), http://www.winfield.demon.nl
11. MySQL Database Server (2006), http://dev.mysql.com
12. Hajič, J.: Disambiguation of Rich Inflection (Computational Morphology of Czech). Karolinum, Charles University Press, Prague, Czech Republic (2004)
13. Hajič, J.: Morphological Tagging: Data vs. Dictionaries. In: 6th ANLP Conference / 1st NAACL Meeting. Proceedings, Seattle, Washington, pp. 94–101 (2000)
14. Byrne, W., Hajič, J., Ircing, P., Jelinek, F., Khudanpur, S., McDonough, J., Peterek, N., Psutka, J.: Large Vocabulary Speech Recognition for Read and Broadcast Czech. In: Matoušek, V., Mautner, P., Ocelíková, J., Sojka, P. (eds.) TSD 1999. LNCS (LNAI), vol. 1692, pp. 235–240. Springer, Heidelberg (1999)
15. Young, S., Kershaw, D., Odell, J., Ollason, D., Valtchev, V., Woodland, P.: HTK Book. Entropic Research Laboratory Ltd. (1999), http://htk.eng.cam.ac.uk

# Setting Layout in Dialogue Generating Web Pages

Luděk Bártek, Ivan Kopeček, and Radek Ošlejšek

Faculty of Informatics, Masaryk University
Botanická 68a, 602 00 Brno
Czech Republic
{bartek, kopecek, oslejsek}@fi.muni.cz

**Abstract.** Setting layout of a two-dimensional domain is one of the key tasks of our ongoing project aiming at a dialogue system which should give the blind the opportunity to create web pages and graphics by means of dialogue. We present an approach that enables active dialogue strategies in natural language. This approach is based on a procedure that checks the correctness of the given task, analyses the user's requirements and proposes a consistent solution. An example illustrating the approach is presented.

## 1 Introduction

Giving blind people access to the Internet, which is nowadays one of the most important sources of information for them, is supported by various assistive information technologies - screen readers, special web browsers [2,20,21], standards of accessible web [1,6,7,15], etc. Having one's own web presentation is for blind and visually impaired people more and more useful, advantageous and to some extent also a matter of prestige. However, for most blind users, creating their own web presentation is a not simple task and they usually need the help of a specialist. This is motivation for developing a dialogue system which gives blind users the opportunity to create their own web presentation by means of dialogue. This project also involves developing procedures enabling the blind user to create and manipulate computer graphics, at least to some extent.

In this paper, we present the concept and outline of the procedure that enables the user to put their requirements on the layout of a web page or graphical composition in a natural language without priory limitations. We also outline the architecture of the system and present a simple example of generating a web page layout.

## 2 Structure and Architecture of the System

The basic procedure of generating a web presentation within our system [3] consists of the following steps:

1. Selecting the type of the presentation: the system distinguishes several types of different presentations, such as personal page, blog, photo slide show, product

V. Matoušek and P. Mautner (Eds.): TSD 2007, LNAI 4629, pp. 613–620, 2007.

presentation, etc. Each kind of presentation specifies a set of elements that are included within the presentation (e.g. name, address, hobbies, links to relevant sites, etc.)

2. Defining the presentation content: the user specifies the content of the elements included in the presentation. The specific form of the dialogue used to perform this step depends on the presentation type selected in the previous step.
3. Selecting the web page layout: the user can choose either an existing predefined layout or create a new layout as described later in the paper.

Technically, the layouts are represented in the form of XML Stylesheets [14] that are used to transform the XML data that describe the web site into a set of HTML/XHTML files and Cascading Stylesheets [5] used to add a style (e.g. fonts, colors, spacing, positioning, etc.) to a document.

The whole dialogue system for generating web-based presentations is based on VoiceXML technology [16]. Its subsystem for layout generation consists of the following modules:

- Dialogue interface: this module is implemented using VoiceXML, Speech Recognition Grammar Specification (SRGS) and Semantic Interpretation for Speech Recognition (SISR) standards [18].
- Database of layouts: this database can be embedded as well as distributed over the Internet. The embedded database is advantageous for standalone computers. The distributed solution allows layouts to be shared among a large group of users. The only limitation of the database engine is that the JDBC [10] driver for the particular database must exist.
- Description of a data structure for each website type: this information is stored in the form of an XML Schema [19] that allows the description of a web site to be validated as well as extracting the information about the structure of the web site. Both actions can be easily performed using standard Java classes [9].

The system itself is implemented as a combination of Java Server Pages, Java Beans and servlets [8]. This solution offers robustness and enables distributed storage of the existing presentation layouts. The schema of the system is shown in Fig. 1.

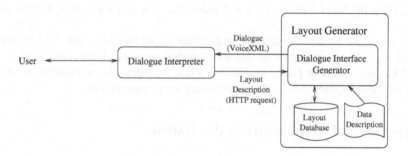

**Fig. 1.** Schema of the system

# 3    Generating a Layout

We assume that our goal is to generate a layout within a fixed rectangular area, which is referred to in this paper as a *composition frame*. Composition frame can represent a web page, an internal frame of a page, a canvas for image composition, etc. Its size can be defined either absolutely, in pixels for example, or relatively, as an aspect ratio. Composition frames contain elements representing pictures, text blocks, tables and subframes, etc. These elements are used to create the layout and are considered to have also a fixed rectangular size defined either absolutely in pixels or relatively in percentages of the composition frame.

To restrict ourselves to rectangular areas and elements is a reasonable simplification; if some elements are not rectangular, we can always construct a boundary rectangle around it to approximate the original non-rectangular shape to the rectangular one.

The users can choose several strategies to set the elements in the composition frame. They can specify the position of the elements either exactly or approximately. The layout generator has to compute an acceptable layout taking into consideration space restrictions appointed by individual elements. For example, if the user specifies that element $A$ is on the left of the element $B$, then the system has to compute a convenient configuration satisfying this space requirements. Approximate placement can be given in one of the following forms:

- The approximate position is specified in the context of the whole frame. For instance, the user requires a location near the lower-left corner, close to the center, etc.
- The user defines the mutual position of two elements (e.g. element $A$ is above element $B$). As a result, we get a system of spatial constraints. The mutual position can also take a more "fuzzy" form, e.g. elements $A$ and $B$ are placed side-by-side. In this case, no exact order is defined and the positioning system has more possibilities for configuring the final layout.

All approaches of the layout description can be mixed, i.e. some elements can be fixed in exact positions, others approximately located in the frame, and some of them linked by relations. The goal of the system is to solve this complex task. As we will show in what follows, the core of this task can be solved by an optimization process based on given requirements and simple formatting principles. Let us present here a brief outline of this approach.

Assume that set $R = \{R_i, i = 1..n\}$ of $n$ rectangles is given, each having width $w_i$ and height $h_i$, and the object is to allocate them orthogonally, without overlapping, to a given rectangle $S$, representing a composition frame, by respecting a given set of conditions $F$ describing their absolute or relative position, and satisfying a global condition $G$. It is assumed that the rectangles have a fixed orientation, i.e., they cannot be rotated. In other words, our goal is to find coordinates $(x_i, y_i)$ of the centers of the given rectangles so that the above-mentioned conditions hold.

For our purposes, we propose the following three-stage algorithm. First we try, by some heuristics, to directly eliminate the cases that are not solvable. For instance, we can check whether the sum of the areas of the rectangles belonging to $R$ does not exceed the area of $S$, or check the antisymmetry property of the binary relation "on-the-left" is not violated, etc.

Second, we try to get one solution that satisfies the set of conditions $F$. If such a solution does not exist, the problem is not solvable. This stage is similar to the two-dimensional knap-sack problem [12], but differs in the presence of the conditions $F$. Because the number of the rectangles can be assumed to be small, we can adopt a version of the exhaustive search in which most of the possible cases to be examined are eliminated by the conditions belonging to $F$. Even though this algorithm is potentially exponential, the computation is for most practical cases a real-time problem, and some possible exceptions can be eventually handled by requiring the user to add some more pieces of information.

Finally, we use the resulting solution as a starting point for an optimization procedure, aiming to slightly rearrange the obtained configuration of the rectangles in such a way that the conditions from $F$ stay true while condition $G$, formulated as an objective function, will be reached in the sense of finding at least a local optimum for $G$. Generally, the meaning of the function $G$, for our purposes, reads as "find a reasonable configuration for the rectangular objects fulfilling the user's wishes about their relative or absolute position". Such a condition is necessary, otherwise we have no criterion for determining the final positions of the objects. Of course, we could also simply omit this and, as the final configuration, take the starting solution - this would, however, ignore basic formatting and aesthetic principles.

To keep the problem as simple as possible, we choose $G$ in the form that can be expressed by the following requirement: minimize the sum of squares of the distances between the boundary lines of the neighboring rectangles, respecting the given requirements. In other words, the minimization procedure keeps the distances between the boundary lines of the neighboring rectangles as uniform as possible, but it does not affect the distances in the cases in which it would influence the position of the rectangles that is determined by the user's requirements.

Once the layout is generated, the blind user can check it by means of *navigation grid*. Navigation grid [11,13] represents a simplified navigation concept providing the user a sufficient feedback via a dialogue.

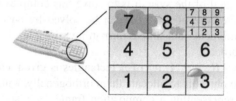

**Fig. 2.** Recursive navigation grid

Navigation grid divides the composition frame into the nine identical rectangular sectors. Each sector can be subdivided recursively. The user can traverse the grid and be informed about the situation in any chosen sector. The user can also select the scale of the navigation and set one or more sectors as a working space. Smaller sectors at the lower levels provide more detailed information than the higher-level ones. Sectors of the navigation grid can be numbered lexicographically for rapid reference, i.e. the sector 6 has sub-sectors 6.1,6.2, etc.

## 4   Example

Instead of going into more technical details, we present a very simple example. This example demonstrates the process of the web page layout generation via a dialogue.

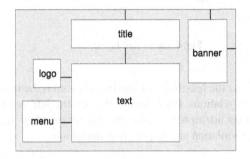

**Fig. 3.** Scheme of the test page with named elements and relations between them

Assume that we have five elements, each element uniquely named and having proportions as shown in Fig. 3. First, the user specifies a spacial constraints of the future layout. These constrains are suggested as lines between elements in Fig. 3. In what follows, $U$ is the user, $S$ is the system:

```
U: Put the title on the top of the page, the banner in
   the upper right corner, the text below the title,
   the menu and logo on the left of the text.
S: Accepted. Location of the logo and menu remains ambiguous.
U: Put the menu below the logo.
S: Accepted. The layout is unambiguous now. Generate the layout?
U: Yes.
S: Page layout created successfully.
```

The system checks the spatial constraints permanently, searching possible configurations. If the system discovers that some constraint is ambiguous then the user is immediately informed. On the contrary, as soon as the constraints make the layout unambiguous the system should offer the automatic layout generation. In the previous dialogue, the logo and menu are ambiguous because they are both defined to be on the left of the text but their mutual position is not specified. Therefore, either the menu can be located above the logo or vice versa. Final layout produced by the dialogue is shown in Fig. 4.

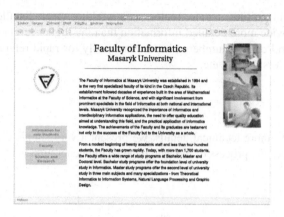

**Fig. 4.** Generated layout

Another way of ensuring the feasibility of requested layout is the inspection of properties of the related binary relations. For example, the "on-the-left" relation is transitive and asymmetric. If the user attempts to violate the asymmetry then the system should inform the user about the violation and require new definition:

```
U: Put the logo on the left of the text, the text on the
   left of the banner.
S: Accepted. The layout remains ambiguous.
U: Put the banner on the left of the logo.
S: Invalid location: The banner is already on the right
   side of the logo. Location ignored.
...
```

The banner is located on the right of the logo due to the transitivity. The attempt to locate it on the left therefore violates the asymmetry.

Automatic generation of the layout optimizes spreading of the elements on the web page area by refining spaces between the elements. The menu and logo in Fig. 4 are therefore uniformly distributed in horizontal direction, which makes them lying in the middle of the page. To make them located on the top of the page, the user has to adjust their location:

```
U: Put the title on the top of the page, the banner in
   the upper right corner, the text below the title,
   the menu and logo on the left of the text.
S: Accepted. Location of the logo and menu remains ambiguous.
U: Put the menu below the logo.
S: Accepted. The layout is unambiguous. Generate the layout?
U: No. Put the logo in the upper left corner and the menu
   close to the logo.
S: Accepted. Layout is unambiguous. Generate the layout?
U: Yes.
S: Page layout created successfully.
```

**Fig. 5.** Adjusted layout with menu and logo on the top

Bold text in the dialogue emphasizes the part where the user moves the logo and menu upward. The "close" relation instructs the system to omit the relevant space from the optimization algorithm distributing the elements uniformly. Instead, the space between the elements is kept small. Final layout is shown in Fig. 5.

## 5   Conclusions and Future Work

This paper presents the problem of generating the layout of web pages and graphical compositions by means of dialogue while briefly discussing a proposed procedure to manage this task.

Our future work involves thoroughly testing the system and enhancing the efficiency of its modules. As for testing, we are collaborating with the Teiresias Centre (Support Centre for Students with Special Needs) of Masaryk University. A major task for future research is to apply aesthetic criteria for the generated web presentations.

## Acknowledgement

The authors are grateful to James Thomas for proofreading a draft of this paper and to the students and staff of the Support Centre for Students with Special Needs of Masaryk University for their advices, support and collaboration. This work has been supported by Grant Agency of Czech Republic under Project No. 201/07/0881.

## References

1. Authoring Tool Accessibility Guidelines 1.0, available at
   http://www.w3.org/TR/ATAG10/
2. Bártek, L., Kopeček, I.: Adapting Web-based Educational Systems for the Visually Impaired. In: Proceedings of the International Workshop on Combining Intelligent and Adaptive Hypermedia Methods/Techniques in Web-Based Education Systems, Salzburg, pp. 11–16 (2005)

3. Bártek, L., Kopeček, I.: Web Pages for Blind People — Generating Web-based Presentations by means of Dialogue. In: Proceedings of 10th International Conference on Computers Helping People with Special Needs, Linz, Austria, pp. 114–119 (July 2006)
4. Brewer, J.: Web accessibility highlights and trends. In: ACM International Conference Proceeding Series. In: Proceedings of the 2004 international cross-disciplinary workshop on Web accessibility (W4A), New York City, pp. 51–55 (2004)
5. Cascading Style Sheets, available at http://www.w3.org/Style/CSS/
6. Engelen, J., Evenepoel, F., Wesley, T.: The World Wide Web: A Real Revolution in Accessibility. In: Proceedings of ICCHP1996, Linz, Austria, pp. 17–22 (July 1996)
7. HTML Techniques for Web Content Accessibility Guidelines 1.0, available at http://www.w3.org/TR/WCAG10-HTML-TECHS/
8. Java EE at a Glance, available at http://java.sun.com/javaee
9. Java SE at a Glance, available at http://java.sun.com/javase
10. Java SE - Java Database Connectivity (JDBC), available at http://java.sun.com/javase/techno-logies/database/index.jsp
11. Kamel, H.M., Landay, J.A.: Sketching images eyes-free: a grid-based dynamic drawing tool for the blind. In: Proceedings of the fifth international ACM conference on Assistive technologies, pp. 33–40. ACM Press, New York (2002)
12. Kellerer, H., Pferschy, H., Pisinger, U.: Knapsack Problems. Springer, Heidelberg (2004)
13. Kopeček, I., Ošlejšek, R.: The Blind and Creating Computer Graphics. In: Proceedings of the Second IASTED International Conference on Computational Intelligence, Anaheim, Calgary, pp. 343–348 (2006)
14. The Extensible Stylesheet Language Family (XSL), available at http://www.w3.org./Style/XSL/
15. User Agent Accessibility Guidelines 1.0, available at http://www.w3.org/TR/UAAG10/
16. Voice Extensible Markup Language (VoiceXML) 2.0 (2004), available at http://www.w3.org/TR/,/REC-voicexml20-20040316/
17. W3C HTML Home Page, available at http://www.w3.org/MarkUp/
18. W3C Voice Browser Activity, available at http://www.w3.org/Voice
19. W3C XML Schema, available at http://www.w3.org/XML/Schema
20. Walshe, E., McMullin, B.: Browsing Web Based Documents Through an Alternative Tree Interface: The WebTree Browser. In: Proceedings of 10th Internationtional Conference on Computers Helping People with Special Needs, Linz, Austria, July, pp. 106–113 (2006)
21. Webbie and the Accessible Programs, available at http://www.webbie.uk/

# Graph-Based Answer Fusion
# in Multilingual Question Answering[*]

Rita M. Aceves-Pérez, Manuel Montes-y-Gómez, and Luis Villaseñor-Pineda

Laboratorio de Tecnologías del Lenguaje,
Instituto Nacional de Astrofísica, Óptica y Electrónica, México
{rmaceves, mmontesg, villasen}@inaoep.mx

**Abstract.** One major problem in multilingual Question Answering (QA) is the combination of answers obtained from different languages into one single ranked list. This paper proposes a new method for tackling this problem. This method is founded on a graph-based ranking approach inspired in the popular Google's PageRank algorithm. Experimental results demonstrate that the proposed method outperforms other current techniques for answer fusion, and also evidence the advantages of multilingual QA over the traditional monolingual approach.

## 1 Introduction

Question Answering (QA) has become a promising research field whose aim is to provide more natural access to textual information than traditional document retrieval techniques. In essence, a QA system is a kind of search engine that responds to natural language questions with concise and precise answers.

One major challenge that currently faces this kind of systems is the multilinguality. In a multilingual scenario, it is expected for a QA system to be able to: (*i*) answer questions formulated in various languages, and (*ii*) look for the answers in several collections in different languages.

Evidently, multilingual QA has some advantages over standard monolingual QA. In particular, it allows users to access much more information in an easier and faster way. However, it introduces additional challenges caused by the language barrier.

A multilingual QA system can be described as an ensemble of several monolingual systems [5], where each system works over a different –monolingual– document collection. Under this schema, two additional tasks are of great importance: first, the translation of questions to the target languages, and second, the combination or fusion of the extracted answers into one single ranked list.

The first problem, namely, the translation of the questions from one language to another, has been widely studied in the context of cross-language QA[1] [1, 10, 11, 14]. In contrast, the second task, i.e., the fusion of answers obtained from different

---

[*] Work done under partial support of CONACYT (Project grand 43990).
[1] Cross-language QA is a special case of multilingual QA. It addresses the situation where questions are formulated in a language different from that of the (single) target collection.

V. Matoušek and P. Mautner (Eds.): TSD 2007, LNAI 4629, pp. 621–629, 2007.
© Springer-Verlag Berlin Heidelberg 2007

languages, has only recently been addressed [2]. Nevertheless, it is important to mention that there is considerably work on combining lists of monolingual answers extracted by different QA systems [4, 6, 12] as well as on integrating lists of documents in several languages for cross-lingual information retrieval applications [13].

In this paper, we propose a new method for tackling the problem of answer fusion in multilingual QA. The proposed method is founded on a graph-based ranking approach inspired in the popular Google's PageRank algorithm [3]. Similar to previous related approaches, this method also centers on the idea of taking advantage of the redundancy of several answer lists. However, it models the fusion problem as a kind of recommendation task, where an answer is recommended or "voted" by similar answers occurring in different lists. In this model, the answer receiving the greatest number of votes is the one having the greatest relevance, and therefore, it is the one selected as the final answer.

The rest of the paper is organized as follows. Section 2 shows the architecture of a multilingual QA system and describes some previous works on answer fusion. Section 3 describes the proposed method. Section 4 shows some experimental results. Finally, section 5 presents our conclusions and outlines future work.

## 2  Related Work

Figure 1 shows a common architecture for a multilingual QA system. This architecture includes, besides the set of monolingual QA systems, a stage for question translation and a module for answer fusion.

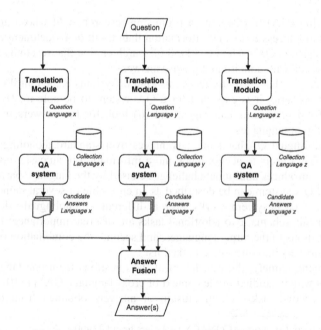

**Fig. 1.** General architecture of a multilingual QA system

As we previously mentioned, the problem of question translation has already been widely studied. Most current approaches rest on the idea of combining the capacities of several translation machines. They mainly consider the selection of the best instance from a given set of translations [1, 11] as well as the construction of a new query reformulation by gathering terms from all of them [10, 14].

On the other hand, the problem of answer fusion in multilingual QA has only very recently been addressed by [2]. This work compares a set of traditional ranking techniques from cross-language information retrieval in the scenario of multilingual QA.

In addition, there is also some relevant related work on combining lists of monolingual answers. For instance, [6] proposes a method that performs a number of sequential searches over different document collections. At each iteration, this method filters out or confirms the answers calculated in the previous step. [4] describes a method that applies a general ranking over the five-top answers obtained from different collections. They use a ranking function that is inspired in the well-known RSV technique from cross-language information retrieval. Finally, [12] uses various search engines in order to extract from the Web a set of candidate answers for a given question. It also applies a general ranking over the extracted answers, nevertheless, in this case, the ranking function is based on the confidence of search engines instead that on the redundancy of individual answers.

The method proposed in this paper is similar in spirit to [2, 4] in that it also applies a general ranking over the answers extracted from different languages, and it is comparable to [6] in that it performs an iterative evaluation process. However, our method uses a novel graph-based approach that allows taking into consideration not only the redundancy of answers in all languages but also their original ranking scores in the monolingual lists.

## 3  Proposed Method

The aim of the answer fusion module is to combine the answers obtained from all languages into one single ranked list. In order to do that, we propose using a graph-based ranking approach. In particular, we decide adapting the Google's PageRank algorithm[2] [3].

In short, a graph-based ranking algorithm allows deciding on the importance of a node within a graph, by taking into account global information recursively computed from the entire graph, rather than relying only on local node-specific information. In other words, this kind of ranking model put into practice the idea of voting or recommendation, where the node having the greatest number of votes is considered the most relevant one, and therefore, it is selected as the system's final output.

The application of this approach to the problem of multilingual answer fusion consists of the following steps:

1. Construct a graph representation from the set of extracted answers.
2. Iterate the graph-based ranking algorithm until convergence.

---

[2] It is important to mention that this algorithm has recently been used in other text processing tasks such as text summarization and word sense disambiguation [7, 8].

3. Sort nodes (answers) based on their final score and select the top-ranked as the system's response.

The following sections describe in detailed these steps. In particular, section 3.1 explains the proposed graph representation, and section 3.2 presents the graph-based ranking function.

## 3.1 Graph Representation

Formally, a graph $G = (V, E)$ consists of a set of nodes $V$ and a set of edges $E$, where $E$ is a subset of $V \times V$.

In our case, each node represents a different answer. This way, we will have as many nodes as answers obtained from the different languages (that is, $|V| = |A|$).

Each node $v_i \in V$ contains a set of content words $\{w_1,...,w_n\}$ that describes an specific answer $a_i \in A$. In particular, we consider two levels of representation for nodes.

***Direct representation:*** In this case, the set of content words is directly extracted from the corresponding answer. For instance, given the Spanish answer $a_j =$ "*1 de enero de 1994*", its related node will be $v_j = \{1,$ enero, 1994$\}$.

***Extended representation:*** In order to make comparable the answers obtained from different languages, we extend the node representations by considering the answer's translations to all languages. For instance, if we are working with Spanish, French and Italian, then answer $a_j$ will be represented by the node $v_j = \{1,$ enero, 1994, janvier, gennaio$\}$.

The initial weight $s_\pi$ of a node $v_i$ is calculated in accordance with the ranking position of answer $a_i$ in its original answer list ($r(a_i)$):

$$s_\pi^{\ 0}(v_i) = 110 - (10 \times r(a_i)) \tag{1}$$

Using this formula, the answers at the first positions –of each language– will have a weight of 100, the second ones a weight of 90, and so on.

On the other hand, the edges of the graph establish a relation between two different answers. They mainly indicate that the answers are associated, i.e., that they share at least one content word. Obviously, the greater the number of common words between them, the greater their association value. Based on the last consideration, the weight $s_\sigma$ of an edge $e_{ij}$ between the nodes $v_i$ and $v_j$ is calculated as follows:

$$s_\sigma(e_{ij}) = \frac{|v_i \cap v_j|}{|v_i \cup v_j|} \tag{2}$$

where $|v_i \cap v_j|$ indicates the number of common content words of nodes $v_i$ and $v_j$, and $|v_i \cup v_j|$ is the number of different words in both nodes.

Figure 2(a) shows the graph representation of the set of answers for the question "*When did the NAFTA come into effect?*". In particular, this graph includes answers in three different languages (Spanish, Italian and French), and uses the extended representation of nodes.

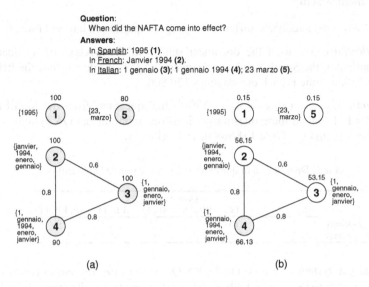

**Question:**
When did the NAFTA come into effect?
**Answers:**
In Spanish: 1995 **(1)**.
In French: Janvier 1994 **(2)**.
In Italian: 1 gennaio **(3)**; 1 gennaio 1994 **(4)**; 23 marzo **(5)**.

(a)                    (b)

**Fig. 2.** Example of a graph representation for answer fusion

## 3.2 Ranking Function

The ranking algorithm computes the scores of nodes in line with: (*i*) the number of their neighbor nodes, (*ii*) the initial weight of these neighbors (refer to formula 1), and (*iii*) the strength of their links (refer to formula 2). Therefore, the idea behind this algorithm is to reward the answers that are strongly associated to several other top-ranked responses.

Formula 3 denotes the proposed ranking function. As can be noticed, it defines the ranking algorithm as an iterative process, that –following the suggestions by Mihalcea [8]– must break off when the change in the score of one single node be less than a given specified threshold.

$$s_\pi^m(v_i) = (1-d) + d \times \left( \sum_{v_j \in adj(v_i)} \frac{s_\sigma(e_{ij})}{\sum_{v_k \in adj(v_j)} s_\sigma(e_{jk})} s_\pi^{m-1}(v_j) \right) \tag{3}$$

In this formula, $s_\pi^m(v_i)$ is the score of the node $v_i$ after $m$ iterations, $s_\sigma(e_{ij})$ is the weight of the edge between nodes $v_i$ and $v_j$, and $adj(v_i)$ is a function that indicates the set of adjacent nodes to $v_i$.

Figure 2(b) shows the final state of the example graph after performing the ranking process. In this case, the selected answer (top-ranked node) for the question *"When did the NAFTA come into effect?"* is *"1 gennaio 1994"*.

## 4 Experimental Evaluation

### 4.1 Experimental Setup

*Languages.* We considered three different languages: Spanish, Italian and French.

*Search Collections.* We used the document sets from the QA@CLEF evaluation forum. In particular, the Spanish collection consists of 454,045 documents, the Italian one has 157,558, and the French one contains 129,806.

*Test questions.* We selected a subset of 170 factual questions from the MultiEight corpus of CLEF. From all these questions at least one monolingual QA system could extract the correct answer. Table 1 shows their distribution.

**Table 1.** Distribution of questions (by answer source language)

|  | \multicolumn{7}{c}{Answers in:} | | | | | | |
|---|---|---|---|---|---|---|---|
|  | SP | FR | IT | SP, FR | SP, IT | FR, IT | SP, FR, IT |
| *Questions* | 37 | 21 | 15 | 20 | 25 | 23 | 29 |
| *Percentage* | 21% | 12% | 9% | 12% | 15% | 14% | 17% |

*Monolingual QA system.* We used the TOVA QA system [9]. Its selection was supported on its competence to deal with all considered languages. It obtained the best precision rate for Italian and the second best ones for Spanish and French in the CLEF-2005 evaluation exercise.

*Translation Machine.* For all translation combinations we used Systran[3].

*Evaluation Measure.* In all experiments we used the precision as evaluation measure. It indicates the general proportion of correctly answered questions. In order to enhance the analysis of results we show the precision at one, three and five positions.

*Baseline.* We decided using as a baseline the results from the best monolingual system (the Spanish system in this case). This way, it is possible to conclude about the advantages of multilingual QA over the standard monolingual approach. In addition, we also present the results corresponding to other fusion techniques[4] [2].

### 4.2 Results

In order to evaluate the usefulness of the proposed method, we considered the top-ten ranked answers from each monolingual QA system. Table 2 shows the results obtained when using the direct and extended graph representations. The conclusions from these results are the following.

1. Combining answers extracted from different languages sources makes possible to respond a large number of questions. In other words, multilingual QA allows improving the performance of the standard monolingual approach.

---

[3] www.systranbox.com
[4] This comparison is possible because both experiments used the same set of questions as well as the same target document collections.

2. The proposed approach is pertinent for the task of multilingual answer fusion. In particular, using the extended representation leads to a better performance (14% of improvement over the baseline), since it allows better capturing the redundancy of answers across different monolingual answer lists.

**Table 2.** Precision achieved by the proposed graph-based approach

| Method's configuration | Precision at: | | |
|---|---|---|---|
| | 1st | 3rd | 5th |
| Using the direct node representation | 0.45 | 0.62 | 0.72 |
| Using the extended node representation | **0.48** | **0.68** | **0.78** |
| Best Monolingual Run (baseline) | 0.45 | 0.57 | 0.64 |

On the other hand, table 3 compares the results of the proposed method with those obtained by a set of traditional ranking techniques from cross-language information retrieval (for details on these techniques refer to [2]). This table indicates that the graph-based method outperforms all previously used techniques for answer fusion in multilingual QA. We believe this is mainly because our graph-based ranking approach not only takes into consideration the redundancy of the answers into the different languages, but also makes a better use of their original ranking scores in the monolingual lists.

**Table 3.** Comparison of several ranking techniques in multilingual QA

| Method | Precision at: | | |
|---|---|---|---|
| | 1st | 3rd | 5th |
| Graph-based approach | **0.48** | **0.68** | **0.78** |
| RSV | 0.44 | 0.61 | 0.69 |
| RoundRobin | 0.45 | **0.68** | 0.74 |
| CombSum | 0.42 | 0.66 | 0.75 |
| CombMNZ | 0.42 | 0.62 | 0.70 |

Finally, table 4 shows the evaluation results corresponding to the set of questions having their answers in more than one collection. As it was expected, the answer fusion approach had a greater impact on this subset. It is also important to notice that for this particular subset the extended node representation was much better than the direct one (10% of improvement at five positions). We consider this is because the extended representation betters capture the redundancy of answers in different languages.

**Table 4.** Precision on questions with answers in more than one collection

| Method's configuration | Precision at: | | |
|---|---|---|---|
| | 1st | 3rd | 5th |
| Using the direct node representation | 0.52 | 0.71 | 0.79 |
| Using the extended node representation | **0.54** | **0.79** | **0.89** |

# 5  Conclusions

This paper proposed a new method for tackling the problem of answer fusion in multilingual QA. This method is founded on a graph-based ranking algorithm that allows combining the answers obtained from different languages into one single ranked list. The algorithm takes into consideration not only the redundancy of answers but also their original ranking scores in the monolingual lists.

Experimental results showed that the proposed method is pertinent for this task. It outperforms the best monolingual performance as well as the results obtained using other current techniques for answer fusion.

As noticed from table 4, the precision at five positions is considerably greater than that for the first ranked answer. We believe this behavior is consequence of having several incorrect answer translations. In order to reduce the errors on these translations we plan to apply, as future work, some techniques for combining the capacities of several translation machines [1]. It is important to point out that the proposed scheme allows easily integrating several translations for each answer, and therefore, incrementing the possibility of retrieving the correct one.

# References

1. Aceves-Pérez, R., Montes-y-Gómez, M., Villaseñor-Pineda, L.: Enhancing Cross-Language Question Answering by Combining Multiple Question Translations. In: Gelbukh, A. (ed.) CICLing 2007. LNCS, vol. 4394, Springer, Heidelberg (2007)
2. Aceves-Pérez, R., Montes-y-Gómez, M., Villaseñor-Pineda, L.: Fusión de Respuestas en la Búsqueda de Respuestas Multilingüe. Procesamiento de Lenguaje Natural, 38 (2007)
3. Brin, S., Page, L.: The Anatomy of a Large-Scale Hypertextual Web Search Engine. Computer Networks and ISDN Systems, 30(1-7) (1998)
4. Chu-Carroll, J., Czuba, K., Prager, A.J., Ittycheriah, A.: Question Answering, Two Heads are Better than One. In: Conference of the North American Chapter of the Association for Computational Linguistics on Human Language Technology (vol. 1). Canada (2003)
5. García-Cumbreras, M.A., Ureña-López, L. A., Martínez-Santiago, F., Perea-Ortega, J.M.: BRUJA System: The University of Jaén at the Spanish Task of CLEFQA 2006. Working Notes of CLEF 2006. Alicante, España (2006)
6. Jijkoun, V., Mishne, G., Rijke, M., Schlobach, S., Ahn, D., Muller, K.: The University of Amsterdam at QA@CLEF 2004. In: Peters, C., Clough, P.D., Gonzalo, J., Jones, G.J.F., Kluck, M., Magnini, B. (eds.) CLEF 2004. LNCS, vol. 3491, Springer, Heidelberg (2005)
7. Mihalcea, R.: Graph-Based Ranking Algorithms for Sentence Extraction Applied to Text Summarization. In: 42$^{nd}$ Annual Meeting of the Association for Computational Linguistics (ACL-2004). Barcelona, Spain (2004)
8. Mihalcea, R., Tarau, P.: TextRank: Bringing Order into Texts. In: Conference on Empirical Methods in Natural Language Processing (EMNLP-2004). Barcelona, Spain (2004)
9. Montes-y-Gómez, M., Villaseñor-Pineda, L., Pérez-Coutiño, M., Gómez-Soriano, J.M., Sanchis-Arnal, E., Rosso, P.: INAOE-UPV Joint Participation in CLEF 2005: Experiments in Monolingual Question Answering. In: Peters, C., Gey, F.C., Gonzalo, J., Müller, H., Jones, G.J.F., Kluck, M., Magnini, B., de Rijke, M., Giampiccolo, D. (eds.) CLEF 2005. LNCS, vol. 4022, Springer, Heidelberg (2006)

10. Neumann, G., Sacaleanu, B.: DFKI's LT-lab at the CLEF 2005 Multiple Language Question Answering Track. In: Peters, C., Gey, F.C., Gonzalo, J., Müller, H., Jones, G.J.F., Kluck, M., Magnini, B., de Rijke, M., Giampiccolo, D. (eds.) CLEF 2005. LNCS, vol. 4022, Springer, Heidelberg (2006)
11. Rosso, P., Buscaldi, D., Iskra, M.: Web-based Selection of Optimal Translations of Short Queries. Procesamiento de Lenguaje Natural, 38 (2007)
12. Sangoi-Pizzato, L.A., Molla-Aliod, D.: Extracting Exact Answers using a Meta Question Answering System. In: Australasian Language Technology Workshop. Australia (2005)
13. Savoy, J., Berger, P.Y.: Selection and Merging Strategies for Multilingual Information Retrieval. In: Peters, C., Clough, P.D., Gonzalo, J., Jones, G.J.F., Kluck, M., Magnini, B. (eds.) CLEF 2004. LNCS, vol. 3491, Springer, Heidelberg (2005)
14. Sutcliffe, R., Mulcahy, M., Gabbay, I., O'Gorman, A., White, K., Slatter, D.: Cross-Language French-English Question Answering using the DLT System at CLEF 2005. In: Peters, C., Gey, F.C., Gonzalo, J., Müller, H., Jones, G.J.F., Kluck, M., Magnini, B., de Rijke, M., Giampiccolo, D. (eds.) CLEF 2005. LNCS, vol. 4022, Springer, Heidelberg (2006)

# Using Query-Relevant Documents Pairs
# for Cross-Lingual Information Retrieval*

David Pinto[1,2], Alfons Juan[1], and Paolo Rosso[1]

[1] Department of Information Systems and Computation,
Polytechnic University of Valencia, Spain
[2] Faculty of Computer Science,
B. Autonomous University of Puebla, Mexico
{dpinto, ajuan, prosso}@dsic.upv.es

**Abstract.** The world wide web is a natural setting for cross-lingual information retrieval. The European Union is a typical example of a multilingual scenario, where multiple users have to deal with information published in at least 20 languages. Given queries in some source language and a target corpus in another language, the typical approximation consists in translating either the query or the target dataset to the other language. Other approaches use parallel corpora to obtain a statistical dictionary of words among the different languages. In this work, we propose to use a training corpus made up by a set of Query-Relevant Document Pairs (QRDP) in a probabilistic cross-lingual information retrieval approach which is based on the IBM alignment model 1 for statistical machine translation. Our approach has two main advantages over those that use direct translation and parallel corpora: we will not obtain a translation of the query, but a set of associated words which share their meaning in some way and, therefore, the obtained dictionary is, in a broad sense, more semantic than a translation one. Besides, since the queries are supervised, we are working in a more restricted domain than that when using a general parallel corpus (it is well known that in this context results are better than those which are performed in a general context). In order to determine the quality of our experiments, we compared the results with those obtained by a direct translation of the queries with a query translation system, observing promising results.

## 1 Introduction

The fast growth of the Internet and the increasing multilinguality of the web poses an additional challenge for language technology. Therefore, the development of novel techniques for managing of data, especially when we deal with information in multiple languages, is needed. There are sufficient examples in which users may be interested in information which is in a language other than their own native language. A common language scenario is where a user has some comprehension ability for a given language but s/he is not sufficiently proficient to confidently specify a search request in that language. Thus, a search engine that may deal with this cross-lingual problem should be of a high benefit.

* This work has been partially supported by the MCyT TIN2006-15265-C06-04 research project, as well as by the BUAP-701 PROMEP/103.5/05/1536 grant.

V. Matoušek and P. Mautner (Eds.): TSD 2007, LNAI 4629, pp. 630–637, 2007.
© Springer-Verlag Berlin Heidelberg 2007

In Cross-Language Information Retrieval (CLIR), the usual approach is to first translate the query into the target language and then retrieve documents in this language by using a conventional, monolingual information retrieval system. The translation system might be of any type (rule-based, statistical, hybrid, etc.). For instance, in [1] and [2], a statistical machine translation system is used, but it had to be previously trained from parallel texts. See [3], [4], and [5] for a survey on CLIR.

Since our perspective, the above two-step approach is too sensitive to translation errors produced during the first step. In fact, even if we have a very accurate retrieval system, translation errors prevent correct retrieval of relevant documents. To overcome this drawback, we propose to use a set of queries with their respective set of relevant documents as an input training set for a direct probabilistic cross-lingual information retrieval system which integrates both steps into a single one. This is done on the basis of the IBM alignment model 1 (IBM-1) for statistical machine translation [6]. Probabilistic approaches which use parallel corpora in order to translate the input queries by means of a statistical dictionary in CLIR have been used from many years ago (see [2]). However, our aim is *not* to translate queries but to obtain a set of associated words for a given query. Therefore, a parallel corpus does not have sense for our purpose, since we need to find a possible set of relevant documents for each query given. To our knowledge, this novel approach has not been presented earlier in literature.

We carried out some experiments by using a subset of the EuroGOV corpus [7] which was first used in the bilingual English to Spanish subtask of WebCLEF 2005 [8]. A document indexing reduction was also proposed in order to improve precision of our approach and to diminish its storing space. The corpus reduction was based on the use of a technique for selecting mid-frequency terms, named the Transition Point (TP), which was used in other research works with the same purpose [9,10]. We evaluated four different percentages of TP observing that it is possible to improve precision by reducing the number of terms for a given corpus.

Section 2 and 3 describe the query-relevant document pairs model in detail. Section 4 introduces the corpus used in the experiments, and explains the way we implemented the reduction process. The results obtained after the evaluation are illustrated in Section 5 and discussed in Section 6.

## 2   The QRDP Probabilistic Model

Lex $x$ be a query text in a certain *input (source)* language, and let $y_1, y_2, \cdots, y_W$ be a collection of $W$ web pages in a different *output (target)* language. Let $\mathcal{X}$ and $\mathcal{Y}$ be their associated input and output vocabularies, respectively. Given a number $k < W$, we have to find the $k$ most relevant web pages with respect to the input query $x$. To do this, we have followed a probabilistic approach in which the $k$ most relevant web pages are computed as those most probable given $x$, i.e.,

$$\{y_1^*(x), \cdots, y_k^*(x)\} = \operatorname*{argmax}_{\substack{S \subset \{y_1, \cdots, y_W\} \\ |S|=k}} \min_{y \in S} p(y \mid x) \tag{1}$$

In the particular case of k=1, Equation (1) is simplified to

$$y_1^*(x) = \operatorname*{argmax}_{y=y_1,\cdots,y_W} p(y \mid x) \qquad (2)$$

In this work, $p(y \mid x)$ is modelled by using the well-known IBM alignment model 1 (IBM-1) for statistical machine translation [6,11]. This model assumes that each word in the web page is *connected to exactly one word* in the query. Also, it is assumed that the query has an initial "null" word to which words in the web page with no direct connexion are linked. Formally, a hidden variable $a = a_1 a_2 \cdots a_{|y|}$ is introduced to reveal, for each position $i$ in the web page, the query word position $a_i \in \{0, 1, \ldots, |x|\}$ to which it is connected. Thus,

$$p(y \mid x) = \sum_{a \in \mathcal{A}(x,y)} p(y, a \mid x) \qquad (3)$$

where $\mathcal{A}(x, y)$ denotes the set of all possible alignments between $x$ and $y$. The *alignment-completed* probability $p(y, a \mid x)$ can be decomposed in terms of individual, web page position-dependent probabilities as:

$$p(y, a \mid x) = \prod_{i=1}^{|y|} p(y_i, a_i \mid a_1^{i-1}, y_1^{i-1}, x) \qquad (4)$$

$$= \prod_{i=1}^{|y|} p(a_i \mid a_1^{i-1}, y_1^{i-1}, x) p(y_i \mid a_1^{i}, y_1^{i-1}, x) \qquad (5)$$

In the case of the IBM-1 model, it is assumed that $a_i$ is uniformly distributed

$$p(a_i \mid a_1^{i-1}, y_1^{i-1}, x) = \frac{1}{|x| + 1} \qquad (6)$$

and that $y_i$ only depends on the query word to which it is connected

$$p(y_i \mid a_1^{i}, y_1^{i-1}, x) = p(y_i \mid x_{a_i}) \qquad (7)$$

By sustitution of (6) and (7) in (5); and thereafter (5) in (3), we may write the IBM-1 model as follows by some straighforward manipulations:

$$p(y \mid x) = \sum_{a \in \mathcal{A}(x,y)} \prod_{i=1}^{|y|} \frac{1}{(|x| + 1)} p(y_i \mid x_{a_i}) \qquad (8)$$

$$= \frac{1}{(|x| + 1)^{|y|}} \prod_{i=1}^{|y|} \sum_{j=0}^{|x|} p(y_i \mid x_j) \qquad (9)$$

Note that this model is governed only by a *statistical dictionary* $\Theta = \{p(w|v),$ for all $v \in \mathcal{X}$ and $w \in \mathcal{Y}\}$. The model assumes that the order of the words in the query is not important. Therefore, each position in a document is equally likely to be connected to each position in the query. Although this assumption is unrealistic in machine translation, we consider the IBM-1 model is particularly well-suited for our approach.

## 3  Maximum Likehood Estimation

It is not difficult to derive an Expectation-Maximisation (EM) algorithm to perform maximum likelihood estimation of the statistical dictionary with respect to a collection of training samples $(X, Y) = \{(x_1, y_1), \ldots, (x_N, y_N)\}$. The *(incomplete)* log-likelihood function is:

$$L(\Theta) = \sum_{n=1}^{N} \log \sum_{a_n} p(y_n, a_n | x_n) \tag{10}$$

with

$$p(y_n, a_n | x_n) = \frac{1}{(|x_n| + 1)^{|y_n|}} \prod_{i=1}^{|y_n|} \prod_{j=0}^{|x_n|} p(y_{ni} | x_{nj})^{a_{nij}} \tag{11}$$

where, for convenience, the alignment variable, $a_{ni} \in \{0, 1, \ldots, |x_n|\}$, has been rewritten as an indicator vector in (11), $a_{ni} = (a_{ni0}, \ldots, a_{ni|x_n|})$, with 1 in the query position to which it is connected, and zeros elsewhere.

The so-called *complete* version of the log-likelihood function (10) assumes that the hidden (missing) alignments $a_1, \ldots, a_N$ are also known:

$$\mathcal{L}(\Theta) = \sum_{n=1}^{N} \log p(y_n, a_n | x_n) \tag{12}$$

The EM algorithm maximises (10) iteratively, through the application of two basic steps in each iteration: the E(xpectation) step and the M(aximisation) step. At iteration $k$, the E step computes the expected value of (12) given the observed (incomplete) data, $(X, Y)$, and a current estimation of the parameters, $\Theta^{(k)}$. This reduces to the computation of the expected value of $a_{nij}$:

$$a_{nij}^{(k)} = \frac{p(y_{ni} | x_{nj})^{(k)}}{\sum_{j'} p(y_{ni} | x_{nj'})^{(k)}} \tag{13}$$

Then, the M step finds a new estimate of $\Theta$, $\Theta^{(k+1)}$, by maximising (12), using (13) instead of the missing $a_{nji}$. This results in:

$$P(v|w)^{(k+1)} = \frac{\sum_n \sum_{i=1}^{|y_n|} \sum_{j=0}^{|x_n|} a_{nij}^{(k)} \delta(y_{ni}, w) \delta(x_{nj}, v)}{\sum_{w'} \sum_n \sum_{i=1}^{|y_n|} \sum_{j=0}^{|x_n|} a_{nij}^{(k)} \delta(y_{ni}, w') \delta(x_{nj}, v)} \tag{14}$$

$$= \frac{\sum_n \frac{p(w|v)^{(k)}}{\sum_{j'} p(w|x_{nj'})^{(k)}} \sum_{i=1}^{|y_n|} \sum_{j=0}^{|x_n|} \delta(y_{ni}, w) \delta(x_{nj}, v)}{\sum_{w'} \left[ \sum_n \frac{p(w'|v)^{(k)}}{\sum_{j'} p(w'|x_{nj'})^{(k)}} \sum_{i=1}^{|y_n|} \sum_{j=0}^{|x_n|} \delta(y_{ni}, w') \delta(x_{nj}, v) \right]} \tag{15}$$

for all $v \in \mathcal{X}$ and $w \in \mathcal{Y}$; where $\delta(a, b)$ is the Kronecker delta function; i.e., $\delta(a, b) = 1$ if $a = b$; 0 otherwise.

An initial estimate for $\Theta$, $\Theta^{(0)}$, is required for the EM algorithm to start. This can be done by assuming that the translation probabilities are uniformly distributed; i.e.,

$$p(w \mid v)^{(0)} = \frac{1}{|\mathcal{Y}|} \tag{16}$$

for all $v \in \mathcal{X}$ and $w \in \mathcal{Y}$.

## 4  The EuroGOV Corpus

We have used a subset of the EuroGOV corpus for the evaluation of the QRDP model. This subset was made up by a set of Spanish Internet pages, originally obtained from European government-related sites and particularly used in the WebCLEF track of the Cross-Language Evaluation Forum[1] (CLEF) [8]. A better reference to this corpus can be seen in [7].

We refined the evaluation corpus, with those documents automatically identified as in the "Spanish" language, by using the TexCat language identification program [2]. For the evaluation of this corpus, a set of 134 supervised queries in the "English" language was used. The pre-processing step was applied to both, the web pages and the queries, and consisted of the elimination of punctuation symbols, Spanish and English stopwords, numbers, html tags, script codes and cascading style sheets codes.

For convenience, we built a training corpus comprising pairs of query and target web page. We observed that a possible improvement in time indexing and search engine precision may be obtained by reducing the size of this corpus. Therefore, we applied a term selection technique, named transition point, in order to obtain only the mid-frequency terms which will represent every document (see [9] and [10] for further details).

For this purpose, a term frequency value of the web page vocabulary is selected as the transition point, and then a neighbourhood of TP is used as threshold for determining those terms which will be selected. After using four different thresholds (10%, 20%, 40%, and 60%), we obtained five corpora for the evaluation. Table 1 shows the size of every test corpus used, as well as the percentage of reduction obtained for each of them. As can be seen, the TP technique obtained a high percentage of reduction (between 75 and 89%), which also implied a time reduction for constructing the statistical dictionary.

**Table 1.** Test corpora

| Corpus | Size ($\approx$ Kb) | Reduction (%) |
|--------|--------|--------|
| Full   | 117 | 0 |
| TP60   | 29  | 75.37 |
| TP40   | 20  | 82.55 |
| TP20   | 19  | 83.25 |
| TP10   | 13  | 89.25 |

---

[1] http://www.clef-campaign.org/

[2] http://www.let.rug.nl/~vannoord/TextCat/

## 5   Evaluation of the Results

In the experiments, we used the leave-one-out procedure which is a standard procedure in predicting the generalisation power of a classifier, both from a theoretical and empirical perspective [12].

Table 2 shows the results for every run executed by applying only 10 iterations in the EM algorithm. The first column indicates the name of the run carried out for each corpus. The last column shows the Mean Reciprocal Rank (MRR) obtained for each run. Additionally, the Average Success At (ASA) different number of documents retrieved is shown. As can be seen, an improvement by using an evaluation corpus was obtained employing the TP technique with a neighbourhood of 40%, which is exactly the same percentage used in other research works (see [10] and [13]). We consider that this improvement is derived from the elimination of noisy words, which helps to rank better the web pages.

**Table 2.** Evaluation results

| Run | ASA | | | | | MRR |
| --- | 1 | 5 | 10 | 20 | 50 | |
| FULL | 0.0000 | 0.0299 | 0.0970 | 0.2687 | 0.3955 | 0.0361 |
| TP10 | 0.0149 | 0.0522 | 0.0672 | 0.0970 | 0.4030 | 0.0393 |
| TP20 | 0.0149 | 0.0299 | 0.0448 | 0.0746 | 0.4030 | 0.0323 |
| TP40 | 0.0149 | 0.0448 | 0.1045 | 0.1940 | 0.3881 | **0.0470** |
| TP60 | 0.0000 | 0.0448 | 0.1269 | 0.2164 | 0.4030 | 0.0383 |

Three teams participated at the bilingual "English to Spanish" subtask at WebCLEF in 2005. Every team submitted at least one run [14,10,15]. A comparison among the results obtained by each team and our best results can be seen in Table 3. In this case, we are presenting the results obtained with the TP40 corpus and by applying 100 iterations in the EM algorithm. Each of these teams translated each query from English to Spanish and thereafter they used a traditional monolingual information retrieval system for carrying out the searching process. Particularly, the UNED team reported two results (UNED_FULL and UNED_BODY) which are related with the information of each web page used; their first aproximation makes use of information stored in html fields or tags identified during the preprocessing, like *title, metadata, heading, body, outgoing links*. Their second aproximation (UNED_BODY) only considered the information in the *body* field. We also considered only the information inside the *body* html tag and, therefore, the UNED_BODY run can be used for comparison. On the other hand, the ALICANTE's team has used a combination of three translation systems for obtaining the best translation of a query. Thereafter, they used a passage retrieval-based system as a search engine, indexing in the documents all the information except html tags.

We may observe that by using the same information from a web page, we have slightly outperformed the results obtained by other approaches, even when we have trained our model with only 3 target web pages in average per query, and executing 100 iterations on the Expectation-Maximization model.

**Table 3.** Comparison results over 134 topics

| Run name | ASA | | | | | |
|---|---|---|---|---|---|---|
| | 1 | 5 | 10 | 20 | 50 | MRR |
| OurApproach | 0.0672 | 0.1045 | 0.1418 | 0.2164 | 0.4403 | **0.0963** |
| UNED_FULL | 0.0821 | 0.1045 | 0.1194 | 0.1343 | 0.2090 | 0.0930 |
| BUAP/UPV40 | 0.0597 | 0.0970 | 0.1119 | 0.1418 | 0.2164 | 0.0844 |
| UNED_BODY | 0.0224 | 0.0672 | 0.1045 | 0.1716 | 0.2612 | 0.0477 |
| BUAP/UPVFull | 0.0224 | 0.0672 | 0.1119 | 0.1418 | 0.1866 | 0.0465 |
| ALICANTE | 0.0299 | 0.0522 | 0.0597 | 0.0746 | 0.0970 | 0.0395 |

## 6  Conclusions

We have described a query-relevant document pairs based model for cross-language information retrieval. The QRDP model uses a statistical dictionary of associated words directly to rank documents according to their relevance with respect to the query. We consider that inaccuracies of query translation have a negative effect on document retrieval and, therefore, using the probabilistic values of association should help to overcome this problem.

The application of statistical machine translation for CLIR may be often seen in literature, but what we proposed in this paper is to study the derivation of the translation (association) dictionary from query-relevant document pairs. The probabilistic model assumes that the order of the words in the query is not important. Therefore, each position in a document is equally likely to be connected to each position in the query. Although this assumption is unrealistic in machine translation, we consider the IBM-1 model to be particularly well-suited for our approach.

We have used a term selection technique in order to reduce the size of the training corpus with good findings. For instance, by using a 82.5% of reduction, the results can improve those of using the complete corpus.

Last but not least, we would emphasize that the QRDP probabilistic model is language independent and, therefore, it can be employed to model cross-language query-document pairs in any language.

## References

1. Franz, M., McCarley, J.S., Roukos, S.: Ad-hoc and multilingual information retrieval at ibm. In: Proceedings of the TREC-7 Conference, pp. 157–168 (1998)
2. Kraaij, W., Nie, J.Y., Simard, M.: Embedding web-based statistical translation models in cross-language information retrieval. Computational Linguistics 29(3), 381–419 (2003)
3. Fuhr, N.: Probabilistic models in information retrieval. The Computer Journal 35(3), 243–255 (1992)
4. Rijsbergen, C.J.V.: Information Retrieval, 2nd edn. Dept. of Computer Science, University of Glasgow (1979)
5. Baeza-Yates, R., Ribeiro-Neto, B.: Modern Information Retrieval. ACM Press, New York, Addison-Wesley (1999)

6. Brown, P.F., Cocke, J., Pietra, S.A.D., Pietra, V.J.D., Jelinek, F., Lafferty, J.D., Mercer, R.L., Roossin, P.S.: A statistical approach to machine translation. Computational Linguistics 16(2), 79–85 (1990)
7. Sigurbjörnsson, B., Kamps, J., de Rijke, M.: Eurogov: Engineering a multilingual web corpus. In: Peters, C., Gey, F.C., Gonzalo, J., Müller, H., Jones, G.J.F., Kluck, M., Magnini, B., de Rijke, M., Giampiccolo, D. (eds.) CLEF 2005. LNCS, vol. 4022, pp. 825–836. Springer, Heidelberg (2006)
8. Sigurbjörnsson, B., Kamps, J., de Rijke, M.: Overview of webclef 2005. In: Peters, C., Gey, F.C., Gonzalo, J., Müller, H., Jones, G.J.F., Kluck, M., Magnini, B., de Rijke, M., Giampiccolo, D. (eds.) CLEF 2005. LNCS, vol. 4022, pp. 810–824. Springer, Heidelberg (2006)
9. Pinto, D., Jiménez-Salazar, H., Rosso, P.: Clustering abstracts of scientific texts using the transition point technique. In: Gelbukh, A. (ed.) CICLing 2006. LNCS, vol. 3878, pp. 536–546. Springer, Heidelberg (2006)
10. Pinto, D., Jiménez-Salazar, H., Rosso, P.: Buap-upv tpirs: A system for document indexing reduction on webclef. In: Peters, C., Gey, F.C., Gonzalo, J., Müller, H., Jones, G.J.F., Kluck, M., Magnini, B., de Rijke, M., Giampiccolo, D. (eds.) CLEF 2005. LNCS, vol. 4022, pp. 873–879. Springer, Heidelberg (2006)
11. Civera, J., Juan, A.: Mixtures of ibm model 2. In: Proceedings of the EAMT Conference, pp. 159–167 (2006)
12. Vapnik, V.N.: The Nature of Statistical Learning Theory. Springer, Heidelberg (1995)
13. Rojas-López, F., Jiménez-Salazar, H., Pinto, D.: A competitive term selection method for information retrieval. In: Gelbukh, A. (ed.) CICLing 2007. LNCS, vol. 4394, pp. 468–475. Springer, Heidelberg (2007)
14. Artile, J., Peinado, V., Peñas, A., Verdejo, F.: Uned at webclef 2005. In: Peters, C., Gey, F.C., Gonzalo, J., Müller, H., Jones, G.J.F., Kluck, M., Magnini, B., de Rijke, M., Giampiccolo, D. (eds.) CLEF 2005. LNCS, vol. 4022, pp. 888–891. Springer, Heidelberg (2006)
15. Martínez, T., Noguera, E., noz, R.M., Llopis, F.: University of alicante at the clef2005 webclef track. In: Peters, C., Gey, F.C., Gonzalo, J., Müller, H., Jones, G.J.F., Kluck, M., Magnini, B., de Rijke, M., Giampiccolo, D. (eds.) CLEF 2005. LNCS, vol. 4022, pp. 865–868. Springer, Heidelberg (2006)

# Detection of Dialogue Acts Using Perplexity-Based Word Clustering

Iosif Mporas, Dimitrios P. Lyras, Kyriakos N. Sgarbas, and Nikos Fakotakis

Artificial Intelligence Group, Wire Communications Laboratory,
Electrical and Computer Engineering Department, University of Patras,
26500 Rion, Patras, Greece
Tel.: +30 2610 996496; Fax.: +30 2610 997336
{imporas, d.lyras, sgarbas, fakotaki}@wcl.ee.upatras.gr

**Abstract.** In the present work we used a word clustering algorithm based on the perplexity criterion, in a Dialogue Act detection framework in order to model the structure of the speech of a user at a dialogue system. Specifically, we constructed an n-gram based model for each target Dialogue Act, computed over the word classes. Then we evaluated the performance of our dialogue system on ten different types of dialogue acts, using an annotated database which contains 1,403,985 unique words. The results were very promising since we achieved about 70% of accuracy using trigram based models.

## 1 Introduction

In the area of language technology the development of spoken Dialogue Systems (DSs) is an essential task. DSs are widely used in many applications, such as speech enabled systems and voice portals for security, customer services and information retrieval. Because of the complexity of the task, DS architectures exploit different interconnected modules, each one dedicated to more specific tasks [1]. In this modular structure any mistakes during the primary stages of the process, consisting of speech recognition and understanding, affect greatly the subsequent modules. In order to enhance the performance of these primary modules, statistical language modeling units for the prediction of the next Dialogue Act (DA), have been employed.

A DA represents the meaning of an utterance and can be thought of as a set of tags classifying utterances according to a combination of pragmatic, semantic and syntactic criteria [2]. Usually DA categories are specifically constructed to match each particular application; i.e. they are not domain-independent. DA prediction seems useful for several subtasks of a DS. For example a conversational agent needs to know whether it has been asked a question or ordered to do something. Moreover, DA labels would enrich the available input for higher-level processing of spoken words. Finally, DA information can provide feedback to lower-level processing, such as the speech recognition module, in order to constrain the potential recognition hypotheses and to reduce the word recognition error rate improving thus the speech recognition accuracy [3], [4]. Exploitation of this information can lead to more natural and fluent dialogues. The detection of DAs can also be very useful in dialogue

V. Matoušek and P. Mautner (Eds.): TSD 2007, LNAI 4629, pp. 638–643, 2007.

management [5], where the knowledge of the intention of the user is required for the estimation and generation of the machine response [6], [7], or even in shallow parsing applications – i.e. recognizing DAs is equivalent to understanding the utterance on a more general level [8].

There are a number of previous efforts on DA modeling and prediction. The most popular approach is the use of *n-grams* for modeling the probabilities of the corresponding DA sequences [9]. N-gram modeling has been extended using standard techniques from statistical language processing [7]. Approaches on prosodic modeling of DAs have used various speech parameters, such as duration, pitch, and energy patterns, to disambiguate utterance types [10].

In the present study we evaluate a word clustering algorithm based on perplexity measures, for the needs of DA detection. Section 2 describes the architecture of the DA detection module as well as the clustering algorithm that has been used. Section 3 presents the data used for our evaluation. Section 4 shows the experimental setup and results achieved, followed by concluding remarks in section 5.

## 2 Dialogue Act Detection

The task of Dialogue Act Recognition involves finding an interpretation of a given utterance in the context of a dialogue in terms of a dialogue act with type and semantic content.

DA modeling and detection is based on the observation that different DAs use characteristic word strings. There are distinguishing correlations between particular phrases and DA types. In order to exploit this information, statistical language models that model the word sequences $w_1...w_N$ related with each DA type are used. For each DA word sequence likelihoods are computed.

$$P_r(w_1...w_N) = \prod_{n=1}^{N} P_r(w_n \mid w_1...w_{n-1}) \qquad (1)$$

For large vocabulary applications the conditional probability is practically restricted to the immediate $m-1$ words and is referred as *m-gram*. These m-gram models are estimated from the text corpus during training, however there are word sequences that never occur in the training corpus. For this reason general smoothing distributions are used, in order to cover events not apparent during training. The approach we describe computes m-grams over word classes $g$, instead of words $w$. This algorithm, proposed by Matrin et al. [11], maps each word to a class and so even if a certain m-gram did not appear in the training corpus, it is possible that the class m-gram appeared and can be computed.

### 2.1 Word Clustering Algorithm

The algorithm maps every word to exactly one word class. Since the number of word classes is smaller than the number of words, the number of model parameters is reduced so that each parameter can be estimated more reliably. On the other hand, reducing the number of model parameters makes the model prediction less precise. To balance these two extremes, the statistical criterion of maximum likelihood or equivalently minimum perplexity $PP$ is used. The perplexity of a probability model $q$ is defined as:

$$PP = 2^{-H(p,q)} \tag{2}$$

The exponent is the cross-entropy of the empirical distribution of the test sample $p$.

$$H(p,q) = -\sum_x p(x) \log_2 q(x) \tag{3}$$

The algorithm maps each word $w$ of the vocabulary to a word class $g_w$, using a mapping function $G:w \rightarrow g_w$. The number of classes is a fixed number $G$ and for initialization the G-1 most frequent words are mapped to one class, while all the rest words are mapped to the remaining class. For each class the transition probability function $p_1(g_w|g_u)$ from class $g_w$ to class $g_u$ and the membership probability function $p_0(w|g)$ for word $w$ and class $g$ are computed. The log likelihood probability for the bigram model will be

$$p(w|u) = p_0(w|g_w) \cdot p_1(q_w|g_u) \tag{4}$$

For the trigram model the log likelihood probability will be

$$p(w|u,v) = p_0(w|g_w) \cdot p_2(q_w|g_u,g_v) \tag{5}$$

As a side effect all words with a zero unigram count $N(w)$ remain at the last word class, since they do not affect perplexity. The exchange algorithm employed between the word classes is outlined below.

```
-> Start with some initial mapping w→gw
  ->> for each word w of the vocabulary do
    ->>> for each class k do
      -tentatively exchange word w from class gw to
                     class k and update counts
      -Compute perplexity for the tentative exchange
    ->>exchange word w from class gw to class k with
                     minimum perplexity
-> do until stopping criterion is met
```

The stopping criterion is a prespecified number of iterations. In addition, the algorithm stops if no words are exchanged any more.

## 2.2 Detection of Dialogue Acts

In order to detect the DA of a current utterance two factors are computed. The first factor is the probability $p(u|DA_i)$ of an utterance $u$ to belong to each DA. On the other hand, the second factor is the probability $p(DA_k|DA_{k-1}...DA_1)$ of a DA to occur after its predecessor DAs. The detected DA is considered as the one with the maximum probability

$$DA = \arg\max_{i} \left\{ p(u \mid DA_i) \cdot p(DA_{k=i} \mid DA_{k-1}...DA_1) \right\} \tag{6}$$

The computation of the probability for each class to belong to a DA is performed using n-gram models. The models are trained on the word classes, as computed by the perplexity minimization criterion. N-gram models are also employed to compute the probability functions of appearance of each DA after its n predecessor DAs. For the first utterance unigram probability functions are used.

## 3 Database Description

The evaluation of our experiments was carried out using the 2000 Communicator Dialogue Act Tagged Database [12], (catalog number: LDC2004T15), produced by the Linguistic Data Consortium (LDC). This database contains annotations on the transcriptions of the system and user utterances as taken from the 2000 Communicator Evaluation corpus. The database includes the ability of a user, during one 10-minute session, to plan a three-leg trip, with the three flights/legs on three different days, with rental car and hotel in each of the two "away" cities, plus dictating/sending a voice-mail message. Dialogue Act annotations are provided for system utterances in the dialogues. The total number of dialogues is 648. There are 314,223 words (tokens) and 1,403,985 unique words.

The Dialogue Act dimension captures distinctions between distinct communicative goals such as requesting information (REQUEST-INFO), presenting information (PRESENT-INFO) and making offers (OFFER) to act on behalf of the caller. The types of DAs are specified and tabulated in Table 1.

**Table 1.** Types of the Dialogue Acts and examples of them

| Dialogue Act | Example |
| --- | --- |
| REQUEST-INFO | *And, what city are you flying to?* |
| PRESENT-INFO | *The airfare for this trip is 390 dollars.* |
| OFFER | *Would you like me to hold this option?* |
| ACKNOWLEDGMENT | *I will book this leg.* |
| STATUS-REPORT | *Accessing the database; this might take a few seconds.* |
| EXPLICIT-CONFIRM | *You will depart on September 1st. Is that correct?* |
| IMPLICIT-CONFIRM | *Leaving from Dallas.* |
| INSTRUCTION | *Try saying a short sentence.* |
| APOLOGY | *Sorry, I didn't understand that.* |
| OPENINGS/CLOSINGS | *Hello. Welcome to the CMU Communicator.* |

## 4 Experimental Setup and Results

For the task of evaluating our DA detection framework, we conducted experiments using bigram and trigram based models. The detection accuracy on different training set sizes has been measured also.

The statistical technique chosen for our experiments was the ten-fold cross validation. In each dialogue there were about 20 dialogue acts that were to be predicted. The experimental results over the fixed tag set of 10 DAs of Communicator Dialogue Act Tagged Database are shown in Table 2.

**Table 2.** DA detection results

| Number of Dialogues | Accuracy | |
|:---:|:---:|:---:|
| | Bigrams | Trigrams |
| 150 | 42.85% | 44.98% |
| 200 | 48.92% | 51.57% |
| 250 | 52.38% | 55.67% |
| 300 | 55.76% | 58.69% |
| 350 | 59.52% | 61.90% |
| 400 | 65.31% | 69.53% |

Based on the experimental results, we may conclude that the accuracy of the dialogue system highly depends on the number of dialogues used and that the trigram models performed slightly better than the bigrams, because of the bigger size of observation sequence over which they compute probabilities of occurrence. The most important conclusion to be drawn is that although the amount of training data is not big comparable to other approaches [2], a satisfactory accuracy was achieved. This is explained through the fact that by the use of the perplexity-based word clustering algorithm, word-based trigrams can be computed, in contrast to word trigrams.

## 5   Conclusion

In the present work we evaluated a DA detection module using a word clustering algorithm based on the perplexity criterion. Thus, we achieved the mapping of every word to exactly one word class and by implementing the statistical criterion of maximum likelihood we reduced the number of model parameters, so that each parameter can be estimated more reliably.

DA detection was performed on the 2000 Communicator Dialogue Act Tagged Database using n-gram based statistical models for different sizes of training corpus and different size of n-grams. The experimental results suggest that the trigram based models provided a slightly better accuracy than the bigrams. It is also important to mention that the use of the word clustering algorithm enables the computation of trigrams that do not appear in the training data. This extracted knowledge may be useful to several DA detection applications such as machine translation in dialogue systems, conversational analysis, information retrieval etc. DA recognition is essential for modeling the structure of a user's speech when operating with a DS as it can be used as a feedback for a speech recognizer of a DS, in order to choose between the N-best of hypothesized words.

# References

1. Gianchin, E., Mc Glashan, S.: Corpus-based Methods in Speech Processing. Kluwer Academic, Dordrecht (1997)
2. Stolke, A., Coccaro, N., Bates, R., Taylor, P., van Ess-Dykema, C., Ries, K., Shriberg, E., Jurafsky, D., Martin, R., Meteer, M.: Dialogue Act Modeling for Automatic Tagging and Recognition of Conversational Speech, Computational Linguistics, 26(3)
3. Alshawi, H.: Effective Utterance Classification with Unsupervised Phonotactic Models. In: Proceedings of the 2003 Conference of the North American Chapter of the Association for Computational Linguistics on Human Language Technology, Edmonton, Canada, vol. 1, pp. 1–7 (2003)
4. Grau, S., Sanchis, E., Castro, M.J., Vilar, D.: Dialogue Act Classification Using a Bayesian Approach. In: Proceedings of the 9th International Conference Speech and Computer, pp. 495–499 (2004)
5. Nagata, M., Morimoto, T.: First steps toward statistical modeling of dialogue to predict the speech act type of the next utterance, Speech Communication, 15 (1994)
6. Fernandez, R., Ginzburg, J., Lappin, S.: Using Machine Learning for Non-Sentential Utterance Classification. In: Proceedings of the 6th SIGdial Workshop on Discourse and Dialogue, Lisbon, Portugal, pp. 77–86 (2005)
7. Reithinger, N., Engel, R., Kipp, M., Klesen, M.: Predicting Dialogue Acts for a speech to speech translation system. In: Proceedings of the International Conference on Spoken Language Processing, Philadelphia, vol. 2, pp. 654–657 (1996)
8. Lendvai, P., van den Bosch, A., Krahmer, E.: Machine Learning for Shallow Interpretation of User Utterances in Spoken Dialogue Systems. In: Proceedings of the EACL-03 Workshop on Dialogue Systems: Interaction, Adaptation and Styles of Management, Budapest, Hungary, pp. 69–78 (2003)
9. Nagata, M.: Using pragmatics to rule out recognition errors in cooperative task-oriented dialogues. In: Proceedings of the International Conference on Spoken Language Processing, Banff, Canada, vol. 1, pp. 647–650 (1992)
10. Yoshimura, T., Hayamizu, S., Ohmura, H., Tanaka, K.: Pitch pattern clustering of user utterances in human-machine dialogue. In: Proceedings of the International Conference on Spoken Language Processing, Philadelphia, vol. 2, pp. 837–840 (1996)
11. Martin, S., Liermann, J., Ney, H.: Algorithms for bigram and trigram word clustering, Speech Communication 24 (1998)
12. Prasad, R., Walker, M.: 2000 Communicator Dialogue Act Tagged, Linguistic Data Consortium, Philadelphia (2002)

# Dialogue Management for Intelligent TV Based on Statistical Learning Method

Hyo-Jung Oh, Chung-Hee Lee, Yi-Gyu Hwang, and Myung-Gil Jang

ETRI, 161 Gajeong-dong, Yuseong-gu, Daejeon, 305-700, Korea
{ohj, forever, yghwang, mgjang}@etri.re.kr

**Abstract.** In this paper, we introduce a practical spoken dialogue interface for intelligent TV based on goal-oriented dialogue modeling. It uses a frame structure for representing the user intention and determining the next action. To analyze discourse context, we employ several statistical learning techniques and device an incremental dialogue strategy learning method from training corpus. By empirical experiments, we demonstrated the efficiency of the proposed system. In case of the subjective evaluation, we obtained 73% user satisfaction ratio, while the objective evaluation result was over 90% in case of a restricted situation for commercialization[1].

## 1 Introduction

During the last decade, research in the field of spoken dialogue systems has experienced increasing growth. The success of these researches lets the machine interact with users to provide ubiquitous and personalized services. For instance, technological advances in spoken language have resulted in a new generation of automated contact center services that offer callers the flexibility to speak their request naturally using their own words as opposed to the words dictated to them by the machine [1].

This paper proposes a voice-enabled TV interface for Electronic Program Guide (EPG). In Korea's current TV broadcasting environment there are over 70 channels and about 1,000 programs are broadcasted every day. Thus, the TV itself has more multifunctional controls and its operation has become more complex and difficult. *Intelligent TV* provides advanced functions that endure users get convenience, quality, and enjoyment from the TV. However, an ordinary remote control has many limitations for the intelligent TV interface.

In the EPG domain, a voice-based commercial TV remote control was announced in December 2001 [2]. Its major functions are channel selection, genre search, volume control, recording and playing video by voice. But it can only control the TV by voice commands and cannot handle an actual conversation with the user. Another research area for EPG is the generating of TV program recommendations based on information

---

[1] This work was supported by the IT R&D program of MIC/IITA. [2006-S036-01, Development of large vocabulary/interactive distributed/embedded VUI for new growth engine industries].

V. Matoušek and P. Mautner (Eds.): TSD 2007, LNAI 4629, pp. 644–652, 2007.

garnered from program and channel descriptions and the user's profile [3]. By the way, while most personalized EPG systems use the user's conversation, they are not integrated with spoken language.

In this paper, we introduce a practical dialogue management model for intelligent TV which uses a semantic frame for representing the user intention and meaning. To analyze discourse context, we employ several statistical learning techniques. Conditional random fields (CRF) classifier is used for anaphoric expression recognition and maximum entropy (ME) is for discourse structure analysis. The context information is stored by history manager during the ongoing dialogue. Contrary to the rule-based system, we devise an incremental learning algorithm for dialogue strategies from the training corpus. To verify the efficacy of our proposed system, we conduct a set of experiments.

## 2  System Overview

As shown in Figure 1, our spoken dialogue system broadly consists of four parts; one each for input, output, control, and data. The input part receives a speech signal from TV, converts the signal to text using Automatic Speech Recognition (ASR), and then captures meaningful concepts and intentions in the user utterance with Spoken Language Understanding (SLU). The control part manages dialogues between the system and a user and determines responding actions appropriate to user intentions. The data part manages internal databases with Question Answering (QA). Finally, the output part renders communicative responses as audio with Text-To-Speech (TTS) or TV screen to the user.

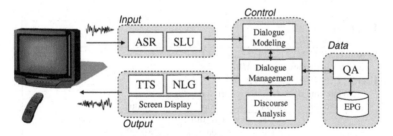

**Fig. 1.** System Architecture

We used ESTk (ETRI Speech Toolkit, [4]) for ASR. ESTk is a large vocabulary continuous speech recognizer, based on tree/flat lexicon search and token-passing algorithm. It is also featured as the multiple pronunciations dictionary and static recognition network. For SLU, we utilized a concept spotting approach [5]. This approach originates from the idea of information extraction (IE) in text processing. In our concept spotting approach, several essential factors for understanding spoken language and dialog management are extracted based on pre-defined domain-dependent meaning representation. The meaning representation directly reflects the task structure for dialog management by including goal-oriented dialog acts and the possible slot structures for dialog acts.

To obtain answers from the EPG database, the QA module takes a question frame analyzed by dialogue manager, and then converts it into SQL query for the relational database. This paper is focused on the control part which contains dialogue management module, after input part.

# 3 Dialogue Management for Intelligent TV

Dialogue management module maintains discourse context history, decides what action to take, and determines the system response appropriate to the user utterance.

## 3.1 Dialogue Modeling

We utilized frame-based semantic representation [6] as the internal data structure in dialogue modeling. Figure 2 shows the frame for the utterance, "What time does *Joo-Mong* start, today?" For the EPG domain, there are 20 frame slots such as *start_time, program, genre, actor*, and so on. The spotting results of SLU are interpreted by a set of slot-value pairs.

This semantic representation also includes discourse analysis result, such as history inheritance tags – CUR and PREV. They represent whether the slot value is inherited from the previous context or not. Action tags denote the next action system has to process. The action tag of User2 in Figure 2 indicates that user wants to know the start time of the program (marked with the "?" tag).

| • | User1: *MBC*로 돌려봐. | | | |
|---|---|---|---|---|
| | Turn on *MBC*⁺ | | | |
| • | System1: *MBC* 9시 뉴스데스크입니다. | | | |
| | This is 9 *news-desk* on *MBC*. | | | |
| • | User2: 오늘 주몽 언제 하지? | | | |
| | What time does *Joo-mong* start, today? | | | |
| | Action | History | Slot | Value |
| | ? | CUR | start_time | |
| | | CUR | date | 2007.02.12 |
| | | CUR | program | Joo-Mong |
| | | PREV | channel | MBC |
| • | System2: *MBC*에서 오늘 밤 10시에 방송될 예정입니다. | | | |
| | It will be played at 10:00 pm on *MBC*. | | | |
| | Action | History | Slot | Value |
| | * | CUR | start_time | 22:00 |
| | | CUR | date | 2007.02.12 |
| | | CUR | program | Joo-Mong |
| | * | PREV | channel | MBC |

⁺: *Italic words* indicate key concepts such as channel, program, etc.

**Fig. 2.** An example of dialogue and internal semantic frame

## 3.2 Discourse Analysis

We implemented a discourse analysis procedure encoding four functions: (1) anaphora resolution, (2) time normalization, (3) discourse structure analysis, and (4) history management.

The anaphora resolution [7] consists of two processes: anaphoric expression recognition and antecedent assignment. Anaphoric expression recognition process is to find the boundary of anaphoric expression and assign one of anaphora tags, such as *N_program, N_channel,* and so on. We solve the task by conditional random fields (CRF) classifier using lexical word, suffix, POS, dictionary, and previous dialog acts as features [5]. The equation of CRF is as follows:

$$P_{CRF}(\mathbf{y} \mid \mathbf{x}) = \frac{1}{Z(\mathbf{x})} \exp\left(\sum_{t=1}^{T}\sum_{k} \lambda_k f_k(y_{t-1}, y_t, \mathbf{x})\right) \tag{1}$$

where $x$ is the word sequence and $y$ is the anaphora tag.

All antecedents appearing in dialogue are stored in anaphora stack according to its types. If a user utterance has some anaphora, we link the anaphora with recently appeared antecedent by a plain first-candidate search. For example, in User1 and System1 of Figure 3, there are four antecedents: *SBS* (channel), *February 13* (date), *10:00 pm* (time), and *Matrix2* (program). They are stored in anaphora stack, respectively. By the stack information, the anaphora *"that movie"* with tag *"N_program"* from User2 is resolved as the *Matrix2*.

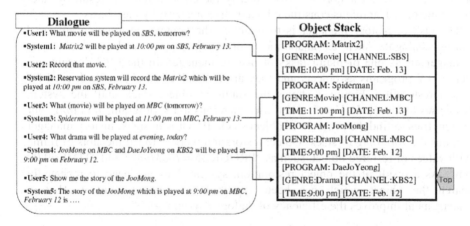

**Fig. 3.** A dialogue example for discourse analysis and object stack

Time normalization task of our system is to assign absolute value to relative concept of time. For example, the value of time word, *"today"* in Figure 3 is variable according to the date when the utterance happens. The time normalization module finds the appropriate value of it. The next table shows some examples of time normalization.

**Table 1.** Examples of time normalization

| Relative time expression | Intermediate result | Absolute value |
|---|---|---|
| 오늘(today) | SystemDate | 20070309* |
| 내일(tomorrow) | SystemDate+1 | 20070310 |
| 다음 주(next week) | (SystemDayMon+7) ~ (SystemDaySun+7) | 20070312~20070318 |

\* The absolute value is retained on the assumption that the dialog happen at 2:23 pm in March 9.

Most frame-based dialogue systems have a set of frames. For deciding the next action of system, they map the speech recognizer output onto a sequence of semantic frames and select one frame in focus. But our frame-based dialogue system maintains only one focus frame and changes the focus according to the condition of current utterance and previous context. Dialogue system need to make a decision whether previous slot values is inherited or not. We use the maximum entropy (ME) [5] method for discourse structure analysis in terms of a binary classification problem, 'clear' or 'not clear'. The used features are morpheme, POS, concept sequence, result of anaphora resolution, current dialog acts, and previous dialog acts. If more than one slot has to be inherited then the result is 'not clear'. In User3, values of genre and date slot are inherited from previous context because they are omitted. At that time, the history tag denoting inheritance status is 'PREV' as shown in Figure 2. On the other hand, User4 does not refer any previous context and so all history tags of the fame are 'CUR'. Not all of the ME result is correct, so we rectify the error of the ME result by rule-based method. The rule is constructed manually on a case-by-case basis.

The task of history manager is to maintain the context information appeared during dialogue, such as anaphora information, object, and so on. Object management is one of major tasks of history manager. In the EPG domain, we define TV programs as objects and they are maintained in object stack used for object restoration. Figure 3 shows the status of object stack until the system responses to User4. In User5, when user mentions a TV program, *JooMong*, our system tries to find the program in object stack. If that is in stack, history manager restores the *JooMong* of stack. Object restoration work is to fill the slots of frame with values of restored object, such as "PROGRAM=*JooMong*" and "TIME=9:00 pm". As a result of the object restoration, system does not need to repeatedly search the object which has been appeared in previous context. History management improves the efficiency of dialogue system.

### 3.3 Dialogue Control

Given the context, dialogue manager selects an appropriate strategy that controls the system response. Depending on the dialogue strategy, system performs QA module, controls the TV, or reserves recording for a particular program. In EPG domain, there are 40 dialog acts such as *search_program, search_channel*, or *change_channel*. The dialog act for the User1 in Figure 4 is assigned as *change_channel*, so the system should control the TV to change channel, "11."

Contrary with rule-based system which is constructed by manual, our system automatically learns dialogue strategies from training corpus. The training corpus consists of a set of pairs: user utterance and system response. Each utterance is tagged by a frame as shown in Figure 4. However, adequate training data is an expensive and time consuming process, thus we developed an incremental learning method to construct dialogue strategy. This model is based on a weighted majority voting algorithm [8] which is proposed for improving the performance of weak classifiers from small training set.

Our training model is able to learn additional information from new dialogue. If it finds some different system responses from existing strategies, then our model tries to expand dialogue control rules, while at the same time to preserve previously acquired knowledge. In example of *change_channel* in Figure 4, the primary dialogue strategy from initial corpus just finds the channel number, and then gives the signal to TV for changing the channel. However, the second corpus shows a more complex situation. The channel which user wants is not unique, so system must confirm which channel the user exactly wants to change, before controls the TV. Finally, the advanced dialogue strategy is incrementally expanded with interactive clarification.

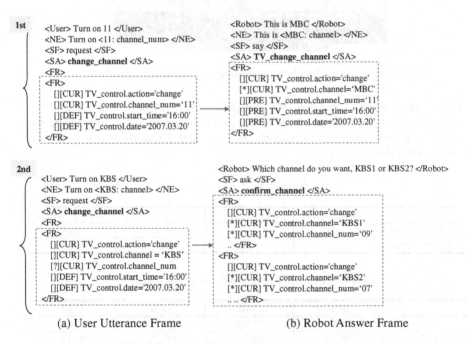

(a) User Utterance Frame                    (b) Robot Answer Frame

**Fig. 4.** Learning dialogue strategy from corpus

Our system generates system response utterance using templates, which are associated with answer frames. The dialogue manager marks entries which have to be delivered to user with "*" as shown Figure 2 and 4. The final system response might be sent to the TTS module for spoken out synthesis or be displayed on the TV screen when answer frames are multiple.

## 4 Empirical Experiments

Our system was implemented in TV set-top box with the remote controller in which a wireless microphone is embedded and it provides multi-modal I/O interface consist of speech, remote control, keyboard, and TV screen. Figure 5 shows a screenshot of our TV.

**Fig. 5.** The screenshot of our proposed spoken dialogue TV interface

For training, we used 597 dialogue scenarios collected from 170 persons. They consist of over 4,100 unique names of TV program and channel. To evaluate the efficiency of our system, 100 persons (80 men and 20 women) examined our system with following test sets:

**Table 2.** Evaluation Set

| Evaluation set | # Scenarios | # Utterances | Average turns | Average words |
|---|---|---|---|---|
| ES1 (freestyle) | 225 | 2,282 | 4.48 | 3.49 |
| ES2 (restricted) | 100 | 642 | 3.21 | 2.67 |

**Table 3.** Evaluation results of the proposed system

| | SRR | DSR | TCR | USR |
|---|---|---|---|---|
| ES1 | 53.5% | 77.8% | 87.7% | 73.0% |
| ES2 | 78.3% | 96.3% | 96.7% | 90.4% |

As the first evaluation (ES1), we conduct the *subjective* user evaluation used to investigate the user friendliness and easy use of end-to-end system [9]. Novice users had free conversations with our system, even including out-of-function, out-of-grammar (OOG), and out-of-vocabularies (OOV). As expected, the ASR result shown in Table 3, the Sentence Recognition Ratio (SRR) of ES1 (53.5%) was too low due to OOG and OOV. Based on the fact that the state-of-the-art in ASR system has limitations in terms of accuracy and response time, we should restrict the user utterance patterns. Nevertheless, the patterns cover most functions which users want to perform with TV.

The *Objective* evaluation measures the system performance on the basis of pre-defined references with respect to the action and reactions taken [9]. ES2 consist of 260 utterance patterns for commercialization. To complement the performance of ASR and to evaluate the dialog system, if an ASR output is not correct, the user correct the error by typing.

We used three metrics for measuring our system. In these metrics, "task" is a collection of utterances which have the same goal. The sentence recognition ratio (SRR) is ASR evaluation result; the dialogue success ratio (DSR) divides the number of correct responses by total user utterances; and the task completion ratio (TCR) divides the number of task successes by the total tasks. The user satisfaction ratio (USR) is an arithmetic mean of DSR, TCR and SRR.

As expected, the performance of ES2 is better than ES1 because it uses restricted utterances. Compared to [2], which uses only 100 controlling words or commands and has a 4.4% word error rate, our system uses 4,100 words for ASR. JUPITER [1] uses 2,000 words and answers nearly 80% of the utterances; The USR of the proposed system is 73.0% with ES1 and 90.4% with ES2. This result shows that the proposed system can be applicable for the EPG domain.

## 5 Conclusion

To develop an intelligent and convenient interface for the EPG domain, we proposed a dialogue-based Intelligent TV system. For dialogue modeling, we utilized frame-based semantic representation as the internal data structure. We also applied statistical learning methods to discourse analysis procedure, encoding four functions: anaphora resolution, time normalization, discourse structure analysis, and history context management. Furthermore, our system automatically learns dialogue strategies from training corpus and is able to learn additional information from new dialogue in terms of incremental learning. Future works include the extension of service domains like news, weather and personal schedules in the TV interface. Moreover, for higher accuracy and commerciality, we will consider including an error recovery scheme in the dialog management module.

## References

1. Zue, V., Seneff, S., Glass, J.R., Polifroni, J., Pao, C., Hazen, T.J., Hetherington, L.: JUPITER: A Telephone-Based Conversational Interface for Weather Information. IEEE Transactions on Speech and Audio Processing 8(1), 85–96 (2000)
2. Fujita, Keiko, Kuwano, Hiroyasu, Tsuzuki, Takashi, Ono, Yoshio, Ishihara, Toshihide.: A New Digital TV Interface Employing Speech Recognition. IEEE Transactions on Consumer Electronics 49(3), 765–769 (2003)

3. Ardissono, L., Kobsa, A., Maybury, M.: Personalized Digital Television: Targeting Programs to Individual Viewers. Kluwer Academic Publishers, Dordrecht (2004)
4. Park, J., Lee, S., Kim, S.: Keyword Spotting for Far-field Speech Input by Categorical Fillers and Speech Enhancement. In: The Proceedings of 22nd Speech Communication and Signal Processing (2005)
5. Lee, Changki, Eun, Jihyun, Jeong, Minwoo, Lee, Geunbae, G., Hwang, Yi-Gyu, Jang, Myung-Gil: A Multi-Strategic Concept-Spotting Approach for Robust Spoken Korean Understanding, ETRI Journal, 29(2) (2007)
6. Chu-Carroll, Jennifer.: MIMIC: An Adaptive Mixed Initiative Spoken Dialogue System for Information Queries. In: Christodoulakis, D.N. (ed.) NLP 2000. LNCS (LNAI), vol. 1835, pp. 97–104. Springer, Heidelberg (2000)
7. Kim, Harksoo, Cho, Jeong-Mi, Seo, Jungyun.: Anaphora Resolution Using an Extended Centering Algorithm in a Multi-modal Dialogue System. In: The proceedings of the Workshop on the Relation of Discourse/Dialogue Structure and reference (1999)
8. Gangardiwala, A., Polikar, R.: Dynamically Weighted Majority Voting for Incremental Learning and Comparison of Three Boosting Based Approaches. In: The proceeding of IJCNN, pp. 1131–1136 (2005)
9. Minker, Wolfgan.: Evaluation Methodologies for Interactive Speech Systems. In: The proceedings of LREC'98, pp. 199–206 (1998)

# Multiple-Taxonomy Question Classification for Category Search on Faceted Information*

David Tomás and José L. Vicedo

Departamento de Lenguajes y Sistemas Informáticos
Universidad de Alicante, Spain
{dtomas,vicedo}@dlsi.ua.es

**Abstract.** In this paper we present a novel multiple-taxonomy question classification system, facing the challenge of assigning categories in multiple taxonomies to natural language questions. We applied our system to category search on faceted information. The system provides a natural language interface to faceted information, detecting the categories requested by the user and narrowing down the document search space to those documents pertaining to the facet values identified. The system was developed in the framework of language modeling, and the models to detect categories are inferred directly from the corpus of documents.

## 1 Introduction

From its very beginning, there have been two main paradigms to information search on the web. The first one, *navigational search* (represented by websites like the *Open Directory Project*), helps people to narrow down the general neighborhood of the information they seek using topical directories or taxonomies. The second paradigm, *direct search* (present in websites like *Google*), allows users to write their queries as a set of keywords in a text box to perform *information retrieval* (IR).

*Faceted search* [9] is a new approach that has recently emerged. This paradigm aims to combine navigational and direct search, allowing users to navigate a multiple-dimensional information space by combining text search with a progressive narrowing of choices in each dimension. Faceted search systems assume that the information is organized into multiple independent facets, rather than a single taxonomy. For instance, we could define for a restaurant guide attributes such as *Cuisine*, *City* or *Features*. These attributes are *facets* that help the users to navigate through them selecting the *values* desired, for instance *Mexican* for *Cuisine*, *Madrid* for *City* or *Online Reservation* for *Features*.

This paradigm is complemented by *category search* [8], which is not a direct search against the information recorded but a search in the space of facet values. While direct search retrieves a set of records[1] that can be further refined using a faceted search approach, category search provides results that are themselves entry points into faceted

---

* This work has been partially supported by the framework of the project QALL-ME (FP6-IST-033860), which is a 6th Framenwork Research Programme of the European Union (EU), and the Spanish Government, project TEXT-MESS (TIN-2006-15265-C06-01).
[1] Although information search covers different formats like images or sounds, our research is focused on textual information.

---

V. Matoušek and P. Mautner (Eds.): TSD 2007, LNAI 4629, pp. 653–660, 2007.

navigation. In the restaurant guide example, a user would query the system with requests such as "Madrid" or "Italian" to restrict the results to restaurants in this *City* or with this *Cuisine*.

Current interfaces to category search are limited to keyword search on facet values (*categories*). In this paper we propose a novel approach to category search. We face the challenge of identifying facet values requested in natural language questions from the user. We tackle this problem from the point of view of *question classification*, which is the task that, given a question, maps it to different semantic classes. This task has been largely used in the context of *question answering* (QA) [3], where it tries to assign a class or category from a fixed taxonomy to the question in order to semantically constrain the space of valid answers.

While traditional question classification systems are limited to single taxonomy categorization, we introduce the idea of *multiple-taxonomy question classification*. In the context of category search, our question classification system accepts a natural language question from the user and detects the different facets (taxonomies) and values (categories) implicitly requested in the query. The values assigned narrow down the set of relevant documents to those pertaining to the categories identified. Following the previous example, a question like "I'm looking for a Tukish restaurant in Madrid", would set the value of facets *Cuisine* to *Turkish* and *City* to *Madrid*, in order to retrieve only restaurants that fulfill these two constraints.

Our system makes use of *language modeling*. A language model is built for each category based on the document set. To identify categories in the question, the probability of generating it is calculated for each category through its language model. Then several heuristics are applied to determine the final classification. Thus, unlike traditional category search systems we do not limit our search to the list of possible values of the facets, but take advantage of the statistical regularities of the documents classified under these categories. We follow the intuition that words occurring in documents assigned to a category are related to it. Going back to the restaurant guide example, documents describing restaurants categorized as *Mexican* would probably contain words like "burrito", "fajita" or "taco". Moreover, in documents describing restaurants with features like *Reservation*, words like "book" or "reserve" would be common. Thus, our system can interpret a request like "I want to book a table to eat a burrito" and infer that the user is asking for *Cuisine* and *Features* facets with values *Mexican* (triggered by "burrito") and *Reservation* (triggered by "book") respectively.

In the rest of this paper, Section 2 reviews related work. Section 3 describes the language modeling framework. Section 4 depicts the sytem architecture, paying attention on the question processing and the identification of facets and their values. Section 5 describes the corpus employed in the experiments carried out, the results obtained and the error analysis derived from these results. Finally, conclusions and future work are discussed in Section 6.

## 2   Related Work

The approach presented in this paper is related to the fields of question classification and faceted information. Question classification has been mainly employed in the field of QA. A majority of systems use hand-crafted rules to identify expected answer types [5]. To overcome the lack of flexibility of these systems, several machine learning approaches have been successfully applied, like *Maximum Entropy* (ME) [1] and *Support Vector Machines* (SVM) [11]. These systems require different levels of linguistic knowledge and tools to learn the classifiers. To avoid this dependence, our approach to question classification is based on statistical language modeling. Unlike other similar approaches [6], we do not need to obtain a training set of questions to build the models as they are built directly from the classified documents.

In [4], Moschitti and Harabagiu presented a novel approach to the task of question classification. Instead of mapping the question to an expected answer type, they assigned document categories from a single taxonomy to questions. They used a set of training questions related to five categories of the *Reuters-21578* text classification benchmark. The idea was to classify questions into these categories in order to filter out all the answers occurring in documents that do not pertain to the category detected. Our system also performs the task of assigning document categories to natural language queries, but extending the classification task to multiple different taxonomies.

In the field of category search on faceted information, current systems [8] perform the task of mapping keywords from the queries to categories. In this sense we go beyond keywords to deal with natural language questions. Moreover, we do not limit the search to category values but take advantage of the faceted documents to infer knowledge to map categories to questions.

## 3   Language Modeling Framework

Our approach follows the ideas described in [7] for statistical language modeling applied to IR, i.e., to infer a language model for each document and to estimate the probability of generating the query according to each of these models. For a query $q = q_1 q_2 \ldots q_n$ and document $d = d_1 d_2 \ldots d_m$ we want to estimate $p(q|d)$, the probability of the query $q$ given the language model of document $d$. One important advantage of this framework over previous approaches is its capability of modeling not only documents but also queries directly though statistical language models. This makes it possible to set retrieval parameters automatically and improve retrieval performance through utilization of statistical estimation methods.

In our experiments we follow a unigram language modeling algorithm based on *Kullback-Leibler* (KL) *divergence* [10]. Documents are ranked according to the negative of the divergence of the query language model from the document language model. We employed *Dirichlet prior* as smoothing method. Previous studies [10] suggest that this method surpasses other smoothing strategies when dealing with short queries. This property is interesting as we estimate probabilities for n-grams in the question (this process is described in detail in Section 4).

In our system, in order to detect the categories that occur in a question, we previously grouped all the documents by the values of their facets, obtaining clusters of documents for every category. Then we infer a language model for each of these clusters to estimate the probability of generating the question according to each of these models. We then rank the clusters of documents according to the negative KL-divergence described above and choose the best candidate following some heuristics further detailed in the next section. In this framework, collection statistics such as term frequency, document length and document frequency are integral parts of the language model and are not used heuristically as in many other approaches. Length normalization is implicit in the calculation of the probabilities and does not have to be done in an *ad hoc* manner. In our experiments all the words in the language were indexed, since we do not want to be biased by any artificial choice of stopwords.

## 4   System Description

Our approach deals with questions on documents that are organized into multiple independent facets, rather than a single taxonomy. The system carries out the task of identifying the set of facet values that occur in a natural language request on faceted documents.

We first group all the documents by the values of their facets obtaining clusters of documents for every category in the corpus. For instance, we would get clusters *Chinese* and *Thai* for facet *Cuisine*, or *Takeout* and *Outdoor Dinning* for facet *Features*. Neither stemming nor stopword removal is carried out in this stage. After this previous process, a language model is derived for each of these clusters. When a question is sent to the system, all the n-grams of the question are extracted and the probability of generating each one according to the models is estimated, following the aforementioned language modeling framework. We use the *Lemur toolkit*[2] to perform this task.

As we said before, the underlying idea of our approach is that content of the documents is related to the categories assigned to these documents. For example, words like "pizza" or "risotto" will commonly appear in the description of restaurants with value *Italian* assigned to facet *Cuisine*. A request like "I want a pizza" will promote the cluster of *Italian* cuisine in the language modeling retrieval framework over other clusters, thus detecting this category in the question. So we do not only search on the values of the facets to detect categories, but also the contents of the documents classified in these categories.

Another assumption is that, as facets are orthogonal sets of values [2], we consider that one n-gram from the question can only determine one facet, i.e., "pizza" can not determine values on different facets as *Cuisine* or *City* at the same time. Otherwise, it could be possible to obtain different values for the same facet (*Italian* and *Greek* for facet *Cuisine*), but in our system hard classification is performed allowing only one possible value per facet.

To process the question in the system and obtain multiple-taxonomy classification, we define the following algorithm:

**Compute similarity:** Compare each n-gram ($n \leq 3$ in our experiments) from the question with each facet value in the corpus. If there is an exact match, the category value is assigned to that n-gram with the maximum ranking value (0, as we employ negative KL-divergence). These first steps perform the classic approach to category search. If no match is found, the similarity with the models estimated for every category is computed on each n-gram, and ranking values obtained are stored.

---

[2] http://www.lemurproject.org

**Stopword removal:** First, In order to detect which n-grams in the query behave as stopwords (and should not be taken into account for the classification task), each unigram is treated in isolation, measuring the variance of the ranking values obtained in the previous step. We consider as stopwords all the unigrams that present close similarity values for every model of the clusters, indicating that they are almost equally distributed through the corpus and cannot discriminate between categories. An empirical variance threshold is established to detect these stopwords. All the remaining n-grams (bigrams and trigrams) that start or end with a unigram that is a stopword are also considered as stopwords.

**Category selection:** For all the n-grams not labeled as stopwords, the best category detected in the ranking process is stored with its corresponding weight. If two n-grams have the same best category, their weights are compared: the n-gram with the greatest weight keeps the category assigned while the other is discarded.

**Category assignment:** Finally, we get a set of n-grams with the assigned categories. These categories are finally associated to the question if the weight given by the similarity of the language model for these n-grams is over a threshold empirically set.

# 5   Experiments

This section describes the corpus of documents and set of questions used in the test carried out, the experiments and results obtained and the error analysis derived from these results.

## 5.1   Dataset

In order to test the system, we created a corpus of documents extracting information from *Lanetro*[3], a website specialized in tourist information about places and events, that offers a faceted search interface for browsing the data. From this web we collected all the data related to restaurants. We obtained 2,146 documents in Spanish, one for each restaurant, with information about the street address, the telephone number and a brief textual description of the restaurant (about 50 words on average). Every document was originally classified in four different facets: *City* (the city where the restaurant is located), *Average price* (average price per person), *Features* (such as *Open Late*, *Romantic*, *Delivery Available...*) and *Cuisine* (*Chinese*, *Italian*, *Seafood...*). There are 77 possible values for facet *City*, 5 for facet *Average price*, 21 for *Features* and 87 for *Cuisine*.

In addition to the corpus of documents, we created a test set of questions in Spanish gathered from potential users outside our project. Eight different users were asked to formulate ten call centre-like free questions. The only restriction imposed to the users was that they must ask questions that should have as answers a restaurant or a list of restaurants. Thus we obtained eighty different natural language questions with a significant variety of utterances, from "Want a Kebab" to "I'm in Alicante. I'm vegetarian and I'm looking for a cheap place that fits my needs". All the questions in the test set

---

[3] http://www.lanetro.com

were labeled by two assessors that assigned values to the facets present in the questions and detected those facets that did not occur.

Users were not informed about the facets involved in the experiments or its possible values, as users of real systems do not necessary know the multiple taxonomies present in this type of systems. This way we also wanted to test the robustness of the approach detecting existing features and values, and also discarding those that do not exist in the taxonomies.

## 5.2   Evaluation and Results

We tested the system on the eighty questions described above. We carried out three different experiments in order to compare the performance of the language modeling framework with other traditional IR empirical methods. For this purpose, we computed the similarity between the questions and the documents also with *TF-IDF* and *Okapi*. In the language modeling experiments, we used Dirichlet prior smoothing with prior parameter $\mu$ empirically set to 3500.

Three different measures of performance are defined. $P$ represents the precision detecting categories in the questions, that is, the number of categories present in the questions that where correctly detected from the total number of categories present. $P_\emptyset$ indicates the precision of the system detecting absent facets in the questions, i.e., if there are restrictions on facets or not. This value is the number of facets not present in the questions that were correctly identified as absent. Finally, $P_T$ is the total number of categories correctly detected in the questions plus the number of facets correctly detected as not present, divided by the total number of facets in the questions.

Table 1 presents the results obtained for the three approaches mentioned above: language modeling, TF-IDF and Okapi. The results are detailed for each of the facets. Language modeling achieved the best overall result in these experiments, with a value of $P_T = 0.6500$. It also achieved the best value of $P$ for all the facets, while best values for $P_\emptyset$ are more distributed through the three approaches.

**Table 1.** Detailed results for *language modeling*, *Okapi* and *TF-IDF*

| Facets | LM | | | Okapi | | | TF-IDF | | |
|---|---|---|---|---|---|---|---|---|---|
| | $P$ | $P_\emptyset$ | $P_T$ | $P$ | $P_\emptyset$ | $P_T$ | $P$ | $P_\emptyset$ | $P_T$ |
| City | **0.8621** | **0.5686** | **0.6750** | 0.5862 | 0.4902 | 0.5250 | 0.1724 | 0.0588 | 0.1000 |
| Average price | **0.4000** | 0.9467 | 0.9125 | **0.4000** | **0.9733** | **0.9375** | **0.4000** | 0.7067 | 0.6875 |
| Features | **0.3462** | **0.7407** | 0.6125 | 0.2692 | 0.8148 | **0.6375** | 0.1923 | 0.6852 | 0.5250 |
| Cuisine | **0.6250** | 0.0625 | 0.4000 | 0.3125 | 0.0313 | 0.2000 | 0.0833 | **1.000** | **0.4500** |
| Overall | **0.6111** | 0.6698 | **0.6500** | 0.3796 | **0.6745** | 0.5750 | 0.1481 | 0.5896 | 0.4406 |

We can conclude from this results that language modeling offers more robustness and precision through all the facets. These values are coherent with the results obtained in [10] for the task of IR. Okapi obtains the second best overall precision ($P_T = 0.5750$) and also obtains the best results for facets *Average price* and *Features*. TF-IDF obtained the worst results with $P_T = 0.4406$.

### 5.3  Error Analysis

The first problem detected is due to verbosity in questions. A request such as "Could you recommend me a restaurant recently opened specialized in Asian cuisine?" presents many n-grams not related to any facet (like "recently opened"). These terms introduce noise in the detection of facet values.

Another problem is sparse facets, which severely harms the performance of statistical methods. There are facets that present many possible values (87 for *Cuisine*) and few documents classified in each value (there are only two restaurants offering *Vietnamese* cuisine in our corpus). This makes necessary the increase of data to predict the models.

As we said before, we do not perform preprocessing of the corpus, not even a stemming process. This way, terms like "burger" and "burgers" are considered completely different. Performing stemming on the data would solve this problem.

Finally, the way the set of test questions was built results a bit risky. We wanted to set a real open domain environment, so that questions were uttered freely with no restriction on facet values. Thus, many requests related to "anniversaries" or "weddings" occur, while no categories in our system match these requests. This results in an increase of the noise introduced into the system.

## 6  Conclusions and Future Work

In this paper we presented a novel approach to multiple-taxonomy question classification. The system proposed receives a natural language question and maps it to categories in different taxonomies. In our experiments we used this system as a natural language interface to category search on faceted data, allowing the user to formulate a free question to narrow down the set of candidate relevant documents to its query. We tested the system on a corpus of faceted documents describing restaurants. We gathered questions from potential users in other to build a corpus of real test questions.

The system was built on a language modeling framework that demonstrated to perform better than other traditional approaches to IR, like Okapi or TF-IDF. All the information to build the models was obtained from the corpus of documents. Thus, we do not need any linguistic knowledge or tools but statistical information. The results obtained are promising for this novel task as free questions made this type of multiple-classification a hard problem. We obtained a best performance $P = 0.6500$ in detecting categories for language modeling.

Error analysis revealed that some improvements must be done. Adding stemming to the preprocessing of the corpus and improving the stopword detection algorithm could easily rise the overall performance. More improvement in the system could be introduced by expanding the original unigram model approach to language modeling with a more complex one based on bigrams or trigrams.

As the system does not employ linguistic tools, there is room for much improvement in this field. For instance, we could use WordNet in order to expand the terms of the question and increase the recall of categories.

# References

1. Blunsom, P., Kocik, K., Curran, J.R.: Question classification with log-linear models. In: SIGIR, ACM, New York (2006)
2. Denton, W.: How to make a faceted classification and put it on the web (November 2003), http://www.miskatonic.org/library/facet-web-howto.html
3. Li, X., Roth, D.: Learning question classifiers. In: International Conference on Computational Linguistics (2002)
4. Moschitti, A., Harabagiu, S.: A novel approach to focus identification in question/answering systems. In: HLT-NAACL Workshop on Pragmatics of Question Answering (2004)
5. Pasca, M.A., Harabagiu, S.M.: High performance question/answering. In: SIGIR, pp. 336–374. ACM, New York (2001)
6. Pinto, D., Branstein, M., Coleman, R., Croft, W.B., King, M., Li, W., Wei, X.: Quasm: Â a system for question answering using semi-structured data. In: 2nd ACM/IEEE-CS Joint Conference on Digital libraries, pp. 46–55 (2002)
7. Ponte, J.M., Croft, W.B.: A language modeling approach to information retrieval. In: SIGIR, pp. 275–281. ACM, New York (1998)
8. Tunkelang, D.: Dynamic category sets: An approach for faceted search. In: SIGIR Workshop on Faceted Search, ACM, New York (2006)
9. Yee, K., Swearingen, K., Li, K., Hearst, M.: Faceted metadata for image search and browsing. In: CHI, ACM, New York (2003)
10. Zhai, C., Lafferty, J.: A study of smoothing methods for language models applied to information retrieval. ACM Trans. Inf. Syst. 22(2), 179–214 (2004)
11. Zhang, D., Lee, W.S.: Question classification using support vector machines. In: SIGIR, ACM, New York (2003)

# Author Index

# Lecture Notes in Artificial Intelligence (LNAI)

Vol. 4451: T.S. Huang, A. Nijholt, M. Pantic, A. Pentland (Eds.), Artifical Intelligence for Human Computing. XVI, 359 pages. 2007.

Vol. 4441: C. Müller (Ed.), Speaker Classification. X, 309 pages. 2007.

Vol. 4438: L. Maicher, A. Sigel, L.M. Garshol (Eds.), Leveraging the Semantics of Topic Maps. X, 257 pages. 2007.

Vol. 4434: G. Lakemeyer, E. Sklar, D.G. Sorrenti, T. Takahashi (Eds.), RoboCup 2006: Robot Soccer World Cup X. XIII, 566 pages. 2007.

Vol. 4429: R. Lu, J.H. Siekmann, C. Ullrich (Eds.), Cognitive Systems. X, 161 pages. 2007.

Vol. 4428: S. Edelkamp, A. Lomuscio (Eds.), Model Checking and Artificial Intelligence. IX, 185 pages. 2007.

Vol. 4426: Z.-H. Zhou, H. Li, Q. Yang (Eds.), Advances in Knowledge Discovery and Data Mining. XXV, 1161 pages. 2007.

Vol. 4411: R.H. Bordini, M. Dastani, J. Dix, A.E.F. Seghrouchni (Eds.), Programming Multi-Agent Systems. XIV, 249 pages. 2007.

Vol. 4410: A. Branco (Ed.), Anaphora: Analysis, Algorithms and Applications. X, 191 pages. 2007.

Vol. 4399: T. Kovacs, X. Llorà, K. Takadama, P.L. Lanzi, W. Stolzmann, S.W. Wilson (Eds.), Learning Classifier Systems. XII, 345 pages. 2007.

Vol. 4390: S.O. Kuznetsov, S. Schmidt (Eds.), Formal Concept Analysis. X, 329 pages. 2007.

Vol. 4389: D. Weyns, H.V.D. Parunak, F. Michel (Eds.), Environments for Multi-Agent Systems III. X, 273 pages. 2007.

Vol. 4386: P. Noriega, J. Vázquez-Salceda, G. Boella, O. Boissier, V. Dignum, N. Fornara, E. Matson (Eds.), Coordination, Organizations, Institutions, and Norms in Agent Systems II. XI, 373 pages. 2007.

Vol. 4384: T. Washio, K. Satoh, H. Takeda, A. Inokuchi (Eds.), New Frontiers in Artificial Intelligence. IX, 401 pages. 2007.

Vol. 4371: K. Inoue, K. Satoh, F. Toni (Eds.), Computational Logic in Multi-Agent Systems. X, 315 pages. 2007.

Vol. 4369: M. Umeda, A. Wolf, O. Bartenstein, U. Geske, D. Seipel, O. Takata (Eds.), Declarative Programming for Knowledge Management. X, 229 pages. 2006.

Vol. 4343: C. Müller (Ed.), Speaker Classification. X, 355 pages. 2007.

Vol. 4342: H. de Swart, E. Orłowska, G. Schmidt, M. Roubens (Eds.), Theory and Applications of Relational Structures as Knowledge Instruments II. X, 373 pages. 2006.

Vol. 4335: S.A. Brueckner, S. Hassas, M. Jelasity, D. Yamins (Eds.), Engineering Self-Organising Systems. XII, 212 pages. 2007.

Vol. 4334: B. Beckert, R. Hähnle, P.H. Schmitt (Eds.), Verification of Object-Oriented Software. XXIX, 658 pages. 2007.

Vol. 4333: U. Reimer, D. Karagiannis (Eds.), Practical Aspects of Knowledge Management. XII, 338 pages. 2006.

Vol. 4327: M. Baldoni, U. Endriss (Eds.), Declarative Agent Languages and Technologies IV. VIII, 257 pages. 2006.

Vol. 4314: C. Freksa, M. Kohlhase, K. Schill (Eds.), KI 2006: Advances in Artificial Intelligence. XII, 458 pages. 2007.

Vol. 4304: A. Sattar, B.-h. Kang (Eds.), AI 2006: Advances in Artificial Intelligence. XXVII, 1303 pages. 2006.

Vol. 4303: A. Hoffmann, B.-h. Kang, D. Richards, S. Tsumoto (Eds.), Advances in Knowledge Acquisition and Management. XI, 259 pages. 2006.

Vol. 4293: A. Gelbukh, C.A. Reyes-Garcia (Eds.), MICAI 2006: Advances in Artificial Intelligence. XXVIII, 1232 pages. 2006.

Vol. 4289: M. Ackermann, B. Berendt, M. Grobelnik, A. Hotho, D. Mladenič, G. Semeraro, M. Spiliopoulou, G. Stumme, V. Svátek, M. van Someren (Eds.), Semantics, Web and Mining. X, 197 pages. 2006.

Vol. 4285: Y. Matsumoto, R.W. Sproat, K.-F. Wong, M. Zhang (Eds.), Computer Processing of Oriental Languages. XVII, 544 pages. 2006.

Vol. 4274: Q. Huo, B. Ma, E.-S. Chng, H. Li (Eds.), Chinese Spoken Language Processing. XXIV, 805 pages. 2006.

Vol. 4265: L. Todorovski, N. Lavrač, K.P. Jantke (Eds.), Discovery Science. XIV, 384 pages. 2006.

Vol. 4264: J.L. Balcázar, P.M. Long, F. Stephan (Eds.), Algorithmic Learning Theory. XIII, 393 pages. 2006.

Vol. 4259: S. Greco, Y. Hata, S. Hirano, M. Inuiguchi, S. Miyamoto, H.S. Nguyen, R. Słowiński (Eds.), Rough Sets and Current Trends in Computing. XXII, 951 pages. 2006.

Vol. 4253: B. Gabrys, R.J. Howlett, L.C. Jain (Eds.), Knowledge-Based Intelligent Information and Engineering Systems, Part III. XXXII, 1301 pages. 2006.

Vol. 4252: B. Gabrys, R.J. Howlett, L.C. Jain (Eds.), Knowledge-Based Intelligent Information and Engineering Systems, Part II. XXXIII, 1335 pages. 2006.

Vol. 4251: B. Gabrys, R.J. Howlett, L.C. Jain (Eds.), Knowledge-Based Intelligent Information and Engineering Systems, Part I. LXVI, 1297 pages. 2006.

Vol. 4248: S. Staab, V. Svátek (Eds.), Managing Knowledge in a World of Networks. XIV, 400 pages. 2006.

Vol. 4246: M. Hermann, A. Voronkov (Eds.), Logic for Programming, Artificial Intelligence, and Reasoning. XIII, 588 pages. 2006.

Vol. 4223: L. Wang, L. Jiao, G. Shi, X. Li, J. Liu (Eds.), Fuzzy Systems and Knowledge Discovery. XXVIII, 1335 pages. 2006.

Vol. 4213: J. Fürnkranz, T. Scheffer, M. Spiliopoulou (Eds.), Knowledge Discovery in Databases: PKDD 2006. XXII, 660 pages. 2006.

Vol. 4212: J. Fürnkranz, T. Scheffer, M. Spiliopoulou (Eds.), Machine Learning: ECML 2006. XXIII, 851 pages. 2006.